Earth Sciences and Archaeology

Earth Sciences and Archaeology

Edited by

Paul Goldberg
Boston University
Boston, Massachusetts

Vance T. Holliday
University of Wisconsin–Madison
Madison, Wisconsin

and

C. Reid Ferring
University of North Texas
Denton, Texas

Kluwer Academic / Plenum Publishers
New York, Boston, Dordrecht, London, Moscow

Library of Congress Cataloging-in-Publication Data

Earth sciences and archaeology/edited by Paul Goldberg, Vance T. Holliday, and C. Reid Ferring.
    p. ; cm.
    Includes bibliographical references and index.
    ISBN 0-306-46279-6
      1. Archaeological geology.   2. Archaeological geology—Methodology.   3. Earth sciences.   4. Earth sciences—Methodology.   I. Goldberg, Paul.   II. Holliday, Vance T.   III. Ferring, C. Reid.

CC77.5 .E2 2000
930.1—dc21
                                                                                                     99-058246

ISBN 0-306-46279-6

©2001 Kluwer Academic / Plenum Publishers, New York
233 Spring Street, New York, New York 10013

http://www.wkap.nl/

10 9 8 7 6 5 4 3 2 1

A C.I.P. record for this book is available from the Library of Congress

All rights reserved

No part of this book may be reproduced, stored in a retrieval system, or transmitted in any form or by any means, electronic, mechanical, photocopying, microfilming, recording, or otherwise, without written permission from the Publisher

Printed in the United States of America

Contributors

Arthur Bettis, Department of Geoscience, University of Iowa, Iowa City, Iowa 52242

Ofer Bar-Yosef, Department of Anthropology, Peabody Museum, Harvard University, Cambridge, Massachusetts 02138

Marie-Agnès Courty, CRA-CNRS, Laboratoire de Science des Sols et Hydrologie, INA-PG, 78850 Grignon, France

Jill Cruise, University of London Guildhall and Greenwich University, London LU78J4, United Kingdom

C. Reid Ferring, Department of Geography, University of North Texas, Denton, Texas 76203

Charles Frederick, Department of Archaeology and Prehistory, University of Sheffield, Sheffield S1 4ET, United Kingdom

Norman Herz, Department of Geology, University of Georgia, Athens, Georgia 30602

Vance T. Holliday, Department of Geography, University of Wisconsin, Madison, Wisconsin 53706

Kenneth L. Kvamme, Department of Anthropology, University of Arkansas, Fayatteville, Arkansas 72701

David Leigh, Department of Geography, University of Georgia, Athens, Georgia 30602

Richard I. Macphail, Institute of Archaeology, University College London, London WC1H 0PY, United Kingdom

Rolfe D. Mandel, Department of Geography, University of Kansas, Topeka, Kansas 66045

Jay Noller, Department of Crop and Soil Science, Oregon State University, Corvallis, Oregon 97331

Lee Nordt, Department of Geology, Baylor University, Waco, Texas 76798

W. Jack Rink, School of Geography and Geology, McMaster University, Hamilton, Ontario, Canada, L8S 4MI

Sarah Sherwood, Department of Sociology and Anthropology, Middle Tennessee State University, Murfreesboro, Tennessee 37132

Julie K. Stein, Department of Anthropology, University of Washington, Seattle, Washington 98195

James Stoltman, Department of Anthropology, University of Wisconsin–Madison, Madison, Wisconsin 53706

Lisa E. Wells, Department of Geology, Vanderbilt University, Nashville, Tennessee 37235

Preface

Over the past several decades, a number of volumes have appeared on the subject of "Archaeological Geology" or "Geoarchaeology" (Davidson and Shackley, 1976; Herz and Garrison, 1998; Rapp and Gifford, 1985; Rapp and Hill, 1998; Waters, 1992). Although the supposed differences between these two endevors continue to be discussed (e.g., Butzer, 1982; Rapp and Hill, 1998), here we are basically concerned with any subject that bridges the interface between the earth sciences and archaeology, with the earth sciences including a wide array of subjects, such as sedimentology, stratigraphy, geomorphology, pedology, geochemistry, geophysics, and geochronology. The number of books that focus on geoarchaeology (and the organization of goups such as the Archaeological Geology Division of the Geological Society of America, and the Geoarchaeology Interest Group of the Society for American Archaeology) demonstrate the increasing amount of and interest in geoarchaeology and the validity of considering geeoarchaeology as a subdiscipline in its own right. These volumes also reflect the interest in geoarchaeology in universities throughout North America; this interest is further demonstrated by the Guide to Geoarchaeology Programs and Departments published by the Archaeological Geology Division of the GSA.

Among the earliest of the volumes on geoarchaeology was a collection of papers from a symposium titled "Sediments in Archaeology" held in England in the early 1970s (Davidson and Shackley, 1976). The papers from this groundbreaking effort were organized into themes that included Techniques, Sedimentary Environments (coastal, lacustrine, and terrestrial environments), and Biological Sediments. Although some of these articles were local in scope, those on methodology encompassed a number of techniques that included stratigraphy, magnetic properties of sediments as applied to prospection, petrography, phosphate chemistry, and cave sediments.

A decade or so later, Rapp and Gifford (1985) produced a multiauthored volume that was very much methodology oriented. It included a broad array of subjects and techniques and their applications to archaeological problems, including the following: geomorphology (including sedimentary and palaeoenvironments), palynology, anthrools, geophysical surveying and archaeomagnetism, isotope and dating studies, and sourcing of materials.

More recently, a number of general geoarchaeology textbooks have appeared. The first in this group is by Waters (1992), which takes on a larger scale perspective, stressing geomorphology and site formation from a North American viewpoint. *Geoarchaeology* by Rapp and Hill (1998) covers most of the topics that are encompassed in modern geoarchaeological studies, ranging from field-based geomorphology/sedimentology to laboratory techniques. Published at about the same time was *Geological Methods for Arrchaeology* by Herz and Garrison (1998), which considers with some detail geological techniques in archaeology from a variety of aspects: geomorphology, sediments and soils, dating techniques, site exploration, and artifact analysis.

What struck us about these earlier collections—but less so with the most recent publications—is that articles tended to describe an earth science technique used in archaeology, provide some theoretical background, and then discuss the results. What seemed to be typically lacking were explicit statements of a number of issues: (1) the type(s) of problem being solved; (2) why a particular technique (or techniques) was being applied in the first place; (3) why this technique was the most suitable to tackle this problem; and (4) the implications of the results to both the archaeological and the earth science communities. Any ramifications directed toward these groups were commonly left to the readers to figure out for themselves.

We developed this book to avoid these shortcomings by making it as didactic as possible. We wanted to present a sampling of a variety of earth science techniques and strategies that can be used to answer problems that are of interest to both archaeologists and earth scientists. We attempted to choose techniques and their practitioners that represent up-to-date thinking and methodology on each subject. Some of these techniques are not widely performed or widely known. We stressed that authors should not present just a summary of technique or types of studies (and that they also avoid simply summarizing old research or publishing new results), but demonstrate how such studies are actually carried out. We wished the authors to convey some of their experiences as means of furnishing some *practical* information, a type of information that rarely gets into press. Finally, we wanted to acquaint different members of the academic community with concrete geoarchaeological problems and their significance. We hope that by explicitly stating the goals, techniques, and implications in each case, archaeologists will be made aware of the value of a particular technique, whereas the practicing or potential geoscientist would be exposed to possible problems that can be attacked along the suture of archaeological and geological research.

We hope this volume will have wide appeal to archaeologists and earth scientists alike. Archaeologists are increasingly making use of modern technologies and are becoming increasingly aware of the role of earth sciences in modern archaeology. The interest of earth scientists as well is much higher than it was in the late 1980s, and many universities are currently offering formal courses in geoarchaeology, when then there were few. The increased collaboration and awareness is clear to those of us who are involved in the production of the journal *Geoarchaeology*, where most of the contributions are joint efforts between archaeologists and earth scientists. Finally, the "explicit approach" we wish to promote in this collection of articles should make clear to scholars in both fields the wealth of opportunity that each community has to offer.

Not surprisingly, we were not able to provide papers on all subjects. First, this was not possible because of space limittions and the ever-increasing breadth of the discipline. For example, we would have liked to include chapters on a broad spectrum of aeolian deposits (both dunes and loess), but this would have been a formidable task for anyone. Similarly, coastal and periglacial environments, as well as colluvial settings, could also merit their own chapters. A discussion of general geoarchaeological problems that repeatedly arise would have been especially useful. Any number of regional or site-specific studies that carefully document the interface and interaction between archaeology and the earth sciences would have also constituted cogent additions. On the other hand, we believed that so much has been written about radiocarbon that we could reasonably forgo its inclusion in this publication. These and similar topics could readily fit into a volume of their own, and we can only hope that continued interest in the subject will bring about publication of such a book.

Second, certain subjects that were included in the original table of contents simply never made it to press. They were promised but not delivered. There was no time to approach other authors without significantly delaying the volume beyond our 2-year schedule. Such delays would also hold up those authors who did provide chapters. So, for example, a timely chapter on the "Stratigraphy and Sedimentology of Caves and Rockshelters" is missing. We had also wished to cover the topic of magnetic susceptibility in soils and in archaeological deposits, a subject well known in Europe, but unfortunately used relatively little in North America. A chapter specificlly on fluvial landscapes in arid environments is also absent, much to our regret. Nevertheless, the subjects that are presented in this volume do constitute a realistic representation of the majority of techniques and themes involving the interaction of earth sciences and archaeology.

The book is organized into a series of sections that share common themes. The articles in Part I furnish background material that reflect broader issues. Holliday's chapter titled "Quaternary Geosciences in Archaeology" considers a number of important issues, such as geologic time, the record of and reason for Quaternary environmental changes, and approaches to reconstructing past environments. These issues are critical in the communication and execution of proper and modern geoarchaeological research. Stein, in Chapter 2, provides a historical background of the study of site formation studies. This theme is critical to correctly documenting and interpreting the archaeological record, and now forms—or should form—the basis for modern archaeological research.

Geomorphological studies are the main focus of Part II. Frederick (Chapter 3) examines the nature of alluvial sequences and discusses in detail the explanations of some of their causes and interpretations, ranging from "natural" ones to those induced by human activities. Chapter 4 by Ferring examines fluvial landscapes from both arid and humid environments; he provides some basic information on this archaeologically significant geomorphic setting and shows how geoarchaeological sequences have been studied in the past. Wells, in Chapter 5, considers the relationship of settlement pattern to geomorphological change, providing examples from both the New and Old Worlds. The second part concludes with a view by Noller (Chapter 6) of a promising new avenue of research along the interdisciplinary junction between archaeology and the earth sciences, archaeoseismology. He demonstrates how the archaeological record provides valuable information in understanding the effects of earthquakes on both past and modern societies.

Soils, sediments, and microstratigraphy are dealt with in Part III. These issues constitute the bulk of geoarchaeological research in North America and somewhat less so in Europe. In this part we present chapters that range from general issues and problems to environmentally and technique-specific ones. So, for example, Mandel and Bettis (Chapter 7) discuss the practical details of soils and landscapes, such as distinguishing soils from sediments, an often daunting problem to archaeologists and earth scientists alike; their discussion on soils and archaeological surveys provides practical information that is timely to field problems associated with modern cultural resource management projects. The chapters by Courty (Chapter 8) and Macphail and Cruise (Chapter 9) provide a somewhat different approach to many of the studies in the volume, with their emphasis on microstratigraphy and site formation dynamics and the methodological means to study them using soil micromorphology. They also furnish a valued European perspective because these types of studies are relatively uncommon in North America. Finally, Leigh (Chapter 10) examines the relationships among artifacts found in sandy contexts. These settings are widespread but not very well documented or understood, and the question of artifact mobilization is of critical importance here.

In Part IV, the shift is to studies involving more specific techniques. Stoltman (Chapter 11), for example, discusses the use and methodologies involved in ceramic petrography, including its application to issues of trade and its relationship to geochemical methodologies. Sherwood (Chapter 12) provides insights into the methodology, use, and interpretation of microartifact studies in archaeology. She shows how the method—one inspired from the earth sciences—can be used to interpret space in the archaeological context. Turning to geophysical methods, Kvamme (Chapter 13) demonstrates how modern remote sensing techniques are conducted. He illustrates the differences in suitability of different geophysical techniques at the same site and thus provides some vauable insights as to the choice of available techniques.

Part V includes three chapters dealing with geochemical methods. Rink (Chapter 14) explains the variety of techniques—including field and laboratory procedures—appropriate for dating artifacts and contexts that are beyond the range of radiocarbon. This information is particularly useful to those working on sites from the Old World, where archaeological records extend well beyond the Holocene. Chapter 15 by Nordt considers isotope analysis of soils, a subject that has proven to provide valuable paleoenvironmental data in Quaternary and geoarchaeological studies in both the New and Old Worlds. Chapter 16 by Herz reveals his extensive experience with the use and application of instrumental analyses in the sourcing of lithic materials.

Part VI concludes the volume with a prehistorian's perspective of the earth scientist and archaeology. It utilizes Bar-Yosef's background in earth science and archaeology, as well as decades of collaboration with earth scientists and environmentalists. It provides a number of lessons learned from his experience in Near Eastern sites on how (geo)archaeological research has been carried out and how it might be better conducted in the future, regardless of locale.

In sum, our ultimate goal here is to provide pragmatic information that is translatable into better field and lab studies, as well as more and better interaction between archaeology and the earth sciences. We hope that practitioners from both disciplines will benefit from the perspectives and talents of our authors.

ACKNOWLEDGMENTS. We would like to gratefully acknowledge the editorial assistance of S. Weinberger for keeping together the infinite pieces of text, references, and correspondence. We also wish to thank those authors who submitted their manuscripts early and on their patience in waiting for this volume to appear.

References

Butzer, K. W., 1982, *Archaeology as Human Ecology*. Cambridge University Press, Cambridge.
Davidson, D. A., and Shackley, M. L., 1976, *Geoarchaeology*. Duckworth, London.
Herz, N., and Garrison, E., 1998, *Geological Methods for Archaeology*. Oxford University Press, New York.
Rapp, G., Jr., and Gifford, J., 1985, *Archaeological Geology*. Yale University Press, New Haven.
Rapp, G. R., Jr., and Hill, C., 1998, *Geoarchaeology*. Yale University Press, New Haven.
Waters, M. R., 1992, *Principles of Geoarchaeology: A North American Perspective*. University of Arizona Press, Tucson.

Contents

I. BACKGROUND

Chapter 1

Quaternary Geoscience In Archaeology 3

    Vance T. Holliday

1. Introduction 3
2. Definitions and Boundaries 5
   2.1. The Pliocene–Pleistocene Boundary 11
   2.2. The Pleistocene–Holocene Boundary 12
   2.3. Stratigraphic Subdivisions 13
3. Glacial–Interglacial Cycles 13
4. Causes of Quaternary Climate Cycles 16
5. Reconstructing Quaternary Environments: Data versus Models . . . 20
6. Discussion and Conclusions 22
   Acknowledgments 27
7. References 28

Chapter 2

A Review of Site Formation Processes and Their Relevance to Geoarchaeology 37

    Julie K. Stein

1. Introduction 37
2. History and Definitions of Formation Processes 38
   2.1. Definition of Site Formation Analysis 39
   2.2. The Unit of Site Formation Analysis 42

3. Examples 45
4. Conclusion 47
   Acknowledgments 48
5. References 48

II. GEOMORPHOLOGICAL STUDIES

Chapter 3

Evaluating Causality of Landscape Change: Examples from Alluviation . 55

Charles Frederick

1. Introduction 55
2. Basic Elements of Alluvial Stratigraphic Sequences 57
   2.1. The Stratigraphic Sequence 57
   2.2. Chronology 58
   2.3. Combination Proxies 61
3. Evidence Used to Link Causal Factors with Alluviation 64
   3.1. Climatic Forcing 65
   3.2. Anthropogenic Alluviation 65
   3.3. Tectonic Activity 67
   3.4. Eustasy 68
   3.5. Internal or Endogenic Factors, Geomorphic Thresholds, and
        Complex Responses 68
4. Investigating Causality in Alluviation: Some Examples 69
   4.1. Southern Illinois 69
   4.2. Southwestern Utah 70
   4.3. Colorado Plateau—Northeastern Arizona 71
   4.4. Comments 71
5. Summary 71
6. References 72

Chapter 4

Geoarchaeology in Alluvial Landscapes 77

C. Reid Ferring

1. Introduction 77
   1.1. Perspectives 78
   1.2. Time, Environments, and Fluvial Systems 78
2. Fluvial Environments, Geology, and Archaeological Implications . . 80
   2.1. General Factors 80
   2.2. Vegetation, Weathering, and Sediment Yield 81
   2.3. Channel Patterns and Stream Load 81
   2.4. Facies, Architecture, and Alluvial Geomorphology 84

	2.5. Alluvial Morphogenesis and Pedogenesis	86
	2.6. Alluvial Response to Climatic Change	88
3.	Geoarchaeological Methods in Fluvial Environments	89
	3.1. Stratigraphy	89
	3.2. Geochronology	91
	3.3. Site Discovery Methods	93
	3.4. Excavations and Formation Analyses	94
	3.5. Landscapes, Change, and Human Settlements	96
4.	Conclusions	98
5.	References	99

Chapter 5

A Geomorphological Approach to Reconstructing Archaeological Settlement Patterns Based on Surficial Artifact Distribution: Replacing Humans on the Landscape 107

Lisa E. Wells

1.	Introduction	107
2.	Techniques and Methods	109
	2.1. Morphostratigraphy and Allostratigraphy	109
	2.2. Chronostratigraphy	113
	2.3. Integration of Geomorphology and Archaeological Survey Data: Questions of Scale	114
3.	Case Studies	115
	3.1. Paleolandscapes of the Andean Foothills, Northern Coastal Peru: Reinterpreting Site-Based Surveys	115
	3.2. Paleolandscapes of the North Troodos Foothills, Cyprus: Toward an Interdisciplinary Framework	126
4.	Conclusions: Replacing Humans on the Landscape	136
	Acknowledgments	137
5.	References	137

Chapter 6

Archaeoseismology: Shaking Out the History of Humans and Earthquakes 143

Jay Stratton Noller

1.	Using Archaeology to Solve a Paleoseismic Problem	143
2.	Approaches and Results of Archaeoseismology	144
	2.1. When Did the Earthquake Occur	145
	2.2. What Did the Earthquake Do?	152
	2.3. When's the Next Earthquake?	156

3. Case Study: Offset of the Seal Cove Archaeological Site by the
   San Gregorio Fault 159
   3.1. Introduction 159
   3.2. Approach: Identify an Archaeological Site on a Fault 160
   3.3. Methods: Excavate and Date 161
   3.4. Results: Reading between the Fault Lines 162
   3.5. Implications of Results from Seal Cove 165
4. Closing 166
   Acknowledgments 166
5. References 167

III. SOILS, SEDIMENTS, AND MICROSTRATIGRAPHY

Chapter 7

Use and Analysis of Soils by Archaeologists and Geoscientists: A North American Perspective 173

Rolfe D. Mandel and E. Arthur Bettis III

1. Introduction 173
2. Distinguishing Soil from Sediment 174
3. Soils and Archaeological Surveys 181
4. Soils and Site Evaluations 185
5. Soils and Site Excavations 190
6. Summary and Conclusions 194
   Acknowledgments 195
7. References 195

Chapter 8

Microfacies Analysis Assisting Archaeological Stratigraphy 205

Marie-Agnès Courty

1. Introduction 205
2. Basic Concepts and Definitions 207
   2.1. Anthropogenic Processes 207
   2.2. Archaeological Facies and Facies Patterns 208
3. Methodology 209
   3.1. Problems 209
   3.2. Research Strategy 211
   3.3. Sampling 212
   3.4. Analytical Procedure 213
   3.5. Synchronization with Other Techniques 215

4.	Formation of Archaeological Strata	217
	4.1. General Principles	217
	4.2. Dynamics of the Soil Interface	220
	4.3. From the Soil Interface to the Archaeological Layer	220
	4.4. Stratigraphic Relationships and Three-Dimensional Reconstruction	229
5.	Implications	232
	5.1. Implications for Archaeology	232
	5.2. Implications for Soil Science	232
	5.3. Implications for Paleoenvironmental Research and Paleoclimatology	234
6.	Conclusion	235
7.	References	236

Chapter 9

The Soil Micromorphologist as Team Player: A Multianalytical Approach to the Study of European Microstratigraphy ... 241

Richard Macphail and Jill Cruise

1.	Introduction	241
2.	Methods	243
	2.1. Getting the Sampling Right	243
	2.2. Multidisciplinary–Analytical Approach	245
	2.3. Numerical/Semi-numerical Data Gathering	245
3.	Research Base	247
	3.1. Experimental Findings	247
4.	Discussion	259
	4.1. A Final Cautionary Tale	263
5.	Conclusions	263
	Acknowledgments	263
6.	References	264

Chapter 10

Buried Artifacts in Sandy Soils: Techniques for Evaluating Pedoturbation versus Sedimentation ... 269

David S. Leigh

1.	Introduction	269
	1.1. Equifinality of Pedoturbation and Sedimenttion	271
2.	Techniques	272
	2.1. Geomorphic Setting	272
	2.2. Sedimentary Structures, Stratigraphy, and Pedology	274
	2.3. Particle Size Analysis	277

        2.4. Distribution and Integrity of Cultural Materials and Features . . 280
        2.5. Micromorphology 284
        2.6. Dating Techniques 284
        2.7. Other Techniques 286
    3. Case Example 286
    4. Conclusions 291
    5. References 291

IV. SPECIFIC TECHNIQUES

Chapter 11

The Role of Petrography in the Study of Archaeological Ceramics . . . 297

   James B. Stoltman

    1. Introduction 297
    2. Basic Principles of Ceramic Petrography 299
        2.1. Qualitative Observations 301
        2.2. Quantitative Observations 305
    3. Archaeological Problems Amenable to Petrographic Analysis 307
        3.1. Ceramic Classification 307
        3.2. Ceramic Engineering/Functional Considerations 309
        3.3. Ceramic Production 312
        3.4. Ceramic Exchange 319
    4. Summary and Conclusions 322
    5. References 323

Chapter 12

Microartifacts **327**

   Sarah C. Sherwood

    1. Introduction 327
    2. Defining Microartifacts 328
        2.1. Theoretical Framework 329
    3. Methods 330
        3.1. Microartifact Identification 331
        3.2. Microartifact Recovery 332
        3.3. Size Distribution 333
        3.4. Microartifact Quantification 335
        3.5. Data Representation 337
    4. Research Questions 338
        4.1. Site-Scale Research 338
        4.2. Landscape-Scale Research 344
    5. Conclusions 346

Acknowledgments 348
6. References 348

Chapter 13

Current Practices in Archaeogeophysics: Magnetics, Resistivity, Conductivity, and Ground-Penetrating Radar 353

Kenneth L. Kvamme

1. Introduction 353
2. Geophysical Prospection Principles 355
3. Field Survey Methods 356
4. Geophysical Methods and Instruments 356
    4.1. Magnetic Methods 356
    4.2. Electrical Resistivity 358
    4.3. Electromagnetic Conductivity 362
    4.4. Ground-Penetrating Radar (GPR) 363
5. Computer Methods 365
6. Case Studies I: Field Methods and Results 366
    6.1. Whistling Elk Village, South Dakota 366
    6.2. Menoken Village, North Dakota 370
    6.3. Sluss Cabin, Kansas 372
    6.4. Breed's Hill, Massachusetts 373
7. Case Studies II: Advanced Geophysical Data Processing 375
    7.1. Navan Fort, Northern Ireland 375
    7.2. 3D Ranch, Kansas 377
    7.3. Whistling Elk Village, South Dakota 378
    7.4. Breed's Hill, Massachusetts 378
8. Conclusions 378
    Acknowledgments 379
9. Glossary 381
10. References 382

V. GEOCHEMICAL METHODS

Chapter 14

Beyond ^{14}C Dating: A User's Guide to Long-Range Dating Methods in Archaeology 385

W. Jack Rink

1. Introduction 385
    1.1. Scope and Current Issues 386
2. How to Choose the Right Dating Methods 386

3. Radiogenic Isotopes for Dating: Physical Basis ... 389
   3.1. Applications of Radiogenic Isotope Dating ... 391
   3.2. ^{40}Ar/^{39}Ar Dating ... 392
   3.3. Closed-System Uranium-Series Dating ... 392
   3.4. Open-System Uranium-Series Dating ... 393
   3.5. Sampling Requirements for Uranium Series and ^{40}Ar/^{39}Ar Dating ... 393
4. Radiation Exposure Dating ... 394
   4.1. Physical Basis of Fission-Track Dating ... 394
   4.2. Applications of Fission-Track Dating ... 395
   4.3. Sampling Requirements for Fission-Track Dating ... 395
   4.4. Physical Basis of ESR and Luminescence Dating ... 396
   4.5. Dosimetry Requirements ... 397
   4.6. Applications of ESR Dating ... 399
   4.7. Applications of Luminescence Dating ... 405
5. Dating Intercomparisons and General Problems with Interpretation of Dating Results ... 408
6. Potential Problems with Various Dating Methods ... 410
7. Summary ... 412
   Acknowledgments ... 412
8. References ... 412

Chapter 15

Stable Carbon and Oxygen Isotopes in Soils: Applications for Archaeological Research ... **419**

Lee C. Nordt

1. Introduction ... 419
2. Theory of Isotope Pedology ... 422
   2.1. Soil Genesis ... 422
   2.2. Stable C Isotopes of Soil Organic Matter ... 422
   2.3. Stable C and O Isotopes of Pedogenic Carbonate ... 423
3. Field Application of Stable C and O Isotopes in Soils ... 425
   3.1. Radiocarbon Dating and δ^{13}C Depth Distributions ... 425
   3.2. Sources of Soil Carbon ... 428
4. Stable Isotope Laboratory Procedures ... 432
   4.1. Sample Collection ... 432
   4.2. Procedures for Soil Organic Matter ... 433
   4.3. Procedures for Pedogenic Carbonate ... 433
   4.4. Laboratory Comparisons ... 434
5. Geoarchaeology Case Studies ... 434
   5.1. Arid Southwest ... 434
   5.2. Southern Great Plains ... 437
   5.3. East Africa ... 440
   5.4. China ... 443

	5.5. Summary of Case Studies	443
6.	Conclusions	444
	Acknowledgments	444
7.	References	445

Chapter 16

Sourcing Lithic Artifacts by Instrumental Analyses **449**

Norman Herz

1.	Introduction	450
2.	Instrumental Analysis	451
3.	Determining Provenance of Lithic Materials	453
	3.1. Obsidian	453
	3.2. Basalt	455
	3.2. Granitic and Other Felsic Igneous Rocks	455
	3.4. Serpentine and Related Rocks	456
	3.5. Marble	456
	3.6. Sandstone and Quartzite	456
	3.7. Chert and Other Siliceous Sediments	459
	3.8. Carbonates	461
	3.9. Amber	463
4.	Summary and Conclusions	464
5.	References	466

VI. A PREHISTORIAN'S PERSPECTIVE

Chapter 17

A Personal View of Earth Sciences' Contributions to Archaeology . . . **473**

Ofer Bar-Yosef

1.	Opening Remarks	473
2.	Open-Air Sites and Their Environments—Are We Doing What Is Needed?	474
3.	What Do We Expect to Learn from Site Formation Processes in Cave and Rockshelters?	476
4.	Geochronology—Is It Simply a Game of Numbers?	
	4.1. Radiocarbon Chronology	480
	4.2. The Preradiocarbon Techniques	481
5.	Conclusions	484
6.	References	485

Index **489**

Background I

Quaternary Geoscience in Archaeology

VANCE T. HOLLIDAY

1. Introduction

The Quaternary is the most recent period of the geologic time scale, spanning roughly the past 2 million years. It is the coolest period of the Cenozoic era and is characterized by dramatic and repeated cycles of global climatic changes, resulting in, among other things, the repeated growth and decay of glaciers worldwide. Of particular interest in archaeology, the Quaternary spans much of human evolution, including the development of the genus *Homo*, the appearance of fully modern human beings, and the evolution of modern cultures and societies. Quaternary environments and environmental changes provided the backdrop for these processes and for most of human prehistory. Moreover, given the impact of Quaternary climate changes on plant and animal communities, on sea level, on rivers, on lakes, and on all other components of the environment, human prehistory must be inextricably linked to Quaternary environmental changes, although the degree and *nature* of the linkages are hotly debated (e.g., Bell and Walker, 1992; Feibel, 1997; Potts, 1996; Sikes and Wood, 1996; Vrba et al., 1995). Coping with or even (more recently) distancing ourselves from Quaternary environments and environmental changes is arguably a key aspect of human prehistory and evolution. Archaeology is thus one of the Quaternary

---

VANCE T. HOLLIDAY • Department of Geography, University of Wisconsin, Madison, Wisconsin 53706.

Earth Sciences and Archaeology, edited by Paul Goldberg, Vance T. Holliday, and C. Reid Ferring. Kluwer Academic/ Plenum Publishers, New York, 2001.

sciences and, therefore, understanding the record of the Quaternary and some basic principles of the Quaternary geosciences is an important part of archaeology and geoarchaeology in both research and teaching.

The Earth sciences historically have been employed in archaeology along traditional subdisciplinary lines, such as stratigraphy, geomorphology, geochronology, sedimentology, pedology, and geophysics (e.g., Davidson and Shackley, 1976; Herz and Garrison, 1998; Rapp and Gifford, 1985; Rapp and Hill, 1998; Waters, 1992), but Quaternary geologists have been key players in the development of geoarchaeology (Rapp and Hill, 1998, pp. 1–17). Rapp and Hill further argue that

> One can consider geoarchaeology as a component of prehistoric archaeology that, in turn, may be considered a part of geoecology or paleogeography, which is an aspect of Quaternary geology.... These research fields and subdisciplines [are part of]...a broader framework of natural history and natural science focused on the evaluation of the complete Late Cenozoic record. (1998, p. 4)

Quaternary geoscience, similar to geoarchaeology, crosscuts the usual subdivisions of the Earth sciences and indeed goes beyond these subdisciplines to include aspects of the biological sciences, atmospheric sciences, oceanography, and geography. Quaternary geosciences are an important component of the "contextual archaeology" so eloquently and forcefully advocated by Butzer (1982). Studies of the Quaternary period and more typically and explicitly the Pleistocene epoch have long been a component of archaeological research. However, discussions of the significance of Quaternary studies or the basic principles of Quaternary research as a body of knowledge similar to, for example, botany or pedology, are rarely addressed in the archaeological literature, including environmental archaeology and geoarchaeology. A sampling of introductory texts on archaeological method and theory, world prehistory, and North American archaeology published since the mid-1980s (Table 1.1) shows that generally one percent or less of text space is devoted to a discussion of the Pleistocene epoch. The Holocene epoch fares worse, which is startling given the archaeological record of the Holocene, and the Quaternary generally is not mentioned at all (Table 1.1). This tendency is not confined to archaeology texts, however. Most of the few comprehensive volumes on geoarchaeology (Herz and Garrison, 1997; Rapp and Gifford, 1985; Rapp and Hill, 1998; Waters, 1992) have no explicit discussion of Quaternary geoscience, though the topics they do cover would fall under this heading. The notable exceptions are the classic volumes by Butzer (1964, 1971, 1982), which deal directly with concepts of Pleistocene research as a component of geoarchaeology. Some, especially older texts on dating methods in archaeology also included discussions of Pleistocene and Holocene stratigraphy and geochronology (e.g., Oakley, 1964; Wagner, 1998; Zeuner, 1958). Volumes on Quaternary geoscience vary in the amount of attention paid to human evolution and archaeology. Those focusing specifically on Quaternary geology (Bowen, 1978; Dawson, 1992; Ehlers, 1996) pay scant attention to these topics, whereas more comprehensive volumes do (Nilsson, 1983; Williams et al., 1993), though not necessarily in any detail (e.g., Andersen and Borns, 1994; Flint, 1971;).

Table 1.1. Discussion of the Quaternary, Pleistocene, and Holocene in archaeology texts.

		Pages devoted to discussion of the		
Book	Total pages of text	Quaternary	Pleistocene	Holocene
General Archaeology				
Fagan 1997	463	1[a]	6	0
Renfrew & Bahn, 1996	538	0	3	1[b]
Thomas, 1989	575	0	0	0
Hester & Grady 1982	465	0	3	0
World Prehistory/Human Evolution				
Price & Feinman, 1997	492	0	4	0
Fagan, 1992	628	1[a]	10	4
Klein, 1989	427	35[c]	0[c]	0
Wenke, 1984	450	0	1	0
North American Archaeology				
Fiedel, 1992	366	0	0	0
Fagan, 1991	458	0	0	0

[a] Mentioned in one sentence.
[b] Mentioned only in the Glossary.
[c] The discussions deal with stratigraphic units and the geologic time scale (4 pp) and late Cenozoic climates (31 pp); the Quaternary Period and Pleistocene Epoch are mentioned only in passing.

This chapter discusses some principles of Quaternary geoscience that have an impact on the study of and understanding of human prehistory. The discussion includes seemingly mundane topics such as the geologic time scale and stratigraphic nomenclature, which are important to understand for precise and accurate communication, similar to an understanding of basic cultural chronologies and associated terminology in archaeology. The chapter also includes a review of current understanding of the Quaternary climate record as revealed in cores from the ocean floors and ice sheets, the mechanisms that forced the climate changes that characterize the Quaternary period, and contemporary approaches to reconstructing Quaternary environments. These issues are important for providing a context for pursuing archaeological questions and for understanding human prehistory.

2. Definitions and Boundaries

The Quaternary period is a subdivision of the geologic time scale. The ordering or grouping of geologic events, rocks, and sediments into a chronological sequence is a key component of "stratigraphy." A discussion of basic stratigraphic principles is beyond the scope of this chapter, but a few key terms and concepts must be introduced. The organization of rocks and sediment based on their age relationships is referred to as "chronostratigraphy." Grouping deposits on the basis of their lithologic characteristics is called "lithostratigraphy." Organizing

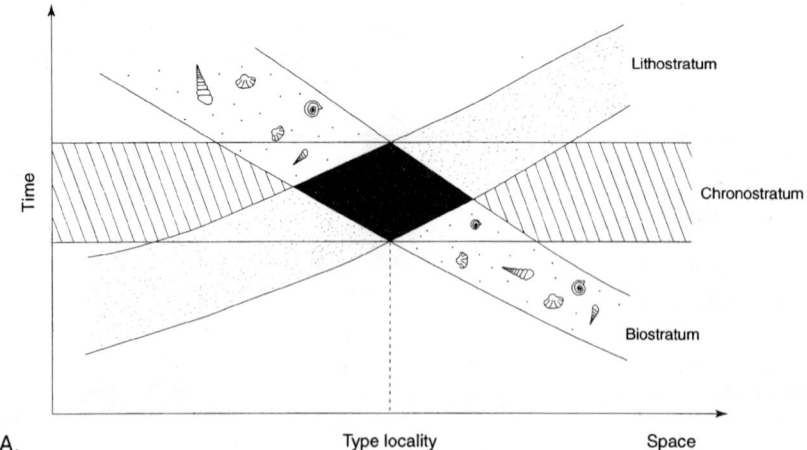

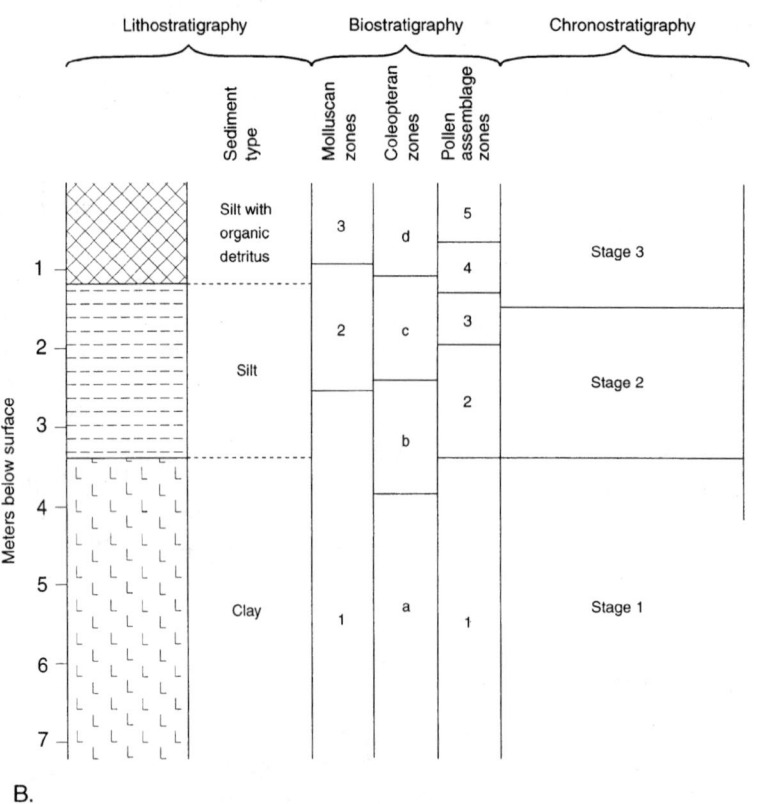

Figure 1.1. Illustrations of the different approaches that can be taken in the grouping of rocks and sediment into stratigraphic units. (A) The three principal kinds of stratigraphic units, illustrating the diachronous nature of boundaries for lithostratigraphic and biostratigraphic units and in comparison to the synchronous boundaries of chronostratigraphic units (modified from Wagner, 1998, Fig. 1). (B) Stratigraphic subdivisions of a sedimentary sequence showing the different positions of stratigraphic boundaries depending on different stratigraphic criteria.

rocks and sediment according to their fossil content is known as "biostratigraphy." Lithostratigraphic units and biostratigraphic units can have diachronous boundaries, but chronostratigraphic units always have synchronous boundaries (Fig. 1.1). Moreover, in a sequence of rocks or sediment, the boundaries of different stratigraphic units may not coincide, that is, the rocks or sediment may be grouped in several different ways depending on the stratigraphic approach taken (Fig. 1.1). The principles of stratigraphy are spelled out by Salvador (1994), and excellent discussions of stratigraphy from the perspective of Quaternary studies are provided by Bowen (1978, pp. 84–104) and by Lowe and Walker (1997, pp. 298–323). Further elaboration of specific components of stratigraphy are presented by Wells, in this volume (Chapter 5).

The Quaternary is the younger of the two periods that comprise the Cenozoic era (Fig. 1.2). The Quaternary traditionally is divided into the Pleistocene and the Holocene epochs, terms that seem to be more widely used in archaeology than the term Quaternary. Beyond these basic relationships, however, definitions of the Quaternary, Pleistocene, and Holocene, and the boundaries of these components of the geologic time scale vary significantly and are quite contentious (e.g., Farrand, 1990; Partridge, 1997b; Van Couvering, 1997a). These issues may seem picayune, but are vital to intra- and interdisciplinary communication and understanding. "Stratigraphic nomenclature, like systematic taxonomy, is a tedious means to interesting ends" (Hopkins, 1975, p. 10).

The key here is that ultimately the terms Quaternary, Pleistocene, and Holocene have their origins in and are still part of a formalized (standard) time scale (Harland et al., 1990; NACOSN, 1983; Salvador, 1994). "A geologic time scale (geochronologic scale) is composed of standard stratigraphic divisions based on rock sequences [a chronostratigraphic scale] and calibrated in years [a chronometric scale]" (Harland et al., 1990:1). Thus the subdivisions of the time scale represent intervals of time and the boundaries for each interval are, by definition, synchronous. As discussed below, a physical representation of the boundaries is established somewhere in the world in a real rock section, but elsewhere boundaries have no necessary physical manifestation. The passage of time marking the Pliocene–Pleistocene boundary or the Pleistocene–Holocene boundary in and of itself left no physical evidence (e.g., Goodyear, 1991, 1993), though clearly the change in environmental conditions of the late Pleistocene to those of the early Holocene can be manifested or documented in a variety of physical systems.

For any standard scale to be useful, the intervals must be universally accepted. "... [S]tandardization is intended to give a convenient and stable [time] scale that will not vary with changing opinion" (Harland et al, 1982:41). For example, a yardstick, a tape measure, or a clock would serve no purpose if we had varying views of what constituted an inch, a yard, a meter, or an hour. Units of weights and measures are standardized by international scientific organizations.This is also the case with the geologic time scale, though its definitions are not so crucial to our daily lives, and it has evolved and will continue to do so as we learn more about the geologic past. Moreover, the definitions and terminology of the geologic time scale are based to some extent on pragmatic considerations and acceptance by convention. "The chronostratic scale is a

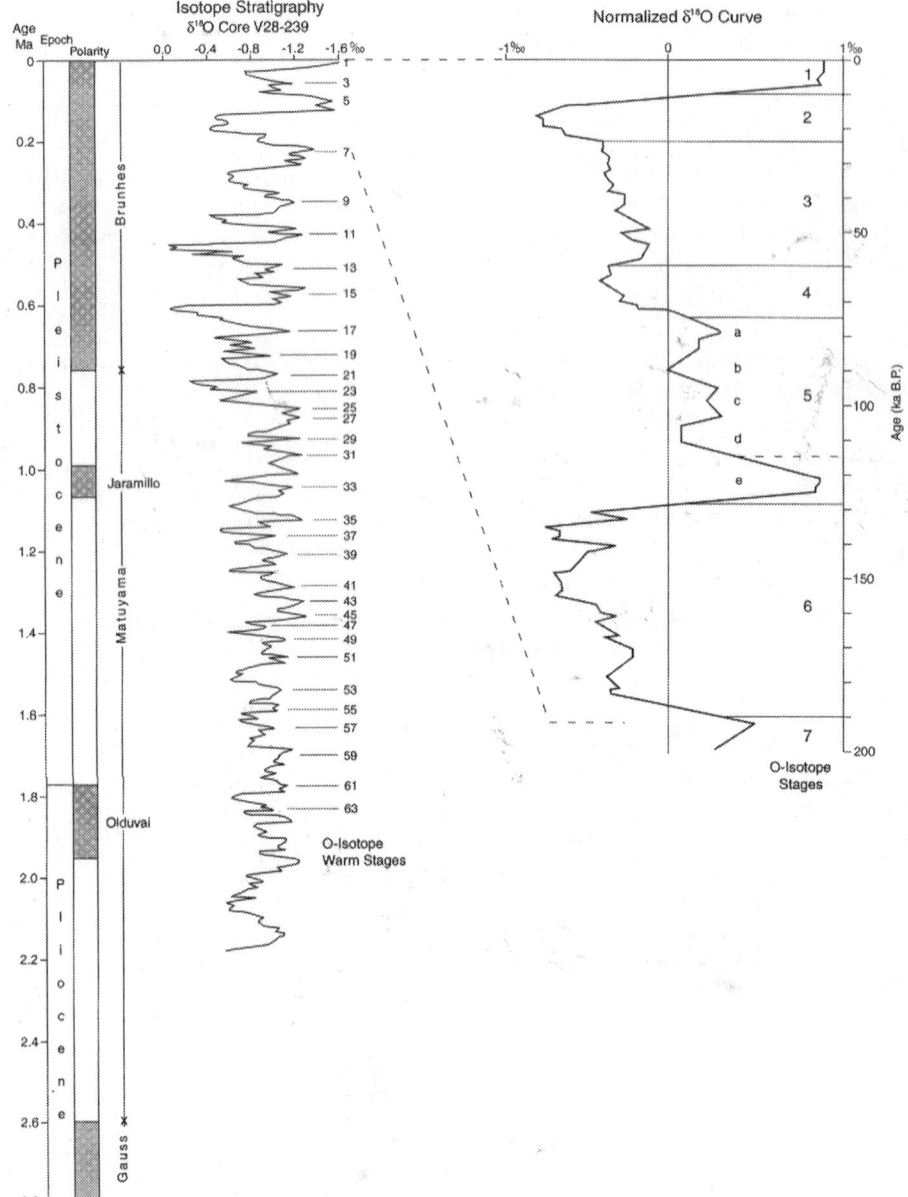

Figure 1.2. Synthesis of Late Cenozoic time stratigraphy, magnetic polarity, and oxygen isotope ($\delta^{18}O$) stratigraphy. The O-isotope diagram at right shows the details of the late Quaternary record. For both O-isotope curves increasing ice volume is to the left. For the long O-isotope record only warm (interglacial) stages are numbered. In the long isotope record note the change in frequency of glacial–interglacial cycles from a period of ∼41,000 years to ∼100,000 years at ∼0.9 my. Dating of magnetic stratigraphy follows Shackleton et al. (1990). The long record of O-isotope stratigraphy is based on Lowe and Walker (1997). The late Quaternary oxygen–isotope stratigraphy at right is after Martinson et al. (1987).

convention to be agreed rather than discovered, while its calibration in years is a matter for discovery or estimation rather than agreement" (Harland et al., 1990:1). The nomenclature of the time scale "provides a convenient set of labels for slices of time whose artificially selected boundaries are inadequately known" (Hopkins, 1975, p. 10). Ultimately, all stratigraphic subdivisions are to some extent arbitrary and were determined in part by the locations of early geologic field work. The basic terminology of the geologic time scale is now generally accepted and standardized, though for the Quaternary period some terminological issues remain (discussed in the following paragraph). Otherwise, the points of contention generally focus on deciding on and dating the boundaries.

At its most basic level, the Quaternary period is a time of relatively cool global climate at the culmination of long-term cooling that characterizes the Cenozoic era. Within this period of relative cooling were dramatic and repeated cycles of climate change, typified by the growth and decay of glaciers. Hence, the Quaternary is often called the "Ice Age." The cycles of glacial expansion and contraction are referred to as "glacial" and "interglacial" stages with superimposed "stadials" and "interstadials" (Lowe and Walker, 1997:8). A glacial stage was a prolonged period of cooling when major expansion of ice sheets and glaciers took place. A stadial, in contrast, was a shorter cold episode when local ice advance occurred. An interglacial stage was a warm interval when temperatures reached or exceeded those of the present time and when ice coverage of the Earth's surface was minimal. An interstadial was a relatively short-lived period of warming within a glacial stage, but temperatures did not reach the levels of today (see also the discussion of Bond cycles that follows).

The Pleistocene epoch comprises all of the glacial–interglacial cycles of the Quaternary except for the current interglacial called the Holocene epoch. Thus, the Pleistocene also has been referred to as the "Ice Age." The Holocene, significantly, is not unique geologically or environmentally relative to any other interglacial stage and, therefore, is an arbitrary subdivision of the Quaternary period. As Sherratt (1997) correctly observes, "Over 90% of [the Quaternary] period has been cooler and dryer than the Holocene, so contemporary conditions are unrepresentative. Archaeologically, though, the Holocene is the time during which food production and complex societies appeared. Archaeologically and anthropologically, therefore, the Holocene is significant.

On a conceptual level, differentiating the Holocene epoch from the Pleistocene epoch has generated considerable discussion (Farrand, 1990; discussed in the following text), but these terms are perhaps more problematic at the level of everyday usage. An unfortunate tendency in archaeology as well as in Quaternary studies is to look on "The Pleistocene" as something distinctly different from "The Holocene" (Fagan, 1997:112-115; Renfrew and Bahn, 1996:120; Wenke, 1984:52). It is not. The Holocene is simply the latest in a long series of interglacial stages (discussed further below). Some investigators also refer to the "Pleistocene–Holocene transition" and infer the passage of some time (several thousand years) (e.g., Walthall, 1998). Certainly the environmental conditions at the end of the Pleistocene (e.g., at the end of the last glacial period) evolved over several thousand years into Holocene (e.g., postglacial environments), but the

Pleistocene–Holocene "transition" is a boundary on the geologic time scale and as such represents a moment in geologic time.

So how are Quaternary, Pleistocene, and Holocene defined? As noted, the most common informal definition of the Pleistocene and Quaternary is "The Ice Age." Another view was that this was "The Age of Man" or "Age of Humanity," referring to the idea that the first appearance of humans coincided with the beginning of the Pleistocene (Farrand, 1990). Vertebrate and invertebrate faunal changes were formally proposed as defining criteria, following long-standing geologic tradition. There are substantial problems with most of these definitions, however. The first hominids preceded the onset of the Pleistocene epoch by 1 to 2 million years (regardless of one's view on dating the beginning of the Pleistocene, discussed in the following text). Vertebrate faunas were evolving throughout this time, and there is no globally significant, clearcut change in faunal assemblages. Scattered, small, high-latitude and high-altitude glaciers appeared in the late Cenozoic long before more substantial lower latitude and lower altitude glaciers. Likewise, environmental cooling was a continuous process through the late Cenozoic. As summarized by Farrand (1990:19) "From the time of Desnoyers and Lyell [both 19th-century geologists who helped 'construct' the geologic time scale] to the present, it has been noted that the passage from the Tertiary to the Quaternary, or from the Pliocene to the Pleistocene, was generally one of transition, not of abrupt change."

Attempts at standardizing the geologic time scale began in the 19th century, but not until after the Second World War did standardization become a reality (Harland et al., 1990, pp. 2–3). This work was and continues to be under the auspices of the International Geological Congress (IGC) and the International Union of Geological Sciences (IUGS), though ultimately the adoption of terms, definitions, and dates is by agreed convention among workers in the field. Subdivisions of the time scale are standardized at their boundaries only. The conceptual basis for each boundary is a marine biostratigraphic sequence. This definition goes back to the roots of the geologic time scale. Lyell (1833, 1839), for example, subdivided the Tertiary period on the basis of percentage of living Mollusca represented in fossil assemblages. For the Pleistocene (initially viewed as part of the Tertiary) Lyell (1839) proposed that this term be applied to deposits having more than 70% modern molluscs (see discussion in Farrand, 1990:17). Significantly, this most enduring of definitions was proposed before and independently of the recognition that the Pleistocene epoch was a time of glaciation. Under the new procedures, a boundary is agreed upon based on the approximate position in a biostratigraphic sequence that best fits existing usage of the particular subdivisions of the time scale (Harland et al., 1990:3).

Once the general biostratigraphic concept of the boundary is agreed on, a stratigraphic section somewhere in the world is selected to be the physical representation of the boundary, called a "boundary stratotype." The boundary is a unique rock section in which a time horizon is defined as a universal standard of reference for dating purposes. Exposed marine sections are usually used because marine invertebrate fauna or faunal assemblages can change rapidly, they can be correlated over wide regions, and because marine sections tend to be more continuous (i.e., have fewer time gaps) than terrestrial sections. Beyond its representativeness, the boundary stratotype also is selected on the basis of its completeness and access (Harland et al., 1990:3).

2.1. The Pliocene–Pleistocene Boundary

The Pliocene–Pleistocene boundary was one of the first boundaries with which the IGC attempted to deal (in 1948). It has proven to be a tough boundary to establish and date (see the detailed discussion in Nikiforova and Alekseev, 1997, and also Harland et al., 1990:68, from which the following is distilled; and Mauz, 1998). Several decades passed following the initial 1948 discussions before a boundary and stratotype were proposed. The first section considered (in 1972) was Le Castella, Italy (first proposed by Grignoux, 1913), but problems with the section soon appeared and in 1977 a section at Vrica, Italy was agreed on. The Vrica section, fully discussed and described in Van Couvering (1997b), still stands as the boundary stratotype for the Pliocene–Pleistocene boundary. Most investigators agreed to define the Pliocene–Pleistocene boundary within the section on the basis of the appearance of certain cold-water marine fauna in the Mediterranean. The idea is that such fauna would provide clear evidence of significant global cooling, including low-latitude and otherwise warm-water settings.

Historically, chronometric estimates for the age of the base of the Pleistocene range from 0.6 to 4 Ma (Haq et al., 1977). Establishment of a boundary stratotype has focused the issue but hasn't necessarily settled it. Within the Vrica section, the boundary was placed at the top of the Olduvai normal polarity zone (Fig. 1.2), dated to 1.64 Ma (Aguirre and Pasini, 1985; Tauxe et al., 1983). The age of 1.6 Ma was widely adopted; it was incorporated, for example, into the standard time scale of the Geological Society of America (Palmer, 1983) and the U.S. Geological Survey (Hansen, 1991, Fig. 15). Reinvestigation of the section and some revision of the geomagnetic polarity time scale resulted in a revised age of 1.796 (1.8) Ma for the boundary (Fig. 1.2; Pasini and Colalongo, 1997).

Selection of the Vrica section as a Pleistocene boundary stratotype and placement of the boundary has never had universal acceptance (Partridge, 1997b). The issues are several (well summarized by Mauz, 1998; see also the debate between Morrison and Kukla, 1998, and Aubry et al., 1998). Some believe that the definition of the base of the Pleistocene should be expressly climatic and should take into account evidence of significant global cooling between 3.0 and 2.0 Ma (Partridge, 1997a), whereas others question the interpretation of the section (e.g., Jenkins, 1987). Several investigators now argue that the boundary should be placed at the Gauss–Matuyama paleomagnetic reversal (Fig. 1.2), that is, moved back to 2.6 Ma (formerly dated at 2.4 Ma e.g., Ding et al., 1997; Ehlers, 1996:3; Morrison and Kukla, 1998; Partridge, 1997a; Suc et al., 1997). The arguments are based on evidence for significant cooling around the G–M boundary in both marine and terrestrial (loess) records along with the ease of correlation of the G–M boundary.

The issue is far from resolved, but most earth scientists, at least in the United States, seem to follow the recommendation to use 1.8 Ma (formerly 1.6 Ma; e.g.,

Berggren et al., 1995a,b; Harland et al., 1990; Woodburne and Swisher, 1995). In any event, investigators working with late Pliocene and early Pleistocene localities should explicitly state their definitions of the boundary.

2.2. The Pleistocene–Holocene Boundary

The Pleistocene–Holocene boundary is one of the most difficult boundaries to deal with because: (1) we know so much about the last glacial–postglacial transition; (2) the nature of the transition varied significantly from region to region, and among different components of the environment (e.g., climate, flora, fauna); and (3) dating resolution is much finer than the length of the transition. Over the years, arguments for the age of the boundary ranged from 20,000 to 4,000 yrs BP (Hopkins, 1975; Morrison, 1969). Indeed, some argue that the Pleistocene–Holocene boundary should be formally recognized as diachronous depending on local records (Watson and Wright, 1980), and others propose that the term should be abandoned altogether because the Holocene simply represents the current interglacial (Flint, 1971:384). These arguments have merit, but neither is likely to be accepted. The general concept of the Holocene epoch and the usage of the term are too well established and we have such a vast amount of data for it compared to any other subdivision of the Quaternary period. The concept of the Holocene epoch has proven especially useful in archaeology because the Pleistocene–Holocene transition was a time of significant changes in the archaeological record, followed soon after by the development of complex societies (e.g., Eriksen and Straus, 1998; Straus et al., 1996). A diachronous boundary or at least the provision for differing boundary criteria from region to region (e.g., Haynes, 1991; Watson and Wright, 1980) is contrary to the concept of chronostratigraphy and a standard time scale. This would be akin to allowing different places to have their own definition of a day because day length varies latitudinally and seasonally.

In 1969 the International Congress for Quaternary Research (INQUA) proposed that the Pleistocene–Holocene boundary be placed at 10,000 yrs BP and that a suitable boundary stratotype be searched for in a fossiliferous marine section in Sweden (Hageman, 1972; Harland et al., 1990:71, sec. 3.21.5). The proposed date represents a good midpoint between full glacial conditions of the terminal Pleistocene and maximum warming of the early Holocene, and, as Hopkins (1975) observed, "we place the onset of the Holocene Epoch at 10,000 B.P. simply because that's a nice round number' (p. 10). The age of 10,000 yr BP for the boundary seems to be gaining general acceptance based on its wide adoption and usage (e.g., Hansen, 1991; Harland et al., 1990, Fig. 15; Palmer, 1983; Roberts, 1989; Woodburne and Swisher, 1995). Perhaps the most noteworthy aspect of the INQUA proposal is that this is the only instance of a boundary stratotype being proposed on the basis of an absolute age and for a suitable section to be found that best documents the date marking the full-glacial–postglacial transition.

2.3. Stratigraphic Subdivisions

The most common subdivision of the Pleistocene time scale is a tripartite ("early," 'middle," and "late") scheme (Fig.1.2). The early–middle Pleistocene boundary is placed at the Brunhes–Matuyama polarity reversal, 788 ka (Fig. 1.2) (Harland et al., 1990:68, sec. 3.21.2). The middle–late Pleistocene boundary is placed at the beginning of marine oxygen isotope stage 5e (Fig.1.2; see the following section) Harland et al., 1990:68–69, sec. 3.21.2), which represents the beginning of the last interglacial period before the Holocene, dated to ca. 125 ka (following Winograd et al. 1997, though interpretations vary on the dating of this boundary). A formal nomenclature for a tripartite subdivision of the Quaternary has not been proposed but seems appropriate given, for example, the wide use of the term "Late Quaternary" (e.g., Bell and Walker, 1992; Bryant and Holloway, 1985; Dawson, 1992; Wright, 1983). The same standards that apply to the Pleistocene probably can apply to the Quaternary.

An important distinction here is the difference between "early," "middle," and "late" and "lower," "middle," and "upper." The former are geologic time terms, used to refer to subdivisions of real time (e.g., the genus *Homo* appeared in early Pleistocene time). In contrast, the latter trio of terms refer to subdivisions of the chronostratigraphic record itself, that is, a real rock sequence (lithostratigraphy) on which time subdivisions are based (e.g., the bones of *Homo* are found in lower Pleistocene rocks).

3. Glacial–Interglacial Cycles

Since the beginning of the 20th century, geologists have recognized that the Pleistocene epoch was characterized by a series of dramatic environmental changes, most famously represented by the advance and retreat of glaciers. For much of the 20th century, the evidence for these environmental cycles was used to subdivide the Pleistocene. The best known subdivisions were the glacial–interglacial stages of the Alps (Günz-Mindel-Riss-Würm, oldest to youngest) and the Midwestern United States (Nebraskan-Kansan-Illinoian-Wisconsin, oldest to youngest). These two glacial stratigraphic sequences in particular, along with several others from Europe, came to dominate much of the thinking about the Quaternary record worldwide such that most Pleistocene deposits, soils, and landforms were usually somehow "fit" into or otherwise correlated with these schemes. This situation is well described by Bowen (1978:10–56) and by Ehlers (1996:323–325, 355–357), and is illustrated in Wright and Frey (1965) and in Flint (1971).

With the expansion of academic and governmental research into the Quaternary following the Second World War, and with the development of numerical dating methods, many investigators began to realize that the simple fourfold stratigraphic sequences for the Midwest and for Europe were grossly oversimplified. For example, studies of glacial stratigraphy and intercalated volcanic ashes on the Great Plains and the Midwestern United States clearly showed that there were more than four glacial–interglacial cycles represented and that the classic terminology for all pre-Illinoian deposits should be abandoned (Boellstorff, 1978; Hallberg, 1980a,b). In eastern Europe, a classic investigation of loess stratigraphy showed that there were perhaps 17 glacial–interglacial cycles in the past 1.7 million years (Fink and Kukla, 1977; Kukla, 1975, 1977).

The most far-reaching advances in understanding and classifying the climate cycles of the Quaternary, however, came from studies of sediments on the floors of the ocean and from ice locked in glaciers. The basic assumption and condition that allowed the deep ocean research is that parts of the ocean basins have been the sites of essentially continuous sedimentation during much or all of the Quaternary (and earlier) in contrast to terrestrial settings with stratigraphically incomplete and regionally discontinuous records. On glaciers, very high elevations have extremely low rates of ablation so that snow accumulation has been essentially continuous. Among other contributions, the deep sea work demonstrated (and continues to show) the lengthy, cyclic, and complex nature of Quaternary environmental changes, and it also provided an exceptional standard scale for comparison and correlation of all other Quaternary stratigraphic records. The glacial ice, though not spanning as much time as the ocean sediments, provides a high resolution record of precipitation amounts, air temperature, atmospheric composition, and explosive volcanic activity, among other kinds of proxy indicators.

Beginning in the 1950s, cores recovered from the ocean basins have been analyzed for their isotopic, paleontological, and magnetic characteristics. These studies revolutionized thinking about Quaternary climates. For the purposes of this discussion the most pertinent issue is the variation in composition of oxygen isotopes in marine fossils. At scales of tens to hundreds of thousands of years, the ratio of ^{18}O to ^{16}O varies significantly (Fig. 1.2). Initially, this variation (expressed as $\delta^{18}O$) was believed indicative of fluctuations in global temperatures (Emiliani, 1955; Epstein et al., 1953; Urey, 1947), but subsequent studies indicated that most of the changes were due to changes in the global extent of glaciers (e.g., Shackleton, 1967). Regardless, the $\delta^{18}O$ curves illustrate worldwide environmental changes throughout the Quaternary. The usual approach taken for interpreting the cores is as an indicator of ice volume (these issues are well outlined and explained by Bowen, 1978:61–69, and by Bradley, 1985:178–189). Because it is the lighter of the two isotopes, ^{16}O is preferentially removed from the oceans by evaporation. Water or ice subsequently precipitated will be enriched in ^{16}O. As glaciers build up, therefore, they become enriched in ^{16}O, while ocean water becomes depleted in ^{16}O relative to ^{18}O. When glaciers melt at the onset of an interglacial or interstadial cycle, the ^{16}O-enriched waters return to the oceans, altering the $^{18}O/^{16}O$ ratio.

Studies of hundreds of cores from across all the world's ocean basins show that these curves can be correlated and are synchronous globally, thus providing a worldwide standard for comparison of Quaternary environmental changes (Bradley, 1985:178–189). "It is highly unlikely that any superior stratigraphic subdivision of the Pleistocene will ever emerge" (Shackleton and Opdyke,

1973:48). Indeed, this seems to be the case and the curves are widely used as a standard in correlation of both marine and terrestrial stratigraphic records. The fluctuations in the oxygen isotope curves were assigned numbered "Stages" beginning at the top (most recent, Fig. 1.2). Periods of increased ice volume (glacial periods) have even numbers and periods of decreased ice volume (interglacial periods) have odd numbers (with the exception of Stage 3, which is an interstadial period). Boundaries between stages are at the midpoints between maximum and minimum values (Fig. 1.2). The Holocene epoch, the present interglacial, correlates to oxygen–isotope Stage 1, for example. The last full glacial period (e.g., the late Wisconsinan in the midwest United States, the late Wurmian in the Alps), is Stage 2, the penultimate glaciation (the Illinoian in the Midwest) is Stage 6, and so on.

A glance at the oxygen–isotope curve clearly shows that there were many more than four glacial–interglacial cycles (Fig. 1.2). The cycles have a period of about 100,000 years for the past 1 million years, and have a period of 41,000 years prior to that back to about 2.5 million years (Ruddiman et al., 1986; Shackleton et al., 1990), yielding strong evidence for approximately 30 glacial–interglacial cycles during the Quaternary period.

In recent decades, and especially in the 1990s, cores from glaciers (mostly but not exclusively from Greenland and from Antarctica) have begun to provide an unprecedented record of a variety of characteristics of the atmosphere spanning much (and in some cases most) of the late Quaternary (see Bradley, 1999:125–190, for an excellent summary of ice core studies). The ice core studies brought about yet another revolution in our understanding of Quaternary paleoclimatology "by providing high resolution records of many different parameters, recorded simultaneously at each location" (Bradley, 1999:153). Ice cores are not as numerous as ocean cores and ice cores do not provide the time depth of deep-sea cores. Less than two dozen ice cores go back to the last glaciation; a few extend back to the penultimate glaciation (oxygen isotopes Stage 6 Bradley, 1999:126, Table 5.2). However, the layers of ice in the ice cores tend not to be subjected to the mixing that characterizes the deep ocean cores, and thus some provide an annual record of Holocene and even late Pleistocene atmospheric variation (e.g., Thompson et al., 1985). In addition, a wider variety of dating techniques can be applied to ice cores.

The cores from glaciers have yielded a number of significant results. For example, levels of atmospheric dust (which can affect the global energy balance) were much higher during glacial periods, and the composition of gases in the atmosphere (which can influence radiation) changed over glacial–interglacial cycles (e.g., Mayewski et al., 1993). The cores also provide a record of explosive volcanic eruptions that can influence climate. Perhaps the most significant result to come out of the ice core research, however (certainly from an archaeological viewpoint), is evidence that Holocene climates were relatively stable whereas climates during cold (glacial) stages fluctuated rapidly between two or more modes (e.g., Adams et al., 1999; Taylor et al., 1993). For example, Dansgaard et al. (1993) identified 24 interstadial episodes between 12,000 and 110,000 yrs BP. This issue is discussed further in the following section.

4. Causes of Quaternary Climate Cycles

The question of what caused Quaternary climate changes has been with us for as long as the concepts of the Quaternary and Pleistocene have. Until the 1970s, no satisfactory, generally acceptable theories were available, and the issue remained a fundamental problem in Quaternary research (Flint, 1971). Flint (1971:788–809) summarizes the theories prevalent at the beginning of the 1970s. Most broadly they fall into six categories: (1) variation in solar emissivity, (2) veils of cosmic dust, (3) geometric variations in the Earth's orbital parameters, (4) variations in the optical depth (transmissivity and absorptivity) of the Earth's atmosphere (due, for example, to volcanic eruptions), (5) lateral and vertical movements in the Earth's crust, and (6) changes in the system of ocean/atmosphere circulation. At the time the biggest obstacle to sorting out and evaluating these ideas was lack of data. That situation began to change dramatically shortly after Flint's (1971) volume was published, and more and more data have become available at an ever-increasing rate.

Several of the processes listed previously appear to play roles in changing the earth's climate, but at different time scales. Cooling of the earth's surface to the point where glaciers can exist and grow probably is linked to tectonic movement of continental plates to high (northern) latitudes (late Mesozoic to today), the opening of the North and South Atlantic and the connection of the North Atlantic to the Arctic Ocean, the opening of the Bering Strait (late Mesozoic to today), the closing of the Isthmus of Panama (connecting North and South America and significantly affecting ocean circulation in the late Tertiary), and tectonic uplift of mountain ranges, especially the Tibetan Plateau (late Tertiary; summarized by Bell and Walker, 1992:65–66; Ehlers, 1996:7–8; Maslin et al, 1998; Partridge et al., 1995; Williams et al, 1993:15–25). These processes do not account for the cyclicity of climate change within the Quaternary, however.

The key to the "pacemaker of the Ice Ages" came with the discoveries of the long and complex record of climate changes and ice volume combined with an old idea that climate cycles were due to variations in the geometry of the Earth's orbit around the Sun (a fascinating detective story well told by Imbrie and Imbrie, 1979). The basic idea of the "astronomic theory" was proposed in 1842 by James Croll, but the person who provided the most detailed and elaborate calculations and whose name is most closely associated with it is Milutin Milankovitch, who worked on the idea from 1910 to 1940 (Imbrie and Imbrie, 1979). His ideas generated considerable interest, but few data were available to support them. By the mid-1970s, however, a variety of new dating methods were applied to the deep-sea cores and the global significance of the deep-sea record became clear, whereas the loess studies in central Europe showed that the terrestrial glacial record was comparable to and correlated with the ocean record. The results of this research created a revolution in thinking about Quaternary history (Imbrie and Imbrie, 1979). The impact on the Quaternary sciences was comparable to the impact of plate tectonic theory in geology or of Darwinian evolution in biology.

Changes in the earth's orbit around the Sun affect the seasonal and hemispheric distribution of the insolation, which is the radiation influx or the amount of solar radiation reaching the earth's surface (Fig. 1.3). These changes appear to influence climate though the linkages remain unclear. Three orbital factors are involved (well described and discussed by Bell and Walker, 1992; Dawson, 1992; Ehlers, 1996; Imbrie and Imbrie, 1979; and Lowe and Walker, 1997; among others); (Fig. 1.3). *Eccentricity* refers to elongation in the elliptical path of the Earth's orbit around the Sun. Variations in the shape of the path change in 100,000-year cycles. *Obliquity* refers to the tilt of the Earth's axis of rotation relative to its orbital plane. Today the tilt is 23.5° but it varies between 24.5° and 22.0° in 41,000-year periods. Increasing tilt increases the amount of solar radiation reaching the poles in the summer and increases seasonal contrasts. *Precession* refers to the "wobble" of the Earth's axis as it spins. This wobble describes a circle with a period of 19,000 to 23,000 years. As noted previously, the 100,000-year cycles seem to have dominated glacial–interglacial cycles in the later half of the Quaternary, and 41,000-year cycles dominated the first half (Fig. 1.3). Variations in insolation due to changes in eccentricity appear weak so why the 100,000-year period is so strong remains a question as does the reason why the cyclicity changed from 41,000 years to 100,000 years (see Raymo, 1998, and Bradley, 1999:35–46 for further discussion).

Other factors such as volcanic emissions, ice sheet instability, and characteristics of ocean circulation likely play a role in changing the climate, particularly at millennial time scales. Volcanic eruptions and the injection of aerosols into the atmosphere are known to have dramatic short-term effects on climate, and some investigators have suggested that fluctuations in climate at scales shorter than Milankovitch cycles may be the result of increases in volcanic eruptions (Hammer et al., 1981; Lamb, 1982; Porter, 1986). Others have suggested that volcanic eruptions can modulate the effects of Milankovitch forcing (Bryson, 1989; Rampino and Self, 1993) and that rapid climatic changes can force volcanism (McGuire et al., 1997; Zielinski et al., 1996).

Evidence for cyclic instability of ice sheets at time scales shorter than the Milankovitch periodicities recently has been recognized in cores from the Greenland ice sheet and from North Atlantic sediments (summarized by Bradley, 1999; Broecker, 1995, Lowe and Walker, 1997, and). The climate of the last glaciation was composed of a series of cooling cycles of 10,000 to 15,000-year duration (Bond cycles; Fig. 1.4) (Adams et al., 1999; Bond et al., 1993). The Bond cycles were composed of a series of cold stadials, culminating in the coldest stadial of approximately 1000 years duration (Fig. 1.4), associated with the massive release of icebergs into the North Atlantic (Heinrich events; Adams et al., 1999; Andrews, 1998; Bond and Lotti, 1995; Broecker et al., 1992; Heinrich, 1988). The Bond cycles included periods of relatively warmer climatic conditions (Dansgaard–Oeschger events) lasting 2000 to 3000 years (Fig. 1.4; Adams et al., 1999; Dansgaard et al., 1993; Johnsen et al., 1992). Following the Heinrich event that culminated a Bond cycle was an abrupt shift to a warmer climate (an interstadial; Bond and Lotti, 1995; Bond et al. 1993) and then gradual cooling that marked the beginning of the next Bond cycle (Adams et al., 1999). The causes of these various cycles are unclear; they may have been due to internal

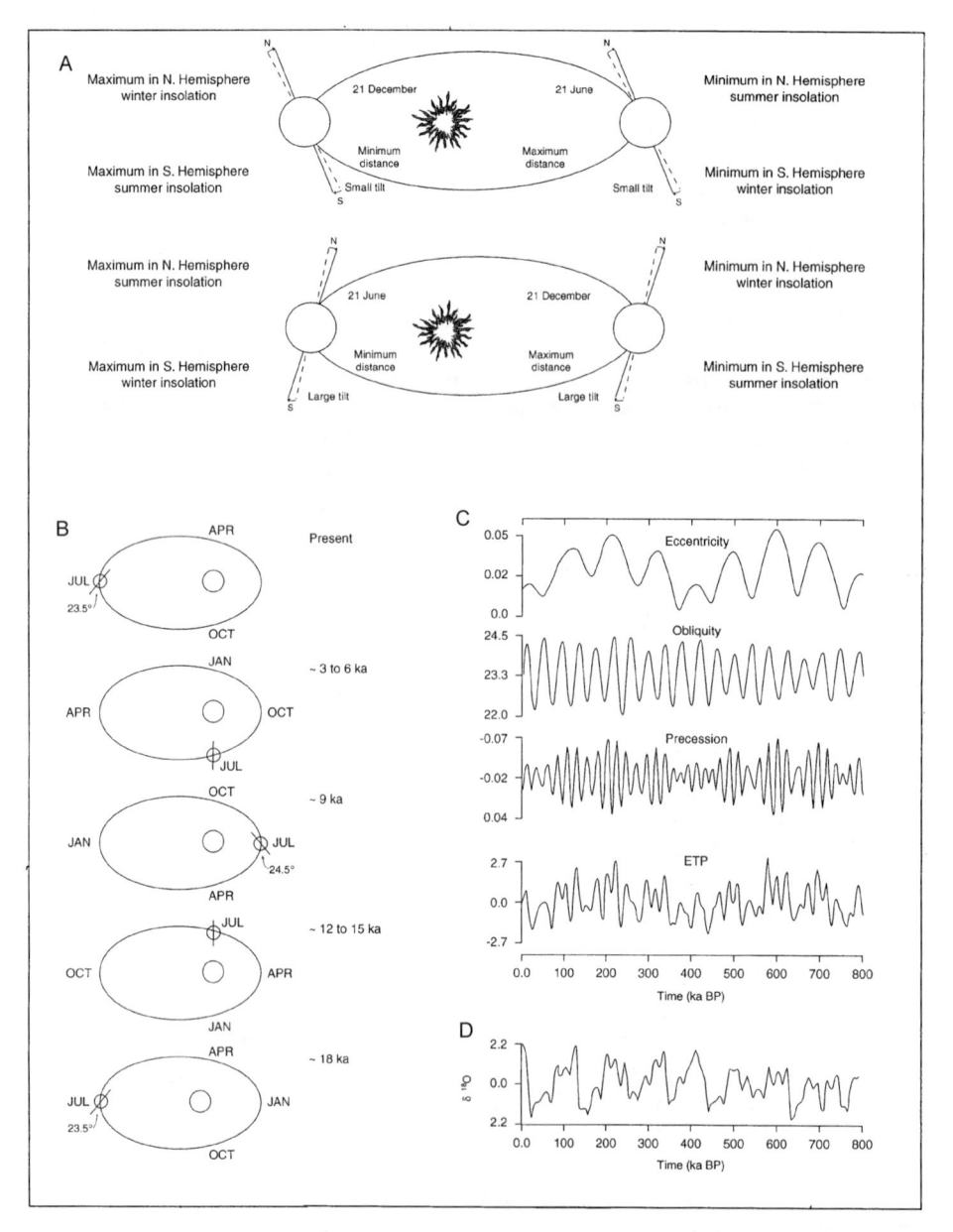

Figure 1.3. Key components of the variations in Earth's orbit around the Sun (Milankovitch cycles) and their effect on Quaternary climate. (A) Schematic representation of Earth's orbital elements (modified from Ruddiman and Wright, 1987, Fig. 4). (B) Earth–Sun geometry for ca. 18,000, ca. 15,000–12,000, ca. 9000, and ca. 6000–3000 yrs BP, and the present. Perihelion (minimum distance from the sun) is in January at present and in July ca. 9000 yrs BP, and tilt is greater at ca. 9000 yrs BP than at ca. 18,000 yrs BP and at present (modified from Kutzbach and Webb, 1993, Fig. 2.1). (C) Variations in eccentricity, obliquity, and the precessional index over the past 800,000 years. ETP is a composite curve constructed by normalizing and adding the three time series. The scale for obliquity is in degrees and for ETP in standard deviation units. (D) The $\delta^{18}O$ curve showing the similarity between it and the ETP curve in (C, C and D modified from Lowe and Walker, 1997, Fig. 1.8).

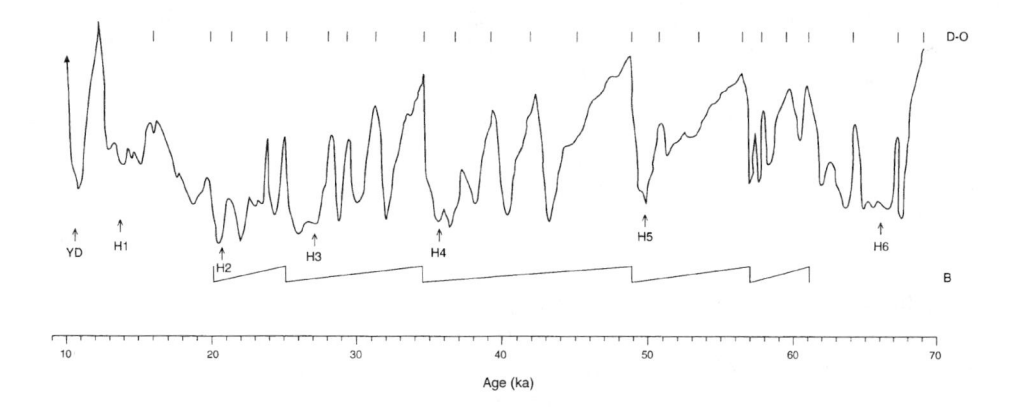

Figure 1.4. Generalized diagram of temperature fluctuations (warm is up) during the last full glacial in the North Atlantic illustrating the relationship of Bond cycles (B), Dansgaard-Oeschger events (D–O), Heinrich events (H1–H6), and the Younder Dryas (YD, modified from Lowe and Walker, 1997, Fig. 7.13).

oscillations of the ice sheets to external climatic forcing, or more likely, to a complex combination of factors (Adams et al., 1999; Andrews, 1998; Berger, 1990; Bradley, 1999:261–268). The Heinrich events, however, probably resulted from ice sheet instability and collapse. The effects of these cycles beyond the Greenland ice sheet and the North Atlantic are unclear, but at least some of the cycles may be globally significant (e.g., Adams et al., 1999; Allen and Anderson, 1993; Bradley, 1999; Lowe et al, 1994; Lowell et al., 1995;).

Regardless of the causes of these different cycles, the apparently rapid changes in environmental conditions have significant archaeological implications because the rate of change may have approached the scale of human life spans (Adams et al., 1999; Dansgaard et al., 1993; Johnsen et al., 1992). This would mean that prehistoric populations would have to adapt to rapid and dramatic climate shifts. The best known of the rapid cooling cycles is the "Younger Dryas" dated to 11,000 to 10,000 BP, similar to and sometimes characterized as a Heinrich event (Broecker, 1992, 1995; Keigwin and Jones, 1995). A variety of paleoenvironmental and even archaeological events dated to this period from around the world have been correlated to the Younger Dryas (Anderson, 1997; Haynes, 1991; Sherratt, 1997).

Changes in thermohaline circulation of the oceans, the movement of ocean water masses due to temperature and salinity (density) gradients, has been recognized in the 1990s as being linked to late glacial climate changes (e.g., Bond et al., 1993; Boyle, 1995; Broecker, 1991). Proposed causes of the changes include both discharge of massive amounts of glacial meltwater into the oceans (Broecker et al., 1989, 1990) and the periodic release of iceberg "armadas" associated with Heinrich events (Baumann et al., 1995; Broecker, 1994). The linkages between climate, ocean circulation, and ice sheets, however, are probably much more complex than these proposals suggest (e.g., Adams et al., 1999; Anderson, 1997; Bradley, 1999:260–275; Lowe and Walker, 1997:362–365).

These processes do not, however, account for abrupt climate changes and possible changes in ocean circulation during interglacial periods (Stager and Mayewski, 1997).

5. Reconstructing Quaternary Environments: Data versus Models

Traditionally the most direct bearing of Quaternary studies on archaeology probably is the application of methods for reconstructing paleoenvironments. The methods for such reconstructions fall into two broad categories: proxy records and computer simulation models. These approaches provide distinctly different kinds of information, distinctions that are not always fully appreciated by field investigators. Each category also has its advantages and disadvantages.

Proxy records are organic and inorganic remains that provide an indirect indicator of past environments (Bell and Walker, 1992:11). This approach relies on the principal of uniformitarianism, which states that our understanding of geological and biological processes operating today can be used to reconstruct the past (Rymer, 1978). Proxy indicators are derived from glaciological (ice core), geological, biological, and historical records (Table 1.2; Bell and Walker, 1992; Bradley 1985, 1999; Lowe and Walker, 1997). Plant and animal remains are used to reconstruct floral and faunal communities that in turn provide clues to environmental and even climatic conditions. Inorganic remains such as sediments, soils, and stable isotopes can provide information on conditions of formation (e.g., depositional environments and weathering conditions). These methods have been used in archaeology and in other Quaternary sciences for environmental reconstructions for decades. For example, the study of vertebrate remains has been a component of archaeological research since the middle of the 19th century (Grayson, 1983), and palynology was incorporated into archaeology since the inception of pollen studies earlier in the 20th century (Bryant and Holloway, 1983). Indeed, archaeological research has been an important driving force in the development of some paleoenvironmental methods such as phytolith analysis (Piperno, 1988: p. 1–10).

Computer simulation climate models are, at their most basic level, mathematical computations of climate conditions given a set of specified and calculated parameters (see Bradley, 1999, pp. 471–505, for an excellent summary of paleoclimate models). The best known models are General Circulation Models (GCMs), which attempt to simulate the three-dimensional structure and flow of the atmosphere (Kutzbach, 1985; Street-Perrott, 1991). Their development resulted from the availability of fast and powerful computers. In the models, Earth's surface is represented by a grid, usually 4 to 10°. For each grid cell, a number of values can be prescribed or calculated at various levels from the surface up into the atmosphere (usually 10–20 levels, with more near the surface).

One of the better known applications of a GCM for reconstructing Quaternary climates is COHMAP (the Cooperative Holocene Mapping Project; COHMAP, 1988; Webb, 1998; Wright et al., 1993). This project modeled global

Table 1.2. Sources of Proxy Data for Paleoclimate Reconstructions[a]

Glaciological (Ice Cores)
    Oxygen and hydrogen isotopes, major ions
    Gas content in air bubbles
    Trace elements and microparticle concentrations (e.g., dust)
    Physical characteristics

Geological
  Marine (ocean sediment cores)
    Microfossils
    Oxygen isotopes (of Foraminifera)
    Sediment mineralogy and geochemistry
    Eolian dust and pollen
    Ice-rafted debris
    Clay mineralogy
  Terrestrial
    Glacial landforms and deposits
    Periglacial and other mass-wasting landforms and deposits
    Eolian landforms and deposits
    Fluvial landforms and deposits
    Lacustrine landforms and deposits
    Cave deposits
    Mire and bog deposits
    Soils and other weathering characteristics

Biological
    Tree rings
    Pollen
    Phytoliths
    Corals
    Plant macrofossils
    Invertebrate fossils
    Vertebrate fossils

Historical
    Weather records
    Weather-dependent phenomena
    Phenological records

[a]Modified from Bradley (1985, 1999), following Lowe and Walker (1997).

late-Quaternary climates at 3000-year intervals from 18,000 yrs BP (full glacial time) to the present. Their GCM, described by Kutzbach (1987), by Kutzbach and Webb (1993), and by Kutzbach and Ruddiman (1993) incorporated principles of atmosphere dynamics based on equations of fluid motion, radiative and convective processes, and condensation and evaporation; and prescribed levels of incoming solar radiation (based on orbital parameters), atmospheric CO_2 concentration, sea-surface temperature, sea-ice limits, snow cover, land albedo, effective soil moisture, sea level, ice sheet height, volume, and extent, and continental topography. The model has nine vertical levels at a grid scale of $4.4° \times 7.5°$ and was calculated for January 16 and July 16 for each of the 3000-year "snapshots." A newer version of the model (Kutzbach et al, 1998) includes interactive components for soil moisture, snow hydrology, sea-ice, and mixed-layer ocean temperature, and is configured to simulate the full seasonal cycle.

The results of the COHMAP modeling include maps of each time slice illustrating large-scale patterns of circulation and surface climate, jet stream locations, and regional averages of temperature, precipitation, and precipitation minus evaporation. The latest version of the model (Kutzbach et al., 1998) also simulates major biomes. The models are tested by producing a model of modern conditions for comparison with known conditions and by comparing model results for each time slice with proxy evidence from that time. The numerical simulations of GCMs have been able to reproduce aspects of paleoclimate indicated by proxy records (Street-Perrott, 1991; Wright et al., 1993) and, moreover have provided insights into the significance of factors such as Milankovitch forcing, ice sheet coverage, the effects of seasonality and vegetation feedbacks, and explanations of regional linkages of climate such as the evolution of monsoon climates and the effects of tectonic uplift (e.g., Kutzbach et al., 1993; Ruddiman and Kutzbach, 1989; TEMPO, 1996; Webb et al., 1993).

Another approach to climate modeling of particular relevance in archaeology is the "archaeoclimate model" developed by Bryson and Bryson (1997a,b). It is the inverse of and a complement to the GCM. The model presents computations of climate for a specific locality (such as an archaeological site) over time, whereas the GCMs yield climate simulations for a point in time around the globe. Archaeoclimatic modeling is specifically for archaeologists working on sites with lengthy records. The model includes the heat budget of Earth, incoming solar radiation flux through time as determined by Milankovitch forcing and modulated by volcanic eruptions, and a model of surface ice cover. The locations of major climate systems such as the jet stream and the intertropical convergence zone are calculated for each hemisphere at 200 year intervals back to 14,000-BP and then at 500-year intervals back to 40,000 yrs BP. Synoptic climatology provides the link between the past locations of major circulation features and the climate of a particular site. The results provide estimates of mean temperature and precipitation by month at the 200- and 500-year intervals.

6. Discussion and Conclusions

Archaeology is a Quaternary science, arguably one of the most fundamental ones. The Quaternary period is unique in Earth history in having anatomically modern humans or their predecessors present throughout the period. This characteristic of the period has driven much of the interest and research into the Quaternary as scientists seek to understand human biological and cultural evolution, the relationship of this evolution to environmental changes, and the potential effects of and adaptations to future environmental changes. The Quaternary sciences, therefore, should be an intrinsic component of archaeological teaching and training. Quaternary science encompasses a wide array of disciplines and subdisciplines, similar to archaeology, and archaeologists cannot be expected to be familiar with, much less master, all of these areas. There are several key aspects of Quaternary studies that are fundamental, however, and should be familiar to all of those studying the history of the Earth and its occupants over the past 2 million or so years.

The basic terminology of the geologic time scale is a fundamental issue of communication, although terminological problems can easily become exercises in semantics, if not nit-picking. Understanding the history, basic concepts, and current conventions regarding the time scale in general, and the Quaternary, Pleistocene, and Holocene in particular, are as important as understanding the meaning of, for example, Paleoindian and Archaic or the Upper Paleolithic and Mesolithic. Archaeologists using the time-scale terminology must be aware of this just as other Quaternary scientists involved in archaeology should know and understand the basic cultural chronologies they deal with. As in archaeological chronologies, the details of the time scale (e.g., specific definitions of time periods, ages of the boundaries) may not be agreed on. In such cases the criteria used or the conventions followed should be clearly spelled out. Depending on one's view of the Plio–Pleistocene boundary, miscommunication could make a difference of 800,000 ka or more (1.6 my vs. 2.4 my vs. 2.6 my for the age of the boundary), spanning a significant interval of human evolution (the end of the Australopithecines and the appearance of *Homo*; Kimbel, 1995). For the Pleistocene–Holocene boundary, most definitions vary by only a few thousand years at most (e.g., 11,000 yrs vs. 10,000 yrs), but those years span significant changes in environment and in culture worldwide (e.g., Eriksen and Straus, 1998; Straus et al., 1996).

Among the most distinctive characteristics of the Quaternary are global cooling and dramatic, cyclic environmental change. The most obvious and best known physical manifestation of these characteristics is the formation of huge ice sheets and the evidence for their repeated growth and decay. The record of Quaternary environments in general, and of glacial–interglacial cycles in particular, were long viewed fairly simplistically in the absence of extensive, dateable field evidence, but with the post–World War II growth of Quaternary research and methodology, this view changed. The deep-ocean record of ice volume and environmental changes revolutionized thinking about the Quaternary. Moreover, the ocean record provided a standard scale and terminology for correlation and comparison of Quaternary events.

The record from the ocean and the complexities of the terrestrial stratigraphic records have now been known for over a quarter century. There is no need to perpetuate the myth of the fourfold glacial sequences for the Midwest and for Europe and they should be excised from texts, archaeology and otherwise (Fagan: 1992, Table 3.2; Hester and Grady, 1982, Table 6-1.) Some volumes on archaeology and even a few on geography still present the old glacial schemes, or both the old schemes and the new interpretations, and yet others suggest that the old, named glacial intervals are simply the most recent ones (e.g., Fagan, 1992, 1997:113, Table 3.2; Hester and Grady, 1982:105–107; McKnight, 1996, Table 20–1; Renfrew and Bahn, 1996: 119–120), which is both confusing and misleading. Most of the classic glacial stratigraphic terminology is largely outdated, if not hopelessly muddled, and some of the terminology is now abandoned (Bowen, 1978:10–56; Ehlers, 1996; Sibrava et al., 1986). As noted, for example, in the midwestern United States only the terms for the more recent intervals are still retained (Wisconsin, Sangamon, and Illinoian; e.g., Hallberg, 1986; Johnson, 1986).

Working with proxy indicators presents several potential problems. A continuing problem is the lack of modern analogues for a number of plant and animal communities and for some geologic and geomorphic conditions found in Quaternary records (Rymer, 1978). This is particularly a problem for glacial periods and for the late glacial–postglacial transition. For example, certain mixed-conifer hardwood forests that existed in eastern North America in late glacial times have no modern analogues (Delcourt and Delcourt, 1987). Among mammal communities during full-glacial times, heterogeneity was greater, differing environmental gradients resulted in dissimilar species composition, and individual species shifted at different times, in different directions, and at different rates in response to late Quaternary environmental changes (FAUNMAP Working Group, 1996). Fluvial systems were significantly modified by greatly expanded ice sheets and permafrost regions, by different patterns of atmospheric circulation, and by drainage of huge quantities of glacial meltwater (Knox, 1995). Immense glacial lakes that once existed at the southern margin of the Laurentide and Cordilleran ice sheets produced huge catastrophic floods at scales unknown historically (Baker, 1997). The magnitude of the discharges were such that modern hydrological parameters may be insufficient for flood reconstructions (Baker, 1978). A related problem is that of evolution and extinction. "The frequency of climatic and environmental change during the Quaternary may have resulted in an acceleration in rates of evolution and of morphological characteristics compared with preceding geological periods" (Lowe and Walker, 1997: p. 230). This is particularly a problem with older fossil assemblages due to the higher proportion of extinct species and the increasing evolutionary distance from the modern equivalents (Kurten, 1968; Lundelius, 1976).

Interpretation of individual proxy indicators also have limitations and usually some problematic issues. None provides "The Truth." Generally these issues fall into one of three categories: preservation, representativeness, and interpretation. Differential preservation can provide a biased sample. Pine pollen is much more resistant to degradation than pollen from other species, so high levels of pine pollen in a sample are not necessarily meaningful (Bryant and Hall, 1993). Likewise, pine trees produce pollen in massive quantities that can be distributed for tens to hundreds of miles downwind (Bryant and Holloway, 1983). What, therefore, do high levels of pine pollen in a sample mean? In contrast, phytoliths are deposited very close to the plants that produced them (Piperno, 1988:142–146). Dispersion, therefore, is not as problematic as in pollen studies, but phytoliths from plants in the bottom of an alluvial valley subject to fluctuating hydrological conditions may not be representative of regional conditions. Finally, what do biological and geological records and changes in the records mean? Are they indicators of effective precipitation or of absolute rainfall? Are there lag effects in geological and biological systems following a climate change, that is, do plant and animal communities and geomorphic systems respond to climate changes at different rates (e.g., Cole, 1990; FAUNMAP Working Group, 1996; Knox, 1972)? If the changes are synchronous, does the direction of change vary depending on local and regional factors (e.g., Knox, 1983; FAUNMAP Working Group, 1996)? Ultimately, the best approach to solving or avoiding these problems is to use an array of proxy indicators in environmental reconstruction.

A number of problems and limitations warrant considerable caution in the utilization of outputs from GCMs, especially at archaeological scales of time and space. Most of the available results are from models that did not incorporate factors such as cloud cover or interactive ocean dynamics (Kutzbach et al., 1993:60–61; Street-Perrott, 1991). Cloud cover and ocean circulation have significant impacts on climate but for a variety of reasons are not yet part of most models. Street-Perrott (1991) also correctly pointed out another major criticism and concern: "a serious danger of circular reasoning if the same palaeoecological data are used both to specify the surface-boundary conditions and to test the model output" (p. 78). Street-Perrott (1991) also discusses several instances of "glaring discrepancies...between modeled climates and geological data" (p. 78; see also critiques of GCMs by Shackley et al., 1998, and Trenberth, 1997). For example, the models did not reproduce paleoclimatic conditions clearly evident from a variety of proxy data (e.g., Grayson, 1998). Also, because of their coarse spatial (continental to subcontinental) scale and because the experiments have been run for only a few simulated months, "it is unwise to place too much reliance on the regional details of the experiments" (Street-Perrott, 1991, p. 74). "GCMs...cannot be expected to predict what happened at a particular site. Instead, comparisons between model predictions and palaeoenvironmental data are best made by considering the general pattern of changes across a continent" (Harrison et al., 1991:236). Most field-based questions about past environments in archaeology and in other Quaternary sciences deal with much smaller spatial scales than can be modeled by GCMs (e.g., at the scale of an archaeological site or group of sites). Investigators must consider more seriously the applicability of a GCM to a single site or a small area such as a watershed, especially if these study areas are some distance from the grid points used in the model. Part of the answer will depend on the significance of local environmental factors such as slope, aspect, soil and rock types, and the local surface and subsurface hydrology.

Archaeoclimatic modeling has been able to reproduce records of climate change based on proxy indicators for a number of localities around the world with reasonable results (Bryson and Bryson, 1996, 1997b). Because the models are site specific and are intended for archaeological applications, they should and have generated considerable interest among archaeologists. The model results provide no linkages or explanations for climate trends (either in phase or out of phase, and based on either computer computations or proxy data) observed among a set of sites, however. This may limit some applications of this approach because contemporary archaeology tends to deal with problems beyond the level of individual sites. As with other climate models, the output is intended to provide testable hypotheses for further research. A continuing problem with all model results is reconciling differences between the field data and the model results. A rule of thumb applicable to any comparison of field or proxy data with data generated by laboratory analyses or computer-generated output is to rely on the field or proxy data.

A fundamentally important distinction between paleoclimatic reconstructions based on proxy records and those based on computer simulations (including both GCMs and archaeoclimatic models) is that the former are based directly on data whereas the latter are based on mathematical computations, that is, simplifications and generalizations. Too often the tendency seems to be to assume that because computer-generated models are expensive and run on very powerful computers, they must provide some sort of mathematical truth (e.g., Shackley et al., 1998). They do not. What they do provide are hypotheses regarding climate changes and climate patterns that can be tested using proxy indicators. Reconstructions based on proxy indicators are not tested against climate models; the proxy data are used to test the model results.

Several mechanisms appear to drive late Cenozoic environmental changes, and each operates at a different scale of time and space. Most of these probably had some effect on human prehistory. Sherratt (1997) is emphatic on this point: "Environmental change is not simply a backdrop to evolution: it is a principal reason for major episodes of biological change. It is no coincidence that successive species of hominid made their appearance during the Quaternary period, with its rapid pace and massive scale of environmental alteration" (p. 283) The links between cause and effect are unclear, however, and in any event are not agreed on. The longer term changes such as the overall trend in late Cenozoic cooling and in glacial–interglacial cycles probably contributed to early hominid evolution. For example, Kimbel (1995) noted "a general correlation...between the late Pliocene onset of global climatic change and a net increase in hominid taxonomic and adaptive diversity over the period 3.0-2.0 myr." deMenocal and Bloemendal (1995) were somewhat more specific, arguing that "If climate had a role in determining hominid evolution, the most parsimonious interpretation of the available data is that it was a change in mode of subtropical climate *variability* rather than a wholesale, stepwise change in climate that prompted evolutionary response." (p. 284). Kimbel (1995) also noted, however, that "Of the three million years of hominid evolution..., about one-half of that time remains virtually undocumented" (p. 435). That is, the record of hominid evolution and hominid environments is substantially coarser, and for some time intervals is altogether lacking in comparison to paleoclimatic data from other regions and sources. In any event, "the hominid data base must improve considerably before we can move from proposing correlations to the meaningful testing of causally specific hypotheses" (p. 436).

The databases for the paleoclimates and the archaeology of later Quaternary time are significantly more complete and better documented than for the earlier Quaternary. A striking feature of late Quaternary human evolution is the "activity" in physical and cultural development compared to earlier times. Having passed through 30 or so glacial–interglacial cycles, the first "clear morphological evidence of cold adaptation" comes with the Neanderthals (Stringer, 1995, p. 530), and the diversity of modern humans comes only in the most recent cycle (e.g., Eriksen and Straus, 1998; Straus et al, 1996).

Probably one of the most archaeologically significant developments in Quaternary research is the discovery of evidence for very rapid (at the scale of human generations) and significant environmental changes during the last glacial interval (oxygen isotope Stage 2) and perhaps in the Holocene as well. These changes are well documented in the Greenland ice sheet and in sediments from the floor of the North Atlantic, but their manifestation in terrestrial stratigraphic and archaeological records around the globe remains unclear. Nevertheless, the general correlation between the environmental "flip-flops" of this time and the dramatic changes in human dispersion, settlement, subsistence, technology, and art that took place as fully modern *Homo sapiens* lived during and emerged from the last full-glacial episode have generated considerable archaeological and paleoecological interest and research (e.g., the collections of papers assembled by Eriksen and Straus, 1998; Gamble and Soffer, 1990b, Soffer and Gamble, 1990, and Straus et al., 1996).

Several significant issues must be resolved in order to understand better the relationship between the various environmental changes of the late Pleistocene and Holocene and human physical and cultural evolution, however. Adequate age control is generally not available from most localities to document environmental shifts occurring at the same rate as the onset of the Younger Dryas and other rapid environmental shifts, and several "plateaus" occur in the radiocarbon calibration curves for this time (Stuiver and Braziunas, 1993), further hampering correlation (e.g., Bar-Yosef, 1995:517–519). Moreover, a remark by Kimbel (1995) with regard to early hominid studies is equally applicable here: the "data base must improve considerably before we can move from proposing correlations to the meaningful testing of causally specific hypotheses" (p. 436).

More broadly, some human adaptations probably were directly related to Quaternary environmental fluctuations, some undoubtedly were not, and many probably were complexly interwoven with both environmental and nonenvironmental forces (e.g., Butzer, 1982; Gamble and Soffer, 1990a; Sherratt, 1997; Wobst, 1990). An important step in dealing with this issue is the recognition of the varying spatial and temporal scales of environmental change (Butzer, 1982:23–32; Gamble and Sofer, 1990a:8–12). A related issue is assessing the linkage, if any, between specific environmental changes and the archaeological record (e.g., Meltzer, 1991; Sherratt, 1997). "[A]lthough the biophysical evidence leaves no doubt as to repeated environmental changes of different amplitudes and wavelengths, there is no archaeological case for causally related technological or behavioral readjustments" (Butzer, 1982:301). "Human adaptations, at whatever wavelength of climate cycle, need to be assessed in terms of organization rather than simply as a set of technological solutions" (Gamble and Soffer, 1990b:12). Searching for and sorting out these various relationships and linkages is one of the most important, daunting, and exciting areas of research for archaeologists and for other Quaternary scientists.

ACKNOWLEDGMENTS. The idea for this chapter first came out of some conversations I had with D. Gentry Steele some years ago when we were colleagues at Texas A&M University. Additional and more recent encouragement for preparing this discussion came from Bill Farrand, Paul Goldberg, and Don Grayson, who also provided very helpful commentary on earlier drafts. Caspar Ammann patiently helped me sort out Bond cycles, Heinrich events, and so forth. The manuscript also benefitted from reviews by Reid Bryson, Reid Ferring, Eileen Johnson, David Meltzer, and James Stoltman.

References

Adams, J., Maslin, M., and Thomas, E., 1999, Sudden Climate Transitions during the Quaternary, *Progress in Physical Geography* 23:1–36.
Aguirre, E., and Pasini, G., 1985, The Pliocene–Pleistocene Boundary, *Episodes* 8:116–120.
Allen, B. D., and Anderson, R. Y., 1993, Evidence from Western North America for Rapid Shifts in Climate during the Last Glacial Maximum, *Science* 260:1920–1923.
Andersen, G. G., and Borns, H. W., Jr., 1994, *The Ice Age World: An Introduction to Quaternary History and Research with Emphasis on North America and Northern Europe during the Last 2.5 Million Years*, Scandinavian University Press, Oslo.
Anderson, D. E., 1997, Younger Dryas Research and its Implications for Understanding Abrupt Climate Change. *Progress in Physical Geography* 21:230–249.
Andrews, J. T., 1998, Abrupt Changes (Heinrich events) in Late Quaternary North Atlantic Marine Environments: A History and Review of Data and Concepts. *Journal of Quaternary Science* 13:3–16.
Aubry, M.-P., Berggren, W. A., Van Couvering, J. A., Rio, D., and Castradori, D., 1998, The Pliocene–Pleistocene Boundary Should Remain at 1.81 Ma, *GSA Today*, 8:22.
Baker, V. R., 1978, Paleohydraulics and Hydrodynamics of Scabland Floods. In *The Channeled Scabland*, edited by V. R. Baker and D. Nummedal, pp. 59–79. National Aeronautics and Space Administration, Washington, DC.
Baker, V. R, 1997, Mega-floods and Glaciation. In *Late Glacial and Postglacial Environmental Changes*, edited by I. P. Martini, pp. 98–108. Oxford University Press, New York.
Bar-Yosef, O., 1995, The Role of Climate in the Interpretation of Human Movements and Cultural Transformations in Western Asia. In *Paleoclimate and Evolution, with Emphasis on Human Origins*, edited by E. S. Vrba, G. H. Denton, T. C. Partridge, and L. H. Burckle, pp. 507–523. Yale University Press, New Haven, CT.
Baumann, K.-H., Lackschewitz, K. S., Mangerud, J., Spielhagen, R. F., Wolf-Welling, T. C. W., Heinrich, R., and Kassens, H., 1995, Reflection of Scandinavian Ice Sheet Fluctuations in Norwegian Sea Sediments During the Last 150,000 Years. *Quaternary Research* 43:185–197.
Bell, M., and Walker, M. J. C., 1992, *Late Quaternary Environmental Change: Physical and Human Perspectives*, Longman/Wiley, New York.
Berger, W. H., 1990, The Younger Dryas Cold Spell—A Quest for Causes, *Palaeogeography, Palaeoclimatology, Palaeoecology* 89:219–237.
Berggren, W. A., Kent, D. V., Swisher, C. C., III, and Aubry, M.-P., 1995a, A Revised Cenozoic Geochronology and Chronostratigraphy. In *Geochronology, Time Scales and Global Stratigraphic Correlation*, edited by W. A. Berggren, D. V. Kent, M.-P. Aubry, and T Haudenbal, pp. 129–212. SEPM Special Publication 54.
Berggren, W. A., Hilgen, F. J., Langereis, C. G., Kent, D. V., Obradovich, J. D., Raffi, I., Raymo, M. E., Shackleton, N. J. 1995b, Late Neogene Chronology: New Perspectives in High-Resolution Stratigraphy. *Geological Society of America Bulletin* 107:1272–1287.
Boellstorff, J., 1978, North American Pleistocene Stages Reconsidered in Light of Probable Pliocene–Pleistocene Continental Glaciation, *Science* 202:305–307.
Bond, G. C., and Lotti, R., 1995, Iceberg Discharge into the North Atlantic on Millennial Time Scales during the Last Glaciation, *Science* 267:1005–1010.
Bond, G. C., Broecker, W., Johnsen, S., McManus, J., Labeyrie, L., Jouzel, L., and Bonani, G., 1993, Correlations Between Climate Records from North Atlantic Sediments and Greenland Ice, *Nature* 365:143–147.
Bowen, D. Q., 1978, *Quaternary Geology*, Pergamon, Oxford, UK.
Boyle, E. A., 1995, Last-Glacial-Maximum North Atlantic Deep Water: On, Off, or Somewhere in Between? *Philosophical Transactions of the Royal Society*, London B348:243–253.
Bradley, R. S., 1985, *Quaternary Paleoclimatology: Methods of Paleoclimatic Reconstruction*, Allen and Unwin, Boston.
Bradley, R. S., 1999, *Paleoclimatology: Reconstructing Climates of the Quaternary* 2nd ed., Academic Press, New York.
Broecker, W. S., 1991, The Great Ocean Conveyer, *Oceanography* 4:79–89.
Broecker, W. S., 1992, Defining the Boundaries of the Late Glacial Isotope Episodes. *Quaternary Research* 38:135–138.

Broecker, W. S., 1994, Massive Iceberg Discharges as Triggers for Global Climate Change. *Nature* 372:421–424.
Broecker, W. S., 1995, *The Glacial World According to Wally*, Eldigio Press, Palisades, NY.
Broecker, W. S., Bond, G., McManus, J., Klas, M., and Clark, E., 1992, Origin of the Northern Atlantic's Heinrich Events. *Climate Dynamics* 6:265–273.
Broecker, W. S., Kennett, J. P., Teller, J., Trumbore, S., Bonani, G., and Wolfli, W., 1989, The Routing of Laurentide Ice-Sheet Meltwater during the Younger Dryas Cold Event, *Nature* 341:318–321.
Broecker, W. S., Bond, G., and Klas, M., 1990, A Salt Oscillator in the Glacial Atlantic: 1. The Concept, *Palaeoceanography*. 5:469–477.
Bryant, V. M, and Hall, S. A., 1993, Archaeological Palynology in the United States: A Critique, *American Antiquity* 58:277–286.
Bryant, V. M., and Holloway, R. G., 1983, The Role of Palynology in Archaeology. *Advances in Archaeological Method and Theory* 6:191–224.
Bryant, V. M., and Holloway, R. G. (eds), 1985, *Pollen Records of Late-Quaternary North American Sediments*, American Association of Stratigraphic Palynologists Foundation, Dallas.
Bryson, R. A., 1989, Late Quaternary Volcanic Modulation of Milankovich Climate Forcing, *Theoretical Applied Climatology* 39:115–125.
Bryson, R. A., and Bryson, R. U., 1996, *Site-Specific High-Resolution Archaeoclimatic Modeling for the Great Plains*. Paper presented at the 54th Annual Plains Anthropological Conference, Laramie, Wyoming.
Bryson, R. A., and Bryson, R. U., 1997a, High Resolution Simulation of Iowa Holocene Climates. *Journal of the Iowa Archeological Society* 44:121–128.
Bryson, R. A., and Bryson, R. U., 1997b, High Resolution Simulations of Regional Holocene Climate: North Africa and the Near East. In *Third Millennium BC Climate Change and Old World Collapse*, edited by H. N. Dalfes, G. Kukla, and H. Weiss, pp. 565–593. Springer-Verlag, Berlin.
Butzer, K. W., 1964, *Environment and Archaeology: An Introduction to Pleistocene Geography*, Aldine, Chicago.
Butzer, K. W., 1971, *Environment and Archaeology: An Ecological Approach to Prehistory*, Aldine, Chicago.
Butzer, K. W., 1982, *Archaeology as Human Ecology*. Cambridge University Press, Cambridge, UK.
COHMAP Members, 1988, Climatic Changes of the Last 18,000 years: Observations and Model Simulations, *Science* 241:1043–1052.
Cole, K. L., 1990, Late Quaternary Vegetation Gradients through the Grand Canyon. In *Packrat Middens: The Last 40,000 Years of Biotic Change*, edited by J. L. Betancourt, T. R. Van Devender, and P. S. Martin, pp. 240–258. University of Arizona Press, Tucson.
Dansgaard. W., Johnsen, S. J., Clausen, H. B., Dahl-Jensen, D., Gundestrup, N. S., Hammer, C. U., Hvidberg, C. S., Steffensen, J. P., Sveinbjornsdottir, A. E., Jouzel, J., and Bond, G., 1993, Evidence for General Instability of Past Climate from a 250-kyr Ice-Core Record, *Nature* 364:218–220.
Davidson, D. A., and Shackley, M. L., 1976, *Geoarchaeology: Earth Science and the Past*, Westview Press, Boulder, CO.
Dawson, A. G., 1992, *Ice Age Earth: Late Quaternary Geology and Climate*, Routledge, New York.
Delcourt, H. R., and Delcourt, P. A., 1987, Late-Quaternary Dynamics of Temperate Forests: Applications of Paleoecology to Issues of Global Environmental Change. *Quaternary Science Reviews* 6:129–146.
deMenocal, P. D., and Bloemendal, J., 1995, Plio–Pleistocene Climatic Variability in Subtropical Africa and the Paleoenvironment of Hominid Evolution: A Combined Data-Model Approach. In *Paleoclimate and Evolution, with Emphasis on Human Origins*, edited by E. S. Vrba, G. H. Denton, T. C. Partridge, and L. H. Burckle, pp. 262–288. Yale University Press, New Haven, CT.
Ding, Z., Rutter, N. W., and Liu, T., 1997, The Onset of Extensive Loess Deposition around the G/M Boundary in China and its Palaeoclimatic Implications. In "The Plio-Pleistocene Boundary" edited by T. C. Partridge, *Quaternary International* 40:53–60.
Ehlers, J., 1996, *Quaternary and Glacial Geology*, Wiley, New York.
Emiliani, C., 1955, Pleistocene Temperatures, *Journal of Geology* 63:538–578.
Epstein, S., Buchsbaum, R., Lowenstam, H. A., and Urey, H. C., 1953, Revised Carbonate-Water Isotopic Temperature Scale, *Geological Society of America Bulletin* 64:1315–1326.

Eriksen, B. V., and Straus, L. G. (eds), 1998, As the World Warmed: Human Adaptations across the Pleistocene/Holocene Boundary. *Quaternary International* 49/50:1–199.

Fagan, B. M., 1991, *Ancient North America: The Archaeology of a Continent*, Thames and Hudson, New York.

Fagan, B. M., 1992, *People of the Earth*, 7th ed., HarperCollins, New York.

Fagan, B. M., 1997, *In the Beginning: An Introduction to Archaeology*, 9th ed., Longman, New York.

Farrand, W. R., 1990, Origins of Quaternary–Pleistocene–Holocene Stratigraphic Terminology. In *Establishment of a Geologic Framework for Paleoanthropology*, edited by L. Laporte, pp. 15–22. Geological Society of America Special Paper 242. Boulder, CO.

FAUNMAP Working Group, 1996, Spatial Response of Mammals to Late Quaternary Environmental Fluctuations, *Science* 272:1601–1606.

Feibel, C. S., 1997, Debating the Environmental Factors in Hominid Evolution, *GSA Today* 7(3):1–7.

Fiedel, S. J., 1992, *Prehistory of the Americas*, Cambridge University Press, New York.

Fink, J., and Kukla, G., 1977, Pleistocene Climates in Central Europe: At Least 17 Interglacials after the Olduvai Event. *Quaternary Research* 7:363–371.

Flint, R. F., 1971, *Glacial and Quaternary Geology*, Wiley, New York.

Gamble, C., and Soffer, O., 1990a, Introduction: Pleistocene Polyphony: The Diversity of Human Adaptations at the Last Glacial Maximum. In *The World at 18,000 BP*, Volume 1: *High Latitudes*, edited by C. Gamble and O. Soffer, pp. 1–23. Unwin Hyman, London.

Gamble, C., and Soffer, O. (eds), 1990b, *The World at 18,000 BP*, Volume 2: *Low Latitudes*. Unwin Hyman, London.

Goodyear, A. C., 1991, *Geoarchaeological Criteria for the Recognition of the Pleistocene–Holocene Boundary in the Southeastern United States*. Paper presented at the Southeast Archaeological Conference, Jackson, Mississippi.

Goodyear, A. C., 1993, The Stratigraphic Significance of Paleosols at Smiths Lake Creek, 38AL135, for the Study of the Pleistocene–Holocene Transition in the Savannah River Valley. In *Proceedings of the First International Conference on Pedo-Archaeology*, edited by J. E. Foss, M. E. Timpson, and M. W. Morris, pp. 27–40. The University of Tennessee Agricultural Experiment Station, Knoxville, TN.

Grignoux, M., 1913, Les formations marines pliocènes et quaternaires de l'Italie du sud et de la Sicilie. *Annales de l'Université de Lyon*, n.s. 1.

Grayson, D. K., 1983, *The Establishment of Human Antiquity*, Academic Press, New York.

Grayson, D. K., 1998, Moisture History and Small Mammal Community Richness during the Latest Pleistocene and Holocene, Northern Bonneville Basin, Utah, *Quaternary Research* 49: 330–334.

Hageman, B. P., 1972, *Reports of the International Quaternary Association Subcommission on the Study of the Holocene*, Bulletin 6.

Hallberg, G. R., (ed.), 1980a, *Illinoian and Pre-Illinoian Stratigraphy of Southeast Iowa and Adjacent Illinois*, Iowa Geological Survey, Technical Information Series No. 11.

Hallberg, G. R., 1980b, *Pleistocene Stratigraphy in East-Central Iowa*, Iowa Geological Survey, Technical Information Series No. 10. Iowa City, IA

Hallberg, G. R., 1986, Pre-Wisconsin Glacial Stratigraphy of the Central Plains Region in Iowa, Nebraska, Kansas, and Missouri. In "Glaciations in the Northern Hemisphere" edited by V. Sibrava, D. Q. Bowen, and G. M. Richmond, *Quaternary Science Reviews* 5:11–15.

Hammer, C. U., Clausen, H. B., and Dansgaard, W., 1981, Past Volcanism and Climate Revealed by Greenland Ice Cores, *Journal of Volcanology and Geothermal Research* 11:3–11.

Hansen, W. R., 1991, *Suggestions to Authors of the Reports of the United States Geological Survey*, U.S. Government Printing Office, Washington, DC.

Haq, B. U., Berggren, W. A., and Van Couvering, J. A., 1977, Corrected Age of the Pliocene/Pleistocene Boundary, *Nature* 269:483–488.

Harland, W. B., Cox, A. V., Llewellyn, P. G., Pickton, C. A. G., Smith, A. G., and Walters, R., 1982, *Geologic Time Scale*. Cambridge University Press, Cambridge.

Harland, W. B., Armstrong, R. L., Cox, A. V., Craig, L. E., Smith, A. G., and Smith, D. G., 1990, *A Geologic Time Scale, 1989*, Cambridge University Press, Cambridge.

Harrison, S. P., Kutzbach, J. E., and Behling, P., 1991. General Circulation Models, Palaeoclimatic Data and Last Interglacial Climates. *Quaternary International* 10–12:231–242.

Haynes, C. V., Jr., 1991, Geoarchaeological and Paleohydrological Evidence for a Clovis-Age Drought in North America and its Bearing on Extinction, *Quaternary Research* 35:438–450.

Heinrich, H., 1988, Origin and Consequences of Cyclic Ice-rafting in the Northeast Atlantic Ocean During the Past 130,000 Years, *Quaternary Research* 29:142–152.

Herz, N., and Garrison, E. G., 1998, *Geological Methods for Archaeology*, Oxford University Press, New York.

Hester, J. J., and Grady, J., 1982, *Introduction to Archaeology* 2nd ed., Holt, Rinehart and Winston, Austin.

Hopkins, D. M., 1975, Time-Stratigraphic Nomenclature for the Holocene Epoch, *Geology* 3:10.

Imbrie, J., and Imbrie, K. P., 1979, *Ice Ages: Solving the Mystery*, Enslow Publishers, Hillside, NJ.

Jenkins, D. G., 1987, Was the Pliocene–Pleistocene Boundary Placed at the Wrong Stratigraphic Level? *Quaternary Science Reviews* 6:41–42.

Johnsen, S. J., Clausen, H. B., Dansgaard, W., Fuhrer, K., Gundestrup, N., Hamer, C. U., Iversen, P., Jouzel, J., Stauffer, B., and Steffensen, J. P., 1992, Irregular Glacial Interstadials Recorded in a New Greenland Ice Core. *Nature* 359:311–313.

Johnson, W. H., 1986. Stratigraphy and Correlation of the Glacial Deposits of the Lake Michigan Lobe Prior to 14 ka BP. In "Glaciations in the Northern Hemisphere" edited by V. Sibrava, D. Q. Bowen, and G. M. Richmond, *Quaternary Science Reviews*, 5:17–22.

Keigwin, L. D., and Jones, G.A., 1995, The Marine Record of Deglaciation from the Continental Margin off Nova Scotia. *Paleoceanography* 10: 973–985.

Kimbel, W. H., 1995, Hominid Speciation and Pliocene Climatic Change. In *Paleoclimate and Evolution, with Emphasis on Human Origins*, edited by E. S. Vrba, G. H. Denton, T. C. Partridge, and L. H. Burckle, pp. 425–437. Yale University Press, New Haven, CT.

Klein, Richard G., 1989, *The Human Career: Human Biological and Cultural Origins*. University of Chicago Press, Chicago.

Knox, J. C., 1972, Valley Alluviation in Southwestern Wisconsin. *Annals of the Association of American Geographers* 62:401–410

Knox, J. C., 1983, Responses of River Systems to Holocene Climate. In *Late-Quaternary Environments of the U.S.*, Volume 2, edited by H. E. Wright, pp. 26–41. University of Minnesota Press, Minneapolis.

Knox, J. C., 1995, Fluvial Systems since 20,000 Years BP. In *Global Continental Palaeohydrology*, edited by K. J. Gregory, J. L. Starkel, and V. R. Baker, pp. 87–108. Wiley, New York.

Kukla, G. J, 1975, Loess Stratigraphy of Central Europe. In *After the Australopithecines*, edited by K. W. Butzer and G. Isaac, pp. 99–188. Morton Publishers, The Hague.

Kukla, G. J., 1977, Pleistocene Land–Sea Correlations. I. Europe, *Earth Science Reviews* 13:307–374.

Kurten, B., 1968, *Pleistocene Mammals of Europe*, Aldine Publishing, Chicago.

Kutzbach, J. E., 1985, Modeling of Paleoclimates. *Advances in Geophysics* 28A:159–196.

Kutzbach, J. E., 1987, Model Simulations of the Climatic Patterns during the Deglaciation of North America. In *North America and Adjacent Oceans During the Last Deglaciation*, edited by W. F. Ruddiman and H. E. Wright, Jr., pp. 425–446. Geological Society of America, The Geology of North America K-3, Boulder, Colorado.

Kutzbach, J. E., and Webb, T., III, 1993, Conceptual Basis for Understanding Late-Quaternary Climates. In *Global Climates since the Last Glacial Maximum*, edited by H. E. Wright, Jr., J. E. Kutzbach, and W. F. Ruddiman, 1993, Model Description, External Forcing, and Surface Boundary Conditions. In *Global Climates since the Last Glacial Maximum*, edited by H. E. Wright, Jr., J. E. Kutzbach, T. Webb, III, W. F. Ruddiman, F. A. Street-Perrott, and P. J. Bartlein, pp. 5–11. University of Minnesota Press, Minneapolis.

J. E. Kutzbach, T. Webb, III, W. F. Ruddiman, F. A. Street-Perrott, and P. J. Bartlein, pp. 12–23. University of Minnesota Press, Minneapolis.

Kutzbach, J. E., Guetter, P. J., Behling, P. J., and Selin, R., 1993, Simulated Climatic Changes: Results of the COHMAP Climate-Model Experiments. In *Global Climates since the Last Glacial Maximum*, edited by H. E. Wright, Jr., J. E. Kutzbach, T. Webb, III, W. F. Ruddiman, F. A. Street-Perrott, and P. J. Bartlein, pp. 24–93. University of Minnesota Press, Minneapolis.

Kutzbach, J., Gallimore, R., Harrison, S., Behling, P., Selin, R., and Laarif, F., 1998, Climate and Biome Simulations for the Past 21,000 Years, *Quaternary Science Reviews* 17: 473–506.

Lamb, H. H., 1982. *Climate, History and the Modern World*, Methuen, London.

Lowe, J. J., and Walker, M. J. C., 1997, *Reconstructing Quaternary Environments*, 2nd ed. Longman, Essex, UK.

Lowe, J. J., Amann, B., Birks, H. H., Bjorck, S., Coope, G. R., Cwynar, L. C., De Beaulieu, J.-L., Mott, R. J., Peteet, D. M., and Walker, M. J. C., 1994, Climatic Changes in Areas adjacent to the North Atlantic during the Last Glacial–Interglacial Transition (14–9 ka BP). *Journal of Quaternary Science* 9:185–198.

Lowell, T. V., Heusser, C. J., Andersen, B. G., Moreno, P. I., Hauser, A., Heusser, L. E., Schluchter, C., Marchant, D. R., and Denton, G. H., 1995, Interhemispheric Correlation of Late Pleistocene Glacial Events, *Science* 269:1541–1549.

Lundelius, E. L., Jr., 1976, Vertebrate Paleontology of the Pleistocene: An Overview. *Geoscience and Man* 13:45–59.

Lyell, C., 1833, *Principles of Geology*, Volume 3 (Reprinted in 1969), Johnson, New York.

Lyell, C., 1839. *Nouveaux éléments de géologie*, Pitois-Levrault, Paris.

Maslin, M. A., Li, X. S., Loutre, M.-F., and Berger, A., 1998, The Contribution of Orbital Forcing to the Progressive Intensification of Northern Hemisphere Glaciation. *Quaternary Science Reviews* 17: 411–426.

Mauz, B., 1998, The Onset of the Quaternary: A Review of New Findings in the Pliocene–Pleistocene Chronostratigraphy, *Quaternary Science Reviews* 17: 357–364.

Mayewski, P. A., Meeker, L. D., Twickler, M. S., Morrison, M. C., Alley, R. B., Bloomfield, P., and Taylor, L., 1993, The Atmosphere during the Younger Dryas, *Science* 261: 195–197.

Martinson, D. G., Pisias, N. G., Hays, J. D., Imbrie, J., Moore, T. C., and Shackleton, N. J., 1987, Age dating and the orbital theory of the ice ages: Development of a high resolution 0–300,000 year chronostratigraphy. *Quaternary Research* 27: 1–29.

McGuire, W. J., Howarth, R. J., Firth, C. R., Solow, A. R., Pullen, A. D., Saunders, S. J., Stewart, I. S., and Vita-Finzi, C., 1997, Correlation between Rate of Sea-Level Change and Frequency of Explosive Volcanism in the Mediterranean, *Nature* 389: 473–476.

McKnight, T. L., 1996, *Physical Geography: A Landscape Appreciation*, 5th ed. Prentice Hall, Upper Saddle River, New Jersey.

Meltzer, D. J., 1991, Altithermal Archaeology and Paleoecology at Mustang Springs, on the Southern High Plains of Texas, *American Antiquity* 56: 236–267.

Morrison, R. B., 1969, The Pleistocene–Holocene Boundary. *Geologie En Mijnbouw* (Journal of the Royal Geological and Mining Society of the Netherlands) 48:363–371.

Morrison, R, B., and Kukla, G., 1998. The Pliocene–Pleistocene (Tertiary–Quaternary) Boundary Should be Placed at about 2.6 Ma, not at 1.8 Ma! *GSA Today* 8:9.

NACOSN (North American Commission on Stratigraphic Nomenclature), 1983, North American Stratigraphic Code, *American Association of Petroleum Geologists Bulletin* 67: 841–875.

Nikiforova, K. V., and Alekseev, M. N., 1997, International Geological Correlation Program, Project 41: Neogene/Quaternary Boundary. In *The Pleistocene Boundary and the Beginning of the Quaternary*, edited by J. A. Van Couvering, pp. 3–12. Cambridge University Press, Cambridge, UK.

Nilsson, T., 1983, *The Pleistocene: Geology and Life in the Quaternary Ice Age*, D. Reidel, Boston.

Oakley, K. P., 1964, *Frameworks for Dating Fossil Man*, Aldine, Chicago.

Palmer, A. R. (compiler), 1983, The Decade of North American Geology 1983 Geologic Time Scale, *Geology* 11:503–504.

Partridge, T. C., 1997a, Reassessment of the Position of the Plio–Pleistocene Boundary: Is there a Case for Lowering it to the Gauss–Matuyama Paleomagnetic Reversal? In "The Plio–Pleistocene Boundary," edited by T. C. Partridge, *Quaternary International* 40:5–10.

Partridge, T. C., (ed.). 1997b, The Plio–Pleistocene Boundary, *Quaternary International* 40:1–100.

Partridge, T. C., Bond, G. C., Hartnady, C. J. H., deMenocal, P. D., and Ruddiman, W. F., 1995, Climatic Effects of Late Neogene Tectonism and Volcanism. In *Paleoclimate and Evolution, with Emphasis on Human Origins*, edited by E. S. Vrba, G. H. Denton, T. C. Partridge, and L. H. Burckle, pp. 8–23.Yale University Press, New Haven, CT.

Pasini, G., and Colalongo, M. L., 1997, The Pliocene–Pleistocene Boundary-Stratotype at Vrica, Italy. In *The Pleistocene Boundary and the Beginning of the Quaternary*, edited by J. A. Van Couvering, pp. 15–45. Cambridge University Press, Cambridge, UK.

Piperno, D. R., 1988, *Phytolith Analysis: An Archaeological and Geological Perspective*, Academic Press, New York.

Porter, S. C., 1986, Pattern and Forcing of Northern Hemisphere Glacier Variations during the Last Millennium. *Quaternary Research* 26:27–48.
Potts, R., 1996, Evolution and Climate Variability, *Science* 273:922–923.
Price, T. D., and Feinman, G. M., 1997, *Images of the Past*, 2nd ed., Mayfield Publishing, Mountain View, CA.
Rampino, M. R., and Self, S., 1993, Climate–Volcanism Feedback and the Toba Eruption of 74,000 Years Ago, *Quaternary Research* 40:269–280.
Rapp, G., Jr., and Gifford, J. A., 1985, *Archaeological Geology*. Yale University Press, New Haven, CT.
Rapp, G., Jr., and Hill, C. L., 1998. *Geoarchaeology: The Earth-Science Approach to Archaeological Interpretation*. Yale University Press, New Haven, CT.
Raymo, M., 1998, Glacial Puzzles, *Science* 281:1467–1468.
Renfrew, C., and Bahn, P., 1996, *Archaeology: Theories, Methods, and Practice*, 2nd ed., Thames and Hudson, London.
Roberts, N., 1989, *The Holocene: An Environmental History*, Basil Blackwell, London.
Ruddiman, W. F., and Kutzbach, J. E., 1989, Forcing of Late Cenozoic Northern Hemisphere Climate by Plateau Uplift in Southern Asia and the American West, *Journal of Geophysical Research* 94:18, 409–418, 427.
Ruddiman, W. F., McIntyre, A. F., and Raymo, M. E., 1986, Matuyama 41,000-Year Cycle: North Atlantic Ocean and Northern Hemisphere Ice Sheets. *Earth and Planetary Science Letters* 80:117–129.
Rymer, L., 1978, The Use of Uniformitarianism and Analogy in Palaeoecology. In *Biology and Quaternary Environments*, edited by D. Walker and J. C. Guppy, pp. 245–258. Australian Academy of Science, Canberra.
Salvador, A. (ed.), 1994, *International Stratigraphic Guide: A Guide to Stratigraphic Classification, Terminology, and Procedure*, 2nd ed., Geological Society of America, Boulder, CO.
Shackleton, N. J., 1967, Oxygen Isotope Analyses and Pleistocene Temperature Re-assessed, *Nature* 218:15–17.
Shackleton, N. J., and Opdyke, N. D., 1973, Oxygen Isotope and Paleomagnetic Stratigraphy of Equatorial Pacific Core V28-238: Oxygen Isotope Temperatures and Ice Volumes on a 10^5 and 10^6 Year Scale. *Quaternary Research* 3:39–55.
Shackleton, N. J., Berger, A., and Peltier, W. A., 1990, An Alternative Astronomical Calibration of the Lower Pleistocene Timescale Based on ODP Site 677. *Transactions of the Royal Society of Edinburgh: Earth Sciences* 81:251–261.
Shackley, S., Young, P., Parkinson, S., and Wynne, B., 1998, Uncertainty, Complexity and Concepts of Good Science in Climate Change Modelling: Are GCMs the Best Tools? *Climatic Change* 38:159–205.
Sherratt, A., 1997, Climatic Cycles and Behavioural Revolutions: The Emergence of Modern Humans and the Beginning of Farming. *Antiquity* 272:271–287.
Sibrava, V., Bowen, D. Q., and Richmond, G. M. (eds.), 1986, Glaciations in the Northern Hemisphere. *Quaternary Science Reviews* 5:1–14.
Sikes, N. E., and Wood, B. A., 1996, Early Hominid Evolution in Africa: The Search for an Ecological Focus, *Evolutionary Anthropology* 4:155–159.
Soffer, O., and Gamble, C. (eds.), 1990, *The World at 18,000 BP*, Volume 1: *High Latitudes*, Unwin Hyman, London.
Stager, J. C., and Mayewski, P. A., 1997, Abrupt Early to Mid-Holocene Climatic Transition Registered at the Equator and Poles, *Science* 276:1834–1836.
Straus, L. G., Eriksen, B. V., Erlandson, J. M., and Yesner, D. R. (eds.), 1996, *Humans at the End of the Ice Age: The Archaeology of the Pleistocene–Holocene Transition*, Plenum Press, New York.
Street-Perrott, F. A., 1991, General Circulation (GCM) Modelling of Palaeoclimates: A Critique, *The Holocene* 1:74–80.
Stringer, C. B., 1995, The Evolution and Distribution of Later Pleistocene Human Populations. In *Paleoclimate and Evolution, with Emphasis on Human Origins*, edited by E. S. Vrba, G. H. Denton, T. C. Partridge, and L. H. Burckle, pp. 524–531. Yale University Press, New Haven, CT.
Stuiver, M, and Braziunas, T. F., 1993, Modeling Atmospheric ^{14}C Influences and ^{14}C Ages of Marine Samples Back to 10,000 BC. *Radiocarbon* 35:137–189.

Suc, J. P., Bertini, A., Leroy, S. A. G., and Suballyova, D., 1997, Toward the Lowering of the Pliocene/Pleistocene Boundary to the Gauss–Matuyama Reversal. In *The Plio–Pleistocene Boundary*, edited by T. C. Partridge. *Quaternary International* 40:37–42.

Tauxe, L., Opdyke, N. D., Pasini, G., and Elmi, C., 1983, Age of the Plio–Pleistocene Boundary in the Vrica Section, Southern Italy. *Nature* 304:125–129.

Taylor, K. C., Alley, R. B., Doyle, G. A., Grootes, P. M., Mayewski, P. A., Lamorey, G. W., White, J. W. C., and Barlow, L. K., 1993, The "Flickering Switch" of Late Pleistocene Climate Change, *Nature* 361:432–436.

TEMPO, 1996, Potential Role of Vegetation Feedback in the Climate Sensitivity of High-Latitude Regions: A Case Study at 6000 Years BP. *Global Biogeochemical Cycles* 10:727–736.

Thomas, D. H., 1989, *Archaeology* 2nd Ed., Holt, Rinehart and Winston, Orlando, FL.

Thompson, L. G., Mosley-Thompson, E., Bolzan, J. F., and Koci, B. R., 1985, A 1500 Year Record of Tropical Precipitation in Ice Cores from Quelccaya Ice Cap. *Science* 229:971–973.

Trenberth, K. E., 1997, The Use and Abuse of Climate Models, *Nature* 386:131–133.

Urey, H. C., 1947, The Thermodynamic Properties of Isotopic Substances. *Journal of the Chemical Society* 152:190–219.

Van Couvering, J. A., 1997a, Preface: The New Pleistocene. In *The Pleistocene Boundary and the Beginning of the Quaternary*, edited by J. A. Van Couvering, pp. xi–xvii. Cambridge University Press, Cambridge, UK.

Van Couvering, J. A. (ed.), 1997b, *The Pleistocene Boundary and the Beginning of the Quaternary*, Cambridge University Press, Cambridge UK.

Vrba, E. S., Denton, G. H., Partridge, T. C., and Burckle, L. H. (eds.), 1995, *Paleoclimate and Evolution, with Special Emphasis on Human Origins*. Yale University Press, New Haven, CT.

Wagner, G., 1998, *Age Determination of Young Rocks and Artifacts: Physical and Chemical Clocks in Quaternary Geology and Archaeology*, Springer-Verlag, New York.

Walthall, J. A., 1998, Rockshelters and Hunter–Gatherer Adaptation to the Pleistocene/ Holocene Transition, *American Antiquity* 63:223–238.

Waters, M. R., 1992, *Principles of Geoarchaeology: A North American Perspective*, University of Arizona Press, Tucson.

Watson, R. A., and Wright, H. E., 1980, The End of the Pleistocene: A General Critique of Chronostratigraphic Classification, *Boreas* 9:153–163.

Webb, T., III (ed.), 1998, Late Quaternary Climates: Data Synthesis and Model Experiments. *Quaternary Science Reviews* 17:465–688.

Webb, T., III, Ruddiman, W. F., Street-Perrott, F. A., Markgraf, V., Kutzbach, J. E., Bartlein, P. J., Wright, H. E., Jr., and Prell, W. L., 1993, Climatic Changes During the Past 18,000 Years: Regional Syntheses, Mechanisms, and Causes. In *Global Climates Since the Last Glacial Maximum*, edited by H. E. Wright, Jr., J. E. Kutzbach, T. Webb, III, W. F. Ruddiman, F. A. Street-Perrott, and P. J. Bartlein, pp. 514–535. University of Minnesota Press, Minneapolis.

Wenke, R. J, 1984, *Patterns in Prehistory: Humankind's First Three Million Years*, 2nd ed., Oxford University Press, New York.

Williams, M. A. J., Dunkerley, D. L., De Deckker, P., Kershaw, A. P., and Stokes, T., 1993, *Quaternary Environments*, Edward Arnold, New York.

Winograd, I. J., Landwehr, J. M., Ludwig, K. R., Coplen, T. B., and Riggs, A. C., 1997, Duration and Structure of the Past Four Interglaciations. *Quaternary Research* 48:141–154.

Wobst, H. M., 1990, Afterword: Minitime and Megaspace in the Palaeolithic at 18 K and Otherwise. In *The World at 18,000 BP*, Volume 1: *High Latitudes*, edited by O. Soffer and C. Gamble, pp. 331–343. Unwin Hyman, London.

Woodburne, M. O., and Swisher, C. C., III, 1995, Land Mammal High-Resolution Geochronology, Intercontinental Overland Dispersals, Sea Level, Climate, and Vicariance. In *Geochronology, Time Scales and Global Stratigraphic Correlation*, edited by W. A. Berggren, D. V. Kent, C. C. Sivisher, III, and M.-P. Aubry, pp. 335–364. SEPM Special Publication 54, Tulsa, OK.

Wright, H. E., Jr. (ed.), 1983, *Late Quaternary Environments the United States* (2 Volumes). University of Minnesota Press, Minneapolis.

Wright, H. E., Jr., and Frey, D. (eds.), 1965, *The Quaternary of the United States*, Princeton University Press, Princeton, New Jersey.

Wright, H. E., Jr., Kutzbach, J. E., Webb, T., III, Ruddiman, W. F., Street-Perrott, F. A., and Bartlein, P. J. (eds.), 1993, *Global Climates Since the Last Glacial Maximum*, University of Minnesota Press, Minneapolis.

Zeuner, F. E., 1958, *Dating the Past: An Introduction to Geochronology*, 4th ed., Methuen, London.

Zielinski, G. A., Mayewski, P. A., Meeker, L. D., Whitlow, S., and Twickler, M. S., 1996, A 100,000-yr Record of Explosive Volcanism from the GISP2 (Greenland) Ice Core. *Quaternary Research* 45: 109–118.

A Review of Site Formation Processes and Their Relevance to Geoarchaeology

JULIE K. STEIN

1. Introduction

Someone unfamiliar with the history of archaeological methods and theory who read the title of this chapter would guess that the content addresses the processes responsible for the formation of archaeological sites. That guess would be based on the English use of the word "formation"; an act of giving form or shape to something, or of taking form. The word formation in archaeology, however, has a connotation that goes well beyond this English definition.

    A chapter concerning site formation processes is included in this book about earth sciences and archaeology because the study of site formation processes in the discipline of archaeology means more than just analyzing the processes responsible for the formation of archaeological sites. Formation processes are crucial to the discipline because archaeologists use the patterns of artifacts in the ground to infer behaviors. They identify patterns that are created by ancient behaviors and separate those patterns from the ones created by later cultural and natural processes. Earth science methods are required to decipher any natural

---

JULIE K. STEIN • Department of Anthropology, University of Washington, Seattle, Washington 98195.

Earth Sciences and Archaeology, edited by Paul Goldberg, Vance T. Holliday, and Reid Ferring. Kluwer Academic/Plenum Publishers, New York, 2001.

process that may have disturbed the original patterns created by behaviors, and is, therefore, an integral part of site formation analysis. Rather than just a concern with formation, site formation analysis focuses on a broad array of theoretical and methodological issues.

In many ways site formation analysis links archaeology and the earth sciences as no other concept in archaeology ever has. In archaeology, site formation studies grew out of a theoretical approach involving the identification of behaviors common to all people that was called "New Archaeology" (S. R. Binford and Binford, 1968; Clarke, 1968; Reid et al., 1975; Schiffer, 1976, 1995; Taylor, 1948; Watson, 1986; Watson et al., 1971; Willey and Sabloff, 1993; Wylie, 1989). Although many people suggest that the new archaeology did not contribute significantly to changes in theory (Dunnell, 1986), everyone agrees that it had a powerful impact on fields ancillary to archaeology. The systematic and scientific methods advocated by the new archaeologists became the focus of such subdisciplines as geoarchaeology, archaeometry, taphonomy, experimental archaeology, ethnoarchaeology, and lithic analysis. The new focus on formation of sites, depositional history of artifacts, and reconstruction of paleoenvironments linked archaeology and earth sciences. This focus took hold because archaeologists wanted to infer behavior from artifacts and new studies of site formation processes demonstrated that the placement of artifacts shifted since they were deposited (Goldberg et al., 1993; Nash and Petraglia, 1987) and required an earth science approach to be understood.

In this chapter I discuss the history and kinds of formational processes in archaeology, in the hope of clarifying their various connotations and the potentials for misunderstanding. My approach is admittedly Americanist in orientation, as has been my training, but it should provide a starting point for a more international synthesis. The emphasis on formational processes in the decade of the 1970s influenced greatly the subdiscipline of geoarchaeology and the interactions of archaeologists and earth scientists. To clarify this impact I discuss the history of site formation studies and point out the theoretical orientation of such studies as well as the methodological ones. I emphasize the effects of shifting from the artifact assemblage for empirical observations to the deposit as the appropriate unit for analyzing formational processes. In the end, I note a number of successful research examples where site formation analysis contributed to the archaeological research, each example chosen for its strongly contrasting approach.

2. History and Definitions of Formation Processes

The history of collaboration between archaeologists and earth scientists most certainly began early in the history of the discipline (e.g., Butzer, 1964; Pyddoke, 1961), but the person given most credit for championing the particular emphasis of site formation analysis is Michael Schiffer (1972). He was likely influenced by the research of many influential archaeologists and colleagues (e.g., Binford, 1962; Isaac, 1967; Rathje, 1974, 1979; Reid, 1985; Taylor, 1948) but he seems to be the one archaeologist who wrote in such a way as to capture the attention of the archaeological community.

In the 1970s, Michael Schiffer was one of the developers of the behavioral theoretical approach that included this new wrinkle of formation processes. Like many others during this decade (e.g., Binford, 1962, 1964; S. R. Binford and Binford, 1968), Schiffer was interested in gleaning more from the archaeological record than just the classification and ordering of artifacts. Schiffer (1972) asked the questions "How is the archaeological record formed by behavior in a cultural system?" (p. 156), and how is that behavior obscured by later natural and cultural processes? He was trying to create theory and laws to account for characteristics of the archaeological record. He emphatically pointed out that archaeologists had to recognize that both cultural and natural processes operated on the artifacts while they lay in the archaeological record. These natural and cultural processes had to be identified and accounted for to obtain the behavioral information that was of real interest.

Formation processes analysis requires consideration of three kinds of processes, each of which encompasses different fields of study. First, the cultural processes (the behaviors) are the ones of primary interest to the archaeologists. These are responsible for the formation of the archaeological record, including the manner in which objects are procured, used, maintained, and discarded. These behaviors create patterns of artifacts over various locations. They involve the activities that create and use the artifacts, as well as those that result in their deposition. Second, the cultural processes that alter, or obscure, the original behavioral signatures are analyzed. These cultural processes include the actions of people contemporary with deposition, as well as the actions of the archaeologists well after deposition. These actions can either create their own patterns, alter them slightly, or even obliterate them. Third, the natural processes are those non-cultural events that alter, obscure, or preserve the original behavioral signatures. These fall mostly in the realm of the earth sciences, and include a wide variety of environmentally produced actions.

Some people consider formation processes to be only the second and third cultural and natural processes discussed previously. Confusion occurs because most archaeologists do not explicitly differentiate between these three kinds of processes or understand which cultural processes are the target of theoretical and behavioral questions and which relate to formation processes.

2.1. Definition of Site Formation Analysis

Schiffer, in 1972, devoted much of the article to a "simple flow model with which to view the life history of any element, and account behaviorally for the production of the archaeological record" (1972:157). Simply put, an element within a behavioral system goes through states. First is the *systemic context* that "labels the condition of an element which is participating in a behavioral system" (1972:157). Second is the *archaeological context* that "describes materials which have passed through a cultural system, and which are now the objects of investigation of archaeologists" (1972:157).

The systemic context is defined as the activities in which an element participates during its life, and Schiffer (1972) suggested five processes for durable elements: procurement, manufacture, use, maintenance, and discard. Each process consists of stages, and stages can consist of one or more activities. The most important notion of systemic context is that the processes take place in specific locations, and the locations are places where the probability of finding an element is high (1972:160).

Discard is the termination of an element's use-life; it is then considered refuse. Three kinds of refuse can be identified, if location is considered along with discard. *Primary refuse* is material discarded at its location of use, and *secondary refuse* is material discarded in a location different from its location of use. Elements that reach archaeological context without the performance of discard activities are called *de facto refuse* (Schiffer:1972:160).

The 1972 introduction of formation processes was embedded in an introduction to behavioral archaeology, called a life history model of artifactual elements. Schiffer did not even offer a definition of formation processes in this article. He merely supplied a statement introducing a nameless concept as "the conceptual system that explains how the archaeological record is formed" and stated, "the cultural aspect of *formation process concepts* has not been appreciably developed" (Schiffer, 1972:156). In subsequent publications (e.g., Schiffer 1976) the term "formation processes" and the definition come together, as "the factors that create the historic and archaeological records are known as formation processes" (Schiffer, 1987:7).

From the beginning, as the new archaeologists and behavioral archaeologists recognized patterns in assemblages of artifacts, their behavioral conclusions met with criticism almost immediately. The patterns they recognized were not the result of behavior but rather were produced by decomposition, by sampling biases, and (even more problematic) by cultural and noncultural processes (see discussion in Schiffer, 1983:676–678). One had to be sure that the patterns noted in the artifact distributions were really the result of the behavior of the original makers and users and were not identified mistakenly as the result of behaviors associated with the systemic use of the artifacts.

Recognizing that patterns introduced by cultural and post depositional processes alter the original (behavioral) patterns became known as a transformational view of the archaeological record (Schiffer, 1976; Schiffer and Rathje, 1973). The remains in an archaeological site "experienced successive transformations from the time they once participated in a behavioral system to the time they are observed by the archaeologist" (Schiffer, 1975:838). These transformations could be *cultural formation processes*—proccesses of human behavior that affect or transform artifacts after their initial period of use in a given activity (Schiffer, 1987:7)—or *natural (noncultural) formation processes*—processes of the natural environment that impinge on artifacts and archaeological deposits in systematic or archaeological contexts.

Schiffer assumed in 1972 that natural processes were the realm of study of the earth scientist (e.g., geoarchaeologists and archaeometrists) and that these processes are governed by laws already identified within the earth sciences, and therefore were in need of little more research. By 1983, Schiffer's assumptions about natural formation processes changed (Schiffer, 1983). He no longer assumed that all natural processes relevant to archaeology had been studied adequately by other scientists (a fact pointed out early by earth scientists such as Butzer [1964, 1971, and 1982]), and he encouraged a wide variety of research related to pattern recognition. This research and the patterns are described in great detail in a book devoted to the subject (Schiffer, 1987).

Almost immediately following the appearance of these early publications, and for at least a decade, archaeologists and earth scientists offered anecdotal examples of patterns within artifact distributions produced by processes other than the original behavior (Bocek, 1986; Erlandson, 1984; Gifford and Behrensmeyer, 1977; Johnson, 1989; Rick, 1976; Stein, 1983; Stockton 1973; Thomas, 1971; Villa, 1982; Villa and Courtin, 1983; Wood and Johnson, 1978). Experimental research replicated in the lab the patterns observed in artifact distributions of archaeological sites but which were imposed by noncultural processes. This research threw in the face of every archaeologists the fact that if they wanted to identify patterns of artifacts in the archaeological record for the purpose of inferring behaviors, then the transformations (i.e., all the other pattern-producing processes) had to be identified first and taken into account.

> From the standpoint of inference, then, the behavioral and organizational properties that interest archaeologists are reflected—sometimes redundantly and often in complex or subtle ways—in artifacts. However, except in ethnoarchaeological settings and modern material culture studies, we do not deal with items in systemic context. Artifacts recovered archaeologically have been deposited by adaptive systems and subjected to other cultural and natural processes. Thus, in order to infer the *systemic* properties of interest, the archaeologist must identify *and take into account* these formation processes. (Schiffer, 1983:676; emphasis in original)

Many researchers consider wrongly that these other pattern-producing processes are the entire purview of site formation processes. The archaeologists central to the development of site formation analysis perceived site formation studies as central to the building of theory about cultural formation processes. These archaeologists (increasing in numbers in the late 1980s and early 1990s) are mostly associated (as faculty or students) with the University of Arizona. Their theoretical approach is called behavioral archaeology, and their focus is primarily on (but not exclusively) artifacts and sites within the southwestern United States. They did not restrict themselves to the identification of cultural and natural processes that produce or alter patterns of artifacts in assemblages (that was the work to be done by others), but they examined behavioral inferences that affected material objects to identify behavioral laws operating within systemic contexts. To them, formation processes are behavioral processes important for theory building in archaeology.

Thus, the term formation processes describes research about the transformations of the record, but also describes the original behavior surrounding the artifacts. Formation processes connote cultural and natural processes transforming the archaeological record and the cultural processes inferred in theories and laws.

The multiple aspects of site formation analysis is important to note. Those archaeologists not interested in theory building consider formation processes primarily as the identification of processes that create the archaeological record (e.g., Bar-Yosef, 1993; Nash and Petraglia, 1987; Straus, 1993). The different kinds of cultural or natural processes and their relationship to theory is not relevant to them. There exist, therefore, nontheoretical archaeologists who ignore those archaeologists with theoretical concerns who are using the phrase "cultural formation processes" in their search for the laws that dictate behavior, and there are theoretical archaeologists who ignore those archaeologists with earth science concerns who are using the term in their search for natural processes that alter the spatial distribution of objects in the archaeological record.

Site formation analyses are not mutually exclusive—all archaeologists and earth scientists can continue to pursue their search for theory or their search for natural processes that create the archaeological record. The differences are important to understand, however, because each reads the literature of the other, resulting in the potential for confusion.

2.2. The Unit of Site Formation Analysis

As site formational studies increased in popularity in the 1980s, the unit of analysis in the discipline began to change. Previous inferences about behavior (and the preoccupation with chronology), lead archaeologists to focus on the "artifact" as the unit of analysis, or the assemblage. The emphasis on patterns, and recognition of all the forces acting on those patterns, required that attention be shifted to the "deposit." Schiffer (1983) stated that the analytical level at which the identification of formation processes occurs is the "deposit." He expanded this idea later when he stated that

> the perspective elaborated in this chapter leads us to view deposits themselves as peculiar artifacts, the characteristics of which must be studied in their own right. Deposits are the packages containing evidence that might be relevant to one's research questions. Establishing such relevance, however, requires that the genesis of deposits be determined, in terms of both cultural and noncultural formation processes. (Schiffer, 1987:302–303).

Clearly, archaeologists had recognized the importance of the artifact, the assemblage, and the deposit.

Schiffer did not invent this idea independent from other events in archaeology. Karl Butzer had been advocating regional approaches, paleoenvironmental reconstructions, and a focus on deposition through his research in Egypt, in Europe, in South Africa, and in North America (e.g., Butzer, 1960, 1965, 1973, 1976, 1977, 1978, 1981). Others, such as Francois Bordes (1961, 1972) in France, Vance Haynes (1964) in North America, George Rapp Jr. (1975) in Greece, and Myra Shackley (1975) in Great Britain, added volume to the advocacy. As archaeologists shifted their emphasis to patterns and transformations, the deposit as described by earth scientists became the necessary analytical unit.

Immediately a problem appeared, associated with the shift in analytical units. Archaeologists knew how to describe deposits during excavation and in wall profiles (they had been grouping artifacts into assemblages for years; Browman and Givens, 1996), but they were not as comfortable describing or analyzing the deposit itself (the deposit from which the assemblage came). Most archaeologists could not make the leap from describing deposits for the purpose of grouping artifact classes (assemblages) and telling relative time to the purpose of describing them to infer depositional histories and complex postdepositional alterations. These archaeologists had to either attempt the description and interpretations themselves (and make mistakes) or hire earth scientists (who were often untrained in archaeology and make other mistakes).

The collaboration that developed out of this descriptive and interpretive necessity contributed to the growth and emergence of the subdiscipline called geoarchaeology. Geoarchaeology, as a name, was coined by Renfrew in 1973 in his introduction to a conference later published as a book of the same name (Davidson and Shackley, 1976). His definition indicates that on both sides of the Atlantic concern for the archaeological record, patterns, and transformations had lead to an increase awareness of the need for interdisciplinary cooperation between archaeologists and earth scientists. Formation processes and geoarchaeology gained in popularity at about the same time.

> This discipline employs the skills of the geological scientist, using his concern for soils, sediments, and landforms to focus these upon the archaeological "site," and to investigate the circumstances which governed its location, its formation as a deposit and its subsequent preservation and life history. This new discipline of geoarchaeology is primarily concerned with the context in which archaeological remains are found. And since archaeology, or at least prehistoric archaeology, recovers almost all its basic data by excavation, every archaeological problem starts as a problem in geoarchaeology. (Renfrew, 1976:2)

Now archaeologists had reason to believe they needed geoarchaeology to accomplish their basic study of artifacts and assemblages. Because the pattern (context) of artifacts had been transformed, they absolutely had to utilize earth science methods. Geoarchaeology was not a set of methods used only by archaeologists who excavated very old sites or by those who witnessed massive climatic or geomorphological changes. Everyone had to face the concerns of formational processes, which included geoarchaeology and the deposit. Earth scientists had been offering assistance to archaeologists for over a decade, but as site formation analysis gained popularity, numerous articles were added that explained basic tenets and emphasized its potential (Butzer, 1982; Cornwall, 1958; Davidson and Shackley, 1976; Gladfelter, 1977, 1981; Hassan, 1978; Haynes, 1964, 1971; Limbrey, 1975; Pyddoke, 1961; Rapp, 1975). These articles were now being read by a larger proportion of the archaeological community.

Evidence of the problems associated with the shifting of analytical unit from artifacts and assemblages to the deposit is expressed clearly in criticisms of the deposit: the severe shortcoming of the "deposit," as the appropriate unit for archaeological analysis, is that it represents a single depositional episode, which is too simplistic as an organizing principle in archaeology. Schiffer (1987) articulated this concern that a minimal unit of deposition means a *single* event of deposition with the implication that a single activity was responsible for the deposition of the single deposit (or assemblage). This view follows archaeologists' expressed interest in defining specific patterns of artifacts in particular locations. The pattern of artifacts (assemblages) is the information that they investigate to infer behavior, and the minimal unit of deposition had been described in terms of a single activity and behavior that results in single sets of artifacts (assemblages). The grouping (archaeologists refer to it as an assemblage, but here it is referred to as a deposit) is thought to be based on the activity, and because a deposit would likely contain other objects not related to a single activity, it is problematic.

Traditionally in archaeology, the level is a unit created to group artifacts into assemblages (Phillips et al., 1951), and assemblages are grouped because they are considered one kind of behavior (component) or period of time. Given this perspective, a deposit is problematic for the following reasons: a single depositional process can give rise to materials in different deposits, items originally deposited together by one process can be divided up subsequently among several deposits, and a single deposit can contain the products of many different depositional processes (Schiffer, 1987:266). Although articulated most clearly by Schiffer, this list reflects the long tradition of considering artifact deposition as behavior and a heterogeneous deposit only as artifact assemblages.

Earth scientists consider the deposit differently, emphasizing the history of all particles in the deposit through analysis of their source, transport agent, and environment of deposition (Birkeland, 1999; Krumbien and Sloss, 1963; Pettijohn, 1957; Selly, 1988). If the phrase "depositional processes" in each of the three concerns is replaced with the phrase "behavioral processes" or "cultural activities," then the difference between the sedimentologist's view of a deposit and an archaeologist's view becomes clear: a single *cultural activity* can give rise to materials in different deposits, items originally deposited together by one *behavioral process* can be divided up subsequently among several deposits, and a single deposit can contain the products of many different *cultural activities*. The concept of deposit or depositional processes as seen by earth scientists includes more kinds of particles than just artifacts. There are differences between earth scientists' use of depositional processes and the use by some theoretical archaeologists. To study site formation processes, the differences should be understood.

The point here is that earth scientists (and geoarchaeologists) define deposits by their lithology for the ultimate purpose of inferring the sedimentological history of the depositional event (Hassan, 1978; Reading and Levell, 1996; Salvador, 1994; Stein, 1987, 1990). They realize that a depositional event is defined on the basis of what they see before them—the group of particles that look alike. Archaeologists call such a group a *natural level*. The processes involved in the creation of the depositional event (as defined by the analyst in the field) are diverse, bringing particles from many sources, many transport agents, and at variable rates. The deposit is an organizational unit (defined at any scale) in which to decipher the depositional event that created it. It is not only a unit useful for recognizing behaviors or patterns in the artifacts produced by one cultural activity. It is a unit that can provide information about cultural or natural processes at a scale different from that of a single artifact. These groups of objects can add to information derived from single artifacts by adding the contextual relationship of all grains. Deposits are units to study groups of objects brought together through deposition (Stein, 1987, 1990, 1992). For example, analyzing groups of bones (referred to as bone beds) led archaeologists such as Joe Ben Wheat (1972) and George Frison (1974) to innovative conclusions concerning kill sites. Examining individual bones could not provide the depositional perspective that the group of bones provided. Earth scientists examine all grains in a deposit to analyze the depositional history of the entire entity.

This historic review of site formational analysis, and its relationship to earth science, indicates the variety of meanings for the term "site formation"—meanings that go far beyond the English use of the word. Schiffer captured the attention of Americanist archaeologists by using the term to describe both a theoretical approach and a methodological one. The methodological approach emphasized the need to collaborate with earth scientists, and it is the one with which most readers of this book will be familiar. The theoretical approach, however, is found in the literature under formational processes and should be recognized for what it was intended.

3. Examples

Some examples may illustrate the multiple uses of site formation analysis in archaeology. I conducted a library database search of anthropological and earth science literature published since 1984 using the key words "formation processes." This search generated a list of hundreds of entries, most of them using the English definition of the words "formation processes" (e.g., titles of articles such as: "Different Formation Processes of the Moon, the Earth and Meteorites," "Analysis of Reformation and Formation Processes in the course of Karst Development", and "Formation Processes of Gallstones: with special reference to Cholesterol Gallstones"). These examples aside, about 40 of the entries used formation processes in a manner specific to archaeology and illustrate the multiple uses of the term.

Michael Barton and Geoffrey Clark entitled a chapter "Cultural and Natural Formation Processes in Late Quaternary Cave and Rockshelter Sites of Western Europe and the Near East" (Barton and Clark, 1993). The publication summarizes how material is incorporated into rockshelters, separating the transport agents (or processes) into those that are natural and those that are cultural. The list is informative and of interest to anyone attempting to infer cultural events taking place in a rockshelter. The authors, however, did not differentiate the cultural formation processes that are behaviors involving the use and manufacture of artifacts, and those that are responsible for altering the pattern of artifacts after deposition. The authors instead provided an excellent synthesis of the various kinds of cultural and natural transport agents affecting rockshelters.

Stein (1996) summarized research in an article titled "Geoarchaeology and Archaeostratigraphy: view from a Northwest Coast Shell Midden" that explained the origin of a stratigraphic sequence observed frequently in shell midden sites on the Northwest coast. Two layers (shell midden with light colored matrix overlying shell midden with dark-colored matrix) had been interpreted as the result of two different cultural activities representing different cultural components in each site. Stein's research suggested that the color of the matrix is related to the differential water content held within the organic matter and clay minerals. More water is held in the lower layer because a 1m rise in sea level has effectively "raised" the water table of sites near the shore to the point where lower portions of sites are affected. The upper layer is too porous to hold water percolating from the surface, and only the water pulled upward through capillary action hydrates the matrix and produces the dark color. She identified a natural formation process that altered the site to produce a pattern where none had existed previously. The natural pattern has been misinterpreted as cultural. This is a classic example of identifying natural formation processes that affects behavioral interpretations.

In the article "Pollen-Record Formation Processes, Interdisciplinary Archaeology, and Land Use by Mill Workers and Managers: The Boott Mills Corporation, Lowell, Massachusetts 1836–1942" Gerald Kelso (1993) discussed the pollen record in terms of how pollen is naturally deposited on the surface of the site and moved into the subsurface through translocation associated with soil development. Kelso also discussed how cultural site formation processes distort this natural pollen "formation" pattern. These cultural events produce other patterns that are characteristic of broad classes of human behavior and not of natural pollen rain. The pollen was transported by natural processes but was moved post depositionally by cultural processes. This research differentiates between natural and cultural processes, but with a twist. The target event is natural and the cultural processes are those that are altering it. Again, however, this use of formation processes is an excellent study of how the earth's deposits can be altered by a variety of processes. As in the previous examples, this article does not include the theoretical concern associated with formation studies.

Charles Miksicek wrote a synthesis entitled "Formation Processes of the Archaeobotanical Record" (Miksicek, 1987). After reviewing many topics related to plants as part of archaeology (e.g., preservation environments of plants, methods of extraction, and uses of plants), the cultural transformations of the archaeobotanical record were described. These "transformations" were in turn used to predict how plant remains could be used to infer one kind of discard (refuse), and when a charred plant remain was *de facto*, primary, or secondary refuse (terms introduced by Schiffer [1972] as part of his model to describe different kinds of behaviors). Miksicek describe how archaeobotanical objects are transported and altered within the archaeological record, but he went further than other archaeologists toward theoretical concerns and inferring behaviors that might be "laws" applicable to human actions across time and space.

Masakazu Tani (1995) presented "Beyond the Identification of Formation Processes: Behavioral Inference Based on Traces Left by Cultural Formation Processes," in which he clearly addresses behaviors that affect the life history of artifacts in the hopes of defining behavioral laws. Tani's publication flows directly from the original theoretical goals of Schiffer. Tani targeted only cultural depositional processes—those involving refuse disposal. He articulates the depositional formation processes, the refuse deposits, and the theoretical components within behavioral archaeology. He attempts to infer site functions and changes in occupation over time. This study follows faithfully the research advocated by Schiffer toward the goal of understanding behavior through material culture. Tani recognized that the transformational processes had to be considered before the behaviors could be inferred.

These examples demonstrate a range of research conducted under the rubric formation processes. Most researchers use site formation processes to study the creation or alteration of the archaeological record. Many of them focus on geoarchaeological studies of how artifact patterns are destroyed and rearranged after deposition. But other archaeologists use cultural formation processes as synonymous with the search for material culture's use, procurement, or other behavioral inferences. How many archaeologists realize that the term formation process really contains all these connotations?

4. Conclusion

The analysis of formation processes, as defined by Schiffer in the early 1970s and 1980s, required that archaeologists take the giant step beyond just interpreting the histories of artifacts. In addition to the artifacts, archaeologists had to focus on the deposit as a whole and interpret its depositional history. That is a difficult thing to ask of archaeologists, especially ones who were interested essentially in the systemic context (the behavior behind the artifact) that Schiffer originally proposed. The systemic context, however, cannot be determined until the formation processes (transformations) are identified. The artifact patterns that allow us to interpret behavior are altered, obliterated, and sometimes replaced with new patterns, created by processes that are in no way related to the behavior of interest. Whether we like it or not, we must study these processes.

The most appropriate unit for studying these formation processes is the deposit, not the individual artifact. The deposit requires that the archaeological record be thought of in terms less dichotomous than cultural and natural, but rather as one depositional event. Many attributes traditionally used by earth scientists are appropriate to study this event and can be effectively incorporated into archaeology. For example, studies of grain size distributions (Stein and Teltser, 1989), micromorphology (Courty et al., 1989), and sequence stratigraphic methods (Laville et al., 1980; Farrand, 1975, 1993) have all made significant contributions. Archaeologists are not necessarily familiar with these earth science methods, because they are not taught traditionally as part of archaeology. More important, these methods did not become important until archaeologists became interested in questions involving artifact patterns rather than strictly their chronology. Once that step was taken, the new approaches became the appropriate ones.

Geoarchaeology gained popularity in archaeology about the same time as did formational studies. This coincidence is perhaps related. Archaeologists have often concerned themselves with the regional environment, the burial of their sites, or the preservation conditions of organic remains (concerns identified as part of geoarchaeology). The concern blossomed, however, when processes were identified that could affect the patterns of artifacts. Artifacts are the central concern of archaeology and their interpretation is the focus of the discipline. Geoarchaeologists could really assist in the identifications of these processes, and such contributions became crucial.

The discipline of archaeology now recognizes that formational processes are integral, and most excavations include a watchful eye toward identifying transformational events that could distort artifact patterns. Not everyone, however, is equally equipped to identify these processes or to make the correct measurements of deposit-based attributes, archaeologists and earth scientists alike. The archaeological community must make an effort to provide students with training in the methods necessary to succeed in identifying formation processes correctly.

One step that will lead us closer to this goal is to recognize that the phrase formation processes means different things to different people. To behavioral archaeologists it includes the search for behavioral laws; to geoarchaeologists it means the search for processes of formation associated with the archaeological context. These are not mutually exclusive goals, just connotations that could cause confusion.

ACKNOWLEDGMENTS. The author thanks Michael Schiffer for many years of conversation about formation processes, and Patrice Teltser for acting as translator between the two of us. This manuscript was greatly improved by the comments of Michael Schiffer, Debora Kligmann, Vance Holliday, Paul Goldberg, and Reid Ferring, the indexing of Christopher Lockwood, and the bibliographic assistance of Jennie Deo.

6. References

Bar-Yosef, O., 1993, Site Formation Processes from a Levantine Viewpoint. In *Formation Processes in Archaeological Context*, edited by P. Goldberg, D. T. Nash, and M. D. Petraglia, pp. 11–32. Monographs in World Archaeology No. 17, Prehistory Press, Madison, WI.

Barton, C. M. and Clark, G. A., 1993, Cultural and Natural Formation Processes in Late Quaternary Cave and Rockshelter Sites of Western Europe and the Near East. In *Formation Processes in Archaeological Context*, edited by P. Goldberg, D. T. Nash, and M. D. Petraglia, pp. 33–60. Monographs in World Archaeology No. 17, Prehistory Press, Madison, WI.

Binford, L. R., 1962, Archaeology as Anthropology. *American Antiquity* 28:217–225.

Binford, L. R., 1964, A Consideration of Archaeological Research Design. *American Antiquity* 29:425–441.

Binford, S. R. and Binford, L. R., 1968, *New Perspectives in Archeology*. 2nd ed., Aldine-Atherton, Chicago.

Birkeland, P. W., 1984, *Soils and Geomorphology*. Oxford University Press, New York.

Birkeland, P. W., 1999, *Soils and Geomorphology*, 3rd ed.. Oxford University Press, Oxford, U.K.

Bocek, B., 1986, Rodent Ecology and Burrowing Behavior: Predicted Effects on Archaeological Site Formation. *American Antiquity* 51:589–603.

Bordes, F., 1961, Mousterian Cultures in France. *Science* 134:803–810.

Bordes, F., 1972, *A Tale of Two Caves*. Harper & Row, New York.

Browman, D. L. and Givens, D. R., 1996, Stratigraphic Excavation: The First "New Archaeology" *American Anthropologist*. 98:80–95.

Butzer, K., 1960, Archeology and Geology in Ancient Egypt. *Science* 132:1617–1624.
Butzer, K., 1964, *Environment and Archaeology: An Introduction to Pleistocene Geography*. Aldine, Chicago.
Butzer, K., 1965, Acheulian Occupation Site at Torralba and Ambrona, Spain: Their Geology. *Science* 150:1718–1722.
Butzer, K., 1971, *Environment and Archaeology: An Ecological Approach to Prehistory*, Aldine-Atherton, Chicago.
Butzer, K., 1973, Geology of Nelson Bay Cave, Robberg, South Africa. *South African Archaeological Bulletin* 28:97–110.
Butzer, K., 1976, *Early Hydraulic Civilization in Egypt*. University of Chicago Press, Chicago.
Butzer, K., 1977, *Geomorphology of the Lower Illinois Valley as a Spatial–Temporal Context for the Koster Archaic Site*. Illinois State Museum, Reports of Investigations No. 34, Springfield, Illinois.
Butzer, K., 1978, Changing Holocene Environments at the Koster Site: A Geoarchaeological Perspective. *American Antiquity* 43:408–413.
Butzer, K., 1981, Cave Sediments, Upper Pleistocene Stratigraphy and Mousterian Facies in Cantabrian Spain. *Journal of Archaeological Science* 8:133–182.
Butzer, K., 1982, *Archaeology as Human Ecology*. Cambridge University Press, Cambridge, UK.
Clarke, D. L., 1968, *Analytical Archaeology*. Methuen, London.
Cornwall, I. W., 1958, *Soils for the Archaeologist*, Phoenix House, London.
Courty, M. A., Goldberg, P., and Macphail, R., 1989, *Soils and Micromorphology in Archaeology*. Cambridge University Press, New York.
Davidson, D. A., and Shackley, M. L., 1976, *Geoarchaeology*. Westview Press, Boulder, CO.
Dunnell, R. C., 1986, Five Decades of American Archaeology. In *American Archaeology: Past and Future*, edited by D. J. Meltzer, D. D. Fowler, and J. A. Sabloff, pp. 23–49. Smithsonian Institution Press, Washington, DC.
Erlandson, J. M., 1984, A Case Study in Faunalturbation: Delineating the Effects of the Burrowing Pocket Gopher on the Distribution of Archaeological Materials. *American Antiquity* 49:785–790.
Farrand, W. R., 1975, Sediment Analysis of a Prehistoric Rockshelter: The Abri Pataud. *Quaternary Research* 5:1–26.
Farrand, W. R., 1993, Discontinuity in the Stratigraphic Record: Snapshots from Franchthi Cave. In *Formation Processes in Archaeological Context*, edited by P. Goldberg, D. T. Nash, and M. D. Petraglia, pp. 85–96. Monographs in World Archaeology No. 17, Prehistory Press, Madison, WI.
Frison, G. C., 1974, *The Casper Site: A Hell Gap Bison Kill on the High Plains*, Academic Press, San Diego.
Gifford, D., and Behrensmeyer, A. K., 1977, Observed Formation and Burial of a Recent Human Occupation Site in Kenya. *Quaternary Research* 8:245–266.
Gladfelter, B. G., 1977, Geoarchaeology: the Geomorphologist and Archaeology. *American Antiquity* 42:519–538.
Gladfelter, B. G., 1981, Developments and Directions in Geoarchaeology. *Advances in Archaeological Method and Theory* 4:344–364.
Goldberg, P., Nash, D. T., and Petraglia, M. D., 1993, *Formation Processes in Archaeological Context*. Monographs in World Archaeology No. 17, Prehistory Press, Madison, WI.
Hassan, F. A., 1978, Sediments in Archaeology: Methods and Implications for Paleoenvironmental and Cultural Analysis. *Journal of Field Archaeology* 5:197–213.
Haynes, C. V., Jr., 1964, The Geologists's Role in Pleistocene Paleoecology and Archaeology. In *The Reconstruction of Past Environments*, edited by J. J. Hester and J. Schoenwetter, pp. 61–64. Publication of the Fort Burgwin Research Center, No. 3, Taos, N.M.
Haynes, C. V., Jr., 1971, Time, Environment, and Early Man. *Arctic Anthropology* 8:3–14.
Isaac, G. L., 1967, Towards the Interpretation of Occupation Debris: Some Experiments and Observations. *Kroeber Anthropological Society Papers* 37:31–57.
Johnson, D. L., 1989, Subsurface Stone Lines, Stone Zones, Artifact–Manuport Layers, and Biomantles Produced by Bioturbation via Pocket Gophers (Thomomys bottae). *American Antiquity* 54:370–389.
Kelso, G. K., 1993, Pollen-Record Formation Processes, Interdisciplinary Archaeology, and Land Use by Mill Workers and Managers: The Boott Mills Corporation, Lowell, Massachusetts, 1836–1942. *Historical Archaeology* 27:70–94.
Krumbein, W. C., and Sloss, L. L., 1963, *Stratigraphy and Sedimentation*. W. H. Freeman, San Francisco.

Laville, H., Rigaud, J.-P., and Sackett, J., 1980, *Rock Shelters of the Perigord: Geological Stratigraphy and Archaeological Succession*. Academic Press, New York.
Limbrey, S., 1975, *Soil Science and Archaeology*, Academic Press, London.
Miksicek, C. H., 1987, Formation Processes of the Archaeobotanical Record. *Advances in Archaeological Method and Theory* 10:211–247.
Nash, D. T., and Petraglia, M. D., 1987, *Natural Formation Processes and the Archaeological Record*. BAR International Series 352, London.
Pettijohn, F. J., 1957, *Sedimentary Rocks*, 2nd ed. Harper & Bros., New York.
Phillips, P., Ford, J. A., and Griffin, J. B., 1951, *Archaeological Survey in the Lower Mississippi Valley, 1940–1947*. Papers of the Peabody Museum of Archeology and Ethnology No. 25, Harvard University, Cambridge.
Pyddoke, E., 1961, *Stratification for the Archaeologist*. Phoenix House, London.
Rapp, G., Jr., 1975, The Archaeological Field Staff: The Geologist. *Journal of Field Archaeology* 2:229–237.
Rathje, W. L., 1974, The Garbage Project: A New Way of Looking at the Problems of Archaeology. *Archaeology* 27:236–241.
Rathje, W. L., 1979, Modern Material Culture Studies. *Advances in Archaeological Method and Theory* 2:1–37.
Reading, H. G., and Levell, B. K., 1996, Controls on the Sedimentary Rock Record. In *Sedimentary Environments: Processes, Facies and Stratigraphy*, 3rd ed., edited by H. G. Reading, pp. 5–36. Blackwell Science, Cambridge.
Reid, J. J., 1985, Formation Processes for the Practical Prehistorian. In *Structure and Process in Southeastern Archaeology*, edited by R. S. Dickens, Jr., and H. T. Ward, pp. 11–13. University of Alabama Press, University, Alabama.
Reid, J. J., Schiffer, M. B., and Rathje, W. L., 197 , Behavioral Archaeology: Four Strategies. *American Anthropologist* 77:864–869.
Renfrew, C., 1976, Archaeology and the Earth Sciences. In *Geoarchaeology: Earth Science and the Past*, edited by D. A. Davidson and M. L. Shackley, pp. 1–5. Westview Press, Boulder, Colorado.
Rick, J. W., 1976, Downslope Movement and Archaeological Intrasite Spatial Analysis. *American Antiquity* 41:133–144.
Salvador, A., 1994, *International Stratigraphic Guide: A Guide to Stratigraphic Classification, Terminology, and Procedure*, 2nd ed. The Geological Society of America, Boulder, Colorado.
Schiffer, M. B., 1972, Archaeological Context and Systemic Context. *American Antiquity* 37:156–165.
Schiffer, M. B., 1975, Archaeology as Behavioral Science. *American Anthropology* 77:836–848.
Schiffer, M. B., 1976, *Behavioral Archaeology*. Academic Press, New York.
Schiffer, M. B., 1983, Toward the Identification of Formation Processes. *American Antiquity* 48:675–706.
Schiffer, M. B., 1987, *Formation Processes of the Archaeological Record*. University of New Mexico Press, Albuquerque.
Schiffer, M. B., 1995, *Behavioral Archaeology: First Principles*, University of Utah Press, Salt Lake City.
Schiffer, M. B., and Rathje, W. L., 1973, Efficient Exploitation of the Archeological Record: Penetrating Problems. In *Research and Theory in Current Archeology*, edited by C. L. Redman, pp. 169–179. Wiley, New York.
Selley, R. C., 1988, *Applied Sedimentology*, Academic Press, San Diego.
Shackley, M. L., 1975, *Archaeological Sediments*. Wiley, New York.
Stein, J. K., 1983, Earthworm Activity: A Source of Potential Disturbance of Archaeological Sediments. *American Antiquity* 48:277–289.
Stein, J. K., 1987, Deposits for Archaeologists. In *Advances in Archaeological Method and Theory*, Volume 11, edited by M. B. Schiffer, pp. 337–393. Academic Press, Orlando, Florida.
Stein, J. K., 1990, Archaeological Stratigraphy. In *Archaeological Geology of North America*, edited by N. P. Lasca and J. Donahue, pp. 513–523. Geological Society of America, Centennial Special Volume 4, Boulder, Colorado.
Stein, J. K., 1992, Interpreting Stratification of a Shell Midden. In *Deciphering a Shell Midden*, edited by J. K. Stein, pp. 71–93. Academic Press, San Diego.
Stein, J. K., 1996, Geoarchaeology and Archaeostratigraphy: View from a Northwest Coast Shell Midden. In *Case Studies in Environmental Archaeology*, edited by E. J. Reitz, L. A., Newson and S. J. Scudder, pp. 35–54. Plenum Press, New York.

Stein, J. K. and Teltser, P. A., 1989, Size Distributions of Artifact Classes: Combining Macro- and Micro-Fractions. *Geoarchaeology* 4:1–30.

Stockton, E., 1973, Shaw's Creek Shelter: Human Displacement of Artifacts and its Significance. *Mankind* 9:112–117.

Straus, L. G., 1993, Hidden Assets and Liabilities: Exploring Archaeology from the Earth. In *Formation Processes in Archaeological Context*, edited by P. Goldberg, D. T. Nash, and M. D. Petraglia, pp. 1–10. Monographs in World Archaeology No. 17, Prehistory Press, Madison, Wisconsin.

Tani, M., 1995, Beyond the Identification of Formation Processes: Behavioral Inference Based on Traces Left by Cultural Formation Processes. *Journal of Archaeological Method and Theory* 2:231–252.

Taylor, W. W., 1948, *A Study of Archaeology*. American Anthropological Association, Memoir 69, Washington DC.

Thomas, D. H., 1971, On Distinguishing Natural from Cultural Bone in Archaeological Sites. *American Antiquity* 36:366–371.

Villa, P., 1982, Conjoinable Pieces and Site Formation Processes. *Amercian Antiquity* 47:276–290.

Villa, P., and Courtin, J., 1983, The Interpretation of Stratified Sites: A View from the Underground. *Journal of Archaeological Science* 10:267–281.

Watson, P. J., 1986, Archaeological Interpretation, 1985. In *American Archaeology Past and Future: A Celebration of the Society for American Archaeology 1935–1985*, edited by D. J. Meltzer, D. D. Fowler, and J. A. Sabloff, pp. 439–457. Smithsonian Institution Press, Washington, DC.

Watson, P. J., LeBlanc, S. A., and Redman, C. L., 1971, *Explanation in Archaeology: An Explicitly Scientific Approach*. Columbia University Press, New York.

Wheat, J. B., 1967, A Paleo-Indian Bison Kill. *Scientific American* 216:43–52.

Wheat, J. B., 1972. The Olsen–Chubbuck Site: A Paleo-Indian Bison Kill. Society for American Archaeology Memoir 26, Washington, DC.

Willey, G. R., and Sabloff, J. A., 1993, *A History of American Archaeology*, 3rd ed. W. H. Freeman, New York.

Wood, W. R., and Johnson, D. L., 1978, A Survey of Disturbance Processes in Archaeological Site Formation. In *Advances in Archaeological Method and Theory*, Volume 1, edited by M. B. Schiffer, pp. 315–381. Academic Press, Orlando, Florida.

Wylie, A., 1989, The Interpretive Dilemma. In *Critical Traditions in Contemporary Arcaheology: Essays in the Philosophy, History and Socio-Politics of Archaeology*, edited by V. Pinsky and A. Wylie, pp. 18–27. Cambridge University Press, Cambridge.

Geomorphological Studies II

Evaluating Causality of Landscape Change

Examples from Alluviation

CHARLES FREDERICK

1. Introduction

Compiling and summarizing regional and local evidence of geomorphic change are among the more common tasks of field geoarchaeologists in North America, where they have become almost routine aspects of archaeological survey. The scale of geoarchaeological projects in America is often regional, owing to a considerable number of large development projects such as reservoirs, open cast mines, and highways. In Europe, the scale of development is often (but clearly not always) of smaller scale, and it is my impression that the majority of geoarchaeological projects are performed at the scale of individual sites. The principal exceptions to this generalization are often academic surveys, which in some regions of Europe routinely incorporate geoarchaeological studies. Perhaps the best example of this may be found in Greece (e.g., Pope and van Andel, 1984; Wells *et al.*, 1990; Zangger, 1994; Zangger *et al.*, 1997).

The rationale of investigating geomorphic change as part of archaeological research is clear where the archaeological record under investigation has no

---

CHARLES FREDERICK • Department of Archaeology and Prehistory, University of Sheffield, Sheffield S1 4ET, United Kingdom.

Earth Sciences and Archaeology, edited by Paul Goldberg, Vance T. Holliday, and C. Reid Ferring. Kluwer Academic/Plenum Publishers, New York, 2000.

significant architectural components. In such situations, deciphering multiple prehistoric occupations from stratigraphically compressed scatters of burnt rock and debitage on ancient landforms is often problematic. In an attempt to find clearer separation between occupations, archaeologists in some regions (e.g., Texas) have turned to excavating sites in dynamic or rapidly aggrading portions of the landscape, such as river valleys, where occupations of short duration are found in stratigraphic isolation (see Collins, 1995). The effects of aggradation on the stratigraphic separation of cultural deposits has been illustrated by Ferring (1986). In general terms, as the sedimentation rate increases, so does the thickness of sediment between occupations of different age. Clearly, the degree to which this occurs is a function of the rate of sedimentation and the periodicity, intensity, and duration of occupation. Sites with multiple, stacked, but stratigraphically separated components have been referred to as sites with "isolable components" or as *gisements* (cf. Collins, 1995:374). This stratigraphic situation facilitates interpretation of the archaeological record, sometimes allowing us to refine artifact seriations as well as simply providing a much clearer image of cultural activity areas.

In this context, which is common in North American archaeology, stratigraphic and geomorphic data are used as a proxy for climate and vegetation change, to search strategically for archaeological sites of a particular age, or to establish age estimates for sites discovered during the survey work. Although it is rarely articulated explicitly (e.g., Xu, 1998:67), most researchers assume that climatic variation is principally responsible for erosion and sedimentation, except where tectonic activity is suspected or in proximity to the modern coast where sea level variation may affect stream response (e.g., Blum, 1991; Blum and Price, 1994, 1998). Exceptions to this generalization include regions where significant sedentary populations were present during the prehistoric period and/or deposits that postdate European settlement. In both cases, anthropogenic factors are argued to have caused soil erosion and concomitant valley alluviation.

Where sites with earthworks and/or earth or stone buildings are common, as in Europe and in Mesoamerica, the majority of field archaeology occurs on stable upland surfaces that have relatively high archaeological visibility and may comprise graphic palimpsests. In these regions, the role of human activity in environmental change is a prominent research theme. Studies addressing this issue are not often performed in site context, because most archaeological excavations in these landscapes are not suitable. Researchers must seek out settings where evidence of environmental change is likely to be preserved, such as lakes, bogs, mires, slopes, and alluvial valleys. Sites in bogs and in mires are suitable for compiling records of past vegetation change, but are not necessarily suitable for evaluating ancient soil erosion and sediment mobilization. Lake basins, slopes and river valleys are the preferred landscape elements for compiling records sensitive to soil erosion. Lakes may yield high resolution and potentially less problematic records, but they are uncommon in arid and in semiarid landscapes and as such are not universally applicable. Alluvial and colluvial deposits, on the other hand, occur much more frequently and are therefore more useful for evaluating the origins of prehistoric and historic erosion and sedimentation.

In Europe and in North America, emphasis is placed on evaluating the causes of sedimentation, especially where human agency is concerned. Although this process is not critical to most archaeological projects, the issue is nonetheless one which has caused much interest among archaeologists, geographers, and Quaternary geologists. A myriad of factors influence erosion and sedimentation in rivers (see discussions in Bull, 1991; Schumm, 1977 for an overview). The most frequently cited factors in the literature are: climate, land use, tectonic activity, changes in base level (sea level in most cases), and internal factors such as geomorphic thresholds. How researchers support these inferences is not often clear; this calls their conclusions into question. Hence the goal of this chapter is to explicate how researchers have constructed their arguments, identify which lines of evidence are considered indicative of each factor, and illustrate where the problems may lie.

2. Basic Elements of Alluvial Stratigraphic Sequences

Three main elements are necessary to evaluate the origins of alluviation: (1) a stratigraphic sequence, (2) a chronology for this sequence, and (3) one or more dated proxy records that document change of an external variable that may influence alluviation. Many studies of Holocene alluviation are lacking on the second and third counts, which significantly reduces their utility and therefore the validity of their arguments.

2.1. The Stratigraphic Sequence

Alluvial chronology is based on stratigraphy. Examples of how such sequences are constructed are provided elsewhere in this volume (see Chapter 4, this volume) and are only briefly reviewed here. Initially, the reaches of a valley are examined for geomorphic variability. A generalized sequence of geomorphic surfaces (i.e., terraces, floodplain, and alluvial fans) is compiled and is used as a guide for siting stratigraphic excavations. Where there are few natural exposures, mechanical excavation or coring aids this work. In arid and semiarid landscapes, a relatively complete sequence may be compiled from observation of natural exposures such as cut banks and arroyo walls. Sections must be obtained from the valley axis and the valley margins, the latter of which are often where older deposits may be found at depth. Once exposures are found or made, they are described. Observations of the lithology, morphology, and stratigraphic occurrence of soils are compiled and are used to construct a model of the alluvial architecture. Most sequences found in the North American Quaternary literature are based on either lithologic observations (a lithostratigraphic sequence) or the occurrence of unconformities (an allostratigraphic sequence; cf. Autin, 1992; North American Commission of Stratigraphic Nomenclature, 1983).

Once the stratigraphic framework has been compiled, samples are collected for dating in order to temporally define each major stratigraphic unit. Although geoarchaeological investigations may share goals with the archaeological fieldwork, constructing an alluvial stratigraphic sequence necessitates gathering stratigraphic information from outside site contexts in a variety of settings. This includes even the most recent depositional environments, which may seem irrelevant to some archaeologists. Rarely can site excavations be used as the exclusive source of exposures for this process, as site excavations are chosen for different reasons (the presence of cultural material rather than stratigraphic potential). To use them exclusively risks missing significant stratigraphic units. Also, some cultural activities (e.g., creation and natural infilling of reservoirs and borrow pits, earthworks, etc.) result in deposits that are only of local significance. This may be recognized only if stratigraphic work is performed in various settings. Furthermore, human settlements often do not occur in settings that are typical of the broader alluvial environments.

A final point is worthy of consideration when examining stream response to external or internal variables: recent works suggest that small-order stream systems are more sensitive to changes in external factors. Therefore they are likely to yield less ambiguous results when compared to large streams, which tend to average results of many small tributary basins (Arbogast and Johnson, 1994; McFadden and McAuliffe, 1997:306). The other advantage of small basins is that the linkage between slope erosion and valley aggradation is much more direct. The disadvantage of small basins is that major geomorphic events such as channel entrenchment may not extend into the upper reaches of the drainage network, resulting in different architectural and stratigraphic sequences. Hence the stratigraphic sequences in such settings may not be representative of the entire basin.

2.2. Chronology

Perhaps the most critical element of stratigraphic studies is the chronology. Several dating methods are applicable to slope and fluvial systems (see Brown, 1997; Smart and Frances, 1991 for an overview). The most widespread and useful is radiocarbon (Brown, 1997:48–52), although other dating techniques such as dendrochronology (e.g., Becker, 1975; Karlstrom, 1988) and luminescence methods (Bailiff, 1992; Nanson and Young, 1987; Nanson et al., 1991; Preusser, 1998) are increasingly common. Lack of adequate chronological control has been recognized as a problem with Holocene alluvial chronologies for more than 20 years (cf. Butzer, 1980), and many of the comments made by Butzer are as pertinent today as they were then. Without an adequate chronology, it is nearly impossible to draw correlations with either archaeological or other paleoenvironmental information (Butzer, 1980; Knox, 1983:38; Macklin, 1995:179). Poor chronologies arise from at least four distinct problems: (1) lack of financial resources, (2) a lack of datable material from stratigraphically significant locations, (3) poor sample precision (Table 3.1), and (4) taphonomic factors.

In terms of financial resources, there are several identifiable problems. Stratigraphic studies funded through archaeological projects often compete with archaeological goals and in my experience, the archaeology nearly always prevails. This is not as much a criticism as it is a pragmatic reality. In compliance-

Table 3.1. Materials Used in Radiocarbon Dating of Alluvial Stratigraphic Sequences and Their Limitations

Type of material and context	Limitations
Wood or charcoal from *in situ* trees	Few significant limitations. Significant errors may be incurred depending on the portion of the tree that is dated. Best used for dendrochronology if possible.
Wood or charcoal from redeposited trees	Same as previous, except significant errors may be obtained for material that has been repeatedly reworked. Date is a maximum age for the deposit.
Charcoal from hearths	Relatively few. Dates will vary according to parts of tree selected (samples collected from the interior of long-lived trees can cause significant errors). Likewise, cultural selection of fuel source can bias results (e.g., driftwood vs. recently fallen branches). Can be quite accurate if annual plant remains are dated. Resulting date is a maximum age for the deposit.
Charcoal from cultural contexts (nonhearth)	Similar to hearths, except there is a greater probability of errors associated with redeposition. Resulting date is a maximum age for the deposit.
Organic sediments	Plants growing in aquatic environments may incorporate ancient carbon dissolved in the water, thereby incurring an age error, often referred to as a "hard water error." Organic sediments also may be redeposited in cultural contexts.
Paleosol organic matter	Samples suffer a mean residence error that results in a date older than the time of burial. May provide a maximum age for unit that buries soil.
Bulk sediment (unmodified by pedogenesis)	Assumed to approximate the time of deposition but where they have been checked, they have been found to nearly always date older than time of sedimentation due to inclusion of older organic matter (e.g. Frederick, 1993; Abbott, 1995; Quigg and Peck, 1995). This error is generally less than 1,000 years but can be as much as several thousand years in some instances (see Abbott and Valastro, 1995, for an example of this). The only occasions they will date younger is if contaminated with younger carbon through processes like pedogenesis, pedoturbation, or recent cultural practices such as the use of borrow pits for silage.
Scattered charcoal	Unpredictable as charcoal can be reworked over long periods yieding unreliable dates. Otherwise, same limitations as wood or charcoal from redeposited trees.

related archaeological projects, the goal is to advance the understanding of archaeological issues, not geological ones, even though the two are often complementary. Furthermore, if the stratigraphic work is done in the right sequence, that is before or concomitant with survey, it is probable that the work was funded from the survey budget, which is often highly competitive. If the level of dating is not specified in the contract, dating budgets are all too often cut in order to make bids more competitive. The best dated projects I have seen were those that specified the level of geoarchaeological effort for the initial phases of fieldwork (survey) or alternately, advocated the geoarchaeology as a separate phase of work altogether. Academic projects do not suffer from the same financial restraints as contract archaeology and are often better dated. However, the number of dates is often a function of the ease with which money for radiocarbon dates or other forms of geochronology may be obtained. In general, European projects appear to have a fewer number of Carbon-14 dates than American ones.

A fieldwork axiom is that one rarely encounters datable material where it is desired most, and therefore nearly all studies suffer from opportunistic dating. Clearly, this situation affects each sequence to different degrees and there are few ways around it. Mixing dating techniques may be one way to circumvent this problem. For example, using luminescence methods to directly date sandy channel and proximal overbank deposits where charcoal or other suitable organic matter is not available, and then employing radiocarbon for the fine grained facies, temporally brackets each depositional unit.

Another factor that influences the quality of the chronology, poor sample precision, is largely a function of the method used and the sample dated. Luminescence methods, for example, often yield dates with large standard deviations (ca. 10–15%). With radiocarbon dating, the significance of each dated sample must be evaluated separately, as the geologic implications vary widely depending on the context and nature of the dated material (see Table 3.1). Unfortunately, the choice of material for dating is often severely limited and so we have to rely on less accurate materials. In most contract archaeological studies, limitations of field and total project time cause researchers to choose less reliable materials for dating even when more reliable materials are present but difficult to find. Where field time and access are less limited, it is possible to be more selective of material, which results in more accurate chronologies.

A classic example of this may be found in comparing several recent alluvial stratigraphic studies from Texas. Nordt's (1992) work at Fort Hood, Texas, is an example of a study performed over a period of years that allowed the author to select charcoal from secure stratigraphic contexts, such as cultural hearths, thereby increasing the precision of the alluvial chronology. Similar kinds of studies performed in conjunction with reservoir projects in the same state (e.g., Blum and Valastro, 1992; Frederick, 1993) were compiled from as little as 20 days in the field, which severely limited the options of material that could be selected and sampled for dating. In the two studies cited here, the majority of the radiocarbon ages were on bulk sediment organic matter as opposed to charcoal. Dates for this kind of organic matter are assumed to approximate the time of deposition, but in reality they include an age error that is variable and difficult to quantify except on a per case basis. Where these age errors have been investigated they have been found to range from a few hundred years to as much as several thousand years.

Some of the most precisely dated alluvial chronologies are from dendrochronological dating. The use of this method is limited to saturated or very dry settings where wood is preserved, so it is not widely applicable. In fact, the lack of well-established dendrochronologies, in conjunction with poor preservation of wood, are the primary factors that limit the use of this dating method. The best examples are from the southwestern United States (e.g., Hereford et al., 1996; Karlstrom, 1988; Plog et al., 1988), although this technique has been used in Europe as well (Becker, 1975; Becker and Schirmer, 1977; Bell and Walker, 1992). Dendrochronology may also be used in studies of very young deposits, where germination dates of modern trees provide minimum ages for geomorphic surfaces and deposits. Examining the development of adventitious roots that form anew following burial of trees by alluvium may yield information on sedimentation that occurred during the life of the tree (Orbock Miller et al., 1993).

Luminescence dating is still under evaluation with respect to alluvial sediments. The method as it applies to alluvial deposits has been summarized by Bailiff (1992). Luminescence methods, especially optically stimulated luminescence (OSL), are more frequently employed in Europe than in North America. Although the use of luminescence techniques is increasing in the New World, the popularity of this method in Europe is probably related to the limited funds available for radiocarbon dating, especially in Britain. Optical luminescence dating has been successfully applied to alluvial deposits in northeast Spain (e.g., Macklin et al., 1994), in Switzerland (Preusser, 1998), in the Netherlands (Duller and Törnqvist, 1998) and in eastern Germany (Mol, 1997). From a theoretical perspective, OSL dates from fluvial deposits may yield erroneously old ages owing to inadequate bleaching during transportation. This is most likely to occur in streams with very cloudy water columns. Other forms of luminescence dating have been used with alluvial deposits as well. For instance, impressive sedimentary chronologies have been compiled using thermoluminescence in Australia (e.g., Nanson and Young, 1987; Nanson et al., 1991).

In the end, even with adequate funding, plenty of field time, adequate sampling, and presumably precise methods, a number of factors may lead to dating irregularities and to less than useful chronologies. The nature of these irregularities is difficult to generalize as many are particular to the dating techniques employed (e.g., partial bleaching in OSL, mean residence times in soil radiocarbon dates). Nevertheless, taphonomic processes such as pedoturbation are also capable of leading to erroneous ages.

2.3. Combination Proxies

In order to argue causality successfully, one must employ one or more proxy data sets. These permit comparison and correlation of the timing of alluvial events with factors known to influence alluvial sedimentation (Tipping, 1992). In general terms, useful factors include the following: variation in the frequency, seasonality and magnitude of precipitation; relative differences in past precipitation; changes in vegetation; tectonic and eustatic variation; and past land use and demography. Ideally, an alluvial chronology is compared with dated proxy records for all of the major external factors. In practice, not all of the factors are warranted in every case. In the American midcontinent, prehistoric land use and tectonic activity are commonly discounted because neither are considered to have been significant enough influences on Holocene alluviation.

In southern Europe (e.g., Greece and Italy), information on all of the aforementioned factors would be useful given the active tectonics and long period of settlement. The problems associated with Mediterranean environments have been well articulated over the last few years, and they illustrate the problems of establishing causality in dominantly arid to semiarid landscapes with dynamic climatic, vegetation, tectonic, and land use histories (see papers in Lewin et al., 1995; Bottema et al., 1990 for examples addressing these issues). Rarely are all of the useful proxies available. When they are present, many lack sufficient resolution or sensitivity to be of much use.

When working with proxy records it is important to keep in mind how far removed the interpretations are from the value they represent. Caran (1998) provided a nice illustration of how this can affect the integrity of paleoenvironmental arguments. Bell and Walker (1992) provided the example of a climatic interpretation derived from a pollen sequence. To make such an interpretation, one must first reconstruct plant communities and vegetation change from the data and then use those changes to infer variations in paleoclimate. Both of these inferential steps are potentially erroneous, and the more removed the proxy record, the more likely the error.

2.3.1. Vegetation Change

2.3.1.a. Pollen Perhaps the most commonly employed proxy records are studies of past vegetation change inferred from stratigraphic sequences of fossil pollen. Both climate and land use may be discerned from pollen records and in many regions this form of paleoenvironmental evidence is the primary source of proxy palaeoclimatic data. Pollen records may be sensitive or complacent and therefore are not always fruitful sources of information. They are particularly salient records for geomorphic studies in that they inform on the nature and character of vegetation that buffer and protect the surface from erosion. The mechanics of how vegetation protects the soil surface from erosion are particularly well known today (e.g., Thornes, 1987a, 1987b) but extrapolating this kind of response into the past is often a difficult proposition.

Unfortunately, palynology is limited to locations that favor preservation of pollen. In more arid environments, pollen preservation is generally poor and knowledge of past vegetation change may be fuzzy at best. Pollen records have been derived from alluvial sequences (e.g., Hall, 1977), but there is some debate concerning the degree of hydrodynamic sorting that occurs in pollen redeposited by streams. Evidence of anthropogenic interference is easier to ascertain in some landscapes than in others and is not always a cut and dried proposition.

2.3.1.b. Isotopes A method for building proxy vegetation records is to document changes in the chemical remains of former vegetation through stable isotopic ratio analysis (SIRA) of organic carbon ($^{13}C/^{12}C$) in soil, sediment, and biotic systems (Humphrey and Ferring, 1994; Monger et al., 1993; Nordt et al., 1994; see also Nordt, Chapter 15, this volume). The most common application is to examine the isotopic variation in soil epipedons (modern or fossil) and in subsoil concretions such as carbonate nodules. Theoretical considerations supporting SIRA studies of soils have been published elsewhere (e.g., Boutton, 1991; Cerling et al., 1989) and are not described here in detail. In essence, the stable carbon isotopic composition of organic matter in a soil reflects the source of the carbon from which it was derived. Plants serve as the principal source of soil organic matter and the stable isotopic composition of modern soils generally reflect the standing biomass (Cerling et al., 1989; Kelly et al., 1991). Most plants fall into one of three major photosynthetic pathways (C_3, C_4, and cassulacean acid metabolism [CAM]), each of which fractionates carbon in a different fashion. This fractionation effect permits chemical (isotopic) discrimination of organic matter produced through the two major photosynthetic pathways (C_3 and C_4) and therefore permits estimation of the composition of the vegetation that contributed organic matter to the deposit in question. This method is becoming increasingly common where other proxy vegetation records such as pollen are not available.

As with radiocarbon dating of sediments and soils, understanding the origin of the organic matter that is being analyzed is critical in interpreting such records. Most stable isotopic studies involve soils because it can be assumed that the majority of the organic matter present was formed *in situ* from vegetation growing on the site. However, if the amount of organic matter in the soil is low, interpreting such records can be difficult, owing to problems of contamination with allochthonous organic matter that is irrelevant with respect to paleoenvironmental change.

2.3.2. Discharge and Precipitation: Dendrohydrology and Dendroclimatology

In addition to providing a dating tool, fossil trees can be used to reconstruct prehistoric variations of discharge and precipitation. This is accomplished by correlating variations in ring width with one or more climatic variables, such as mean annual precipitation or discharge. Several studies in the southwestern United States have incorporated time series of paleodischarge or paleoprecipitation estimates derived from tree ring records (known as dendrohydrology and dendroclimatology) in order to interpret the timing and influence of climate on fluvial system response (e.g., Hereford et al., 1996; Karlstrom, 1988). This is a unique means of retrodicting detailed paleoclimatic variability that may then be correlated with known stratigraphic variability in order to evaluate the linkage between precipitation and alluvial aggradation and incision.

2.3.3. Land Use

In landscapes that have a long history of agriculture (e.g., Europe, Mesoamerica, parts of the Andes), evidence of past land use is often employed as an explanatory tool in evaluating the origins of alluviation. Information useful in this context includes paleodemographic data based on archaeological survey and direct archaeological evidence of agricultural practices, such as terracing and irrigation. Some of the best examples of this type of data include the number of archaeological sites associated with each archaeological period (e.g., Barker and Hunt, 1995; Pope and van Andel, 1984), and the number of sites from different periods from which paleodemographic estimates are calculated (e.g., Sanders et al., 1979). In most cases, evidence of population change is only very basic and is not temporally refined. This is because such estimates rely on survey data that is tied to artifact seriations that may have poor chronological resolution, be inadequately dated, or both. Furthermore, such paleodemographic estimates rely on surface survey data and nearly always underestimate actual populations owing to sites that lie buried and undetected in alluvial environments.

In an ideal situation, regional studies of alluviation would be performed in tandem with archaeological survey, permitting direct comparison of periods of alluviation, stability, and erosion with temporal and spatial variations in settlement. The principal limitation to this approach is the assumption that the geographic distribution of past settlements is a good proxy for land use. In reality this may not be true and people in the past (and today in some cultures) often travel long distances from their homes in order to cultivate land. Hence, just because a basin does not have much settlement does not mean that it was not affected by prehistoric agricultural activity.

2.3.4. Eustasy

In regions close to a coast, local records of eustatic variation help researchers evaluate the potential role of base-level change in alluviation. In tectonically active regions, it is important to have a locally derived relative sea-level curve because tectonic activity may significantly alter the temporal and spatial effect (see Kraft et al., 1987:185 for an example).

Although other proxy records may be useful in evaluating causality, these are some of the most common in the literature. For proxy records to be of any use, they too must be dated. As with the alluvial chronology, the number and precision of dates/ages will determine the utility of the proxy record. Undated sequences are functionally useless. Others, such as past records of tectonic activity, are generally not available and must be inferred on a case-by-case basis.

3. Evidence Used to Link Causal Factors with Alluviation

Once a stratigraphic sequence has been described and dated, and as many dated proxy records as possible are compiled, interpretation of causality is evaluated. The following sections summarize the kinds of evidence that have been used to infer a causal relationship in alluviation for each of the major mechanisms. My goal is to illustrate the manner in which these arguments are compiled rather than to provide an exhaustive list.

3.1. Climatic Forcing

Geomorphic response to climate change can be complex and variable. The magnitude of this complexity is well illustrated by Bull (1991), who explicates how one may disentangle fluvial response to climate and tectonic activity. Despite detailed treatises such as Bull's, the inference of climatic causality in many stratigraphic studies is often much simpler than he describes.

In tectonically stable regions, many authors assume that climate is the driving mechanism behind alluviation. Some state this explicitly (albeit for specific temporal periods; e.g., Xu, 1998:67), but most do not. Classic examples of this abound in the literature and can be exemplified by two alluvial stratigraphic studies from central Texas published about the same time (Blum and Valastro, 1992; Nordt, 1992). Both of these works evaluate their sequences in light of the meager Late Quaternary paleoclimatic data for this region and evoke climatic variation to explain the timing and changes observed in Late Pleistocene and Holocene alluvial sedimentation. This is reasonable in that the region is tectonically stable, it is assumed to have had negligible anthropogenic activity that could have influenced alluviation, and both areas are relatively far from the coast. Nevertheless, neither study attempted to compare the chronology of alluviation with factors other than climate.

Where there are well-documented paleoclimatic anomalies, interpretations of climatic causality have been made by comparing the temporal correlation between significant alluvial events and the documented climatic anomalies (excessively wet or dry periods, or periods of extreme variation; e.g., Hereford et al., 1996; Plog et al., 1988).

Another commonly cited attribute of climatically stimulated behavior is regionally synchronous aggradation or incision (e.g., Hall, 1990; Knox, 1983:36; McFadden and McAuliffe, 1997:320; Vita Finzi, 1978). A good example of this kind of argument is Hall's work from the Southern Great Plains in the United States. In this paper Hall argued that the widespread occurrence of stream incision observed in numerous streams east of the Llano Estacado around 1,000 years BP is indicative of a synchronous change in regional climatic to more arid conditions. One problem with this approach is deciding how much temporal variation between different streams is "synchronous" behavior, given that there are always minor and often major differences between the alluvial chronologies of streams between regions and that the quality of the chronology varies between streams. In the case of Hall's study, recent work has found that one of the streams cited in his study entrenched several hundred years earlier (e.g., Frederick, 1996) and others in the same region incised much (400 years) later (Nordt, 1992) than the 1,000 years BP date to which he drew attention.

3.2. Anthropogenic Alluviation

Arguments favoring anthropogenic causality generally take several forms. Where vegetation records are available, anthropogenic causality is generally inferred where periods of sedimentation correspond with palynological evidence indicative of cultivation and/or clearance. Conversely, some authors have argued in favor of anthropogenic alluviation where phases of aggradation do not correspond with evidence of climatic change observed in pollen records (e.g., Collier et al., 1995:42) or general paleoclimatic trends (e.g., Hunt and Gilbertson, 1995). Others have evoked increases in the apparent rate of sedimentation or sediment flux, either alone or in combination with other evidence to support a case of anthropogenic alluviation (e.g., Collier et al., 1995; Mei-e and Xianmo, 1994; Orbock Miller et al., 1993; Xu, 1998).

When paleodemographic data are available it is possible to compare periods of alluviation with temporal and spatial variations in population (e.g., Pope and van Andel, 1984). An anthropogenic cause has been asserted where alluvial sedimentation occurred during periods of population expansion (e.g., Barker and Hunt, 1995:155–156) or where there is a reasonably close correlation between historical information of increased land use and patterns of alluvial activity (e.g., Coltori, 1997). In some instances, an anthropogenic cause has been employed as a ruling hypothesis and alluvial aggradation is attributed to both population expansion and contraction (see comments in Endfield, 1997). Alluviation has been attributed to erosion that followed a decline in agriculture or in population owing to the collapse and failure of slope features such as terraces (Brückner, 1986:15; Pope and van Andel, 1984), as well as to periods of deforestation and population expansion. Conceptually, both may occur, but rarely are alluvial chronologies sufficiently well dated and other proxies such as climatic variation and demography well enough known to adequately support this degree of interpretive detail.

Combinations of these factors have also been employed. Frederick (1995, 1997) compared alluvial chronologies in separate regions of central Mexico that differed principally in the nature and degree of prehistoric land use. Where no sedentary prehistoric populations were present, alluviation was assumed to be climatically stimulated and the stratigraphic sequence was compared to paleoclimatic records that generally supported this assumption. Subsequently, the stratigraphic records of several streams in the densely settled and cultivated Basin of Mexico were compared to climate proxies, the paleodemographic record, the pollen record, and the alluvial chronology presumably controlled by climate. An anthropogenic cause was evoked in the Basin of Mexico region because the chronology of alluviation between the two regions did not agree as could be expected if climate were the sole mechanism, alluviation in the Basin of Mexico exhibited a prominent correlation with periods of demographic expansion and varied between basins with different demographic histories, and in at least one period, the Formative, rapid alluviation and catastrophic slope failure coincided with the first phase of pronounced forest clearance.

Where historical data concerning land use are available, it is sometimes possible to directly link erosion in uplands with sedimentation in lowlands (e.g., Orbock Miller et al., 1993). In a similar fashion it is possible to use time series photography to link soil erosion and other geomorphic changes that may be correlative with stratigraphic information (e.g., Hereford et al., 1996; Orbock Miller et al., 1993). These sources of evidence may be useful in very recent periods and provide a clearer image of recent geomorphic and stratigraphic change.

Another line of inference are studies that link alluvial sediments with source areas or provenance studies. In order to link sediments and source areas, a wide range of sediment properties have been employed to form a "fingerprint" for a source. The most common properties are mineralogy (Klages and Hsieh, 1975), chemical or elemental composition (Peart and Walling, 1996), mineral magnetics (Slattery et al., 1995), and particle size (e.g., Langedal, 1997). Using any one of these, however, has proven less that completely satisfactory, and recent efforts typically employ more than one form of characterization in order to identify source area fingerprints (Collins et al., 1997; Langedal, 1997; Woodward et al., 1995). In terms of anthropogenic alluviation, provenance studies are most compelling when they are used to link alluvial sediments to a source that is available only through human agency such as mining. For example, Knox (1989) used the concentration of elements such as lead and zinc to identify sediments derived from mining. Provenance studies have also been used to examine sediment storage and routing through drainage basins, as in the case of Graf's (1994) study of the Río Grande in the southwestern United States that employed radioactive elements such as plutonium derived from environmental contamination associated with the development of nuclear weapons. Where the sediment source cannot be tied to anthropogenic activity, provenance studies may only identify which parts of the landscape are contributing sediment, and this may not necessarily be linked to any cultural process.

3.3. Tectonic Activity

It is widely recognized that tectonic activity (as expressed as uplift, subsidence, and lateral crustal movements) may influence the temporal and spatial pattern of erosion and sedimentation. It is often difficult to provide detailed information about past tectonic activity except at the most basic temporal and spatial scales. Indeed, Lewin (1995:30) noted that the timing and rate of tectonic activity is often not known and that interpretation of tectonic event histories from alluvial chronologies may lead to circular arguments. Tectonic activity is most often discussed in studies concerned with long-term (Pleistocene or older) adjustments of the fluvial system although it clearly influences Holocene erosion and sedimentation.

Evidence of tectonic activity in alluvial sequences takes a variety of forms. Bull's (1991:26) tripartite genetic division of terraces considers strath terraces cut into bedrock as the principal tectonic stream landform. In some instances the presence of landslide deposits and geographically restricted dewatered and deformed sediments have been used as evidence of seismic activity occurring contemporaneously with Pleistocene alluviation (e.g., Mather et al., 1995:86). This study also attributed a shift in sedimentation from fluvial to palustrine conditions behind a fault as evidence of tectonic activity. Mather and Harvey (1995:73) and Wenzens and Wenzens (1995:65) attributed the existence of asynchronous series of terraces on rivers in the same region to local variations in tectonics. Deeply entrenched rivers are frequently cited as evidence of tectonic uplift (e.g., Kuzucuoglu, 1995:49) and the accumulation of excessively thick sequences of sediment without any evidence of downcutting has been cited as indicative of subsidence (Starkel, 1988). Other evidence of tectonic activity is the arching of terrace fragments as viewed in longitudinal profiles (Harvey and Wells, 1987:693).

3.4. Eustasy

Base level is the lowest elevation to which erosion may lower a landscape (Bull, 1991:4). Sea level is the ultimate base level for most streams, although in internally drained basins a lake serves in this capacity. Changes in base level may influence alluvial erosion and aggradation by changing the channel gradient. Experiments that examined the effects of lowering base level suggest that this process promotes channel entrenchment throughout basins (Schumm, 1977). Conversely, raising base level may promote sedimentation, at least in proximity to the coast. There is debate concerning the applicability of these models to fluvial response in continental settings where it has been argued that climatic events are more influential than base level in controlling alluvial activity (cf. Blum, 1991; Blum and Price, 1994; Durbin et al., 1997).

I encountered few examples that explicitly interpreted alluviation to be the result of changes in base level in any but the most simplistic terms. The classic example is Fisk's (1944) study of the Mississippi River where he argued that the Mississippi was entrenched during the late Pleistocene owing to the lower sea level during the last glacial period, and this entrenched valley was subsequently infilled as the valley slope was decreased by the subsequent rise in sea level in the Holocene. Similar interpretations have been advanced for rivers in the Mediterranean. Brückner (1986:7), for instance, argued that rising sea levels of the Holocene promoted alluvial aggradation following erosion that was associated with the Würm regression.

3.5. Internal or Endogenic Factors, Geomorphic Thresholds, and Complex Responses

Changes in fluvial erosion and deposition are also controlled by internal factors. Schumm (1977) demonstrated that in some instances, a simple change in an external condition may evoke a complex series of events that arise from the new equilibrium established following the initial change.

Identification of a complex response in the historical record is much less common than the recognition of a change in external influences. Numerous criteria have been used to exclude or support a complex response as an explanatory mechanism. In one case, McFadden and McAuliffe (1997:327) argued on the basis of Schumm's (1977) work that a complex response is likely to produce suites of unpaired terraces or terraces that cannot be mapped or correlated over large areas. These features form during progressive incision, initiated by relative uplift because of tectonic activity or valley floor aggradation

(Bull, 1991). Hence, the presence of alluvial sequences with paired terraces that are of mappable extent are unlikely to be the product of such fluvial response. Alternately, Bettis and Autin (1997) argued that the Holocene stratigraphic sequence of Mud Creek (southeastern Iowa, United States) is a historical, long-term example of a complex response originating with changes in sediment yield which occurred at the close of the Pleistocene. Their argument relies on three principal interpretations:

1. Channel entrenchment that did not correlate with either a change in vegetation (as derived from regional pollen data) or flooding was due to a complex response originating with valley floor over-steepening;
2. When a known change in climate occurred but no change in stream behavior or stratigraphy was observed, they assumed "the fluvial system had adjusted to sediment supply in the late Holocene to the degree that large changes in extrinsic factors were needed to produce stratigraphically significant change in system" (Bettis and Autin, 1997:745);
3. The activity of Mud Creek appears to decrease through time, which Schumm (1977:73) suggested is illustrative of a complex response to one perturbation, which in the case of Mud Creek Bettis and Autin argued is the end of loess deposition and slope sedimentation in the upper midwestern United States at the close of the Pleistocene.

4. Investigating Causality in Alluviation: Some Examples

As with many geologic processes, our ability to resolve historical events becomes increasingly difficult with the passage of time. As might be expected, some of the most interesting studies addressing causality of alluviation examined the alluvial record of the last several hundred years, when various forms of proxy records not traditionally applicable to prehistoric studies may be employed. The following examples illustrate how one may test hypotheses concerning which external variables (e.g., climate or land use) best explain variations in alluvial sedimentation.

4.1. Southern Illinois

Orbock Miller et al. (1993) examined alluviation in Illinois during the Historic period (since A.D. 1860). The basic framework of the study was a stratigraphic sequence in which the four different deposits (Qa1, Qa2, Qa2b, Qa3) were identified on the basis of texture, internal structures, clast composition, topographic and stratigraphic position, and the occurrence of buried soils. Live trees and tree stumps found in growing position were dated by dendrochronlogical methods in order to temporally constrain the two major stratigraphic units (Qa1 and Qa2). Historic records delineated the major phases of land use and provided graphic descriptions of soil erosion. Using aerial photographs the researchers evaluated changes in land use between 1938 and 1980. Climatic variation was evaluated from six stations, which provided an 89-year record for the period from 1901 to 1990. Four parameters were examined: (1) total annual precipitation, (2) number of rain days per year, (3) number of days per year in which rainfall exceeded 2.5 cm, and (4) number of days per year that received more than 5 cm of precipitation.

Using these data, Orbock Miller et al. (1993) concluded that deposition of the Qa2 unit, a floodplain deposit, was primarily attributable to land use that resulted in soil erosion during the period between 1860 and 1940. The sedimentation rate for this deposit was compared with the presettlement sedimentation rates elsewhere in the central United States and was found to be one or two orders of magnitude greater. This was not considered to be attributable to climate. Because Qa2 was a floodplain deposit emplaced primarily by vertical accretion, they reasoned that it formed during flooding associated with two phases of above average precipitation. Although they lacked sufficient temporal resolution to firmly support this assertion, it is supported in general terms by the presence of two sets of adventitious roots that formed after phases of excessive floodplain sedimentation. Dendrogeomorphological criteria were also used to evaluate a period of channel entrenchment. This phase of downcutting (between Qa2 and Qa3) was estimated to occur in the 1940s, as about this time the rate of Qa2 sedimentation decreased, a variety of trees requiring a relatively stable floodplain began to grow on the valley floor, and sycamore trees in the valley began to form a new series of adventitious roots. Orbock Miller et al. (1993) speculated that this phase of entrenchment was during a period of intense rainfall that occurred between 1945 and 1951, which also corresponded with the introduction of soil conservation measures and revegetation of some watersheds in the study.

4.2. Southwestern Utah

Hereford et al. (1996) examined the evolution of alluvial stratigraphy and geomorphology of the Virgin River in southeastern Utah near Zio National Park during the last 300 years. Deposits within the valley were mapped, described, and dated. Prehistoric deposits were dated with archaeological material and the historic deposits were dated by dendrochronological methods. Climatic information (in this case the proxy paleodischarge) was deduced from a dendrohydrological study where tree-ring chronologies from two localities were calibrated with the gauged streamflow record in order to reconstruct discharge for a 303-year period between A.D. 1690 and 1992. Historic information primarily consisted of records of flood damage to structures and farm land and comparison of modern and historic photographs.

Interpretation of the changes in the alluvial history were primarily deduced from comparison of the timing of alluviation and arroyo formation with variations in paleodischarge. This comparison revealed a prominent correlation between periods of entrenchment and above average discharge and extremely variable precipitation, whereas periods of sedimentation apparently coincided with periods of below average discharge. That the peak in grazing pressure coincided with a period of above average discharge and precipitation complicated unraveling the potential role of grazing in the major phase of arroyo formation and a climatic hypothesis was favored.

4.3. Colorado Plateau — Northeastern Arizona

One of the best examples of how to evaluate causality of prehistoc alluviation has been presented by Plog et al. (1988) for the Colorado Plateau region of the southwestern United States. This study integrated a detailed and well-dated alluvial chronology compiled by Karlstrom (1988) and a dendroclimatological study by Dean (1988), as well as palynological and archaeological evidence of paleodemography in order to examine in detail how streams responded to climate and to generalized land use. The alluvial chronology was anchored by more than 200 ages determined by means of dendrochronological and radiocarbon methods on subfossil trees. By cross-correlating all these variables, Plog et al. (1988) were able to illustrate that alluvial activity (deposition) correlates relatively well with palynological estimates of relative effective moisture as represented by the proportions of arboreal to nonarboreal pollen. Periods of greater arboreal pollen were inferred to represent periods of higher effective moisture. They also noted the apparent correspondence of alluvial entrenchment with periods of high temporal climatic variability. No correlation between alluviation and land use was noted.

4.4. Comments

The common theme in these examples is the use of tree-ring data to provide both dates and climate proxy records. They are idealized examples that are not applicable to many regions. That said, they also illustrate how multiple lines of evidence may be used to test specific hypotheses regarding the cause of alluviation. In the case of Orbock Miller et al. (1993), it is apparent that even with relatively good temporal control, a perfect correlation could not be made for the timing of alluviation or the channel entrenchment. The argument made by Hereford et al. (1996) convincingly illustrates a causal relationship between climate and alluviation, although one wonders how different their conclusions may have been if their data concerning land use was less focused on flood damage and provided more information on the change in land use through time. Finally, Plog et al. (1988) illustrate how the full integration of these methods may permit a sound retrodiction of the causes of alluvial aggradation and incision, as well as providing a rich data set for addressing archaeological issues.

5. Summary

This chapter has examined the manner in which arguments of causality have been constructed to support historic and prehistoric changes in fluvial systems. In particular, I have tried to show, by using selected papers, the basic approach necessary to construct an argument of causality, and then list some of the lines of evidence that have been used to support the major external and internal factors that influence alluvial activity. Experimental work with river erosion and sedimentation casts doubt on our ability to retrodict the factor(s) responsible for past episodes of erosion or sedimentation. Reading a few of Schumm's works gives us a good idea of the potential errors we may incur in this process (e.g., Schumm, 1977, 1985, 1991). Nevertheless, researchers still try to understand how and why fluvial systems change. And, as in some of the examples I cited, they are able to advance convincing arguments of causality.

Clearly, the most persuasive cases rely on multiple proxy records and a well-dated alluvial chronology. As I have suggested, these situations are rare. Indeed, the only completely convincing cases concern recent alluviation. They involve high-precision dating of deposits, high-precision climate proxy records, and detailed evidence of past land use. The historic age example of Orbock Miller et al. (1993) is considered a good example of how such arguments are presented. In studies where there is an inadequate chronology for the alluvial record, and few or no proxy records, statements of causality should be considered critically. The prehistoric example provided by Plog et al. (1988) shows what kinds of controls are necessary to better understand how geomorphic systems respond to external and internal controls. This should be a model for future studies.

Although the majority of this discussion concerned alluvial deposits, the general approach to establishing causality applies to lake and colluvial settings as well. Lakes are perhaps the ideal setting for this kind of work, as it is often possible to compile one or more proxy records from the lacustrine sequence alone. To date, however, the processual concerns that pervade the literature on fluvial sedimentation appear to be unknown or of minimal concern to paleolimnologists. Rarely are autogenic responses considered in the interpretation of lacustrine sequences, but then, unlike alluvial or colluvial records, there is usually much better temporal correlation of geomorphic, climatic, and vegetation events when they are all reconstructed from the same core.

6. References

Abbott, James T., 1995. Geomorphic Context of the Barton Site (41HY202) and the Mustang Branch Site (41HY209). In *Archaic and Late Prehistoric Human Ecology in the Middle Onion Creek Valley, Hays County, Texas*, edited by Robert A. Ricklis and Michael B. Collins, pp. 353–380. Studies in Archeology 19, Texas Archeological Research Laboratory, The University of Texas at Austin.

Abbott, James T., and Valastro, Salvatore, Jr., 1995. The Holocene Alluvial Records of the Chorai of Metapontum, Basilicata, and Croton, Calabria, Italy. In *Mediterranean Quaternary Environments*, edited by John Lewin, Mark G. Macklin, and Jamie C. Woodward, pp. 195–206. A. A. Balkema, Rotterdam, Netherlands.

Arbogast, A., and Johnson, W. C., 1994, Climatic Implications of the Late Quaternary Alluvial Record of a Small Drainage Basin in the Central Great Plains. *Quaternary Research* 41:298–305.

Autin, W. J., 1992, Use of Alloformations for Definition of Holocene Meander Belts in the Middle Amite River, Southeastern Louisiana. *Geological Society of America Bulletin* 104:233–241.

Bailiff, I. K., 1992, Luminescence Dating of Alluvial Deposits. In *Alluvial Archaeology in Britain*, edited by S. Needham and M. G. Macklin, pp. 27–36. Oxbow Monograph 27, Oxford, Oxbow Books.

Barker, G. W., and Hunt, C. O., 1995, Quaternary Valley Floor Erosion and Alluviation in the Biferno Valley, Molise, Italy: The Role of Tectonics, Climate, Sea Level Change and Human Activity. In *Mediterranean Quaternary River Environments*, edited by J. Lewin, M. G. Macklin, and J. C. Woodward, pp. 145–158. A.A. Balkema, Rotterdam.

Becker, B., 1975, Dendrochronological Observations on the Postglacial River Aggradation in the Southern Part of Central Europe. *Biuletyn Geologiczn* 19:127–136.

Becker, B., and Schirmer, W., 1977. Paleoecological Study on the Holocene Valley Development of the River Main, Southern Germany. *Boreas* 6:303–321.

Bell, M., and Walker, M. J. C., 1992, *Late Quaternary Environmental Change: Physical and Human Perspectives*. Addison, Wesley Longman, Essex.

Bettis, E. A., III, and Autin, W. J., 1997, Complex Response of a Mid-continent North America Drainage System to Late Wisconsinan Sedimentation. *Journal of Sedimentary Research* 67:740–748.

Blum, M. D., 1991, Climatic and Eustatic Controls on Gulf Coast Plain Fluvial Sedimentation: An Example from the Late Quaternary of the Colorado River, Texas. In *Sequence Stratigraphy as an Exploration Tool: Concepts and Practices in the Gulf Coast*, edited by J. Armentrout and B. S. Perkins, pp. 71–83. Proceedings of the 11th Annual Research Conference, Gulf Coast Section, Society of Economi Paleontologists and Mineralogists, Houston.

Blum, M. D., and Price, D. M., 1994, Glacio-Eustatic and Climatic Controls on Pleistocene Alluvial Plain Deposition, Texas Coastal Plain. *Transactions of the Gulf Coast Association of Geological Societies*. 44; 85–92.

Blum, M. D., and Price, D. M., 1998, Quaternary Alluvial Plain Construction in Response to Interacting Glacio-Eustatic and Climatic Controls, Texas Gulf Coastal Plain. In *Relative Role of Eustasy, Climate, and Tectonism in Continental Rocks.*, edited by K. Shanley and P. McCabe, pp. 31–48. SEPM Special Publication 59.

Blum, M. D., and Valastro, S., Jr., 1992, Quaternary Stratigraphy and Geoarchaeology of the Colorado and Concho Rivers, West Texas, *Geoarchaeology* 7(5):419–448.

Bottema, S., Entjes-Nieborg, G., and van Zeist, W., (eds.), 1990, *Man's Role in the Shaping of the Eastern Mediterranean Landscape*. A. A. Balkema, Rotterdam.

Boutton, T. W., 1991, Stable carbon Isotope Ratios of NaturMaterials: II. Atmospheric, Terrestrial, Marine and Freshwater Environments. In *Carbon Isotope Techniques*, edited by D. C. Coleman and B. Fry, pp. 173–185. Academic Press, New York.

Brown, A. G., 1997. *Alluvial Geoarchaeology: Floodplain Archaeology and Environmental Change*. Cambridge University Press, Cambridge.

Brückner, H., 1986, Man's Impact on the Evolution of the Physical Environment in the Mediterranean Region in Historical Times. *GeoJournal* 13:7–17.

Bull, W. L., 1991, *Geomorphic Responses to Climatic Change*. Oxford University Press, Oxford.

Butzer, K. W., 1980, Holocene Alluvial Sequences: Problems of Dating and Correlation. In *Timescales in Geomorphology*, edited by R. Cullingford, D. Davidson, and J. Lewin, pp. 131–141. Wiley, New York.

Caran, S. C., 1998, Quaternary Paleoenvironmental and Paleoclimatic Reconstructions: A Discussion and Critique, with Examples from the Southern High Plains. *Plains Anthropologist* 43, (164): 111–124.

Cerling, T. E., Quade, J., Wang, Y., and Bowman, J. R., 1989, Carbon Isotopes in Soils and Paleosols as Ecology and Paleoecology Indicators, *Nature* 341:138–139.

Collier, R. E. Ll., Leeder, M. R., and Jackson, J. A., 1995, Quaternary Drainage Development, Sediment Fluxes, and Extensional Tectonics in Greece. In *Mediterranean Quaternary River Environments*, edited by J. Lewin, M. G. Macklin, and J. C. Woodward, pp. 31–44. A.A. Balkema, Rotterdam.

Collins, A. L., Walling, D. E., and Leeks, G. J. L., 1997, Fingerprinting the Origin of Fluvial Sediment in Larger River Basins: Combining Assessment of Spatial Provenance and Source Type, *Geografiska Annaler* 97A:239–254.

Collins, M. B., 1995. Forty Years of Archaeology in Central Texas. *Bulletin of the Texas Archaeological Society* 66:361-400.

Coltori, M., 1997, Human Impact in the Holocene Fluvial and Coastal Evolution of the Marche Region, Central Italy, *Catena* 30:311–335.

Dean, J. S., 1988, Dendrochronology and Palaeoenvironmental Reconstruction on the Colorado Plateaus. In *The Anasazi in a Changing Environment*, edited by G. J. Gummerman, pp. 119–167. Cambridge University Press, Cambridge.

Duller, G. A. T., and Törnqvist, T. E.,1998, Luminescence Dating of Fluvial Deposits. *Flag News* (Newsletter of the Fluvial Archives Research Group, a research group of the Quaternary Researc Association), 1(3):11–12.

Durbin, J. M., Blum, M. D., and Price, D. M., 1997, Late Pleistocene Stratigraphy of the Nueces River, Corpus Christi, Texas: Climatic and Glacio-Eustatic Control on Valley Fill Architecture. *Transactions of the Gulf Coast Association of Geological Societies.* 47:119–130.

Endfield, G. H., 1997, Myth, Manipulation, and Myopia in the Study of Mediterranean Soil Erosion. In *Archaeological Science 1995*, edited by A. Sinclair, E. Slater, and J. Gowlett, pp. 241–248. Oxbow Monograph 64. Oxbow Books, Oxford.

Ferring, C. R., 1986. Rates of Fluvial Sedimentation: Implications for Archaeological Visibility. *Geoarchaeology* 1:258–274.

Fisk, H. N., 1944, *Geological Investigation of the Alluvial Valley of the Lower Mississippi River*. United States Corps of Engineers, War Department, Vicksburg, Mississippi.

Frederick, C. D., 1993, Geomorphology. In *Historic and Prehistoric Data Recovery at Palo Duro Reservoir, Hansford County, Texas*, edited by, J. M. Quigg, C. Lintz, F. M. Oglesby, A. C. Earls, C. Frederick, N. Trierweiler, and D. Owsley, pp. 70-116 Mariah Associates Technical Report No. 485, Austin Texas.

Frederick, C. D., 1995, *Fluvial Response to Late Quaternary Climate Change and Land Use in Central Mexico*. Ph.D. dissertation, The University of Texas at Austin, University Microfilms, Ann Arbor.

Frederick, C. D., 1996, Geomorphic Investigations. In *Early Archaic Use of the Concho River Terraces: Archaeological Investigations at 41TG307 and 41TG309, Tom Green County, Texas*, edited by J. M. Quigg et al., pp. 85–108. Technical Report No. 11058, TRC-Mariah Associates, Inc., Austin.

Frederick, C. D., 1997, *Landscape Change and Human Settlement in the Southeastern Basin of Mexico*, unpublished research reporsubmitted to the Texas Higher Education Coordinating Board.

Graf, W. L., 1994, *Plutonium and the Ro Grande: Environmental Change and Contamination in the Nuclear Age*, Oxford University Press, Oxford.

Hall, S. A., 1977, Late Quaternary Sedimentation and Paleoecologic History of Chaco Canyon, *Geological Society of America Bulletin* 88:1593–1618.

Hall, S. A., 1990, Channel Trenching and Climatic Change in the Southern U.S. Great Plains, *Geology* 18:342–345.

Harvey, A. M., and Wells, S. G., 1987, Response of Quaternary Fluvial Systems to Differential Epeirogenic Uplift: Aguas and Feos River Systems, Southeast Spain, *Geology* 15:689–693.

Hereford, R., Jacoby, G. C., and McCord, V. A. S., 1996, *Late Holocene Alluvial Geomorphology of the Virgin River in the Zion National Park Area, Southwest Utah*. Geological Society of America Special Paper 310, Boulder, Colorado.

Humphrey, J. D., and Ferring, C. R., 1994, Stable Isotopic Evidence for Latest Pleistocene and Holocene Climatic Change in North Central Texas, *Quaternary Research* 41:200–213.

Hunt, C. O., and Gilbertson, D. D., 1995, Human Activity, Landscape Change and Valley Alluviation in the Feccia Valley, Tuscany, Italy. In *Mediterranean Quaternary River Environments*, edited by J. Lewin, M. G. Macklin, and J. C. Woodward, pp. 167–178. A.A. Balkema, Rotterdam.

Karlstrom, T. N. V., 1988, Alluvial Chronology and Hydroloic Change of Black Mesa and Nearby Regions. In *The Anasazi in a Changing Environment*, edited by G. J. Gummerman, pp. 45–91. Cambridge University Press, Cambridge.

Kelly, E. F., Amundson, R. G., Marino, B. D., and Marino, M. J., 1991, The Stable Isotope Ratios of Carbon in Phytoliths as a Quantitative Method of Monitoring Vegetation and Climatic Change, *Quaternary Research* 35:222–233.

Klages, M. G., and Hsieh, Y. P., 1975. Suspended Solids Carried by the Gallatin River of Southwestern Montana II: Using Mineralogy for Inferring Sources, *Journal of Environmental Quality* 4:68–73.

Knox, J. C., 1983, Responses of River Systems to Holocene Climates. In *Late Quaternary Environments of the United States*, Volume 2, *The Holocene*, edited by H.E. Wright, Jr., pp. 26–41. University of Minnesota Press, Minneapolis.

Knox, J. C., 1989, Long- and Short-Term Episodic Storage a Removal of Sediment in Watersheds of Southwestern Wisconsin and Northwestern Illinois. In *Sediment and the Environment*, IAHS Publication No. 184.

Kraft, J. C., Rapp, G., Jr., Szemler, G. J., Tziavos, C., and Kase, E. W., 1987, The Pass at Thermopylae, Greece, *Journal of Field Archaeology* 14:181–198.

Kuzucuoglu, C, 1995, River response to Quaternary Tectonics with Examples from Northwestern Anatolia, Turkey. In *Mediterranean Quaternary River Environments*, edited by J. Lewin, M. G. Macklin, and J. C. Woodward, pp. 45–54. A.A. Balkema, Rotterdam.

Langedal, M., 1997. The Influence of a Large Anthropogenic Sediment Source on the Fluvial Geomorphology of the Knabena-Kvina Rivers, Norway. *Geomorphology* 19:117–132.

Lewin, J., 1995, The impact of Quaternary Tectonic Activity on River Behaviour. In *Mediterranean Quaternary River Environments*, edited by J. Lewin, M. G. Macklin, and J. C. Woodward, pp. 29-30. A.A. Balkema, Rotterdam.

Lewin, J., Macklin, M. G., and Woodward, J. C. (eds.), 1995, *Mediteranian* Quarternary River Environments. A.A. Balkema, Rotterdam.

Macklin, M. G., 1995, Geochronology, Correlation and Controls of Quaternary River Erosion and Sedimentation. In *Mediterranean Quaternary River Environments*, edited by J. Lewin, M. G. Macklin, and J. C. Woodward, pp. 179–181. A.A. Balkema, Rotterdam.

Macklin, M. G., Passmore, D. G., Stevenson, A. C., Davis, B. A., and Benevente, J. A., 1994, Responses of Rivers and Lakes to Holocene Environmental Change in the Alcañez Region, Teruel, Northeast Spain. In *Effects of Environmental Change in Drylands*, edited by A. C. Milligan and K. Pye, pp. 113–130. Wiley, Chichester, New York.

Mather, A. E., and Harvey, A. M., 1995, Controls on Drainage Evolution in the Sorbas Basin, Southeast Spain. In *Mediterranean Quaternary River Environments*, edited by J. Lewin, M. G. Macklin, and J. C. Woodward, pp. 65–76. A.A. Balkema, Rotterdam.

Mather, A. E., Silva, P. G., Goy, J. L., Harvey, A. M., and Zazo, C., 1995, Tectonics Versus Climate: An Example from Late Quaternary Aggradational and Dissectional Sequences of the Mula Basin, Southeastern Spain. In *Mediterranean Quaternary River Environments*, edited by J. Lewin, M. G. Macklin, and J. C. Woodward, pp. 77–87. A.A. Balkema, Rotterdam.

McFadden, L. D. and McAuliffe, J. R., 1997, Lithologically Influenced Geomorphic Responses to Holocene Climatic Changes in the Southern Colorado Plateau, Arizona: A Soil-Geomorphic and Ecologic Perspective, *Geomorphology* 19:303–332.

Mei-e, R., and Xianmo, Z., 1994, Anthropogenic Influences on Changes in the Sediment Load of the Yellow River, China, during the Holocene, *The Holocene* 4:314–320.

Mol, J., 1997, *Fluvial Response to Climate Variations. The Last Glaciation in Eastern Germany*. Thesis Vrije Universiteit Amsterdam, Febodruk BV, Enschede.

Monger, H. C., Cole, D. R., and Giordano, T. H., 1993, Stable Isotopes in Pedogenic Carbonates of Fan-Piedmont Soils as Indicator of Quaternary Climate Change. In *Soil-Geomorphic and Paleoclimatic Characteristics of the Fort Bliss Maneuver Areas, Southern New Mexico and Western Texas*, edited by H. C. Monger, pp. 75–88. Historical and Natural Resources Report No. 10, Cultural Resources Management Program, Directorate of Environment, United States Army Air Defense Artillery Center, Fort Bliss, Texas.

Nanson, G. C., and Young, R. W., 1987, Comparison of Thermoluminescence and Radiocarbon Age Determinations from late-Pleistocene Alluvial Deposits Near Sydney, Australia, *Quaternary Research* 27:263–269.

Nanson, G. C., Price, D. M., Short, S. A., Young, R. W., and Jones, B. G., 1991, Comparative Uranium–Thorium and Thermoluminescence Dating of Weathered Quaternary Alluvium in the Tropics of Northern Australia, *Quaternary Research* 35:347–366.

Nordt, L. C., 1992, *Archaeological Geology of the Fort Hood Military Reservation, Fort Hood, Texas*. U.S. Army Fort Hood, Archaeological Resource Management Series, Research Report Number 25.

Nordt, L. C., Boutton, T. W., Hallmark, C. T., and Waters, M. R., 1994, Late Quaternary Vegetation and Climate Changes in Central Texas Based on the Isotopic Composition of Organic Carbon, *Quaternary Research* 41:109–120.

North American Commission of Stratigraphic Nomenclature, 1983, North American Stratigraphic Code, *American Association of Petroleum Geologists Bulletin* 67:841–875.

Orbock Miller, S., Ritter, D. F., Kochel, R. C., and Miller, J. R., 1993, Fluvial Responses to Land-Use Changes and Climatic Variations Within the Drury Creek Watershed, Southern Illinois, *Geomorphology* 6:309–329.

Peart, M. R., and Walling, D. E., 1986. Fingerprinting Sediment Source: The Example of a Drainage Basin in Devon, UK. In *Drainage Basin Sediment Delivery*, edited by R. F. Hadley, pp. 41–55. IAHS Press, Wallingford. IAHS Publication No. 159.

Plog, F., Gumerman, G. J., Euler, R. C., Dean, J. S., Hevley, R. H., and Karlstrom, T. N. V., 1988, Anasazi Adaptive Strategies: The Model, Predictions and Results. In *The Anasazi in a Changing Environment*, edited by G. J. Gummerman, pp. 230–276. Cambridge University Press, Cambridge.

Pope, K. O., and van Andel, T. H., 1984, Late Quaternary Alluviation and Soil Formation in the Southern Argolid: Its History, Causes, and Archaeological Implications. *Journal of Archaeological Science* 11:281–306.

Preusser, F., 1998, Luminescence Dating of Fluvial Sediments: Case Studies from Switzerland. *Flag News* (Newsletter of the Fluvial Archives Research Group, a research group of the Quaternary Research Association), 1(3):11.

Quigg, J. M. and Peck, J., 1995, *The Rush Site (41TG346):A Stratified Late Prehistoric Locale in Tom Green County, Texas*. Mariah Associates, Inc., Technical Report 816c, Austin, Texas.

Sanders, W. T., Parsons, J. R., and Santley, R. S., 1979, *The Basin of Mexico: Ecological Processes in the Evolution of a Civilization*, Academic Press, New York.

Schumm, S., 1977, *The Fluvial System*, John Wiley, New York.

Schumm, S., 1985, Explanation and Extrapolation in Geomorphology: Seven Reasons for Geologic Uncertainty, *Transactions, Japanese Geomorphological Union* 6(1):1–18.

Schumm, S., 1991, *To Interpret the Earth: Ten Ways to be Wrong*, Cambridge University Press, Cambridge.

Slattery, M. C., Burt, T. P., and Walden, J., 1995. The Application of Mineral Magnetic Measurements to Quantify Within Storm Variations in Suspended Sediment Sources. In *Tracer Technologies for Hydrological Systems*, edited by C. Leibundgut, pp. 143–151. IAHS Press, Wallingford. IAHS Publication No. 229.

Smart, P. L., and Frances, P. D., 1991. *Quaternary Dating Methods—A User's Guide*. Quaternary Research Association, Cambridge.

Starkel, L., 1988, Tectonic, Anthropogenic and Climatic Factors in the History of the Vistula River Valley Downstream of Cracow. In *Lake, Mire and River Environments*, edited by G. Lang and C. Schluchter, pp. 161–170. A.A. Balkema, Rotterdam.

Thornes, J. B., 1987a. Models for Palaeohydrology in Practice. In *Palaeohydrology in Practice*, edited by K. J. Gregory, J. Lewin, and J. B. Thornes, pp. 17–36. John Wiley, London.

Thornes, J. B., 1987b. The Palaeo-Ecology of Erosion. In *Landscape and Culture: Geographical and Archaeological Perspectives*, edited by J. M. Wagstaff, p. 37–55. Basil Blackwell, Oxford.

Tipping, R., 1992, The Determination of Cause in the Generation of Major Prehistoric Valley Fills in the Cheviot Hills, Anglo-Scottish Border. In *Alluvial Archaeology in Britain*, edited by S. Needham and M. G. Macklin, pp. 111–121. Oxbow Monograph 27. Oxbow Books, Oxford.

Vita-Finzi, C., 1978, *Archaeological Sites in Their Setting*, Thames and Hudson, London.

Wells, B., Runnels, C., and Zangger, E., 1990. The Berbati–Limnes Archaeological Survey. The 1988 Season, *Opuscula Atheniensia* 18:207–238.

Wenzens, E., and Wenzens, G., 1995, The Influence of Quaternary Tectonics on River Capture and Drainage Patterns in the Huèrcal-Overa Basin, Southeastern Spain. In *Mediterranean Quaternary River Environments*, edited by J. Lewin, M. G. Macklin, and J. C. Woodward, pp. 55–64. A.A. Balkema: Rotterdam.

Woodward, J. C., Lewin, J., and Macklin, M. G., 1995., Glaciation, River Behaviour and Palaeolithic Settlement in Upland Northwest Greece. In *Mediterranean Quaternary Environments*, edited by J. Lewin, M. G. Macklin, and J. C. Woodward, pp. 115–129. A. A. Balkema, Rotterdam.

Xu, J., 1998, Naturally and Anthropogenically Accelerated Sedimentation in the Lower Yellow River, China, Over the Past 13,000 years, *Geografiska Annaler* 80A:67–78.

Zangger, E., 1994, The Island of Asine: A Paleogeographic Reconstruction, *Opuscula Atheniensia* 20:221–239.

Zangger, E., Timpson, M. E., Yazvenko, S. B., Kuhnke, F., and Knauss, J, 1997. The Pylos Regional Archaeological Project. Part II: Landscape Evolution and Site Preservation, *Hesperia* 66:599–641.

Geoarchaeology in Alluvial Landscapes

C. REID FERRING

1. Introduction

This chapter is a discussion of geoarchaeology in fluvial environments ranging from temperate to arid. This environmental framework is chosen because of the striking differences in fluvial systems that pertain to alluvial records across different environments. The archaeological objectives in these different settings may differ slightly, owing to unique or specialized forms of adaptation in different environments; for example, the use of irrigation. But in the main, it is the characteristics of alluvial landforms, deposits, and soils that vary across bioclimatic clines, rather than the kinds of archaeological questions and problems that are pursued there. In this sense, geoarchaeological methods employed for defining stratigraphic frameworks, dating deposits and sites, and reconstructing environments and site formation histories are not specific to certain bioclimatic settings.

On the other hand, bioclimatic clines correlate with significant differences in the processes that underlie construction of the geoarchaeological records, such that those records look different. So, my purpose here is not to examine the rationale for studying site formation, or for constructing detailed geochronologies, or to demonstrate that we need to establish the effects of geomorphic

---

C. REID FERRING • Department of Geography, University of North Texas, Denton, Texas 76203.

Earth Sciences and Archaeology, edited by Paul Goldberg, Vance T. Holliday, and C. Reid Ferring. Kluwer Academic/Plenum Publishers, New York, 2000.

change on site preservation and site visibility. Rather, I characterize the archaeological significance of some important parameters of fluvial systems and alluvial geology across bioclimatic clines from temperate to arid to summarize and illustrate methods used in the course of geoarchaeological research in those settings. These discussions deal primarily with fluvial environments in midlatitude, low–altitude contexts.

1.1. Perspectives

Most prehistoric archaeologists conduct most of their fieldwork in fluvial environments. This especially true in the United States, where Cultural Resources Management (CRM) investigations are concentrated in alluvial settings. Understanding the fundamental aspects of fluvial geology is essential for producing sound fieldwork. Unfortunately, many archaeologists have not had the opportunity to obtain this familiarity, either through academic preparation or through long interaction with geologists. And, one might add, many geologists have not availed themselves of corresponding archaeological fluency! A number of textbook approaches to fluvial geology are available to archaeologists (e.g., Rapp and Hill, 1997; Waters, 1992). These can provide students of alluvial geology with an introduction to the principles of fluvial systems and also to the methods used to study them.

Processes and factors of alluviation, comprising independent variables in fluvial equations, are critically reviewed by in this volume by Frederick (Chapter 3). He outlines both concepts and methods for relating alluvial sedimentation to factors such as climate change, tectonism, and eustacy. In this chapter, my emphasis is on the dependent variables in those equations, namely the sediments that contain archaeological records. Variations in alluvial sediments are discussed with respect to different environmental settings and over varying temporal ranges.

1.2. Time, Environments, and Fluvial Systems

Variability in alluvial records is discussed here at two temporal scales. The first is that period considered here as the Holocene period. (I intentionally avoid the arbitrarily set date of 10,000 B.P. for the end of the Pleistocene, because the physical and the bioclimatic shift from "glacial to postglacial" environments is the only empirical concern, and that shift is notoriously time transgressive; see Holliday, Chapter 1, this volume). The Holocene encompasses late Paleo–indian and younger occupations of the New World. It begins roughly at the shift from Epipaleolithic to Neolithic in the Near East and from Upper Paleolithic to Mesolithic in much of Europe. The late Holocene, roughly the last five millennia, witnessed the emergence of all of the world's complex cultures. In essence, the Holocene is the period of substantial and lasting human modifications of landscapes, including fluvial systems (cf. Binford, 1968).

Geoarchaeology in Alluvial Landscapes

The Post-Pleistocene is distinctive in regards to fluvial systems as well. It can be generalized that this period is one of common floodplain aggradation, manifesting the lack of macroclimatic change and near–modern sea level (Schumm and Lichty, 1965). Bull (1991) stressed that in all climatic and tectonic settings, the end of the Pleistocene corresponded with "... strong shifts to the aggradational side of the threshold of critical power" (p. 276). At the same time, it is important to note that there were significant climatically driven differences in the tempo of aggradation in many drainages; those include episodic periods of dynamic equilibrium that caused cut–fill cycles. Overall, however, the major variations in post-Pleistocene alluvial records are geographic (interdrainage or interregional) rather than temporal. Comparison of chronostratigraphic columns from valleys in the midwestern to the southwestern United States illustrate regional variability. These patterns illustrate the necessity of local perspectives on alluvial geologic records and their geoarchaeological implications.

The second scale is the Pleistocene. This period encompasses the initial peopling of the New World and practically all of the human occupations in the Old World. Both the magnitude and the qualitative parameters of environmental change, including changes in fluvial environments, are very different for these two time scales (Table 4.1). And, significantly, the Pleistocene was a time of comparatively minimal and/or transitory human impacts on their environments. The Pleistocene fluvial records were created in a context of net long-term entrenchment of rivers that is registered by alluvial deposits associated with terraces. Minimally, terraces are evidence of punctuated hydrogeomorphic evolution of drainage basins. Likewise, the alluvial deposits, including enveloped artifacts and ecofacts, are "punctuated" within and between terraces. Both the

Table 4.1. Comparison of Fluvial Environments at Pleistocene and Holocene Scales

Parameter	Holocene	Pleistocene
Macroenvironment	Quasistable	Macroenvironmental changes common
Fluvial environments	Final succession to modern environments	Cycles of glacial–interglacial or stadial–interstadial environments
Marine base levels	Fluctuations about near-modern levels	Major cycles of change, notably the last transgression
Floodplains	Period dominated by construction of modern floodplains via episodic alluviation; uncommon terraces[a]	Cycles of floodplain construction and abandonment (creation of terraces)
Meltwater effects	Negligible in most cases	Major cycles of meltwater discharge; effect changes with glacier–coast distance
Channel adjustments	Usually minor shifts, controlled by internal, self-regulating mechanisms (e.g., avulsion) and/or by minor cycles of climate change	Major changes in stream class
Soils genesis	Monogenetic soils of subequilibrium age (azonal to weakly zonal)	Polygenetic soils representing serial equilibria

[a]Terraces predominantly in loessal regions or those with neotectonics or glacioeustatic rebound.

completeness of geoarchaeological records and the difficulties in dating and correlating those records are increased in proportion to the age of the deposits.

2. Fluvial Environments, Geology, and Archaeological Implications

This section contains a brief overview of the geologic results of fluvial change at both spatial and temporal scales. The causes of those changes, considered in detail by Frederick (Chapter 3, this volume) are the subject of major works by a number of principals, including, for example, Bull (1991), Knox (1983), Leopold et al. (1964), and Schumm (1977). Alluvial processes and results are integral to geomorphology (e.g., Bloom, 1991; Cook and Warren, 1973; Glennie, 1970; Lewin, 1978; Thorns and Brunsden, 1977). Additionally, the literature in alluvial sedimentology is similarly extensive (e.g., Allen, 1965; Collinson and Lewin, 1983; Miall, 1981, 1992; Reineck and Singh, 1980). The emphasis here is on those aspects of fluvial geology that are most important to archaeologists: landforms, sediments, stratigraphic markers, and associated evidence for past environments.

2.1. General Factors

The general factors of alluvial geology are those that would be considered in virtually any fluvial system. At least ten factors must be controlled in order to quantitatively model the behavior of a graded stream, one that is neither aggrading nor eroding its channel (Bloom, 1991). These first include the resistive framework of the system, including bedrock lithology and structures (Bull, 1991:133; Knox, 1976; Patton and Schumm, 1981). Tectonics (Bull, 1991; Miall, 1981; Summerfield, 1991; Vita-Finzi, 1986) and/or glacio-eustatic rebound (Brakenridge et al., 1988) are endogenic factors of system change. Climatic control on fluvial systems is probably the dominant theme in the process literature. Much too extensive to treat here, that literature considers climatic patterns (e.g., annual precipitation, flood frequency and magnitude, seasonality) as direct factors in dynamic parameters such as sediment yield, stream discharge, stream load, and channel shape, sinuosity, and so on. (e.g., Leopold, *et. al.*, 1964; Schumm and Lichty, 1965). A deficiency in research on alluvial morphogenesis is the shortage of studies on chemical denudation, which, in some humid environments is subequal to mechanical denudation of metamorphic rocks (Cleaves et al., 1970).

Much of the research concerning climatic controls on fluvial systems is based on short-term records within basins (Harlin, 1978, 1980), or on comparison of modern fluvial systems among different climatic regions (Knox, 1983; Schumm and Lichty, 1965). Longer-term records matched with independent climatic proxies are also studied in this respect (Baker and Penteado-Orellana, 1977).

Despite significant risks of circular reasoning, it is also common to use quantitative or semiquantitative climate-response models such as the Langbein–Schumm (1958) sediment yield function as devices for inferring past climates from alluvial geology (e. g., Hall, 1982, 1990a,b; see Knox, 1983). Last, contributions are made by measuring the impacts of human land use on fluvial process (McDowell, 1983). Several parameters of fluvial systems and their geologic records are discussed below, with an emphasis on patterns expressed across temperate to arid environments. The purpose of these discussions is to identify empirical and methodological implications for geoarchaeology.

2.2. Vegetation, Weathering, and Sediment Yield

The interdependent factors of bedrock, climate, vegetation, and rates of weathering are the principal controls on sediment yield (Langbein and Schumm, 1958). The sediment yield in basins has been of great interest to geomorphologists because this factor is diagnostic of soil loss, reservoir filling, and flood properties. The controls on sediment yield are important to the research of Quaternary scientists because sediment yield is a major aspect of landscape evolution. For geoarchaeologists, sediment yield is important for somewhat different reasons. The amount and distribution of sediment yield are factors that largely define where, when, and at what rate sediments will be eroded or deposited. If even moderately accurate, such a sediment yield model would allow identification of places on past landscapes where archaeological materials could be preserved by burial. Indeed, predicting site locations and site contexts in fluvial environments can benefit greatly from a diachronic sediment yield model, coupled with known patterns of settlement choice based on resource distributions or on other cultural factors. Examples of this are presented later.

The Langbein–Schumm (1958) curve (L–S) is the model most frequently invoked in the interpretation of alluvial deposits, despite the fact that the curve is only a fit to a remarkably variable data base. Significantly different patterns of sediment yield are associated with environments different from those used in the L–S model (Walling and Webb, 1983), cautioning against its unpracticed use. The L–S model indicates peak sediment yields ca. 800 mm annual precipitation, corresponding to a grassland environment, and decreasing yields with either less (desert) or more (forest) rainfall. In equilibrium conditions, the majority of the sediment yield is stored on slopes as colluvium, and a fraction is deposited as alluvium, the principal focus of alluvial geoarchaeology.

2.3. Channel Patterns and Stream Load

Schumm and Brakenridge (1987) used variables such as stream power, sediment load, sediment size, and channel patterns to classify channels (Figure 4.1). This well-known scheme associates channel shape and stream load properties with grades of stability and specific modes of change (see Baker and Penteado-Orellena, 1977).

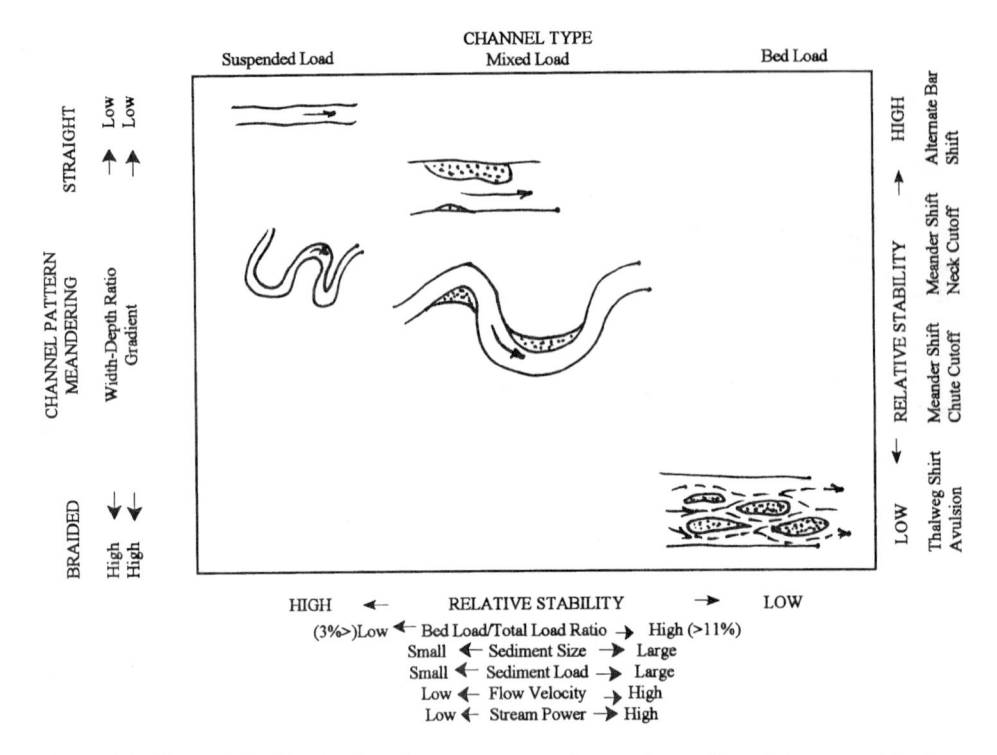

Figure 4.1. Channel classification based on pattern, stability, and type (from Schumm and Brakenridge, 1987).

Two features of their classification are very significant regarding geoarchaeology: channel form and the ratio of bed load to total load. Channel forms are important in that they correlate with key patterns of erosion and deposition on floodplains (Table 4.1). Both the rates and the styles of channel shifting may be inferred from channel patterns. For example, the instability of braided channels denotes low preservation potentials compared to straight ones. Relatively unstable, meandering channels are associated with higher rates of levee and splay construction, which are important factors in constructing floodplain landforms for occupation. They are also associated with rapid site burial. Unstable channels also have higher rates of avulsion, one of the larger scale changes in floodplain depositional morphology (Allen, 1965; Lewin, 1978), and concomitant changes in site formation contexts (Ferring 1986, 1992). Channel stability in meandering streams is also related to meander cutoff style (chute versus neck cutoff), which largely defines site formation processes in the cutoff channels.

The type of stream load is strongly correlated with flow velocity and stream power. Sediment texture, as a correlate to stream load, also has postdepositional significance for archaeological formation processes. Low sand content, for example, promotes more rapid soil development (Birkeland, 1999). In sum, it is possible to deduce from stream class and stream load an array of alluvial features that bear on site formation processes that operated before, during, and after occupations.

Stream classes in temperate environments are predominantly meandering to straight, with perennial discharge. With fine-grained sediment supply, these streams exhibit meanderbelt features including point bars, natural levees, crevasse splays, and cut-off channels/ oxbows (Lewin, 1978). Regional variations are caused by different bedrock substrates and/or tectonism, resulting in changes in discharge, channel migration patterns, or rates of avulsion. Major changes in stream class are well documented in temperate settings, both at Pleistocene and Post-Pleistocene temporal scales. During the Holocene, large midwestern U.S. river channels evolved from meandering to straight (Bettis and Hajic, 1995). In the same region, significant differences in Holocene alluviation are registered for small, medium, and large drainages. Similar variability is also well documented in Central Plains valleys (Johnson and Logan, 1990; Mandel, 1994a,b; 1995; May, 1989). In Poland, river channels shifted from straight or braided to meandering in the early Holocene, yet the timing of this change varied between drainages (Schumm and Brakenridge, 1987).

In semiarid to arid environments drainages may be classified as either allogenic or ephemeral. Through–flowing (allogenic) rivers that cross deserts are maintained by headwaters usually emanating from distant higher elevations. Allogenic rivers are the settings for many of the early civilizations in the Old World. Those rivers in deserts include the Tigris–Euphrates, the Nile, and the Indus, which were all transformed by irrigation systems in the fourth and third millennia B.C.E. Examples in the New World include the Gila River in Arizona that nourished the Hohokam culture (Haury, 1976) and the rivers crossing the coastal deserts of Peru where Archaic peoples constructed some of the first mound-dominated ceremonial complexes in the New World (see Wells, Chapter 5, this volume).

Most archaeological work in arid to semiarid regions is conducted along ephemeral streams (Bull, 1991; Leopold and Miller, 1956; Schumm, 1977:150; Schumm and Hadley, 1957). These are variously called wadis, nahals, draws, or arroyos. The behavior of these streams differs significantly from those in temperate settings (Table 4.2). These desert streams are prone to extremely "flashy" floods resulting from high runoff and brief periods of high peak discharge (Cooke and Warren, 1973:163). These streams are slow to respond to climate change (Bull, 1991) and characteristically exhibit alternating reaches of erosion and aggradation along the same channel (Patton and Schumm, 1981). Rapid changes along an arroyo that do not correspond with climate change are common.

The causes for widespread and rapid development of arroyo systems in the American Southwest over the last two centuries has been an intensely studied and debated phenomenon: "the arroyo problem." Interpretations have invoked a shift to drier climates (Antevs, 1952; Bryan, 1925; Melton, 1965), an onset of larger summer storms, or a threshold response to overgrazing. These and other internal dynamics result in interdrainage variations in alluvial records, even for adjacent tributaries. Quite commonly, then, it is difficult or impossible to extend stratigraphic correlations over multiple valleys, or even for appreciable distances down a single valley.

Table 4.2. Generalized Relations between Stream Class, Equilibrium State, and Alluvial Activity

State	Braided	Bedload	Suspended load
Erosional	Braidplain widening and incision strath development	Channel widening, shift toward braided class	Entrenchment (terrace genesis) shift to bedload, possibly braided
Stable	Limited braidplain shifting, lateral and downstream bar migration	Periodic changes in channel shape (w × d) and meander geometry	Lateral shifting of depositional geomorphic features (point bars, levees, oxbows); periodic avulsion (local shifting of meanderbelt)
Aggradational	Shift to bedload class	Shift to suspended load class	Floodplain aggradation within limits of geomorphic thresholds

In contrast to streams in temperate regions, another well-documented pattern of arroyo behavior is the increased lag time between climatic–vegetational change and the fluvial response (Bull, 1991:118). This suggests that cultural responses to environmental change may have been out of phase with stream response, adding further cautionary flags for interpreting geological–archaeological records in arid lands. When extended to the Pleistocene time scale, even terrace genesis may not correlate closely with the timing and magnitude of climatic–environmental change (Bull, 1990). But the world of arid fluvial systems is not one of total chaos at all scales.

Despite the highly variable character of arid lands streams just described, cases of regional change in fluvial systems are well documented. The late Quaternary phases of wadi aggradation along the Nile Valley is an excellent example (Butzer and Hansen, 1968; Wendorf and Schild, 1980).

2.4. Facies, Architecture, and Alluvial Geomorphology

The major components of alluvial records include geomorphic features, such as terraces, and also the packages of alluvium contained below the surfaces of terraces or floodplains. In this context, the focus on alluvial deposits is for purposes of defining sedimentary environments pertinent to archaeological inquiry. Stratigraphic study of the same sediments requires different approaches, as discussed in the following section.

2.4.1. Alluvial Facies

Sedimentary facies are bounded sediment packages that are distinct from contiguous packages in their lithology, primary structures, and sometimes biotic remains (Walker, 1992). For alluvial sediments, Miall's (1992) facies classification is well suited for use in Quaternary sediments, yet it lacks soils features of alluvial sediments. Because of the dependent relations between alluvial parent materials and alluvial soil formation, the concept of soils facies (Birkeland, 1999) or pedofacies (Krause and Bown, 1988) is necessary and very useful.

Miall's (1992) hierarchical classification of facies is based primarily on sediment textures and then on bedforms. Architectural elements are recurrent facies assemblages associated with various channel and overbank environments (Allen, 1965). Miall recognized eight architectural elements that are common in ancient alluvial deposits. Alternate definitions of elements appears useful to convey sedimentary environments pertinent to Quaternary deposits, especially those of mixed load streams (Ferring, 1993).

Facies are the constituent features that permit recognition of fluvial environments such as floodbasins, levees and splays, and point bars (Allen, 1965; Lewin, 1978). Facies description and study is necessary for defining both the ecological setting of archaeological sites and context-specific formation processes.

Alluvial fans, formed by mudflows and by ephemeral stream deposition, are especially common in arid lands (Bull 1972, 1977; Hooke, 1967). Compared to ancient, including Pleistocene fans, modern fans are small, having radii between 1.5 to 7.0 km (Schumm, 1977). These contrast with the much larger fluvial fans that form in humid environments by perennial streams. Coalesced fans, or bajadas, often occupy major areas of arid lands basins and often contain archaeological records, especially of agricultural groups (Waters, 1988, 1998). Small alluvial fans are also common in temperate settings; those fans differ from arid fans in morphology and in process (Kochel and Johnson, 1984). Because of climatic patterns and more complete vegetation cover, temperate fans tend to have gentler topography, less diverse sedimentary facies, and more common buried soils. Fans were commonly favored locations for Archaic and Woodland occupations in the midwestern United States, where upland loess supplied sediment for fan growth during the Holocene (Bettis and Hajic, 1995; Stafford et. al., 1992; Wiant et al., 1983).

2.4.2. Associated Facies

Nonalluvial sediments are commonly associated with alluvium. Colluvium and alluvial fan sediments accumulate at floodplain margins and frequently contain archaeological sites. Dunes form along the lee side of braid plains or broad sandy channels; they may also form on terrace surfaces. In contrast to other regions in the Great Plains, draws on the Southern High Plains contain thick Middle and Late Holocene eolian sand that buried older alluvial and lacustrine and palustrine sediments (Holliday, 1995).

Incorporation of loess into aggrading alluvial deposits is more difficult to recognize, especially if it is the thinner and finer loess deposited farther from the source. This situation can be revealed with precise textural analyses afforded by electronic or laser–based instruments (Autin, 1992). Quantifying, or in some cases simpy recognizing loess-alluvium contacts is necessary for both sediment and soils analysis (Ruhe and Olsen, 1980). Loess mantles are common on both interfluves and alluvial terraces in the Pleistocene loessal plains of North America, Europe, and Asia (Ruhe, 1983; van Andel and Tzedakis, 1996).

2.5. Alluvial Morphogenesis and Pedogenesis

With or without changes in external factors including climate, tectonics, or base level, alluvial systems will undergo geomorphic change as a result of internal factors (Knox, 1976; Schumm, 1977). On floodplains, these morphogenic changes range from lateral channel shifting to buildup of depositional geomorphic features such as levees or splays (Allen, 1965; Lewin, 1972). Even floodplain abandonment and terrace formation can ensue development of geomorphic instability through reduction in floodplain gradient (Knox, 1976, 1983). So common are these kinds of change that these processes need to be first ruled out before invoking climate or other factors as forcing agents in alluvial morphogenesis (Ferring, 1986). Cleaves et al. (1970) demonstrated the significant role of geochemical weathering geomorphic change in alluvial basins.

Formation of soils in alluvial environments is intimately related to internal morphogenesis, because soil development is usually contingent on local rates of deposition (Ferring, 1986). Because of different rates of deposition across a floodplain, associated with different sedimentary environments, soils genesis will proceed with very different rates, on different parent materials, and in different biotic and hydrologic settings (see Mandel and Bettis, Chapter 7, this volume).

It is always true, however, that an alluvial soil is evidence of local geomorphic change at some scale, ranging from periodic internal adjustments to major geomorphic response to extrinsic factors. Field recognition and study of soils is therefore intimately and necessarily integrated into alluvial geomorphology (Birkeland, 1999; Bull, 1990, 1991; Gerrard, 1987; Gile et al., 1981). These relationships are discussed below, first as they pertain to floodplain development and second as they relate to terrace morphogenesis.

2.5.1. Floodplains: Depositional Geomorphic Features

On floodplains, depositional geomorphic features are those landforms constructed through differential patterns and rates of channel migration and sedimentation (Lewin, 1978). These include natural levees, crevasse splays, oxbows, and alluvial ridges. Avulsion of a meanderbelt is the abandonment of an alluvial ridge and creation of a new meanderbelt reach in a lower part of the floodplain (Allen, 1965). The potential scale of the process is illustrated by the 200-mile shift in the Yellow River channel in 1851 (Schumm, 1977:297). After avulsion, both depositional and pedogenic processes change, and different sequences of pedofacies form in the position of the old and new meanderbelts (Ferring, 1986, 1992). Avulsion has had dramatic impacts on cultures, for example the site abandonment accompanied by an 18th century A.D. avulsion of the Middle Ucayali River in the Amazon Basin (Prässinen et al., 1996). Avulsion has been related to both occupation site selection and later site preservation processes along meandering rivers (Guccione et al., 1998) and alluvial fans (Stafford et al., 1992).

2.5.2. Floodplains: Soils

The formation of soils on floodplains is largely controlled by the patterns of depositional geomorphic change (Ferring, 1992). Because all parts of a floodplain receive sediment at least every few years, all floodplain soils are to some degree cumulic soils. This term denotes a soil that is "overthickened" through deposition of parent material concurrently with pedogenesis. However, floodplain soils are "cumulic" at different scales for two related reasons. Both the rate of deposition and the texture of the alluvium vary across floodplain environments (facies), and both of these factors influence the rate of pedogenesis. In many cases, rates of deposition and sediment texture are correlated, such as in meandering suspended load systems, in which clay-rich floodbasin sediments are deposited more slowly than sandier levee or splay facies. For this reason it is important that alluvial soils should be compared within facies, not between them, for purposes of stratigraphic correlations based on floodplain histories. Soils in floodbasin or vertical accretions deposits will yield a much more accurate record of overall aggradation (Knox, 1983), whereas soils in channel–associated facies will yield irregular if not misleading patterns. However, Borchardt and Hill (1985) analyzed clay minerals, especially smectite, to establish relative ages of Holocene soils in wetter environments. That study also is an excellent example of parent material analysis as a control on rates of pedogenesis.

In arid environments, aside from the climatic controls on weathering, patterns of deposition on floodplains, and indeed the styles of deposition and erosion in general, are such that soils genesis is substantially different than in temperate regions. Cumulic soils are rare in arid settings, and climatic factors usually override sedimentary controls on soil morphology as seen in temperate settings.

Flood patterns and groundwater hydrology are major factors in formation of floodplain soils. In temperate regions, overbank flooding is common and streams are generally effluent. Thus, floodplain soils are subjected to high throughput of water and are saturated at least seasonally. This promotes leaching, shrink-swell of clay, and gleying of the lower parts of soil profiles. (Hayward and Fenwick, 1983; Veneman et al., 1976; Walker and Coventry, 1976).

In arid climates, overbank flooding is rare, the streams are influent, and both leaching and groundwater removal of soluble salts is reduced. Floodplain soils are rarely gleyed, although organic-rich Cienega soils form in locally saturated and grassy areas near springs or arroyo heads (Cooke and Warren, 1973:171). Seasonal drying of clay-rich floodplain soils often results in Vertisols that may also exhibit silt coats on crack surfaces.

2.5.3. Terraces: Alluvial Morphogenesis and Soils

Floodplain abandonment immediately changes the environment of soil formation, principally through cessation of flooding and deposition, but also through changes in vegetation and groundwater (Bull, 1990). Terrace soils manifest the state of the floodplain prior to abandonment, as well as the sum of conditions since that event. Through time, the original floodplain soil features will be lost, but sedimentary facies are generally preserved. If soil parent materials are similar, chronosequences of terrace soils can be defined, but the environmental significance of those soils is lost to compounded changes in different environments over time (Birkeland, 1999). Bryan and Albritton (1943) identified "polygenetic soils" as those that formed over a period with distinct changes in climate.

Terrace soils in temperate environments are usually Alfisols, Mollisols, or Vertisols. The development of soil morphology over time may be expressed through different indices of soil development (Ponti, 1985). Later addition of loess to terrace surfaces complicates analysis of the original soil that formed in alluvium by creating "welded" soils (Ruhe and Olsen, 1980). In mesic environments, where calcic soils are not dominant, calcic horizons form by the redistribution of primary (inherited or allogenic) carbonate (Machette, 1985).

In arid environments, terrace soils are usually Aridisols, Entisols, or Vertisols. These are largely formed by infiltration and illuviation of dust and by buildup of soil carbonates (Gile et al., 1981). Rates of calcic horizon development in arid environments are generally assessed for soils formed on stable surfaces (Machette, 1985). Overall, rates of weathering and development of soils morphology are so slow that chronosequences are useful for surfaces more than 500,000 years old (Gile et al., 1981).

2.6. Alluvial Response to Climatic Change

The relationships between climate change and fluvial responses is a dominant theme in fluvial research (Bull, 1991; Knox, 1983; Schumm and Brakenridge, 1987). In geoarchaeology, neither actualistic nor theoretical studies of these relations are sufficient to justify detailed reconstruction of past climates based on sediments and soils alone. This conclusion is based on the detail of records cited below that could not have been identified from either of those avenues of pure research. Moreover, Quaternary research has shown that the high degree of variability within and between drainages provides ample warnings against simplistic interpretations. General relations between specific patterns of climate change and fluvial responses such as aggradation and erosion have been defined for both arid and temperate settings (Knox, 1983; Bull, 1991). Yet the invitations of those authors to use these relations to deduce geologic records for a specific drainage are noticeably lacking. In addition to the spatial complexity of responses, the deficiencies of independent records of climate commonly deplete the left side of the "climate = response" equations. We are simply in a better position to define the response in most cases.

For example, the semiarid cycle of erosion (Graf et al., 1987; Schumm, 1977; Schumm and Hadley, 1957) applies to some, but not all desertic settings. This cycle includes processes that accommodate simultaneous erosion and deposition in the same channel system. Feedback relations between aggradation, channel floor instability and knickpoint migration. Patton and Schumm (1981) extended the semiarid cycle's implications to multiple drainages, demonstrating that among ephemeral streams on the Colorado Plateau, there was no correlation in the timing of erosion and deposition. Similarly, in the Rolling Plains of Texas, remarkable interdrainage variability in the architecture of late Quaternary sediments has been documented (Gustafson, 1986).

These situations contrast with patterns of allometric change within drainage systems, where change in one part of the system correlates well with a related change in another part of the same system (Bull, 1991). A good example of this is the patterns of erosion and deposition within Central Plains drainage systems in Kansas (Mandel, 1994b). There, erosion of small tributaries in the middle Holocene led to development of alluvial fans in midsized valley reaches downstream. Entrainment of sediment in those positions inhibited middle Holocene alluviation in the associated large valleys. In the late Holocene, multiple cycles of alluviation and pedogenesis are registered in all stream orders, suggesting storage of sediments that were delivered from valley margins.

An important point is that in arid to semiarid settings, intra- and interdrainage alluvial variability is typically high; in temperate settings it is also high, but better understood in terms of allometric change. Thus stratigraphic correlations are best established within major depositional environments, such as alluvial fans (Bettis and Hajic, 1995), or where very similar controls on sedimentation prevailed, such as the Southern High Plains (Holliday, 1995, 1997). Where differential sediment supply or other factors exist, inter-drainage variability will be enhanced. For example, late Holocene alluvial records in North Texas differ markedly between areas with easily eroded bedrock (Ferring, 1990, 1995; Ferring and Yates, 1997) and nearby valleys with resistant limestone bedrock (Nordt, 1995).

Geoarchaeologists must rely heavily on empirical evidence in all these contexts and must approach each drainage or drainage segment inductively. This approach will ensure that substantial contributions to archaeology are not compromised by high-risk assumptions, and that the important results of research may accumulate as unbiased contributions to those larger issues in fluvial geology.

3. Geoarchaeological Methods in Fluvial Environments

Geoarchaeology in fluvial settings shares goals with work in any environmental context, and to some degree the methods employed are the same as well. The following discussions stress the contexts and the methods that bear most particularly on work in fluvial environments, although there are brief allusions to some general issues. As much as possible the subjects are treated with reference to published research and by means of brief reviews of selected case studies.

3.1. Stratigraphy

Stratigraphic investigation of late Quaternary sediments in geoarchaeology is conducted with methods and databases that are largely the same as those in purely geologic investigations, although the advantage of artifacts as stratigraphic markers is notably more common in geoarchaeology. Archaeological stratigraphy is distinctive in the role it plays in excavation of sites and also in that it is exclusively the domain of cultural stratigraphy. What's more, despite the evolutionary (almost genealogical at times) terms bestowed on stone tools by their analysts, culture change as a chronometer requires much more stringent assumptions concerning succession than does biostratigraphy.

3.1.1. Morphostratigraphy

Establishing a sequence of landforms has been a defining aspect of geomorphology since its inception. Today the importance of this approach to geoarchaeology is largely seen at the Pleistocene scale, but Post-Pleistocene terrace sequences may be readily achieved as well (Borchardt and Hill, 1985; McDowell, 1983). In arid environments there is also an emphasis on dating and correlation of alluvial fans (Bull, 1972).

Landform sequences are integral to prehistoric archaeology in the Old World by virtue of the time depth of occupations. Correlation with presumed glacial stages has given way to independent dating. For many alluvial terrace sequences, long-term net incision results in flights of terraces that may contain sediments and archaeology of great age. Terraces in the Elbe–Saale region in Germany contain records of occupations dating back to the middle Pleistocene (Mania, 1996). Spring-lacustrine marls and also loess deposits complicate terrace correlations with other valleys. However, the travertines provided an opportunity for U–series dating (Schwarcz et al., 1988), yielding ages of ca. 350 to 400 ka. For the Lower Paleolithic occupations and *Homo erectus* fossils at the Bilzingsleben II locality (Mania, 1996).

In late Pleistocene and Holocene contexts, relative surface ages usually cannot be defined without ancillary evidence such as alluvial stratigraphy, soils, or cobble weathering rinds. The possible age relations of alluvial terraces that may be either cut or fill surfaces is a classic illustration of this problem (Leopold, et al., 1964:460). Additionally, as mentioned previously, the weathering-based methods for dating surfaces have much shorter applicable ranges in temperate environments than in arid ones (Bull, 1991:272), although these restrictions apply mainly to Pleistocene-scale investigations. Geomorphic–soils analysis of exposed and buried aggradation surfaces in the Duck River Basin (Tennessee) by Turner and Klippel (1989) revealed archaeological implications of differential rates of deposition and surface stability, building on Brakenridge's (1984) study of this "ingrown meandering" stream.

3.1.2. Lithostratigraphy

Allostratigraphic units are the fundamental divisions of an alluvial stratigraphic sequence (Walker, 1992; NACSN, 1983:865). These units are defined on the basis of "bounding discontinuities" including straths, erosional disconformities, soils, and surfaces of either terraces or floodplains. Autin's (1992) application of allostratigraphic principles to Quaternary alluvium in the lower Mississippi Valley is exemplary. Allostratigraphic units may incorporate any variety of alluvial facies, because the units are defined by boundaries rather than contents. The boundaries connote a significant change in depositional regimes, caused by changes in climate, tectonics, or base level. Thus defining an allostratigraphic sequence sets the stage for explaining the mechanisms that forced the changes.

A variety of methods are available for constructing and correlating stratigraphic sequences. Debusschere et al. (1989) made field descriptions of sediments and soils within an Archaic site and also in off-site sections. These were correlated using granulometry, clay mineralogy by X-ray diffractometry and total element analysis using atomic absorption. The value of these multivariate approaches is alo illustrated by correlation of anthropogenic sediments in valleys in northern China (Jing et al., 1995). To refine definitions and correlations among lithostratigraphic units, they used various statistical parameters of sediment texture, soil chemistry, and magnetic parameters including low-field magnetic susceptibility and anhysteretic remanent magnetization (ARM; Jing and Rapp, 1998). Their analyses demonstrated the effects and geoarchaeological implications of dramatic post-Neolithic alluviation. Neolithic sites were buried by as much as 10 m of alluvium, whereas subsequent predynastic site distributions register major readaptations to the altered floodplains.

3.1.3. Pedostratigraphy

Use of soils for stratigraphic correlations is one of the more common approaches in Quaternary geology (Morrison, 1978) and geoarchaeology (Holliday, 1997). Holliday (1989, 1995) intensively studied Quaternary soils on the High Plains of Texas and used them as components of his broader approach to developing interdrainage stratigraphic correlations. His work is fundamental to his geoarchaeological analysis of Paleoindian sites, which is a detailed application of regional geoarchaeology. In Holocene contexts, soils morphology can be difficult to use in correlations because local differences in parent material or in patterns of alluviation yield greater variability than does the interval of soil formation (Figure 4.2). Radiocarbon dating of soil horizons is commonly employed (e.g., Mandel, 1994b; Martin and Johnson, 1995). Correlation of fan, alluvial, and colluvial sediments based on soils morphology and radiocarbon ages has been extensively applied in geoarchaeological research in the midwestern United States (Artz, 1995; Bettis and Hajic, 1995; Bettis and Littke, 1987; Wiant et al., 1983) and in the Great Plains (Mandel, 1994a, 1995; Johnson and Logan, 1990).

3.2. Geochronology

Methods of dating sediments are broadly applicable (see Frederick, Chapter 3, and Rink, Chapter 14, this volume). In alluvial contexts, several guidelines are notable. Radiocarbon dating of humate fractions from floodplain soils entails risks of dating inherited organics, potentially from older sediments upstream, thereby obtaining anomalously old ages for the soil horizon. Nonetheless, dates on bulk soil can yield accurate chronologies of sediment accumulation and pedogenesis (e.g., Blum et al., 1992). Another example is the dating of bulk

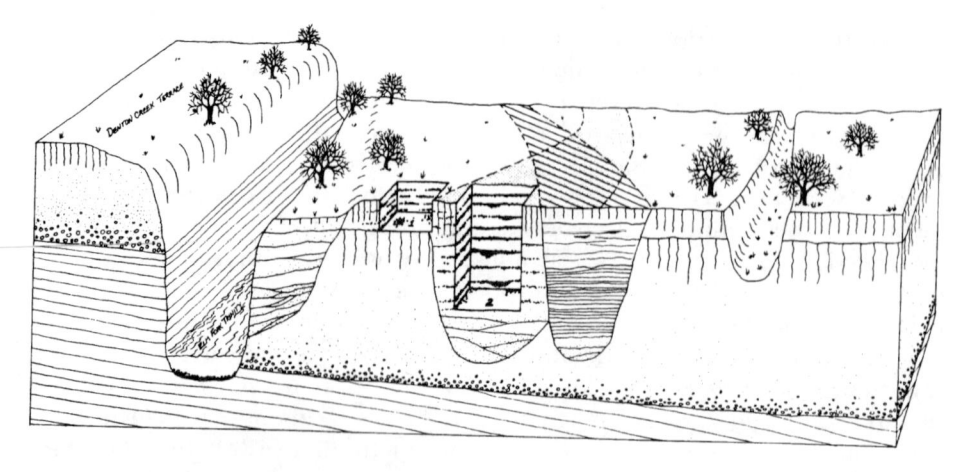

Figure 4.2. Complex series of filled channels containing stratified archaeological materials at Elm Fork Trinity River, Texas (from Ferring and Yates, 1997). Channel 1 (chute cutoff) filled with seven stratified late Archaic living surfaces over 900-year period. Channel 2 (neck cutoff) filled initially with laminated oxbow lake silts, then with subaerial clays, with cumulic soil and late Prehistoric occupation materials.

samples of floodplain clays at the Aubrey Clovis Site, Texas (Ferring, 1994). The distal floodplain sediments there had been deposited slowly enough that the 7 to 9 m section was essentially a stacked series of cumulic soils.

In all cases, careful field techniques of collection for dating, whether of particulate organic material or of bulk samples, is critical. All features that could betray displacement or translocation of the dating material must be recognized and avoided during sampling (Rink, this volume, Chapter 14). Common crayfish burrows in the sediments at the Aubrey site (Ferring, 1994) required very careful exposure of the sediments in three dimensions so that the samples would not include any of the younger burrow fill. The importance of proper cleaning of a profile prior to sample collection is highlighted by the work of Haas et al. (1986), at the Lubbock Lake locality (Texas). They measured the rate of organic matter degradation after sediments are exposed (in either natural cuts or excavated trench walls), yielding ages that are too young.

Dating alluvium that is beyond the range of radiocarbon dating can be achieved using a variety of methods (Rink, this volume, Chapter 14; Aitken, 1995). Paleomagnetism may be used for sediments older than Bruhnes (Easterbrook, 1988). Pope et al. (1984) used Uranium-series dating on pedogenic carbonates and also on groundwater carbonate crusts on Mousterian artifacts to date middle Paleolithic sites in alluvium in Greece. Their dates range from ca. 270 ka to 33 ka. Van Andel (1998) also established a U-series chronology for alluvial deposits in Greece that contain Middle Paleolithic artifacts. Nanson et al. (1991) employed U/Th and TL dating of alluvium in Australia and applied U/Th methods not only to carbonates but also to Fe/Mn oxyhydroxides/oxides.

3.3. Site Discovery Methods

The discovery of archaeological sites in alluvial contexts is a principal objective of geoarchaeology. Over the history of archaeology, sites have been discovered with little or no consideration of geology, yet this process is rife with missed opportunities as well as accumulated biases in interpretations. The main contributions from geoarchaeology are in the area of buried site discovery. This entails both prediction of site contexts and their physical detection. In this endeavor, both site construction and site preservation are important, requiring routine analysis of postoccupational alluvial histories.

3.3.1. Site Prediction

Buried site prediction models require integration of geomorphic, sedimentary, and stratigraphic data (Gladfelter, 1985; Hassan, 1985). The utility of those models is, therefore, measured by the detail of geologic research in a given setting. At the risk of sounding circular, the strength of predictive models is also a measure of how much patterning there is in alluvial geologic records. This can only be measured with reference to fluvial dynamics, as considered earlier in this chapter. A summary opinion on this matter would have to be that there are few cases, either in temperate or arid environments, where simplistic models of alluvial geology will assist in site prediction. Variability down valleys, among streams of different rank, and between drainage systems is best considered the rule. Thus the most useful site prediction models are those that are based on extensive, sound research over whole drainage systems.

Site prediction models should incorporate both contextual and site preservation components (Guccione et al., 1998). Potential site contexts must be summarized not only stratigraphically but also with consideration of sedimentary environments as controls on site locations (Kauflulu, 1990; Mooers and Dobbs, 1993; Phillips and Gladfelter, 1983; Putnam, 1994). Thus sedimentary, soils, and stratigraphic data are required. Successful (i.e., useful) examples of this are found in the detailed records established in the Midwest, including work cited previously by Bettis, Hajic, Artz, and others. Likewise, the work of Mandel, Johnson, and others has provided similarly detailed models for the Central Plains. Their models incorporate sufficient geologic data to discuss the known and probable contexts for sites of different ages, as well as the factors that have led to differential preservation and loss within different parts of a drainage system. Blum et al. (1992) used similar approaches and extensive radiocarbon dating in central Texas.

3.3.2. Surveys and Subsurface Exploration

Discovery of buried sites requires implementation of all available means to examine the target sediments. Careful inspection of natural outcrops is the cheapest method, but it is successful only when substantial natural exposures are available. This situation is more common in arid settings where vertical channel banks are common (Waters, 1986).

Where natural exposures are inadequate, which is generally the rule, artificial exposures are made by backhoe trenches, auger tests, cores, and for shallow contexts, shovel tests. Auguring or coring can be used to refine strategies for more serious site-discovery efforts. An excellent example of this is a survey done on the South Platte River in Colorado by McFaul, et al. (1994). They recovered 150 cores, each 3 in in diameter, with a Giddings drilling rig, and used sediment–soils descriptions to define areas for subsequent exploration, which was accomplished with excavation of 50 backhoe trenches. They defined patterns of geomorphic change and the age and character of alluvium and eolian deposits.

Successful use of coring has also been done in the intensive work of Holliday (1995) in the draws of the Southern High Plains. To actually locate sites, recovery methods must be adequate. Microscreening of core samples is effective but broadly underutilized (Sherwood, Chapter 2, this volume). A very successful program of vibracoring, diver surveys, and dredge excavations has led to discovery of sites associated with stream channels, now marine inundated, on the shallow Gulf of Mexico shelf of Florida (Faught and Donoghue, 1997). Their survey methods included prior definition of channels using a shallow seismic reflection system.

Remote sensing is more commonly used in terrestrial settings to define subsurface stratigraphy, to detect archaeological features, or both (Kvamme, Chapter 13, this volume). Resistivity profiling can reveal major lithologic changes (Darwin et al., 1990), as will seismic profiling (Noller, Chapter 6, this volume). Ground-penetrating radar can reveal shallow sedimentary–soils features, whereas magnetometry is useful for defining shallowly buried archaeological features (Wynn, 1990).

In arid lands, surveys have always been facilitated by sparse vegetative cover. However, burial of ancient deposits by sand sheets has inhibited surveys. Implementation of shuttle imaging radar (SIR) images to survey the "radar channels" of the Egyptian Sahara has been remarkably successful in revealing Plio–Pleistocene drainages (Wendorf et al., 1987). Follow-up surveys assisted by backhoe trenching and hand excavations resulted in discovery of Middle to Late Acheulian sites (ca. 0.15–0.500 Ma) associated with the buried paleodrainages (McHugh et al., 1988). Remote sensing, coupled with geographic information systems (GIS) analysis should enjoy more success in the future, especially in arid lands (Linse, 1993).

3.4. Excavations and Formation Analyses

Geoarchaeology in fluvial settings does not stop after site discovery but rather turns to different yet important goals during site excavations. These goals include documentation of site contexts, collection of dating samples, and integration of natural and cultural records. The latter is the domain of site formation processes, which are important in any geologic context (Stein, Chapter 2, this volume; Butzer, 1982; Schiffer, 1987; Wood and Johnson, 1978). Most processes of site formation are not unique to fluvial environments, but certain processes of formation and methods of investigation are encountered commonly in alluvial sites.

Transport of artifacts and ecofacts is a problem commonly encountered in alluvial sites. This is not simply a question of finding "in-place" sites, but rather is one in critical assessment of natural and cultural deposits (Hanson, 1980; Petraglia and Nash, 1987; Schick, 1987). Actualistic studies have yielded approaches for investigating sites by documenting patterns of bone transportation and reorientation at early East African sites (Behrensmeyer 1982, 1987; Irving et al., 1989). These kinds of studies require close integration of sedimentology, bone positioning data, and taphonomic observations (Johnson and Holliday, 1989; Koster, 1987; Kreutzer, 1988; Steele and Carlson, 1989).

Abrasion of lithic artifacts during fluvial transport has received similar attention (Shackley, 1974, 1978). A good example is Shea's (1999) analysis of artifact abrasion and site formation processes at 'Ubeidiya, the well-known lower Paleolithic site in the Jordanian Rift.

All channel contexts are not necessarily poor site formation settings. In suspended load meandering systems, cutoff channels were sometimes favored occupation locations and were ideal preservation contexts as well. Multiple superposed occupation surfaces at the Gemma site in North Texas were found in a chute cutoff channel (Ferring and Yates, 1997). Their analysis of sediment texture and soils development illustrated that episodic deposition promoted burial of successive occupation surfaces while vegetation apparently served to baffle potentially erosive currents (Figure 4.3). At this site occupation surfaces were separated by sterile flood deposits. These ideal formation contexts may be difficult to find unless exposed by later channel migration, but they are also very amenable to detection using geophysical techniques including seismic and resistivity methods (see Noller, this volume, Chapter 6; Darwin et al., 1990).

Surface stability, soil formation, and site formation processes are intimately related in fluvial environments (Ferring, 1992; Ferring and Peter, 1987). Analysis of soil morphology, supplemented by radiometric dating, is essential to quantify those processes as they relate to the character and density of archaeological materials (Holliday, 1990). An excellent example of methods is the study of a sequence of buried soils in an alluvial terrace of the Susquehanna River in Pennsylvania by Cremeens et al., (1998). Their work revealed the syndepositional and postdepositional (overprinting) implications of pedogenesis on a series of stratified occupation surfaces.

Facies analysis, as described earlier, is essential for defining the primary contexts of archaeological site formation as well as the environmental settings of the sites. Walker et al. (1997) employed detailed facies study to define changes in depositional environments, controls of Lake Erie base-level fluctuations, and the resulting implications for site formation in a context of initial horticultural settlements in Ontario. Similar approaches work equally well in ancient deposits, such as the investigation of formation processes at a Lower-Middle Pleistocene site in the Malawi Rift, where Kauflulu (1990) used microstratigraphic facies analysis to define processes of site formation in near-channel overbank deposits.

Figure 4.3. Late Holocene alluvium with buried soils at Delaware Canyon, Oklahoma. Figure is pointing to cumulic soil that formed ca. 1950–900 BP and contains Plains Woodland archaeological materials. Above that is a soil with Plains Village materials. Note that these are weakly developed A-C soils, which can be difficult to correlate between drainages based on morphology.

3.5. Landscapes, Change, and Human Settlements

Fluvial landscapes have been settings for human adaptive systems ranging from mobile forager–collectors to complex societies practicing intensive agriculture. This range of adaptive systems is broadly correlative with increasing population densities and also with human modification of the alluvial landscape. In settlement-pattern and land-use analyses therefore, geoarchaeological methods have been brought to focus on a wide array of archaeological contexts and research questions. To illustrate these methods, several case studies are described below.

3.5.1. Hunter–Gatherer Settlement Systems

Hunter–gatherer sites are essentially "embedded" in natural geologic contexts; those systems exhibited both a negligible array of constructed features and an equally small impact on alluvial environments. Geoarchaeology can improve study of hunter–gatherer settlement systems in every stage of research, from implementing site discovery models to the establishing the contexts and character of the sites. These tasks are complicated in settings where significant environmental change has occurred.

Settlement systems in the American Midwest evolved dramatically through the late Archaic, Woodland, and Mississippian periods. Because of intensive exploitation of fluvial environments by all those populations, geoarchaeology has been essential in recent decades of research along the midwestern rivers. A foundation for geoarchaeology in the Illinois Valley was set by Butzer's (1977) geomorphic protocol for defining the context of the famous Koster site, located near St. Louis. His conclusions were explicit statements concerning the relevance of paleogeography to subsistence and settlement models. All students should consider Butzer's stepwise approach toward those conclusions.

Locations of Archaic sites relative to paleomeanderbelts on the Mississippi "American Bottom" were studied by Phillips and Gladfelter (1983). Similarly, Bettis and Hajic (1995) used geomorphic, stratigraphic, and facies data to reconstruct changing site locations of Archaic occupations. In both of these cases, the paleomeanderbelts were associated with channels that were more sinuous than today, and sites were found near former channels as well as near oxbow lakes. Holocene landscape change and settlement patterns in the Upper Mississippi Valley have also been studied by Stafford et al. (1992).

In arid settings, similar objectives are met with methods appropriate for the settings, which frequently involve heterogeneous fluvial, colluvial, lacustrine, and eolian sediments (Waters, 1990). These records are especially important in Paleolithic studies, where long-term, significant environmental change has shaped desert landscapes and has left complex geoarchaeological records (Gladfelter, 1990; Goldberg, 1986; Pappu, 1999; Schuldenrein and Clark, 1994). Goldberg and Bar-Yosef (1995) summarized sedimentary records from the Southern Levant related to settlement patterns and adaptive strategies by Paleolithic through Iron Age cultures. Quite similar approaches to diverse sedimentary contexts in semiarid settings are illustrated by the late Quaternary records of the streams on the Southern Plains (Blum et al., 1992; Hester, 1972; Holliday, 1997) and the subhumid Central Plains (Johnson and Logan, 1990), and the Pleistocene rift valleys of East Africa (Helgren, 1997).

3.5.2. Agricultural Settlement Systems

All early complex societies, which emerged only after successful, large-scale agricultural systems had been developed, were situated in fluvial environments. These systems were initiated by much older shifts toward intensive plant collection and village life that did not entail significant landscape modification (Bar-Yosef and Belfer-Cohen, 1989; Roberts, 1991). Later, both irrigation and the shear size of the agricultural systems began the processes of human modification of fluvial systems that continue today. Moreover, both early and contemporary agriculturalists have demonstrated repeatedly that their control is less than complete and that farmers have always been subject to nature's will. Sufficient evidence is found in MacKay's (1945) demonstration that avulsion of the Euphrates left the cities of Ur and Lagash far from the river, stripping those locations of their long-enjoyed importance. By contrast, avulsion is not possible in the Nile Valley, helping to explain the persistence of major centers in the same locations for millennia. Geoarchaeology has been successfully applied to studying those systems and to deciphering complex patterns of cultural and geologic change.

Impacts of humans on their fluvial landscapes include changing patterns and rates of erosion. Joyce and Mueller (1992, 1997) used coring, soils analysis, and radiocarbon dating of soils to reconstruct the causes and results of increased erosion caused by intensified agriculture in the Rio Verde Basin, Oaxaca. In this case, downstream sedimentation increased agricultural potentials, and it may have prompted population growth. In the humid lowlands of Veracruz, Hebda et al., (1991) conducted detailed facies analysis, stratigraphy, and soils analysis as part of their study of late preclassic and Classic agricultural systems. Both natural changes in sedimentation and the construction of extensive canals and raised field complexes were documented.

Geomorphology and facies analysis were used by Huckleberry (1995) to define late Holocene patterns of alluvial change in the Gila River basin, Arizona, the setting for the complex Hohokam culture. Although irrigation technology had been established earlier, consolidation of the canal networks and also redistribution of settlements was necessary to adjust to adverse affects of braid-plain expansion resulting from larger scale floods after A.D. 1000.

One of Courty's (1995) principal approaches to investigating complex patterns of landscape change associated with Harappan settlements in the Ghaggar Plain, India, was micromorphology. In addition to fieldwork that defined major lithofacies and soil morphology, she documented soils–hydrologic factors that influenced site locations, documented natural versus human causes for eolian activity, and assessed the role of climatic change in the decline of the Harappan civilization.

4. Conclusions

Fluvial environments are common settings for archaeology, and therefore application of geoarchaeological approaches can be of widespread utility. The complexity of alluvial records, including local variation in sedimentary environments as well as larger scale, interdrainage patterns of variation, can be and has been dealt with by Quaternary scientists. Their methods should be studied well by geoarchaeologists. Studying alluvial sediments that contain archaeological materials allows geoarchaeologists to draw on the broad foundation of Quaternary geology for concepts and methods, yet these endeavors also require careful consideration of the archaeological implications of those geologic records. Different approaches are required in different environments, ranging from temperate to arid, and at different temporal scales. Also, different approaches are required for varying archaeological problems. Hunter–gatherer studies, for example, entail emphasis on site locational strategies and environmental (habitat) reconstructions that support formational analyses (Stein, Chapter 2, this volume). Finally, studying complex social systems and their impact on past landscapes is aided by methods that are able to deal with those research challenges. The researchers cited in this chapter have all contributed to these geoarchaeological goals, either as Quaternary scientists or as archaeologists or as persons or teams who apply geoarchaeology to those problems.

5. References

Aitken, M. J., 1995, Chronometric Techniques for the Middle Pleistocene. In *The Earliest Occupation of Europe*, edited by W. Roebroeks, and T. V. Kolfschoten, pp. 269–278. University of Leiden, Leiden, The Netherlands.

Allen, J. R. L., 1965, A Review of the Origin and Character of Recent Alluvial Sediments *Sedimentology* 5:89–191.

Antevs, E., 1952, Arroyo-Cutting and Filling, *Journal of Geology* 60:375–385.

Artz, J. A., 1995, Geologic Contexts of the Early and Middle Holocene Archaeological Record in North Dakota and Adjoining Areas of the Northern Plains. In *Archaeological Geology of the Archaic Period in the United States*, edited by E. A. Bettis III, pp. 67–86. Geological Society of America Boulder, CO. Special Paper 297.

Autin, W. J., 1992, Use of Alloformations for Definition of Holocene Meander Belts in the Middle Amite River, Southeastern Louisiana, *Geological Society of America Bulletin* 104:233–241.

Baker, V. B., and Penteado-Orellana, M. M., 1977, Adjustment to Quaternary Climate Change by the Colorado River in Central Texas, *Journal of Geology* 85:395–422.

Bar-Yosef, O., and Belfer-Cohen, A., 1989, The Origin of Sedentism and Farming Communities in the Levant, *Journal or World Prehistory* 3 (4):447–498.

Behrensmeyer, A. K., 1982, Time Resolution in Fluvial Vertebrate Assemblages. *Paleobiology* 8:211–227.

Behrensmeyer, A. K., 1987, Miocene Fluvial Facies and Vertebrate Taphonomy in Northern Pakistan. In, *Recent Developments in Fluvial Sedimentology*, edited by F. G. Ethridge, R. M. Flores and M. D. Harvey, pp. 169–176. Society of Economic Paleontologists and Mineralogists, Special Publication No. 39. Tulsa, OK.

Bettis, E. A., III, and E. R., Hajic, 1995, Landscape Development and the Location of Evidence of Archaic cultures in the Upper Midwest. In *Archaeological Geology of the Archaic period in the United States*, edited by E. A. Bettis III, pp. 87–113. Geological Society of America, Boulder, CO. Special Paper 297.

Bettis, E. A., III, and Littke, J. P., 1987, *Holocene Alluvial Stratigraphy and Landscape Development in Soap Creek Watershed, Appanoose, Davis, Monroe, and Wapello Counties, Iowa*. Iowa Department of Natural Resources, Iowa City.

Binford, L. R., 1968, Post-Pleistocene Adaptations. In *New Perspectives in Archaeology*, edited by S. R. Binford and L. R. Binford, pp. 313–341. University of Chicago Press, Chicago.
Birkeland, P. W., 1999, *Soils and Geomorphology*, Oxford University Press, New York.
Bloom, A. L., 1991, *Geomorphology, A Systematic Analysis of Late Cenozoic Landforms*, Prentice-Hall, Englewood Cliffs, NJ.
Blum, M. D., Abbott, J. T. and Valastro, S. Jr., 1992, Evolution of Landscapes on the Double Mountain Fork of the Brazos River, West Texas: Implications for Preservation and Visibility of the Archaeological Record, *Geoarchaeology, An International Journal* 7(4):339–370.
Borchardt, G., and Hill, R. L., 1985, Smectite Pedogenesis and late Holocene Tectonism along the Raymond Fault, San Amrino, California. In *Soils and Quaternary Geology of the Southwestern United States*, edited by D. L. Weide, pp. 65–78. Geological Society of America Special Paper 203. Boulder, CO.
Brakenridge, G. R., 1984, Alluvial Stratigraphy and Radiocarbon Dating along the Duck River Tennessee: Implications Regarding Flood-plain Origin. *Geological Society of America Bulletin* 95:9–25.
Brakenridge, G. R., Thomas, P. A. Conkey, L. E. and Schiferle, J. C., 1988, Fluvial Sedimentation in Response to Postglacial Uplift and Environmental Change, Missisquoi River, Vermont, *Quaternary Research* 30:190–203.
Bryan, K., 1925, Date of Channel Trenching (Arroyo Cutting) in the Arid Southwest, *Science* 62:338–344.
Bryan, K., and Albritton, C. C., Jr., 1943, Soil Phenomena as Evidence of Climatic Changes, *American Journal of Science* 241:469–490.
Bull, W. B., 1972, Recognition of Alluvial Fans in the Stratigraphic Record, *Society of Economic Paleontologists and Mineralogists*, Special Publication No. 16., pp. 63–83.
Bull, W. B., 1977, The Alluvial Fan Environment, *Progress in Physical Geography* 1:222–270.
Bull, W. B., 1990, Stream–Terrace Genesis, Implications for Soil Development, *Geomorphology* 3:351–367.
Bull, W. B., 1991, *Geomorphic Responses to Climatic Change*, Oxford University Press, New York.
Butzer, K. W., 1977, *Geomorphology of the Lower Illinois Valley as a Spatial-Temporal Context for the Koster Archaic Site*, Reports of Investigations No. 34, Illinois State Museum, Springfield, Illinois.
Butzer, K. W., 1982, *Archaeology as Human Ecology*, Cambridge University Press, Cambridge, UK.
Butzer, K. W., and Hansen, C. L., 1968, *Desert and River in Nubia*. University of Wisconsin Press, Madison.
Cleaves, E. T., Godfrey, A. E., and Bricker, O. P., 1970, Geochemical Balance of a Small Watershed and its Geomorphic Implications, *Geological Society of America Bulletin* 81: 3015–3032.
Collinson, J. D. and Lewin, J. (eds.), 1983, *Modern and Ancient Fluvial Systems*, Special Publication No. 6, International Association of Sedimentologists. Blackwell, Oxford.
Cooke, R. U., and Warren, A., 1973, *Geomorphology in Deserts*, University of California Press, Berkeley.
Courty, M.-A., 1995, Late Quaternary Environmental Changes and Natural Constraints to Ancient Land Use (Northwest India), In *Ancient Peoples and Landscapes*, edited by E. Johnson, pp. 105–126. Museum of Texas Tech University, Lubbock.
Cremeens, D. L., Hart, J. P., and Darmody, R. G., 1998, Complex Pedostratigraphy of a Terrace Fragipan at the Memorial Park Site, Central Pennsylvania, *Geoarchaeology: An International Journal* 13:339–360.
Darwin, R., Ferring, C. R., and Ellwood, B., 1990, Geoelectric Stratigraphy and Subsurface Evaluation of Quaternary Stream Sediments at Cooper Basin, NE Texas, *Geoarchaeology. An International Journal* 5 (1):53–79.
Debusschere, K., Miller, B. J., and Ramenofsky, A. F., 1989, A Geoarchaeological Reconstruction of Cowpen Slough, A Late Archaic Site in East Central Louisiana, *Geoarchaeology. An International Journal* 4(3):251–270.
Easterbrook, D. J., 1988, Paleomagnetism of Quaternary Deposits. In *Dating Quaternary Sediments*, edited by D. J. Easterbrook, pp. 111–122. Geological Society of America Special Paper 227. Boulder, CO.
Faught, M. K., and Donoghue, J. F., 1997, Marine Inundated Archaeological Sites and Paleofluvial Systems: Examples from a Karst-Controlled Continental Shelf Setting in Apalachee Bay, Northeastern Gulf of Mexico, *Geoarchaeology, An International Journal* 12(5):417–458.

Ferring, C. R., 1986, Rates of Fluvial Sedimentation: Implications for Archaeological Variability. *Geoarchaeology: An International Journal* 1(3):259–274.
Ferring, C. R., 1990, Archaeological Geology of the Southern Plains. In *Archaeological Geology of North America*, edited by N. P. Lasca and J. Donahue, pp. 253–266 Geological Society of America, Centennial Special Volume 4. Boulder, CO.
Ferring, C. R., 1992, Alluvial Pedology and Geoarchaeological Research. In *Soils in Archaeology*, edited by V. Holliday, pp. 1–39. Smithsonian Institution Press, Washington, DC.
Ferring, C. R., 1993, *Late Quaternary Geology of the Upper Trinity River Basin, Texas*. Ph.D. Dissertation, University of Texas at Dallas.
Ferring, C. R., 1994, The Role of Geoarchaeology in Paleoindian Research. In *Method and Theory for Investigating the Peopling of the Americas*, edited by R. Bonnichsen, and G. Steele, pp. 52–72. Center for the Study of the First Americans, Corvallis, Oregon.
Ferring, C. R., 1995, Middle Holocene Environments, Geology and Archaeology in the Southern Plains. In, *Archaeological Geology of the Archaic Period in the United States*, edited by E. A. Bettis, III, pp. 21–35. Geological Society of America, Special Paper 297, Boulder, CO.
Ferring, C. R. and Peter, D. E., 1987, Geoarchaeology of the Dyer Site, A Prehistoric Occupation in the Western Ouachitas, Oklahoma, *Plains Anthropologist* 32(118):351–366.
Ferring, C. R., and Yates, B. C., 1997, *Holocene Geoarchaeology and Prehistory of the Raz Roberts Lake Area, North Central Texas*. Institute of Applied Science, University of North Texas, Denton, Texas.
Gerrard, A. J. (ed.), 1987, *Alluvial soils*. Van Nostrand, New York.
Gile, L. H., Hawley, J. W., and Grossman, R. B., 1981, *Soils and Geomorphology in the Basin and Range Area of Southern New Mexico, Guidebook to the Desert Project: Socorro, Memoir 39*. New Mexico Bureau of Mines and Mineral Resources.
Gladfelter, B. G., 1985, On the Interpretation of Archaeological Sites in Alluvial Settings. In *Archaeological Sediments in Context*, edited by, J. K. Stein and W. R. Farrand, pp. 41–52 Center for the Study of Early Man, University of Maine, Orono.
Gladfelter, B. G., 1990, The Geomorphic Setting of Upper Paleolithic Sites in Wadi el Sheikh, Southern Sinai, *Geoarcheology: An International Journal*, 5(2):99–119.
Glennie, K. W., 1970, *Desert Sedimentary Environments*, Elsevier, Amsterdam.
Goldberg, P., 1986, Late Quaternary Environmental History of the Southern Levant. *Geoarchaeology: An International Journal*, 1(3):225–244.
Goldberg, P., and Bar-Yosef, O., 1995, Sedimentary Environments of Prehistoric Sites in Israel and the Southern Levant. In *Ancient Peoples and Landscapes*, edited by E. Johnson, pp. 29–49. Museum of Texas Tech University, Lubbock.
Graf, W. L., Hereford, R., Laity, J., and Young, R. A., 1987, Colorado Plateau, In *Geomorphic Systems of North America*, edited by W. L. Graf, pp. 259–302. Geological Society of America, Boulder, CO.
Guccione, M. J., Sierzchula, M. C., Lafferty, R. H., III, and Kelley, D., 1998, Site Preservation along an Active Meandering and Avulsing River: The Red River, Arkansas, *Geoarchaeology: An International Journal* 13(5):475–500.
Gustafson, T. C. (ed.), 1986, *Geomorphology and Quaternary Stratigraphy of the Rolling Plains, Texas Panhandle*, Bureau of Economic Geology, University of Texas, Austin.
Haas, H., Holliday, V., Stuckenrath, R., 1986, Dating of Holocene Stratigraphy with Soluble and Insoluble Organic Fractions at the Lubbock Lake Archaeological Site. In *Twelfth International Radiocarbon Conference*, edited by, M. Stuiver, and R. Kra, pp. 473–485. *Radiocarbon* 28 (2A). American Journal of Science, New Haven.
Hall, S. A., 1982, Late Holocene Paleoecology of the Southern Plains, *Quaternary Research* 17:391–407.
Hall, S. A., 1990a, Channel Trenching and Climatic Change in the Southern U.S. Great Plains, *Geology* 18:342–345.
Hall, S. A., 1990b, Holocene Landscapes of the San Juan Basin, New Mexico: Geomorphic, Climatic and Cultural Dynamics. In *Archaeological Geology of North America*, edited by, N. P. Lasca and J. Donahue, pp. 323–334. Geological Society of America Centennial Special Volume 4, Boulder, CO.
Hanson, C. B., 1980, Fluvial Taphonomic Processes: Models and Experiments. In *Fossils in the Making*, edited by, A. K. Behrensmeyer, and A. P. Hill, pp. 156–181. University of Chicago Press, Chicago.

Harlin, J. M., 1978, Reservoir Sedimentation as a Function of Precipitation Variability. *Water Resources Bulletin* 14(6):1457–1465.

Harlin, J. M., 1980, The Effect of Precipitation Variability on Drainage Basin Morphometry. *American Journal of Science* 28:812–825.

Hassan, F. A., 1985, Fluvial Systems and Geoarchaeology in Arid Lands: With Examples from North Africa, the Near East and the American Southwest. In *Archaeological Sediments in Context*, edited by J. K. Stein, and W. R. Farrand, pp. 53–68. Center for the Study of Early Man, University of Maine, Orono.

Haury, E. W., 1976, *The Hohokam, Desert Farmers and Craftsmen*, University of Arizona Press, Tucson.

Hayward, M., and Fenwick, I., 1983, Soils and Hydrological Change. In *Background to Paleohydrology*, edited by, K. J. Gregory, pp. 167–187. Wiley, London.

Hebda, R. J., Siemens, A. H., and Robertson, A., 1991, Stratigraphy, Depositional Environment, and Cultural Significance of Holocene Sediments in Patterned Wetlands of Central Veracruz, Mexico, *Geoarchaeology: An International Journal*, 6(1):61–84.

Helgren, D. A., 1997, Locations and Landscapes of Paleolithic Sites in the Semliki Rift, Zaire. *Geoarchaeology, An International Journal* 12(4):337–361.

Hester, J. J., 1972, Paleoarchaeology of the Llano Estacado. In *Late Pleistocene Environments of the Southern High Plains*, edited by F. Wendorf, and J. J. Hester, pp. 247–256. Southern Methodist University, Dallas, TX.

Holliday, V. T., 1989, Paleopedology in Archaeology, *Paleopedology* 16:187–206.

Holliday, V. T., 1995, *Stratigraphy and Paleoenvironments of Late Quaternary Valley Fills on the Southern High Plains*. Geological Society of America, Memoir 186. Boulder, CO.

Holliday, V. T., 1990, Pedology in Archaeology. In *Archaeological Geology of North America*, edited by N. P. Lasca and J. Donahue, pp. 525–540. Geological Society of America, Centennial Special Vol. 4, Geological Society of America, Boulder, CO.

Holliday, V. T., 1997, *Paleoindian Geoarchaeology of the Southern High Plains*, University of Texas Press, Austin.

Hooke, R. LeB., 1967, Processes on Arid-Region Alluvial Fans. *Journal of Geology* 75: 438–460.

Huckleberry, G. A., 1995, Archaeological Implications of Late-Holocene Channel Changes on the Middle Gila River, Arizona, *Geoarchaeology, An International Journal* 10(3):159–182.

Irving, W. N., Jopling, A. V. and Kritsch-Armstrong, I., 1989, Studies of Bone Technology and Taphonomy, Old Crow Basin, Yukon Territory, In *Bone Modification*, edited by R. Bonnichsen and M. H. Sorg, pp. 347–379. Center for the Study of the First Americans, Orono, Maine.

Jing, Z., and Rapp, G., Jr., 1998, Environmental Magnetic Indicators of the Sedimentary Context of Archaeological Sites in the Shangqiu Area of China. *Geoarchaeology: An International Journal*, 13(1):37–54.

Jing, Z., and Rapp, G., Jr., and T. Gao, 1995, Holocene Landscape Evolution and Its Impact on the Neolithic and Bronze Age Sites in the Shangqiu Area, Northern China. *Geoarcheology: An International Journal*, 10(6):481–513.

Johnson, E., and Holliday, V. T., 1989, Lubbock Lake: Late Quaternary Cultural and Environmental Change on the Southern High Plains. *Journal of Quaternary Science* 4(2):145–165.

Johnson, W. C., and Logan, B., 1990, Geoarchaeology of the Kansas River Basin, Central Great Plains. In *Archaeological Geology of North America*, edited by N. P. Lasca, and J. Donahue, pp. 267–300. Centennial Special Volume 4, Geological Society of America, Boulder, CO.

Joyce, A. A., and Mueller, R. G., 1992, The Social Impact of Anthropogenic Landscape Modification in the Rio Verde Drainage Basin, Oaxaca, Mexico, *Geoarchaeology, An International Journal* 7(6):503–526.

Joyce, A. A., and Mueller, R. G., 1997, Prehispanic Human Ecology of the Rio Verde Drainage Basin, Mexico, *World Archaeology* 29(1):75–94.

Kauflulu, Z. M., 1990, Sedimentary Environments at the Mwanganda site, Malawi. *Geoarchaeology. An International Journal*, 5(1):15–27.

Knox, J. C., 1976, Concept of the Graded Stream. In *Theories of Landform Development*, edited by, W. N. Melhorn, and R. C. Flemal, pp. 169–198. State University of New York at Binghamton Publications pin Geomorphology, Binghamton.

Knox, J. C., 1983, Responses of River Systems to Holocene Climates. In *Late Quaternary Environments of the United States*, Volume 2, *The Holocene*, edited by H. E. Wright, pp. 26–41. University of Minnesota Press, Minneapolis.

Kochel, R. C., and Johnson, R. A., 1984, Geomorphology and Sedimentology of Humid-Temperate Alluvial Fans, Central Virginia. In *Sedimentology of Gravels and Conglomerates*, edited by E. H. Koster and R. J. Steel, pp. 109–122. Canadian Society of Petroleum Geologists, Memoir 10. McAra Printing, Calgary.

Koster, E. H., 1987, Vertebrate Taphonomy Applied to the Analysis of Ancient Fluvial Systems. In *Recent Developments in Fluvial Sedimentology*, edited by F. G. Ethridge, R. M. Flores, and M. D. Harvey, pp. 159–168. Society of Economic Paleontologists and Mineralogists, Special Publication No. 39, Tulsa, OK.

Kraus, M. J., and Bown, T. M., 1988, Pedofacies Analysis: A New Approach to Reconstructing Ancient Fluvial Sequences. In *Paleosols and Weathering Through Geologic Time: Principles and Applications*, edited by J. Reinhardt and W. R. Sigleo, pp. 143–152. Special Paper 216, Geological Society of America, Boulder, CO.

Kreutzer, L., 1988, Megafaunal Butchering at Lubbock Lake, Texas: A Taphonomic Reanalysis, *Quaternary Research* 30:221–231.

Langbein, W. B., and Schumm, S. A., 1958, Yield of Sediment in Relation to Mean Annual Precipitation. *American Geophysical Union Transactions* 39:1076–1084.

Leopold, J. P., Wolman, M. G., and Miller, J. P.,1964, *Fluvial Processes in Geomorphology*. Freeman, San Francisco.

Leopold, L. B., and Miller, J. P., 1956, *Ephemeral Streams—Hydraulic Factors and Their Relation to the Drainage Net*. U.S. Geological Survey Professional Paper 352, Washington DC.

Lewin, J., 1972, Late-stage Meander Growth. *Nature; Physical Science* 240 (101):116

Lewin, J., 1978, Floodplain Geomorphology, *Progress in Physical Geography* 2:408–437.

Linse, A. R., 1993, Geoarchaeological Scale and Archaeological Interpretation: Examples from the Central Jornada Mogollon. In *Effects of Scale on Archaeological and Geoscientific Perspectives*, edited by, J. K. Stein, and A. R. Linse, pp. 11–28. Geological Society of America Special Paper 283, Boulder, CO.

Machette, M. N., 1985, Calcic Soils of the Southwestern United States. In *Soils and Quaternary Geology of the Southwestern United States*, edited by D. L. Weide, pp. 1–22. Geological Society of America, Boulder, Colorado. Special Paper 203.

MacKay, D., 1945, Ancient River Beds and Dead Cities, *Antiquity* 19:135–144.

Mandel, R. D., 1994a, Geoarchaeology of the Lower Walnut River Valley at Arkansas City, Kansas, *Kansas Anthropologist* 15(1):46–49.

Mandel, R. D., 1994b, Holocene Landscape Evolution in the Pawnee River Valley, Southwestern Kansas. Kansas Geological Survey Kansas State Historical Society, *Bulletin* 236.

Mandel, R. D., 1995, Geomorphic Controls of the Archaic Record in the Central Plains of the United States. In *Archaeological Geology of the Archaic Period in the United States*, edited by E. A. Bettis III, pp. 37–66. Geological Society of America Boulder, CO. Special Paper 297.

Mania, D., 1996, The Earliest Occupation of Europe: The Elbe–Saale Region (Germany). In *The Earliest Occupation of Europe*, edited by W. Roebroeks, and T. Van Kolfschoten, pp. 85–101. University of Leiden, Leiden.

Martin, C. W., and Johnson, W. C., 1995, Variation in Radiocarbon Ages of Soil Organic Matter Fractions from Late Quaternary Buried Soils, *Quaternary Research* 43:232–237.

May, D. W., 1989, Holocene Alluvial Fills in the South Loup Valley, Nebraska, *Quaternary Research* 32(1):117–120.

McDowell, P. F., 1983, Evidence of Stream Response to Holocene Climatic Change in a Small Wisconsin Watershed, *Quaternary Research* 19:100–116.

McFaul, M., Traugh, K. L., Smith, G. D., Doering, W., and Zier, C. J., 1994, Geoarchaeologic Analysis of South Platte River Terraces: Kersey, Colorado, *Geoarchaeology. An International Journal* 9(5):345–374.

McHugh, W. P.,McCauley, J. F., Haynes, C. V., Breed, C. S., and Schaber, G. G., 1988, Paleorivers and Geoarchaeology in the Southern Egyptian Sahara, *Geoarchaeology: An International Journal*, 3(1):1–40.

Melton, M. A., 1965, The Geomorphic and Palaeoclimatic Significance of Alluvial Deposits in Southern Arizona, *Journal of Geology* 73:1–38.

Miall, A. D. (ed.), 1981, *Sedimentation and Tectonics in Alluvial Basins*, Geological Association of Canada, Special Paper 23. Waterloo.

Miall, A. D., 1992, Alluvial Deposits. In *Facies Models, Response to Sea Level Change*, edited by, R. G. Walker, and N. P. James, pp. 119–142. Geological Association of Canada, St. Johns.

Mooers, H. D., and Dobbs, C. A., 1993, Holocene Landscape Evolution and the Development of Models for Human Interaction with the Environment: An Example from the Mississippi Headwaters Region, *Geoarchaeology: An International Journal*, 8(6):475–492.

Morrison, R. B., 1978, Quaternary Soil Stratigraphy—Concepts, Methods and Problems. In *Quaternary Soils*, edited by W. C. Mahaney, pp. 77–108. Geological Abstracts, Norwich.

Nanson, G. C., Price, D. M., Short, S. A., Young, R. W., and Jones, B. G., 1991, Comparative Uranium-Thorium and Thermoluminescence Dating of Weathered Quaternary Alluvium in the Tropics of Northern Australia. *Quaternary Research* 35:347–366.

Nordt, L. C., 1995, Geoarchaeological Investigations of Henson Creek: A Low-Order Tributary in Central Texas, *Geoarchaeology: An International Journal* 10(3):205–221.

North American Commission on Stratigraphic Nomenclature, 1983, North American Stratigraphic Code, *American Association of Petroleum Geologists Bulletin*. 67:841–875,

Pappu, S., 1999, A Study of Natural Site Formation Processes in the Kortallayar Basin, Tamil Nadu, South India, *Geoarchaeology: An International Journal* 14(2):127–150.

Pärssinen, M. H., Salo, J. S., and Räsänen, M. E., 1996, River Floodplain Relocations and the Abandonment of Aborigine Settlements in the Upper Amazon Basin: A Historical Case Study of San Miguel de Cunibos at the Middle Ucayali River, *Geoarchaeology: An International Journal* 11(4):345–359.

Patton, P. C., and Schumm, S. A., 1981, Ephemeral-Stream Processes: Implications for Studies of Quaternary Valley Fills, *Quaternary Research* 15:24–43.

Phillips, J. L., and Gladfelter, B. G., 1983, The Labras Lake Site and the Paleogeographic Setting of the Late Archaic in the American Bottom. In *Archaic Hunters and Gatherers in the American Midwest*, edited by J. L. Phillips and J. A. Brown, pp. 197–218. Academic Press, New York.

Ponti, D. J., 1985, The Quaternary Alluvial Sequence of the Antelope Valley, California. In *Soils and Quaternary Geology of the Southwestern United States*, edited by D. L. Weide, pp. 79–96. Geological Society of America, Boulder, CO. Special Paper 203.

Pope, K. O., Runnels, C. N. and Ku, T-L., 1984, Dating Middle Paleolithic Red Beds in Southern Greece, *Nature* 312 (5991):264–266.

Petraglia, M. D., and Nash, D. T. ,1987, The Impact of Fluvial Processes on Experimental Sites. In *Natural Formation Processes and the Archaeological Record*, edited by D. T. Nash and M. D. Petraglia, pp. 108–130. BAR International, Series 352, Oxford.

Putnam, D. E., 1994, Vertical Accretion of Flood Deposits and Deeply Stratified Archaeological Site Formation in Central Maine, USA. *Geoarchaeology: An International Journal* 9(6):467–502.

Rapp, G. R., and Hill, C., 1997, *Geoarchaeology*, Yale University Press, New Haven.

Reineck, H.-E., and Singh, I. B., 1980, *Depositional Sedimentary Environments*, Springer-Verlag, New York.

Roberts, N., 1991, Late Quaternary Geomorphological Change and the Origins of Agriculture in South Central Turkey, *Geoarchaeology: An International Journal* 6(1):1–26.

Ruhe, R. V., 1983, Depositional Environment of Late Wisconsin Loess in the Midcontinental United States. In *Late Quaternary Environments of the United States*, Volume 1, *The Late Pleistocene*, edited by S. C. Porter, pp. 130–137. University of Minnesota Press, Minneapolis.

Ruhe, R. V., and Olson, C.G., 1980, Soil Welding, *Soil Science* 130 (3):132–139.

Schick, K. D., 1987, Experimentally-Derived Criteria for Assessing Hydrologic Disturbance of Archaeological Sites. In *Natural Formation Processes and the Archaeological Record*, edited by D. T. Nash and M. D. Petraglia, pp. 86–107. BAR International, Series 352, Oxford.

Schiffer, M. J., 1987, *Formation Processes in the Archaeological Record*, University of Arizona Press, Tucson.

Schuldenrein, J., and Clark, G. A., 1994, Landscape and Prehistoric Chronology of West-Central Jordan, *Geoarchaeology: An International Journal* 9(1):31–55.

Schumm, S. A., 1977, *The Fluvial System*. Wiley, New York.

Schumm, S. A., and Brakenridge, R., 1987, River Responses. In *North America and Adjacent Oceans during the Last Deglaciation*, pp. 221–240. *The Geology of North America*, Volume K-3. Geological Society of America, Boulder, CO.

Schumm, S. A. and Hadley, R. F., 1957, Arroyos and the Semiarid Cycle of Erosion, *American Journal of Science* 225:164–174.

Schumm, S. A., and Lichty, R. W., 1965, Time, Space and Causality in Geomorphology, *American Journal of Science* 263:110–119.

Schwarcz, H. P, Grün, R., Latham, A. G., Mania, D., and Brunnacker, K., 1988, The Bilzingsleben Archaeological Site: New Dating Evidence, *Archaeometry* 30:5–17.

Shackley, M. L., 1974, Stream Abrasion of Flint Artifacts, *Nature* 248(5448):501–502.

Shackley, M. L., 1978, The Behavior of Artifacts as Sedimentary Particles in a Fluviatile Environment, *Archaeometry* 20(1):55–61.

Shea, J., 1999, Artifact Abrasion, Fluvial Processes, and Living Floors from the Early Paleolithic Site of 'Ubeidiya (Jordan Valley, Israel), *Geoarchaeology: An International Journal* 14(2):191–207.

Stafford, C. R., Leigh, D. S., and Asch, D. L., 1992, Prehistoric Settlement and Landscape Change on Alluvial Fans in the Upper Mississippi River Valley, *Geoarchaeology: An International Journal* 7(4):287–314.

Steele, D. G., and Calson, D. L., 1989, Excavation and Taphonomy of Mammoth Remains from the Duewall–Newberry Site, Brazos County, Texas. In *Bone Modification*, edited by R. Bonnichsen and M. H. Sorg, pp. 413–430. Center for the Study of the First Americans, Orono, ME.

Summerfield, M. A., 1991, *Global Geomorphology, An Introduction to the Study of Landforms*. Wiley, New York.

Thornes, J. B., and Brunsden, D., 1977, *Geomorphology and Time*. Wiley, New York.

Turner, W. B., and Klippel, W. E., 1989, Hunter–Gatherers in the Nashville Basin: Archaeological and Geological Evidence for Variability in Prehistoric Land Use, *Geoarchaeology: An International Journal*, 4(1):43–67.

van Andel, T. H., 1998, Paleosols, Red Sediments, and the Old Stone Age in Greece, *Geoarchaeology: An International Journal*, 13(4):361–390.

van Andel, T. H., and Tzedakis, P. C., 1996, Paleolithic Landscapes of Europe and Environs, 150,000–25,000 Years Ago: An Overview. *Quaternary Science Reviews*, 15:481–500.

Veneman, P. L. M., Vepraskas, M. J., and Bouma, J., 1976, The Physical Significance of Soil Mottling in a Wisconsin Toposequence. *Geoderma* 15:103–118.

Vita-Finzi, C., 1986, *Recent Earth Movements, An Introduction to Neotectonics*. Academic Press, London.

Walling, D. E., and Webb, B. W., 1983, Patterns of Sediment Yield. In *Background to Paleohydrology*, edited by K. J. Gregory, pp. 69–99. Wiley, London.

Walker, I. R., Levesque, A. J., Cwynar, L. C., and Lotter, A. F., 1997, An Expanded Surface–Water Paleotemperature Inference Model for Use with Fossil Midges from Eastern Canada, *Journal of Paleolimnology* 18(2):165–178.

Walker, P. H., and Coventry, R. J., 1976, Soil Profile Development in Some Alluvial Deposits of Eastern New South Wales. *Australian Journal of Soil Research*. 14(3):305–317.

Walker, R. G., 1992, Facies, Facies Models and Modern Stratigraphic Concepts. In *Facies Models, Response to Sea Level Change*, edited by R. G. Walker, and N. P. James, pp. 1–14. Geological Association of Canada, St. Johns.

Waters, M. R., 1986, The Sulphur Spring Stage and Its Place in New World Prehistory, *Quaternary Research* 19:373–387.

Waters, M. R., 1988, The Impact of Fluvial Processes and Landscape Evolution on Archaeological Sites and Settlement Patterns Along the San Xavier Reach of the Santa Cruz River, Arizona, *Geoarchaeology: An International Journal* 3(3):205–219.

Waters, M. R., 1992, *Principles of Geoarchaeology, A North American Perspective*, University of Arizona Press, Tucson.

Waters, M. R., 1998, The Effect of Landscape Hydrologic Variables on the Prehistoric Salado: Geoarchaeological Investigations in the Tonto Basin, Arizona, *Geoarchaeology: An International Journal* 13(2):105–160.

Wendorf, F., And Schild, R., 1980, *Prehistory of the Eastern Sahara*, Academic Press, New York.

Wendorf, F., Close, A. E., and Schild, R., 1987, A Survey of the Egyptian Radar Channels: An Example of Applied Archaeology, *Journal of Field Archaeology* 14:43–63.

Wiant, M. D., Hajic, E. R., and Styles, T. R., 1983, Napoleon Hollow and Koster Site Stratigraphy: Implications for Holocene Landscape Evolution and Studies of Archaic Period Settlement

Patterns on the Lower Illinois River Valley. In *Archaic Hunters and Gatherers in the American Midwest*, edited by J. L. Phillips and J. A. Brown, pp. 147–164. Academic Press, New York.

Wood, W. R., and Johnson, D. L., 1978, A Survey of Disturbance Processes in Archaeological Site Formation. In *Advances in Archaeological Method and Theory*, edited by M. B. Schiffer, pp. 315–381. Academic Press, New York.

Wynn, J. C., 1990, Applications of High–Resolution Geophysical Methods to Archaeology. In *Archaeological Geology of North America*, edited by N. P. Lasca and J. Donahue, pp. 603–618. Geological Society of America Centennial Special Volume 4, Boulder, CO.

5

A Geomorphological Approach to Reconstructing Archaeological Settlement Patterns Based on Surficial Artifact Distribution

Replacing Humans on the Landscape

LISA E. WELLS

1. Introduction

Physical and cultural landscapes evolve in response to a combination of natural and anthropogenic factors. Therefore any attempt to understand landscape evolution requires the integration of anthropology and the physical sciences. As such, both landscape archaeology and geomorphology are interdependent and interdisciplinary endeavors. Unfortunately, we are hampered in our re-creation of past landscapes by the set of assumptions and the scientific language we bring to this endeavor. This chapter attempts to present a geomorphological approach

---

LISA E. WELLS • Department of Geology, Vanderbilt University, Nashville, Tennessee, 37235.
Earth Sciences and Archaeology, edited by Paul Goldberg, Vance T. Holliday, and C. Reid Ferring. Kluwer Academic/Plenum Publishers, New York, 2001.

to re-creating physical and cultural landscapes and in this process elucidates the strengths and weaknesses of this approach.

Regional archaeological survey commonly attempts to trace social and economic relationships as they are manifest in changes in settlement pattern through time. Survey data collection is therefore often focused toward the creation of maps or images that show the physical relationships between a network of anthropogenic features and how these relationships change through time and space (Binford, 1992; Dunnell, 1992). To this end, a variety of methods have been developed by which archaeologists collect data representing the distribution of anthropogenic features across the physical landscape. The maps that result from the survey represent the spatial distribution of artifacts across the modern landscape. Many levels of interpretation and assumption are required to reconstruct settlement patterns from these representations of artifact distribution. At the most fundamental level, the reconstruction of settlement patterns requires a geomorphological understanding of landscape evolution because each artifact or feature is reworked, at least to some degree, by geological processes after its original emplacement.

Without an appreciation of landscape evolution and geomorphic change, the potential for misinterpretation of archaeological survey data is immense. A morphostratigraphic understanding of the landscape provides a context for determining the spatial relevance of artifacts. Unless archaeologists incorporate this fundamental geologic information into their data analysis, they cannot hope to provide a secure interpretation of the human landscape. To be specific, knowing the geologic context of an artifact provides the survey team a framework for determining if the artifact is *in situ* (i.e., anthropogenically deposited) or was reworked by geologic processes; establishing the age of surfaces relative to the times of land use allows us to more accurately interpret the archaeological relevance of sterile surfaces (be they depositional, erosional, or never occupied); reconstruction of the physical landscape at the time of occupation allows for the understanding of the environmental context of sites and features; and our experience suggests that once the morphostratigraphy and the chronostratigraphy of the physical landscape have been established, the success rate in finding artifacts increases dramatically.

The goal of this chapter is to describe the methods by which geomorphology can be integrated into an archaeological survey to facilitate survey sampling strategies, prioritize survey regions, reconstruct paleolandscapes, and provide an environmental framework for survey data interpretation. To this end I describe two projects conceived and executed in very different physical and cultural landscapes and the lessons learned from each. What the two survey regions do have in common is that they are steep mountainous terrains whose landscapes are predominantly erosional rather than aggradational. This means that the potential for site burial is less important than in predominantly aggradational settings such as the plains of North America (e.g., Holliday, 1995; Bettis, 1995). The case studies presented here serve to illustrate the process of a geomorphologic study and how regional archaeological interpretations are impacted by landscape evolution and the strength of integrating geomorphology from the onset of an archaeological survey.

In coastal Peru the archaeology and geomorphology have been fully independent endeavors. The archaeological studies are largely site based and are focused on understanding the cultural evolution of the region. Although cultural artifacts are thinly scattered across the landscape, discrete peaks in artifact density and architectural features are relatively easy to define in terms of individual sites *per se*. I mapped archaeological features in the process of the investigation, but the original geoarchaeologic goal was not a settlement pattern study, but rather reconstruction of the impacts of long-term land use on a hyperarid landscape. The integration of data collected by different projects with different research agendas results in strong limitations in developing a coevolutionary landscape model.

The second study is the Sydney–Cyprus Survey Project (SCSP), a siteless survey of the northern piedmont fringe of the Troodos Mountains, Cyprus, that was completed in the year 2000. Being a siteless survey, data collection focused on the quantity and nature of individual artifacts in the field; the standard concept of "site" was not considered during data collection. Unlike the Peruvian desert, discrete peaks in artifact density and architectural features are poorly defined here, and thus the geomorphological context of individual artifacts is centrally important to any subsequent data analysis. As a result, geomorphology became an integral part of SCSP during the second year of survey, after the sampling methods were established. Geomorphology was included during all subsequent stages of planning and data collection. This has resulted in an improved method of data collection and an enhanced understanding of landscape evolution. Ideally, geomorphologists should be involved in survey project planning from the original stages of project conception.

2. Techniques and Methods

The techniques described as follows are a primer for the nonspecialist on the process of creating a geomorphologic map and on how geomorphology can be incorporated in statistically stratifying the landscape for survey. These are largely apprenticed skills learned in geology field camps, and although there are good manuals of field geology (e.g., Compton, 1985; Lahee, 1961), the techniques are not well described in the literature. In addition, Quaternary geology and geomorphology are poorly represented in the technique-based literature. As part of an interdisciplinary research team, I have found that teaching these skills is most effective on joint walking transects across the landscape. Ideally, a geomorphologist is assigned to work with each survey crew to collect geological data temporally and spatially coincident with the archaeological survey data.

2.1. Morphostratigraphy and Allostratigraphy

One of the initial lessons learned during the SCSP was that when the archaeologists requested a geomorphologic map of the landscape, they were not envision-

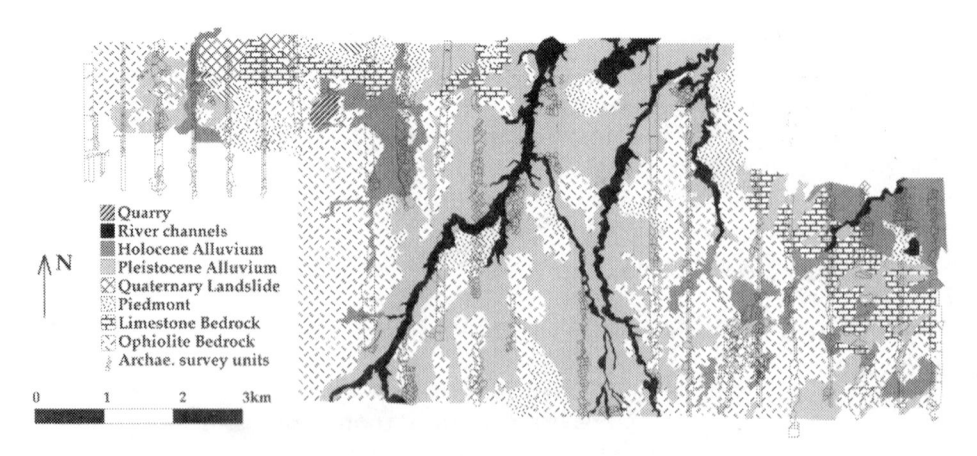

Figure 5.1. Map showing the geomorphology of the Sydney Cyprus Survey Project field area. The superposed N/S "souvlaki" lines represent the individual archaeological survey units.

ing the same end product as the geomorphologists. The result of the first geomorphology field season at the SCSP was a map that subdivided the landscape into standard geomophologic units classified by relative age (Figure 5.1). This map was created from aerial photographs and from field transects across key areas of the survey region. The geomorphologic units read like the chapters in a geomorphology text: river terraces, alluvial fans, and colluvial surfaces were identified. Each piece of the landscape was assigned a geomorphologic category based on surface age and on the dominant process that formed the surface at a place. Unfortunately, to an archaeologist untrained in geomorphology, this map had little meaning or direct relevance in answering the foremost question: what is the context of this artifact? Thus the geomorphology was reinterpreted into a surface stability map in a format that was useful in answering the primary questions and helpful in survey strategy and priority.

Although the field of geomorphology is well established, morphostratigraphy is not recognized in the North American Stratigraphic Code (North American Commission on Stratigraphic Nomenclature 1983). It is, however, a useful tool in surficial studies and has been defined as follows: "A morphostratigraphic unit is defined as comprising a body of rock that is identified primarily on the basis of the surface form it displays; it may or may not be distinctive lithologically from contiguous units; it may or may not transgress time throughout its extent" (Frye and William, 1960, 1962). Surface-based morphostratigraphy is an important component of surface archaeological survey, as it is focused on identifying the forms and processes resulting in the surface that is actually being surveyed. Standard morphostratigraphic mapping entails classifying the landscape into surfaces whose form indicates a dominant depositional or erosional process for a discrete period of time. These surfaces generally have a constant average slope and direction defined by the process of formation (e. g., wind, waves, river water, gravity), the substrate materials (e.g., bedrock, sand, clay), and the hydrology of the region. The contacts, or boundaries between morphostratigraphic units, commonly occur at the places in the landscape where slope changes abruptly. For example, at the contact zone between colluvium and alluvium the slope will generally decrease and the direction of the slope may change from being parallel to the gravitational gradient (i.e., the shortest distance to the base of the hillslope) to being parallel to the hydrologic gradient (i.e., parallel to the direction of river flow). Preliminary mapping of these surfaces is ideally done on the largest available scale topographic map or directly onto aerial photographs. Topographic map or aerial photograph interpretations are then field checked by observations of fine-scale surface morphology, grain size and shape distributions, surface sediment texture, and soils characteristics. The contacts between units are adjusted where necessary. The result is a map that classifies that landscape by surface type and by relative position. These morphostratigraphic units can be considered the substrate on which humans impose a land-use pattern.

An example of a high-resolution morphostratigraphic map of a small isolated mountain in the Troodos foothills is provided in Fig. 5. 2. Rectified aerial photographs (Fig. 5.2a) served as the base map for the morphostratigraphic map (Fig. 5.2b). I classify this as a morphostratigraphic map because it relies almost exclusively on surface information with only minor field descriptions of exposed soil horizons or sedimentary characteristics. The map was made by interpreting stereo aerial photographs and entering the observations into a GIS as an overlay on the rectified aerial photographs. The morphostratigraphy is comprised of four primary units: Anthropogenic, Holocene, Pleistocene, and Cretaceous rocks. The Cretaceous limestone unit is comprised of surface exposure of stratigraphically in–place limestone bedrock. The Pleistocene morphostratigraphic unit is subdivided into two subunits: piedmont and landslide. Large landslides are present on the northern and western slope of the mountain because the limestone bedrock fails along bedding planes that dip to the north-northwest. The antiquity of the Pleistocene landslide is based on the degree of surface weathering and the infilling of the swales within the landslide. Pleistocene piedmont is the remnant of a colluvial drape that once surrounded the southern and eastern faces of the mountain. It is preserved as only as narrow spur ridges, and it has a well-developed carbonate soil horizon that attests to its antiquity. The Holocene unit is subdivided into five subunits: gully, younger alluvium, older alluvium, colluvium, and landslide. Holocene landslides have oversteepened slopes along their margins (note the shadows in the aerial photograph) and the low areas within the landslide mass have yet to be filled with sediment. A younger colluvial drape, set physically below the Pleistocene piedmont, is present on slopes both near the top of the mountain and at its base. The older and younger alluvium units are deposits of an ephermeral stream running along the base of the mountain; the older alluvium is a small alluvial terrace set about 1 m above the younger active floodplain. The Holocene gully units are small channels with basalt exposed below an intermittent gravel lag of basalt and limestone boulders. Active erosion in these gullies is cutting down into the bedrock and upward and laterally into colluvial sediments. Anthropogenic units are comprised of mine tailings and mine works where the morphology of the mountain has been directly impacted by intentional movement of rock and soil by humans.

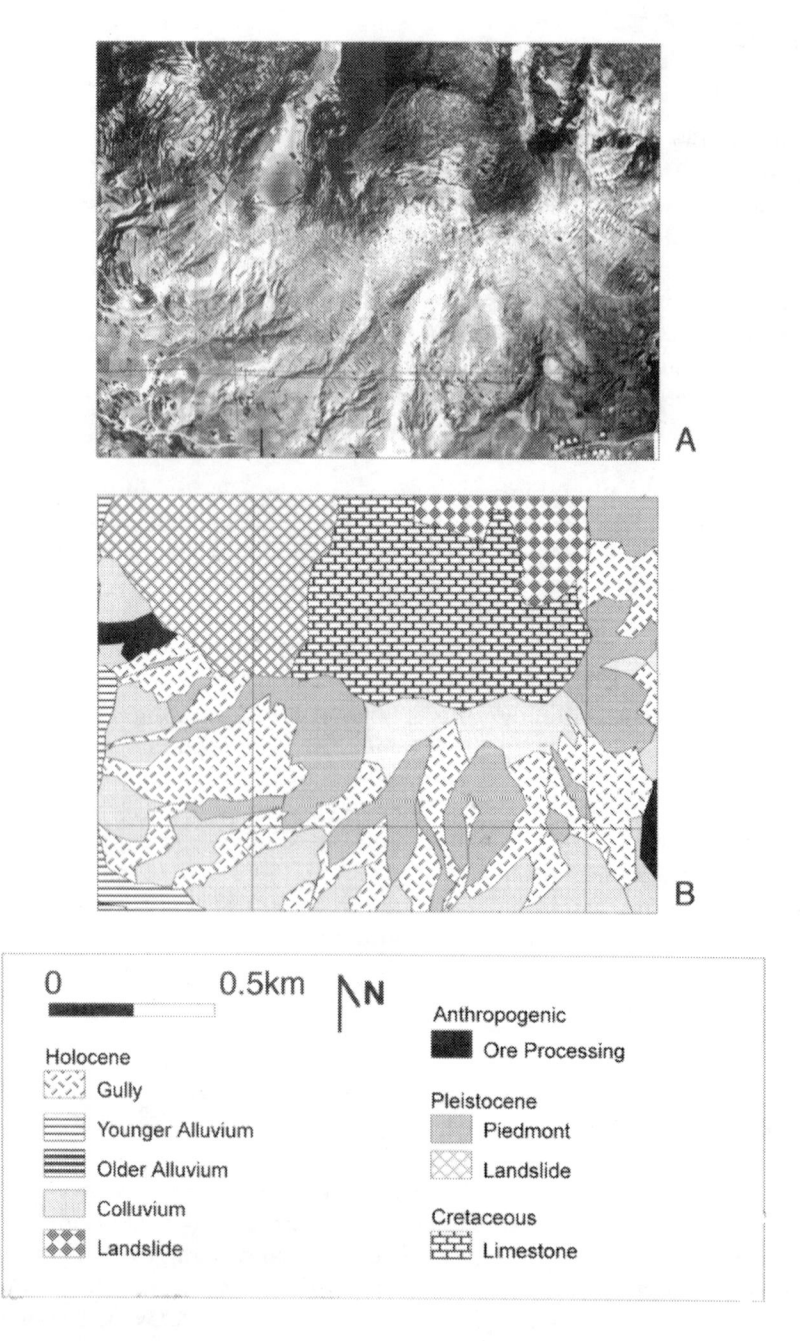

Figure 5.2. (A) Montage of rectified aerial photographs of Mount Kreatos, Cyprus. (B) Morphostratigraphic map of Mount Kreatos. A 1 km grid is superposed over Map B. The morphostratigraphic map was created as a layer in a GIS data base allowing a variety of surface characteristics to be included with each mapped unit. The map represents a 2.1 km² area that lies 2 km east of the northwestern corner of SCSP as represented in Fig. 5.1.

Allostratigraphic units provide a method of combining the surface morphology with the subsurface stratigraphy. Allostratigraphic units are mappable bodies of sediment or sedimentary rock identified and defined by their bounding discontinuities (North American Commission on Stratigraphic Nomenclature, 1983). Thus, in surficial sediments, the unit is commonly defined as the sedimentary package between a lower erosive contact and the upper nondepositional surface. For example, a single uninterrupted sequence of sedimentary fill associated with an alluvial terrace may be defined as an allostratigraphic unit. Each morphostratigraphic unit may coincide with an individual allostratigraphic unit or may be underlain by a number of allostratigraphic units. The additional stratigraphic control added by the recognition of unconformities allows for the development of a relative chronology between the units. The classification of allostratigraphic units requires field description of outcrops or core samples such that the bounding discontinuities will be recognized and described.

In environments where deposition has occurred during or subsequent to the time of occupation, a significant percentage of the artifacts and sites may have been buried or reworked by geological processes. In these environments, archaeological survey must include subsurface investigation if the results are to be meaningful or relevant. Subsurface investigations may include direct sampling by coring or shovel probes, or they may rely on geophysical techniques such as ground-penetrating radar or electrical resistivity. Even in the best of circumstances, it is difficult to argue that a statistically significant sample depicting artifact distribution can be generated by these subsurface methods for any broad area of the landscape. These methods are most useful for investigations at the scale of a few square meters, or at most tens of square meters. Settlement pattern studies in aggradational environments are equivocal at best.

2.2. Chronostratigraphy

The presence of relict geomorphologic units implies that the landscape has changed through time and the placement of these units relative to the active modern surfaces, and to one another, allows us to establish a relative chronology. Both geologists and archaeologists are familiar with the basic rules of stratigraphy (e.g., Steno's Laws and cross-cutting relationships) that allow us to establish the relative ages of stratigraphic units. There are, however, complications in both anthropogenic and natural Quaternary sections that can make the application of these laws difficult. The Harris (1989) matrix, developed specifically for archaeological stratigraphy, is also useful in deconvolving complicated Quaternary sections. The strength of the Harris matrix lies in the equal weight given to strata and their interfaces (boundaries or unconformities). The scale-free box diagrams provide a powerful method of visualizing and correlating allostratigraphy and removing the apparent conflict created when erosion results in the stranding of older stratigraphic units at higher elevations than younger units.

In addition to these basic stratigraphic techniques, a geomorphologist has other tools that can be used to distinguish the relative age of a surface. Weathering and soil development combine to transform a depositional surface through time, and the extent to which the surface has been altered can provide an estimation of the relative age of the surface (Birkeland et al., 1991). In the best of circumstances, these methods can provide correlative ages, that is, they can establish age equivalency for deposits for which there is no numeric age estimate (Colman and Pierce, 1998). Surficial processes will cause any extant, exposed surface to transform with time in the following ways: soil horizons will become more strongly developed (Birkeland 1984; Birkeland et al., 1991), surfaces will lose their depositional morphology (Birkeland and Noller, 1997; McFadden et al., 1989), mean clast size will decrease, overall surface roughness will decrease, and rock varnish may form and become more intense. Description of the results of these processes can be used to provide qualitative or quantitative indicators of where a surface fits within the regional stratigraphic framework. Studies of the formation of soil with time are referred to as soil chronosequences, and they can be used to correlate surfaces of the same relative age over a region with constant climate, biota, and geology. Catenas are studies of the distribution of soil across changes in topography, and they help us to constrain how much of soil formation at a place is derived from hillslope rather than pedogenic processes. When the rate of soil formation is established by chronosequence and catena studies, then soil characteristics can provide us with ballpark estimates ($\pm 20\%$) of the age of a stable surface.

The relative and correlative ages of landscape elements can be further constrained by numeric age estimates. Numeric-age methods include any method that produces a quantitative estimate of the age of a deposit or surface based on ratio or absolute time scale and with a known uncertainty (Colman and Pierce, 1998). Cosmogenic nuclide analysis is the most versatile quantitative technique for estimating the ages of late Pleistocene surface materials (Trumbore, 1998; Zreda and Phillips, 1998). Cosmogenic nuclides include the commonly used radiocarbon technique for organic materials as well as newer methods that utilize both radioactive (^{36}Cl, ^{10}Be, ^{26}Al, ^{14}C) and stable (^{3}He, ^{21}Ne) elements that accumulate in surficial inorganic materials. Archaeological artifacts found within a stratigraphic section can be very effective fossils in developing a numeric chronology. In the best circumstances, the artifact stratigraphy will have a much smaller error than any geochronologic method. Knowledge of the regional distribution of sites—again used as marker fossils—can provide a framework for subdividing the landscape into units that predate or postdate various phases of occupation.

Once the regional stratigraphic framework is developed, we can determine which geomorphic surfaces are most likely to have *in situ* archaeological surface materials (remnant stable surfaces), buried deposits (aggradational surfaces), or lag deposits (lightly eroded surfaces), or be sterile for geologic reasons (highly eroded surfaces).

2.3. Integration of Geomorphology and Archaeological Survey Data: Questions of Scale

At the first stages of analysis, the features of interest to archaeologists and geomorphologists are rarely presentable at the same scale. As geomorphologists, we generally attempt to create a continuous map of surficial geology across the entire survey area. Archaeological survey commonly requires compromises between intensive discontinuous data collection or extensive continuous data collection. For a working survey, the best case scenario is that a preliminary small-scale geomorphologic map is constructed of the region and that this map serves as a basis for an initial sampling stratification of the landscape. Although the geomorphological observations need to be field tested, the subdivision of landscape elements by surface stability allows for the most productive use of survey field time.

Additionally, surficial geomorphological studies are useful in helping construe working units of the landscape. Cognitive geography has shown that the Cartesian space that we statistically analyze has little relationship to the way in which individuals conceptualize or utilize the space they inhabit (Downs and Stea, 1977; Gould and White, 1986). The establishment of watershed or viewshed boundaries, of regions with coherent soil and biological characteristics, or of juxtaposed resource bases may be more reasonable units for the collection of survey data. Simple gridded collection strategies have the possibility of missing areas where the geologic information tells us there is a higher likelihood of finding materials based on temporal or environmental constraints. Including environmental information in a survey data collection strategy can greatly improve results relative to purely random collection strategies.

When regional archaeological survey begins, geomorphological data should be collected at the same scale as the archaeological data. A geomorphologist should be included with each survey team. As this was not possible in the SCSP, field crews were trained at the beginning of the field season to make primary observations of landscape position, bedrock geology, and surface sediment and soil characteristics for each survey unit. Because the field crews had little prior experience with geomorphological observation, a geomorphologist rotated through the crews, helping them to refine their skills. The geomorphologists then focused on mapping individual places of special interest (POSIs) and special interest areas (SIA) at a scale compatible for statistical comparison with the archaeological data.

3. Case Studies

3.1. Paleolandscapes of the Andean Foothills, Northern Coastal Peru: Reinterpreting Site-Based Surveys

3.1.1. Environmental Setting of Coastal Peru

Northern coastal Peru is a land of extreme contrasts: the adjacent ocean is one of the world's most productive ecosystems whereas the adjoining desert is nearly barren of life except for the verdant valley oases that drain the front range of the Andes. The boundary between the vegetated and barren land surface is anthropogenically determined, located at the edge of outermost modern irrigation canal. This canal most commonly traces the contact between trunk stream alluvium and bedrock, dune, or tributary alluvium. To add to the contrasts, once every 3 to 10 years El Niño brings torrential rains and the collapse of the cold-water marine ecosystem. Mean annual rainfall, inclusive of El Niño years, ranges from about 5 to 10 mm/yr on the north coast (SENAHMI, unpublished data).

The environmental limitations of life in this harsh desert landscape have long colored the interpretation of the archaeological history (e.g., Moseley, 1975, 1983, 1987, 1992; Shimada, 1994). Unfortunately, long-standing misconceptions of the evolution of this landscape in response to its tectonic setting and climatic history flavor these interpretations. The resulting view is often of a nearly static desert landscape impacted exclusively by intermittent catastrophic events, either earthquakes or El Niño flooding. There appears to be little appreciation for the evolution of the landscape in response to changes in climate, sea level, or sediment flux. The Holocene stratigraphy provides the basis for reinterpretation of the archaeological stratigraphy within the context of landscape evolution.

The description that follows focuses on the Rio Casma Valley where I completed a detailed survey of geomorphology as well as a cursory survey of the archaeology. A more detailed settlement pattern study of Casma is in process (Wilson, 1998). Subsequent reconnaissance geomorphology in the Santa, Supe, Moche, Chicama, and Chiclayo valleys supported the general finds of geoarchaeological survey and the predictive strength of using geomorphology to locate sites.

3.1.2. The Casma Region

3.1.2a. History of the Research The original goal of my research in Peru was to document the impacts to the physical landscape of some 7000 years of agriculture. To that end, the river valley and adjacent desert and coastal zones were mapped using aerial photographs and 1:10,000 scale planimetric maps as a base. Both surficial geology and archaeological structures and cemeteries were recorded, and a morphostratigraphic framework was developed. Only a few of the archaeological structures have age estimates, based on either published information (Malpass, 1983; S. Pozorski and Pozorski, 1986; Tello, 1956; Wilson, 1998), or field collaboration with Tom and Shelia Pozorski (personal communication).

The study region extends along the coast from Huaynuna in the north to Quebrada Rio Seco in the south (Fig. 5.3) and upstream along the Casma and Sechin Valleys for a distance of about 22 km. The focus of the mapping was a strip roughly 2 km wide adjacent to the hydrologic resources (coastline, river channel). The only open desert region mapped was that between the Sechin and Casma Rivers.

3.1.2b. Geomorphic History The surficial geology is subdivided into five categories: Santa Alloformation (Holocene beach and nearshore deposits), Sechin Alloformation (Holocene trunk stream alluvium), Colorado Alloformation (tributary alluvium), eolian deposits, and bedrock. Each is described briefly as follows; more detail is available in Noller (1993) and in Wells (1988).

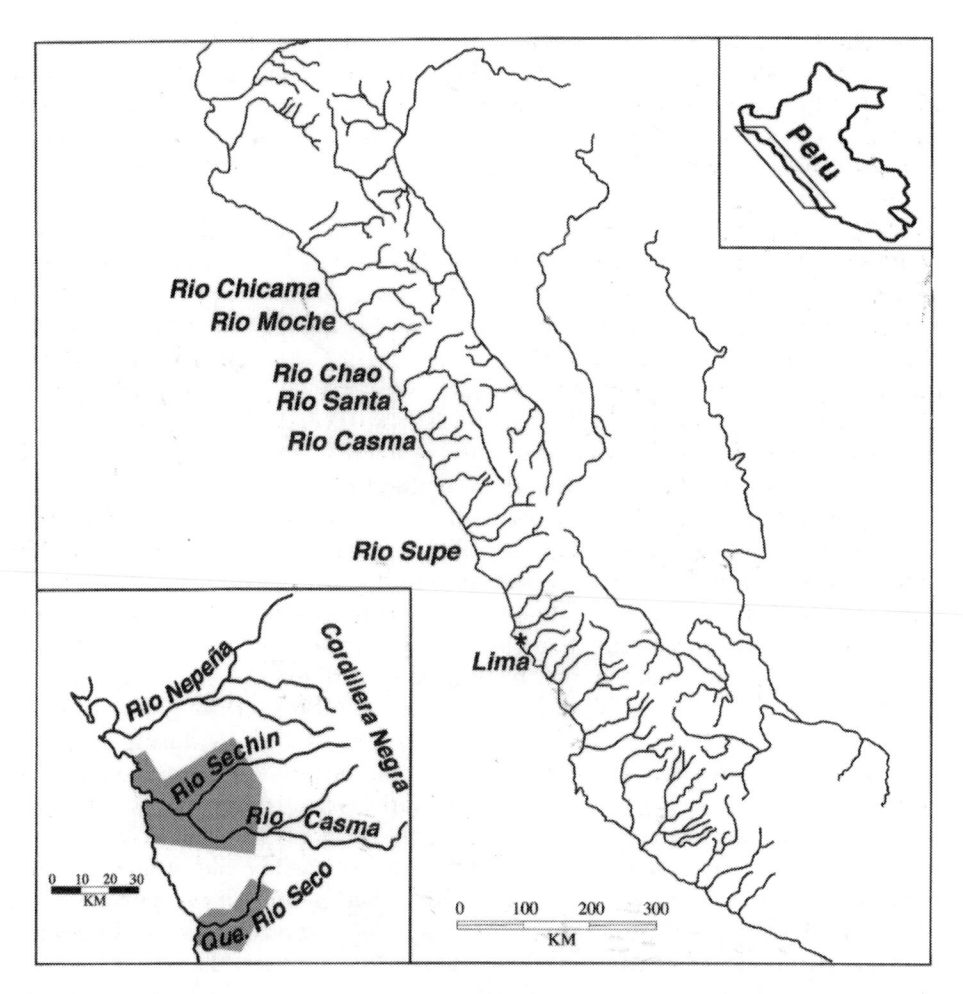

Figure 5.3. Map of the northern coastal Peru and the study area. Shaded regions in the inset map of the Rio Casma–Rio Sechin region show the area mapped by Wells (1988).

Rocky eroding shorelines make up over 90 percent of the coast. Both steep sea cliffs and marine platforms are common. The sea cliffs separate protected embayments where littoral sediment accumulates at valley mouths. The Santa Alloformation is comprised of dunes, beach ridges and littoral drift deposits, fan delta, and salinas (evaporative salt pans) that were deposited between the early Holocene transgressive sea cliff and the modern shoreline. The size and earliest age of the littoral complexes is directly related to sediment discharge from the source river, as the Holocene shift from marine transgression to coastal progradation occurred earlier adjacent to rivers that yielded higher sediment flux to the coastline. Within the Casma embayment, the Holocene maximum transgressive shoreline ranges from 4 to 6 km inland of the modern shoreline. Beach ridges and littoral deposits are concentrated on the north side of the coastal complex, indicating that there has been no change in the predominant direction of littoral drift. The fan delta that fills the central portion of the bay has a radius of about 5 km, a mean slope of about 0. 23°, and classic conical shape with the fan apex where the Rio Casma emerged into the bay. The southernmost margin of the fan delta abuts against the Cerro Santa Christina massif, and a narrow estuary is trapped between the fan delta and the alluvial fans that drape the northeast face of the massif.

The Sechin Alloformation is comprised of the deposits of the main channel of Rio Casma and its large tributary the Rio Sechin, fifth- and fourth-order streams, respectively. These valleys are filled with channel gravels, overbank fluvial sediments, and intercalated eolian sand. Deposits of the 1982 to 1983 El Niño were described in detail to characterize an El Niño deposit (Wells, 1988): they are predominantly well-sorted sands, silts, clays, and occasional poorly sorted, matrix-supported gravels that result from overbank sedimentation and debris flow events. These flood deposits commonly overly well-sorted gravels deposited in the channels under normal or flood conditions. Similar sediments were observed to comprise the majority of the floodplain deposits. Flood layers are commonly separated by incipient soil horizons (Noller, 1993) and/or thin eolian sands, both of which reflect the hiatus in floodplain deposition between El Niño events. Detrital ceramic fragments are common within the gravel fraction of the Sechin Alloformation. These detrital materials provide a maximum age constraint for the sediments in which they are incorporated.

The sediments of the Colorado Alloformation fill first- through third-order valleys. Today, the streams within these valleys flow exclusively during major El Niño years when the coastal zone receives precipitation. The sediments, although confined to the valleys, are similar in form and facies to alluvial fan deposits typical in other arid environments. Six morphostratigraphic subdivisions were recognized within the alloformation (Qt1 through Qt6). The oldest terrace, Qt1, is preserved only as isolated remnants of a dissected pediment on spur ridges of the local peaks. This strath terrace is generally located at elevations of between 1000 and 1500 m, some 700 to 1000 m above the modern thalweg. Qt2 is an aggradational terrace that marks a significant fill event and comprises the bulk of the landscape in these valleys. It is incised up to 50 m at its apex but is buried at its distal end by younger sediments. Qt3 through Qt6 are small strath and fill terraces, inset within the channels on the proximal fan or burying older deposits on the distal fan. Some of these small young terraces change nature, from erosional straths to aggradational terraces down valley. The distal ends of these alluvial sediments form depositional lobes that prograde into the trunk valleys and the thalwegs of the trunk streams are commonly diverted around them. With increasing age, the terraces

1. are stranded higher and higher above the most recent thalweg of the proximal fan,
2. have increasingly darker rock varnish (Qt6: 2. 5Y6/4; Qt2: 10R4/3; Qt1: 5YR3/2),
3. have increased development of desert pavement (except for Qt1, which is actively eroding and thus is also losing its pavement),

4. have decreased surface roughness (again except Qt1, where new roughness is imposed by erosion), and
5. as a result of (3) and (4) gradually lose their fine-scale depositional morphology (i.e. channels, bars, and floodplain regions lose their distinctiveness).

Both active and ancient (stable) dunes are common. The dunes take many forms, including small erg, longitudinal, transverse, barchan, climbing, and falling dunes (see also Grolier et al., n.d.). Active dunes overlie landforms of all types, and eolian sand sheets are intercalated in all depositional units. Very thick old inactive dunes make up much of the coastal plain and are commonly exposed in the modern sea cliff. These ancient stable dunes range from a few meters to more than 300 m in thickness and are commonly covered with a thin lag gravel or the remains of lomas (fog-drip) vegetation. In the smaller valleys near the coast, eolian processes are generally more active than fluvial processes and the landforms are transitional in character. The sediments are dominantly eolian although their surface form is closer to that of an alluvial fan, that is, a cone of sediment filling a first-order valley. Stable discrete climbing and falling dunes are also common in these small valleys.

Finally, much of the desert outside the river valleys is exposed Cretaceous bedrock: it is comprised of the coastal batholith and an overlying sequence of volcanic rocks intercalated with shale and quartzite (Myers, 1976). Colluvial cover over the bedrock is generally thin, especially on the batholithic rocks that weather directly to sand. The strong persistent winds quickly remove any fine-grained colluvium.

3.1.2c. Settlement History For this study, the distribution of obvious architectural remains, roads, mines, and cemeteries were mapped from aerial photographs (Wells, 1988). Because this study was not meant to be a comprehensive archaeological survey, no attempt was made to map the distribution of individual artifacts in the field. Discrete loci of surface artifacts, with or without obvious architecture, were conceptualized as sites for this survey. The archaeological chronology was developed from published sources and was occasionally supplemented by field visits with T. and S. Pozorski. Only in a few instances was archaeological chronology determined by Wells, and those cases focused on the recognition of the coastal lithic sites that have a very distinctive tool kit (predominantly unifacial choppers, cobble cortex flakes, and simple knives) and in a few locations where surficial charcoal was sampled for radiocarbon analysis.

There have been a number of archaeological surveys of the Casma Valley, beginning with Tello (1956), followed by M. Malpass (1983), S. Pozorski and Pozorski (1986, 1987) and most recently by Wilson (1998). A summary archaeological chronology is presented in Table 5.1. Malpass' survey focused on finding lithic sites in a small areas outside the valley proper. Malpass' Mongoncillo sites are predominantly located around the Lomas Las Aldas, some 15 km south of Casma. Wilson's (1998) comprehensive archaeological survey documented the distribution of roads and architecture throughout the lower Casma/Sechin Valleys. Wilson recognizes an order of magnitude more occupations than Wells

Table 5.1. Site Distribution as a Function of Chronology, Casma

Casma Period[a]	Estimated chronology (years × 1000)	Central Andrean period	Sites Malpass	Occupations Wells	Wilson[c]
			[d]	[b]	
Early Lithic[b]	9 to 8 BP	Paijan	3		
Mongoncillo[d], Tortugas[c] or Coastal Lithics[b]	8 to 5 BP	Preceramic	38	6	
Cotton Preceramics[b]	5 to 4 BP	Cotton Preceramic	4		10
Moxeke[c]	4 to 3.1 BP	Initial period	8		65
Pallka[c]	3.1 to 2.3 BP	Early Horizon/Chavin	8		45
Patazca[c]	2.3 to 2 BP	Early Intermediate period			196
Cachipampa[c]	2 to 1.5 BP	Middle EIP/Gallinazo			194
Nivin[c]	1.5 to 1.3 BP	Late EIP/Moche	2		31
Choloque[c]	1.3 to 1.1 BP	Early Middle Horizon/Wari	3 3		249 249
Casma[c]	1.1 to 0.6 BP	Late Middle Horizon			399
Manchan[c]	0.6 to 0.4 BP	Late Intermediate/Late Horizon	7		154
		No temporal information			68
	total			108	1343

[a]This terminology used here is a synthesis of that used by Wilson (1998; those followed by a [c]) and Wells (1988; those followed by a [b]). Because of differences in their schemes for lumping or subdividing the chronology, not every period is represented in each of the surveys. For example, Wilson does not subdivide the Preceramic periods, and Wells did not subdivide the early Intermediate period.
[b]Wells (1988)
[c]Wilson (1998)
[d]Malpass (1983)

(1988) does for a number of reasons: the area surveyed was larger, many sites were recognized to have multiple occupations, and sites were subdivided more discretely. Wilson (1998) also revised the chronology of many of the sites; for example, many of the sites that Wells classified as Early Horizon (based predominantly on Tello, 1956) Wilson considers to be Patazca (Early Intermediate period) occupations.

The Casma region has had nearly continuous occupation for probably the last 9 BP, and artifacts are ubiquitous across the landscape. The 9 BP date is based on two very early lithic workshops south of the valley (Malpass, 1983), but we have no other information about the earliest occupants. The earliest dated occupation is from a coastal lithic period shell midden on the north side of the Casma paleoembayment. Charcoal from the midden's surface suggest occupation around 7.0 to 7.7 ka BP (SMU–1915, 7670 ± 60 Cal. yr BP; SMU–1916, 6980 ± 240 Cal. yr BP). The valley has a very rich early prehistory and probably one of the largest concentrations of early agricultural sites in the region. Based on Wilson's (1998) preliminary data, we can make the very general observation that population in the Casma Valley increased with time with the apparent exceptions of decreased occupations during the Nivin (Moche) and possibly the Manchan (Chimu/Inca) periods.

3.1.3. Distribution of Sites on a Geomorphologically Stratified Landscape

As expected, the great bulk of archaeological sites (79%) in the Casma Valley is located on older (>0.5 BP) stable geomorphic surfaces: alluvial fans, dunes, coastal deposits, or the older floodplain (Table 5.2). Only 5 percent of the known sites have been found in a buried or partially buried context on the active floodplain. A small but significant number (14%) are located on erosional bedrock surfaces. The fact that the sites on bedrock directly overly bedrock suggests that significant erosion predated occupation. The higher percentage of sites on stable alluvial surfaces as compared to those on exposed bedrock surfaces in part reflects differences in slope: alluvial surfaces generally have slopes less than 4°; bedrock surfaces have slopes commonly steeper than 25°. Sites are rare on active aggradational surfaces (5%). One small pyramid was found partially buried by overbank sediments on the aggrading floodplain of the Rio Sechin in 1985. This pyramid was entirely removed by erosion during flooding caused by the El Niño event of 1997 to 1998. Detrital artifacts and buried agricultural soils are common in the aggradational floodplain sediments, indicating that the floodplain has been utilized to some degree throughout the Holocene. Therefore, we must conclude that the lack of architectural sites in the floodplain reflects reworking and erosion rather than nonoccupation. It is, at this point, impossible to imply what the actual occupational history of this active aggradational environment was.

The choice of site locations not only reflects landscape stability but also environmental parameters. The majority of sites (67%) are located on the depositional fill of the small tributary channels or on the higher fluvial terraces (Table 5.3). Collectively these sites are constructed on the surfaces adjacent to, but largely outside, the irrigable floodplain. It appears that prehistoric Casmeños chose not to build large structures on viable agricultural land. Wilson's (1998) preliminary maps confirm these observations.

A priori, sand dunes seem like an unlikely location for habitation sites. And indeed, half of the surveyed sites (11% of the total) are cemeteries in stable

Table 5.2. Site Distribution as a Function of Surface Stability, Casma, Peru

Stability class	N[a]	Percent	Geomorphic units
Stable	84	79	Alluvium, dunes, nearshore, older floodplain
Stripped	15	14	Bedrock
Aggrading	5	5	Fan delta, floodplain
Eroding	2	2	Alluvial fans in 3° drainage

[a]Number of sites cataloged by Wells (1988).

Table 5.3. Site Distribution as a Function of Geomorphology, Casma, Peru

Substrate	N[a]	Percent	Comments
1986 thalweg	0	0	Channel eroded to bedrock or into older alluvium.
1982–1983 alluvial deposits	0	0	Overbank floodplain surface was aggradational during this event.
Trunk valley active alluvium <0.5 BP	1	1	One partially buried pyramid was removed by erosion during the floods of 1997–1998.
Trunk channel alluvium >0.5 BP	27	25	Incision has left this Holocene terrace well above the active floodplain.
Alluvial fans in tributary channels	44	42	
Early Holocene shoreline	4	4	Preceramic sites located just inland of the Early Holocene seacliff.
Holocene fan-delta	3	3	
Holocene nearshore deposits	1	1	
Sand dunes and ramps	12	11	Six of these sites are cemeteries located in the lower river valley.
Cretaceous bedrock	14	13	

[a]Number of sites cataloged by Wells (1988).

falling/climbing dunes adjacent to the lower valley regions. Looting of the cemeteries results in a characteristic "pockmarked" texture on aerial photographs that allows them to be easily identified. Consequently, cemeteries are probably overrepresented in the survey. Presumably, the dunes were chosen as burial grounds due to the ease of excavation and perhaps due to the low moisture of the soils. Eling (1987) suggested that the dunes were also used as storage areas for food during prehistory but no relict dune storage sites are reported. Architectural sites in eolian sediments are located on the southern Casma valley margin and include the site of Manchan. These sites were partially buried by sand blown into the valley by the persistent southwestern winds. The famous sites of Huaca del Sol and Huaca de la Luna are located in a similar context in the Moche Valley; their abandonment has, in part, been blamed on dune incursion from the southwest (Moseley et al., 1983).

Ridged-field agriculture was a local adaptation in the Casma Valley that resulted from the unique geography of the lower river valley. Raised fields are found elsewhere in Peru and in South America, but in northern coastal Peru they are only found in the lower Casma Valley (Moore, 1988). This occurrence is the result of the geography of the lowermost Casma valley. Coastal mountains north and south of the delta restrict direct wave energy to westerly swells. The Casma shoreline is thus extremely well protected from both the predominant southwest swell and the rarer north or northwest swell (Fig. 5.4). Although littoral drift built a small beach ridge complex on the northern side of the delta, the cone shape of the central delta is well preserved. This cone morphology results in low-elevation

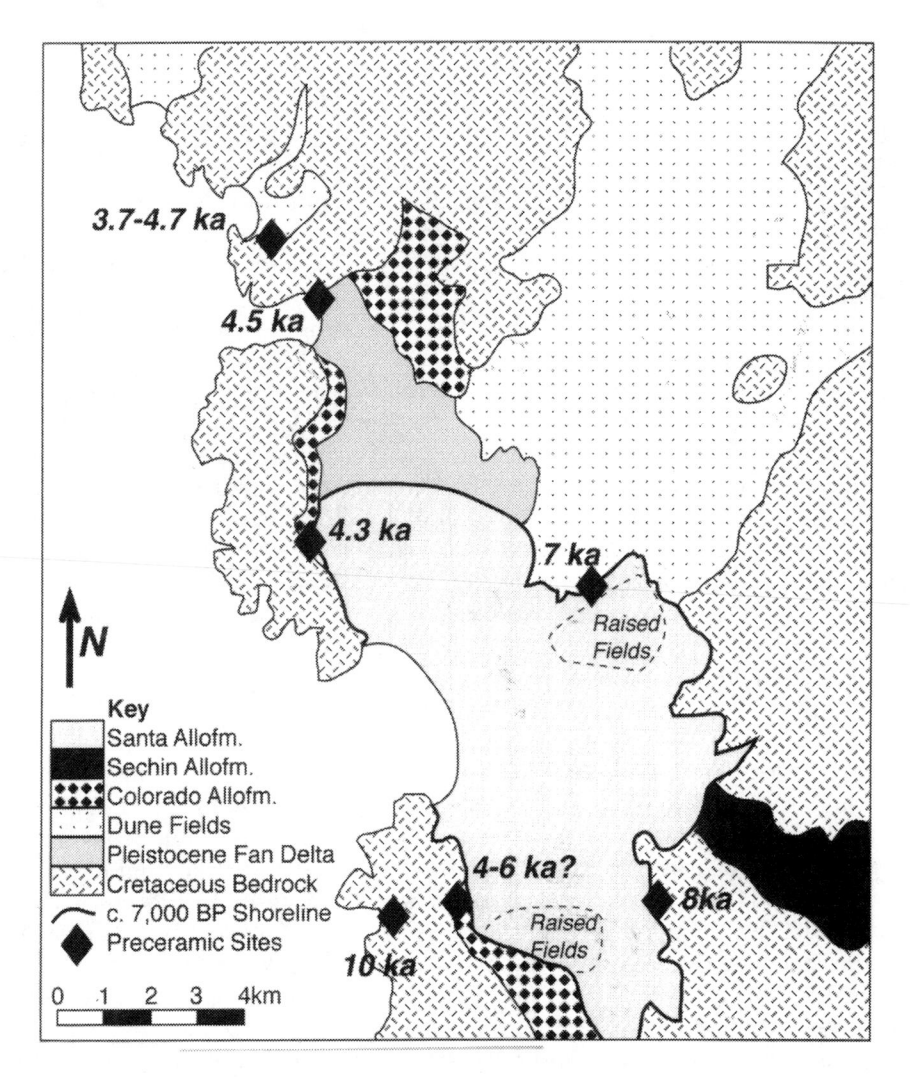

Figure 5.4. Map showing the distribution of morphostratigraphic units, coastal sites, and ridged field complexes in the area of the Casma fan delta.

areas at the far northern and southern edges of the delta, and estuaries or salinas are trapped in these depressions where the water table is very near the surface. Today the areas are either nonagricultural or are used exclusively for salt-tolerant crops. During the late Chimu Period (Late Intermediate Period—14th century A.D.) raised-field techniques were employed to make use of the depression on the southern fringe (Moore, 1991); no dates are available for the raised fields on the northern side of the delta. Moore (1991) suggested that raised-field agriculture was an adaptation in the response to the loss of arable land during the aftermath of a 14th-century A.D. El Niño event.

3.1.4. An Integrated Prehistory of the Region

Combining the geomorphological history with the archaeological history at Casma allows us to paint a more accurate picture of what the settlement might have looked like during specific periods of prehistory (Wells and Noller, 1999). Additionally, including the archaeological stratigraphy with the geomorphologic history improves our chronological control on landscape evolution. What follows is a brief description of changes in human habitat and regional environment through time.

During the early Holocene, sea level rise rapidly flooded the nearshore coastal zone until ca. 7 ka BP when the maximum Holocene transgression was reached. No sites are reported from the valley at this time (the Lithic Period: 10–8 ka BP). However, in the course of the geomorphic survey, a small site with Paijan–like blades was found on a Pleistocene alluvial fan in a Quebrada Rio Seco south of Casma; Malpass (1983) described similar sites from the coastal headlands south of the Casma Valley. If these sites indeed correlate with the Paijan sites to the north (Chauchat, 1975, 1978, 1988; Ossa, 1973, 1978; Ossa and Moseley, 1972), they record a small occupation of the region prior to the stabilization of sea level. The paucity of artifact finds of this period indicates that either sites from the early Holocene are submerged offshore or are buried under Holocene sediment, or that there was only minor occupation of the region prior to 8 ka BP.

Sometime around 8 ka BP the rate of sedimentation outpaced the rate of sea level rise such that coastal progradation began. In the window of time surrounding this shift from transgression to progradation, the shoreline was occupied by Coastal Lithic communities that left their remains in shell middens. On the easternmost and earliest edge of the bayshore are the earliest middens with radiocarbon ages indicating occupation between 8 ka and 7 ka BP (Pozorski and Pozorski, 1995, in press; Wells, 1988). These are small deposits; Almejas the southeastern site (Fig. 5.4) is 0.5 ha in surface area and 1.2 m thick (Pozorski and Pozorski, 1995, in press). Later and larger Preceramic occupations (5 to 3.5 ka BP) are located farther to the west, adjacent to younger shoreline segments or the modern shoreline (Pozorski and Pozorski, 1992). This pattern suggests a westward migration coincident with coastal sedimentation. It has long been clear that the primary protein resource for these early occupants was nearshore and estuarine fauna (Moseley, 1975), but the inhabitants of these sites also made use of riparian plant resources (Pozorski and Pozorski, 1990, 1992). The absence of local freshwater sources or of paleo-spring deposits suggests that they must have traveled to the river valleys for fresh water. However, like the preceding period, no valley sites are known to correspond to this time. If there was occupation in the valley, it must either be buried below the Holocene floodplain or have been eroded. The faunal subsistence assemblages from this time have led to some controversy surrounding the water temperature of the adjacent oceans and the associated climate at the time of occupation (DeVries and Wells, 1990; DeVries et al., 1996; Sandweiss et al., 1996a, 1996b). The most recent paleoclimatic studies (Andrus et al., 1998; Keefer et al., 1998; Riedinger et al., 1998; Rodbell et al., 1999; Steinitz-kannan et al., 1997; Wells and Noller, 1999) all concur that the climate was arid and that the frequency of El Niño events was probably different from that of today.

Shortly after sea level stabilized (ca. 5 ka BP) the coastal valleys began to aggrade and backfill. Sedimentation in these valleys changed the substrate of the valley floor from a gravel fill to finer grained sand and silt (Wells, 1988). Coincident with this change in fluvial substrate was an increased reliance on agricultural resources. Over the course of the next two millennia, as the rivers backfilled and the floodplain grew, there was concomitant increase in reliance on agricultural resources. By sometime between 3 and 4 ka BP the population had largely moved inland, and although marine protein continued to be an important resource, the focal point for occupation was clearly centered around agricultural resources. Thus, the technological developments that resulted in a change from a maritime focus to an agricultural focus occurred coincidentally with an evolution of the river valley such that an appropriate substrate was available on which agriculture could flourish (see also Stanley and Warne, 1997).

For the subsequent 3 ka BP of prehistory, the preferred location to build monumental architecture has been on the alluvial fan surfaces in the first- to third-order tributary valleys directly adjacent to the floodplains of the larger rivers (Table 5.2, see also, Wilson, 1998). The distribution of agricultural soils, canals, and related artifacts suggests that the entire floodplain was agricultural throughout this period of prehistory. The paucity of monumental architecture on the floodplain does not imply that there was no occupation there, as the waddle and daub style construction typical of small dwellings is unlikely to be preserved in the active floodplain environment.

Humans have cleared the desert pavement from alluvial fan surfaces for tens or hundreds of generations. Although not as spectacular as the Nasca geoglyphs, numerous generations of roads, geoglyphs, and other cleared features are present in the desert surrounding the Casma (e.g., T. Pozorski et al., 1982). Cross-cutting relationships show that most of the prehistoric roads were cleared sometime between the deposition of Qt3 and Qt4, although some postdate Qt4. Scattered ceramic fragments along the roads implies that their earliest use dates to the Early Horizon, and that the Casma–Sechin valleys were an important transportation corridor between the coast and highlands during this time. The Casma Valley was undoubtedly the most heavily occupied valley in this region of the coast during the Initial Period and Early Horizon (S. Pozorski and Pozorski, 1987; Strong and Evans, 1952; Willey, 1953; Wilson, 1988, 1998) and this may in part explain the early development of a transportation network here.

In summary, cultural change occurred hand-in-hand with environmental change in coastal Peru. The distribution and variety of natural resources changed through time in response to landscape evolution as sea level first rose rapidly and then stabilized, as rivers valleys once only incising began to aggrade in their lower reaches, and as the regolith continued to weather and to be eroded from the hillsides. Technological innovation appears to have quickly allowed local inhabitants to adapt to changes in the nature and distribution of natural resources. It has been suggested that the local population was preadapted to agriculture (Patterson, 1983). Local inhabitants have also developed strategies that allowed them to weather the recurrent, dramatic shifts in climate that result from El Niño events (Moore, 1988, 1991). The predominance of archaeological sites outside the active river floodplain may, in part, be just such an adaptive response wherein the inhabitants chose to build in areas that were not likely to be destroyed by normal El Niño-induced flooding. However, we must remember that site distribution in the floodplain is also a result of the poor preservation potential of materials on the floodplain and that the record of floodplain occupation is necessarily incomplete. Further, extreme environmental and economic stress associated with the largest El Niño events may have caused short periods of more rapid cultural and technological change or precipitated cultural decline (Moore, 1991; Satterlee, 1993; Wells and Noller, 1999).

3.2. Paleolandscapes of the North Troodos Foothills, Cyprus: Toward an Interdisciplinary Framework

3.2.1. The Sydney Cyprus Survey Project (SCSP)

History, Motivation, and Methods Islands have limited environmental variability due to their small size, narrow climatic range, and the physical constraint on species dispersal (MacArthur and Wilson, 1967). On the semiarid island of Cyprus, the limitations of the environment are believed to have directly controlled the human carrying capacity of the island (Knapp et al., 1994). In this insular landscape, strong correlations should exist between environmental evolution, biologic productivity, human land use, soil erosion rates, and human carrying capacity.

Chroniclers of the Cypriot landscape, from Strabo (ca. 0 A.D.) to Thirgood (1987) have commented on the dramatic deforestation of the island. The impact of agriculture and grazing on this deforested landscape was recognized in the early part of the 20th century (Reid, 1908), and the ubiquity of agricultural terraces and check dams suggests that soil conservation measures have been in effect for millennia. At the time of British colonialism (1878 A.D.) the desolation of the forests was likened to that of a war zone.

The goal of the SCSP is to better understand the relationship between metallurgical resources and production and trade, on the one hand, and the emergence and development of complex social systems on the other. To this end, methodology was developed to examine patterns in site location and settlement networks in relation to the distribution of natural resources. Geomorphologic studies provide the framework to interpret the context of all archaeological finds. The survey region (Fig. 5.5) is located at the contact between the uppermost ophiolite sequence (ore–bearing rocks) and the overlying marine limestones and marls (arable soils) on the northern Troodos Piedmont (e.g., Robertson, 1977). As this contact is transverse to the drainage network, the survey region crosses drainages with a wide variety of catchment and arable land area. The archaeological survey consisted of field crews walking roughly 50 m wide transects at 5 m spacing on north/south lines every half kilometer across the study area. For each survey unit, field crews recorded environmental conditions (topography, surface character, sediment cover, erosional pattern, land use, slope, and visibility),

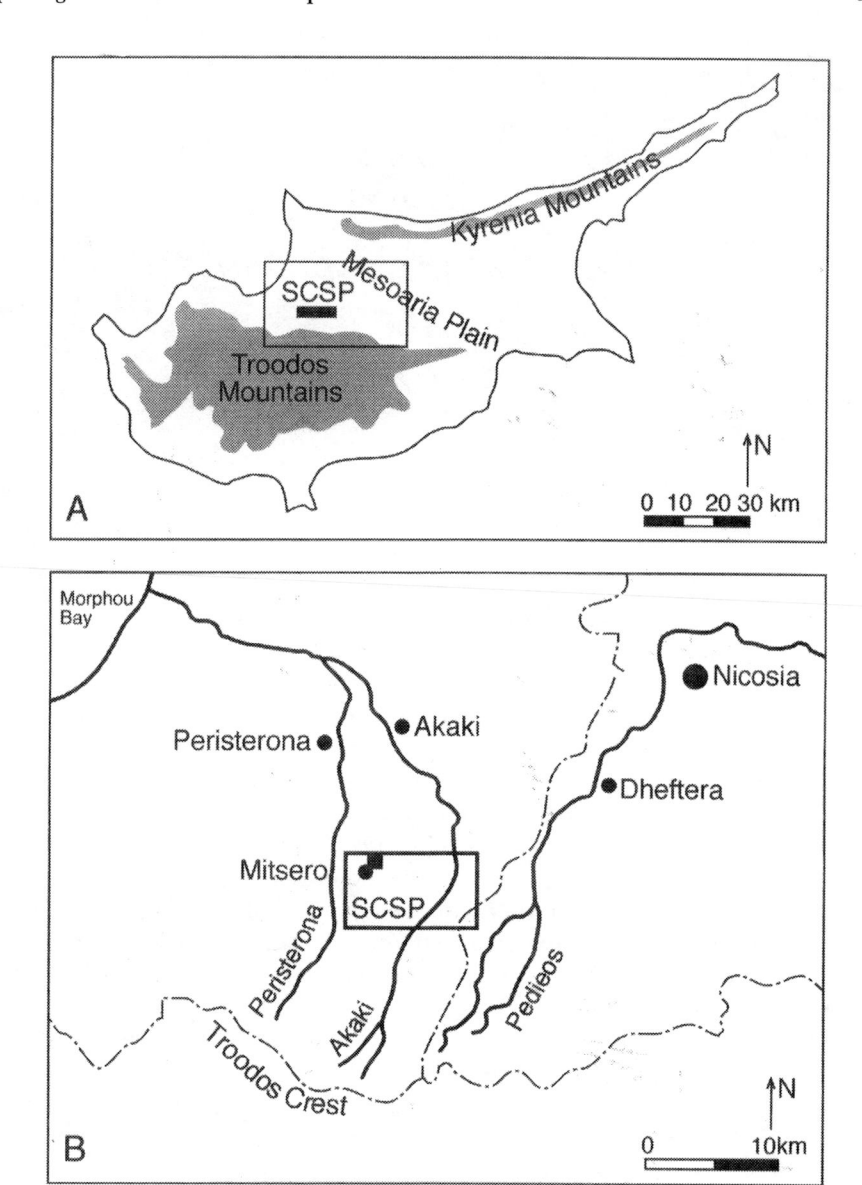

Figure 5.5. (A) Map showing the location of the Sydney Cyprus Survey Project on Cyprus. The area shown in the larger rectangle is enlarged in (B). The survey project boundaries are shown by the inset rectangle in B and the thick black bar in A. The small black square within project boundaries on map B shows the location of the larger scale map presented in Fig. 5.2.

collected diagnostic cultural materials (pottery, lithics, groundstone, metals, slag, ores and fluxes, glass, and tiles), and counted all nondiagnostic cultural material. The survey information was recorded daily in a relational database, and survey unit extent was recorded in a GIS.

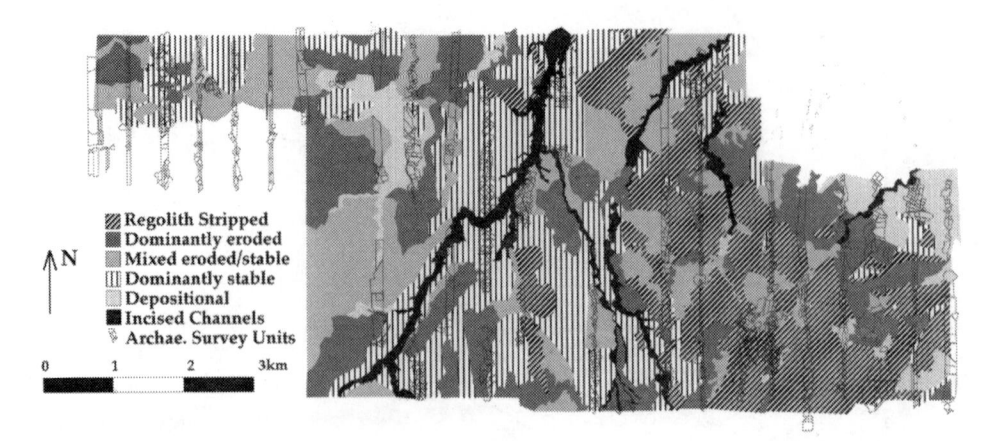

Figure 5.6. Map showing the stability analysis of the SCSP field area. The superposed N/S "souvlaki" lines represent the individual archaeological survey units.

Because large portions of the survey area have suffered from erosion, it is essential to understand the geomorphological context of the cultural material. The geomorphological study began with the development of a standard geomorphological map (e.g., Figs. 5.1, 5.2), created as a layer within the GIS by interpreting stereo–aerial photographs. The landscape was divided into polygons, each with constant geomorphological characteristics over areas as small as 500 m^2. Seven descriptive fields define each polygon: (1) surface cover type (geomorphic descriptor and relative age), (2) surface cover modification, (3) surface stability, (4) incision type, (5) soil presence or absence, (6) agricultural terrace presence or absence, and (7) surface area. A map of surface stability was then created by combining information from the various fields: depositional (surface type = younger alluvium), dominantly stable (stability = stable, surface type ≠ younger alluvium, soil = present), mixed eroded and stable (stability = mixed, soil = present), dominantly eroded (stability = eroded, soil = absent), incised (stability = incised, these are deeply cut river channels), stripped bedrock (surface cover = bedrock, soil = absent), constructed (surface type = constructed, spoils heaps and open pits mines; Fig. 5.6). During subsequent field investigations, the small-scale map was fine tuned and large-scale geomorphologic maps of special interest areas were imbedded in the geographic information system. Fill sequences were described wherever natural exposures were available. Soil chronosequences and catenas provided stratigraphic control allowing us to correlate alluvial terraces between drainage basins. Numeric age estimates on alluvial fill events are based on using artifacts as "fossils" and on a few radiocarbon analyses.

3.2.1b. Geologic and Environmental History The island of Cyprus is a piece of Cretaceous oceanic crust that has been wedged upward during later Neogene time by convergence between Africa and Europe (Kempler and Ben-Avraham, 1987; Lort, 1971; McKenzie, 1970; Robertson, 1977). The rate of uplift reached its peak sometime during late Pliocene or early Quaternary time (Poole et al., 1990; Robertson, 1977). vita Finzi (1993) suggested that the slowing of uplift might be due to transition from a subduction to a strike-slip tectonic regime. Estimates of active vertical motion range from slightly negative to 0.4 mm/yr (Poole, 1992, vita Finzi, 1993). Documented Holocene faulting is limited to very small-scale normal faults (Poole, 1992).

An extensive piedmont drape is preserved along the northern flank of the Troodos Mountains. Poole (1992) mapped four extensive Pleistocene erosion surfaces, each capped with fanglomerate deposits and deep red soils. The fanglomerates thicken down slope and grade into the alluvial fill sequences of the Mesoaria plain. These fanglomerate surfaces attest to a gradual downcutting and incision of the Troodos piedmont, presumably due to the long term uplift of the Troodos. Holocene sediments are rare, and bedrock is exposed in the stream thalwegs.

There is little direct evidence to reconstruct the early Holocene environment of Cyprus (Bintliff and van Zeist, 1982; Butzer, 1975; Knapp, 1994). There was a limited Pleistocene megafauna, (an endemic Pygmy hippopotamus and two species of Pygmy elephant), but otherwise the Pleistocene fauna is very similar to the modern one (Reese, 1995). Holocene pollen data from Cyprus is limited to that presented in Gifford (1978) and this record lacks temporal control. The only clear pattern that emerges from Gifford's palynology is a resurgence of pine pollen in surface samples, likely a result of recent (post-1878) afforestation of the island.

At a regional scale, late Quaternary paleoclimatic data from the Eastern Mediterranean have been interpreted to record marked climatic shifts. Baruch (1994) argued that the pollen data are not extensive enough or adequately correlated to allow detailed regional interpretations. He does, however, suggest that the late Pleistocene/early Holocene forest reached its peak extent earlier in the southern Levant than in the north. Roberts and Wright (1993) interpreted the regional pollen and lake level data to indicate that Cyprus lies near a climatic transition zone that separates the northern Mediterranean (Greece north of the Peloponnese, Turkey, northeastern Syria, northern Iraq, and Iran) from the southern Levant and Arabia. They interpret the data to suggest that the northeastern Mediterranean area was dryer than present between 9 and 6 ka BP, whereas the southerly regions were wetter than present during this time period. Geyh (1994) looked more intensively at the paleohydrological data and suggested that significant climatic variance (dry, wet, warm, and cool phases) occurs on millennial time scales throughout the Holocene. Geyh (1994) attributed these changes to shifts in the position of the ITCZ across the region. Goldberg (1994) saw similar temporal shifts in Israel with some interesting changes in the spatial patterning of climatic zones over the last 24 ka BP. Rossignol-Strick (1993) based an alternate interpretation solely on the pollen data. She chose to reintrepret the geochronology of the sediment cores, dismissing individual radiocarbon dates based on stratigraphic correlations of regional pollen zones and sapropels as well as what she considers reasonable variability in sedimentation rates. Based on a revised chronology, Rossignol-Strick (1993) suggested that early Holocene climate change proceeded in phase throughout the eastern Mediterranean area as follows: the period from 11 to 10 ka BP was arid and cold and was followed by a climatic optimum that peaked between 9 and 7.5 ka BP when summer drought was absent and the lowlands were frost free during the winter. El-Moslimany (1994) agreed that summers during the early Holocene were moister in the continental Middle East than they are today. Most authors agree that there was a marked shift in climate near the Pleistocene–Holocene transition (from moister to dryer conditions), that the Younger Dryas event probably had a significant signature in the region (perhaps very dry and cold), and that the modern vegetation and dry climate, had largely established itself by 6,000 BP. The data do not as yet have sufficient spatial or temporal resolution to yield a clear picture of higher resolution regional climate variability. Additionally, the specifics of the Pleistocene–Holocene climatic transition and the Holocene climate of Cyprus await a local pollen record or other paleoclimatic proxy information.

3.2.1c. Settlement History The regional land use history of Cyprus is reasonably well known (Gifford, 1978; karageorghis, 1968; Knapp et al., 1994). This section provides a brief synopsis of the regional archaeology (Table 5.4) based on the summaries presented in Knapp (1993) and in Knapp et al. (1994), unless otherwise stated.

A single rock shelter site on the southernmost tip of the island contains evidence for a short early island colonization that lasted only a few hundred years around 10.6 ka BP. The stratigraphy and faunal remains at the site indicate that the group changed their reliance from megafauna (pygmy elephant and pygmy hippopotamus) to avifauna and marine resources with time. This change in subsistence base may be due to a drop in the mammalian population of the island resulting from either environmental change or anthropogenic predation (Held, 1989; Knapp et al., 1994; Mandel and Simmons, 1997; Simmons, 1991). The stratigraphic gap following this occupation may indicate a local human extirpation followed by a wholly new colonization (Knapp et al., 1994).

By 9,000 BP the island was settled with a more extensive aceramic population (33 known sites) whose subsistence was based on a combination of hunting, farming, and herding. The pattern of the occupation suggests that initial settlements were located near the coast but that the inland areas of the island were soon occupied as well. Faunal assemblages indicate early reliance on wild species (megafauna and deer) being gradually replaced by domesticated species during early Prehistoric time. Ovicaprid and pig were replaced by cattle, and sheep were replaced by goats gradually through the Bronze Age. These faunal transitions, combined with the evidence for dramatic expansions and contractions of the population base, suggest that environmental constraints limited population growth well into the Bronze Age.

The population increased dramatically with the emergence of mining and copper technology in the Chalcolithic and early Bronze Age. Native copper was probably present as sheets and flakes within soil of the pine forest (Constantinou, 1982; Merrillees, 1982). The earliest mining *per se* was probably direct exploitation of the solum. Technology and cultural change resulted in the gradual evolution of the Bronze Age from the Chalcolithic. Bronze Age settlements were predominantly located along the interface between the agricultural plains and the

Table 5.4. Archaeological Chronology for Cyprus

Date (BP) years (×1000)	Date A.D./B.C.	Knapp '94 revision	Traditional[a]	Subsistence and cultural information
12.0–4.4	10,000 to 2400 B.C.	Early Prehistoric	Neolithic-Chalcolithic	Insular economies; no trade with mainland
12.0–?	10,000 to ? B.C.	Akotiri Phase	"pre-Neolithic"	Megafauna, avifauna; marine fauna
9.0–7.8	7000 to 5800 B.C.	Khirokitia Culture	Aceramic Neolithic	Goats, sheep, grains, wild nuts/fruit, deer
7.5–7.0	5800 to 5000 B.C.	Geochronologic gap		
7.0–5.9	5000 to 3900 B.C.	Sotira Culture	Ceramic Neolithic	Goats, sheep, grains, wild nuts/fruit, deer
5.9–4.4	3900 to 2400 B.C.	Erimi Culture	Chalcolithic	Agropastoralism; deer hunting; initial copper use
4.4–3.0	2400 to 1000 B.C.	Late Prehistoric	Late Chalcolithic to Late Cypriot Bronze Age	Mycenean trade; mixed agropastoral; no coastal settlements
4.4–4.0	2400 to 2000 B.C.	Pre BA 1	Philia Culture	Forest clearance; mixed agrarian economy
4.0–3.7	2000 to 1700 B.C.	Pre BA 2	Early/Middle Cypriot Bronze Age	Cattle and plough complex; milk producing ovicaprines
3.7–3.0	1700 to 1000 B.C.	Pro BA	Middle/Late Cypriot Bronze Age	Cattle, ovicaprid, pig herding; equid draft animals; domesticated cereals, pulses, nuts, fruits
3.0–0	1050 B.C. to present	Historic Period		Subsistence emphasis varies through time; metal production wanes by the Medieval period
3.0–2.7	1050 to 700 B.C.		Cypro-Genometric	
2.7–2.5	709–546 B.C.		Cypro-Archaic	
2.5–2.3	546 to 322 B.C.		Persian	Assyrian domination
2.3–2.0	322 to 58 B.C.		Hellenistic	Egyptian then Persian domination
2.0–1.55	58 B.C.–A.D. 395		Roman	Freed by Alexander the Great Roman Period A.D. 45, conversion to Christianity
1.55–0.75	A.D. 395–1191		Byzantine	
0.75–0.46	A.D. 1192–1489		Lusignan	
0.46–0.38	A.D. 1489–1571		Venetian	
0.38–0.07	A.D. 1571–1878		Ottoman	
	A.D. 1878–1960		British	
	A.D. 1960–1974		Independence	
	A.D. 1974–present		Island split	

[a] From Karageorghis, 1968; Gifford, 1978; Knapp et al., 1994.

mining zones of the Troodos foothills. Population may no longer have been limited by the island's environmental constraints as trade allowed for the purchase of goods from elsewhere.

During the second millennium B.C., the culture became intertwined with that of the eastern Mediterranean and Aegean through trade networks. Textual evidence from surrounding countries confirms the importance of the Cypriot copper trade (Knapp, et al., 1994; Muhly, 1982). Land use intensified during this time with increased pressure on agricultural lands and extensive mining. The direct landscape impacts of mining were probably local to the mines themselves; however, broader landscape impacts were the result of fuel acquisition for smelting. Constantinou (1982) estimated that the volume of wood necessary to smelt copper for 3000 years required clear-cutting the entire forested region of Cyprus 16 times over. If correct, there must have been multiple phases of forest regeneration during prehistory.

The SCSP has produced sound evidence that mining continued from the Bronze Age at least until the fall of the Roman Empire and most probably into the Medieval period; this evidence contradicts all standard histories on the subject. In the time following the Bronze Age, Cyprus has been variously controlled by the powers reigning in the Eastern Mediterranean. With each different occupation of the island came differences in land use practices that may have changed erosion and sedimentation patterns.

3.2.2. Distribution of Sites on a Geomorphologically Stratified Landscape

The primary geomorphologic task for the SCSP was reconstructing how geologic processes filter the distribution of artifacts across the landscape. To this end, a geographic information system was used to statistically compare the landscape classification maps with artifact distribution and density using a GIS (Fig. 5.5, Tables 5.5, 5.6, 5.7). From the comparison of stability class with artifact density we observed the following: the majority of survey units with high artifact densities are located on either stable or mixed eroded and stable surfaces; highly eroded parts of the landscape are commonly devoid of artifacts; only minor parts of the landscape (9%) are aggradational such that older sites might be buried. Areas classified as "mixed" are largely areas where gullies have eroded into an otherwise stable surface. These regions have the highest artifact densities in part because artifacts are being released from shallow burial by surface erosion. Care must be taken in interpreting artifact distributions from the "mixed" regions as the artifacts result from a mixture of primary and secondary contexts. The frequency of artifacts in stable environments and their relative absence from depositional or erosional environments provides confidence that most of the artifactual material derives directly from cultural contexts. Although only a small part of this landscape is aggradational, it is in these small settings that we have found *in situ* late Chalcolithic and early Bronze Age artifacts (Knapp et al., 1998). Thus, we must consider that although the archaeological record preserved is largely in context, it is quite likely that a large part of the record, particularly the early record, is missing due to erosion or burial.

Table 5.5. Calculations of Area Distribution for Various Land Cover Categories within the Confines of SCSP, Cyprus

Category	Area (km²)	Percent of mapped area
Geomorphologic		
Exposed bedrock	19.05	43
River channels	2.81	6
Piedmont	2.85	6
River terraces	15.03	34
Holocene alluvium	4.04	9
Constructed[b]	0.15	<1
Stability Measures		
Stripped	9.81	22
Incised	2.57	6
Eroded	9.66	22
Mixed[c]	8.03	18
Stable	10.66	24
Depositional	4.04	9
Constructed[b]	0.15	<1
Soil present	29.6	67
Soil absent	14.4	33
Gullies present	38.6	88
Gullies absent	1.5	3
No gully data	3.8	9
Agricultural terraces present	27.2	62
Agricultural terraces absent	5.2	12
No agricultural terrace data	11.2	25
Total mapped area[a]	44.9	75% of SCSP

[a]Areas computed using MapInfo™GIS. Only 75% of the entire SCSP survey region has been mapped at this time.
[b]"Constructed" includes quarries and large slag piles.
[c]Some areas could not be differentiated into eroded or stable at this scale, thus resulting in the subcategory "mixed."

Table 5.6. Find Density as a Function of Stability Class, SCSP, Cyprus

| | | Artifact density (finds/km²) | | |
Land cover category	Total artifacts in class	Average	Maximum	Median
All data	57,658	37	1128	1
Depositional	10,064	33	378	0
Stable	20,726	46	671	2
Mixed	17,220	61	1128	2
Eroded	5,874	17	368	1
Incised	2,470	51	394	3
Stripped	1,304	3	73	0

Table 5.7. Find Density as a Function of Geomorphology, SCSP, Cyprus

		Artifact density (finds/km^2)		
Land cover category	Total artifacts in class	Average	Maximum	Median
Bedrock	4,432	5	142	0
River channels	2,291	47	394	3
Piedmont	5,763	115	671	0
River terraces	32,847	53	1128	2
Holocene alluvium	12,325	27	378	0

Pleistocene river terraces yielded both the highest total number of observed artifacts and highest artifact densities. These gently sloping stable surfaces appear to have been the preferred regions for both settlement and agriculture throughout prehistory. Peak artifact densities (1128 finds/km^2) are found in plowed agricultural fields on these surfaces and represent either very long-term composting or settlements that have been well plowed. The limited areas of Holocene deposition yielded 21 percent of the counted artifacts. These areas are also commonly agricultural fields near modern and ancient settlements. Whereas artifact densities tend to be lower in piedmont and bedrock contexts, these areas have yielded the highest frequency of mining related materials.

In summary, these data suggest that stability classification is a strongly predictive tool for artifact density and therefore is useful for statistically stratifying the landscape prior to an archaeological survey. Such an analysis would allow for field time to be focused in environments (stable, mixed, or depositional) appropriate to the questions being asked. The addition of relative age information allows for survey to focus on surfaces that were extant during the occupational period of interest. To be specific, the geomorphology can inform the survey by providing information on where on the landscape to concentrate surface survey efforts (extant surfaces that have been predominantly stable subsequent to the period of interest), subsurface efforts (areas that have undergone aggradation subsequent to the period of interest), or where survey effort are likely to be unproductive (highly eroded areas). The geomorphological analysis of site context is more meaningful to postsurvey interpretations of the collected data.

3.2.3. An Integrated Prehistory of the Region

The SCSP is in the preliminary stages of postsurvey data analysis. At this point, it is clear that the landscape has been used for a combination of agriculture and mining purposes intermittently from the late Chalcolithic until the modern period. Given that economic uses of the landscape have not changed dramatically during this time, much of the patterning we have observed resulted from changes in stratigraphic position of occupations through time. The Quaternary section is dominated by broad Pleistocene erosion surfaces capped with fanglomerates and four Holocene river terraces inset along the modern drainage that were left as the streams incised to their present depth. Each terrace surface may be covered by the remains of occupation younger than the formation age of the surface, and thus the oldest occupation on any surface yields a minimum age estimate. Early Bronze Age or late Chalcolithic period artifacts are found almost exclusively on the oldest of the Holocene terraces. Soil development and surface weathering features confirm that this surface was extant at least until the late Bronze Age. This surface defines an extensive paleolandscape now stranded 4 to 10 m above the modern channel. Roman period occupations are found on the penultimate terrace that is now located 2 to 4 m above the modern channel. Thus it is estimated that stream incision was on the order of 2 and 6 m between the Bronze Age and the Roman period. Using the minimum age for late Prehistoric occupation (1,000 B.C.) and a median age for Roman occupation (170 A.D.) we estimate an average vertical incision rate somewhere between 2 and 5 mm/yr. Occupations and agriculture have thus followed the river as it incised an increasingly deeper valley.

Stone checkdams and hillslope terraces are ubiquitous within the SCSP survey region and most depositional areas are located upslope of these features. Presently, only a small percentage of the older stone-walled terraces are maintained and productive; this very much mirrors a similar situation with terrace walls in mainland Greece (van Andel and Zangger, 1990). The stone walls of the terraces are commonly covered with extensive lichen growth and olive trees with ages in excess of 1,000 years grow on the catchments behind the terraces. Up to seven generations of walls and associated buried soils have been observed in a single checkdam sequence in the Mitsero catchment. Preliminary analysis of radiocarbon age estimates for charcoal from tilled soil behind one such terrace indicate that soil conservation was in use by ca. 1300 cal. yr. BP (radiocarbon age on basal soil of 1292 ± 46 BP; AA–25265). Artifacts included in sediments trapped behind other check dam and hillslope terrace sequences suggest that these features date at least to Roman times and that they have been variously maintained up until the period of British occupation.

As climatic change and tectonic uplift over the past 6,000 years have apparently been minimal (Poole,1992; Roberts and Wright, 1993; Rossignol-Strict, 1993), we attribute much of the recent surface erosion of to anthropogenic causes. Preliminary estimates of hillslope erosion rates within the field area (0. 5 to 4.7 kg^2/yr; Wells et al., 1998) are at the lower end of what is globally considered accelerated erosion (Inbar, 1992; Kosmas et al., 1997). Given what we know of the land-use history, the erosion was most likely initiated by mining practices, enhanced by deforestation as fuel wood was gathered for smelting, and finally exacerbated by overgrazing. The relative importance of agriculture versus pastoralism is believed to have changed with the cultural evolution of the island and thus rates of erosion may not have been constant through time. Cattle were important during the middle Bronze Age but virtually disappeared during the Historic Period. The dependence on goat grazing combined with the deforestation is believed to be the predominant cause of continued soil erosion (Christodoulou, 1959; Thirgood, 1987). The multigenerational checkdams and hillslope terraces attest to a very long history of soil conservation in response to this intensive erosion. The result of some 6,000 years of intense land use and soil abuse is a landscape for which at least 50 percent is classified as stripped, incised, or eroded, and at least 14.5 percent of which has no remaining soil.

4. Conclusions: Replacing Humans on the Landscape

If the goal of a regional archaeological survey is to understand the evolution of social and economic relationships as manifest in paleo-settlement patterns, then regional survey must be placed in a framework that allows for the interpretation of the environmental and geomorphological contexts of the sites or artifacts of interest. Three distinct kinds of geomorphologic data are needed for the re-creation of ancient landscapes: stability, chronology, and paleoenvironment. Without this information it is impossible to deconvolve which changes result from natural as compared to social or cultural factors.

The determination of the relative stability of a geomorphologic surface allows us to determine whether artifacts on that surface are *in situ* or have been reworked by geologic processes. Reworked artifacts should not be included in the data used to reconstruct ancient settlement patterns. Additionally, sterile surfaces may result from either nonoccupation or from postoccupation erosion and nonoccupied space is just as important in interpreting settlement patterns as is occupied space. Regions that have undergone substantial sedimentation or erosion will appear sterile but this sterility may be an artifact of geologic processes. Subsurface investigations can be used to test sterility in depositional areas. It must be recognized that it is impossible to resolve whether settlement ever occurred in regions that have been stripped of surficial sediment.

The determination of the relative age of a geomorphologic surface allows us to project what parts of the landscape were extant during any particular period of occupation. If we are attempting to understand the concepts of space during earlier times, we must also attempt to reconstruct the land surface during the period of interest. Changes in the position of shorelines or river channels or the migration of dunes or glaciers across a landscape may have dramatically changed the appearance of a place through time. The combination of archaeological survey data with a morphostratigraphic analysis of the landscape is a powerful tool for reconstructing which parts of the landscape were extant during the past. If we combine this temporal view of the landscape with a paleoenvironmental interpretation, we can then begin to reconstruct the environmental context of archaeological sites and reconstruct what part of a settlement pattern was driven by the distribution of natural resources. It is the combination of these kinds of data that allows us to determine that a shoreline dotted with small shell midden occupations was four kilometers inland of its modern position or that an ancient smelter was built on the banks of stream channel four meters higher than the modern channel. The richness of our interpretations of ancient landscapes is clearly enhanced by this interdisciplinary technique.

If the geomorphologic study is begun prior to the establishment of a survey methodology then the geomorphologic data can be incorporated into the sampling design. The most useful information for survey design is an analysis of surface stability and relative age, as it allows for a stratification of the landscape based on the highest likelihood of artifact discovery. Some level of random sampling is clearly necessary to avoid prejudicing the survey. However, once the impacts of geomorphologic processes on artifact distribution have been resolved, then the survey may focus on those areas of the landscape that were extant during the period of interest and that have been relatively stable subsequent to the period of interest. This method will result in the most efficient use of valuable field time.

Archaeological settlement patterns are reflections of how humans have used the landscape through time, and these patterns reflect a combination of environmental, economic, cultural, and aesthetic choices. By integrating an understanding of the physical evolution of the landscape into settlement pattern analysis, we are able to better understand which of these choices are predicated by the distribution of natural (hydrologic, biologic, geologic) resources and changes in the distribution of these resources through time. When the impacts of the physical changes have been clearly accounted for, then, perhaps, the patterns resulting from social and cultural evolution will become clear.

ACKNOWLEDGMENTS. There are many people who have contributed intellectually to the concepts presented here. These ideas were first stimulated by discussions with my early advisors: Martin Stout, Tjeerd van Andel, and John Rick. Field work in Peru was supported by the National Science foundation and the National Center for Atmospheric Research. Field excursions with Tom and Shelia Pozorski helped me to visualize archaeological evolution there. Building on these earlier experiences, field work in Cyprus and in Greece facilitated by A. Bernard Knapp and Tim Gregory has been invaluable. I am grateful to the hard work of the SCSP field crews and to A. Bernard Knapp for making the data available. Discussions with and review by the SCSP staff, particularly Nathan Meyer and Michael Given, have helped to formulate ideas. The critical reviews of Paul Goldberg, Reid Ferring, and Vance Holliday greatly improved the manuscript. Finally, this work would not be the same without the intellectual partnership of Jay S. Noller. Although I owe much to all these fine minds, this work and all errors therein are fully the responsibility of the author.

5. References

Andrus, C. F. T., Crowe, D. E., and Sandweiss, D. H., 1998, Mid-Holocene El Niño Variation Recorded in Otoliths from Preceramic Peruvian Middens, *Geological Society of America, Abstracts with Programs* 30:A–16.

Baruch, U., 1994, The Late Quaternary Pollen Record of the Near East. In *Late Quaternary Chronology and Paleoclimates of the Eastern Mediterranean*, edited by O. Bar-Yosef and R. S. Kra, pp. 103–119. *Radiocarbon* 1994, Tucson, AZ.

Bettis, E. A., III, 1995, *Archaeological Geology of the Archaic Period in North America*. Geological Society of America Special Paper 297, Geological Society of America, Boulder, CO.

Binford, L. R., 1992, Seeing the Present and Interpreting the Past—and Keeping Things Straight. In *Space, Time, and Archaeological Landscapes*, edited by J. Rossignol and L. Wandsnider, pp. 43–59. Plenum Press, New York.

Bintliff, J. L., and van Zeist, W., 1982, *Palaeoclimates, Palaeoenvironments and Human Communities in the Eastern Mediterranean Region in Later Prehistory*. British Archaeological Reports, International Series, 133.

Birkeland, P. W., 1984, *Soils and Geomorphology*, Oxford University Press, Oxford.

Birkeland, P. W., Machette, M. N., and Haller, K. M., 1991, *Soils as a Tool for Applied Quaternary Geology*. Miscellaneous Publication 91-3, Utah Geological and Mineral Survey.

Birkeland, P. W., and Noller, J. S., 1997, Rock and Mineral Weathering. In *Dating an Earthquake: Applications of Quaternary Geochronology in Paleoseismology*, edited by J. M. Sowers, J. S. Noller and W. R. Lettis, Washington, DC, pp. 2-453-2-281. U.S. Nuclear Regulatory Agency CR5562, Washington, DC.

Butzer, K. W., 1975, Patterns of Environmental Change in the Near East during Late Pleistocene and Early Holocene Times. In *Problems in Prehistory: North Africa and the Levant*, edited by F. Wendorf and A. E. Marks, pp. 389-410. Southern Methodist University Press, Dallas.

Chauchat, C., 1975, The Paijan Complex, Pampa de Cupisnique, Peru, *Nawpa Pacha* 13: 85-96.

Chauchat, C., 1978, Additional Observations on the Paijan Complex, Pampa de Cupisnique, Peru. *Nawpa Pacha* 16:51-65.

Chauchat, C., 1988, Early Hunter-Gatherers on the Peruvian Coast. In *Peruvian Prehistory*, edited by R. W. Keatinge, pp. 41-66. Cambridge University Press, Cambridge, UK.

Christodoulou, D., 1959, *The Evolution of the Rural Landuse Pattern in Cyprus*. The World Land Use Survey, Monograph 2, Geographical Publications Limited, Cyprus.

Colman, S. M., and Pierce, K. L., 1998, Classifications of Quaternary Geochronologic Methods. In *Dating an Earthquake: Applications of Quaternary Geochronology in Paleoseismology*, edited by J. M. Sowers, J. S. Noller, and W. R. Lettis, pp. 2-11 to 2-19. U.S. Nuclear Regulatory Agency CR5562, Washington, DC.

Compton, R. R., 1985, *Geology in the Field*. Wiley, New York.

Constantinou, G., 1982, Geological Features and Ancient Exploitation for the Cupriferous Sulphide Orebodies of Cyprus. In *Early Metallurgy in Cyprus, 4000-500 B.C.*, edited by J. D. Muhly, R. Maddin, and V. karageorghis, pp. 13-24. Pierides Foundation, Department of Antiquities, Republic of Cyprus, Larnaca.

DeVries, T. J., and Wells, L. E., 1990, Thermally-Anomalous Holocene Molluskan Assemblages from Coastal Peru: Evidence for Paleogeographic Not Climatic Change. *Palaeogeography, Palaeoclimatology, Palaeoecology* 81:11-32.

DeVries, T. J., Ortlieb, L., Diaz, A., Wells, L., Hillaire-Mercel, C., 1996, Determining the Early History of El Niño. *Science* 276:965-966.

Downs, R. M., and Stea, D., 1977, *Maps in Minds: Reflections on Cognitive Mapping*. Harper and Row, New York.

Dunnell, R. C., 1992, The Notion Site. In *Space, Time, and Archaeological Landscapes*, edited by J. Rossignol and L. Wandsnider, pp. 21-41. Plenum Press, New York.

Eling, H., 1987, *The Role of Irrigation Networks in Emerging Societal Complexity During Late Prehispanic Times*. Ph. D. dissertation, University of Texas, Austin.

El-Moslimany, A. P., 1994, Evidence of Early Holocene Summer Precipitation in the Continental Middle East. In *Late Quaternary Chronology and Paleoclimates of the Eastern Mediterranean*, edited by O. Bar-Yosef and R. S. Kra, pp. 121-130. Radiocarbon 1994, Tucson, AZ.

Frye, J. C., and William, H. B., 1960, Classification of the Wisconsin Stage in the Lake Michigan Glacial Lobe, *Illinois Geological Survey Circ.* 285:1-16.

Frye, J. C., and William, H. B., 1962, Note 27—Morphostratigraphic Units in Pleistocene Stratigraphy, *Bulletin of the American Association of Petroleum Geologists* 46:112-113.

Geyh, M. A., 1994, The Paleohydrology of the Eastern Mediterranean. In *Late Quaternary Chronology and Paleoclimates of the Eastern Mediterranean*, edited by O. Bar-Yosef and R. S. Kra, pp. 131-145. Radiocarbon 1994, Tucson, AZ.

Gifford, J. A., 1978, Paleogeography of Archaeological Sites of the Larnaca Lowlands, Southeastern Cyprus. Ph.D. dissertation, University of Minnesota, Minneapolis.

Goldberg, P., 1994, Interpreting Late Quaternary Continental Sequences in Israel. In *Late Quaternary Chronology and Paleoclimates of the Eastern Mediterranean*, edited by O. Bar-Yosef and R. S. Kra, pp. 89-102. Radiocarbon 1994, Tucson, AZ.

Gould, P., and White, R., 1986, *Mental Maps*. Allen and Unwin, Boston.

Grolier, M. J., Erickson, G. E., McCauley, J. F., and Morris, E. C., n.d., The Desert Landforms of Peru; A Preliminary Photographic Atlas, Interagency Report: Astrogeology 57, U.S. Geological Survey, 146 pp, Flagstaff, AZ.

Harris, E. C., 1989, *Principles of Archaeological Stratigraphy*. Academic Press, London.

Held, S. O., 1989, *Early Prehistoric Island Archaeology in Cyprus: Configurations of Formative Culture Growth from the Pleistocene/Holocene Boundary to the mid-3rd Millennium B.C.* Ph. D. Dissertation, Institute of Archaeology, University of London.

Holliday, V. T., 1995, *Late Quaternary Valley Fills on the Southern High Plains*. Geological Society of America Memoir 186, Geological Society of America, Boulder, CO.

Inbar, M., 1992, Rates of Fluvial Erosion in Basins with a Mediterranean Type Climate. *Catena* 19:393–409.

karageorghis, V., 1968, *Cyprus*. Archaeologia Mundi Series, Nagel Publishers, Geneva.

Keefer, D. K., deFrance, S. D., Moseley, M. E., Richardson, J. B., III, Satterlee, D. R., and Day-Lewis, A., 1998, Early Maritime Economy and El Niño events at Quebrada Tacahuay, Peru. *Science* 281:1833–1835.

Kempler, D., and Ben-Avraham, Z., 1987, The Tectonic Evolution of the Cyprean Arc. *Annales Tectonicae* 1:58–71.

Knapp, A. B., 1993, Social Complexity: Incipience, Emergence and Development on Prehistoric Cyprus. *Bulletin of the American School of Oriental Research* 292:85–106.

Knapp, A. B., 1994, Emergence, Development and Decline on Bronze Age Cyprus. In *Development and Decline in the Mediterranean Bronze Age*, edited by C. Mathers and S. Stoddart, pp. 271–304. Sheffiel Archaeological Monographs, John Collis Publications, Sheffield.

Knapp, A. B., Held, S. O., and Manning, S. W., 1994, The Prehistory of Cyprus: Problems and Prospects. *Journal of World Prehistory* 8:377–453.

Knapp, A. B., kassianidou, L., & Donnelly, M., 1998, The 1997 Excavations at Politiko Phorades, Cyprus. *Old World Archaeology Newsletter* 21:15–23.

Kosmas, C., Danalatos, N., Cammeraat, L. H., Chabart, M., Diamantopoulos, J., Farand, R., Gutierrez, L., Jacob, A., Marques, H., Martinez Fernandez, J., Mizara, A., Moustakas, N., Nicolau, J. M., Oliveros, C., Pinna, G., Puddu, R., Puigdefabregas, J., Roxo, M., Simao, A., Stamou, G., Tomasi, N., Usai, D., and Vacca, A., 1997, The Effect of Land Use on Runoff and Soil Erosion Rates Under Mediterranean Conditions, *Catena* 29:45–59.

Lahee, F. H., 1961, *Field Geology*. MacGraw Hill, New York.

Lort, J. M., 1971, The Tectonics of the Eastern Mediterranean: A Geophysical View. *Reviews of Geophysics and Space Physics* 9:189–216.

MacArthur, R. H., and Wilson, E. O., 1967, *The Theory of Island Biogeography*. Princeton University Press, Princeton, New Jersey.

Malpass, M. A., 1983, The Preceramic Occupations of the Casma Valley, Peru. Ph.D. dissertation, University of Wisconsin, Madison.

Mandel, R. D., and Simmons, A. H., 1997, Geoarchaeology of the Akrotiri Aetokremnos Rockshelter, Southern Cyprus. *Geoarchaeology* 12:567–605.

McFadden, L. D., Ritter, J. B., and Wells, S. G., 1989, Use of Multiparameter Relative-Age Methods for Age Estimation and Correlation of Alluvial Fan Surfaces on a Desert Piedmont, Eastern Mojave Desert, California, *Quaternary Research* 32:276–290.

McKenzie, D. P., 1970, Plate Tectonics of the Mediterranean Region, *Nature* 226: 239–243.

Merrillees, R. S, 1982, Early Metallurgy in Cyprus 4000–500 B.C., Historical Summary, 0 B.C. In *Early Metallurgy in Cyprus, 4000–500 B.C.*, edited by J. D. Muhly, R. Maddin, and V. karageorghis, pp. 373–376. Pierides Foundation, Department of Antiquities, Republic of Cyprus, Larnaca.

Moore, J. D., 1988, Prehistoric Raised Field Agriculture in the Casma Valley, Peru, *Journal of Field Archaeology* 15:265–276.

Moore, J. D., 1991, Cultural Responses to Environmental Catastrophes: Post-El Niño Subsistence on the Prehistoric North Coast of Peru, *Latin American Antiquity* 2:27–47.

Moseley, M. E., 1975, *The Maritime Foundations of Andean Civilization*. Cummings Publishing Company, Menlo Park, CA.

Moseley, M. E., 1983, The Good Old Days Were Better: Agrarian Collapse and Tectonics. *American Anthropologist* 85:773–799.

Moseley, M. E., 1987, Punctuated Equilibrium: Searching the Ancient Record for El Niño, *Quarterly Review of Archaeology* 8:7–10.
Moseley, M. E., 1992, *The Incas and Their Ancestors, The Archaeology of Peru*, Thames and Hudson, London.
Moseley, M. E., Feldman, R. A., Ortloff, C. R., and Navarez, A., 1983, Principles of Agrarian Collapse in the Cordillera Negra, Peru, *Annals of the Carnegie Museum* 52:299–327.
Muhly, J. D., 1982, The Nature of Trade in the LBA Mediterranean: The Organization of the Metal's Trade and Their Role in Cyprus. In *Early Metallurgy in Cyprus, 4000–500 B.C.*, edited by J. D. Muhly, R. Maddin, and V. karageorghis, pp. 13–24. Pierides Foundation, Department of Antiquities, Republic of Cyprus, Larnaca.
Myers, J. S., 1976, Erosion Surfaces and Ignimbrite Eruption, Measures of Andean Uplift in Northern Peru, *Geological Journal* 11:29–44.
Noller, J. S., 1993, *Late Cenozoic Stratigraphy and Soil Geomorphology of the Peruvian Desert, 3°–18°S: A Long-Term Record of Hyperaridity and El Niño*. Ph.D. dissertation, University of Colorado, Boulder.
North American Commission on Stratigraphic Nomenclature, 1983, North American Stratigraphic Code, *American Association of Petroleum Geologists Bulletin* 67:841–875.
Ossa, P., 1973, *A Survey of the Lithic Preceramic Occupation of the Moche Valley, North Coastal Peru*. Ph.D. dissertation, Harvard University, Cambridge.
Ossa, P., 1978, Paijn in Early Andean Prehistory: The Moche Valley Evidence. In *Early Man in American for a Circum-Pacific Perspective*, edited by A. L. Bryan, pp. 290–295. Department of Anthropology, Occasional Papers No. 1, University of Alberta, Edmonton.
Ossa, P., and Moseley, M. E., 1972, La Cumbre: A Preliminary Report on Research into the Early Lithic Occupation of the Moche Valley, Peru. *Nawpa Pacha* 9:1–17.
Patterson, T. C., 1983, Historical Development of a Coastal Andean Social Formation in Central Peru, 6000 to 5000 B. C. In *Investigations of the Andean Past*, edited by D. H. Sandweiss, pp. 21–31. Cornell Latin American Studies Program, Cornell University, Ithaca, NY.
Poole, A. J., 1992, Sedimentology, Neotectonics and Geomorphology Related to Tectonic Uplift and Sea Level Change: Quaternary of Cyprus. Ph.D. dissertation, University of Edinburgh.
Poole, A. J., Shimmield, G. B., and Robertson, A. H. F., 1990, Late Quaternary Uplift of the Troodos Ophiolite, Cyprus: Uranium Series Dating of Pleistocene Coral. *Geology* 18:894–897.
Pozorski, S., and Pozorski, T., 1986, Recent Excavations at Pampa Las Llamas–Moxeke, a Complex Initial Period Site in Peru. *Journal of Field Archaeology* 13:381–401.
Pozorski, S., and Pozorski, T., 1987, *Early Settlement and Subsistence in the Casma Valley, Peru*. University of Iowa Press, Iowa City.
Pozorski, S., and Pozorski, T., 1990, Huaynuna, a Late Cotton Preceramic Site on the North Coast of Peru. *Journal of Field Archaeology* 17, 17–26.
Pozorski, S., and Pozorski, T., 1992, Early Civilization in the Casma Valley, Peru, *Antiquity* 66:845–870.
Pozorski, S., and Pozorski, T., 1995, *Paleoenvironment at Almejas, a Mid-Holocene Site in the Casma Valley, Peru*. Paper presented at 60th Annual Meeting, Society for American Archaeology, Minneapolis, Minnesota.
Pozorski, S., and Pozorski, T., in press, Un Entierro Preceramico en el Sitio de Almejas en el Valle de Casma, Peru. In *En El Nombre del Padre: Reflexiones sobre los Antepasados en las Culturas del Norte del Peru*, edited by L. Millones and M. Millones. Editorial Horizonte, Lima.
Pozorski, T., Pozorski, S., and Rick, J., 1982, A Bird Geoglyph near Casma, Peru, *Andean Past* 3:165–186.
Reese, D. S., 1995, *The Pleistocene Vertebrate Sites and Fauna of Cyprus*. Geological Survey Department, Bulletin No. 9, Ministry of Agriculture, Natural Resources and Environment, Republic of Cyprus.
Reid, C., 1908, *Report on Water Supply in Cyprus*. Government Printer, Nicosia (Cited in Thirgood, 1987, p. 29).
Riedinger, M., Steinitz-kannan, M., Last, W., and Brenner, M., 1998, A 6100 YR El Niño Record from the Galapagos Islands, *Geological Society of America, Abstracts with Programs* 30:A–161.
Roberts, N., and Wright, H. E., Jr., 1993, Vegetational, Lake Level and Climate History of the Near East and Southeast Asia. In *Global Climates Since the Last Glacial Maximum*, edited by H. E. Wright, J. E. Kutzbach, T. Webb III, W. F. Ruddiman, R. A. Street-Perrott, and P. J. Bartlein, pp. 194–220. University of Minnesota Press, Minneapolis.
Robertson, A. H. F., 1977, Tertiary Uplift History of the Troodos Massif, Cyprus, *Geological Society of America Bulletin* 88:1763–1772.

Rodbell, D. T., Seltzer, G. O., Anderson, D. M., Abbott, M. B., Enfield, D. B., and Newman, J. H., 1999, An ~15,000-Year Record of El Nino-Driven Alluviation in Southwestern Ecuador, *Science*, 283:516–520.

Rossignol-Strick, M., 1993, Late Quaternary Climate in the Eastern Mediterranean Region. *Paleorient* 19:135–152.

Sandweiss, D. H., Richardson, J. B., III, Reitz, E. J., Rollins, H. B., and Maasch, K. A., 1996b, Geoarchaeological Evidence from Peru for a 5000 Year B.P. Onset of El Niño. *Science* 273:1531–1533.

Sandweiss, D. H., Richardson, J. B. III, Reitz, E. J., Rollins, H. B., and Maasch, K. A., 1996a, Determining the Early History of El Niño. *Science* 276:966–967.

Satterlee, D. R., 1993, The Impact of a Fourteenth-Century El Niño Flood on an Indigenous Population Near Ilo, Peru. Ph. D. dissertation, University of Florida, University Microfilms, Ann Arbor.

Shimada, I., 1994, *Pampa Grande and the Mochica Culture*, University of Texas Press, Austin.

Simmons, A. H., 1991, Humans, Island Colonization and Pleistocene Extinctions in the Mediterranean: The View from Akrotiri–Aetokremnos, Cyprus. *Antiquity* 65:857–869.

Stanley, D. J., and Warne, A. G., 1997, Holocene Sea-Level Change and Early Human Utilization of Deltas. *GSA Today* 7:1–7.

Steinitz-kannan, M., Riedinger, M. A., Last, W., Brenner, M., and Miller, M. C., 1997, Un Registro de 6000 Aos de Manifestos del Fenomeno de El Niño en Sedimentos de Lagunas de las Islas Galapagos. In *Seminario Internacional "Consequencias Climaticas e Hidrologicas del Evento El Niño a Escala Regional y Local,"* edited by E. Cadier and R. Galarrage, pp. 79–88. Memorias Tecnicas, Edicion Preliminar, ORSTOM/INAMHI.

Strong, W. D., and Evans, C., 1952, *Cultural Stratigraphy in the Virú Valley, Northern Peru*. Columbia University Studies in Archaeology and Ethnology 4, Columbia University, New York.

Tello, J. C., 1956, *Arqueologia del Valle de Casma, Culturas: Chavin, Santa o Huaylas Yunga, y Sub–Chimu*. Universidad Nacional Mayor de San Marcos, Lima, Peru.

Thirgood, J. V., 1987, *Cyprus: A Chronicle of its Forest, Land and People*. University of British Columbia Press, British Columbia.

Trumbore, S. E., 1998, Radiocarbon Geochronology. In *Dating an Earthquake: Applications of Quaternary Geochronology in Paleoseismology*, edited by J. M. Sowers, J. S. Noller, and W. R. Lettis, pp. 2-69 to 2-98. U.S. Nuclear Regulatory Agency CR5562, Washington, DC.

van Andel, T. H., and Zangger, E., 1990, Landscape Stability and Destabilization in the Prehistory of Greece. In *Man's Role in Shaping of the Eastern Mediterranean Landscape*, edited by S. Bottema, G. Entjes-Nieborg, and W. van Zeist, pp. 139–157. Balkema, Rotterdam.

vita Finzi, C., 1993, Evaluating Late Quaternary Uplift in Greece and Cyprus. In *Magmatic Processes and Plate Tectonics*, edited by H. M. Prichard, T. Alabaster, N. B. W. Harris, and C. R. Neary, pp. 417–424. Geological Society Special Publications No. 76. Geological Society of London, London, UK.

Wells, L. E., 1988, Holocene Fluvial and Shoreline History as a Function of Human and Geologic Factors in Arid Northern Peru. Ph.D. dissertation, Stanford University, Palo Alto, CA.

Wells, L. E., Gerth, R. A., Noller, J. S., and Whitehill, C. S., 1998, Quantifying Anthropogenic Hillslope Erosion on Mount Kreatos, Cyrpus, *Geological Society of America Abstracts with Programs*. 30:A-143.

Wells, L. E., and Noller, J. S., 1999, Holocene Coevolution of the Physical Landscape and Human Settlement in Northern Coastal Peru. *Geoarchaeology* 14:755–789.

Willey, G. R., 1953, *Prehistoric Settlement Patterns in the Virú Valley, Peru*. Smithsonian Institution, Bureau of American Ethnology, Bulletin 155, Smithsonian Institution, Washington, DC.

Wilson, D. J., 1988, *Prehispanic Settlement Patterns in the Lower Santa Valley Peru*. Smithsonian Institution Press, Washington, DC.

Wilson, D. J., 1998, The Casma Valley Project (1989–1995). http://www. smu. edu/~anthrop/dwilson. html#RESEARCHPROGRAMS.

Zreda, M. G., and Phillips, F. M., 1998, Quaternary Dating by Cosmogenic Nuclide Buildup in Surficial Materials. In *Dating an Earthquake: Applications of Quaternary Geochronology in Paleoseismology*, edited by J. M. Sowers, J. S. Noller, and W. R. Lettis, pp. 2-101 to 2-127. U.S. Nuclear Regulatory Agency CR5562, Washington, DC.

6

Archaeoseismology
Shaking Out the History of Humans and Earthquakes

JAY STRATTON NOLLER

1. Using Archaeology to Solve a Paleoseismic Problem

Intuitively, archaeology and geology should be used together to address one of the common and deadly natural hazards—earthquakes. Yet, until recently, the twain did not meet on even ground, and then not always as successful joint ventures (see commentaries of Karcz and Kafri, 1978, 1981; Stiros, 1988; Guidoboni, 1996). Avenues of investigation of earthquakes lie within a continuum between historical archaeology and seismology for 19th and 20th century earthquakes and between archaeology and geology/paleoseismology for earlier events (Pavlides, 1996). Any one of these disciplines may look at the same event with a different view and purpose. What the body of literature reveals is that multidisciplinary studies are rare until the 1990s.

Earthquakes, like related natural hazards, are phenomena that occasionally interrupt the flow of society by changing topography, altering resource availability, damaging structures, and taking lives. A number of these events have been described as significant points of departure for societal change (e.g., Soren, 1966; Armijo et al., 1991; Nur, 1998). Perhaps the best known of these is the 1755 Lisbon Earthquake that is widely credited as making a mark on European history

JAY STRATTON NOLLER • Department of Crop and Soil Science, Oregon State University, Corvallis, Oregon 97331.

Earth Sciences and Archaeology, edited by Paul Goldberg, Vance T. Holliday, and C. Reid Ferring. Kluwer Academic/Plenum Publishers, New York, 2001.

and thought (e.g., Voltaire, 1759) by precipitating the end of the Age of Enlightenment.

Much of the early work involving the archaeological investigation of earthquakes and their damaging effects was that of explaining the direction of toppling of columns, collapse of structures, and deformation of architecture. Inherent strengths and weaknesses of this approach are easy to see (Rapp, 1986; Stiros, 1988). Unfortunately, early studies did not follow supportable interpretations and relied instead on drama or conjecture (Guidoboni, 1996). Yet despite recent advances in the field this tradition seems to continue.

New insights are being obtained through interdisciplinary study of the source of a damaging earthquake, the accompanying seismically disturbed structures and materials of human construction, and the effects on society. Here not only are answers about the cause of building destruction answered, but also the entire human picture of a natural disaster can be set within the context of a natural (geologic) event. The archaeological record, with its remnants of durable materials and anthropogenic strata, provides the human context for information that, when analyzed with the geologic record, ultimately is of direct relevance to modern society. Better models of earthquake activity and impacts are thus described.

Tectonically active regions provide unique niches for human occupation and food resources (King et al., 1994; Noller and Lightfoot, 1997). Active faults and folds impede and hence alter the flow of surface and ground water. Fault scarps locally dam rivers and form lakes providing long-lived lacustrine and related habitats for life (Sangawa, 1989). The accumulation of sedimentary deposits adjacent to faults and in tectonic depressions, along with development of soils thereon, leads to enhanced, and in some areas unique, floral and faunal communities (King et al., 1994). Thus, it is not surprising that we find occupation sites, especially for nonagrarian societies, along fault scarps and other tectonic landforms.

This chapter is not meant to be a thorough review of the field now commonly referred to as "archaeoseismology." Rather, it provides a view of studies that in one fashion or another have set out to purposefully study the parameters of historic and prehistoric earthquakes, the primary pursuit of the field of paleoseismology, through the use of archaeology. The study of destruction levels in archaeological excavations has been an important and significant starting point for most "mainstream" archaeoseismic studies, and the literature on this subject is voluminous. The reader interested in such works is encouraged to refer to the compilation of Stiros and Jones (1996), as much of that body of literature will not be covered here.

2. Approaches and Results of Archaeoseismology

There is no unique approach to applying archaeology and geology in an archaeoseismic study. The toolbox of approaches is large due to the wide range of potential archaeologic and geologic settings across the globe. Also, the spectrum of expertise that can be drawn into such a study adds further latitude in approach. As Guidoboni (1996) pointed out, a successful archaeoseismic study is one in which many disciplines, including archaeology, geology, seismology, and civil engineering participate throughout the investigation. The approach should be designed to answer, in the simplest sense, questions of earthquake dimension: "When?" "What happened?" "Where?" and "How big?" The goal should be to quantify the parameters that are part of a seismic hazard assessment, including the date of the most recent earthquake or time elapsed since the last event, timing of earlier earthquakes, sense (relative direction) of fault slip, maximum displacement, average displacement, surface rupture length, fault slip rate (creep, co-seismic), earthquake recurrence interval, location of past earthquakes (epicenter), and maximum credible earthquake (McCalpin, 1996).

The study of human history has richly increased our understanding of the magnitude and location of past earthquakes (Lee et al., 1988). Typically, workers develop paleointensity maps that show the distribution of known or interpreted seismic damage. The centroid of the mapped seismic zones is taken as the estimate of the earthquake epicenter. The maximum intensity and distribution of intensity are useful for estimating the magnitude and other parameters of earthquakes (Ambraseys, 1996). Much of the current literature in archaeoseismology addresses these two basic questions. These important foci of archaeoseismology require little in the way of earth science study, and hence, as stated earlier, this chapter does not cover these.

Since the early 1990s studies have shown that archaeology can be used to reveal the activity of a fault, including the chronology of earthquake events or tectonic activity, the amount and direction of surface deformation, and the recurrence of events (Table 6.1). In answering questions related to these three, ancillary information relevant to the location of epicentral areas and event magnitude may arise (Table 6.1). The following sections address three of the basic questions of archaeoseismology and focus on the importance, manner of approach, and history of each. Finally, a case study is detailed that provides data for the important western boundary fault of the San Andreas fault system, the San Gregorio fault. As presented by these examples, important seismic hazard assessment parameters can be quantified by a single archaeoseismic study.

2.1. When Did the Earthquake Occur?

2.1.1. Dating an Earthquake

Establishing the age of an earthquake is one very important contribution that archaeology can make. Natural catastrophes, such as earthquakes and floods, are commonly recorded as they have widespread impact on society. The documentation of these catastrophes as markers in human history has the unintended benefit of being a rich record of earthquake events that have been mined by seismologists (e.g., Ambraseys, 1996). But such records do not extend far back in time and the limit of what is "historical" is not everywhere the same. It is important then to make the distinction between a "historical" earthquake and a

Table 6.1. Use of Archaeology in Paleoseismic Settings around the World

Paleoseismic setting	Archaeological setting	Uses[a]	Event age(s)	References
Faults				
Strike-slip				
Dead Sea Transform, Israel	Offset Umayid house (?)	C		Reches and Hoexter, 1981
	Offset Crusader castle	CD	Dawn, May 20, 1202	Marco et al., 1997; Ellenblum et al., 1998
Dead Sea Transform, Jordan	Offset Islamic city walls	C	A.D. 1068	Whitcomb, 1994
	Offset Roman reservoir	C	A.D. 363	T. Niemi, pers. comm., 1999
San Andreas Fault, USA	Offset occupation site	CDR	5+ in last 2,000 years	Noller et al., 1992; Noller and Lightfoot, 1997
San Gregorio Fault, USA	Offset occupation site	CDM	A.D. 620–A.D. 1400	Noller et al., 1995; Simpson et al., 1997; this chapter
Normal				
Hongguozigou Faults, China	Offset Great Wall of China	CDLM	A.D. 1739	Zhang et al., 1986
Weihe Basin, China	Offset artifact-bearing deposits	CDR	2000 B.C.–A.D. 1556	Zhang et al., 1995
Arima-Takatsuki Fault, Japan	Sheared occupation site	CL	3 B.C.–A.D. 3	Umeda et al., 1984
Amorgos Island, Greece	Submerged architecture	C	Post-Roman	Stiros et al., 1994
Korinth Fault, Greece	Submerged port	CDLM	A.D. 400 (1 of 3 since A.D. 100)	Scranton et al., 1978; Noller et al., 1997
Kyparissi, Greece	Deformed (faulted) stoa	CLD	520 B.C.	Stiros, 1988
Sparta Fault, Greece	Written accounts	DMR	464 B.C.	Armijo et al., 1991
Coastal faults, Israel	Offset and submerged Roman port moles	CL	Unknown number	Mart and Perecman, 1996
Egna, Italy	Offset Roman villa	CD	2500 B.C.–≪A.D. 100; A.D. 200	Galadini and Galli, 1999
Fucino, Italy	Artifact-bearing strata	C	>A.D. 6	Michetti et al., 1996
Hierapolis Fault, Turkey	Offset Roman and Byzantine structures	CDLM	Two post-Roman	Hancock and Altunel, 1997
Thrust				
Reelfoot Fault, USA	Artifact scatter	C	pre-800–1000 A.D.	Kelson et al., 1996
Folds				
Epirus, Greece	Occupation sites, rock shelters	D	Paleolithic (200–10 ka)	King et al., 1994

Coseismic Deformation				
Ground Motion				
Global	Damaged architecture, etc.	CLMR	Various	Mainstream archaeoseismology. For a review see Stiros and Jones, 1996
Liquefaction				
New Brunswick, Canada	Disrupted occupation sites	CL	A.D. 1500–1800	Broster et al., 1993
New Madrid Seismic Zone, USA	Disrupted occupation sites	CLMR	4 in last 6,000 years	Saucier, 1991; Tuttle and Schweig, 1995, 1996; Tuttle et al., 1996, 1999a,b
Wabash Valley, USA	Artifact-bearing strata	C		Munson et al., 1992
Tsunami				
Cascadia Subduction Zone, USA	Buried occupation sites, oral tradition	C	17th century A.D.	G. Carter, pers. comm., 1992; J. Stein, pers. comm., 1995
Land/level Change				
Cascadia Subduction Zone, Canada	Occupation site	CD	B.C. 1500; <A.D. 1100	Reinhardt et al., 1996
Korinthia, Greece	Harbors, marine facilities, quarries	CD	Various, Unknown	Vita Finzi and King, 1985; Flemming and Webb, 1996; Stiros, 1988; Stiros and Pirazzoli, 1995; Pirazzoli et al., 1996; Noller et al., in prep.; Papageorgiou and Stiros, 1996; Stiros et al., 1996
Larrisa Plain, Greece	Settlement patterns, written accounts	D	Various, Unknown	Demitrack, 1986; Caputo et al., 1994
Lokris, Greece	Submerged occupation site	D	2 in last 4,000 years	Ganas and Buck, 1998
Aleutian Islands, USA	Occupation sites	CD	Holocene	Winslow and Johnson, 1989
Cascadia Subduction Zone, USA	Occupation sites, fish weirs	CD	post 410–720 A.D.	Grant and Minor, 1991
Coastal Maine, USA	Occupation sites	D	Holocene	Anderson et al., 1984

*The listed studies apply archaeology principally (C) to derive the chronology of earthquake events or tectonic activity; (D) as an indicator of pre- syn- and/or postseismic deformation; (L) to locate event; (M) to estimate intensity and/or magnitude of an earthquake; (R) to estimate recurrence of events.

"paleoearthquake." Historical earthquakes are those events observed and recorded in a written, carved, or crafted work or by oral history, tradition, or legend. Archaeologists, as well as historians, philologists, and ethnographers, use ancient sources, oral histories, and legends to determine the decade, year, month, day, or time of day of a cultural event (Stiros and Jones, 1996). Paleoearthquakes have no such (at least remaining) human testimonial, thus requiring study of the geological and/or archaeological record and a reliance on a less certain dating scheme.

An interesting study that integrated oral history with paleoseismology is that of G. Carter (Personal Communication., 1992). In this study they used radiocarbon-dated leaves in a tsunami deposit to date the ca. 1700 A.D. Cascadia subduction zone event. An oral tradition of a giant wave (tsunami) that arrived during an autumn's full moon allowed them to reach a subannual estimate of the age. Such a cultural tie would make this a historical earthquake. Although the exact calendar year was not obtained with their data, the study does illustrate how study of a natural hazard of great social importance can strengthen its human context.

Archaeological survey and excavation can reveal earthquake histories through the study of destruction levels and other architectural damage, sites* buried by co-seismic deposits, sites disturbed by faults (see case study that follows), sites disturbed by other types of surface deformation, or features constructed or inscribed to commemorate an earthquake. In these contexts, seismically disturbed archaeological structures, materials, and deposits constrain the maximum age for the event. Overlying postevent structures and strata provide a minimum age constraint. Features produced during the event are of natural construction, such as fault scarps and sand blows, but these have been difficult to date until recently (Noller and Forman, 1998; Zreda and Noller, 1998). Because an earthquake occurs so suddenly and lasts for only a matter of seconds, for anyone to create a record while it occurs would be extremely rare. Rather, the creation of a cultural feature marking the event occurs subsequently, such as the inscription of stiles in China (Lin and Wu, 1993).

Dating of archaeological materials and deposits is carried out by methods familiar to geologists (e.g., thermoluminescence; Noller et al., 2000) and those exclusively within the realm of archaeologists (e.g., seriation of sculpture, ceramics/pottery, and stone tools; Aitken, 1990) (Table 6.2). Geochronologic results can be expressed as sidereal (e.g. 324 B.C. May 20, 1270), numerical (e.g., cal. ^{14}C 1250 $\pm$ 40), calibrated (e.g., aminozone 2), and correlative (e.g., marine oxygen isotope stage 3) (Noller et al., 2000). The best dating methods for archaeoseismology are those providing sidereal results, followed by radiocarbon (the "workhorse" dating method for paleoseismology) and its numerical results, other methods providing numerical results, and methods with calibrated results (e.g., ceramic/pottery). Other methods will generally provide useful results, with specific circumstances making some equal in accuracy to those just listed. At present, sidereal results are useful to no less than one year, the smallest unit of time measure used for events in seismic hazard assessment (SHA). Any further

---

Site is informally used here as a place of cultural reference, whether an occupation site, a place of special interest, or a concentration of human deposits

Table 6.2. Geochronological Methods for Use in Archaeoseismology

TYPE OF RESULTS

≡≡≡ NUMERICAL-AGE ≡≡≡≡≡≡≡≡≡[a]
≡≡ Calibrated-Age ≡≡≡≡≡≡≡[a]
‐‐‐‐‐‐ Relative-Age ‐‐‐‐‐‐‐[a]
≡Correlated-Age ≡[a]

TYPE OF METHOD

Sidereal	Isotopic	Radiogenic	Chemical and biologic	Geomorphic	Correlation
Historical records *Dendrochronology* Sclerochronology and other annual growth in other organisms (e.g., mollusks)	*Radiocarbon* Cosmogenic isotopes ^{36}Cl, ^{10}Be, ^{26}Al, ^{14}C, ^{3}He, and others[b] Uranium-series ^{210}Pb	Thermoluminescence Optically stimulated luminescence Infrared stimulated luminescence Electron-spin resonance	Obsidian hydration and tephra hydration Lichenometry Amino-acid racemization Soil chemistry	*Soil-profile development* Rock and mineral weathering *Scarp morphology and progressive landform modification* Rock-varnish development Rate of deposition Rate of deformation Geomorphic position	*Archaeology* *Stratigraphy* Tephrochronology Paleomagnetism Climatic correlation

[a]Triple-dashed line indicates the type of result most commonly produced by the methods below it; single-dashed line indicates the type of result less commonly produced by the methods below it.
[b]Some cosmogenic methods, particularly exposure ages, have some similarities with methods in the "Radiogenic" column.
Note: Methods in italics are discussed in this chapter. This table is based on that in Noller et al. (2000).

refinement in age, such as season, month, or second, might find use in future SHAs.

An age estimate is only as good as the context from which it is derived (Noller et al., 2000). Context in archaeoseismology is provided through Quaternary geologic and archaeological stratigraphy. These branches of stratigraphy significantly differ from what most geologists practice because so much of the record is still active or present at the surface (Morrison, 1991) or acted on by cultural processes (Harris, 1989). Morphostratigraphy, allostratigraphy, pedostratigraphy, ethnostratigraphy, and other means of keeping account of the relationships of deposits, surfaces, intrusions, event horizons, and other stratigraphic units are integral to this field of study. The Harris Matrix (Harris, 1989) in use by archaeologists parallels that used by paleoseismologists in developing event stratigraphies, an important part in assigning ages to paleoearthquakes.

2.1.2. Case Studies

All archaeoseismic studies contribute age control in one form or another on earthquake or tectonic activity. Worldwide, finding a seismically disturbed archaeological site probably means it was a Holocene earthquake. Such a generality is based on the global rise of civilizations in the past ca. 7,000 years and the dramatic increase in human population of the past several centuries. In the latest Holocene we find more people living in seismically active regions, along with the development of complex societies came the development of more artifacts unique to a specific period of human history. Hence the utility of archaeology in establishing the date of an event, especially in the historic past and during periods of high social order and trade.

In most published accounts, archaeology has only provided *ante quem* or *post quem* constraints on an earthquake. I believe that this is not because of our inability to closely constrain age, but is rather an artifact of a discipline in its early stages of development. Prior to the advent of recent developments in paleoseismology, archaeoseismology was more an accidental type of study involving the geologist as consultant after excavation has removed some of the vital context to reveal a potentially seismogenic feature (Rapp, 1986). However, even purposefully setting out to study an archaeological site that lies across a fault does not necessarily guarantee the result of a well-constrained event age. For example, small (<100 m^2), discrete lithic scatters along the San Andreas fault at Fort Ross typically only provide a maximum age for a series of the last two to five earthquakes. In another example, activity on the Hierapolis fault, Turkey, is constrained only by faulted Roman-era structures (Hancock and Altunel, 1997). The lack of an historical record of an event on the Hierapolis fault leaves us with an uncertain minimum age on the youngest of these surface-rupturing events.

The most commonly sought-after use of archaeology for dating an event involves an earthquake feature in direct context with an artifact of well-defined age. Stained-glass windows, still propped against the walls of the basilica in which they were to be placed, were submerged by an earthquake on the Korinth fault in A.D. 400 (Noller et al., 1997). These impressive art pieces and other artifacts found at Kenchreai, Greece, provide the fix of a firm age on the earthquake.

Locally, archaeological materials may only provide an *ante quem* age estimate. In the epicentral zone of the 1940 Khait earthquake, pottery sherds provide information on the maximum age of gravitational collapse of a fault scarp (Nikonov, 1995). In this study a single collapse debris (colluvial wedge?) deposit is interpreted to represent one (scarp-producing) event. Sherds of the 6th to 8th centuries A.D. are present in a widespread debris layer at the base of the scarp. Locally overlying this is a soil with a 16th century A.D. radiocarbon age estimate. Lichenometry and dendrochronology provide additional lines of evidence for the timing of these three most recent events (Nikonov, 1995).

Ground cracks formed in a residential site during an earthquake on the Arima–Takatsuki fault, Japan, and were infilled with artifacts of Yayoi age (3 B.C.–3 A.D.) (Umeda et al., 1984). Such cracks are unlikely to stay open for more than a few weeks or months, hence the Yayoi age (3B.C.–3A.D.) artifacts establish the age of the event. This age estimate is further constrained by an overlying undisturbed occupation horizon of Yayoi Age. Age of this event may be reported as 2000 ± 300 yr BP, so long as the age of the archaeological period is well constrained and accepted. Otherwise, only the period name (in this case Yayoi Age) should be used as that conveys more of the information on age that is based on seriation of cultural materials.

Liquefaction features are one of the most prevalent and telling lines of evidence of paleoearthquakes. Archaeology has found an important place in the study of liquefaction features produced during earthquakes in relatively stable intraplate tectonic settings, where rate of landscape change typically outpaces that of tectonic deformation (Obermeier et al., 1990). Locating archaeological sites in Holocene water-lain deposits is a successful means of identifying liquefaction-producing earthquakes (Sangawa, 1989; Saucier, 1991). Not only do occupation horizons provide evidence for the age of an event (e.g., Broster et al., 1993), it is argued that they actually set up conditions favorable for the formation of liquefaction features (Saucier, 1991; Tuttle and Schweig, 1995, 1996; Tuttle et al., 1999a, b). The compactness of archaeological deposits and floors leads to greater overpressure in seismically accelerated saturated sand, causing it to liquefy and form dikes, sills, and blows. Areas surrounding an archaeological deposit are less likely to develop the overpressured conditions necessary for liquefaction to occur. It is not surprising then that some of the reported occurrences of paleoliquefaction in intraplate settings have an archaeological context. Examples of these include the Wabash Valley seismic zone of Indiana (Munson et al., 1992) and the maritime provinces of Canada (Broster et al., 1993).

Ages of liquefaction-producing earthquakes are constrained using intercalated blow sands and archaeological deposits revealed in archaeological excavations. The occupation horizon is assumed to be exposed at the surface when the earthquake occurs, leaving the site buried beneath erupted sand blow deposits. Artifacts and other datable material within the occupation horizons provide maximum age estimates for the liquefaction feature present above, and minimum age constraints for a feature present below. Four such events were revealed in excavations at three sites near Osaka and Lake Biwa, Japan (Sangawa, 1989, Fig. 6.1). In the epicentral area of the major destructive earthquakes of the

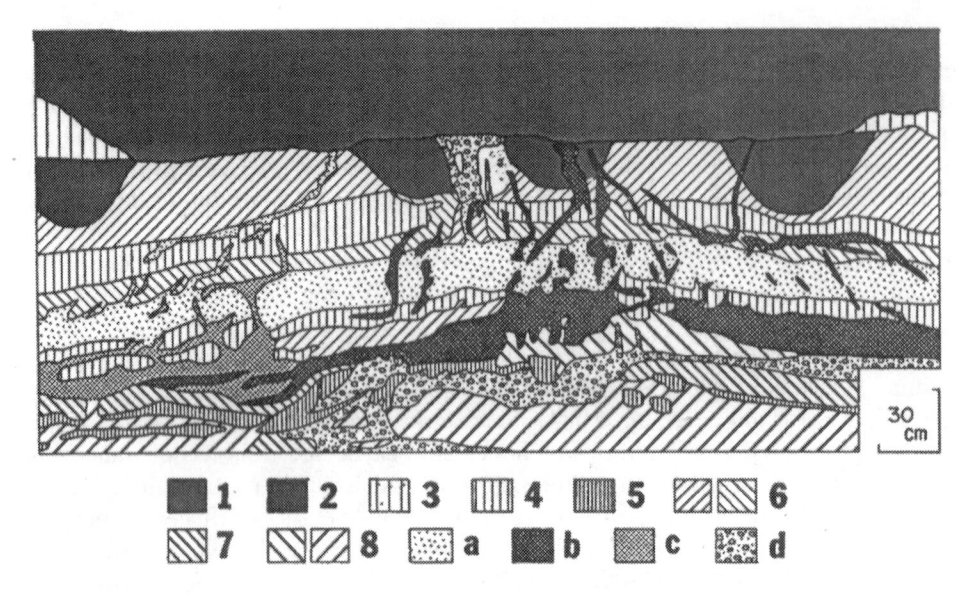

Figure 6.1. Log of excavation wall through archaeological deposits and liquefaction features of four paleoearthquakes. The exposed section is composed of archaeological or artifact-bearing deposits 1–6, and natural deposits 7–8. Events in chronological order are a, b, c, and d. From Sangawa (1989; used with permission of the Quaternary Association of Japan).

1811–1812 New Madrid earthquakes, four events are constrained by archaeological materials and occupation horizons (Tuttle and Schweig, 1995, 1996; Tuttle et al., 1999b).

2.2. What Did the Earthquake Do?

2.2.1. Quantifying Surface Effects

Deformation of the earth's surface occurs as a result of a moderate- to large-magnitude earthquake. Fault slip at depth in the Earth's crust is well represented, although not fully, by surface deformation (Yeats et al., 1996). Rupture on a fault during a >M5 earthquake may be directly or primarily expressed by a fault scarp. Surface deformation may also be indirectly or secondarily produced by processes (e.g., slope failure) accelerated by the earthquake. An important pursuit of paleoseismology is the measurement of the amount, distribution, and short-term and long-term rates of fault slip at the surface because earthquake magnitude and amount of surface deformation are strongly correlated (Wells and Coppersmith, 1994). Even though earthquakes occur at great depth (typically >10 km), the amount of slip at the earthquake focus is well represented (50% or more of it in many cases) by fault slip at the surface (Yeats et al., 1996). Hence, measuring displacement of architecture can provide a good approximation of the energy release and character of an ancient damaging earthquake.

The study of fault displacement using archaeology is outlined by Noller and Lightfoot (1997). This approach applies archaeological and geological stratigraphy within a paleoseismic study. Importantly, the use of a cultural strain marker or "piercing feature" of human production (e.g., wall) can produce equal or better results than the use of a natural feature (e.g., stream; cf. Grant and Sieh, 1994). An important component of applying this approach is in understanding how archaeological deposits will change shape, size, and spatial structure where offset by an active fault. Case studies in which cultural piercing features and other archaeological evidence figured heavily in study results are presented in Section 2.2.3.

The piercing feature is a key concept in assessing the cumulative amount and rate of slip on a fault. In standard paleoseismic practice, a piercing feature is a datable linear geologic unit or its boundary, such as a stream channel (e.g., Grant and Sieh, 1994), that crosses a fault. The rate of slip on a fault is estimated by measuring the distance over which a piercing feature is displaced during a measurable period of time between two earthquake events. This measured period may include more than one event, hence the resultant amount and rate of slip are considered cumulative and mean, respectively.

Employing archaeological deposits to define cultural piercing features has several advantages. First, archaeological deposits may contain discrete linear, planar, or three-dimensional features, such as site boundaries, cultural lenses, or architectural structures, that may be used to measure offset along faults and the rate of displacement. For example, a wall, a house depression, or the outer margin of an artifact scatter that is offset by a fault may be a cultural piercing feature. Second, cultural piercing features may contain archaeological materials that can be directly dated. For example, unlike most geologic deposits, archaeological deposits may be dated by other methods, including obsidian hydration, written and oral histories, and archaeomagnetism. Artifact seriation may be well known and provide reasonable time constraints prior to obtaining numerical age estimates. Finally, the vertical and horizontal stratigraphy of cultural piercing features may be used to reveal complex fault histories.

2.2.2. Tectonic Reshaping of an Archaeological Site

The size and shape of the postearthquake archaeological site will depend on the (1) type of underlying fault; (2) the initial spatial configuration of the human settlement when it was occupied; and (3) the timing and slip of the earthquake starting with the time of the settlement's occupation. In tectonic zones characterized by strike-slip faults, one expects archaeological deposits to be displaced laterally during earthquake events. Along normal faults, deposits are exposed to erosion on one side of the fault and partially or wholly buried on the other. The fate of a site overlying a thrust fault depends on the surface and near-surface expression of the fault. In Quaternary basins, thrust faults commonly do not propagate to the surface. Instead they end at depth (e.g., >1 km) within the core of an anticlinal fold, and because of this character are referred to as "blind thrusts" (Lettis et al., 1997). Hence, a site astride a thrust fault may (1) not experience any deformation, other than change in elevation; (2) be deformed as the hanging wall (upper) block folds and/or ruptures; or (3) be eroded from the overriding block and tectonically and depositionally buried on the footwall (lower) block. To date, many examples of these tectonic settings have been studied, except for cases 1 and 3 of thrust faults (Table 6.1).

Strike-slip faults appear to present the most possible number of outcomes or associations of cultural materials and geologic setting. In many parts of the world human occupation is marked by a deposit of complex stratigraphy and little or no substantial architecture. Without a standing wall, such as the Crusader castle wall at Vadum Jacob, Israel (Marco et al., 1997; Ellenblum et al., 1998), strategies must be followed that will constrain model piercing features. An isoconcentration line (see section 3) is an example. Using the model of circular-shaped sites in the Fort Ross, California, region (Noller et al., 1993), Noller and Lightfoot (1997) proposed four categories of associations of archaeological deposits located above or adjacent to active strike–slip faults (Fig. 6.2). These categories (A–D) are briefly described as follows, along with some of the potential problems.

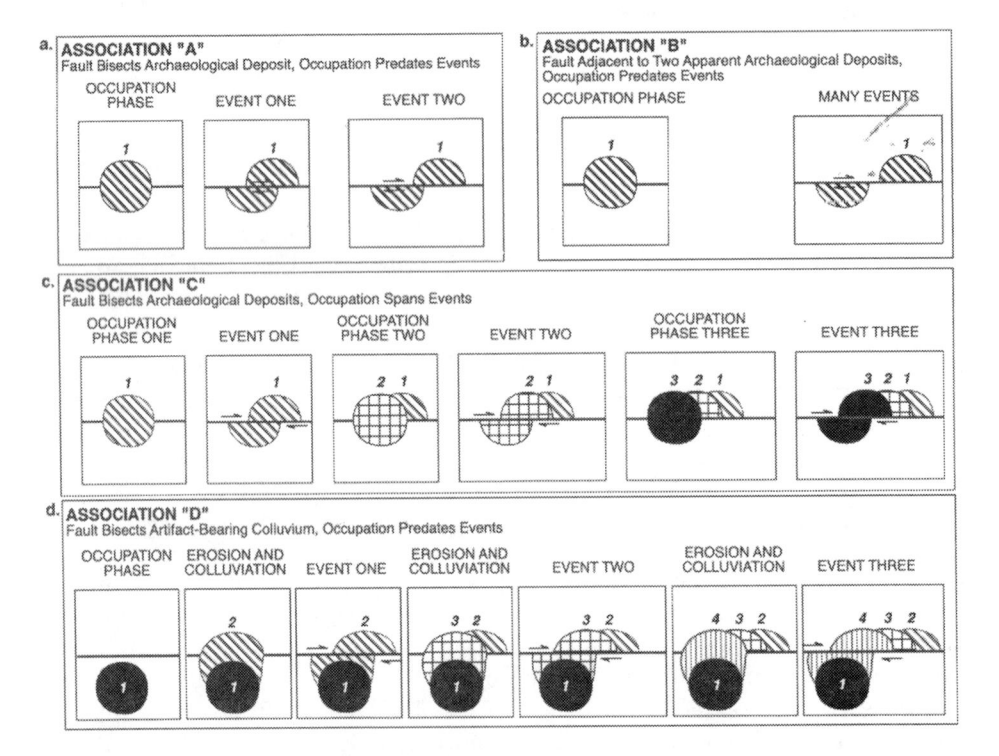

Figure 6.2. Categories of associations of archaeological deposits with an active strike-slip fault include the following: (a) a single architectural feature or deposit belonging to a discrete phase of human occupation predating an earthquake event that is not reused or reoccupied; (b) a single architectural feature or deposit, like association (a), except that the two segments are separated; (c) a site, composed of many architectural features and/or deposits, with long use over a period of multiple earthquakes; (d) a single architectural feature or deposit, or site, located adjacent to a fault forms a point source for colluvium.

A. *A single architectural feature or deposit belonging to a discrete phase of human occupation predating an earthquake event that is not reused or reoccupied.* Following one or more displacements, the feature or deposit is split into two segments that are still touching each other. In the absence of knowledge of faulting, such deposit geometry normally would be interpreted to represent one contiguous and irregularly shaped site. Such is the case at Fort Ross (Noller and Lightfoot, 1997).

B. *A single architectural feature or deposit, like association (A), except that the parts are physically separated.* The two parts could be interpreted to represent two individual sites or discrete occupations, which could certainly be the case if the fault was not recognized. The study team must prove that the two "sites" are parts of one original site. To test this, one may use independently derived fault slip rates, geometry of cryptic compositional patterns in the deposit, and/or deposit age estimates.

C. *A site, composed of many architectural features and/or discrete deposits, with long use over a period of multiple earthquake events.* Repeated offsets of the deposit lead to a pattern of overlapping (A) associations and increased site circumference. The study team must show correspondence between the stratigraphic units and features of the site. A special case of this relationship is where one part of the site is repeatedly reoccupied resulting in a vertical stratigraphy of deposits that can be matched with the lateral stratigraphy of deposits on the other side of the fault.

D. *A single architectural feature or deposit, or site, located adjacent to a fault forms a point source for colluvium.* Archaeological materials are eroded and deposited across the fault. Here, earthquakes are recognized by host geologic deposits that are dated and correlated, in part, by contained artifacts. In a special case, the site is abandoned prior to erosion and is not reoccupied. The zone of redistributed artifacts is stretched along the fault by repeated offsets and is similar in this respect to the special case for association (C).

One or more of these associations may be present at the same site. The presence of one association does not automatically exclude another association. In fact, more than one association may be the consequence of changing environmental conditions as well as changes in site use.

2.2.3. Case Studies

Large, engineered masonry and earthen structures provide ideal piercing features with which to assess timing and amount of fault displacement, sense of slip, distribution of shear, and other important parameters of interest to paleoseismologists and to structural geologists. The Great Wall of China, one of the world's largest engineered structures, crosses a number of active faults, one of which ruptured during the 1739 A.D. Yiuchuan–Pingluo earthquake (Zhang et al., 1986). This event occurred on the Honggouzigou fault that offset the Great Wall by 2.7 m vertically and 3 m right laterally (Zhang et al., 1986). Such precise and accurate measures of displacement on the structure is one of the important and nearly unique contributions of faulted architecture to paleoseismology.

The Crusader castle Vadum Jacob was constructed on the Dead Sea Transform, along the Jordan River in modern Israel, and soon after was offset by an earthquake at dawn, May 20, 1202 (Ellenblum et al., 1998). Masonry castle walls cross the fault at a right angle, presenting ideal cultural piercing features. The trueness of the skillfully constructed wall is unequaled by any of the natural piercing features, (e.g., stream channel). Archaeological excavation of the walls provided the exposure with which to measure with high confidence its displacement (Fig. 2 of Ellenblum). Of the 2.1 m of left-lateral offset, 1.6 m is tied directly to the event in 1202. Earthquakes of 1759 A.D. and/or 1837 added another 0.5m to the total offset of the wall. This study is a good example of the type of results that a coordinated interdisciplinary archaeoseismological study can provide.

Earthquakes on the northern San Andreas fault at Fort Ross offset archaeological and colluvial deposits (Noller et al., 1992, 1993, 1994; Simpson et al., 1997). The last event at this site, the great San Francisco earthquake of 1906, produced 3.7m of offset on a nearby fence. Three cultural piercing features (Fig. 6.3) were used to measure slip rate on the San Andreas fault and include the northern and southern boundaries of the Emergent (500–1812 A.D.) archaeological deposit (Features 1 and 2, respectively) and the loci of Middle to Upper Archaic (3000B.C.–500A.D.) artifacts within the northern part of the midden deposit (Feature 3). Features 1 and 2 are offset by 22.9 ± 2.6 m and 25.6 ± 1.6 m, respectively. The location of Feature 3 is best constrained with an offset of 26 ± 4m. The Archaic deposit could be offset by as much as 6 m more than the Emergent deposit, given the uncertainty in their locations, implying that one to two 1906-size ruptures occurred between the deposition of the two units. This would be an example of association (C) (Fig. 6.2). The amount of offset of the cultural piercing features is consistent with that estimated using recurrence intervals at other sites on the San Andreas fault (Noller and Lightfoot, 1997).

2.3. When's the Next Earthquake?

2.3.1. Guessing the Future Based on the Past

Having some sense of when and how often earthquakes have occurred in the past is valuable information for assessing seismic risk in the future (Wyss and Dmowska, 1997). Predicting earthquakes remains an elusive quest, with hope for new methods ever on the horizon (cf. Geller, 1996; Varotsos et al., 1996). Much of our understanding of the history of humans and earthquakes comes from ancient sources and from archaeological investigations. The former is fraught with problems of communicating exactly what happened as we can get only snippets of an event that may, or in some cases, may not have occurred (Lee et al., 1988; Ambraseys, 1996). Yet resourceful catalogs of events can be assembled (Toppozada et al., 1981; Mart and Perecman, 1996). The five volume "Compilation of Historical Materials of Chinese Earthquakes" stands out among these catalogs because it hosts an impressive record dating from the 23rd century B.C. to A.D. 1980 (from Lin and Wu, 1993).

There is as yet no unifying concept or theory for fault behavior and the occurrence of earthquakes (Yeats et al., 1996). There are concepts that seem to work for some faults and not others. One concept that has seen widespread application is that of the *characteristic earthquake* (Schwartz and Coppersmith,

Archaeoseismology

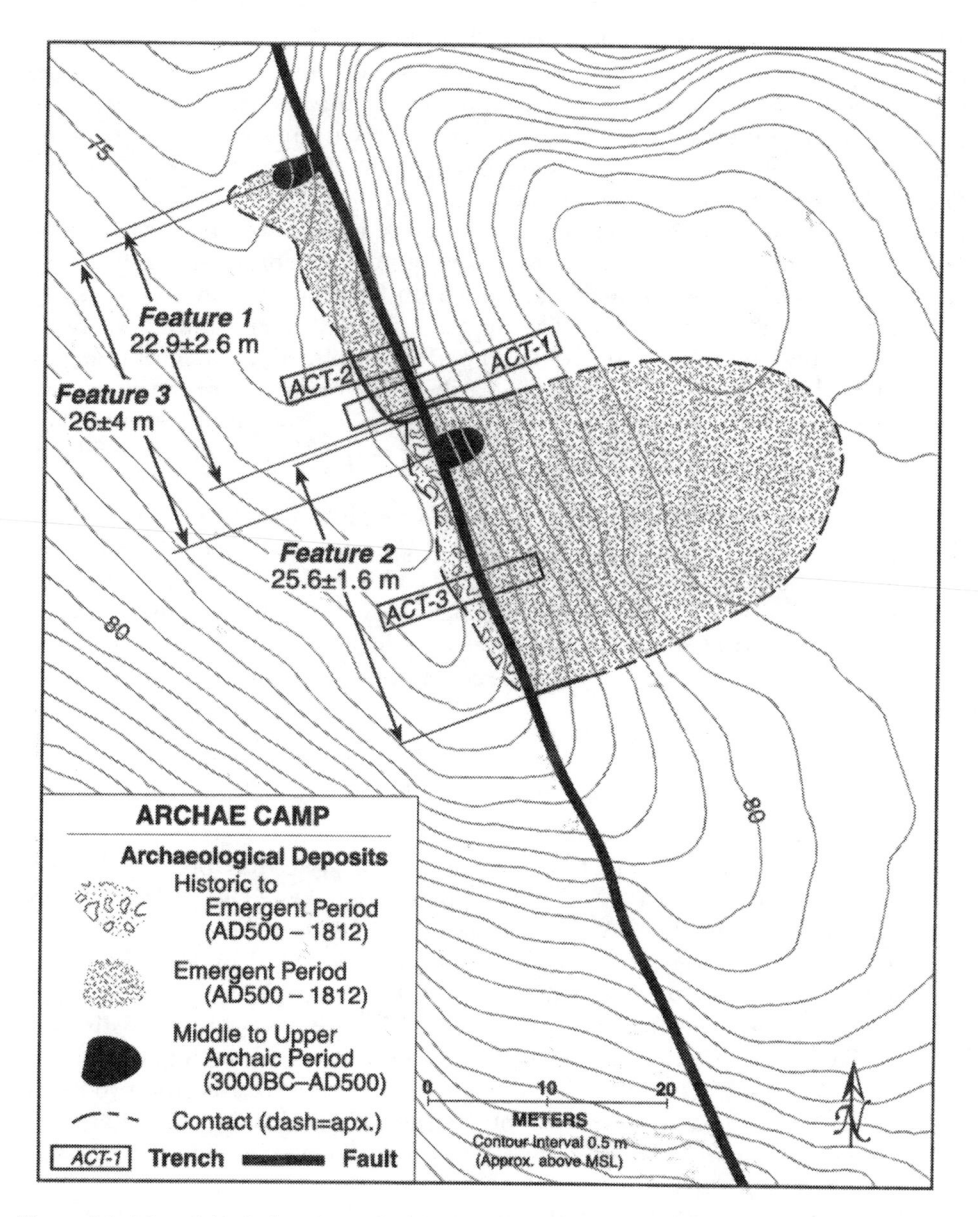

Figure 6.3. Map of the Archae Camp site showing distribution of archaeological deposits and trench locations. The Emergent and Archaic deposits are displaced along the San Andreas fault. Emergent and Archaic deposits west of the fault and northwest of trench ACT-3 are buried beneath fault-scarp-derived colluvium. A less than 25-cm-thick surficial deposit containing Historic to Emergent period materials drapes from the main site across the fault and into the swale. Paleoseismic trenches were excavated across the prominent southwest-facing fault scarp.

1984), in which large-magnitude earthquakes occur with same release of energy at a regular frequency. Another, and not necessarily competing, model is the *clustered earthquake* concept in which a number earthquakes occur followed by a long period of quiesence on the fault. Estimates of recurrence interval depend on the length and resolution of the record of past events. Taking the record on the Hierapolis fault, Turkey (Hancock and Altunel, 1997), as an example, the two events occurred within a period of about a thousand years, or on average every 500 years. If one uses the characteristic earthquake concept as their basis then this would seem to be a reasonable approach. In fact, most current estimates of earthquake recurrence and, by extension, predictions of when future events will occur are based on such short-term, imperfectly defined event chronologies (McCalpin, 1996). Using the clustered earthquake concept, some workers will argue that there is the strong possibility that, despite a geologically recent series of earthquakes, a fault may have entered a period of quiesence and hence will not quake again for another 100,000 or so years.

The historic record of earthquakes, such as that in China (from Lin and Wu, 1993), typically does not provide enough detailed information on earthquakes for recurrence to be calculated for individual faults. Linking historic felt and damage records to the causative fault requires further study in the field. However, such databases of historical seismicity are invaluable in determining the recurrence interval for damaging earthquakes in a region and as the basis for probabilistic seismic hazard studies.

2.3.2. Case Studies

The long use or occupation of tectonically active areas has presented ideal situations for revealing the long history of earthquakes. The chronology of paleoseismic events in the New Madrid Seismic Zone, central United States, is well supported by a sequence of liquefaction sand beds and prehistoric Native American occupation horizons (Saucier, 1991; Tuttle and Schweig, 1995, 1996; Tuttle et al., 1996). The abundance of lignite coal in deposits of the Mississippi River make radiocarbon dating of identical appearing charcoal highly suspect. The seriation of artifacts, although not completely removed from the problems associated with charcoal, provides a robust means of establishing ages of archaeological deposits and hence the liquefaction (earthquake) events that disrupt them (Tuttle et al. 1999b).

Sherds (pottery pieces) provide late Holocene age constraints on a thrice-faulted section of sediments at the Shama Gully sites along the southern border fault of the Weihe Graben, China (Zhang et al., 1995). Artifacts were found throughout the approximately 8 m section and thus bracket the surface fault ruptures at this site. The ages on two early events were established by ^{14}C and pottery seriation. Artifacts were collected from the base of depositional units, yet curiously not from the upper part of the units. The ages of the Weihe Graben earthquakes would be bolstered by collecting artifacts and radiocarbon samples from immediately below each event (stratigraphic) boundary. These samples would provide maximum constraints on the earthquake age estimates.

3. Case Study: Offset of the Seal Cove Archaeological Site by the San Gregorio Fault

3.1. Introduction

Archaeology was used to assess the recency of fault activity and fault slip on the San Gregorio fault at Seal Cove, California (Fig. 6.4; Noller et al., 1995; Simpson et al., 1997). The geological goal of this study was to obtain direct evidence of the date of the most recent earthquake, age estimates for earlier earthquakes, sense of fault displacement, and slip rate. The archaeological goal was to reveal the function and use of the site and its relation to local and regional trade and social

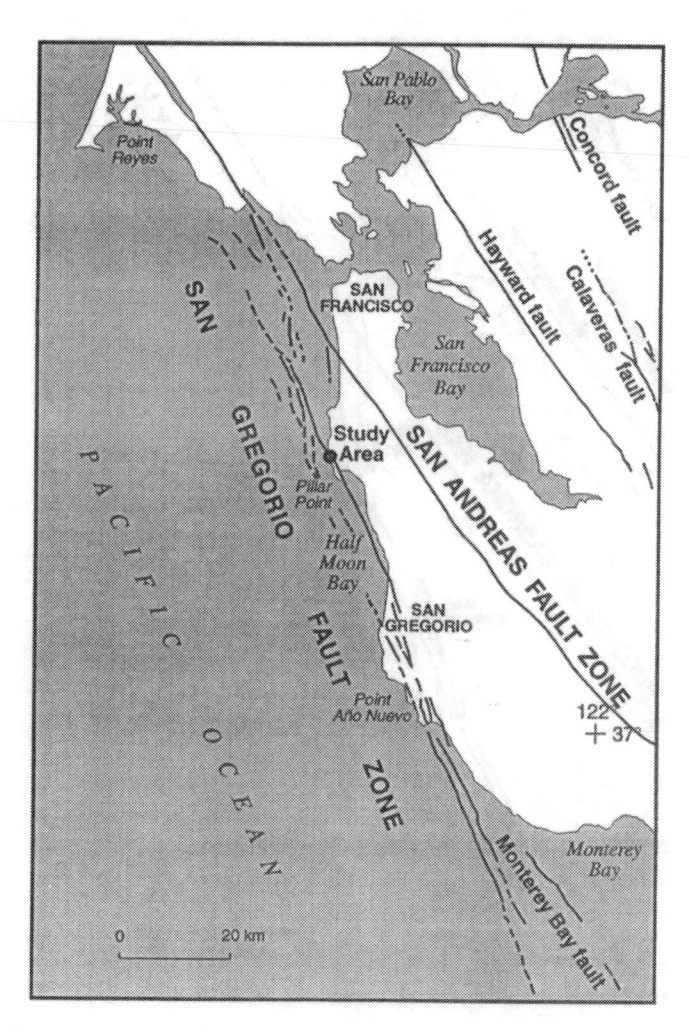

Figure 6.4. The San Gregorio fault lies offshore the California coast for much of its length, coming onshore at Seal Cove, about 20 km south of San Francisco.

organization. Most of these goals were met, providing valuable data for estimating earthquake parameters for the San Gregorio fault, specifically, and for seismic hazards assessments of central coastal California, in general.

The San Gregorio fault is westernmost of a series of faults that subparallel the San Andreas fault in central California. As such it is one of the largest potentially significant seismic source in the San Francisco and Monterey Bay areas. Compared to other faults in the San Andreas system, the San Gregorio fault has received little attention in the way of detailed paleoseismic studies (Weber and Lajoie, 1979) because of the offshore location of much of the fault and the paucity of suitable onshore sites with geologic piercing features (Weber et al., 1995).

The San Gregorio fault is part of a system of laterally continuous, late Pleistocene and Holocene active dextral slip faults, including (from south to north) the Hosgri, San Simeon, and Sur faults. The fault is geomorphically well expressed by features such as sag ponds, linear streams, fault scarps, and offset Pleistocene marine terraces. However, there is no record of surface rupture on this fault since the Spanish first occupied this part of the coast in A.D. 1770 (Monterey) to 1775 (San Francisco Santa María, 1775). Two historic earthquakes of moderate magnitude or greater have occurred that are attributed to the San Gregorio fault: an M $5\frac{1}{4}$ to $5\frac{1}{2}$ event near Pillar Point in 1856 and an M $5\frac{3}{4}$ to 6 event southeast of Point Año Nuevo in 1884 (Toppozada et al., 1981).

3.2. Approach: Identify an Archaeological Site on a Fault

Suitable geological sites were lacking for paleoseismic study of the San Gregorio fault. Having met recent success in locating and studying an archaeological site astride the San Andreas fault (Noller et al., 1993), I thought that the similar location of the San Gregorio fault along the Pacific coast might just yield a suitable archaeological site for investigation. The Seal Cove site was selected by overlaying a map of known archaeological sites in San Mateo County on a geological map showing active traces of the San Gregorio fault. Bill Lettis and I then went out to the site where, despite being a few hundred meters off the officially mapped trace of the fault, we found the site (California Registered Site CA-SMA-134) on a degraded scarp and bordering a sag pond.

The approach used in this study involves detailed mapping of the surface and subsurface distribution of archaeological features, concentrating on those features that are linear or have well-defined margins and therefore can provide strain gauges for evaluating fault displacement (Noller and Lightfoot, 1997). The distribution of ethnostratigraphic units and distinct archaeological features is determined by surface mapping and by a program of subsurface mapping involving shovel probes, hand-auger holes, test pits, and trenches. Cultural artifacts and other site materials also provide valuable information for assessing the age of stratigraphic units.

3.3. Methods: Excavate and Date

Field work consisted of preparing detailed geomorphic and archaeological maps of the Seal Cove site and vicinity, complemented by subsurface exploration consisting of shovel-probe and hand-auger surveys, archaeological test pits, and exploratory trenching (Fig. 6.5). Initially, several dozen random shovel probes were excavated to define the general limits of the archaeological deposits. Following the shovel-probe survey, 145 shallow (1 to 3 m) hand-auger holes were excavated over a 5 to 1.25m grid spacing to better define the margin of the site (Fig. 6.4). Samples were collected at 10 to 20 cm intervals until refusal. Test pits were hand excavated in the archaeological deposit to reveal stratigraphy and materials for dating (Fig. 6.5). The test pits were excavated along two transects

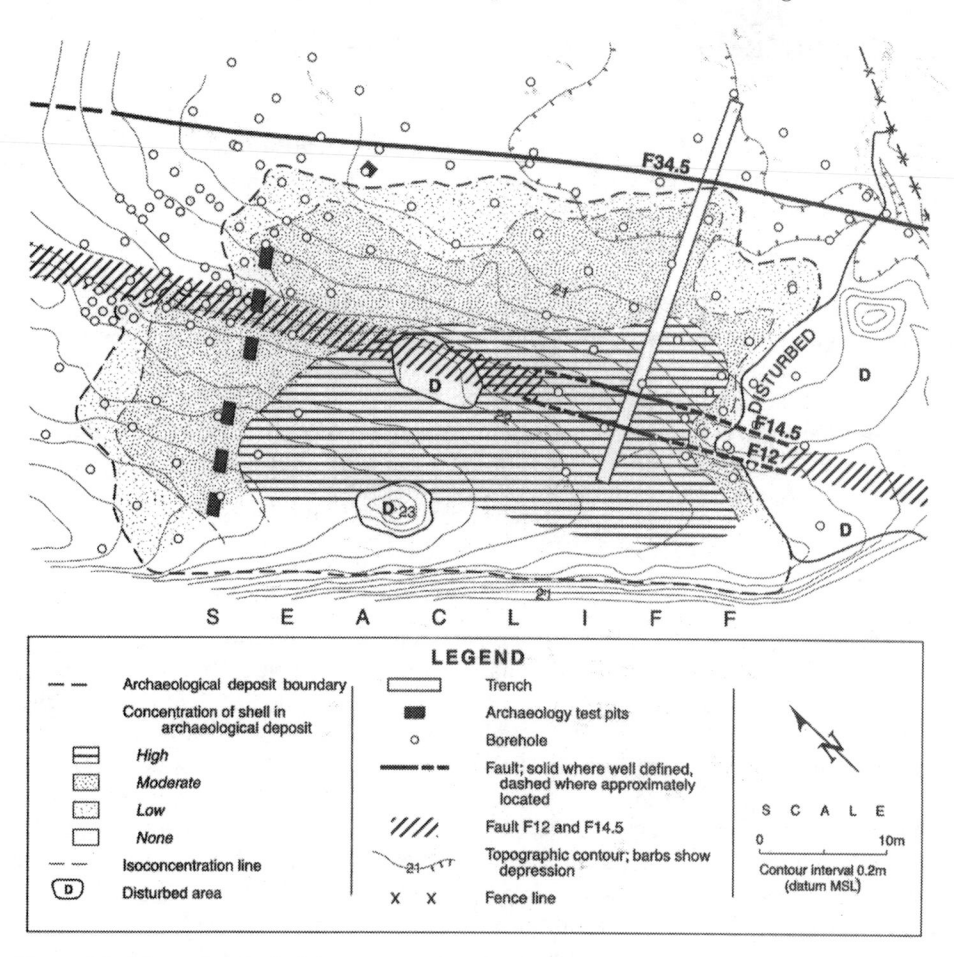

Figure 6.5. The archaeological deposit at Seal Cove is transected by three traces of the San Gregorio fault that underlie a northeast-facing scarp and bound a sag pond.

(northern and southern) across the archaeological deposit orthogonal to the fault. The three-dimensional position of each fire-cracked rock (FCR), bone, and artifact were mapped in the test units. A paleoseismic trench (1 m × 3 m × 36 m) was excavated and described along the southern transect of test pits (Fig. 6.5).

In addition to charcoal, we selected large (>10 cm) mussel shells (*Mytilus californianus*) from the archaeological deposit for radiocarbon dating (Noller et al., 1995). The large shells should maintain original stratigraphic context even if bioturbation (unrecognized) occurred, and thus their context would be better than charcoal sampled from the matrix of the archaeological deposit. Obsidian artifacts were analyzed for source, use, amount of wear and reuse, and measurement of hydration band width (Hylkema et al., 1995).

3.4. Results: Reading between the Fault Lines

The San Gregorio fault extends onshore near Seal Cove, California, where it transects an extensive, prehistoric archaeological deposit (CA-SMA-134) from Native Californian occupation of the site (Fig. 6.4). At the study site, the fault forms a distinct 1.5- to 3-meter-high east-facing scarp along the western margin of a closed depression (Fig. 6.4).

The archaeological deposit at Seal Cove covers 2,500 m², with a circumference of about 200 m around a roughly ellipsoidal boundary (Fig. 6.5). The deposit is about 1 m thick in the center and gradually thins towards its boundary. Where the northern part of the deposit crosses the fault, the boundary has an irregular shape (Fig. 6.5). This part of the deposit was chosen for further study. The southern part of the deposit was historically removed for earthfill, and thus was unsuitable for our study. An unknown amount of the western part of the deposit was removed by retreat of the sea cliff.

3.4.1. Ethnostratigraphic Units

Excavation of the archaeological deposit yielded a voluminous assemblage of stratified prehistoric Native Californian artifacts and refuse that provide a late period date of A.D. 1270–1400 (Table 6.3; Hylkema et al., 1995). Archaeological excavations, sea-cliff exposures, and paleoseismic trenching provide three dimensional control on the distribution of the units and included artifacts throughout much of the deposit. Four facies or ethnostratigraphic units (units A to D) were defined within the archaeological deposit on the basis of concentrations and depositional fabrics of shell, bone, stone handtools, and FCR in matrix materials consisting of black, organic sandy silt (Fig. 6.6). Unit A is shell and artifact poor and overlies an irregular basal and presumably anthropic surface with the underlying sterile soil and host deposits. Unit B is locally shelly, with many of the *Mytilus* fragments ventral side downward as they might have been deposited. Unit C contains a lens of abundant FCR, shell, bone, and stone tools that define a relic cooking hearth at least 7 m × 1 m in size. The base of unit D, a diffuse stone and bone line, is gradational with unit C.

Archaeoseismology

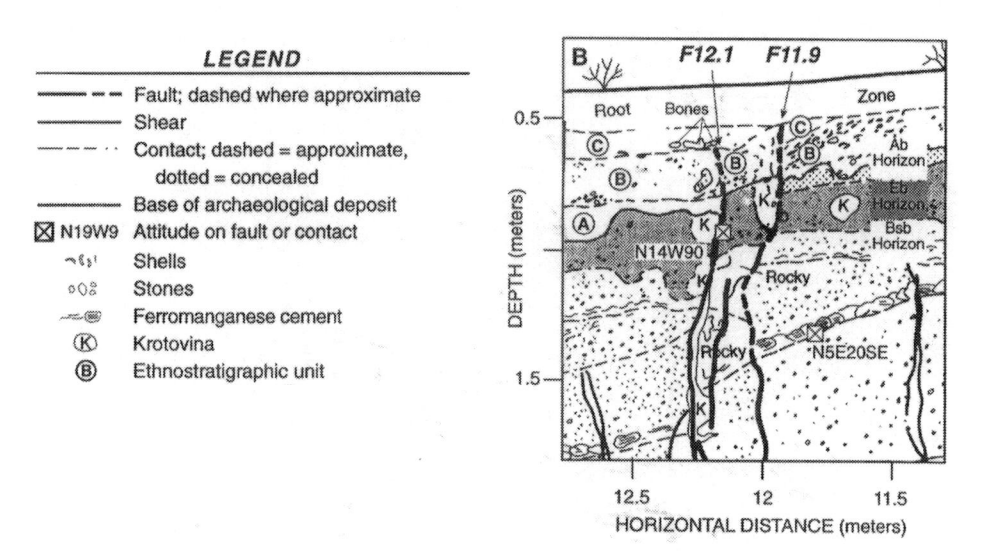

Figure 6.6. Ethnostratigraphic units and their relationship to strand 12.1 of the Seal Cove fault are shown in the south wall of the paleoseismic trench.

The upper 10 to 15 cm of site materials consist of black, organic-rich shelly sediment with few FCR and stone hand tools. Historic artifacts are common in this unit. This surface unit is bound by abundant roots that make calls on stratigraphic boundaries and faults uncertain. Charcoal and shell radiocarbon ages show that the deposit formed quickly. One charcoal fragment and one mussel shell from the core of the hearth feature were submitted for radiocarbon analysis (Noller et al., 1995). The age difference between that of coeval charcoal and shell is about 680 years and is within the range of reservoir ages (∂R) (100–800 yr BP) for marine mussel (*M. californianus*) shell along this coast (Berger et al., 1966; Hylkema, 1991; Robinson and Thompson, 1981). Correcting all shell ages using a ∂R value is not necessary because of the close correspondence of the charcoal-paired shell age with that of the other shells. Hence, the charcoal age is used as the age for all shell.

The archaeological deposit at Seal Cove is coherent in character, artifact assemblage, and age throughout. The uniformity of character and artifact assemblage, limited extent and depth of the deposit, and apparent single use of the site suggest many short periods of occupation. Nearly all of the bones within the deposit are from migrating animals (Table 6.3). On the basis of their historic migratory patterns, the temporal coincidence of these species in the Seal Cove region is April, plus or minus one month. Calibrated radiocarbon ages (2σ, calendric) overlap between A.D. 1270 to 1400, within the range of A.D. 960 to 1660. Given the volume and age range of the deposit it is permissible to suggest that Native Californians deposited the site materials within a period of one generation (20 years) of spring hunts.

Table 6.3. Inventory of Recovered Artifacts at the Seal Cove Site, Ca (CA-SMA-134)

Item	Number of specimen (or pieces)
Fire-cracked rock	2,028
Faunal bone	1,575
Bone tools	6
Pitted stones	152
Handstone/mano	3
Milling slab/metate	4
Incised cobble	2
Hand axe	2
Knapped cobbles	2
Obsidian debitage	11
Non-obsidian debitage	224
Obsidian points/bifaces	6
Chipped stone crescent	2
Olivella bead	1
Dietary shell	N/A

3.4.2. Faults in the Archaeological Deposit

It is vitally important that the discussion of the evidence of rupture of the surface by a fault (surface fault rupture) conform to the modes of description used by paleoseismologists (McCalpin, 1996). In this way, the results are of more immediate import to seismic hazard studies. The following description of stratigraphy and first-order interpretation should serve as an example of the discussion of a tectonically disturbed archaeological site. Other examples include Noller and Lightfoot (1997) and Ellenblum et al. (1998).

Faulted archaeological deposits were observed in the exploratory trench at Seal Cove (ethnostratigraphic units A, B, & C; Fig. 6.6). The northeast–southwest-trending trench crosses a 30-meter-wide zone of Holocene-active dextral faults that strike N15–20W. Three of the five Holocene fault strands in the trench (F4.5, F2.1, and F11.9) can be traced upward to or into the archaeological deposit.

The fault strand at meter 12.1 (F12.1 in Fig. 6.6) extends up to the base of, or possibly through, unit C. Beds of the Pleistocene sediments and the base of the overlying soil Eb horizon are vertically displaced by about 15 cm. Unit A is juxtaposed against the buried A/E/Bs soil profiles and unit B steps down to the east by less than 5 cm. Fault strand F12.1 is weakly expressed in unit B by rotated shell and juxtaposition of shell-rich (west) and shell-poor (east) facies, and is not evident in unit C. The upward termination of F12.1 is at the top of unit B, which is overlain by a floor level with unrotated bones. Because of the decrease in vertical throw going up section, more than one event on this strand may be inferred. It is plausible that this relationship is due to the penultimate event.

On the basis of these stratigraphic relations, one and possibly a second older event are inferred for faults transecting the archaeological deposit. The most recent surface-rupturing earthquake is identified by rotated clasts and changes in shell density in units B and C across F11.9. This earthquake is younger than unit C, which is dated to A.D. 1270–1400, yet must be older than the time of European contact and record-keeping A.D. 1770–1775.

3.4.3. Using the Archaeological Deposit to Determine Fault Slip

The northern boundary of the archaeological deposit is offset in a right lateral sense along the western fault zone (Fig. 6.5) and is used to constrain slip during the most recent, and possibly the penultimate, earthquake. The boundary of the archaeological deposit is diffuse and hence difficult to draw between areas of "low" and "no" shell detritus. The zone of low shell concentration is 2 to 5 m wide along the perimeter of the deposit, except along the fault scarp where the zone is less than 2 m wide (Fig. 6.5). The boundary of the archaeological deposit appears to have a cumulative offset of 8 to 9.5 m across the western fault zone (strands F14.5 and F12) (Fig. 6.5).

Confidence in the use of the northern site boundary is dependent on two key assumptions. First, it is assumed that the offset boundary was not deposited in its present shape by its occupants. Second, it is assumed that this boundary has undergone little or no postdepositional disturbance (e.g., erosion) since the time of fault offset. Disturbance of the northern boundary is ruled out because such disturbance is detectable by the methods employed.

3.5. Implications of Results from Seal Cove

Traces of the San Gregorio fault are identifiable in a 1270 to 1400 A.D. archaeological deposit at Seal Cove. Offset ethnostratigraphy in the deposit provides the basis for estimating the timing of events on this historically quiescent tectonic structure. The sense and magnitude of throw along these traces during the late Holocene are consistent with right lateral slip on the San Gregorio fault.

One and possibly two late Holocene earthquakes on the San Gregorio fault are evidenced in the archaeological deposit at Seal Cove. The most recent earthquake is constrained to between A.D. 1270 and 1775, and it quite possibly occurred while the site was occupied during a spring hunting season prior to A.D. 1400. This and possibly an earlier event produced a (cumulative) maximum right lateral offset of 8 to 10.5 m across the western traces of the San Gregorio fault. Unless more convincing evidence is found of previous displacement of the fault traces, a slip rate cannot be determined based solely on this offset archaeological deposit. Estimates of the timing, sense of throw, and amount of slip for the most recent earthquake were elusive at other sites (Weber and Lajoie, 1979; Weber et al., 1995), and thus there is no comparison for the results from Seal Cove.

These results demonstrate that archaeology can be used to complement standard geology-based paleoseismic studies. Clearly, after repeated unsuccessful attempts at a purely geologic site, for example, one in which stream channels are offset, it was the excavation of an archaeological site and its associated geologic context that provide the best and most complete results to date. The offset hearth was not recognized as such by the geological phase. Also, the block excavation of the exploratory trench did not fully reveal the number and amounts of offset on the other fault strands. Hence, the results also show that the wedding of the two disciplines and their methods are fully complementary: the unique view of one enhances the unique view of the other and vice versa.

4. Closing

Use of archaeology in studies of earthquake history is in its initial phase of development. The fields of paleoseismology and archaeoseismology are in the early stage of definition, as is demonstrated by the recent appearance of textbooks and reviews on these subjects (e.g., Stiros and Jones, 1996; Yeats et al., 1996; McCalpin, 1996). Archaeological evidence and sources have long been used to develop and refine the record of historical seismology (Ambraseys, 1996). The use of archaeology to characterize paleoearthquakes, including age and amount and direction of movement on a fault, began more as a function of describing oddities in archaeological excavations as well as explaining the accidental finds of cultural materials in geologic excavations. Not until the 1990s do we see a concerted effort to conduct truly interdisciplinary studies that provide a wealth of information on paleoearthquakes and fault behavior, as well as the significant contributions to archaeology. The breadth of archaeology-related studies covers the range of tectonic settings, types of active faults, and primary and secondary earthquake effects. The excavation of the Seal Cove site is one example of this new breed of archaeoseismological study. Archaeoseismology can answer the questions basic to seismology and paleoseismology: When did the earthquake occur? What did the earthquake do? Where did the earthquake occur? How big was the earthquake? When is the next earthquake? Archaeoseismologic studies provide additional sites for earthquake research as anthropogenic landscape change is altogether quite different from that due to other surficial processes. Finally, archaeoseismology, unlike purely geologic or geophysical studies, directly relates past human interaction with tectonic forces to that of present and future generations.

5. ACKNOWLEDGMENTS. Research at Seal Cove was supported by the U.S. Geological Survey, Department of the Interior, NEHRP Award 1434-94-G-2275 to the author. The author wishes to thank colleagues Gary Simpson, Stephen Thompson, Mark Hylkema, and Bill Lettis, who worked on the Seal Cove site. Certainly not least, my inspiration for this study is owed to years of discussion with Lisa Wells on the use of archaeology in geologic studies.

6. References

Aitken, M. J., 1990, *Science-based Dating in Archaeology*, Longman, New York.
Ambraseys, N. N., 1996, Material for the Investigation of the Seismicity of Central Greece, in *Archaeoseismology*, edited by S. C. Stiros and R. E. Jones, British School at Athens Fitch Laboratory Occasional Paper 7, pp. 23–36.
Anderson, W. A., Kelley, J. T., Thompson, W. B., Borns, H. W., Jr., Sanger, D., Smith, D. C., Tyler, D. A., Anderson, R. S., Bridges, A. E., Crossen, K. J., Ladd, J. W., Andersen, B. G., and Lee, F. T., 1984, Crustal Warping in Maine. *Geology* 12:677–680.
Armijo, R., Lyon-Caen, H., and Papanastassiou, D., 1991, A Possible Normal-Fault Rupture for the 464 B.C. Sparta Earthquake. *Nature* 351:137–139.
Berger, R., Taylor, R. E., and Libby, W. F., 1966, Radiocarbon Content of Marine Shells from the California and Mexican West Coast. *Science* 153:864–866.
Broster, B. E., Allen, P., and Burke, K. B. S., 1993, The Assessment of Postglacial Neotectonics; An Example of a Multidisciplinary Approach Involving Archaeology, Sedimentology and Historical Seismicity, Miramichi Area, New Brunswick. *Biennial Meeting of the Canadian Quaternary Association*, A8.
Caputo, R., Bravard, J.-P., and Helly, B., 1994, The Pliocene–Quaternary Tecto-Sedimentary Evolution of the Larissa Plain (Eastern Thessaly, Greece). *Geodinamica Acta (Paris)* 7:219–231.
Demitrack, A., 1986, The Late Quaternary Geologic History of the Larissa Plain, Thessaly: Tectonic, Climatic, and Human Impact on the Landscape. Ph.D. diss., Stanford University. University Microfilms, Ann Arbor.
Ellenblum, R., Marco, S., Agnon, A., Rockwell, T., and Boas, A., 1998, Crusader Castle Torn Apart by Earthquake at Dawn, 20 May 1202. *Geology* 26:303–306.
Flemming, N. C., and Webb, C. O., 1986, Tectonic and Eustatic Changes During the Last 10,000 Years Derived from Archaeological Data. *Zeitschrift für Geomorphologie* 62:1–29.
Galadini, F., and Galli, P., 1999, *Faulting of a Roman Building in Northern Italy: Echo of an Ancient Earthquake from a Silent Area*, U.S. Geological Survey Paleoseismology Page http://geo.cr.usgs.gov/paleoseis/PPMain/Gall:FEM.
Ganas, A., and Buck, V. A., 1998, A Model for Tectonic Subsidence of the Allai Archaeological Site, Lokris, Central Greece. *Bulletin of the Geological Society of Greece* 32:181–187.
Geller, R. J., 1996, Earthquake Prediction; A Critical Review. *Geophysical Journal International* 131:425–450.
Grant, W. C., and Minor, R., 1991, Paleoseismic Evidence and Prehistoric Occupation Associated with Late Holocene Sudden Submergence, Northern Oregon Coast. *Eos, Transactions, American Geophysical Union* 72:313.
Grant, L. B., and Sieh, K., 1994, Paleoseismic Evidence of Clustered Earthquakes on the San Andreas Fault in the Carrizo Plain, California. *Journal of Geophysical Research* 99B4:6819–6841.
Guidoboni, E., 1996, Archaeology and Historical Seismology; the Need for Collaboration in the Mediterranean Area. In *Archaeoseismology*. edited by S. C. Stiros and R. E. Jones, British School at Athens Fitch Laboratory Occasional Paper 7:129–152.
Hancock, P. L., and Altunel, E., 1997, Faulted Archaeological Relics at Hierapolis (Pamukkale), Turkey. *Journal of Geodynamics* 24:21–36.
Harris, E. C., 1989, *Principles of Archaeological Stratigraphy*. Academic Press, New York.
Hylkema, M. G., 1991, Prehistoric Native American Adaptations along the Central California Coast of San Mateo and Santa Cruz Counties. M. A. thesis, San Jose State University, San Jose, CA.
Hylkema, M. G., Noller, J. S., and Thompson, S. C., 1995, The Archaeological Excavation of Prehistoric Site CA-SMA-134, Fitzgerald Marine Reserve, San Mateo County, California: *Final Technical Reports of the National Earthquake Hazards Reduction Program*, #1434-92G-2220, Reston, VA, U. S. Geological Survey, pp. 45–72.
Karcz, I., and Kafri, U., 1978, Evaluation of Supposed Archaeoseismic Damage in Israel. *Journal of Archaeological Science* 5:237–253.
Karcz, I., and Kafri, U., 1981, Studies in Archaeoseismicity of Israel: Hisham's Palace, Jericho. *Israel Journal of Earth Sciences* 30:12–23.
Kelson, K. I., Simpson, G. D., VanArsdale, R. B., Haraden, C. C., and Lettis, W. R., 1996, Multiple Late Holocene Earthquakes Along the Reelfoot Fault, Central New Madrid Seismic Zone. *Journal of Geophysical Research* 101B:6151–6170.

King, G., Bailey, G., and Sturdy, D., 1994, Active Tectonics and Human Survival Strategies. *Journal of Geophysical Research* 99B:20,063–20,078.

Lee, W. H. K., Myers, H., and Shimazaki, K. (eds.), 1988, *Historical Seismograms and Earthquakes of the World*, Academic Press, New York.

Lettis, W. R., Wells, D. L., and Baldwin, J. N., 1997, Empirical Observations Regarding Reverse Earthquakes, Blind Thrust Faults, and Quaternary Deformation: Are Blind Thrust Faults Truly Blind? *Bulletin of the Seismological Society of America* 87:1171–1198.

Lin, R., and Wu, Y., 1993, An Overview of Archaeo-Seismological Studies in China. In *Geophysical Exploration of Archaeological Sites. Theory and Practice of Applied Geophysics*, edited by A. Vogel and G. N. Tsokas, pp. 181–191. Vieweg, Bremen.

Marco, S., Agnon, A., Ellenblum, R., Eidelman, A., Basson, U., and Boas, A., 1997, 817-Year-Old Walls Offset Sinistrally 2.1 m by the Dead Sea Transform, Israel, *Journal of Geodynamics* 24:11–20.

Mart, Y., and Perecman, I., 1996, Neotectonic Activity in Caesarea, the Mediterranean Coast of Central Israel. *Tectonophysics* 254:139–153.

McCalpin, J. P., 1996, *Paleoseismology*, Academic Press, New York.

Michetti, A. M., Brunamonts, F., Serva, L., and Vittori, E., 1996, Trench Investigations of the 1915 Fucino Earthquake Fault Scarps (Abruzzo, central Italy): Geological Evidence of Large Historical Events. *Journal of Geophysical Research* 101B:5921–5936.

Morrison, R. B., 1991, Introduction, in *Quaternary nonglacial geology: Conterminous U. S. The Geology of North America, Volume K-2*. edited by R. B. Morrison, pp. 1–12 Geological Society of America. Boulder, CO.

Munson, C. A., Munson, P. J., and Garniewicz, R. C., 1992, Geoarchaeological Dating of Holocene-Age, Earthquake-Induced Paleoliquefaction Features in Southwestern Indiana. *Abstracts with Programs, Geological Society of America Annual Meeting* 24:162.

Nikonov, A. A., 1995, The Stratigraphic Method in the Study of Large Past Earthquakes. *Quaternary International* 25:47–55.

Noller, J. S., and Forman, S. L., 1998, Luminescence Geochronology of Liquefaction Features Near Georgetown, South Carolina, in *Dating and Earthquakes: Review of Quaternary Geochronology and its Application to Paleoseismology* edited by J. M. Sowers, J. S. Noller, and W. R. Lettis, pp. 4-49–4-57. U. S. Nuclear Regulatory Commission, NUREG/CR 5562. Washington, DC.

Noller, J. S., and Lightfoot, K., 1997, An Archaeoseismic Approach and Method for the Study of Active Strike-Slip Faults. *Geoarchaeology* 12:117–135.

Noller, J. S., and Simpson, G. D., Late Holocene Record of Large-Magnitude Earthquakes on the Northern San Andreas Fault, at a Prehistoric Native Californian village, Archae Camp, Fort Ross, California. *Journal of Geophysical Research*, in preparation.

Noller, J. S., Lettis, W. R., Kelson, K. I., and Lightfoot, K., 1992, Holocene Activity of the Northern San Andreas Fault as Determined by Paleoseismic and Archaeologic Methods. *EOS, Transactions, American Geophysical Union* 73:590.

Noller, J. S., Lightfoot, K. Wickens, K. A., Kelson, K. I., Wake, T., and Parkman, E. B., 1993, Preliminary Results of Geoarchaeological Investigations along the Northern San Andreas Fault Zone, Fort Ross State Historic Park, California. *Proceedings of the Society for California Archaeology* 6:249-256.

Noller, J. S., Lettis, W. R., and Simpson, G. D., 1994, Seismic Archaeology: Using Human Prehistory to Date Paleoearthquakes and Assess Deformation Rates of Active Fault Zones. *U.S. Geological Survey Open-File Report* 94–568:138–140.

Noller, J. S., Sowers, J. M., and Lettis, W. R., eds., 2000, *Quaternary Geochronology: Methods and Applications*, American Geophysical Union Reference Shelf Series 4, Washington, DC.

Noller, J. S., Thompson, S. C., Hylkema, M. G., Simpson, G. D., and Lettis, W. R., 1995, Use of Archaeology to Assess Holocene Activity on the San Gregorio Fault, Seal Cove, California. *Final Technical Reports of the National Earthquake Hazards Reduction Program*, #1434-92G-2220, U.S. Geological Survey, Reston, VA.

Noller, J. S., Wells, L. E., Reinhardt, and E., Rothaus, R. M., 1997, Subsidence of the Harbor at Kenchreai, Saronic Gulf, Greece, During the Earthquakes of A.D. 400 and A.D. 1928. *EOS, Transactions, American Geophysical Union* 78:636.

Noller, J. S., Wells, L. E., Reinhardt, E., and Rothaus, R., *Archaeoseismologic Evidence of Recent Episodic Subsidence at Kenchreai, Greece*, manuscript submitted for publication.

Nur, A. M., 1998, The Catastrophic End of the Bronze Age: Large Earthquakes of Sea People? *American Geophysical Union* 78:636.

Obermeier, S. F., Jacobson, R. B., Smoot, J. P., Weems, R. E., Gohn, G. S., Monroe, J. E., and Powars, D. S., 1990, *Earthquake-Induced Liquefaction Features in the Coastal Setting of South Carolina and in the Fluvial Setting of the New Madrid Zone*. U.S. Geological Survey Professional Paper 1504. Washington, DC.

Papageorgiou, S., and Stiros, S. C., 1996, The Harbour of Aigeira (North Peloponnese, Greece); an Uplifted Ancient Harbour. In *Archaeoseismology*. edited by S. C. Stiros and R. E. Jones, 211–214. British School at Athens Fitch Laboratory Occasional Paper 7.

Pavlides, S. B., 1996, Paleoseismology: A Branch of Neotectonics Linking Geological, Seismological and Archaeological Data—An Introduction. In *Archaeoseismology*, edited by S. Stiros and R. E. Jones. pp. 15–19. British School at Athens Fitch Laboratory Occasional Paper.

Pirazzoli, P. A., Stiros, S. C., Arnold, M., Laborel, J., Laborel-Deguen, F., and Papageorgiou, S., 1994, Episodic Uplift Deduced from Holocene Shorelines in the Perachora Peninsula, Corinth Area, Greece. *Tectonophysics* 229:201–209.

Rapp, G., Jr., 1986, Assessing Archaeological Evidence for Seismic Catastrophes. *Geoarchaeology* 1:365–379.

Reches, Z., and Hoexter, D. F., 1981, Holocene Seismic and Tectonic Activity in the Dead Sea Area. *Tectonophysics* 80:235–254.

Reinhardt, E. G., Easton, N. A., and Patterson, T. R., 1996, Foraminiferal Evidence of Late Holocene Sea-Level Change and Amerindian Site Distribution at Montague Harbour, British Columbia. *Geographie Physique et Quaternaire* 50:35–46.

Robinson, S. W., and Thompson, G., 1981, Radiocarbon Corrections for Marine Shell Dates with Application to Southern Pacific Northwest Coast Prehistory. *Syesis* 14:45–57.

Sangawa, A., 1989, Point of Contact Between Archaeology and Paleo-Earthquake Research. *The Quaternary Research (Japan)* 27:241–252.

Santa María, V., 1775, Diario de lo Acaecido en el Nuevo Descubrimiento del Puerto de S[a]n Fran[cis]co. In *The First Spanish Entry into San Francisco Bay* edited by J. Galvin, 1972, pp. 11–76. John Howell Books, San Francisco.

Saucier, R. T., 1991, Geoarchaeological Evidence of Strong Prehistoric Earthquakes in the New Madrid (Missouri) Seismic Zone. *Geology* 19:296–298.

Schwartz, D. P., and Coppersmith, K. J., 1984, Fault Behavior and Characteristic Earthquakes: Examples from the Wasatch and San Andreas Faults. *Journal of Geophyscial Research* 89:5681–5698.

Scranton, R., Shaw, J., and Ibrahim, L., 1978, *Kenchreai Eastern Port of Korinth I. Topography and Architecture*: E. J. Brill, Leiden.

Simpson, G. D., Thompson, S. C., Noller J. S., and Lettis, W. R., 1997, The Northern San Gregorio Fault Zone: Evidence for the Timing of Late Holocene Earthquakes near Seal Cove, California. *Bulletin of the Seismological Society of America* 87:1158–1170.

Soren, D, 1988, The Day the Earth Ended at Kourion. *National Geographic* 174:30–53.

Stiros, S. C., 1988, Archaeology; A Tool to Study Active Tectonics; the Aegean as a Case Study. *Eos, Transactions, American Geophysical Union* 69: 1633, 1639.

Stiros, S. C., and Jones, R. E. (eds.), 1996, *Archaeoseismology*. British School at Athens Fitch Laboratory Occasional Paper 7.

Stiros, S. C., and Pirazzoli, P. A., 1995, Paleoseismic Studies in Greece: A Review. *Quaternary International* 25:57–63.

Stiros, S. C., Marangou, L., and Arnold, M., 1994, Quaternary Uplift and Tilting of Amorgos Island (Southern Aegean) and the 1956 Earthquake. *Earth and Planetary Science Letters* 128:65–76.

Stiros, S. C., Pirazzoli, P. A., Rothaus, R., Papageorgiou, S., Laborel, J., and Arnold, M., 1996, On the Date of Construction of Lechaion, Western Harbor of Ancient Corinth, Greece. *Geoarchaeology* 11:251–263.

Toppozada, T. R., Real, C. R., and Parke, D. L., 1981, *Preparation of Isoseismal Maps and Summaries of Reported Effects for Pre-1900 California Earthquakes*. Open-File Report 81-11, California Division of Mines and Geology, Sacramento.

Tuttle, M. P., and Schweig, E. S., 1995, Evidence of Recurrent Large Earthquakes in the New Madrid Seismic Zone. *Geology* 23:253–256.

Tuttle, M. P., and Schweig, E. S., 1996, Recognizing and Dating Prehistoric Liquefaction Features: Lessons Learned in the New Madrid Seismic Zone, Central United States. *Journal of Geophysical Research* 101B:6171–6178.

Tuttle, M. P., Lafferty, R. H., Guccione, M. J., Schweig, E. S., Lopinot, N., Cande, R., Dyer-Williams, K., and Haynes, M., 1996, Use of Archeology to Date Liquefaction Features and Seismic Events in the New Madrid Seismic Zone, Central United States. *Geoarchaeology* 11:451–480.

Tuttle, M., Chester, J., Lafferty, R., Dyer-Williams, K., and Cande, R., 1999a, *Paleoseismology Study Northwest of the New Madrid Seismic Zone*. U.S. Nuclear Regulatory Commission, NUREG/CR-5730. Washington, DC.

Tuttle, M. P., Collier, J., Wolf, L. W., and Lafferty, R. H., 1999b, New evidence for a large earthquake in the New Madrid seismic zone between A.D. 1400 and 1670. *Geology* 27:771–774.

Umeda, Y., Murakami, H., Iio, Y., Cho, A., Ando, M., and Daicho, A., 1984, Earthquake Traces Impressed on the Yayoi-Age Dwelling Site. *Jishin* 37:465–473.

Varotsos, P., Eftaxias, K., Vallianatos, F., and Lazaridou, M., 1996, Basic Principles for Evaluating an Earthquake Prediction Method. *Geophysical Research Letters* 23:1295–1298.

Vita Finzi, C., and King, G. C. P., 1985, The Seismicity, Geomorphology and Structural Evolution of the Corinth Area of Greece. *Philosophical Transactions of the Royal Society of London A* 314:379–407.

Voltaire, F. -M., 1759, *Candide ou l'optimisme*.

Weber, G. E., and Lajoie, K. R., 1979, Evidence for Holocene Movement on the Frijoles Fault near Point Año Nuevo, San Mateo County, California. In *Coastal Tectonics & Coastal Geologic Hazards in Santa Cruz & San Mateo Counties, California, Field Trip Guide*, edited by G. E. Weber, K. R. Lajoie, and G. B. Griggs, pp. 92–100. Cordilleran Section, Geological Society of America, Boulder, Colorado.

Weber, G. E, Nolan, J. M., and Zinn, E. N., 1995, *Determination of Late Pleistocene–Holocene Slip Rates along the San Gregorio Fault Zone, San Mateo County, California*. Final Technical Report, U.S. Geological Survey National Earthquake Hazards Reduction Program. Reston, VA.

Wells, D. L., and Coppersmith, K. J., 1994, New Empirical Relationships Among Magnitude, Rupture Length, Rupture Width, Rupture Area, and Surface Displacement. *Bulletin of the Seismological Society of America* 84:974–1002.

Whitcomb, D., 1994, *Ayla: Art and Industry in the Islamic Port of Aqaba*. Oriental Institute Museum Publications, Chicago.

Winslow, M. A., and Johnson, L. L., 1989, Prehistoric Human Settlement Patterns in a Tectonically Unstable Environment: Outer Shumagin Islands, Southwestern Alaska. *Geoarchaeology* 4:297–318.

Wyss, M., and Dmowska, R., 1997, *Earthquake Prediction — State of the Art*. Birkhauser Verlag, Boston.

Yeats, R. S., Sieh, K., and Allen, C. R., 1996, *The Geology of Earthquakes*. Oxford University Press, Oxford, UK.

Zhang, B., Liao, Y., Guo, S., Wallace, R. E., Bucknam, R. C., and Hanks, T. C., 1986, Fault Scarps Related to the 1739 Earthquake and Seismicity of the Yinchuan Graben, Ningxia Huizu Zizhiqu, China. *Bulletin of the Seismological Society of America* 76:1253–1287.

Zhang, A., Yang, Z., Zhong, J., and Mi, F., 1995, Characteristics of Late Quaternary Activity Along the Southern Border Fault Zone of the Weihe Graben Basin. *Quaternary International* 25:25–31.

Zreda, M., and Noller, J. S., 1998, Ages of Prehistoric Earthquakes Revealed by Cosmogenic Chlorine-36 in a Bedrock Fault Scarp at Hebgen Lake. *Science* 282:1097–1099.

III

Soils, Sediments, and Microstratiography

III

Soils, Sediments and
Micromorphology

Use and Analysis of Soils by Archaeologists and Geoscientists

A North American Perspective

ROLFE D. MANDEL and
E. ARTHUR BETTIS III

1. Introduction

Archaeologists generally recognize that there is a relationship between cultural deposits and associated soils and landforms. However, their understanding of what a soil is, as well as what soils can reveal about site formation processes, landscape development, and environments of the past varies greatly. Although archaeologists should not be expected to have a complete grasp of pedology, they should be capable of recognizing and interpreting soils in an archaeological context in order to fully comprehend the record of the human past.

A large body of literature addresses the use of soils in archaeological research. Some of it provides general discussions of applications of pedology to archaeology (e.g., Butzer, 1971, 1982; Cornwall,1958; Evans, 1978; Foss et al., 1995; Holliday, 1990; Limbrey, 1975; Lotspeich, 1961; Rapp and Hill, 1998; Shackley, 1975; Tamplin, 1969; Waters, 1992:40–60). In geoarchaeological studies, the focus on soils is often their use as stratigraphic markers (e.g., Artz, 1985; Bettis and Thompson, 1981, 1982; Cremeens and Hart, 1995; Ferring, 1986, 1990, 1992; Gvirtzman et al., 1999; Hajic, 1990; Holliday, 1985a,1985b,1905c,1989; Holliday and Meltzer, 1996; Hoyer, 1980; Mandel, 1992, 1995; Nordt, 1992, 1995; Reider, 1980, 1982, 1990; Styles, 1985; Thoms and Mandel, 1992; Wiant et al., 1983). There is also substantial literature that considers soil chemistry, especially phosphorus, as an indicator of the presence and/or intensity of human occupation and for interpreting use areas within sites (e.g., Ahler, 1973; Brinkmann, 1996; Eidt, 1977, 1985; Gordon, 1978; Groenman-van Waateringe and Robinson, 1988; Jacob, 1995; Kolb et al., 1990; Sandor, 1992; Schuldenrein, 1995; Sjoberg, 1976; White, 1978; Woods, 1984). Soil micromorphology is used in an increasing number of investigations as a tool for interpreting the geoarchaeological context of deposits (e.g., Courty, 1992; Courty et al., 1989, 1991; Fisher and Macphail, 1985; Goldberg, 1983, 1987, 1992; Holliday, 1985c; Holliday et al., 1993; Macphail, 1986, 1992; Macphail and Goldberg, 1995; Macphail et al., 1990, 1994; Mikkelsen and Kangohr, 1996). Other studies have used broader scale archaeological applications of pedology to landscape and climatic reconstruction (e.g., Bettis and Hajic, 1995; Haynes and Grey, 1965; Holliday, 1997; Johnson and Logan, 1990; Mandel, 1992, 1995; Monger, 1995; Paulissen and Vermeersch, 1987; Ranov and Davis, 1979; Reeves and Dormaar, 1972; Reider, 1980, 1982, 1990; Thompson and Bettis, 1980) and have demonstrated the utility of soils as dating tools in archaeological research (e.g., Anderton, 1999; Bischoff et al., 1981; Frink, 1995; Holliday, 1988).

Faced with this vast amount of useful, but often highly technical information from disciplines not often considered in their academic training, archaeologists are frequently left with the following questions: What does one need to know about soils, in a practical sense, when conducting field investigations? What types of soil data are critical to the resolution of archaeological problems, and how does one go about collecting these data? Our chapter addresses these questions. Specifically, we (1) explain why it is important for archaeologists to understand the difference between soil and sediment, (2) describe how to distinguish soil from sediment, and (3) discuss the type of soil information needed at different stages of an archaeological investigation, that is, surveys versus site evaluations versus excavations. Although approaches for collecting soil data are considered, detailed discussions of soil-forming factors and processes, physical and chemical properties of soils, and soil profile nomenclature are not presented here; the reader is directed to other sources for this information (see Birkeland, 1999; Buol et al., 1997; Catt 1986; Soil Survey Division Staff, 1993; Soil Survey Staff, 1996).

2. Distinguishing Soil from Sediment

The Soil Survey Staff (1996: 1) provided the following definition of soil: "soil" is a term used for "the natural bodies, made up of mineral and organic materials, that cover much of the earth's surface, contain living matter and can support vegetation out of doors, and have in places been changed by human activity."

This all-encompassing definition provides for a soil continuum across the present land surface but is inadequate for the four-dimensional (horizontal, vertical, time, environment) landscape analysis necessary in archaeological investigations. Archaeologists and soil geomorphologists alike must distinguish between characteristics of deposits that are a product of sedimentary processes and those that are indicative of secondary alterations related to weathering at a land surface (soil formation). This distinction is important because in most cases a soil signifies a break in deposition; hence it is an indicator of a relatively stable land surface and a zone that represents more time to record human activities than adjacent nonsoil.

Many factors interplay to produce the characteristics of soils, but one characteristic all soils have in common is a vertical sequence of genetically related horizons produced by soil-forming processes acting on geologic materials over time (Fig. 1). The concept that soil horizons require some amount of time to form is probably one of the most significant aspects of pedology in archaeology (Holliday, 1990:530). Another important consideration is that soil development requires not only time but also a relatively stable landscape, one that is neither rapidly aggrading or eroding (Catt, 1986:166–167). Exceptions are bogs and marshes, where organic soils (Histosols) form as a result of accumulation of organic matter. Also, soils on floodplains and on toeslopes may receive influxes of parent material while pedogenesis is underway; that is, soil formation and deposition proceed concurrently (Birkeland, 1999). Disregarding these exceptions, the presence of a soil indicates that the landscape has been relatively stable for a period of time. In general, landscapes that have been stable for long periods have soils that are better developed than those that have been stable for short periods, all other soil-forming factors being equal.

A buried soil in a stratigraphic sequence indicates a hiatus between depositional events. Holliday (1992) stressed that in such a sequence "the sediment, which is the parent material for the soil, may have accumulated rapidly or slowly, but a significant period of nondeposition had to occur for the soil to form (p. 103)." Although large volumes of sediment may be deposited instantaneously, as with a debris flow, soil formation usually requires at least a century or several centuries and commonly millennia (Birkeland, 1999; Holliday, 1992).

In archaeological investigations, consideration of landscape stability, as indicated by soil development, is important in locating cultural deposits, interpreting artifact associations and contexts, defining site stratigraphy, reconstructing the depositional and landscape history, and establishing cultural chronologies. The first step is identifying which parts of the sedimentary deposits present in the site or study area have been modified by soil formation. Although this can be a fairly straightforward task, it often becomes complicated when pedogenically unaltered deposits have properties that mimic some properties of soils.

The first step in the recognition of a soil is the identification of soil horizons and the development of the ability to differentiate between soil horizons and sedimentary deposits with soil-like properties. Soil horizons often parallel a land surface, and they have distinctive properties that result from the complex interactions among a variety of physical, chemical, and biological processes acting on surficial materials over time. The nature and magnitude of influence of these

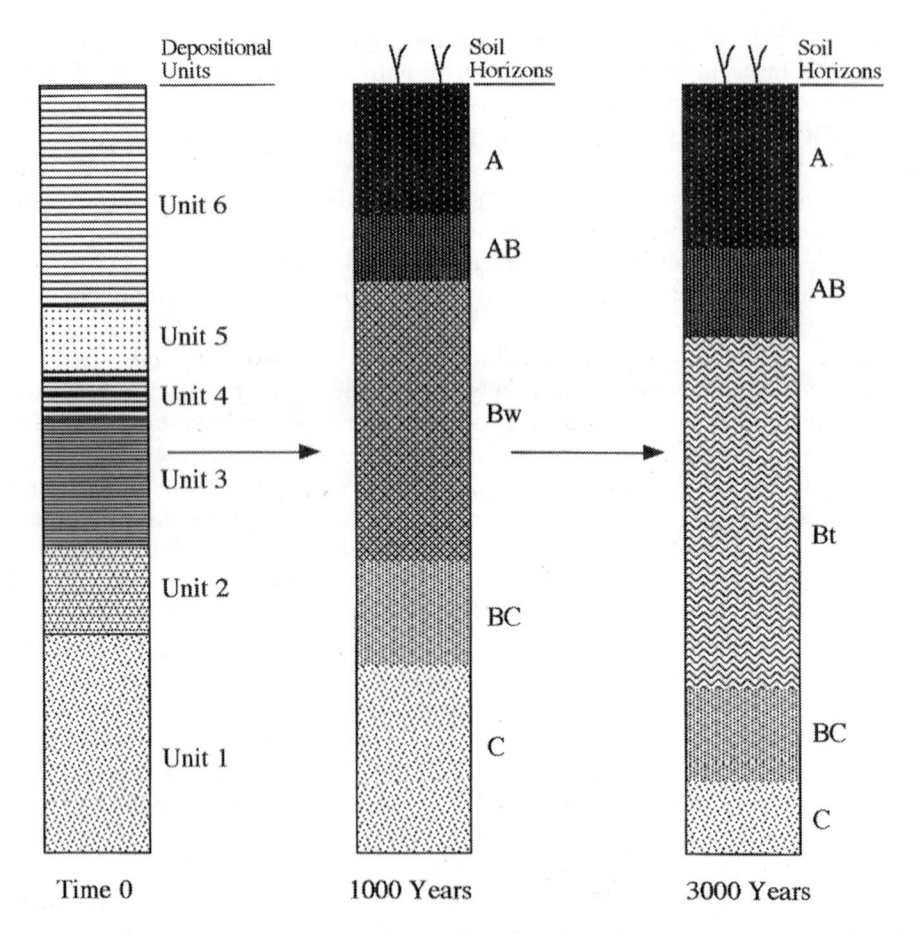

Figure 7.1. The transformation of pedogenically unmodified sediment (parent material) into a soil. At time 0, the sediment is contained in six distinct stratigraphic units that are separated by abrupt boundaries. After 1,000 years of landscape stability, a hypothetical soil with an A-AB-Bw-BC profile has developed in the units, obliterating primary sedimentary features, such as bedding. After 3000 years of landscape stability, the hypothetical soil is thicker and much better developed (A-AB-Bt-BC profile). Note that the soil horizons bear little resemblance to the original stratigraphy.

processes, and the resulting properties of soils, is in large part controlled by various environmental factors. Both soils and unlithified geologic materials that are not soils can be described according to the following properties: color, texture, consistence, soil structure (or lack thereof), cutans (coatings), nodules or concretions, voids, reaction to hydrochloric acid, boundary characteristics, and horizon continuity. Specifics on the nomenclature for describing soil horizons and surficial materials are provided by Birkeland (1999), Birkeland et al. (1991), Buol et al. (1997), Hallberg et al. (1980), and the Soil Survey Staff (1996).

Distinguishing soil from sediments on the basis of the presence or absence of horizons is not always a simple procedure. For example, organic- or clay-rich

Figure 7.2. Stratified alluvium exposed in an archaeological test pit at site 13HA385 in north-central Iowa. A dark, organic-rich flood drape near the bottom of the pit could be confused with a buried A horizon. However, note the abrupt lower boundary of the flood drape. The photo scale is 20 cm long.

alluvium that has not been modified by pedogenesis sometimes exhibits properties that mimic those of A and B horizons of soils, respectively (Fig 7.2). Several criteria may be used to discriminate a soil horizon from a depositional unit in an alluvial setting. The lower boundary of an A or B horizon, for example, is usually clear, gradual, or diffuse, rather than abrupt or wavy (Fig 7.3). Pedogenically unaltered clay- and organic-rich depositional layers often have abrupt and wavy boundaries or graded bedding produced by sedimentary processes. As the deposits are affected by bioturbation and other soil-forming processes, the abrupt boundaries that separate individual beds are obliterated. Micromorphological studies may help distinguish soil horizons from sedimentary zones in these situations by identification of fabrics indicative of mixing, soil-forming processes, or sedimentary processes (Courty, 1992; Courty et al., 1989).

Soil structure, which is the aggregation of primary soil particles into distinctively shaped compound particles (peds), can be used to distinguish organic-rich sedimentary deposits from A horizons of soils. Flood drapes, which tend to be clayey and enriched with organic matter, often contain clay minerals that shrink and swell with drying and wetting. The shrinking and swelling of the these minerals gives the dark alluvium an angular blocky "structure" that may be misinterpreted as having been formed by processes associated with a relatively stable land surface during a hiatus in sedimentation. An A horizon, however,

Figure 7.3. Buried soils developed in alluvium at the Alum Creek site in central Kansas (Mandel, 1992). There is a thin buried A horizon immediately above the handle of the shovel, and the shovel is leaning against a dark, overthickened A horizon. Note the gradual lower boundaries of the buried A horizons, and compare them with the abrupt lower boundary of the organic-rich flood drape in Fig. 7.1.

typically has granular structure produced by the activity of worms and other soil organisms in this biologically active zone of the soil. In buried soils, the granular structure may be transformed to blocky types of structure during compaction, but evidence for the former A horizon may include greater porosity or granular aggregates that can only be detected in soil thin sections (Courty, 1992; Courty et al., 1989). Patterns of rooting and burrowing also provide useful information for identifying former land surfaces and for distinguishing soils and unaltered sediments. Unaltered sediments are usually capable of supporting plant life and often become burrowed, but surfaces stable enough for soils to form are subject to more intensive rooting and potentially more burrowing activity simply because they represent more time per volume than unaltered deposits. The upper horizon of a soil, therefore, should be more heavily rooted, especially with fine roots, and should contain more evidence for bioturbation than unaltered deposits. Because of the dark color of many soil surface horizons, and the rate of bioturbation and the small scale of many bioturbators, such as worms and ants, much of the conclusive rooting- and burrowing-pattern evidence for distinguishing soil horizons from unaltered deposits is at the microscopic scale (Courty, 1992).

In some areas of North America, especially arid and semiarid regions, deposits that have been affected by nonpedogenic accumulations of calcium carbonate can be easily confused with calcic soils. For example, laterally flowing, $CaCO_3$-rich groundwater often forms calcretes that are misidentified as soils with petrocalcic horizons (Machette, 1985). The development of a groundwater calcrete is a nonpedogenic process that requires calcium-charged groundwater to either discharge onto a stream bottom or reach a near-surface position where calcium is concentrated by evaporation. Supersaturation of calcium causes precipitation of $CaCO_3$ and subsequent cementation of relatively porous sands and gravels (Machette, 1985). Further complicating matters, surface runoff may add or redistribute the $CaCO_3$ and produce laminar zones that resemble laminar petrocalcic horizons (Bachman and Machette, 1977). Nevertheless, with careful field observations and micromorphological analyses, it is possible distinguish groundwater calcretes from calcic soils. For example, groundwater calcretes are often strongly indurated to depths of 10 m or more (Machette, 1985). Petrocalcic horizons, however, are rarely more than several meters thick. When viewed in thin section, groundwater calcretes typically consist of simple cement fills enclosing clasts with grain-to-grain contact, whereas the cement of petrocalcic horizons is micritic and replaces or surrounds scattered detrital grains that appear to float in a matrix of carbonate (Arakel, 1986; Jacobsen et al., 1988; Machette, 1985). Also, groundwater calcretes generally lack the horizonation and morphological structures common in calcic soils (Allen, 1986; Bachman and Machette, 1977; Machette, 1985; Wright, 1982).

In an archaeological investigation, it is important to distinguish between a ground-water calcrete and a petrocalcic horizon because of the temporal, soil-stratigraphic, and environmental implications. Ground-water calcretes form in tens to hundreds of years, they are not products of landscape stability, and they develop in a wide range of environments (humid to arid). Petrocalcic horizons, however, require thousands of years of carbonate accumulation in relatively stable arid or semiarid soil-forming environment (Birkeland, 1999; Gile and Hawley, 1966; Gile et al., 1979).

In some situations, sandy C horizons may be confused with albic (E) horizons of soils. This usually occurs where a light-colored sandy C horizon is beneath an A horizon and above a truncated Bt horizon. Such confusion may seriously compromise the interpretation of the soil-stratigarphy, paleoenvironment, and age of an archaeological site. An E horizon is a product of intensive leaching (eluviation) that typically spans thousands of years, whereas a C horizon consists of slightly weathered parent material. The best approach to distinguishing a C horizon from an E horizon is to examine the morphology of the soil. For example, all E horizons have soil structure, which involves a bonding together of individual soil particles. In most E horizons, the particles are arranged about a horizontal plane, forming platy structure (Birkeland, 1999). In contrast, sandy C horizons are single grain or massive; hence, they are structureless. If the material is single grain, the individual particles are easily distinguishable and do not adhere to each other. When the material is massive, individual particles adhere closely to each other, but the mass lacks planes of weakness (Buol et al., 1997). Also, the boundary between an E and a Bt horizon is usually gradual and wavy or irregular, and tongues of the E horizon often extend down into a Bt horizon. In contrast, where a C horizon overlies a truncated Bt horizon, the boundary is abrupt and smooth or wavy and lacks tongues.

Soils, unlike most individual beds of sediment, are laterally extensive across the landscape. They extend across various landforms and underlying geologic deposits and exhibit predictable variations in their properties related to changes in drainage, vegetation, and relative age of the geomorphic surface on which they occur. Hence, surface and buried soils can be mapped in three dimensions over varying topography. In contrast, individual beds of sediment tend to be restricted to a certain depositional environment and will pinch out away from that area (Fig 7.4).

Although identification and study of soils begin in the field with careful observations, it is sometimes necessary to support these observations with laboratory analysis in order to clearly differentiate soil from pedogenically unaltered sediment. One of the most powerful laboratory methods for distinguishing soil from nonsoil is micromorphological analysis (see Fisher and MacPhail, 1985). This procedure, which requires the use of thin sections and petrographic equipment, greatly enhances the ability of the soil scientist to identify micro-

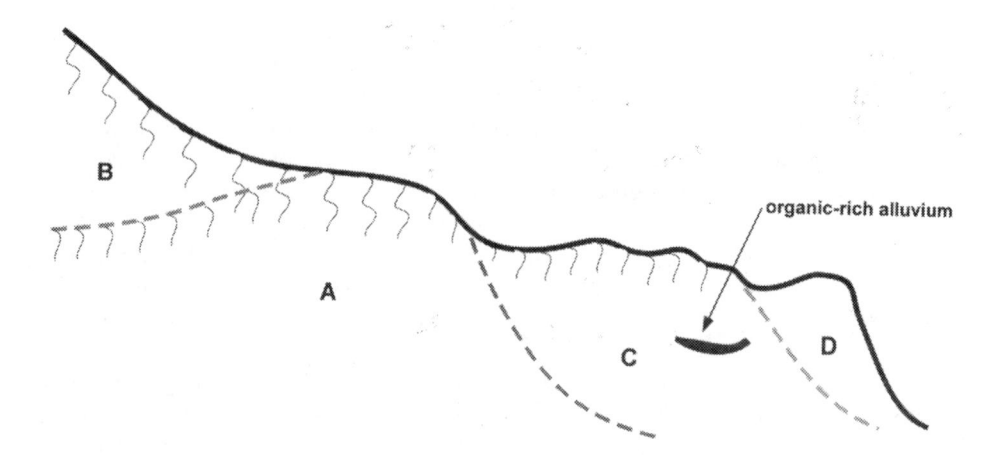

Figure 7.4. Soils form a continuum across the landscape and cross landform and parent material boundaries. The diagram shows a surface soil developed across several landforms of different ages. Organic-rich alluvium sometimes has properties similar to soils, but does not extend across the landscape to the degree that soils do. In this example, the organic-rich alluvium is restricted to one depositional unit (C) and does not cross landform and parent material boundaries.

pedological features, such as soil plasma, voids, skeletal grains, and cutans (Brewer, 1976; Bullock et al., 1985; Courty et al., 1989). However, as Wilding and Flach (1985) pointed out, soil micromorphology is a powerful tool to extend macromorphology and should not be used alone to distinguish soil from sediment. Because many soil properties form by the vertical movement and accumulation of some material, such as clay, iron, and calcium carbonate, a wide variety of laboratory methods, such as grain-size and chemical analyses, can be helpful in differentiating soil from pedogenically unaltered sediment (Buol et al., 1997; Hesse, 1971; Soil Survey Staff, 1996).

3. Soils and Archaeological Surveys

A major aspect of archaeological research involves surveying the landscape for evidence of the human past. In the United States, most archaeological surveys are related to government-mandated cultural resource management (CRM) projects (Green and Doershuk, 1998). Archaeological surveys are also common outside the United States, though they tend to be associated with projects sponsored by research institutions and foundations (e.g., Hassan, 1978; Simmons and Mandel, 1986; Wendorf and Schild, 1980). Regardless of the driving forces, archaeological surveys often encompass many different elements of the modern landscape. Because these landscape elements may be underlain by deposits with varying ages and depositional histories, archaeological surveys should be supported by geomorphological investigations in order to provide archaeologists with important information regarding the effectiveness of various sampling strategies in the survey areas.

With archaeological surveys, the problem of locating sites of a particular cultural period is one of determining where in the landscape sediments of that period are preserved (Artz, 1985). In other words, knowledge of the distribution of the various-age deposits that comprise present landscapes is essential in order to evaluate those landscapes for evidence of past human occupation (Bettis, 1992:132). In most areas, soil variability across the landscape is an important indicator for estimating the age of landforms and sediment assemblages that underlie them. Many studies have demonstrated that the degree of soil development provides important clues to the relative ages of geomorphic surfaces and underlying deposits (e.g., Bettis, 1992; Birkeland, 1999; Birkeland and Burke, 1988; Dethier, 1988; Gile et al., 1981; Harden, 1982; Holliday, 1992; Karlstrom, 1988; McFadden and Weldon, 1987; Yaalon, 1971). Because surface soils can provide relative time control and are mappable, they may be used to devise "quick and dirty" strategies for assessing the cultural resource potential of a survey area. This approach was employed in several major archaeological surveys, including ones in Texas (Lake Creek valley: Mandel, 1987), Oklahoma (Copan Lake area: Reid and Artz, 1984), and Iowa (central Des Moines River valley: Benn and Bettis, 1985; Bettis and Benn, 1984). The survey of the central Des Moines River valley is considered in the following discussion.

A detailed soil-geomorphic investigation was undertaken prior to an intensive archaeological survey and testing program in the Des Moines Valley (Bettis and Benn, 1984). This study demonstrated that certain properties of surface soils developed in Holocene alluvium of midwestern streams are age diagnostic (Bettis, 1992). For example, surface soils formed in early and middle Holocene alluvium are Mollisols with argillic (Bt) horizons or Alfisols. These soils exhibit moderate grade structure, brown or dark brown Bt horizons, few to common argillans, and are well horizonated. In contrast, surface soils developed in late Holocene alluvium are Mollisols or Inceptisols with cambic (Bw) horizons, or Mollisols that lack B horizons. These soils tend to be dark colored throughout, exhibit weak to moderate grade structure, and have weak horizonation. Finally, surface soils developed in Historic deposits are Entisols with thin, weakly expressed A–C profiles. This soil information was combined with other geomorphic and stratigraphic data to construct maps that show the distribution of landform sediment assemblages dating to different periods of the Holocene. These maps provided valuable information used during subsequent archaeological surveys to devise sampling strategies for locating the valley's archaeological deposits. This information also provided archaeologists with estimates of past landscapes removed by river processes. These estimates proved invaluable for interpreting the pattern of known sites in the survey area in terms of past land use patterns (Benn and Bettis, 1985).

Although soils data should be collected by the geoscientist(s) involved in an archaeological survey, soil surveys published by the U.S. Department of Agriculture provide some information useful for planning the archaeological investigation (Almy, 1978; Saucier, 1966). These surveys are available at county and state offices of the Natural Resources Conservation Survey (formerly the Soil Conservation Service). Each survey covers one or two counties and consists of a brief text accompanied by a series of photomosaic or photomap base sheets on which the soil series are outlined in red or in black. The photomaps vary in scale from 1:7,920 to 1:31,680, but most are at a scale of 1:20,000. Individual sheets cover an area of approximately 12 square miles and measure about 9.5 by 12.5 inches. The Natural Resources Conservation Survey (NRCS) is presently digitizing these maps so they can be electronically accessed. Because the relative age and geomorphic stability of surfaces are reflected in some of the properties of soils, the NRCS soil maps may indicate, within broad limits, the relative ages of sediments and sites to be expected in a given area (Artz, 1985). Also, soil drainage may be inferred from these maps, which in turn can be used to isolate poorly drained landscapes with low potential for archaeological sites from well-drained areas with high archaeological potential. However, the soil surveys are generalized and should only be used to establish initial impressions of soil–geomorphic relationships in the study area. Detailed field investigations are required to confirm and elaborate on those impressions and to refine them to the scale (both spatial and temporal) needed by most archaeologists (Artz, 1985).

As noted earlier, buried soils represent previous land surfaces that were exposed for sufficient periods of time to develop recognizable soil profile characteristics. Hence, they represent former stable land surfaces. If one assumes that the probability of cultural utilization of a particular landscape position is equal for each year, it follows that the surface that remains exposed for the longest time would represent those with the highest probability of containing cultural remains (Hoyer, 1980:61). Because buried soils represent former stable surfaces, evidence for human occupation would more likely be associated with them. This reasoning also implies that a soil that had the most time to develop before it was buried would have the highest potential for containing cultural deposits at any given location. Thus, buried soils are also useful indicators for locating archaeological deposits and for assessing an important aspect of the geologic potential for buried cultural deposits.

Knowledge of the temporal and spatial pattern of buried soils in a landscape provides archaeologists with a powerful tool for identifying areas with high potential for buried cultural deposits and for assessing prehistoric cultural patterns. Although there are a number of good examples of how this knowledge can be applied in an archaeological survey (e.g., Bettis and Benn, 1984; Bettis and Littke, 1987; Mandel, 1992, 1994, 1996, 1997, 1999; Mandel et al., 1991; May, 1986; Thompson and Bettis, 1980; Ferring, 1992), only one example is presented here: the Phase II geoarchaeological survey of the Pawnee River Watershed in southwestern Kansas (Mandel, 1992).

The Pawnee River Survey was preceded by an intensive, basinwide study of Holocene alluvial deposits and associated buried soils (Mandel, 1992). Two distinct buried soils, the Hackberry Creek Paleosol and the Buckner Creek Paleosol, were identified beneath the first terrace (T–1) in small (<fourth-order) drainage elements (Fig 7.5) (Mandel, 1992). The T–1 surface stands about 3 m above the surface of the modern floodplain (T–0). A suite of radiocarbon ages suggested that the Hackberry Creek Paleosol formed from about 2,800 B.P. to ca. 2,000 B.P., and that the Buckner Creek Paleosol developed from ca. 1,700 to 1,000 B.P. These two paleosols are superposed and have thick, moderately expressed Ak–Bk profiles. Other buried soils were identified in the late Holocene alluvium that composes the T–1 fill, but they all have thin, weakly expressed A–C profiles that, individually, reflect only tens of years of pedogenesis (Mandel, 1992). Stratified Late Archaic and Plains Woodland cultural deposits were found in the Hackberry Creek and Buckner Creek paleosols, respectively, whereas buried soils with A–C profiles were consistently sterile of archaeological materials (Mandel, 1992). This finding underscores the axiom that the longer the period over which a soil has developed, the higher its potential for containing cultural deposits at any given location. The Hackberry Creek and Buckner Creek paleosols are important stratigraphic markers and were targeted for exploration during the archaeological survey that followed the geomorphological investigation of the Pawnee River basin.

Archaeologists doing surveys have become increasingly aware of the fact that buried soils often harbor much of the archaeological record in areas that were affected by episodic sedimentation throughout the Holocene. However, buried soils often occur at considerable depths; hence, they cannot be detected, much less assessed for cultural resources, using most conventional discovery methods. Faced with this dilemma, archaeologists have turned to sampling and discovery methods that have usually not been employed in past surveys. For example, engine-driven coring devices, such as hydraulic soil probes, can be used to

Figure 7.5. The Buckner Creek site (14HO306) on the east bank of Buckner Creek in southwestern Kansas (from Mandel, 1992, Fig. 2-13). The site was discovered in 1986 during a systematic geoarchaeological survey of the Pawnee River basin (Mandel, 1988). Plains Woodland materials, including a hearth, are associated with the Buckner Creek Paleosol (BCP). The Hackberry Creek Paleosol (HCP) contains stratified Late Archaic deposits at depths of 3.9 to 5.3 m below land surface. The shovel is approximately 1.5 m long. Courtesy Smithsonian Institution Press.

recover intact sediment cores to great depths. This type of mechanical coring is a very efficient method for determining the presence or absence of buried soils in a depositional environment—an important step in determining the geologic potential for buried cultural materials. When buried soils are encountered, they can be easily described from the cores and, if sufficient organic carbon is present, can be dated by radiocarbon analysis. Recent studies have also suggested that fine screening of soil recovered from the cores, and examination of the residue for microdebitage, may aid the discovery of deeply buried archaeological deposits (Stafford, 1995). Once the temporal and spatial patterns of buried soils are known for an area, strategies for subsurface exploration, such as trenching, auguring, and/or inspection of cut banks, can be developed. Even preliminary soil–stratigraphic frameworks that lack absolute time control greatly improve the efficiency and scientific outcome of archaeological surveys (Ferring, 1992).

Since the early 1980s, it has become obvious that surface soils developed on "stable" landscape positions underlain by "old" deposits sometimes contain buried archaeological materials (Artz, 1993, 1995b; Bettis and Hajic, 1995; Van Nest, 1993). This realization has come about primarily because of the implementation of shovel testing during archaeological surveys. However, the theoretical foundations for understanding the burial of cultural deposits in stable upland settings are rooted in pedology and have recently been articulated by Johnson (1990) and by Johnson and Watson-Stegner (1990). Upbuilding of the soil through biomechanical processes may result in the burial of artifacts and cultural features. Some of the primary biological agents of soil upbuilding are ants, termites, and earthworms, all of which can quickly bury surface-occurring materials beneath their mounds or castings (Balek, 1998; Darwin, 1881; Johnson, 1990; Johnson and Watson-Stegner, 1990; Rolfsen, 1980; Stein, 1983; Wood and Johnson, 1978; Van Nest, 1998). For example, in regions where earthworms are extremely active, items may be buried to depths as great as 45 cm below ground surface in about 5 years (Rolfsen, 1980). Hence, worm castings, combined with soil brought to the surface by burrowing animals and tree-throws, form a biomantle that may conceal artifacts on stable uplands (Johnson 1990; Schaetzel et al., 1986; Van Nest, 1993, 1998). The major obstacles preventing a better understanding of site formation processes in these landscape positions are a paucity of datable material, and mixed artifact assemblages because of natural and anthropogenic pedoturbation. A lack of creative approaches to upland archaeological survey is also a major hindrance (Bettis and Hajic, 1995). However, soils can provide information regarding the relative degree of biological activity or soil disturbance across an upland area and can thus provide clues to the potential for burial or site mixing.

4. Soils and Site Evaluations

The discovery of an archaeological site is sometimes followed by an investigation that focuses on the significance of the find. In the United States, site evaluation is a standard procedure with CRM archaeology because of the need to develop and provide information for thoughtful decision making on the disposition of cultural resources that are threatened and/or being considered for National Register eligibility (see Green and Doershuk, 1998). Not all site evaluations, however, are related to a CRM project. For example, foundations and institutions that fund archaeological excavations in the United States and elsewhere often recommend or demand evaluations to determine whether a site warrants the effort and expenditure of resources needed for an excavation.

In most cases, site evaluations must resolve the following questions: What is the depth, horizontal extent, and stratigraphic context of the archaeological materials at the site? Does the site have good vertical and horizontal integrity (i.e., are the archaeological materials *in situ*)? How old are the cultural deposits? Information gleaned from soils at a site may contribute significantly to answering these questions.

Although the boundaries of a site that consists of archaeological materials at or very near the land surface can be easily determined by traditional methods, such as surface collection and excavation of shallow test units, deeply buried cultural deposits can make this procedure difficult. However, if the buried materials are on former stable surfaces (associated with buried soils), soil stratigraphy, combined with deep testing, may be used to determine the spatial limits of cultural deposits.

A good example of a soil stratigraphic approach to site evaluation is a recent investigation that focused on deeply buried cultural deposits associated with an alluvial fan (McNeal Fan) in the Mississippi River valley near Muscatine, Iowa. The proposed construction of a highway across the McNeal Fan initiated an intensive evaluation to assess the significance of the cultural resources. Artz (1995a) determined that three major depositional units underlie the surface of the fan, and that the thickest body of sediment, Unit II, consists of multiple upward-fining sequences (Fig 7.6). Soils are developed at the top of each upward-fining sequence, and archaeological materials are on and within the buried soils. Given the need to determine the lateral extent of the cultural deposits, the McNeal Fan presented a formidable challenge because the cultural deposits are at depths ranging from 0.5 to 8.5 m. Nevertheless, Artz (1995a) was able to define the spatial limits of each buried cultural zone by using a deep auguring device and a backhoe to trace the artifact-bearing soils. In addition, he demonstrated that the fan's soil stratigraphy was very complex because some of the buried soils merge downslope to form welded soil complexes, and locally throughout the site, buried soils are truncated by channels or sheetwash erosion. His findings were critical to the development of excavation strategies that targeted some areas of the site for investigation while avoiding other areas.

Soil evidence is especially useful for determining how and to what extent cultural deposits have been affected by post occupational disturbance processes, such as tree throws, animal burrowing, frost heaving, and erosion. Many disturbance processes result in characteristic soil features that can be recognized with field observations. For example, soils that have been affected by tree throws will have inverted profiles; that is, B-horizon material overlies A-horizon material (Johnson and Watson-Stegner 1990; Schaetzel et al, 1986). Where soils have been

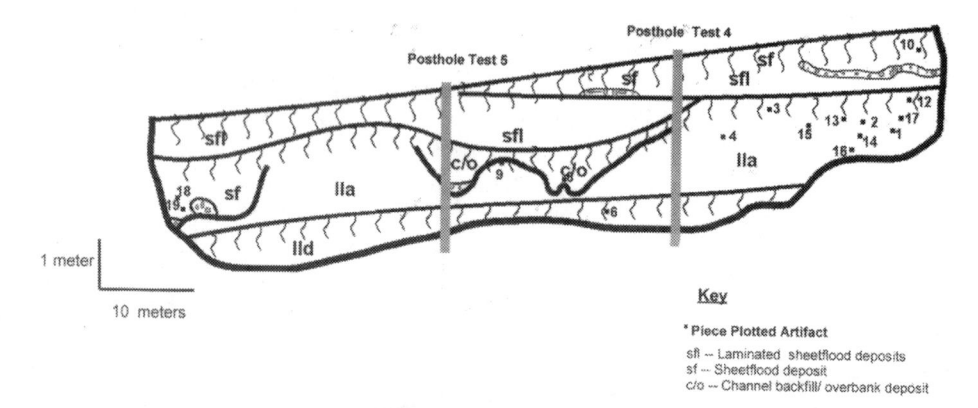

Figure 7.6. Profile of the east wall of Trench 2, Southwest Area, at the McNeal Fan (site 13MC15) in eastern Iowa (from Artz, 1995a, Fig. 19). Soils are developed at the top of each upward-fining sequence, and archaeological materials are within the buried soils. Courtesy Office of the State Archaeologist, University of Iowa.

disturbed by burrowing animals (faunalturbation), krotovina and other biogenic features are likely to be common in the mixing zone, and boundaries between soil horizons will tend to be diffuse (Buol et al., 1997). Cryoturbation, which is soil mixing by seasonal freezing and thawing of the ground, is often indicated by sand wedges, soil deformation, stone polygons and/or stripes, solifluction lobes, and by diagnostic micromorphological features (Van Vliet-Lanoe, 1985; Van Vliet-Lanoe et al., 1984; Wood and Johnson, 1978). Sheetwash erosion may be indicated by truncated soil horizons and rills. Isolated sand- or gravel-filled channels cut in the surface of a buried soil are good evidence of gully erosion. When the magnitude of soil disturbance processes is determined at a site, the potential impact of these processes on the vertical and/or horizontal integrity of the cultural deposits can be assessed (Schiffer, 1987; Wood and Johnson, 1978).

Simply recognizing buried soils in a stratigraphic sequence at an archaeological site has important implications in the evaluation of site formation and preservation processes. For example, it is likely that artifact densities will be greater on and within the buried soils compared to the zones of pedogenically unmodified sediment, but the impacts of pedoturbation on artifacts and on occupation zones are usually greater in the buried soils (Holliday, 1990). This does not imply that archaeological materials associated with buried soils will lack integrity. Some buried soils will harbor cultural deposits that are relatively undisturbed, whereas other buried soils in similar depositional environments will contain archaeological materials that have been greatly displaced by biological and geological processes. Ferring (1992) stressed that in order to determine which buried soils will have well-preserved cultural deposits, one must consider the nature of soil development in a depositional environment. He compared cumulative and noncumulative soils to make his point. Cumulative soils have parent material continuously added to their surfaces while pedogenesis is occurring (Birkeland, 1999), whereas noncumulative soils form during periods of nondeposition. Burial of cultural deposits in a cumulative soil protects artifacts and features from erosional disturbance and active, near-surface bioturbation. In contrast, archaeological materials that were left on the surface of a noncumulative soil were subject to more intense, adverse modifications before they were buried.

It is important to note that a cumulative soil profile is not a requisite for the preservation of cultural deposits. For example, at the Cherokee Sewer site in northwestern Iowa, most of the buried soils have thin, weakly developed profiles (A–C or A–Bw horizonation) that are products of brief episodes of pedogenesis on the rapidly aggrading surface of an alluvial fan (Hoyer, 1980). Archaic cultural horizons that represent short periods of human occupation on the fan are well preserved in poorly developed buried soils that are far below the modern surface soil. The primary context of the archaeological materials was maintained because the soils were not exposed to erosion or pedoturbation processes for long periods before they were buried. In addition, the soils were isolated from postdepositional modifications because of rapid, deep burial. Hence, the depositional environment of the fan (i.e., nearly continuous sedimentation interrupted by brief episodes of stability and pedogenesis) allowed excellent preservation of the material remains from short-term occupations. A similar relationship between the duration of pedogenesis and integrity of archaeological materials is reported for buried soils developed in alluvial fan deposits in the lower Illinois River valley (Wiant et al., 1983), the central Des Moines River valley (Bettis and Benn, 1984), and the upper Mississippi River valley (Artz, 1995a; Bettis et al., 1992). As Ferring (1992) pointed out, " to understand preserved soil-stratigraphic positioning of archaeological materials, it is necessary to document patterns of pedogenesis and sedimentation" (p. 19).

Determining the age of cultural deposits also is a very important aspect of a site evaluation. However, some sites that appear to be significant may lack time-diagnostic artifacts and radiocarbon-datable cultural material. Where this occurs, soils may be used to determine chronologic relationships of the archaeological materials. There are three approaches to using soils for dating archaeological sites: (1) determine the numerical age of organic carbon in soils; (2) correlate soils that are well dated with soils that lack temporal control; and (3) use data derived from a soil chronosequence to estimate the age of other soils in similar setting.

The most common method for determining the numerical age of soil is radiocarbon (^{14}C) dating of soil humates. A ^{14}C age determined on humates is the mean residence time for all organic carbon in the soil sample (Birkeland, 1999; Matthews, 1985). Although mean residence time does not provide the absolute age of a soil, it does give a minimum age for the period of soil development, and it provides a limiting age on material overlying a buried soil (Birkeland 1999:150; Geyh et al. 1975; Haas et al., 1986; Martin and Johnson, 1995; Matthews, 1985; Scharpenseel, 1971). When ^{14}C ages are determined on organic carbon from superposed buried soils, the relative age of cultural deposits bounded by the dated soils can be reasonably estimated (Mandel, 1992).

A relatively new method for determining the numerical age of soil is the Oxidizable Carbon Ratio (OCR) dating procedure (Frink, 1995). This procedure measures the specific rate of biodegradation of organic carbon, either as soil humic material or as charcoal. In general, as the total amount of organic carbon decreases through time due to biological recycling, the relative percentage of readily oxidizable carbon increases (Frink, 1992). This ratio is called the Oxidizable Carbon Ratio, or OCR. Although not fully tested, the OCR dating procedure provides a mean age of the total carbon in the sample (Frink, 1995).

The timing of pedogenesis has been firmly established for some late Quaternary soils through intensive ^{14}C dating of charcoal from cultural deposits (e.g, Artz, 1985; Ferring, 1986, 1990, 1992, 1995; Holliday, 1985a; Holliday et al., 1983, 1985; Mandel, 1992, 1994, 1995; Reid and Artz, 1984; Thoms and Mandel, 1992). By correlating soils, this temporal information can be used to date archaeological deposits at localities where absolute time control is absent. For example, in northeastern Oklahoma and in southeastern Kansas, a buried cumulative soil with a distinct overthickened A horizon is developed in late Holocene alluvium. This soil, often referred to as the Copan paleosol (Artz, 1985; Hall, 1977; Mandel, 1993a, b), developed between ca. 1,900 and 900 yrs B.P. Because the Copan paleosol is well-dated and easy to recognize in late Holocene alluvial sections, it is a time-stratigraphic marker and can be used to estimate the age of cultural deposits contained in its horizons.

At some archaeological sites, the properties of surface and/or buried soils may be used to infer the age of cultural deposits associated with the soils. This inference is possible because the degree of development of a soil profile or specific pedological features in a profile are indicators of time elapsed after deposition of parent material and, in some situations, as an approximate indicator of age (Holliday, 1990). Pedologic features that are time dependent have been summarized by Holliday (1990), and they include overall profile morphology, as determined by soil indices (Bilzi and Ciolkosz, 1977; Harden, 1982); profile thickness (Birkeland, 1999); illuvial clay content and reddening of the B horizon (Birkeland, 1999; Gile, 1979, 1985; Gile et al., 1981; McFadden et al., 1986); calcium carbonate accumulation (Birkeland, 1999; Gile et al., 1981; Machette, 1985; McFadden et al., 1986); clay mineralogy (Birkeland, 1999; McFadden and Hendricks, 1985; Shroba and Birkeland, 1983); and alteration or translocation of certain forms of iron, aluminum, and phosphorous (Birkeland, 1999; Birkeland et al., 1979; McFadden et al., 1986; Scott, 1977).

At localities with soil chronosequences and exceptional time control, it is possible to estimate rates for the development of soil profiles and for some pedological features. The Lubbock Lake archaeological site in the Southern High Plains of northwest Texas is an excellent example of such a situation. Holliday (1985c, 1988) defined a late Holocene chronosequence at the site. By combining field and laboratory data with the site's well-dated geochronology, he was able to determine rates of pedogenesis and, more specifically, time requirements for the development of diagnostic horizons in the Southern High Plains (Holliday, 1985c, 1988). This information has proved useful in dating soils and associated cultural deposits at other sites in the region (Holliday, 1989, 1990, 1995, 1997). Holliday (1990:531) cautioned that in comparing soils from site to site for dating purposes, the soils being compared must be in similar landscape positions and parent materials because both of these factors strongly influence soil morphology. He also stressed that stratigraphic relationships and archaeology must be considered because soils with similar morphology can form at different periods. It is also important to note that rates of pedogenesis vary among the bioclimatic regions of the world (see Birkeland, 1999). Hence, the chronosequence at Lubbock Lake cannot be used to estimate the age of soils far beyond the Southern High Plains.

In a recent Phase II testing of eight archaeological sites in the central Upper Peninsula of Michigan, Anderton (1999) used pedological information not only to determine the degree and processes of site disturbance, but to provide temporal context for the cultural deposits. The sites, which are associated with mid- and late Holocene paleoshorelines of the ancestral Great Lakes, contain stone flakes and fire-cracked rock but rarely yield diagnostic artifacts or datable carbon. Because of the association with dated shorelines, many researchers assumed the sites are Archaic occupations dating to ca. 5,000–2,000 B.P. However, later Woodland cultures (ca. 2,000–500 B.P.) may have also used the abandoned shorelines (Anderton, 1999). By considering the expected pedological and archaeological characteristics, Anderton (1999) developed a soil–artifact context model that provided a preliminary means of relative dating. Specifically, sites that were correlative with shoreline development (i.e., Archaic) have artifacts that are deeper within the soil profile, soil horizon boundaries that cut across middens, and some artifacts that are iron stained from Spodic horizon development. In contrast, sites that post date shoreline development (i.e., Woodland) have artifacts that are at or very near the ground surface. Also, Woodland cultural features, if present, cut across soil horizons, and the artifacts tend not to be iron stained.

5. Soils and Site Excavations

Site excavations require fine-scale analysis of soils aimed at addressing issues regarding site formation processes, paleoenvironmental conditions, and stratigraphy. It is usually at the excavation level of investigation that distinctions among pedologists, physical geographers, or geologists and true geoarchaeologists become most evident. The scale of interest in site excavations is much more refined than at the levels of archaeological investigations discussed previously, often being restricted to single landforms or portions of landforms. Archaeologists and pedologists generally feel most comfortable at this scale, whereas geomorphologists usually feel more comfortable addressing issues at the broader landscape scale. Because of the issue of scale, archaeologists must have a good idea of what information they want to obtain from a soils specialist during the excavation, and they must be able to clearly communicate those needs in order to ensure that the information gathered about soils is of sufficient detail to prove useful.

All other information collected during an excavation relies on proper interpretation and documentation of stratigraphic relationships within the site. As discussed previously, soils can provide essential information for recognizing stratigraphic breaks, recognizing periods of landscape stability, and tracing the lateral extent of geomorphic surfaces. The distribution of archaeological deposits, especially features and age-diagnostic artifacts, can contribute significantly to the recognition and tracing of soils, especially in settings where several weakly expressed soils are superimposed.

Excavations at the Main Site (15BL35), in the Cumberland Valley of southeastern Kentucky, provide an example of how soils and archaeological data were integrated during an excavation to provide a detailed stratigraphic framework for understanding a site buried in alluvium. An archaeological survey for a highway right-of-way recorded a portion of the site situated on a low terrace (T2). During the testing phase of this area, trenching revealed a much larger and more deeply buried site area that included adjacent portions of the floodplain (T1; Creasman, 1994). Excavations into T2 revealed a stratified site consisting of Early Archaic through Middle Woodland cultural horizons contained within the upper 1.2 m of the T2 fill. All the cultural horizons were contained within the Ap-Bt-BC profile of the surface soil. Detailed examination of the block excavations and consultations with the site archaeologists regarding the horizontal and vertical disposition of diagnostic artifacts allowed the site geoarchaeologist to recognize and map the distribution of a weakly expressed buried soil that marked a land surface that had existed on T2 from about 6300 to 3000 BP (Bettis, 1994). The buried soil was severely overprinted by subsequent development of the surface soil and was traceable only by combining soil morphology with artifact distribution data.

Below the buried soil, Early Archaic cultural remains were present, whereas the deposits above the buried soil surface contained archaeological deposits that were 3000 years old and younger.

Distinctly different soil and sediment stratigraphy were discovered beneath T1 compared to T2. The floodplain (T1) consists of a low natural levee paralleling the Cumberland River and a lower lying backswamp behind the levee and bordering T2. Archaeological deposits were encountered throughout the upper two meters of the T1 fill but were shallower and more concentrated in the area of the natural levee. Cultural horizons contained in the T1 fill represented early, middle, and late Woodland occupations that had focused on the slightly higher, better drained natural levee adjacent to the river. The Ap-A-Bw profile of the surface soil is developed throughout the upper 1.5 m of the T1 fill, and the Bw horizon is welded into two weakly expressed buried soils. Middle and Early Woodland occupations were associated with these buried soils, whereas the Late Woodland cultural horizon was encountered above the buried soils (Fig 7.7). Recognition and tracing of these buried soils allowed the investigators to correlate soil and natural stratigraphy and associated archaeological deposits between widely separated block excavations and accomplish mitigation of a large site in an efficient manner.

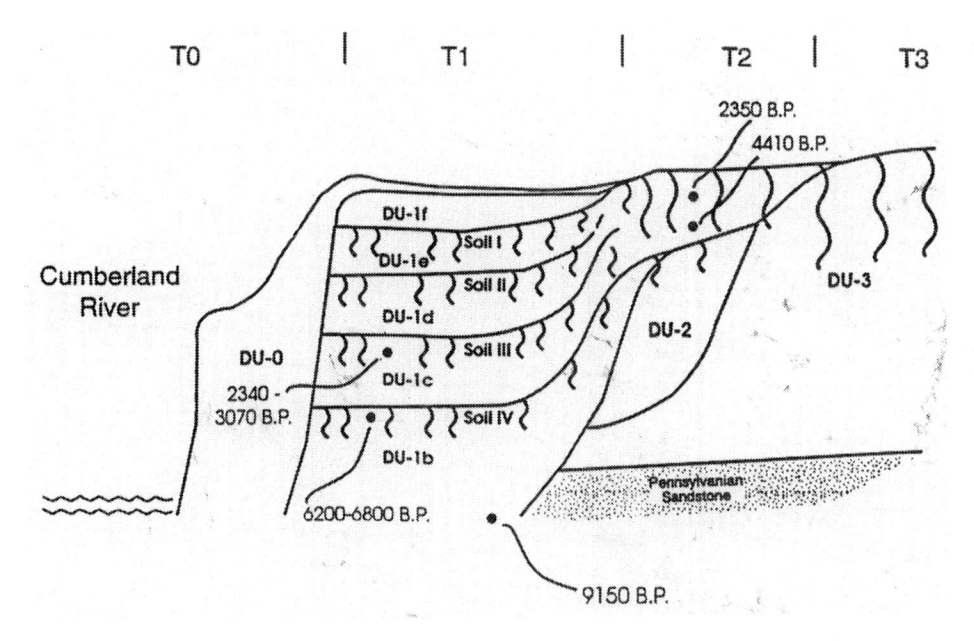

Figure 7.7. Idealized stratigraphic profile at the Main site (15BL35) in the Cumberland River valley of eastern Kentucky (from Bettis, 1994). Several weakly expressed buried soils beneath T_1 merge with the surface soil on T_2. Late Woodland cultural horizons were associated with alluvial deposits above Soil III in the T_1 fill, and Early and Middle Woodland horizons were encountered within and below Soil III. Early Archaic cultural horizons were found within Soil IV beneath T_2, and Middle and Late Archaic as well as Woodland horizons were encountered in the strongly expressed surface soil developed on T_2.

Site formation processes involve a host of natural and anthropogenic processes that combine to produce archaeological deposits (Schiffer, 1983, 1987). Soils can often shed light on the role of natural processes in the formation of archaeological deposits at a site, as well as provide important clues for interpreting the appearance of archaeological deposits in terms of human activity. Mixed soil horizons and thick biomantles point to processes that may have mixed or sorted artifacts and compromised the integrity of some archaeological deposits. Disrupted zones in soils point to major disturbances, such as tree throw, slumping, gullying, or pit digging that mix and destroy the integrity of cultural deposits and, in some cases, produce inverted stratigraphic sequences.

As noted earlier, rates and spatial patterns of sedimentation strongly influence soil morphology and stratigraphy in depositional environments. This is an important consideration during the excavation phase at some archaeological sites. A case in point is the Mahaphy–Akus–Denison (MAD) site (13CF101) in the Boyer River Valley of western Iowa. A Middle and Late Woodland midden was shallowly buried on a low terrace at MAD, whereas diffuse Middle and Late Woodland archaeological deposits were found at much greater depth in the adjacent lower lying floodplain (Benn, 1990). Several A–C soil profiles were present in the floodplain deposits, and these buried soils merged with the more strongly developed surface soil on the adjacent terrace (Fig. 7.8). Based on the soil-stratigraphic record, the late Holocene floodplain was a zone of rapid

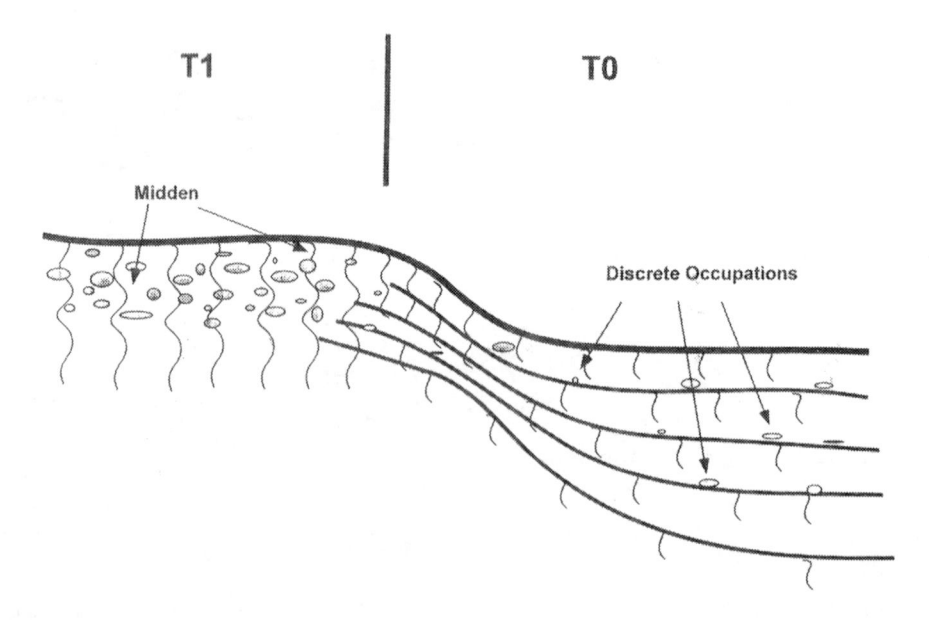

Figure 7.8. At one of the MAD sites (13CF102) in the Boyer River valley of western Iowa, discrete Woodland occupations were associated with weakly expressed buried soils beneath the floodplain (T_0). The Woodland people also inhabited the adjacent terrace (T_1), where sedimentation rates were lower compared to rates on T_0. Consequently, the discrete Woodland occupations observed in weakly expressed buried soils in the T_0 fill are represented by a single midden in a strongly expressed surface soil on T_1 (Bettis, 1990).

sedimentation punctuated by short episodes of landscape stability and pedogenesis. In contrast, the adjacent terrace was a zone of stability and soil formation throughout most of the late Holocene. There is strong evidence suggesting that the rates and spatial patterns of sedimentation during the period of human occupation at the MAD site greatly affected the appearance of the archaeological deposits. For example, the Woodland deposits buried in the floodplain often contained evidence of single residential units that had become subsequently buried, followed by evidence for yet another single residential unit during the next short episode of floodplain stability. Soil-stratigraphic evidence suggesting that the period during which the discrete visits represented in the floodplain area corresponded to the same period represented by the midden on the terrace suggested that the midden represented many episodes of site use by relatively small groups (hamlets) rather than intensive occupation by a large group (a village). Subsequent pottery analyses and radiocarbon ages supported this (Bettis, 1990).

One of the primary goals of archaeological research at a site is the reconstruction of the local and regional environment when the site was occupied or detection of environmental change at the time of site abandonment. Because certain soil properties and soil types are related to climate, archaeologists often turn to pedologists or geoarchaeologists to furnish paleoclimatic information. However, as Holliday (1990) pointed out, soils are probably of limited use for reconstructing climates in archaeological investigations. His reasoning was as follows. In general, soils are not sensitive to discrete climatic changes that may be culturally significant. Also, some climatic changes in the Holocene, the time period that interests most North American archaeologists, were of insufficient magnitude to be detected in the pedological record. Finally, soil properties related to the time factor often resemble those related to climate. It is also important to note that soils change at rates that are usually slower than people respond to a climatic shift.

Despite the limitations described above, soils can be useful in providing some general paleoclimatic and paleoenvironmental information. Soil properties that best reflect the effects of precipitation and temperature on pedogenesis are (1) overall soil morphology, (2) organic-matter and $CaCO_3$ content, (3) depth to leaching, and (4) the depth to the top of the carbonate or salt accumulation zone (Holliday, 1990). Micromorphological features of soils may also be used to infer paleoenvironmental conditions (Fedoroff et al., 1990)

Examples of using soils for paleoclimatic reconstructions at archaeological sites are provided by Reider (1980, 1982, 1990). He noted that soils at terminal Pleistocene archaeological sites in Wyoming and eastern Colorado are mostly Aquolls and Argialbolls with properties indicative of a generally cool, humid environment. In contrast, calcareous or alkaline soils (Calciustolls and Natrargids) at sites dating to the middle Holocene thermal maximum, the Altithermal (ca. 8,000–4,000 B.P.), were interpreted as evidence of significant climatic drying. Reider (1990) suggested that weak development in post-Altithermal soils probably reflects general landscape instability associated with fluctuating Neoglacial climates in the region. Caution should be exercised, however, when inferring regional paleoclimates entirely from soil properties, especially in alluvial settings.

Microenvironmental conditions, such as a perched water table associated with a poorly drained floodplain, can impart soil properties, including gley and carbonate accumulations, that may be incorrectly attributed to regional climatic conditions.

In a recent geoarchaeological investigation, Smith and McFaul (1997) used soil/sediment relationships and supporting geomorphic, paleobotanical, and paleontological data to reconstruct late Quaternary paleoclimates in the San Juan Basin of New Mexico. For example, they documented a strongly expressed buried soil with Stage II to Stage II+ carbonate morphology developed in eolian, alluvial, and playa sediments that were deposited during the late Wisconsinan. Radiocarbon ages suggest that development of this soil was underway by ca. 13,000 B.P. and continued until the soil was buried between ca. 9,300 and 7,800 B.P. Based on the strong morphology of the soil and the regional pollen data, they concluded that the late Pleistocene to early Holocene of the San Juan Basin was relatively moist. In contrast, buried soils that developed in eolian and alluvial sediments towards the end of the Altithermal (ca. 5,000–4,500 B.P.) are weakly expressed and have Bk horizons with Stage I to I+ carbonate morphology. Smith and McFaul (1997) attributed weak soil development during the mid-Holocene to aridity and concomitant landscape instability. They support this interpretation with pollen data and paleohydrological and glacial records for the region.

Soils may also harbor other types of paleoenvironmental information, such as a record of dominant vegetation in the stable carbon and oxygen isotopes (Baker et al., 1998; Cerling et al, 1989; Humphrey and Ferring, 1994; Monger et al., 1998; Nordt et al., 1994) and phytoliths (Fredlund and Tieszen, 1997). In addition, gastropods preserved in soils may yield information on precipitation and temperature trends (Baerreis, 1980), and micromammal remains in soils can provide a wealth of information on the environment as well as human diet (Adovasio et al., 1984; Semken, 1980; Semken and Falk, 1987).

6. Summary and Conclusions

In this chapter, we have shown how soils can and should be used in archaeological investigations. Although many studies have addressed soil science applications in archaeology, pedological information provided to archaeologists is often very technical and, therefore, difficult to apply in the field. This chapter takes a practical approach to soils and archaeology by (1) explaining why it is important for archaeologists to understand the difference between soil and sediment, (2) describing how to distinguish soil from sediment, and (3) discussing the type of soil information needed at different levels of an archaeological investigation.

Distinguishing soil from pedologically unmodified sediment is crucial to all archaeological field investigations. Soils consist of one or more horizons that differ from underlying sediment as a result of the interactions of parent materials, climate, living organisms, and relief through time. Soil horizons are products of pedogenesis, and they develop subsequent to the formation of the body of sediment in which the soil occurs. Hence, soils reflect the passage of time for stable landscapes that supported and recorded human occupations.

During archaeological surveys, it is imperative that information concerning soils geomorphology is gathered early in the investigations. Specifically, knowledge of the temporal and spatial patterns of buried soils enables archaeologists to identify and target areas with high potential for buried cultural deposits. This information also may be used to assess certain aspects of the archaeological record, such as paucity of sites dating to a specific period.

Pedology is also an important component of site evaluations and, as with archaeological surveys, should be implemented during the early stage of investigations. Once the cultural deposits are placed in a soil-stratigraphic context, the potential depth and lateral extent of archaeological materials may be determined. Information gleaned from soils at a site also can be used to assess the integrity of the cultural deposits. Soils are very dynamic, and the various soil properties, factors, and conditions often determine the extent to which archaeological materials are preserved, modified, moved, or destroyed by postdepositional processes. In situations where absolute time control is absent, soils can be used to infer the relative age of cultural deposits. This can be accomplished through direct methods, such as dating the carbon in the artifact-bearing soils, or indirectly through correlation of soils or using data derived from soil chronosequences.

Soils can also serve as key stratigraphic markers for deciphering site stratigraphy and for correlating former land surfaces and human occupation zones. In addition, paleoenvironmental information gleaned from soils can be essential for better understanding the environment during site occupation, and it may provide insights on environmental changes that influenced human subsistence and settlement strategies.

In sum, soils are historical archives that, if interpreted properly, can provide archaeologists with a wide range of information for locating and interpreting archaeological deposits. With a general understanding of the potentials and limitations inherent in interpretation of soils, archaeologists can frame questions and research strategies that will yield important new information for interpreting the record of the human past.

ACKNOWLEDGMENTS. We are grateful to Paul Goldberg, Vance Holliday, and Reid Ferring for their thoughtful comments that helped improve the quality of our manuscript. We also thank Nora Ransom for editing the manuscript.

7. References

Adovasio, J. M., Dohahue, J., Carlisle, R. C., Cushman, K., Stuckenrath, R., and Wiegman, P., 1984, Meadowcroft Rockshelter and the Pleistocene/Holocene Transition in Southwestern Pennsylvania. In *Contributions in Quaternary Vertebrate Paleontology: A Volume in Memorial to John E. Guilday*, edited by H. H. Genoways and M. R. Dawson, pp. 347–369. Carnegie Museum of Natural History, Special Publication No. 8, Pittsburgh, PA.

Ahler, S. A., 1973, A Chemical Analysis of Deposits at Rogers Shelter, Missouri, *Plains Anthropologist* 18:116–131.
Allen, J. R. L., 1986, Pedogenic Calcretes in the Old Red Sandstone Facies (Late Silurian–Early Carboniferous) of the Anglo–Welsh Area, Southern Britain. In *Paleosols: Their Recognition and Interpretation*, edited by V. P. Wright, pp. 58–86. Princeton University Press, Princeton, NJ.
Almy, M. M., 1978, The Archaeological Potential of Soil Survey Reports. *The Florida Anthropologist* 31:75–91.
Anderton, J. B., 1999, The Soil–Artifact Context Model: A Geoarchaeological Approach to Paleoshoreline Site Dating in the Upper Peninsula of Michigan USA, *Geoarchaeology: An International Journal* 14:265–288.
Arakel, A. V., 1986, Evolution of Calcrete in Paleodrainages of the Lake Napperby Area, Central Australia, *Palaeogeography, Palaeoclimatology, Palaeoecology 54:283*-303.
Artz, J. A., 1985, A Soil–Geomorphic Approach to Locating Buried Late-Archaic Sites in Northeast Oklahoma, *American Archaeology* 5:142–150.
Artz, J. A., 1993, The Preservation of Cultural Stratigraphy in Loess-mantled Terrains of Iowa, *Journal of the Iowa Archeological Society* 40:50–62.
Artz, J. A., 1995a, *Archaeology of the Eisele's Hill Locality: Phase II Test Excavations at Six Sites in Muscatine County, Iowa, Primary Roads Project NHS-61-4(55)— 20-70 a.k.a. PIN 92-70040-1*.Project Completion Report Vol. 18, No. 30. Highway Archaeology Program, Office of the State Archaeologist, University of Iowa, Iowa City.
Artz, J. A., 1995b, Geological Contexts of the Early and Middle Holocene Archaeological Record in North Dakota and Adjoining Areas of the Northern Plains. In *Archaeological Geology of the Archaic Period in North America*, edited by E. A. Bettis III, pp. 67-86. Special Paper 297, The Geological Society of America, Boulder, CO.
Bachman, G. O., and Machette, M. N., 1977, *Calcic Soils and Calcretes in the Southwestern United States*. U.S. Geological Survey Open File Report 77-794, Washington, DC.
Baerreis, D. A., 1980, Habitat and Climatic Interpretation Derived from Terrestrial Gastropods at the Cherokee Sewe Site. In *The Cherokee Excavations: Holocene Ecology and Human Adaptations in Northwestern Iowa*, edited by DC. Anderson and H. A. Semken, Jr., pp. 101–122. Academic Press, New York.
Baker, R. G., Gonzalez, L. A., Raymo, M., Bettis, E. A., III, Reagan, M. K., and Dorale, J. A., 1998, Comparison of Multiple Proxy Records of Holocene Environments in the Midwestern United States, *Geology* 26:1131–1134.
Balek, C., 1998, Buried Artifacts in Stable Upland Sites in Illinois: *In situ* or Not? In *Abstracts of the 63rd Annual Meeting of the Society for American Archaeology*, p. 40. Society for American Archaeology, Washington, DC.
Benn, D. W. (ed.), 1990, *Woodland Cultures on the Western Prairies: The Rainbow Site Investigations*. Report No. 18, Office of the State Archaeologist, University of Iowa, Iowa City.
Benn, D. W., and Bettis, E. A., III, 1985, *Archaeology and Landscapes in Saylorville Lake, Iowa*. Fieldtrip Guidebook. Association of Iowa Archaeologists, Iowa City.
Bettis, E. A., III, 1990, *Holocene Alluvial Stratigraphy and Selected Aspects of the Quaternary History of Western Iowa*. University of Iowa, Iowa Quaternary Studies Group Contribution 36, Iowa City.
Bettis, E. A., III, 1992, Soil Morphological Properties and Weathering Zone Characteristics as Age Indicators in Holocene Alluvium in the Upper Midwest. In *Soils in Archaeology*, edited by V. T. Holliday, pp. 119–144. Smithsonian Institution Press, Washington, DC.
Bettis, E. A., III, 1994, Chapter 5. Geology of the Main Site. In *Upper Cumberland Archaic and Woodland Period Archeology at the Main Site (15BL35), Bell County, Kentucky*, Volume 1, edited by S. D. Creasman, pp. 5-1–5-20. Contract Publication Series 94-56. Cultural Resource Analysis, Inc., Lexington, KY.
Bettis, E. A., III, and Benn, D. W., 1984, An Archaeological and Geomorphological Survey in the Central Des Moines River Valley, Iowa, *Plains Anthropology* 29:211–226.
Bettis, E. A., III, and Hajic, E. R., 1995, Landscape Development and the Location of Evidence of Archaic Cultures in the Upper Midwest. In *Archaeological Geology of the Archaic Period in North America*, edited by E. A. Bettis, III, pp. 87–113. The Geological Society of America, Special Paper, 297 Boulder, CO.

Bettis, E. A., III, and Littke, J. P., 1987, *Holocene Alluvial Stratigraphy and Landscape Development in Soap Creek Watershed, Appanoose, Davis, Monroe, and Wapello Counties, Iowa*. Open File Report 87-2. Iowa Department of Natural Resources, Geological Survey Bureau, Iowa City.
Bettis, E. A., III, and Thompson, D. M., 1981, Holocene Landscape Evolution in Western Iowa— Concepts, Methods, and Implications for Archaeology. In *Current Directions in Midwestern Archaeology: Selected Papers from the Mankato Conference*, edited by S. F. Anfinson, pp. 1–14. Occasional Papers in Minnesota Archaeology 9. Minnesota Archaeological Society, St. Paul.
Bettis, E. A., III, and Thompson, D. M., 1982, *Interrelations of Cultural and Fluvial Deposits in Northwestern Iowa*. Fieldtrip Guidebook. Association of Iowa Archaeologists, Iowa City.
Bettis, E. A., III, Baker, R. G., Green, W. R., Whelan, M. K., and Benn, D. W., 1992, *Late Wisconsinan and Holocene Alluvial Stratigraphy, Paleoecology, and Archaeological Geology of East-Central Iowa*. Guidebook Series No. 12. Iowa Department of Natural Resources, Geological Survey Bureau, Iowa City.
Bilzi, A. F., and Ciolkosz, E. J., 1977, A Field Morphology Rating Scale for Calculating Pedogenic Development, *Soil Sciene* 124:45–48.
Birkeland, P. W., 1999, *Soils and Geomorphology*, 3rd ed., Oxford University Press, Oxford, UK.
Birkeland, P. W., and Burke, R. M., 1988, Soil Catena Chronosequences on Eastern Sierra Nevada Moraines, California, U.S.A., *Arctic Alpine Research* 20:473–484.
Birkeland P. W., Machette, M. N., and Haller, K. M., 1991, *Soils as a Tool for Applied Quaternary Geology*. Utah Geological and Mineral Survey Miscellaneous Publications 91-3. Salt Lake City.
Birkeland P. W., Walker, A. L, Benedict, J. B., and Fox, F. B., 1979, Morphological and Chemical Trends in Soil Chronosequences; Alpine and Arctic Environments. In *Agronomy Abstracts*, p. 188. American Society of Agronomy, Madison, WI.
Bischoff, J. L., Shelmon, R. J., Ku, T. L., Simpson, R. D., Rosenbauer, R. J., and Budinger, F. E., 1981, Uranium-series and Soil-geomorphic Dating of the Calico Archaeological Site, California, *Geology* 9:576–582.
Brewer, R., 1976, *Fabric and Mineral Analysis of Soils*, 2nd ed., Krieger, Huntington, NY.
Brinkmann, R., 1996, Pedological Characteristics of Anthrosols in the al-Jadidah Basin of Wadi al-Jubah, and Native Sediments in Wadi al-Ajwirah, Yemen Arab Republic. In *The Wadi Al-Jubah Archaeological Project*, Volume 5, *Environmental Research in Support of Archaeological Investigations in the Yemen Arab Republic, 1982–1987*, edited by W. C. Overstreet and J. A. Blakely, pp. 45–211. American Foundation for the Study of Man, Washington, DC.
Bullock, P., Fedoroff, N., Jongerius, A., Stoops, G., and Tursina, T., 1985, *Handbook for Thin Section Description*, Waine Research Publications, Wolverhampton, UK.
Buol, S. W., Hole, F. D., and McCracken, R. J., 1997, *Soil Genesis and Classification*, 4th ed., Iowa State University Press, Ames.
Butzer, K. W., 1971, *Environment and Archaeology*, 2nd ed., Aldine, Chicago.
Butzer, K. W., 1982, *Archaeology as Human Ecology: Method and Theory for a Contextual Approach*, Cambridge University Press, Cambridge, MA.
Catt, J. A., 1986, Soils and Quaternary Geology; A Handbook for Field Scientists, *Monographs on Soil and Resources Survey 11*. Oxford Science Publications, Oxford, UK.
Cerling, T. E., Quade, J., Wang, Y., and Bowman, R., 1989, Carbon Isotopes in Soils and Paleosols as Ecology and Paleoecology Indicators, *Nature* 341:138–139.
Cornwall, I. W., 1958, *Soils for the Archaeologist*, Phoenix House, London.
Courty, M. A., 1992, Soil Micromorphology in Archaeology, *Proceeding of the British Academy* 77:39–59.
Courty, M. A., Goldberg, P., and Macphail, R. I., 1989, *Soils, Micromorphology and Archaeology*, Cambridge University Press, Cambridge, MA.
Courty, M. A., Macphail, R. I., and Wattez, J., 1991, Soil Micromorphological Indicators of Pastoralism with Special Reference to Arene Candide, Fianle Ligure, Italy, *Rivista di Studi Liguri A*. LVII:127–150.
Creasman, S. D., 1994, *Upper Cumberland Archaic and Woodland Period Archeology at the Main Site (15BL35), Bell County, Kentucky*. Volume I. Contract Publication Series 94-56. Cultural Resource Analysis, Inc., Lexington, KT.
Cremeens, D. L., and Hart, J. P., 1995, On Chronostratigraphy, Pedostratigraphy, and Archaeological Context. In *Pedological Perspectives in Archaeological Research*, edited by M. E. Collins, B. J. Carter,

B. G. Gladfelter, and R. J. Southard, pp. 15-33. SSSA Special Publication No. 44. Soil Science Society of America, Madison, WI.

Darwin, C., 1881, *The Formation of Vegetable Mould through the Action of Worms*. Appleton, New York.

Dethier, D. P., 1988, *The Soil Chronosequence along the Cowlitz River Washington*. U.S. Geological Survey Bulletin 1590-F, Washington, DC.

Eidt, R. C., 1977, Detection and Examination of Anthrosols by Phosphate Analysis, *Science* 197:1327–1333.

Eidt, R. C., 1985, Theoretical and Practical Considerations in the Analysis of Anthrosols. In *Archaeological Geology*, edited by G. Rapp, Jr. and J. A. Gifford, pp. 155–190. Yale University Press, New Haven. CT.

Evans, J. D., 1978, *An Introduction to Environmental Archaeology*, Paul Elek, London.

Fedoroff, N., Courty, M. A., and Thompson, M. L., 1990, Micromorphological Evidence of Paleoenvironmental Change in Pleistocene and Holocene Paleosols, In *Soil Micromorphology (A Basic and Applied Science)*, edited by L. Douglas, pp. 652–665. Elsevier, New York.

Ferring, C. R., 1986, Rates of Fluvial Sedimentation: Implications for Archaeological Variability, *Geoarchaeology: An International Journal* 1:259–274.

Ferring, C. R., 1990, Archaeological Geology of the Southern Plains. In *Archaeological Geology of North America*, edited by N. P. Lasca and J. Donahue, pp. 253–266. The Geological Society of America, Centennial Special Volume 4. Boulder, CO.

Ferring, C. R., 1992, Alluvial Pedology and Geoarchaeological Research. In *Soils in Archaeology*, edited by V. T. Holliday, pp. 1–39. Smithsonian Institution Press, Washington, DC.

Ferring, C. R., 1995, Middle Holocene Environments, Geology, and Archaeology in the Southern Plains. In *Archaeological Geology of the Archaic Period in North America*, edited by E. A. Bettis, III, pp. 21–35. The Geological Society of America, Special Paper 297, Boulder CO.

Fisher, P. F., and Macphail, R. I., 1985, Studies of Archaeological Soils and Deposits by Micromorphological Techniques. In *Palaeoenvironmental Investigations: Research Design, Methods and Data Analysis*, edited by N. R. J. Fieller, D. D. Gilbertson, and N. G. A. Ralph, pp. 93–112. British Archaeological Report (BAR) International Series 258, Oxford, UK.

Foss, J. E., Timpson, M. E., and Lewis, R. J., 1995, Soils in Alluvial Sequences: Some Archaeological Implications. In *Pedological Perspectives in Archaeological Research*, edited by M. E. Collins, B. J. Carter, B. G. Gladfelter, and R. J. Southard, pp. 1–14. SSSA Special Publication No. 44. Soil Science Society of America, Madison, WI.

Fredlund, G. G., and Tieszen, L. L., 1997, Phytolith and Carbon Isotope Evidence for Late Quaternary Vegetation and Climate Change in the Southern Black Hills, South Dakota. *Quaternary Research* 47:206–217.

Frink, D. S., 1992, The Chemical Variability of Carbonized Organic Matter through Time, *Archaeology of Eastern North America*, 20:67–79.

Frink, D. S., 1995, Applications of Oxidizable Carbon Ratio Dating Procedure and its Implications for Pedogenic Research. In *Pedological Perspectives in Archaeological Research*, edited by M. E. Collins, B. J. Carter, B. G. Gladfelter, and R. J. Southard, pp. 95–106. SSSA Special Publication No. 44. Soil Science Society of America, Madison, WI.

Geyh, M. A., Benzler, J.-H., and Roeschmann, G., 1975, Problems of Dating Pleistocene and Holocene Soils by Radiometric Methods. In *Paleopedology: Origin, Nature, and Dating of Paleosols*, edited by D. H. Yaalon, pp. 63–75. Israel University Press, Jerusalem.

Gile, L. H., 1979, Holocene Soils in Eolian Sediments of Bailey County, Texas, *Soil Science Society of America Journal*, 43:994–1003.

Gile, L. H., 1985, *The Sandhills Project Soil Monograph*, New Mexico State University, Rio Grande Historical Collections, Las Cruces.

Gile, L. H., and Hawley, J. W., 1966, Periodic Sedimentation and Soil Formation on an Alluvial-Fan Piedmont in Southern New Mexico, *Soil Science Society of America Proceedings* 30:261–268.

Gile, L. H., Peterson, F. F., and Grossman, R. B., 1979, *The Desert Project Soil Monograph*. U.S. Department of Agriculture, Soil Conservation Service, Washington, DC.

Gile, L. H., Hawley, J. W., and Grossman, R. B., 1981, *Soils and Geomorphology in the Basin and Range Area of Southern New Mexico—Guidebook to the Desert Project*. New Mexico Bureau of Mines and Mineral Resources Memoir 39. Sogiro.

Goldberg, P., 1983, Application of Micromorphology in Archaeology. In *Soil Micromorphology*, edited by P. Bullock and C. P. Murphy, pp. 139–150. AB Academic Publishers, Berkhamsted.

Goldberg, P., 1987, Sediments and Acheulian Artifacts at Berekhat Ram, Golan Heights. In *Micromorphologie des Sols — Soil Micromorphology*, edited by N. Federoff, L. M. Bresson, and M. A. Courty, pp. 583–589. Association Francaise pour /'Etude du Sol (A.F.E.S.), Plaisir, France.

Goldberg, P., 1992, Micromorphology, Soils, and Archaeological Sites. In *Soils in Archaeology*, edited by V. T. Holliday, pp. 145–167. Smithsonian Institution Press, Washington, DC.

Gordon B. C., 1978, Chemical and Pedological Delimiting of Deeply Stratified Archaeological Sites in Frozen Ground, *Journal of Field Archaeology* 5:331–338.

Green, W., and Doershuk, J. F., 1998, Cultural Resource Management and American Archaeology, *Journal of Archaeology Research* 6:121–167.

Groenman-van Waateringe, W., and Robinson, M. (eds.), 1988, *Man-Made Soils*. British Archaeological Report (BAR) International Series 410, Osney Mead, UK.

Gvirtzman, G., Wieder, M., Marder, O., Khalaily, H., Rabinovich, R., and Ron, H., 1999, Geological and Pedological Aspects of an Early-Paleolithic Site: Revadim, Central Coastal Plain, Israel., *Geoarchaeology: An International Journal* 14:101–126.

Haas, H., Holliday, V. T., and Stuckenrath, R., 1986, Dating of Holocene Stratigraphy with Soluble and Insoluble Organic Fractions at the Lubbock Lake Archaeological Site, Texas: An Ideal Case Study, *Radiocarbon* 28:473–485.

Hajic, E. R., 1990, *Koster Site Archeology I: Stratigraphy and Landscape Evolution*. Research Series Vol. 8. Kampsville Archeological Center, Center for American Archeology, Kampsville, IL.

Hall, S. A., 1977, Geology and Palynology of Archaeological Sites and Associated Sediments. In *The Prehistory of the Little Caney River, 1976 Season*, edited by D. O. Henry, pp. 13–41. Contributions in Archaeology 1. University of Tulsa Laboratory of Archaeology, Tulsa, OK.

Hallberg, G. R., Fenton, T. E., and Miller, G. A., 1980, Standard Weathering Zone Terminology for Description of Quaternary Sediments in Iowa. In *Standard Procedures for Evaluation of Quaternary Materials in Iowa*, edited by G. R. Hallberg, pp. 75–109. Technical Information Series No. 8. Iowa Geological Survey, Iowa City.

Harden, J. W., 1982, A Quantitative Index of Soil Development from Field Descriptions: Examples from a Chronosequence in Central California, *Geoderma* 28:1–28

Hassan, F. A., 1978, Archaeological Explorations of the Siwa Oasis Region, Egypt, *Current Anthropology* 19:146–148.

Haynes, C. V., and Grey, DC., 1965, The Sisters Hill Site and its Bearing on the Wyoming Postglacial Alluvial Chronology, *Plains Anthropology* 10:196–207.

Hesse, P. R., 1971, *A Textbook of Soil Chemical Analysis*, Chemical Publishing Company, New York.

Holliday, V. T., 1985a, Archaeological Geology of the Lubbock Lake Site, Southern High Plains of Texas, *Geologica Society of America Bulletin* 96:1483–1492.

Holliday, V. T., 1985b, Holocene Soil-Geomorphological Relationships in a Semi-arid Environment: The Southern High Plains of Texas. In *Soils and Quaternary Landscape Evolution*, edited by J. Boardman, pp. 325–357. Wiley, London.

Holliday, V. T., 1985c, Morphology of Late Holocene Soils at the Lubbock Lake Archaeological Site, Texas, *Soil Science Society of America Journal* 49:938–946.

Holliday, V. T., 1985d, New Data on the Stratigraphy and Pedology of the Clovis and Plainview Sites, Southern High Plains, *Quaternary Research* 23:388–402.

Holliday, V. T., 1988, Genesis of a Late-Holocene Soil Chronosequence at the Lubbock Lake Archaeological Site, Texas, *Annals of the Association of American Geographers* 78:594–610.

Holliday, V. T., 1989, Paleopedology in Archeology. In *Paleopedology: Nature and Applications of Paleosols*, edited by A. Bronger and J. Catt, *Catena Supplement* 16:187–206.

Holliday, V. T., 1990, Pedology in Archaeology. In *Archaeological Geology of North America*, edited by N. P. Lasca and J. Donahue, pp. 525–540. The Geological Society of America, Centennial Special Volume 4, Boulder, CO.

Holliday, V. T., 1992, Soil Formation, Time, and Archaeology. In *Soils in Archaeology*, edited by V. T. Holliday, pp. 101–117. Smithsonian Institution Press, Washington, DC.

Holliday, V. T., 1995, *Stratigraphy and Paleoenvironments of Late Quaternary Valley Fills on the Southern High Plains*. The Geological Society of America, Memoir 186, Boulder, CO.

Holliday, V. T., 1997, *Paleoindian Geoarchaeology of the Southern High Plains*, University of Texas Press, Austin.

Holliday, V. T., and Meltzer, D. J., 1996, Geoarchaeology of the Midland (Paleoindian) Site, Texas, *American Antiquity* 61:755–771.

Holliday, V. T, Johnson, E., Haas, H., and Stuckenrath, R., 1983, Radiocarbon Ages from the Lubbock Lake Site, 1950–1980, Framework for Cultural and Ecological Change in the Southern High Plains, *Plains Anthropology* 28:165–182.

Holliday, V. T, Johnson, E., Haas, H., and Stuckenrath, R., 1985, Radiocarbon Ages from the Lubbock Lake Site: 1981–1984, *Plains Anthropology* 30:277–292.

Holliday, V. T., Ferring, C. R., and Goldberg, P., 1993, The Scale of Soil Investigations in Archaeology. In *Effects of Scale on Archaeological and Geoscientific Perspectives*, edited by J. K. Stein and A. R. Linse, pp. 29–37. The Geological Society of America, Special Paper 283, Boulder, CO.

Hoyer, B. E., 1980, The Geology of the Cherokee Sewer Site. In *The Cherokee Excavations: Holocene Ecology and Human Adaptations in Northwestern Iowa*, edited by D. C. Anderson and H. A. Semken, Jr., pp. 21–66. Academic Press, New York.

Humphrey, J. D., and Ferring, C. R., 1994, Stable Isotope Evidence for Latest Pleistocene to Holocene Climatic Change in North-Central Texas, *Quaternary Research* 41:200–213.

Jacob, J. S., 1995, Archaeological Pedology in the Maya Lowlands. In *Pedological Perspectives in Archaeological Research*, edited by M. E. Collins, B. J. Carter, B. G. Gladfelter, and R. J. Southard, pp. 51–80. SSSA Special Publication No. 44. Soil Science Society of America, Madison, WI.

Jacobsen, G., Arakel, A. V., and Yijian, Y., 1988, The Central Australian Groundwater Discharge Zone: Evolution of Associated Calcrete and Gypdrete Deposits, *Australian J. Earth Sci.* 35: 549–565.

Johnson, D. L., 1990, Biomantle Evolution and the Redistribution of Earth Materials and Artifacts, *Soil Science* 149: 84–102.

Johnson, D. L., and Watson-Stegner, D, 1990, The Soil Evolution Model as a Framework for Evaluating Pedoturbation in Archaeological Site Formation. In *Archaeological Geology of North America*, edited by N. P. Lasca and J. Donahue, pp. 541–560. The Geological Society of America, Centennial Special Volume 4, Boulder, CO.

Johnson, W. C, and Logan, B., 1990, Geoarchaeology of the Kansas River Basin, Central Great Plains. In *Archaeological Geology of North America*, edited by N. P. Lasca and J. Donahue, pp. 267–299. The Geological Society of America, Centennial Special Volume 4, Boulder, CO.

Karlstrom, E. T., 1988, Rates of Soil Formation on Black Mesa, Northeast Arizona: A Chronosequence in Late Quaternary Alluvium, *Physical Geography* 9:301–327.

Kolb, M. F., Lasca, N. P., and Goldstein, L. G., 1990, A Soil-Geomorphic Analysis of the Midden Deposits at the Aztalan Site, Wisconsin. In *Archaeological Geology of North America*, edited by N. P. Lasca and J. Donahue, pp. 199–218. The Geological Society of America, Centennial Special Volume 4, Boulder, CO.

Limbrey, S., 1975, *Soil Science in Archaeology*, Academic Press, London.

Lotspeich, F. B., 1961, Soil Science in the Service of Archaeology. In *Paleoecology of the Llano Estacado*, edited by F. Wendorf, pp. 137–139. Publication 1. Fort Burgwin Research Center, The Museum of New Mexico Press, Santa Fe.

Machette, M. N., 1985, Calcic Soils of the Southwestern United States. In *Soils and Quaternary Geology of the Southwestern United States*, edited by D. L. Weide, pp. 1–21. The Geological Society of America, Special Paper 203, Boulder, CO.

Macphail, R. I., 1986, Paleosols in Archaeology: Their Role in Understanding Flandrian Pedogenesis. In *Paleosols: Their Recognition and Interpretation*, edited by V. P. Wright, pp. 263–290. Blackwell, Oxford, UK.

Macphail, R. I., 1992, Soil Micromorphological Evidence of Ancient Soil Erosion. In *Past and Present Soil Erosion*, edited by M. Bell and J. Boardman, pp. 197–215. Oxford Books, Oxford, UK.

Macphail, R., and Goldberg, P., 1995, Recent Advances in Micromorphological Interpretations of Soils and Sediments from Archaeological Sites. In *Archaeological Sediments and Soils: Analysis, Interpretation and Management*, edited by A. J. Barham and R. I. Macphail, pp. 1–24e. Institute of Archaeology, University College, London.

Macphail, R. I., Courty, M. A., and Goldberg, P., 1990, Soil Micromorphology in Archaeology, *Endeavour, New Series* 14:163–171.

Macphail, R. I., Hather, J., Hillson, S., and Maggi, R., 1994, The Upper Pleistocene Deposits at Arene Candide: Soil Micromophology of Some Samples from the Cardini 1940–42 Excavation. *Quaternary Nova* 4:79–100.

Mandel, R. D., 1987, Geomorphological Investigations. In *Buried in the Bottoms: The Archaeology of Lake Creek Reservoir, Montgomery County, Texas*, edited by L. C. Bement, R. D. Mandel, J. de la Teja, D. Utley, and S. Turpin, pp. 4-1–4-41. Research Report 97. Texas Archaeological Survey, The University of Texas at Austin.

Mandel, R. D., 1988, Geomorphology of the Pawnee River Valley. In *Phase II Archaeological and Geomorphological Survey of the Proposed Pawnee River Watershed, Covering Subwatersheds 3 through 7, Ness, Fort, Lane, and Finney Counties, Southwest Kansas*, edited by R. D. Timberlake, pp. 68–115. Kansas Historical Society, Topeka.

Mandel, R. D., 1992, Soils and Holocene Landscape Evolution in Central and Southwestern Kansas. In *Soils in Archaeology*, edited by V. T. Holliday, pp. 41–117. Smithsonian Institution Press, Washington, DC.

Mandel, R. D., 1993a, Geomorphology. In *Phase II Cultural Resource Survey of High Potential Areas within the Southeast Kansas Highway Corridor*, edited by T. Weston, pp. 44–121. Contract Archeology Publication Number 10. Kansas State Historical Society, Topeka.

Mandel, R. D., 1993b, Geomorphology. In *Cultural Resource Investigations for the U.S. Highway 166 Corridor*, edited by M. F. Hawley, pp. 24–75. Contract Archeology Publication Number 11. Kansas State Historical Society, Topeka.

Mandel, R. D., 1994, Geoarchaeology of the Lower Walnut River Valley at Arkansas City, Kansas, *The Kansas Anthropologist* 15:46–69.

Mandel, R. D., 1995, Geomorphic Controls of the Archaic Record in the Central Plains of the United States. In *Archaeological Geology of the Archaic Period in North America*, edited by E. A. Bettis, III, pp. 37–66. The Geological Society of America, Special Paper 297, Boulder, CO.

Mandel, R. D., 1996, Geomorphology of the South Fork Big Nemaha River Valley, Southeastern Nebraska. In *A Geoarchaeological Survey of the South Fork Big Nemaha Drainage, Pawnee and Richardson Counties, Nebraska*, edited by S. R. Holen, J. K. Peterson, and D. R. Watson, pp. 26–81. Technical Report 96-02. Nebraska Archaeological Survey, University of Nebraska State Museum, Lincoln.

Mandel, R. D., 1997, *Geomorphological Investigation in Support of the Phase I Archaeological Survey of the Highway 60 Corridor, Northwest Iowa*. Report prepared for RUST Environment and Infrastructure, Inc., Waterloo, IA.

Mandel, R. D., 1999, *Geomorphology and Late Quaternary Stratigraphy of the Big Blue River and Lower Beaver Creek Valleys, Southern Nebraska*, Volume 1, *Archeological Investigations of the Lower Beaver Creek and Big Blue Drainages in Furnas, Red Willow, Pawnee and Gage Counties, Nebraska: 1997–1998*. Archeological Contract Series No. 137. Archeology Laboratory, Augustana College, Sioux Falls, SD.

Mandel, R. D., Reynolds, J. D., Williams, B. G., and Wulfkuhle, V. A., 1991, *Upper Delaware River and Tributaries Watershed: Results of Geomorphological and Archeological Studies in Atchison, Brown, Jackson, and Nemaha Counties, Kansas*. Contract Archeology Publication No. 9. Kansas State Historical Society, Topeka.

Martin, C. W., and Johnson, W. C., 1995, Variation in Radiocarbon Ages of Soil Organic Matter Fractions from Late Quaternary Buried Soils, *Quaternary Research* 43:232–237.

Matthews, J. A., 1985, Radiocarbon Dating of Surface and Buried Soils; Principles, Problems, and Prospects. In *Geomorphology and Soils*, edited by K. S. Richards, R. R. Arnett, and S. Ellis, pp. 269–288. Allen and Unwin, London.

May, D. W., 1986, Geomorphology. In *Along the Pawnee Trail: Cultural Resource Survey and Testing at Wilson Lake, Kansas*, edited by D. J. Blakeslee, R. Blasing, and H. Garcia, pp. 72–86. U.S. Army Corps of Engineers, Kansas City, MO.

McFadden, L. D., and Hendricks, D. M., 1985, Changes in the Content and Composition of Pedogenic Iron Oxyhydroxides in a Chronosequence of Soils in Southern California, *Quatern. Res.* 23:189–204.

McFadden, L. D., and Weldon, R. J., 1987, Rates and Processes of Soil Development on Quaternary Terraces in Cajon Pass, California, *Geological Society of America Bulletin* 98:280–293.

McFadden, L. D., Wells, S. G., and Dohrenwend, J. C., 1986, Influence of Quaternary Climate Changes on Processes of Soil Development in Desert Loess Deposits of the Cima Volcanic Field, California, *Catena* 13:361–389.

Mikkelsen, J. H., and R. Kangohr, 1996, *A Pedological Characterization of the Aubechies Soil, a Well Preserved Soil Sequence Dated to the Earliest Neolithic Agriculture in Belgium*. Paper presented at the

Paleoecology XIII International Congress of Prehistoric and Protohistoric Sciences, Colloquium VI, Micromorphology of Deposits of Anthropogenic Origin, Forli, Italy.

Monger, H. C., 1995, Pedology in Arid Lands Archaeological Research: An Example from Southern New Mexico–Western Texas. In *Pedological Perspectives in Archaeological Research*, edited by M. E. Collins, B. J. Carter, B. G. Gladfelter, and R. J. Southard, pp. 35–50. SSSA Special Publication No. 44. Soil Science Society of America, Madison, WI.

Monger, H. C., Cole, D. R., Gish, J. W., and Giordano, T. H., 1998, Stable Carbon and Oxygen Isotopes in Quaternary Soil Carbonates as Indicators of Ecogeomorphic Changes in the Northern Chihuahuan Desert, USA, *Geoderma* 82:137–172.

Nordt, L. C., 1995, Geoarchaeological Investigations of Henson Creek: A Low-Order Tributary in Central Texas, *Geoarchaeology: An International Journal* 10:205–221.

Nordt, L. C., Boutton, T. W., Hallmark, C. T., and Waters, M. R., 1994, Late Quaternary Vegetation and Climate Change in Central Texas Based on Isotopic Composition of Organic Carbon, *Quaternary Research* 41:109–120.

Paulissen, E., and Vermeersch, P. M., 1987, Earth, Man and Climate in the Egyptian Nile Valley During the Pleistocene. In *Prehistory of Arid North Africa: Essays in Honor of Fred Wendorf*, edited by A. E. Close, pp. 29–67. Southern Methodist University Press, Dallas, TX.

Ranov, V. A., and Davis, R. S., 1979, Toward a New Outline of the Soviet Central Asian Paleolithic, *Current Anthropology* 20:249–270.

Rapp, G., Jr., and Hill, C. L., 1998, *Geoarchaeology: The Earth-Science Approach to Archaeological Interpretation*, Yale University Press, New Haven, CT.

Reeves, B. O. K., and Dormaar, J. F., 1972, A Partial Holocene Pedological and Archaeological Record for the Southern Alberta Rocky Mountains, *Arctic and Alpine Research* 4:325–336.

Reid, K. C., and Artz, J. A., 1984, *Hunters of the Forest Edge: Culture, Time, and Process in the Little Caney Basin (1980, 1981, and 1982 Field Seasons)*. Contributions in Archaeology 14. University of Tulsa, Laboratory of Archaeology, Tulsa, OK.

Reider, R. G., 1980, Late Pleistocene and Holocene Soils of the Carter/Kerr-McGee Archaeological Site, Powder River Basin, Wyoming, *Catena* 7:301–315.

Reider, R. G., 1982, Soil Development and Paleoenvironments. In *The Agate Basin Site; A Record of Paleoindian Occupation of the Northwestern High Plains*, edited by G. C. Frison and D. J. Stanford, pp. 331–344. Academic Press, New York.

Reider, R. G., 1990, Late Pleistocene and Holocene Pedogenic and Environmental Trends at Archaeological Sites in Plains and Mountain Areas of Colarado and Wyoming. In *Archaeological Geology of North America*, edited by N. P. Lasca and J. Donahue, pp. 335–360. The Geological Society of America, Centennial Special Volume 4, Boulder, CO.

Rolfsen, P., 1980, Disturbance of Archaeological Layers by Processes in the Soils, *Norwegian Arch. Rev.* 13:110–118.

Sandor, J. A., 1992, Long-Term Effects of Prehistoric Agriculture on Soils: Examples from New Mexico and Peru. In *Soils in Archaeology: Landscape Evolution and Human Occupation*, edited by V. T. Holliday, pp. 217–246. Smithsonian Institution Press, Washington, DC.

Saucier, R. T., 1966, Soil-Survey Reports and Archaeological Investigations, *American Antiquity* 31:419–422.

Schaetzel, R. J., Johnson, D. L., Burns, S. F., and Small, T. W., 1986, Tree Uprooting: Review of Terminology, Process, and Environmental Implications, *Canadian Journal of Forest Research* 19:1–11.

Scharpenseel, H. W., 1971, Radiocarbon Dating of Soils—Problems, Troubles, Hopes. In *Paleopedology: Origin, Nature, and Dating of Paleosols*, edited by D. H. Yaalon, pp. 77–88. Israel University Press, Jerusalem.

Schiffer, M. B., 1983, Toward the Identification of Formation Processes, *American Antiquity* 48:675–706.

Schiffer, M. B., 1987, *Formation Processes of the Archaeological Record*, University of New Mexico Press, Albuquerque.

Schuldenrein, J., 1995, Geochemistry, Phosphate Fractionation, and the Detection of Activity Areas at Prehistoric Sites. In *Pedological Perspectives in Archaeological Research*, edited by M. E. Collins, B. J. Carter, B. G. Gladfelter, and R. J. Southard, pp. 107–132. SSSA Special Publication No. 44. Soil Science Society of America, Madison, WI.

Scott, W. E., 1977, Quaternary Glaciation and Volcanism, Metolius River Area, Oregon, *Geological Society of America Bulletin* 88:113–124.

Shroba, R. R., and Birkeland, P. W., 1983, Trends in Late-Quaternary Soil Development in the Rocky Mountains and Sierra Nevada of the Western United States. In *Late-Quaternary Environments of the United States*, edited by H. E. Wright, Jr., pp. 145–156. University of Minnesota Press, Minneapolis.

Semken, H. A., Jr., 1980, Holocene Climatic Reconstructions Derived from the Three Micromammal Bearing Cultural Horizons at the Cherokee Sewer Site, Northwestern Iowa. In *The Cherokee Excavations: Holocene Ecology and Human Adaptations in Northwestern Iowa*, edited by D. C. Anderson and H. A. Semken, Jr. pp. 67–99. Academic Press, New York.

Semken, H. A., Jr., and Falk, C. R., 1987, Late Pleistocene/Holocene Mammalian Faunas and Environmental Changes on the Northern Plains of the United States. In *Late Quaternary Mammalian Biogeography and Environments of the Great Plains and Prairies*, edited by R. W. Graham, H. A. Semken, Jr., and M. A. Graham, pp. 176–313. Scientific Papers, Vol. 22. Illinois State Museum, Springfield.

Shackley, M. L., 1975, *Archaeological Sediments: A Survey of Analytical Methods*, Butterworths, London.

Simmons, A. H., and Mandel, R. D. (eds.), 1986, *Prehistoric Occupation of a Marginal Environment: An Archaeological Survey near Kharga Oasis in the Western Desert of Egypt*, BAR International Series 303, Oxford, UK.

Sjoberg, A., 1976, Phosphate Analysis of Anthropic Soils, *Journal of Field Archaeology* 3:447–454.

Smith, G. D., and McFaul, M., 1997, Paleoenvironmental and Geoarchaeological Implications of Late Quaternary Sediments and Paleosols: North-Central to Southwestern San Juan Basin, New Mexico, *Geomorphology* 21:107–138.

Soil Survey Division Staff, 1993, *Soil Survey Manual*. Handbook No. 18. U.S. Department of Agriculture, U.S. Government Printing Office, Washington, DC.

Soil Survey Staff, 1996, *Keys to Soil Taxonomy*. U.S. Department of Agriculture, Natural Resources Conservation Service, U.S. Government Printing Office, Washington, DC.

Stafford, C. R., 1995, Geoarchaeological Perspectives on Paleolandscapes and Regional Subsurface Archaeology, *Journal of Archaeological Method and Theory* 23:69–104.

Stein, J. K., 1983, Earthworm Activity; A Source of Potential Disturbance of Archaeological Sediments, *American Antiquity* 48:277–289.

Styles, T. R., 1985, *Holocene and Late Pleistocene Geology of the Napolean Hollow Site in the Lower Illinois Valley*. Research Series Vol. 5. Kampsville Archeological Center, Center for American Archeology, Kampsville, IL.

Tamplin, M. J., 1969, The Application of Pedology to Archaeology Research. In *Pedology and Quaternary Research*, edited by S. Pawluk, pp. 153–161. The University of Alberta Printing Department, Edmonton.

Thompson, D. M., and Bettis, E. A., III, 1980, Archaeology and Holocene Landscape Evolution in the Missouri Drainage of Iowa, *Journal of the Iowa Archeological Society* 27:1–60.

Thoms, A. V., and Mandel, R. D., 1992, The Richard Beene Site (41BX831): A Deeply Stratified Paleoindian Through Late Prehistoric Occupation in South-Central Texas, *Current Research in the Pleistocene* 9:42–44.

Van Nest, J., 1993, Geoarchaeology of Dissected Loess Uplands in Western Illinois, *Geoarchaeology: An International Journal* 8:281–311.

Van Nest, J., 1998, The Good Earthworm: How Natural Processes Preserve Upland Archaic Sites in the Midwest. In *Abstracts of the 63rd Annual Meeting of the Society for American Archaeology*, p. 300. Society for American Archaeology, Washington, DC.

Van Vliet-Lanoë, B., 1985, From Frost to Gelifluction: A New Approach Based on Micromorphology and its Application to Arctic Environment, *Inter-Nord* 17:15–20.

Van Vliet-Lanoë, B., Coutard, J. P., and Pissart, A., 1984, Structures Caused by Repeated Freezing and Thawing in Various Loamy Dediments: A Comparison of Active, Fossil, and Experimental Data, *Earth Surface Processes and Landforms* 9:553–565.

Waters, M. R., 1992. *Principles of Geoarchaeology: A North America Perspective*, The University of Arizona Press, Tucson.

Wendorf, F., and Schild, R., 1980, *Prehistory of the Eastern Sahara*, Academic Press, New York.

White, E. M., 1978, Cautionary Note on Soil Phosphate Data Interpretation for Archaeology, *American Antiquity* 43:507–508.

Wiant, M. D., Hajic, E. R., and Styles, T. R., 1983, Napoleon Hollow and Koster Site Stratigraphy; Implications for Holocene Landscape Evolution and Studies of Archaic Period Settlement Patterns in the Lower Illinois River Valley. In *Archaic Hunters and Gatherers in the American Midwest*, edited by J. L. Phillips and J. A. Brown, pp. 147–164. Academic Press, New York.

Wilding, L. P., and Flach, K. W., 1985, Micropedology and Soil Taxonomy. In *Soil Micromorphology and Soil Classification*, edited by L. A. Douglas and M. L. Thompson, pp. 1–16. SSSA Special Publication No. 15, Soil Science Society of America, Madison, WI.

Wood, W. R., and Johnson, D. L., 1978, A Survey of Disturbance Processes in Archaeological Site Formation. In *Advances in Archaeological Method and Theory*, Volume 1, edited by M. B. Schiffer, pp. 315–381. Academic Press, New York.

Woods, W. L., 1984, Soil Chemical Investigations in Illinois Archaeology; Two Example Studies. In *Archaeological Chemistry-III*, edited by J. D. Lambert, pp. 67–77. American Chemical Society, Advances in Chemistry Series, No. 205, Washington, DC.

Wright, V. P., 1982, Calcrete Paleosols from the Lower Carboniferous Llanelly Formation, South Wales, *Sedentary Geology* 33: 1–33.

Yaalon, D. H., 1971, Soil–Forming Processes in Time and Space. In *Paleopedology*, edited by D. H. Yaalon, pp. 29–40. University of Israel Press, Jerusalem.

Microfacies Analysis Assisting Archaeological Stratigraphy

MARIE-AGNÈS COURTY

1. Introduction

Accurate construction of archaeological stratigraphy has long been recognized as crucial in providing a solid chronocultural framework for discussing past behavioral activities and their linkages with geological processes (Gasche and Tunca, 1984; Harris, 1979). As a consequence, a major effort during excavation has been directed toward the definition of individual strata and their spatial variations. This goal has been accomplished through careful observation of the properties of the sedimentary matrix and its organization in three-dimensional space. The interfering effects of natural agents and human activities on the accumulation of the sedimentary matrix has been considered by some to conform to the principle of stratigraphic succession—as elaborated by earth scientists—and thus conforming to geological laws (Renfrew, 1976; Stein, 1987). Others have strongly argued that the rules and axioms of geological sedimentation cannot be applied to archaeological layers because they are produced by people and thus constitute an entirely distinct set of phenomena (Harris, 1979; Brown and Harris, 1993). Understanding the processes involved in the formation of archaeological stratification has also long been a question of passionate debate, with the views of human or natural deposition being opposed to the theory of biological mixing (Johnson and Watson-Stegner, 1990). These contradictory perceptions have been

---

MARIE-AGNÈS COURTY • CNRS-CRA, UER DMOS, 78850 Grignon, France.

Earth Sciences and Archaeology, edited by Paul Goldberg, Vance T. Holliday, and C. Reid Ferring. Kluwer Academic/Plenum Publishers, New York, 2001.

tentatively reconciled by the recognition of the inherent general complexity of archaeological stratigraphy, that can be isolated into its lithostratigraphic, chronostratigraphic, and ethnostratigraphic components (Barham, 1995; Gasche and Tunca, 1984). Archaeologists have thus been alerted to the difficulty in describing strata objectively—as routinely done by geologists—and the necessity to continuously evaluate the significance of lateral and vertical stratigraphic changes.

The various analytical techniques used in the earth sciences that for a long time have been routinely applied to archaeology (e. g., particle size analysis or geochemistry) have not provided much support for deciphering sediment characteristics and for distinguishing cultural manifestations from natural ones. As a consequence, context analysis, evaluation of site preservation, and reconstruction of past lifeways have not been enhanced greatly above the level of detailed field observations of the sedimentary properties. These analyses are generally debated on the basis of artifact assemblages, including ecofacts and micro artifacts, and the presence of identifiable archaeological features (Bar-Yosef, 1993).

The understanding of the interplay between human activities and natural agents has benefited significantly from the application of microscopic techniques to the study of archaeological sediments— this is known as soil micromorphology (Courty et al., 1989). Similar to sedimentary petrography, soil micromorphology uses thin sections of intact sediments and soils in order to infer their entire depositional and postdepositional history. Originally, the use of microscopic tools was conceived as part of the excavation strategy, and it was expected to refine the criteria traditionally used for establishing archaeological stratigraphy. This technique, however, is still viewed as a difficult and unpractical option; it is conducted by specialists as it is not accessible to every field project, and it is constrained by the lack of a proper methodology for transfering data from the microscope to the field (Barham, 1995; Bar-Yosef, 1993; Matthews et al., 1997). High costs and lack of practitioners have, undoubtedly, constrained the generalized use of microscopic techniques to study archaeological sediments. In addition soil micromorphology—which is derived from petrography—is not a routine procedure used in studying soils. Finally, recognition and interpretation of soil fabrics in thin sections following soil micromorphological principles implies acceptance of the basic concepts of pedology. This latter discipline has given priority to the concept of soil genesis in most soil classification systems (Chesworth et al., 1992). Thus, study of vertical (horizonation) and lateral variations of soil morphology at all analytical levels has focused on properties that reflect the nature of the soil environment and the imprint of the dominant soil-forming processes. However, the complexity of processes involved in the formation of archaeological sediments and soils does not permit simple application of the concepts and methods of soil micromorphology as defined in pedology. Therefore, from its very beginning, the application of microscopic techniques to the study of archaeological soils and sediments was clearly presented as an adaptation of soil micromorphology—as used in pedology—in order to better match the uniqueness of the archaeological context (Courty et al., 1989). Finally, soil micromorphology differs from the approach of soil macromorphology. The latter tends to emphasize descriptive parameters in the field that have genetic overtones (after all, the designation of soil horizons in the field implies a certain realm of pedogenesis); micromorphological criteria are utilized to be as descriptive as possible, with no genetic significance.

The concepts and microscopic methods used in archaeology are in fact more similar to the routine procedures long established in sedimentary geology for studying rocks and sediments. The aim of sedimentary geology is to unravel the comprehensive complex history of depositional environments as based on the classification of sedimentary facies and the sequential analysis of their variability through space and time (Carozzi, 1960). In geology, facies denotes sediments or rocks that are characterized by a unique set of properties related to lithology, texture, structure, and organic remains (Martini and Chesworth, 1992; Miall, 1990; Walker, 1984). This facies concept differs from that of fabric used in soil micromorphology by the absence of any genetic connotation. The idea of facies was introduced in order to help the practical surveyor in the field recognize rocks and sediment (Chesworth et al., 1992). Therefore, facies analysis at the microscopic level has been aimed at providing valuable information for refining criteria visible with the naked eye. In addition, a few sedimentologists concerned with the interplay between sedimentation and pedogenesis introduced the concept of pedofacies in order to better understand how various scales of sediment accumulation across space (particularly in alluvial settings), influence the lateral variability of paleosol morphology (Brown and Kraus, 1987). Facies-based methodologies, however, have been used in archaeology only with timidity (Gilbertson, 1995), although their application to archaeological stratigraphy has been proposed to offer great potential for interpreting the genesis of cultural deposits (Barham, 1995).

The aim of the chapter is to explain how concepts and methods of sedimentary geology — particularly microfacies analysis — can be applied to archaeological settings in order to provide a better integration of microscopic techniques in the excavation. Beyond methodological aspects, we illustrate how analysis at microscopic scales provides important keys to the understanding of the formation of archaeological strata and of their spatiotemporal variability. Knowledge of these factors is an essential requirement for evaluating the integrity of the archaeological record and of site preservation. We intend to emphasize that this methodological orientation is not only important for efficiently assisting archaeological stratigraphy, but is also vital for the future of the discipline, and of geoarchaeology in general. Also considered is the potential of the microscope to illuminate the stratigraphic archaeological record. This strategy impels earth scientists to study geological processes at all temporal and spatial scales compatible with human events. These scales range from daily activities to those of a few generations, and from small habitations, to occupations over a large region.

2. Basic Concepts and Definitions

2.1. Anthropogenic Processes

What makes archaeological strata unique is the interference of humans with natural sedimentation and with pedological processes and their associated post-depositional transformations (Courty et al., 1989). In pedology, anthropogenic processes have been embodied into specific modifications of soil properties that result in the anthropic epipedon (Soil Survey Staff, 1975) without a clear definition of human-related mechanisms. This has lead the pioneers of soil micromorphology to make use of the concepts of cultural processes introduced with the emergence of behavioral archaeology (Schiffer, 1976). This author broadly defined cultural processes at the sedimentary level as the basic physical actions exerted by humans on their living sphere, and more particularly at the surface of occupation: accumulation, transformation, and redistribution. In fact, these new concepts were identical to the one used for artifacts, giving major importance to the intent of past humans in their actions and not to its mechanical nature. The terms *primary*, *secondary*, and *tertiary refuse* that are still in use (Schiffer, 1995) illustrate this anthropological perception of archaeological sediments. We have progressively realized that a definition implying intent is confusing and therefore have given priority to a broad notion referring to all types of actions that are directly or indirectly related to human activities. The diversity of these actions, however, is far from being completely explored (Courty et al, 1994; Matthews et al., 1997). This situation contrasts sharply with most manifestations of sedimentary and pedogenic processes, which are already well known at the microscopic level. Therefore, anthropogenic (i.e., cultural) processes should better be defined as everything not linked to natural factors. This characterization is adopted only to better match the specific objectives of archaeology, and it is not meant to be a philosophical position that counters human agency with "natural" ones. The need to adopt well-defined concepts for anthropogenic processes and cultural layers when studying archaeological stratigraphy should not obscure the fact that any archaeological site has been part of a depositional environment, and that at every moment of its life it has been the stage of natural processes. Therefore, the concept of anthropogenic facies can be theoretically accepted, although they practically never exist in reality, except for materials that suffered irreversible transformations by humans, such as ceramics, baked floors, and bricks.

2.2. Archaeological Facies and Facies Patterns

The definition of facies in sedimentary geology is directly applicable to archaeological layers, although in archaeology lithology is a result of deposition both by natural and anthropogenic processes. Similarly, contacts between facies are not only geogenic or pedogenic (for the pedofacies) limits, but they can also be of anthropogenic origin and complex (Courty et al., 1989). Stratigraphic units identified in the field can be characterized by one or several facies, depending on the homogeneity of each strata and the accuracy of field observations. Therefore, geologist's facies assemblage concepts can be readily extended to archaeological contexts.

Horizontal facies associations refer to lateral variations within each strata, for example, an occupation entity formed of different use areas (e.g., room, court yard, passage, street) or different sub areas within a larger depositional environment (e.g., microdepression, terrace, base of slope). *Vertical facies sequences* relate to the changes of the use of space and site configuration through time in response to occupation dynamics and geomorphic processes. In order to avoid the confusion that commonly occurs in sedimentary geology (Martini and Chesworth, 1992), the terminology used for facies should avoid mixing description and interpretation and should not refer to the product of a process (e.g., "anthropogenic or natural facies") or to a specific activity area (e.g., "courtyard facies").

3. Methodology

3.1. Problems

Many archaeologists often enlist the help of an expert in soil micromorphology when specific problems encountered during excavation or survey remain unsolved. Such problems include the duration and function of a hearth; the presence of questionable archaeological features (e. g., weak traces of combustion suggesting the existence of a poorly preserved fire place); processes producing a certain color of a stratum; difficulty in defining the exact boundary between successive layers; or difficulty in establishing stratigraphic correlations over short distances because of rapid changes in color, texture, or cohesion (Fig. 8.1). After explaining how the aspects of the stratigraphic study performed in the field can benefit from observations at microscopic scales, the specialist can undertake a more comprehensive study. This ideal situation is facilitated when a full-time specialist—whose duties include involvement in the excavation strategies, and responsibility for the gathering of the stratigraphic database—is an integral part of the field project.

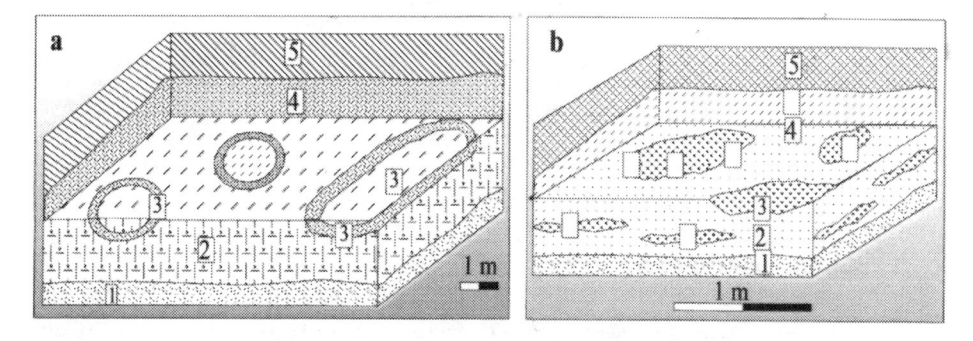

Figure 8.1. Typical situations requiring application of micromorphological study for a particular problem. (a) Circular and elongated ditches evidenced by an uncommon soil material filling with few scattered artifacts suggested to be funerary structures in an open-air site; questions to be solved: origin of the ditch-filling materials, rapidity of the deposition. The small rectangles indicate sampling position. (b) Shelter occupation sequence with four archaeological strata of distinct facies; layer 2 shows diffuse ashy lenses (3) interpreted as poorly structured fireplaces; questions to be solved: mode of deposition and environmental conditions specific to each occupation phase; function of the fireplaces.

Figure 8.2. Example of an archaeological situation where soil micromorphology is exploited to its fullest in the field: Tell Dja'dé, sequence of the Pre-Pottery Neolithic (Middle Euphrates valley, Syria). (a) Topographic map of the site showing the different areas successively excavated and extensively sampled. (b) Mud–brick structures and stone architecture after excavation: samples were taken during the course of excavation to characterize the succession of room fills and the types of construction materials (see Fig. 8.10c, a room fill in thin section). (c) Domestic activity area devoid of massive architecture showing a finely stratified succession of floors and rapid lateral variability. The black rectangles indicate location of samples. (d) Fresh cut of the undisturbed sample displaying clear microstrata that are difficult to identify during excavation (see Fig. 8.10b, the thin section from this sample). (Courtesy of Eric Coqueugniot)

Within the site itself, these duties include the following (Fig. 8.2):

1. Assistance in constructing and correlating stratigraphic sequences (Cammas, 1994)
2. Investigation of the role of human agencies on the production, transport, and transformation of sediments, as well as the effects of humans on bringing about landscape changes (Wattez et al., 1990)
3. Assistance in defining site limits, the spatial configuration of habitation structures and occupation surfaces, and in determining the function of activity areas and their evolution through time (Cammas et al., 1996;

Courty et al., 1991; Matthews et al. 1997; Rigaud et al., 1995; Simpson and Barrett, 1996)
4. Evaluation of the integrity of the archaeological record and the degree of site preservation with respect to the countervailing effects of natural factors (Gé et al., 1993)
5. Reconstructing palaeoenvironmental conditions—along with other environmental specialists—during each phase of occupation and evaluating the evolution of occupations through time in response to climate and environmental changes (Courty, 1989; Wattez and Courty, 1996).

In the off-site context, which is also concerned with archaeological stratigraphy, an important concern of the soil micromorphologist is to better define the exact nature and spatial extent of human-induced landscape transformations. These transformation can be broadly characterized as direct effects, such as changes in soil properties caused by cultivation, or indirect ones that refer to anthropogenic forcing of past geomorphic functioning (Courty and Weiss, 1997; Davidson et al., 1992; Gebhardt, 1988; Macphail et al., 1990)

3.2. Research Strategy

The practice of excavation ideally consists of a strategy of alternating fieldwork and laboratory work in order to continuously evaluate and refine the relevant and easily accessible attributes used to recognize spatial and vertical facies changes in the field. Because most members of a regular archaeological team do not have personal access to the microscope, the specialist is continuously concerned with explaining the information collected under the microscope. Observation of an impregnated, undisturbed block, for example, illustrates how a fresh cut can reveal a much clearer picture of the stratigraphic record than one exhibited by a repeatedly brushed and scraped section (Fig. 8.2d). Excavators thus have a better comprehension of what the specialist is going to look at in thin section. At the same time, they realize why traditional excavation methods—normally quite adequate to properly retrieve artifacts—are not usually adapted to understanding the geometry of stratigraphic layers at all scales. These procedures, for example, if not regularly controlled by closely spaced vertical sections, can easily create an erroneous perception of the surface extent of an occupation.

The specific difficulty that archaeology has in comparison to geology is that the stratigraphic layer is the starting point for multidisciplinary studies and is viewed in different ways depending on the background of each partner. Communication is important to avoid frustration of the excavators and to optimize the use of microscopic data. Accomplishments made by microscopic analysis are strongly dependent on the quality of the sampling, which itself is constrained by archaeological problems that arise from field perceptions. As shown by long experience in the earth sciences, an important difficulty for improving efficiency is to decrease the time lag between fieldwork and laboratory analysis. This lag is one month at a minimum but is generally longer because of a series of practical constraints. The problem is less crucial for long-term excavations where there is the possibility to incorporate microscopic observations produced after each excavation season with the following one. Short-term excavations, particularly rescue projects, face greater difficulty in making efficient and dynamic use of microscopic data.

3.3. Sampling

The iterative field/laboratory strategy offers the possibility of planning sampling strategies over long time periods, which ultimately reduces costs. In general, initial sampling is guided by the need to answer specific problems or to establish the main facies that constitute the stratigraphy in the field. As the excavation progresses, the stratigraphy and associated stratigraphic problems become clearer, and as a consequence it becomes easier to realize how many samples will be needed in order to be certain that the range of stratigraphic variability is covered (Fig. 8.3).

The ideal number of samples to be taken for a successful micromorphological study is a delicate question, often raised by beginners. Those with broader experience, and who are unconstrained by practical limitations, will simply answer, "Take the maximum number of samples. " This query is in fact a very basic principle known to all petrographers: the more samples you take, the more efficient and less speculative are your inferences under the microscope, simply because the "unknown" is reduced to a minimum. However, when working at an archaeological site, it is difficult to adjust sampling to match the spatial complexity of the deposits and the expected level of spatial resolution. For example, understanding a large degree of variability of a stratigraphic layer may often require sampling at <50 cm intervals (see Fig. 8.2 b & 8.2c). Such sampling would undoubtedly create severe damage to the layer and would conflict with the need to retrieve *in situ* artifacts over wide surfaces, which is necessary for understanding intrasite activities. A common solution is to sample a series of sections or pedestals left around the excavated area; this strategy facilitates the extraction of undisturbed blocks with minimum disturbance. It is, however, not ideal for a proper integration of field and microscopic data because the undisturbed samples are taken from a non-excavated area, and, therefore, cannot help test the reality of spatial boundaries or structures as recognized during the excavation.

Sampling is an imperative step, involving a series of choices that can be made only according to the comprehension of the stratigraphy at a certain time, and it requires close collaboration between all field participants. Jointly describing undisturbed, freshly extracted blocks often offers the possibility of confronting different perceptions and of fixing a common view of the field reality that will serve as a solid basis for a joint elaboration of the archaeological stratigraphy.

The novelty of archaeological sites and soil/depositional-related archaeological problems makes irrelevant a standardized sampling procedures for routine soil investigations. To the contrary, sampling requires flexibility, intuition, and the ability to accept the fact that errors will be later revealed under the microscope.

Microfacies Analysis 213

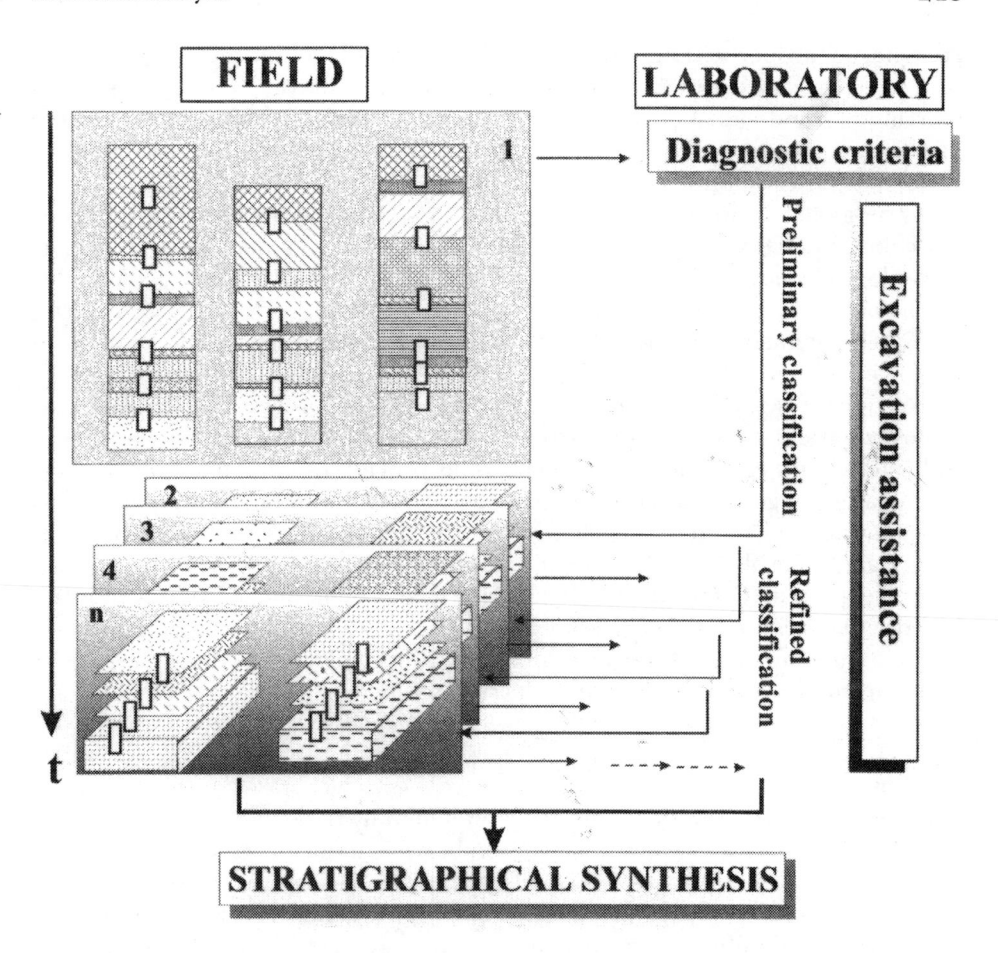

Figure 8.3. Sampling procedure employing an iterative strategy in which excavation and laboratory constraints are continually weighed against each other. Numbers 1 to n represent successive excavation seasons.

3.4. Analytical Procedure

3.4.1. Theoretical Basis

The basic description of depositional environments, associated sediments, and soils has long been recognized to be a difficult task because they are heterogeneous, spatially variable, time dependent, and controlled by nonlinear processes that interact at different spatial and temporal levels (de Marcily, 1996; Perrier and Cambier, 1996). Two scenarios are possible in light of our ability to observe these processes : (1) global changes of the system related to a particular phenomenon can be quantified at a specific operating level, by selecting appropriate indicators even though understanding of causality and the interactions with other phenomena remain speculative; or (2) the measurement of selected "diagnostic" properties does not permit characterization of the global functioning of a system. This situation stems from its great complexity, which necessitates searching for individual processes. This is achieved through objective description of the system coupled with constant elaboration of explanations intended to match our observations.

The global approach is widely applied in soils for quantifying all effects of agricultural practices and is extensively used with isotopic indicators in palaeoclimatological studies of marine, ice, and lake cores. In sedimentary geology this "elementary" approach has long been successfully applied to perform facies analysis of sedimentary basins and to establish a general theory of evolution of geological systems (Bathurst, 1971; Humbert, 1972; Walker, 1984). Due to the interaction of sedimentary, pedogenic, and cultural processes, archaeological depositional environments are extremely complex and it is not yet possible to isolate easily measured indicators that would provide rapid and simple answers to the large range of questions raised in archaeology. Although often presented as a routine analysis in archaeological survey and site interpretation, the exact potential of phosphate analysis for differentiating site function needs to be more deeply explored (Quine, 1995). Therefore, the "disentangled approach" combined with facies analysis is currently the only methodology that can help produce a structural logic in various geomorphic and cultural contexts that are notorious for their unique origin (Brown and Harris, 1993).

3.4.2. Practice

The need to utilize facies analysis as a tool generally depends on the number of samples. This is crucial when a large number of samples is being investigated. Samples used in facies analysis are collected in two basic ways: either extensively across horizontal layers to study the spatial variability of facies and their archaeological significance, particularly for well preserved living floors (see e.g., Fig. 8.2b & 8.2c), or systematically from all strata across the different vertical sections in order to study the evolution through time of their mode of formation.

In both cases, the first stage of the facies analysis requires a systematic comparison of all the thin sections, generally at low levels of magnification. Two objectives should be kept in mind during this stage:

- To establish relationships among field properties and microscopic scales. In particular, to attempt to understand the criteria that the excavators selected for identifying each archaeological strata and its lateral changes;
- To discriminate between: (1) the properties common throughout all the archaeological strata, or at least, a great number of them; these properties are likely to reflect a general trend in the origin, mode and/or conditions of deposition; and (2) the properties encompassed in vertical and lateral changes. This second group can be subdivided into (B1) the ones that change between individual strata, and (B2) those that change within each strata.

Microscopic properties are described according to standard terminology in use for rocks, sediments, soils, and archaeological deposits that are easily available in the most common textbooks such as Bathurst (1975), Bullock et al. (1985), Courty et al. (1989), Humbert (1976), and Pettijohn (1949) Some properties (e.g., clay coatings) have unequivocal morphologies and with reference to published materials it is possible to recognize directly the elementary processes involved. The direct linkage of other attributes (e.g., iron depletion) to basic processes is more ambiguous because their morphology is strongly affected by particular localized circumstances. Their interpretation cannot, therefore, be directly achieved by comparison to extant data. Instead we are required to become familiar with the unique circumstances of each setting.

Three criteria are particularly relevant to reconstructing how archaeological strata were formed and in evaluating the integrity of the archaeological record:

1. The existence of a *soil interface* represented by a distinct boundary either at the contact between two stratigraphic units or within a stratigraphic entity;
2. The *degree of microstratification* as expressed by the thickness of individual laminae, vertical cyclicity, and continuity of the timing of human occupation (seasonal vs continuous), sedimentation, and coeval pedogenesis;
3. The *degree of structural state* that relates to interactions of anthropogenic and natural processes on morphology and arrangement of structural and substructural units (Fig. 8.4).

These criteria can be coupled with ones used in the field by archaeologists for establishing the stratigraphy (Barham, 1995; Brown and Harris, 1993) or for evaluating the degree of site preservation from artifact assemblages (Bar-Yosef, 1993). Progressive awareness of the similarities and differences between and within strata can help to establish an arborescent classification of facies aimed toward revealing the structure and the internal logic of the association of strata toward different levels of organization (Humbert, 1972). The main groups of the classification are defined on the basis of their general properties, whereas different subgroups are distinguished according to the type and intensity of changes in properties. For the classification to be operational in the field, we must be able to choose between unambiguous criteria that can be easily recognized in the field with the naked eye. Facies sharing close morphological similarities at different microscopic levels are assumed to have similar origins and relate to the same mode of deposition. Construction of a strictly descriptive classification implies both the tentative identification of the basic processes as well as their interactions. The classification can be accepted when successive subdivisions in groups and subgroups can be genetically linked (Fig. 8.5). These linkages enable us to interpret sequence of events expressed in the lateral and vertical facies changes.

3.5. Synchronization with Other Techniques

Various analytical techniques performed on bulk samples (e.g., granulometry, organic matter and carbonate contents, pH, phosphorus) have generally been

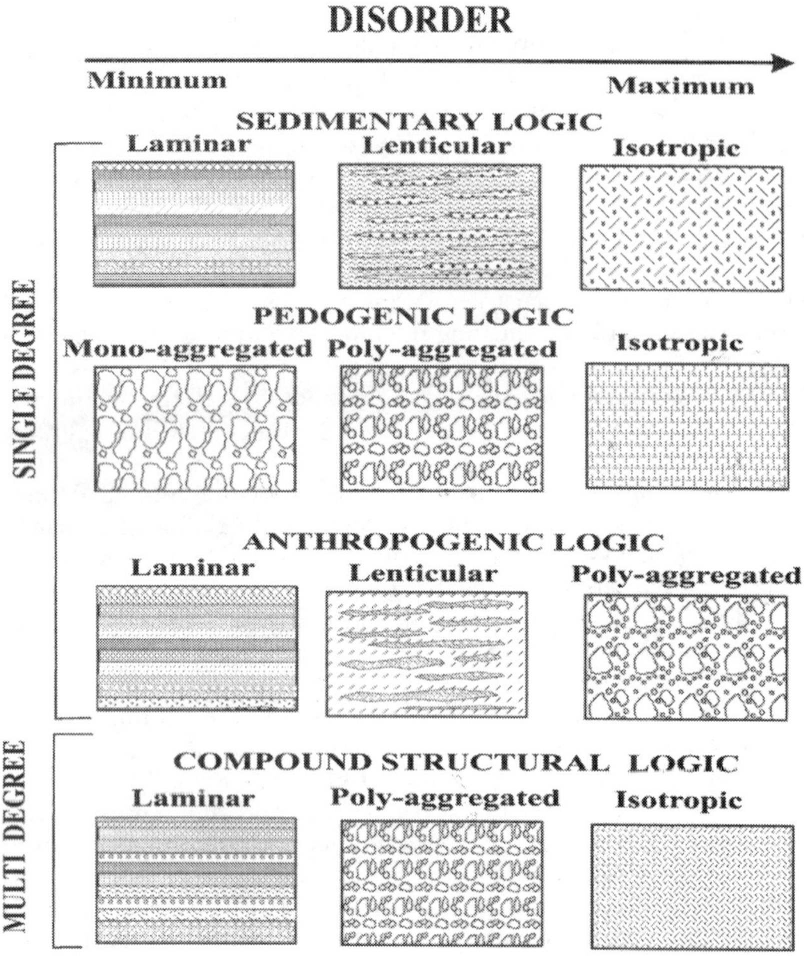

Figure 8.4. Schematic representation of the different types and degrees of structural logic for archaeological strata. This figure illustrates that morphologically similar structures and fabrics can have different origins. For example, for a pedogenic logic the polyaggregated structure can result from both wetting-drying and biological activity, for an anthropogenic logic a similar structure can result from dumping and trampling, and for a compound logic, gravity desegregation, trampling and biological activity can interact.

borrowed from pedology and sedimentology and have been commonly applied to archaeological soils and sediments (for examples see many reports in the archaeological "gray literature"; also Stein and Farrand, 1985; Lasca and Donahue, 1990). These techniques are often presented by the geoarchaeological community as alternative to micromorphological study, and they are often considered to be more rapid, cheaper and statistically reliable to a certain extent (Canti, 1995).

Microfacies Analysis

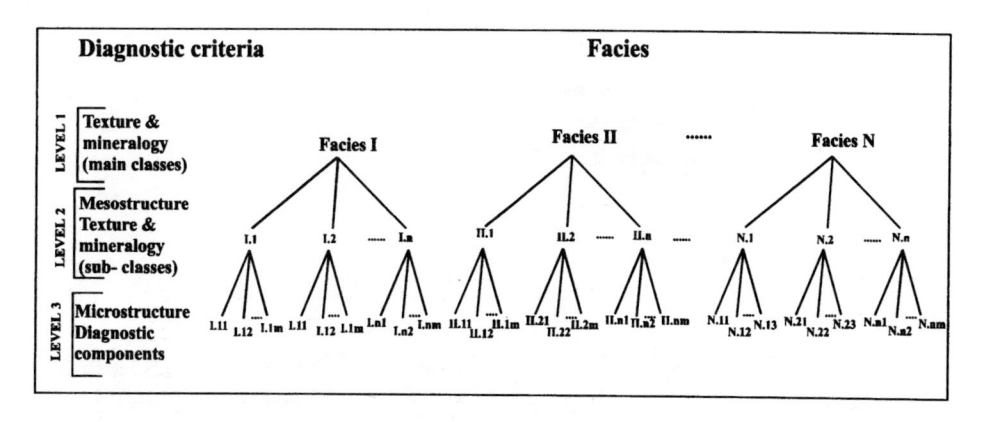

Figure 8.5. Schematic framework of an arborescent classification for interpreting lateral and vertical affiliation between strata: Level 1 groups relate to distinct classes of parent materials used to describe the principle changes in sedimentation or temporal changes of human-used soil materials; Level 2 subgroups detail significant changes of environmental conditions through time or major changes of human activities; Level 3 subdivisions concern lateral spatial variations of local factors or human activities.

The inherent complexity of stratigraphic archaeology and our insufficient understanding of archaeological strata does not, however, allow us to use these techniques on a routine basis independent of microscopic study and to extract relevant conclusions from their statistical analysis. In many cases, the exact significance of the measurement obtained (e.g., particle size data or calcium carbonate content) can be established with the help of microscopic observations.

Choices of analytical techniques may vary widely according to the nature of the problem at hand and the degree of resolution required. However, in order to successfully interpret the data we must adopt a logical progression in the understanding of the levels of organization of the constituent properties. The proposed guideline (Fig. 8.6) emphasizes the difference between two groups of analytical results: (1) easily accessible ones generally obtained from service laboratories and interpreted by the same person who is in charge of the facies study; (2) ones not available on a routine basis that are generally in the hands of a specialist and not the person in charge of the stratigraphic/facies study. The latter, however, keeps a pivotal role in helping link the analytical results with microscopic data and maintaining their relevance to the questions raised in the field.

4. Formation of Archaeological Strata

4.1. General Principles

Application of facies analysis permits the construction of a general model to describe the formation of archaeological strata, including the evaluation of the integrity of the archaeological record. An important aspect of the model lies in

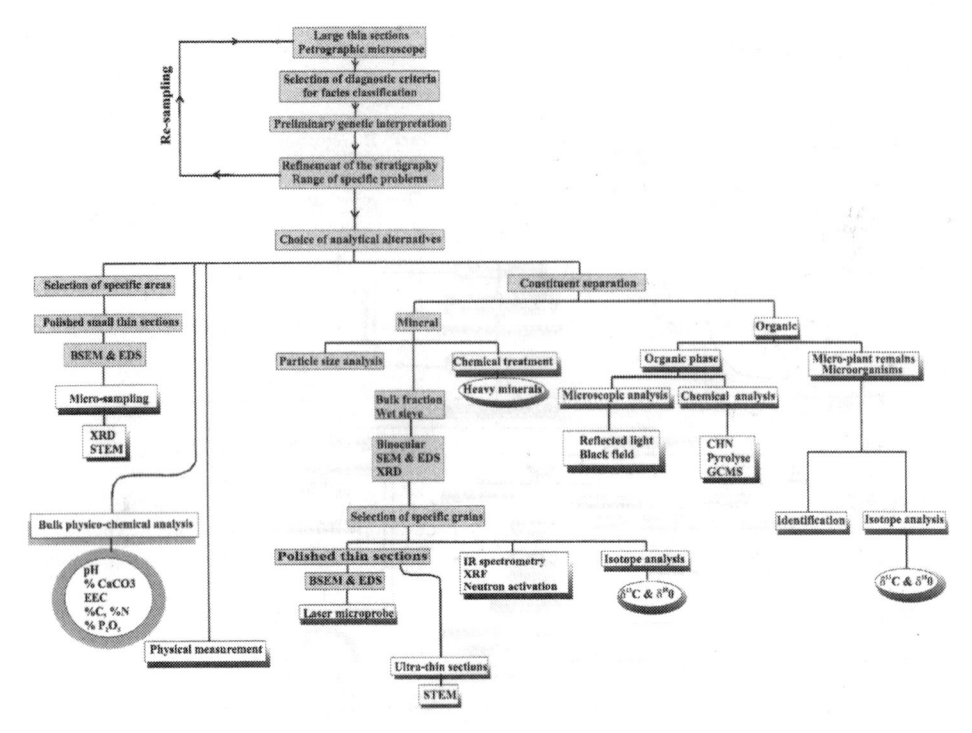

Figure 8.6. Schematic flowchart of a procedure for comprehensive analysis. Gray-filled rectangles indicate the series of investigations performed by the person in charge of the facies study; Gray raised rectangles relate to analyses readily performed by accessible service laboratories; black raised rectangles relate to techniques and analyses that have to be performed by an experienced specialist.

the characterization of the initial state (Phase 0) of the soil interface on which human activities took place. Subsequent stages involve the transformation of a two-dimensional entity into a three-dimensional body. These stages can be subdivided into two phases:

1. Phase 1 represents the period during which anthropogenic processes are dominant, although they interact with sedimentation and pedogenesis. Thus, this phase records human activities in their environmental settings.
2. Phase 2 integrates all the transformations that occurred after site abandonment and includes deterioration of the archaeological signal that was recorded during Phases 0 and 1.

This simplified view aims to illustrate that the evolution of archaeological strata follows general rules that link their morphology, genesis, and their archaeological significance (Table 8.1). Regularly ordered strata display the finest quality temporal signal, as would be the case for Phases 0 and 1 above, with minimal subsequent distortion as (i. e., no Phase 2). They characterize well-preserved strata with good integrity of the archaeological record, and they are ideal for performing sequential analysis.

Table 8.1. Morphological Classification of Archaeological Strata

Group	Subgroup	Basic properties	Interpretation
Ia. Regularly ordered (microstratified)	1. Sedimentary microstratified	• mm thick laminae • Distinct subhorizontal interface	Cyclical sedimentation
	2. Anthropogenic microstratified	• Single degree structural logic	Periodical accumulation
	3. Pedogenic microstratified		Superficial alteration
	4. Compound	• mm thick laminae • Distinct subhorizontal interface • Relict interface	Polycyclical sedimentation/ pedogenesis/ occupation
Ib. Regularly ordered	Anthropogenic	Massive, single degree structural logic	Careful human preparation
II. Weakly ordered	1. Pedosedimentary	Sedimentary microstructures and pedogenic features	Discontinuous sedimentation/ pedogenesis
	2. Anthropogenic	Anthropogenic structural logic. Relict interface	Human accumulation/ reworking
	3. Compound	Multidegree structural logic	Intermittent sedimentation/ pedogenesis/ occupation
III. Randomized	1. Sedimentary	• Single degree structural logic • No sedimentary microstructures	Massive deposition
	2. Pedogenic	Homogeneous Gradual limits Single degree structural logic	High pedogenic maturity
	3. Pedosedimentary	Homogeneous Gradual limits Multidegree structural logic	Redeposited soil horizons
	4. Compound	Homogeneous Gradual limits Erratic structural logic	Human deposition of soil horizons

Weakly ordered strata do not offer a high-quality temporal signal: the record of Phase 0 is generally obscured; that of Phase 1 is present but often not easily accessible, although effects of Phase 2 are moderate. Weakly ordered strata characterize rather well-preserved strata with a medium integrity of the archaeological record.

In randomized strata, the record of Phases 0 and 1 has been totally erased by Phase 2 transformations. Study of their facies cannot be expected to provide information on the original anthropogenic processes and natural events contemporaneous with the occupation, but instead aims at clarifying the nature and environmental significance of the events that strongly obscured the integrity of the archaeological setting.

4.2. Dynamics of the Soil Interface

Archaeologists have long associated high-quality site preservation with the idea of well-preserved living floors, particularly as viewed as the intact record of occupation left after abandonment (Schiffer, 1995). Evidence for minimal vertical dispersion and subhorizontal orientation of artifacts, as well as conjoined pieces, have all been used to recognize these intact surfaces (Bar-Yosef, 1993; David et al., 1973; Leroi-Gourhan and Brézillon, 1966). Analysis at microscopic scales has revealed the physical reality of these occupation surfaces, which appear in the form of inframillimetric layers (Gé et al., 1993) and whose identification provides the opportunity to test their existence independently from field observations and archaeological interpretation. The occurrence of these surfaces is not restricted to special circumstances, such as slow accretion, or rapid burial, as often has been suggested (Bar-Yosef, 1993; Schiffer, 1995), and they appear in all sedimentary contexts (e. g., Fig. 8.7). The high resistance of these surfaces to disturbance is linked to the dynamics of the different processes that originally produced them. Often, intense transit or careful maintenance of habitation and activity areas for prolonged periods has preserved the surfaces from plant colonization, therefore favoring natural agents that are active on bare surfaces (Bresson and Boiffin, 1990; Valentin, 1991). Thus, the induration of ancient surfaces, commonly noticed in the field, appears in thin section to result from hard setting and physical strengthening of interaggregate bonds caused by repeated trampling and alternating wetting–drying. In other contexts, surfaces have remained exposed to splash effects of raindrops at the bare surface and to repeated drying, as expressed by their laminar structure. These processes have induced a surficial compaction that has strongly constrained vertical root penetration and biological mixing, as exhibited by the presence of common very fine subhorizontal channels. Because soils react strongly to atmospheric conditions, well-preserved ancient soil surfaces provide information on the microconditions at the time of occupation.

Living surfaces formed from very resistant materials of concretelike hardness that have been carefully maintained offer a unique situation where the soil interface has remained strictly as a two-dimensional entity, possibly over long periods of time (Fig. 8.8). These exceptions aside, in thin section, soil interfaces of archaeological contexts always express a vertical dimension created by the combined effect of physical actions on the living substrate, production of human microdebris, and the accumulation of dust from various sources. Because these factors operate at very short (i.e., at diurnal) human time scales, viewing the soil interface as a plane surface with irregularities is correct only for instantaneous moments of

Microfacies Analysis

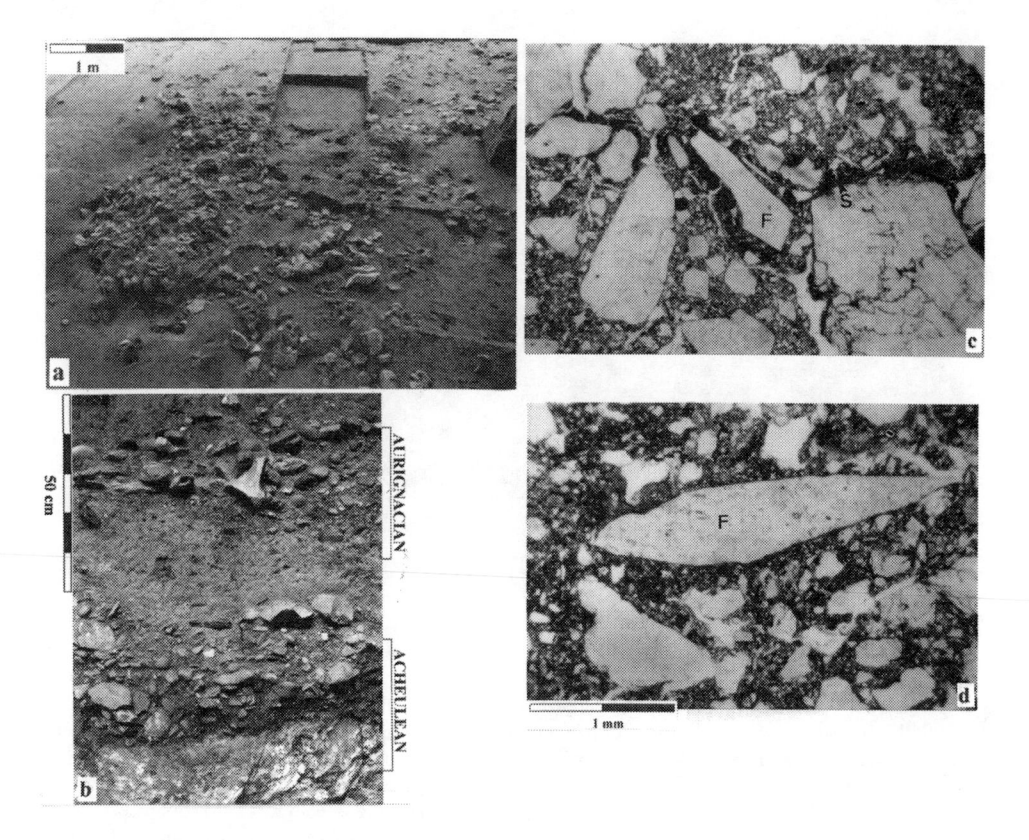

Figure 8.7. The open-air Palaeolithic site of Barbas, Dordogne Valley, southwest France. (a) Field view showing a well-preserved Aurignacian (early Upper Palaeolithic, ca. 35000 yr BP) occupation surface, extensively excavated in one part of the site (Barbas III). (b) Stratigraphic sequence from Barbas I showing evidence of surface flow during deposition of the Acheulean and Aurignacian levels. These flows, however, have not disturbed the archaeological assemblages and interface with associated soil. (c) Microscopic view of the locally well-preserved soil interface (sample 1) on which artifacts are lying; the thin surface illustrates a microstratified pedogenic facies with finely laminated silty clay (S) resulting from water percolation through an impervious material, possibly an animal-skin floor covering, with common microflakes (F) incorporated into the subsoil by trampling. (d) Lateral variations of the same surface (sample 2) showing the slight compaction of a weakly distinct soil interface with subhorizontally layered microflakes. (Photos (a) and (b) courtesy of Eric Boëda)

a few hours. In general, individual ancient soil surfaces are not easily identified during excavation because they are thin and form only patchy physical discontinuities. The field perception of a single, well-preserved living floor appears, in most cases, to correspond to polyphased living floors that represent longer occupation episodes, probably of a few years' duration (Gé et al., 1993).

Human actions have created a unique situation that has nearly no equivalent in natural conditions. An exception occurs in arid regions where soils, weakly protected by an open vegetation cover, have remained exposed to the effects of rain splash and to the effects of surface slaking and crusting. In regions of greater humidity, weakly or nonvegetated surfaces are restricted to zones of active

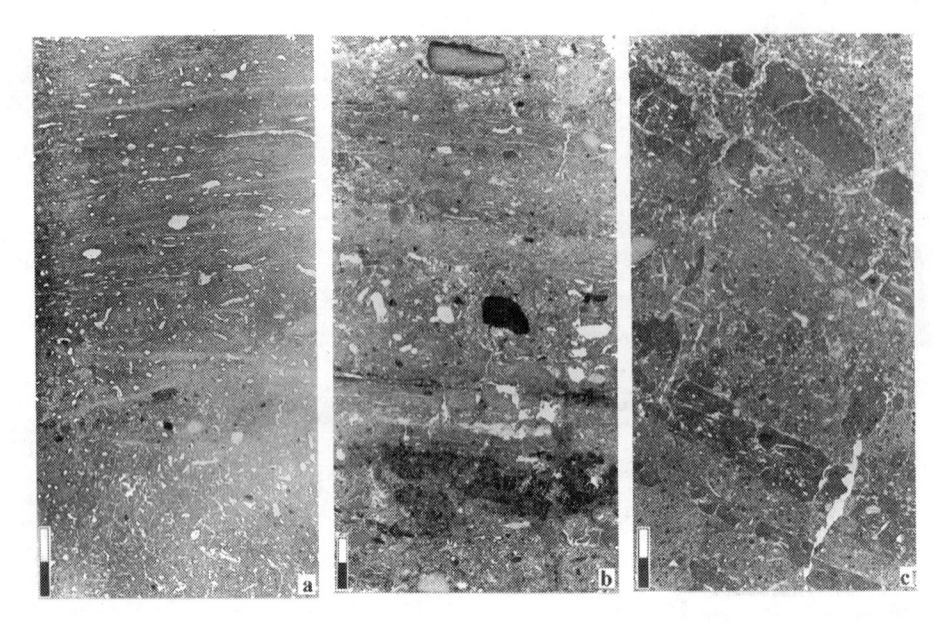

Figure 8.8. Microstratified facies in occupation sequences of urban and proto-urban sites, PPL. (a) Typical anthropogenic microstratified facies resulting from careful maintenance (constant cleaning and seasonal replastering) from a room adjacent to the late 3rd millennium temple at Tell Leilan, Syria. (b) Stratified succession of anthropogenic microstratified facies from a moderately well-maintained room made of light construction materials (thatched roof, thin daub walls); the succession here results from the combined effects of human activity (plastering, domestic debris accumulation and trampling), decay of the walls and roof, and seasonal dripping of water. (c). Stratified succession in a room comprised of anthropogenic massive facies (plastered mud floor) and compound weakly ordered facies that illustrate alternation of seasonal occupation in a well-maintained space and the effects of natural processes (desegregation and insect activity) during phases of nonoccupation. (b) & (c): Pre-pottery Neolithic site of Tell Dja'dé (Middle Euphrates, Syria), see Fig. 8.2. Scale bars equal 1 cm.

sedimentation, where short-term cycles of erosion/sedimentation can maintain instability over long periods. Therefore, the soil surface is permanently refreshed and cannot be morphologically confused with a human-created soil interface. In rare cases, natural conditions can produce a soil interface morphology that can be confused with a human one. For example, an abrupt event (e.g., a wildfire) instantaneously destroying the natural vegetation and accompanied by a heavy rain spell can bring about soil compaction (Weiss et al., 1993). In turn, this compaction can produce a sharp discontinuity, which when buried (Fig. 8.9a) can resemble a human–soil interface (Courty et al., 1998).

4.3. From the Soil Interface to the Archaeological Layer

Archaeological settings display a large range of temporal sequences that document transformations of the initial soil surface into an archaeological layer.

Microfacies Analysis 223

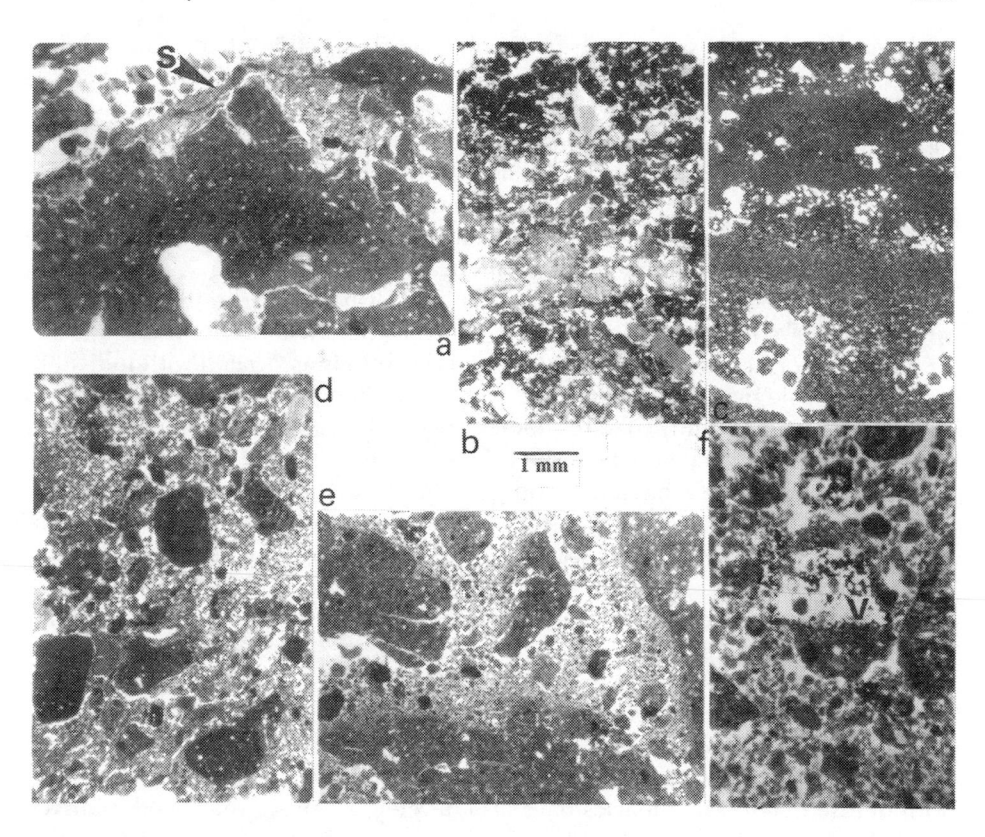

Figure 8.9. (a) View under the microscope (PPL) of a natural soil surface (S) resembling a human soil interface that results from a wildfire contemporaneous with fallout of exogenous dust and strong physical disturbance of the underlying subsoil (Tell Leilan region, NE Syria, burnt surface dated at 3980 yr BP); its wide regional occurrence attests to its nonanthropogenic character. (b) A regularly microstratified pedogenic sequence in the Middle Palaeolithic layer of the cave of Vaufrey (Dordogne, France) resulting from episodic colonization of the cave surface by cryptogamic vegetation and strong weathering of the sparse limestone debris by organic acids; interference with human activity is here marked by repeated burning. PPL. (c) Homogenous massive strata (randomized sedimentary facies) of the middle Paleolithic Cave of Lazaret (Alpes-Maritimes, France): local preservation of this type of clayey silt microlayered accumulation demonstrates slow deposition and weak pedogenic disturbances. PPL. (d), (e), & (f): View in thin section (PPL) of an exceptional event traced over NE Syria that relates to the burnt surface shown in (a); incorporation of exogenous particles (i.e., V in photo f: black vesicular glass) and physical disturbances (here sudden fragmentation of the brick constructions) similar to the ones identified in natural contexts are here identified in occupation sequences and discriminated from anthropogenic processes; (d): Tell Brak, (e): Tell Leilan, (f): Tell Beydar.

Different scenarios are possible depending on the balance between three main dynamics: (1) additions as the result of natural sedimentation and/or anthropogenic inputs; (2) losses as the result of erosion and/or anthropogenic removal; and (3) *in situ* transformation as the result of pedogenesis and/or anthropogenic modification.

Schematically, additions can be understood as exhaustion of the soil interface at different accumulation rates, with differing consequences on the morphology of the soil surface. Losses result in the physical destruction of the soil interface that is, in most cases, instantaneous and generally varies with the microtopography. *In situ* transformations, in theory, leave the soil surface in a stable condition, although its properties may undergo significant changes. The most important modifications involve the layers situated below the supposed stable interface. These changes are governed by two counteracting processes: progressive downward horizonation caused by vertical exchanges between the solid liquid and gas phases and progressive homogenization due to biogenic mixing.

Regularity of the processes involved in the formation of stratified sequences generally allows the estimation of rates of deposition, the duration of exposure of the soil interface, and the environmental conditions associated with individual cycle. Archaeological layers with microstratified sedimentary sequences generally have formed in proximity to water bodies, such as lakes or large meanders, where seasonal flooding has gently refreshed the soil interface, thus preventing vertical disturbance by soil fauna and root growth. In the most regular sequences, occurrences of archaeological strata sandwiched between two depositional episodes provide clear evidence for short occupation of seasonal duration (Fig. 8.10).

The common occurrence of archaeological layers associated with microstratified pedogenetic sequences (although exceptional for natural soils; Soil Survey Staff, 1975) demonstrates the importance of human-induced modifications in the soil microenvironment: reduced seedling and root growth, and preferential colonization of the soil surface by cryptogamic vegetation (mosses and algae), particularly in cave settings (Fig. 8.9b). Microstratified pedogenetic sequences with diffuse horizon boundaries have also developed during the last glacial cycle (and also in postglacial soils) under periglacial conditions as the result of regular aeolian additions, stabilization of the soil surface by a short grass cover, and reduced physical disturbances that are generally restricted to well-drained conditions. Study of the morphology and thickness of microhorizons allows us to evaluate the environmental significance and duration of each pedogenic phase, whereas evaluation of the degree of soil surface alteration by human activities provides indications of the length of occupation.

Anthropogenic microstratified sequences are assumed to occur in well-preserved habitation areas, such as proto-urban and urban sites (Brown and Harris, 1993; Matthews et al., 1997). Microfacies analysis reveals that typical anthropogenic microstratified strata are, in fact, exceptional and restricted to well-maintained habitation areas (Fig. 8.8a). Finely stratified sequences—commonly produced by natural processes such as lacustrine deposition—embody interactions of natural and human processes of similar tempos that are particularly regulated by seasonality (examples in Fig. 8.8b & Fig. 8.10). Cyclical natural sedimentation or surface pedogenic alteration can even be the predominant process involved in the formation of regularly micro stratified sequences of habitation areas (e.g., courtyards and streets) largely open to atmospheric agents (Fig. 8.8c). These sequences offer a unique, high-resolution record to monitor the evolution and environmental conditions with the successive occupation phases (Fig. 8.11). Low-energy deposition and weak pedogenesis that are common to all

Microfacies Analysis

Figure 8.10. The Middle Paleolithic sequence of Umm el Tlel (El Kowm basin, Syria) showing a contrasted succession of well-stratified sedimentary deposits interlayered with a series of short occupation phases. (a) Field view of the stratigraphic section from layers V to VI; rectangles indicate location of samples seen in (b). (b) Detailed field view of the extremely rich V2 Mousterian complex (ca. 42,000 BP) showing the well-preserved V2πa occupation surface and extraction of an undisturbed large-size block during the excavation. (c) Microscopic view (PPL) at low magnification of layer V2β (short Mousterian occupation) showing progressive transition from a regularly microstratified biogenic lacustrine deposit (seasonally wetted pond) to a dark organic-rich facies (swamps episodically affected by wildfires). (d) PPL view at high magnification of the upper part of (b) showing millimeter thick archaeological strata (arrows) sandwiched within the sedimentary sequence. (e) Contrasted view of a disorganized archaeological strata from the coarsely stratified Mousterian VI4a layer (bottom of the sequence shown in photo (a)); reduced vertical dispersion of microartifacts indicates that reworking occurred just after the occupation due to rapid flooding. PPL. (Photos (a) and (b) Courtesy of Eric Boëda)

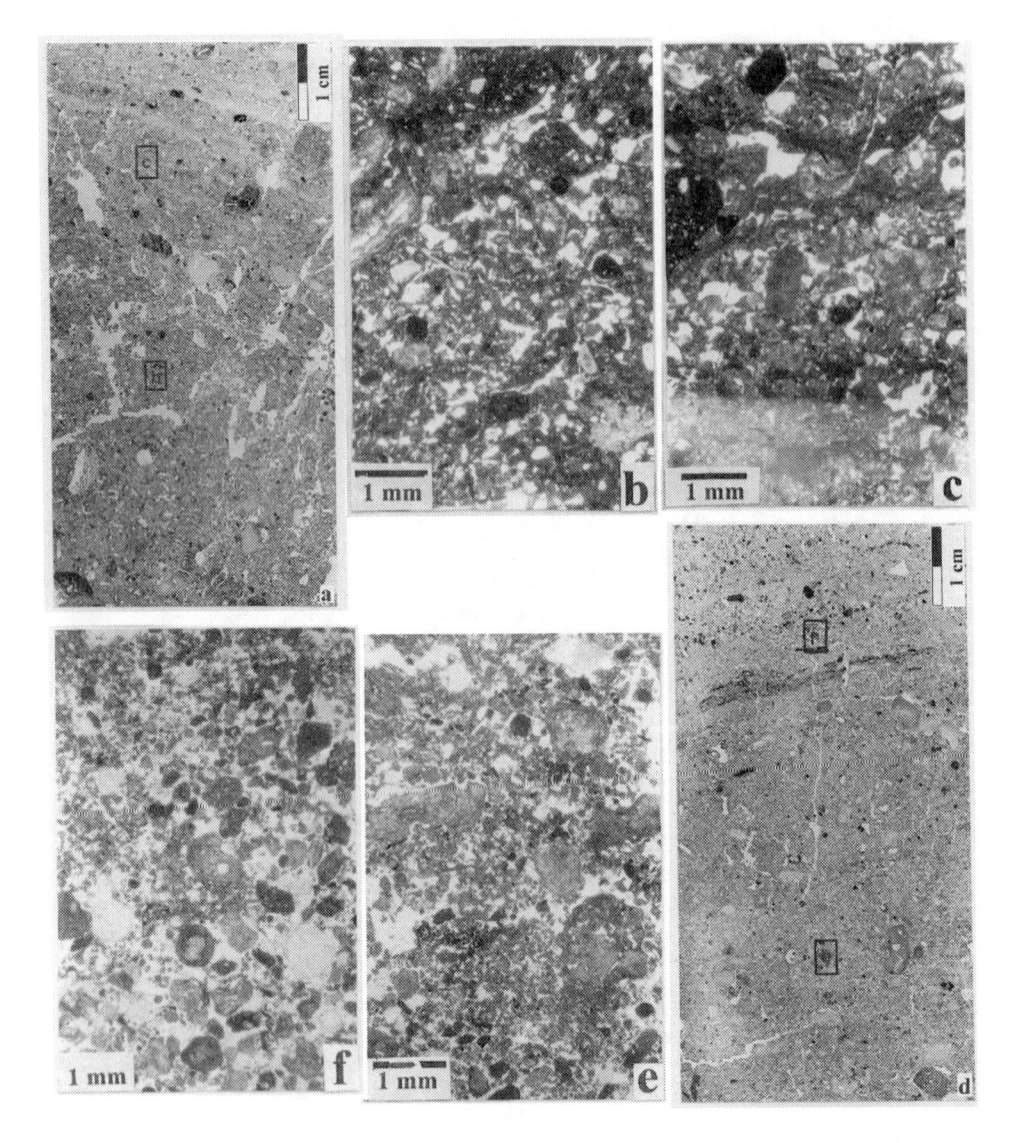

Figure 8.11. Environmental change from wet to dry conditions evidenced from a street sequence, site of Tell Arqa (Northern Lebanon), PPL. (a) Low-magnification view from the bottom part of the street (ca. 2300–2250 B.C.) showing succession of regularly ordered microstratified loose anthropogenic facies. (b) High-magnification view of rectangle 1 from (a): the accumulation here results from slow desegregation of the plastered mud walls adjacent to the street and trampling in well-drained conditions as shown by the lack of compaction. (c) High-magnification view of rectangle 2 from (a): the coarse-textured, loosely packed facies indicates episodic torrential runoff along the street, and an overall maintenance of dry conditions. (d) Low-magnification view from the upper part of the street (ca. 2200–2150 B.C.) showing succession of regularly ordered microstratified dense anthropogenic facies. (e) Massive facies (rectangle 1 from (d)) indicating rapid brick collapse of the adjacent walls and compaction under wet conditions. (f) Microstratified facies formed by the combined action of wall desegregation, human trampling in wet conditions, and surface slaking by mud flow.

microstratified sequences, illustrate that they have hardly suffered from postabandonment transformations (Phase 2), and thus offer a high-quality archaeological record as deduced from excavations.

Weakly ordered strata generally appear in the field as essentially homogeneous deposits with wide variations in properties, such as color, cohesion, and texture, that often give the impression of having gradual boundaries. Three subgroups can be recognized (Table 8.1) that exhibit distinct interactions between anthropogenic and contemporaneous natural processes (Phase 1), each with different archaeological implications.

For subgroup II.1, the environmental conditions contemporaneous with discontinuous deposition, as reconstructed from their pedosedimentary properties, permits identification of two types of archaeological records:

1. In this case, chemically aggressive conditions at the soil surface and intense biological mixing or repeated flooding have strongly erased traces of anthropogenic influence, leaving only the most resistant debris and occasionally weakly preserved structures identified during excavation (Fig. 8.12); the archaeological record is thus concluded to have been strongly altered but stratigraphically coherent.
2. This case is a nonaggressive environment characterized by moderate biological mixing and lack of resistant microdebris, which negates the alteration of anthropogenic properties that are concluded to not have existed, suggesting weak human influence.

Strata from sub-groups II.2 and II.3, although generally similar in the field, present subtle structural differences that reflect on specific modes of formation. For example, the structural logic of aggregation permits differentiation between slow desegregation by natural processes (e.g., insect burrowing and dripping), by contemporaneous human activities, and by rapid destruction of earth-made constructions (see examples in Fig. 8.8c, Fig. 8.9d, e & f, & 8.11). For the latter, however, similarity of physical actions exerted by humans through dumping or through instantaneous natural collapse explains that the distinction between the two scenarios cannot be solely deduced from microscopic observations but requires consideration of the overall excavation context.

The randomized group corresponds to archaeological strata that in the field show an overall homogeneity, with gradual boundaries or sharp erosional contacts, and that are interpreted either as occasional occupations or reworked sites, depending on artifact patterns. At microscopic scales the observed homogeneity appears to relate to postoccupation events (Phase 2), such as rapid sedimentation (III.1), long pedogenic development (III.2) or a combination of both (III.3). Randomized strata have, therefore, lost the memory of their early stages (Phases 0 and 1) and microfacies analysis is unable to restore a reliable image of the original context. The pedosedimentary properties are, however, sufficiently informative to allow recognition the mode of deposition and the nature of pedogenic alterations and their environmental significance. Thus it is possible to evaluate the impact of these transformations on the original configuration of the artifact assemblage. Compound randomized strata (III.4) form a

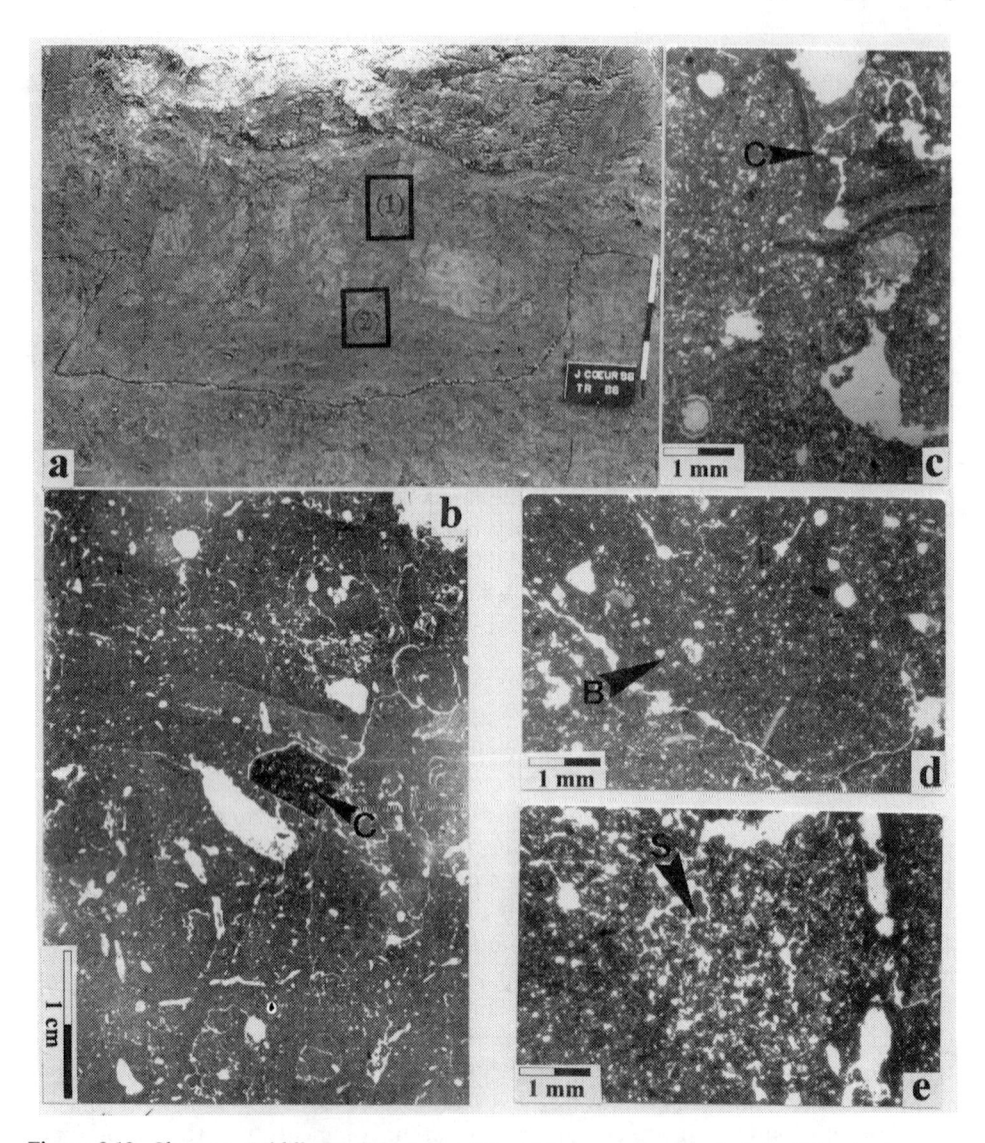

Figure 8.12. Chassean (Middle Neolithic) site of Port-Marianne, Lez flood plain (Hérault, France). (a) Anthropogenic structure with rare sherds recognized during test trenching, interpreted as a foundation pit. (b) View at low magnification of thin section from sample (2) (see location in photo (a)) showing a typical, weakly ordered pedosedimentary facies with massive calcareous loam strongly reworked by biological activity and ceramic fragment (c). Evidence at high magnification of fragmented slaking crusts (S) as shown in photo (c), dense subangular millimetric aggregates identified as brick fragments (B) as shown in photo (d) and concentrations of loamy sand (S) in the packing porosity of aggregates as shown in photo (e) helps to demonstrate that the pedosedimentary facies corresponds to collapse of mud–brick constructions and their strong reworking by biological activity and flooding. PPL. The severe alteration of anthropogenic facies explains the difficulty to identify geometry and function of the archaeological structures suspected in the field. (Photo (a) courtesy of Luc Jalot)

separate subgroup in terms of genetic significance, although they also present an overall homogeneity at microscopic scales and a lack of an anthropogenic signal. In most cases, their anthropogenic origin can be established by excavation. However, human intervention, restricted to transportation, has not significantly modified the original properties of the soil materials quarried for various purposes, such as construction of an earth platform.

4.4. Stratigraphic Relationships and Three-Dimensional Reconstruction

The earth sciences attempt to reconstruct former landscapes, and similarly the ultimate goal of archaeological facies analysis is to restore the three-dimensional image of a human-related space at a given time and to describe its evolution. The precision of this three-dimensional reconstruction is strongly constrained by the quality of the stratigraphic record. Theoretically, the finest reconstruction can be achieved for the sites dominated by regularly ordered sequences, particularly the ones offering well preserved microstratified anthropogenic strata (I.2). In some exceptional situations, sites predominantly made of weakly ordered anthropogenic strata (II.2) offer a record sufficiently coherent to achieve a detailed reconstruction of the geometry of the site and its evolution since its abandonment (Fig. 8.13).

Restoring the configuration of the site contemporaneous with the microstratified signal depends on the ability to accurately control in the field the lateral facies variability for each individual laminae as recognized under the microscope. Reasonably, the sampling interval cannot be smaller than 50 cm in order to preserve the coherence of artifact assemblages and the spatial continuity of occupation surfaces needed for archaeological purposes. Archaeological and radiometric dating are not capable of improving the fine stratigraphic correlation provided by such a high-resolution signal. In some exceptional situations, distinct microstratigraphic markers that are spatially invariable can facilitate the correlation between contemporaneous individual microstrata of different facies (Fig. 8.8d, e, f). Practically, the most efficient method is to correlate segments of vertical sequences with occupation phases defined in the field as based on artifact assemblages, chronological indications, and general information on site configuration. Laminae with distinct properties, and those with more easily recognized with the naked eye, can be tentatively used as stratigraphic markers to match the vertical sequence seen under the microscope with field observation.

Changes in the quality of construction material through time are also helpful indicators for refining stratigraphic correlation between habitation areas. For each segment of a sequence, facies of the successive micro laminae are interpreted in terms of human activities, the nature of the habitation unit, and the local environmental conditions. Spatial variability of the vertical sequence from contemporaneous segments thus provide an evolution of the use of space and contemporaneous environmental conditions for the different sampled areas.

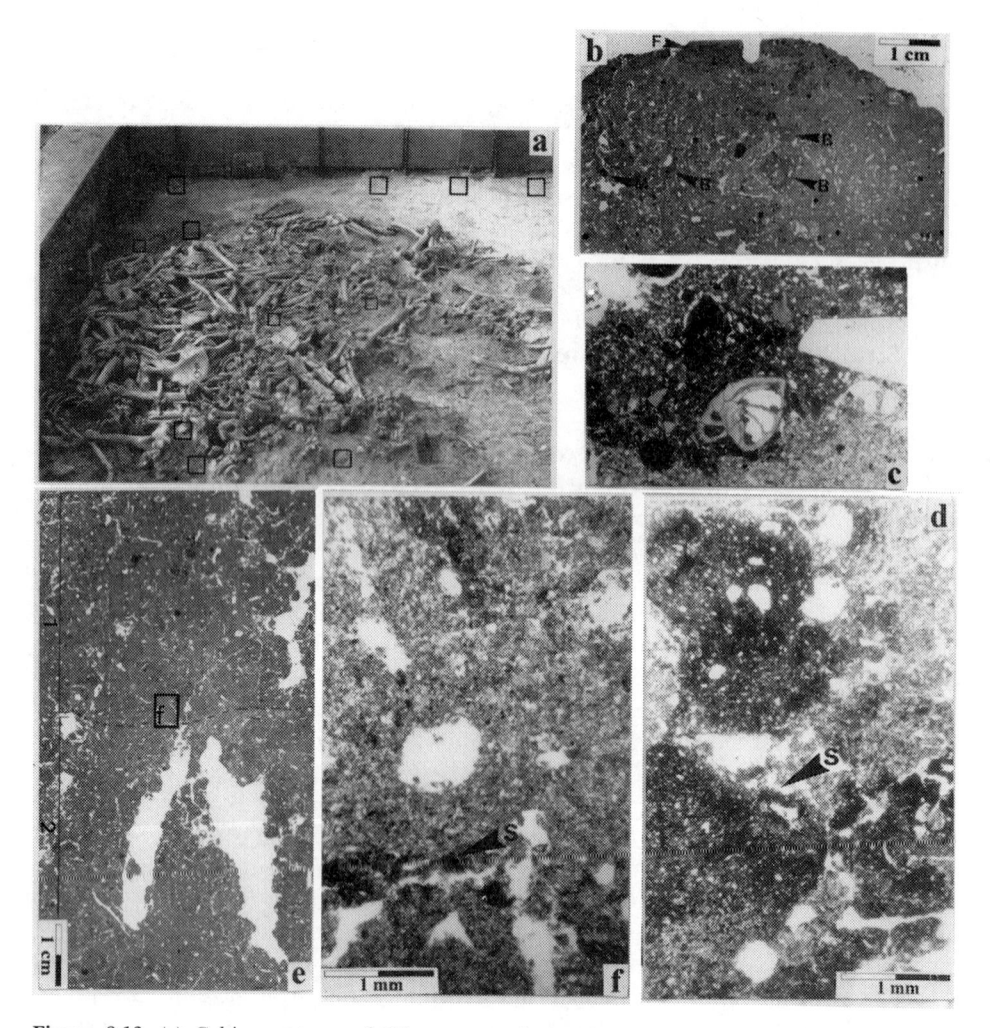

Figure 8.13. (a) Celtic sanctuary of Ribemont-sur-Ancre (Somme, France): rich accumulation of human bones (mostly articulated) that raised questions about preparation of the area and mode of deposition of the corpses; according to ancient texts and archaeological data, the corpses are supposed to have remained exposed for a time while hanging above a ritual platform and then to have fallen down (rectangles indicate sampling location). (b) Low-magnification view of the strata just below large articulated bone assemblage, displaying a weakly ordered pedosedimentary facies with evidence of intense bioturbation, dispersion of mortar (M) and mud–brick fragments (Br), and an upper compacted surface with ferruginous staining (F). (c) View at high magnification of mortar fragments showing the calcitic fine mass, chalk fragments and calcareous coarse grains; their sharp limit with the juxtaposed matrix indicates that the mortar did not suffer dissolution. (d) View at high magnification of the bottom part of the archaeological strata below the bones showing a partly desegregated soil interface (S) marked by ferruginous staining and organic impregnation. (e) Lateral variation of the archaeological strata down the southern slope where bone density rapidly decreases: view at low magnification showing an homogenous, weakly ordered pedosedimentary facies with two weakly distinct subunits (1 & 2). (f) View at high magnification of the contact between subunits 1 & 2 showing relicts of a soil interface (S) with slaking crust sealing a surface horizon that was not disturbed by human activities; also shown is the deposition by gentle runoff of the upper subunit, which is derived from reworking of the bone-rich archaeological strata. The funerary platform is thus concluded to have been carefully prepared with human-made mud–brick and mortar constructions, and then covered with an impervious carpet. This remained exposed to atmospheric agents for a substantial period before the corpses fell on the floor; subsequent gentle runoff along the slope has significantly changed the original configuration as constructed by humans. (Photo (a) courtesy of Jean-Louis Brunaux)

Access to a large diversity of habitation units, (roofed rooms, open courtyards, alleys, streets) offers the possibility to control effects of local factors on the record of environmental conditions and therefore obtain a high-resolution record of short time climate shifts of regional significance (e.g., Fig. 8.11). The most regularly microstratified sequences often can be used to provide an annual record, assuming that seasonal variability constrains the rate of desegregation. An annual rhythm is, however, not always recognizable, particularly when low rates of accumulation and intense trampling obscure the distinction between individual laminae.

Only a partial three-dimensional reconstruction can be achieved for sites offering a juxtaposition between high resolution sequences (group I and II.2) and medium- to low-resolution ones (groups II.1, II.3, and III.1, 2, and 3). The differential preservation of the stratigraphic signal generally expresses the influence of the microtopography on the record of human-induced structural changes at the soil interface and their preservation during the subsequent fossilization (Fig. 8.12). These lateral discontinuities are generally well identified in the field, although they are often confusing due to the difficulty in distinguishing lateral changes related to different functional areas from those caused by subsequent natural variations. The three-dimensional reconstruction based on facies analysis helps to determine whether the natural variations originally present have only affected preservation of the stratigraphic signal after abandonment or whether they have also influenced the spatial patterns of the occupation units. Thus, the common match between subtle microtopographic situations that offer greater protection from natural hazards (e.g., runoff, erosion, flooding, water stagnation) and the mosaiclike distribution of high resolution stratigraphic signals suggests that humans might have preferentially settled at microlocations offering the most suitable living conditions, particularly for occupation on floodplains and universally unstable piedmonts.

For sites dominantly formed of randomized strata, the three-dimensional reconstruction cannot be expected to portray the spatial configuration of a site at the time of occupation due to the lack of high-quality signal related to anthropogenic events and contemporaneous natural incidents. However, an extensive microfacies study of an apparently homogeneous stratum very often draws attention to subtle spatial changes of certain pedosedimentary properties needed to decipher the paleogeography of the site at the time of occupation and subsequent transformations. This is well illustrated, for example, in the Middle Palaeolithic layers of Western European caves, which consist of massive and homogeneous strata traditionally interpreted to relate from episodic colluviation of *terra rossa* soils and/or *in situ* pedogenic weathering (Laville, 1975; Miskovsky, 1974). In thin section, the fine-scale lateral changes in carbonate content of the fine and coarse fraction, and their degree of dissolution, as well as the degree of cohesion of the fine mass help elucidate spatial variations in the configuration of the cave at the time of occupation. They attest to a slow rate of deposition and moderate pedogenic alteration (Fig. 8.9c). These subtle lateral changes, therefore, provide an independent line of evidence that reinforces the impression of good preservation of the archaeological record as deduced from the nature of the artifact assemblage.

It should be stressed that a high-resolution stratigraphic sequence speaks for itself and should not be dismissed by the fact that it is preserved only locally. As long demonstrated in paleogeographical studies (Rat, 1969), the question is not to debate the degree of representation of local observations—particularly ones obtained at microscopic scales, as often confused in archeology and in geoarchaeology (Barham, 1995, Canti, 1995; Glassner, 1994)—but to achieve a comprehensive understanding of the locally preserved high-resolution stratigraphic signals for extrapolating according to the principles of sedimentology and pedology.

5. Implications

Several general comments can be made that have relevance to broad issues in archaeology and in the earth sciences, from site formation processes to soil genesis and paleoclimatology. These remarks should provide direction for future studies.

5.1. Implications for Archaeology

Similar to the achievements in the various branches of petrography, the systematic use of microscope techniques in archaeology—particularly through the development of facies analysis—should help to elaborate a coherent body of stratigraphic theory fully adapted to describe and understand the originality of archaeological stratigraphy. The close similarity in the character of occupation deposits from a large diversity of sociocultural contexts attests to the overall uniformity of most physical actions exerted by humans on their living substrate. Thus, a research priority should be given in archaeology to improve the general classification of anthropogenic facies. Additionally, adoption of a standardized terminology would help unify the different perceptions of archaeological strata. This strategy would reinforce the cohesion of the scientific community involved in stratigraphic archaeology by providing all excavation participants with the possibility of becoming familiar with information obtained at microscopic scales.

The opportunity offered by the microscope to give access to the three-dimensional geometry of archaeological strata represents a unique occasion for refining our perception of archaeological contexts, from habitation areas to landscape units. This tactic requires that we rethink the way we try to control the spatial continuity of individual strata, particularly in cases where rapid lateral changes make excavation over large surface areas extremely difficult. A practical alternative would be to combine surface excavation with a series of small vertical sections, which would provide good stratigraphic control. Development of high-resolution three-dimensional modeling is crucial to a better understanding of the vast range of processes that can produce regular fine stratification with subhorizontal accumulation of artifacts and can thus help us discriminate the soil surfaces that are truly well-preserved occupation floors.

The vital interest in archaeology to accurately estimate the duration of occupation will benefit from the ability of microscope studies to define high-resolution, relative chronologies of environmental changes that reflect rapid shifts at different levels, from local to regional and from seconds to millennia. This issue should stimulate soil micromorphology to better understand the tempos of natural processes for providing independent lines of evidence that can challenge interpretations traditionally based on the logic content of artifact assemblages. In addition, recognition of high-quality paleoenvironmental signals preserved in a large diversity of contexts invites archaeologists to better document the record of natural events at spatiotemporal scales significant to past humans. This course of action is needed in order to refresh engraved ideas on the linkages between natural forcing and sociocultural dynamics of the past.

5.2. Implications for Soil Science

The contribution of the microscope to explain the formation of archaeological layers as the temporal transformation of the soil interface provides an original view on soil dynamics. The great diversity of geomorphic contexts in which archaeological sites occur indicates that they are not present at exceptional locations. Moreover, the strong resemblance of any archaeological layer to a soil horizon, a sediment, or a pedosedimentary unit attests that they are not unique sedimentary bodies. However, the common occurrence of archaeological layers as full stratigraphic entities with original interfaces and pedosedimentary fabrics inherited from the time of deposition shows that a great number of them have escaped vertical soil differentiation and do not present the expected ABC horizonation that should theoretically be developed during slow burial. Only very few sites have benefited from instantaneous burial by rapid and nondestructive sedimentation or human accumulation, and this is clearly not the general explanation. As documented in this chapter, the decreasing gradient of preservation of ancient soil interfaces, from ordered archaeological strata to randomized ones, suggests that vertical soil development has been counteracted by regular to discontinuous surface accretion. The role of sedimentary input on soil genesis has long been recognized by soil scientists, giving rise to the concept of cumulic soils (Soil Survey Staff, 1975). However, the ability of sedimentation to compete with horizonation is assumed to be restricted to the geomorphic contexts influenced by seasonal flooding or endemic airborne dust input (Simonson, 1995). The soil-sedimentary record offered by archaeological sites illustrates the fact that surface accretion is a major component of soil dynamics that is operational in all kinds of geomorphic contexts and at various temporal scales, from seconds to millennia. In addition, the frequent coincidence of the pedogenic boundary with the limit between archaeological strata suggests that horizonation is not simply time transgressive but is strongly constrained by a lithological reality. Thus, the theory of soil genesis can no longer provide the simplified view of successive horizons developing gradually through time below an hypothetical stable soil surface. Instead it should better document the reality of soil surface dynamics.

The integrative use of soil micromorphology and physical measurements has recently enabled us to considerably improve the knowledge of short-term dynamics of the soil surface, particularly through intensive human exploitation (Casenave and Valentin, 1989; Perrier and Cambier, 1996). This research effort has not been extended to medium- and long-term perspectives, most probably because of inaccurate dating of the successive stages of soil development and the difficulty to experimentally control the complex processes involved in the aging of soil fabrics. The threat of global change expected to initiate threshold responses with considerable modifications of properties over short time spans (Stewart et al., 1990) urges the reinforcement of long-term studies of soil dynamics. This crucial issue for the future can greatly benefit from the lessons of the past, and with archaeological sites, soil science is offered a profusion of pedogeomorphic situations to study events that have punctuated the history of our planet's surface.

Differences in ancient site preservation, as illustrated by our classification of archaeological strata focuses future research on a better understanding of spatial soil heterogeneity that so far has been neglected due to the need for standard classification systems for agricultural purposes (Cady and Flach, 1997).

5.3. Implications for Paleoenvironmental Research and Paleoclimatology

Quaternary sequences of prehistoric sites — particularly caves and rock-shelters of southern France — have been the privileged stage since the 1970s to the development of multidisciplinary paleoenvironmental studies aimed toward deciphering the periodicity and nature of climate changes during the recurrent glacial-inter-glacial cycles (Bazile et al., 1986; Laville, 1975; Miskovsky, 1974). The flashy progress of paleoclimatology obtained from high-resolution, long-term sequences from peat, lake, ocean, and ice caps have rapidly obscured the ones from prehistoric sites (Dansgaard and Duplessy, 1981; Jouzel et al., 1987; Pons et al., 1989; Sancetta et al., 1973; Woillard, 1978). Prehistoric sites were relegated to particular sedimentary environments in which the paleoenvironmental record would have been strongly biased by local factors, and therefore would not be reliable for paleoclimatic reconstruction (Campy, 1990; Van Andel and Tzediakis, 1996). Results presented in this chapter encourage us to resuscitate the stratigraphic record of archaeological sites for paleoenvironmental research: these records provide a unique source of information for documenting the in-site and inter-regional complexity of past climate changes at fine temporal scales. The challenge is particularly crucial for the Holocene period, now demonstrated to have undergone a series of abrupt climatic fluctuations (Bond and Lotti, 1995; Gasse and van Campo, 1994; Kutzbach and Liu, 1997; Mayevski et al., 1994; Street-Perrot and Perrot, 1990). The presence of these fluctuations refutes the long-accepted notion of overall climate stability during the Holocene. Moreover, it questions the importance given to human landscape transformations with the emergence and prosperity of early agricultural societies, rather than to climate fluctuations (Bottema and Woldring, 1990; Vernet and Thiebault, 1987; Zangger, 1992). Paleoenvironmental research is now invited to better discriminate natural from anthropogenic forcing and revise the mythical view of a humanity portrayed as the main, largely destructive agent of landscape modification (Crowley and Kim, 1994; O'Brien et al., 1995). Microfacies study of archaeological sites and surrounding regions offer the possibility of obtaining a high-resolution sequence of events from which the effects of cultural factors can be disentangled from the ones of natural agents (Fedoroff and Courty, 1995; Hourani and Courty, 1998).

6. Conclusion

A few years ago, Renfrew (1992) declared that the potential impact of soil micromorphology on the practice of excavation was clearly considerable. At the same time we expected that increasing the number of practitioners and improving the dialogue with archaeologists would be sufficient to reinforce this research direction (Courty, 1992). Although both conditions have now been achieved, the full possibilities of the use of microscopic tools to better understand archaeological strata are still underutilized, if not simply ignored, misused, or even refuted.

Most difficulties now encountered should be viewed as indirect consequences of the general evolution of modern science. Following the technological revolution that gave a leading role to empirical sciences, good research in the modern sense is expected to deal with hard reliable data, with measurement of well-known processes, and with the production of simple models to simulate complex phenomena. For a professional scientist who is expected to be efficient, competitive, and rigorous there is no place for ignorance and no possibility to be wrong. At the same time, the developing complexity of techniques has forced the sciences to become segmented into highly specialized research areas, leaving no other alternatives for those dealing with broad aspects than to remain old-fashioned generalists.

The perception of the various applications of earth sciences to archaeology by the lay human scientist simply reflects this recent partition. Measurement and quantification as provided by the most advanced techniques are generally preferred because they upgrade environmental sciences to the rank of a true science according to modern standards. On the contrary, qualitative techniques, such as soil micromorphology, are accused of being obsolete and are urged to become more reliable and less speculative in order to obtain similar scientific recognition. Maintaining the pressure for more research on quantification might push practitioners of microscopic techniques—particularly those just starting out—into a dead end, simply because our understanding of basic processes is not mature enough to properly design meaningful measurement strategies. The future of soil micromorphology in archaeology, as attempted with this chapter on the study of microfacies, simply depends on our ability to no longer view microscopic tools as the specialized technique adapted to solve specific problems, but as the indispensable companion of archaeological stratigraphy that has been missing for too long.

One can only hope that the eve of the third millennium will mark the end of our naive fascination for high technology and for a science entirely directed to producing a verifiable and reproducible truth. Observing thin sections of sediments, soils, and even more archaeological materials have long taught us that the more we learn, the more we realize how little we know, and how much more we have to simply observe. Accepting our ignorance seems to be one of the most refreshing ideas for the science of the future, and we sincerely hope that many will join us "below the microscope" to enjoy a fascinating challenge that can transform our life into a permanently exciting one.

7. References

Barham, A., 1995. Methodological Aproaches to Archaeological Recording: X-radiography as an Example for a Supportive Recording, Assessment and Interpretative Techniques. In *Archaeological Sediments and Soils*, edited by T. Barham, M. Bates, and R. I. Macphail, pp. 145–182. Archetype Books, London.

Bar-Yosef, O., 1993, Site Formation Processes from a Levantine Viewpoint. In *Formation Processes in Archaeological Context*, edited by P. Goldberg, M. Petraglia, and D. T. Nash, Monographs in World Archaeology, 17:11–32. Prehistory Press, Madison, WI.

Bathurst, R. G., 1971, *Carbonate Sediments and their Diagenesis*, Developments in Sedimentology, Volume 12, Elsevier, Amsterdam.

Bazile, F., Bazile-Robert, E., Debard, E., and Guillerault, P., 1986, Le Pléistocène Terminal et l'Holocène en Languedoc Rhôdanien; Domaines Continental, Littoral et Marin, *Rev. Géol. Dyn. et Gog. Phys.*, 27, (2):95–103.

Bond, G. C., and Lotti, R., 1995, Iceberg Discharges into North Atlantic on Millennial Time Scales During Last Glaciation, *Science* 267:1005–1010.

Bottema, S., and Woldring, H., 1990, Anthropogenic Indicators in the Pollen Record of the Eastern Mediterranean. In *Man's Role in the Shaping of the Eastern Mediterranean Landscapes*, edited by S. Bottema, G. Entjes-Nieborg, and W. Van Zeist, pp. 231–264. Baklema, Rotterdam.

Bresson, L.-M., and Boiffin, J., 1990, Morphological Characterization of Soil Crusts Development Stages on an Experimental Field, *Geoderma* 47:301–325.

Brown, M. R., III, and Harris, E. C., 1993, Interfaces in Archaeological Stratigraphy. In *Practices of Archaeological Stratigraphy*, edited by E. C. Harris, M. R. Brown, III, and G. J. Brown, pp. 7–22. Academic Press, London.

Brown, T. M., and Kraus, M. J., 1987, Integration of Channel and Flood Plain Suites, I. Developmental Sequence and Lateral Relations of Alluvial Paleosols, *Journal Sedimentary Petroleum*. 5:587–601.

Bullock, P., Fedoroff, N., Jongerius, A., Stoops, G., Tursina, T., and Babel, U., 1985, *Handbook for Soil Thin Section Description*, Waine Research Publications, Wolverhampton, UK.

Cady, J. G., and Flach, K. W., 1997, History of Soil Mineralogy in the United States Department of Agriculture. In *History of Soil Science*, edited by D. H. Yaalon and S. Berkowicz, *Advances in Geoecology* 29, CATENA, pp. 211–240. Verlag, Germany.

Cammas, C., 1994, Approche Micromorphologique de la Stratigraphie Urbaine à Lattes: Premiers Résultats, *Lattara* 7, pp. 181–202. Lattes, A.R.A.L.O.

Cammas, C., Wattez, J., and Courty, M. A, 1996, L'enregistrement Sédimentaire des Modes d'Occupation de l'Espace. In *The Colloquium of the XIII International Congress of Prehistoric and Protohistoric Sciences. 3-Paleoecology*, edited by L. Castelleti and M. Cremaschi, pp. 81–86. A.B.A.C.O. Edizioni, Forli, Italy.

Campy, M., 1990, L'enregistrement du Temps et du Climat dans les Remplissages Karstiques: l'Apport de la Sédimentologie, *Karstologia*, Mémoires 2:11–22.

Canti, M. G., 1995, A Mixed Approach to Geoarchaeological Analysis. In *Archaeological Sediments and Soils*, edited by T. Bahram, M. Bates, and R. I. Macphail, pp. 183–190. Archetype Books, London.

Carozzi, A. V., 1960, *Microscopic Sedimentary Petrology*, Wiley, New York.
Casenave, A., and Valentin, C., 1989, *Les États de Surface de la Zone Sahé*lienne, Influence sur l'Infiltration, Coll. Didactiques (ORSTOM, ed.).
Chesworth, W., Spiers, G. A., Evans, L. J., and Martini, I. P., 1992, Classification of Earth-Materials: A Brief Examination of Examples. In *Weathering, Soils and Paleosols*, edited by I. P. Martini and W. Chesworth. In, Developments in Earth Surface Processes 2, pp. 567–586. Elsevier, Amsterdam.
Courty, M.-A., 1989, Analyse Microscopique des Sédiments du Remplissage de la Grotte de Vaufrey (Dordogne), La Grotte de Vaufrey. In *Mémoire de la Société Préhistorique Francaise*, edited by J.-Ph. Rigaud pp. 183–209.
Courty, M. A., 1992, Soil Micromorphology in Archaeology. In *New Developments in Archaeological Science*, edited by M. Pollard, pp. 39–62. Oxford University Press, New York.
Courty, M.-A., and Weiss, H., 1997, The Scenario of Environmental Degradation in the Tell Leilan Region (N. E. Syria) during the Late Third Millennium Abrupt Climate Change, *NATO ASI Series*, I 49, pp. 107–148, Spinger-Verlag.
Courty, M. A., Goldberg, P., and Macphail, R. I., 1989, *Soil, Micromorphology and Archaeology*, Cambridge Manuals in Archaeology, Cambridge University Press, Cambridge, UK.
Courty, M.-A., Macphail, R. I., and Wattez, J., 1991, Soil Micromorphological Indicators of Pastoralism; with Special Reference to Arene Candide, Finale Ligure, Italy, *Rivista di Studi Liguri*, A LVII 1-4:128–150.
Courty, M. A., Goldberg, P., and Macphail, R. I., 1994, Ancient People—Lifestyles and Cultural Patterns, 15th World Congress of Soil Science, Symposium B, "Micromorphological Indicators of Anthropological Effects on Soils," *International Society of Soil Science, Mexico*. 6a:250–269.
Courty, M. A., Cachier, H., Hardy, M., and Ruellan, S., 1998, Soil Record of Exceptional Wild-Fires Linked to Climatic Anomalies (Inter-Tropical and Mediterranean Regions). *Actes du XVIème Congres Mondial de Science du Sol*, Montpellier.
Crowley, T. J. and Kim, K. Y., 1994, Towards Development of a Strategy for Determining the Origin of Decadal–Centennial Scale Climate Variability, *Quaternary Science Reviews* iv:375–385.
Dansgaard, W. and Duplessy, J. C., 1981, The Eemian Interglacial and its Termination, *Boreas* 10:219–228.
David, F., Julien, M., and Karlin, C., 1973, Approche d'un Niveau Archéologiqueer. Sédiment Homogène, In *l'Homme, hier et Aujourd'hui*, edited by Cujas, Paris, pp. 65–72. C.N.R.S., Paris.
Davidson, D. A., Carter, S. P., and Quine, T. A., 1992. An Evaluation of Micromorphology as an Aid to Archaeological Interpretation, *Geoarchaeology*, 7(1):55–65.
Fedoroff, N., and Courty, M. A., 1995, Le Rôle respectif des Facteurs Anthropiques et Naturels dans la Dynamique Actuelle et Passée des Paysages Méditerranéens, *XVème Rencontres Internationales d'Archéologie et d'Histoire d'Antibes*, edited by S. van der Leeuw. pp. 115–141. Editions APDCA, Sophia, Antipodis.
Gasche, H., and Tunca, O., 1984, Guide to Archaeostratigraphic Classification and Terminology: Definitions and Principles, *Journal Field Archaeology* 10(3):325–335.
Gasse, F., and van Campo, E., 1994, Abrupt Post-Glacial Events in West Asia and North Africa Monsoon Domains, *Earth and Planetary Science Letters* 126:435–456.
Gé, T., Courty, M. A., Wattez, J., and Matthews, W., 1993, Sedimentary Formation Processes of Occupation Surfaces. In *Formation Processes in Archaeological Context*, edited by P. Goldberg, M. Petraglia, and D. T. Nash, Monographs in World Archaeology 17, pp. 149–163. Prehistory Press, Madison, Wisconsin.
Gebhardt, A., 1988, Evolution du Paysage Agraire au Cours du Sub-Atlantique dans la Région de Redon (Morbihan, France), Apport de la Micromorphologie, *Bull. Ass. Fr. Et. Quat.* 4:197–203.
Gilbertson, D. D., 1995, Study of Lithostratigraphy and Lithofacies: A Selective Review of Research Developments in the Last Decade and Their Applications to Geoarchaeology. In *Archaeological Sediments and Soils*, edited by T. Bahram, M. Bates, and R. I. Macphail, pp. 99–144. Archetype Books, London.
Glassner, J.-J., 1994, La Chute de l'Empire d'Akkadé, les Volcans d'Anatolie et la Désertification de la Vallée du Habur, *Les Nouvelles de l'Archéologie* 56:49–51.
Harris, E. D., 1979, *Principles of Archaeological Stratigraphy*, Academic Press, London, New York.
Hourani, F., and Courty, M.-A., 1998, L'évolution Morpho-Climatique de 10 500 à 5 500 B.P. dans la Valleé du Jourdain, *Paléorient* 23(2):95–105.

Humbert, L., 1972, *Atlas de Pétrographie des Systèmes Carbonatés*, Technip, Paris.
Johnson, D. L., and Watson-Stegner, D., 1990, The Soil-Evolution Model as a Framework for Evaluating Pedoturbation in Archaeological Site Formation. In *Archaeological Geology of North America*, edited by N. P. Lasca and J. Donahue, pp. 541–560. Centennial Special Volume 4, Geological Society of America, Boulder, CO.
Jouzel, J., Lorius, C., Petit, J. R., Genthon, C., Barkov, N. I., Kotlyakov, V. M., and Petrov, V. M., 1987, Vostok Ice Core: A Continuous Isotope Temperature Record Over the Last Climatic Cycle (160,000 Years), *Nature* 329:403–408.
Kutzbach, J. E., and Liu, Z., 1997, Response of the African Monsoon to Orbital Forcing and Ocean Feedbacks in the Middle Holocene, *Science* 278:440–443.
Lasca, N. P., and Donahue, J., (eds.) 1990, *Archaeological Geology of North America*, Geological Society of America, Centennial Special Volume 4, Boulder, Colorado.
Laville H., 1975, Climatologie et Chronologie du Paléolithique en Périgord. Etude Sédimentologique de Dépôts en Grottes et Sous Abris, *Etudes Quaternaires*. Mémoire 4. Université de Provence. p. 407.
Leroi-Gourhan, A., and Brézillon, M., 1966, L'habitation n°1 de Pincevent, Près Montereau (Seine-et-Marne), *Gallia Préhistoire* 9(2):263–385.
Macphail, R. I., Courty, M. -A., and Gebhardt, A., 1990, Soil Micromorphological Evidence of Early Agriculture in North West Europe, *World Archaeology* 22(1):53–69.
Marcily de, G., 1996, De la Validation des Modèles en Sciences de l'Environnement, Actes des Journées du Programme *Environnement, Vie et Société*, pp. 158–163. C.N.R.S., Paris.
Martini, I. P., and Chesworth, W., 1992, Reflections on Soils and Paleosols. In *Weathering, Soils and Paleosols*, edited by I. P. Martini and W. Chesworth, In, *Developments in Earth Surface Processes* 2, pp. 3–16. Elsevier, Amsterdam.
Matthews, W., French, C. A. I., Lawrence, T., Cutler, D. F., and Jones, M. K., 1997, Microstratigraphic Traces of Site Formation Processes and Human Activities, *World Archaeology* 29(2):281–308.
Mayevski, P. A., Meeker, L. D., Whitlow, S., Twickler, M. S., Morrison, M. C., Blomfield, P., Bond, G. C., Alley, R. B., Gow, A. J., Grootes, P. M., Meese, D. A., Ram, D. A., Ram, M., Taylor, K. C., and Wumkes, W., 1994. Changes in Atmospheric Circulation and Ocean Ice Cover Over the North Atlantic during the Last 41 000 Years, *Science* 263:1747–1751.
Miall, A. D., 1990, *Principles of Sedimentary Basin Analysis*, Springer-Verlag, New York.
Miskovsky, J. C., 1974, Le Quaternaire du Midi Méditerranéen, *Etudes Quaternaires*, Mémoire No. 3.
O'Brien, S. R., Mayewski, P. A., Meeker, L. D., Meese, D. A., Twickler, M. S., and Whitlow S. I., 1995, Complexity of Holocene Climates Reconstructed from a Greenland Ice Core, Science 270:1962–1964.
Perrier, E., and Cambier, C., 1996, Une Approche Multi-Agents pour Simuler les Interactions Entre les Acteurs Hétérogènes de l'Infiltration et du Ruissellement d'Eau sur une Surface Sol, Actes des Journées du Programme *Environnement, Vie et Société*, pp. 77–82. C.N.R.S., Paris
Pettijohn, F. J., 1949, *Sedimentary Rocks*, 1st ed., Harper and Row, New York.
Pons, A., Campy, M., and Guiot, J., 1989, The Last Climatic Cycle in France: The Diversity of Records, *Quaternary International* 3(4):49–55.
Quine, T. A., 1995, Soil Analysis and Archaeological Site Formation Studies. In *Archaeological Sediments and Soils*, edited by T. Bahram, M. Bates, and R. I. Macphail, pp. 77–98. Archetype Books, London.
Rat, P., 1969, Esprit et Démarches de la Paléogéographie, Exemples dans le Bassin Parisien, *Bull. Soc. Géol. Fr.* XI:5–12.
Renfrew, A. C., 1992, The Identity and Future of Archaological Science. In *New Developments in Archaeological Science*, edited by M. Pollard, pp. 285–295. Oxford University Press, New York.
Renfrew, C., 1976, Archaeology and the Earth Sciences. In *Geoarchaeology: Earth Science and the Past*, edited by D. A. Davidson and M. L. Shackley. Duckworth, London.
Rigaud, J.-P., Simek, J. F., and Gé, T., 1995, Mousterian Fires from Grotte XVI (Dordogne, France), *Antiquity* 69:902–1012.
Sancetta, C., Imbrie, J. and Kipp, N. G., 1973, Climate Record of the Past 130,000 Years in North Atlantic Deep Sea Core V 23-82: Correlation with the Terrestrial Record, *Quaternary Research* 3:110–116.

Schiffer, M. B., 1995, Is there a "Pompei premise," *Archaeology? Behavioral Archaeology, First principles*, pp. 5–26. University of Utah Press, Salt Lake City.

Simonson, R. W., 1995., Airborne Dust and its Significance to Soils. *Geoderma* 65:1–43.

Simpson, I. A., and Barrett, J. H., 1996, Interpretation of Midden Formation Processes at Robert's Haven, Caithness, Scotland Using Thin Section Micromorphology. *Journal Archaeological Science* 23:543–556.

Soil Survey Staff, 1975, Soil Taxonomy: A Basic System of Soil Classification for Making and Interpreting Soil Surveys, *USDA-SCS Agricultural Handbook* 436. U.S. Government Printing Office, Washington, DC.

Stein, J. K., 1987, Deposits for Archaeologists. *Advances in Archaeological Method and Theory* 11:337–395.

Stein, J. K., and Farrand, W. R., 1985, Archaeological Sediments in Context. In *Peopling of the Americas Series*, Volume 1, p. 147. Center for the Study of Early Man, Institute for Quaternary Studies, Univeristy of Maine at Orono.

Stewart, J. W. B., Anderson, D. W., Elliot, E. T., and Cole, C. V., 1990, The Use of Models of Soil Pedogenic Processes in Understanding Changing Land Use and Climatic Change. In *Soils On a Warmer Earth*, edited by H. W. Sharpenseel, M. Shomaker, and A. Ayoub, In, *Developments in Soil Science 20*, pp. 121–131. Elsevier, Amsterdam

Street-Perrot, A., and Perrot, R. A., 1990, Abrupt Climate Fluctuations in the Tropics: the Influence of Atlantic Ocean Circulation, *Nature* 343:607–612.

Valentin, C., 1991, Surface Crusting in Two Alluvial Soils of Northern Niger, *Geoderma* 48:201–222.

Van Andel, T. H., and Tzedakis, P. C., 1996, Paleolithic Landscapes of Europe and Environs, 150,000-25,000 Years Ago: An Overview, *Quaternary Science Reviews* 15:481–500.

Vernet, J.-L., and Thiébault, S., 1987, An Approach of Northwestern Mediterranean Recent Prehistoric Vegetation and Ecologic Implications, *Journal of Biogeography* 14:117–127.

Walker, R. G., 1984, General Introduction: Facies, Facies Sequences, and Facies Models. In *Facies Models*, edited by R. G. Walker, pp. 1–9. Geoscience Canada, Toronto.

Wattez, J., and Courty, M.-A., 1996, Modes et Rythmes d'Occupation à Tell Halula, Approche Géoarchéologique (Premiers Résultats), In *Tell Halula (Siria): Un Yacimiento Neolitico del Valle Medio del Euphrates, Campanas de 1991 y 1992*, edited by M. Molist Montana, pp. 53–67. Ministerio de Educacion y Cultura, Madrid.

Weiss, H., Courty, M. -A., Wetterstrom, W., Meadow, R., Guichard, F., Senior, L., and Curnow, A., 1993, The Origin and Collapse of Third Millenium North Mesopotamian Civilization, *Science* 261:995–1004.

Woillard, G. M., 1978, Grande Pile Peat Bog: A Continuous Pollen Record for the Last 140,000 Years, *Quaternary Research*. 9:1–21.

Zangger, E., 1992, Neolithic to Present Soil Erosion in Greece, *Past and Present Soil Erosion, Archaeological and Geographical Perspectives*, Oxbow Monograph, 22:87–103.

The Soil Micromorphologist as Team Player

A Multianalytical Approach to the Study of European Microstratigraphy

RICHARD I. MACPHAIL and
JILL CRUISE

1. Introduction

Soil micromorphology is one of the major subdisciplines within soil science, with subcommission status in the International Society of Soil Science since 1978. It held its initial working-meeting in London in 1981, where Goldberg (1983) made the first review of the application of soil micromorphology to archaeology. First developed by Kubiena (1938) as a way of studying undisturbed soil in thin sections, soil micromorphology now encompassess a range of ultramicroscopic techniques such as scanning electron microscopy (SEM) that is often linked to

---

RICHARD I. MACPHAIL • Institute of Archaeology, University College London. London WC1H OPY, United Kingdom.   Jill CRUISE • University of London Guildhall and Greenwich University. London LU78J4, United Kingdom.

Earth Sciences and Archaeology, edited by Paul Goldberg, Vance T. Holliday, and C. Reid Ferring. Kluwer Academic/Plenum Publishers, New York, 2001.

microchemical instrumental analyses (e.g., qualitative energy dispersive X-ray analysis or Energy Dispersion X-ray Analysis (EDXRA) and microprobe; e.g., Courty et al., 1989).

In Europe, "geoarchaeologist" is broad umbrella term under which are grouped a range of specialists. Many are geographers, pedologists, and Quaternary scientists, who on occasion take on an implied geoarchaeological role when studying sites associated with human activity (e.g., Kemp, 1985; Preece, 1992; Preece, et al., 1995). They also commonly combine geophysical techniques, such as magnetic susceptibility, with standard soil and sediment methodologies, especially when studying wetland sites, although now many archaeological soils are studied in this way (Crowther and Barker, 1995; Oldfield et al., 1985; Taylor et al., 1994). Whereas some are soil micromorphologists, others emphasize the analysis of particle size, tephra, phosphate, river gravels, and the X-ray analysis of sediments (see Barham and Macphail, 1995). Geoarchaeologists are grouped with other environmental archaeologists who study microfossils (e.g., pollen, diatoms, nematode eggs), macrofossils (e.g., seeds/grains, charcoal, mollusks, bones, teeth), isotopes, and human remains, and some notable integrated studies have been published (Dockrill et al., 1994; Maggi, 1997; Matthews and Postgate, 1994). Very broadly speaking, soil micromorphologists, like their soil chemist counterparts, are likely to be asked to focus on the on-site and anthropogenic component of a study. Site geologists and geomorphologists, if present, are more likely to take responsibility for the macro geomorphological setting and the off-site studies. For example, paleosols and colluvium may be identified as macrogeological units, but the soil micromorphologist may confirm these identifications and recognize anthropogenic activities that modified or produced these units. This is not to say that soil micromorphologists cannot also act as competent geomorphologists/geologists, and vice versa. Many workers have been trained in all these fields. As stated as follows, the soil micromorphologist works from the field scale to the microscale, and his/her interpretations may well be of relevance to broad models that reconstruct past landscapes and periods (Crowther et al., 1996; Macphail, 1992; Whittle et al., 1993).

Multidisciplinary environmental studies of archaeological sites that contain a soil science component have been carried out for many years in Europe (Cremaschi, 1985; Dimbleby, 1962; Dockrill et al., 1994; Evans, 1972; Iversen, 1964; Macphail, 1987, Tables 13 and 14, 1994). Such investigations, which involved palynological and land mollusk studies, have contributed enormously to our understanding of past soils, their associated environment, and land use. The specific application of soil micromorphology to European archaeological sites spans the period from the 1950s up to the present day (Castelletti and Cremaschi, 1996; Cornwall, 1958). Nevertheless, one of the constraints that emerged since the late 1950s is a frequent lack of coordination of both sampling and analyses among the various specialists working on a single site or project. For example, when the various specialists sample different parts of the site, conflicting interpretations may result that may be impossible to resolve. This situation is further exacerbated when workers are sampling and analyzing at different scales. Increased dissatisfaction with this situation, which ultimately is a waste of energy and talent, has led in recent years to a growing acceptance of the need for closer integration and collaboration. This is certainly the view of soil scientists actively involved in the Archaeological Soil Micromorphology Working Group, an ad hoc group meeting biannually in Europe (Arpin et al., 1998; see http://www.gre.ac.uk/~at05/micro/soilmain/intro1.htm).

For their own scientific peace of mind, the present authors have adopted procedures that approach the ideal of multianalytical approaches. Thus, the chief aim of this chapter is to illustrate ways in which the soil micromorphologist may more effectively work within a multidisciplinary approach to microstratigraphic studies.

Consensus interpretations will always be more convincing than interpretations based on one discipline working in isolation. Soil micromorphology employing thin sections is in the middle of a scale of stratigraphical studies that involve fieldwork at one end and scanning electron microscopy at the other (e.g., Courty et al., 1989; Macphail, 1998; Macphail et al. 1998a). Soil micromorphology itself is multifaceted, in that organic matter, mineral components, pedological activity, and sedimentary processes, for example, can all be identified (Bal, 1982; Bullock et al., 1985; Courty et al., 1989). This technique also lends itself to being integrated with other disciplines, as detailed as follows. Bulk physical and chemical, macro-, and microfossil data can be linked directly with the undisturbed microstratigraphy evident in thin sections. Schematic and numerical/seminumerical data presentations from combined disciplines is seen as a way of integrating more specialists in the process of creating consensus interpretations.

2. Methods

2.1. Getting the Sampling Right

It is all very good having ambitions to combine post-excavation data in a multidisciplinary way, but this can only work if correct and thoughtful sampling, subsampling, and sample preparation are carried out in the first instance. For example, if soil monoliths are impregnated they cannot be subsampled afterward for soil chemistry. If only large bulk samples are taken, these cannot be used for pollen analysis. Also, if the pollen column is distant from the soil micromorphology samples, data correlation is less certain.

In the field, good results come from combining Kubiena boxes (8 × 7 cm) and square section plastic drainpipe cut into convenient lengths (e.g., 10–20–40 cm) for undisturbed monolith sampling. These are taken exactly alongside plastic bag samples of the archaeological units and layers within them (20–50–200–1,000 gm). Needless to say, all the archaeological contexts of interest must be sampled, with adequate coverage of the vertical stratigraphy, alongside lateral controls, according to the needs of the site study. At this time there must also be good communication with the site's director/area supervisor/environmental manager, in order that archaeological sampling for artifacts and biofacts is coordinated across the site. For example, radiocarbon dating, phasing by pottery analysis, and contextual interpretation based on charred seed and/or bone analysis can all become crucial elements during the post-excavation phase.

Examination of monoliths in the laboratory allows a second and more relaxed chance to examine the stratigraphy. Monolith cores can be first subsampled for pollen and small chemical samples before being impregnated for thin section analysis. As emphasized throughout this chapter all investigators should regard all techniques as equal approaches. In some situations, the early findings from pollen analysis, for example, may allow better targeting of specific parts of a core for chemical and soil micromorphological studies.

2.2. Multidisciplinary–Analytical Approach

2.2.1. Chemistry and Palynology

The chemical and palynological methods employed are already well established in the literature (Clark, 1990; Engelmark and Linderholm, 1996; Moore et al., 1991). Within the text we cite proportions of organic and inorganic phosphate as extracted by 2% citric acid, before and after ignition at 550°C, and refer to "P ratios." Several studies demonstrated empirically that soils with P ratios of <1.0 contain inorganic phosphate in the form of neoformed apatite, bone, vivianite, poorly crystallized forms of phosphate and mineralized coprolites, whereas soils with P ratios >1.0 have been manured and/or contain organic herbivore dung (Engelmark and Linderholm, 1996; Macphail et al., in press).

2.2.2. Choosing Techniques

Different archaeological and pedological questions require a flexibility of approach. For example, at the Romano–British site of Folly Lane, St. Albans (UK), it was necessary that archival information from the Soil Survey of England and Wales should be combined with on-site soil micromorphology, microprobe, and diatom studies in order to investigate the composition and archaeological significance of "turf" mound material (Avery, 1964; Macphail et al., 1998b).

During this first stage of soil micromorphological description and identification, some specific features can be analysed by SEM/EDXRA and/or microprobe (see the following sections). It will be seen that such data retrieval then permits the presentation of soil micromorphological data alongside that from other disciplines, such as chemistry and palynology (cf. Preece et al., 1995, Fig. 6).

2.2.3. Soil Micromorphology

In soil micromorphology, descriptive analysis has produced good results (Bullock et al., 1985) in the identification of (1) microfabric types (absolutely essential), (2) structural and porosity features, (3) natural inclusions (e.g., plant remains such as roots, gravel-size flint, and chalk), (4) anthropogenic inclusions (e.g., charcoal, bone, various coprolites, slag, allocthonous stones), and (5) pedofeatures. The presence of fine charcoal, an abundance of phytoliths, or the presence of diatoms, pollen grains, and fungal spores can all be included within the definition of a microfabric type. Pedofeature studies may include the identification of different types of clay coatings, secondary iron, and manganese nodular impregnations, neoformed vivianite, and different types of soil animal excrements.

2.3. Numerical/Semi-numerical Data Gathering

Since 1992, a combination of description of the previously listed components and features and area counting (as opposed to point counting), has been adopted in about 20 studies. The latter can be extremely accurate, and when tested against image analysis of a counted slide from Overton Experimental Earthwork, as little as a 0 to 5 percent difference was found for each of the 13 vertical 0.5 cm deep transects (Acott et al., 1997; Macphail and Cruise, 1996). As the slide was counted at vertical intervals of 0.5 cm, estimates were based on 0.5 cm squares across the slide. Counting of a slide (7.5 × 5.5 cm) at 0.5 cm intervals, however, takes about 8 working hours, and so it is no light undertaking to carry out this kind of analysis where estimates attain numerical validity. On the other hand, where budget and time constraints are factors in a study, area counting may be carried out at a variety of scales, some of which are considerably less time consuming (see the following text). Estimates of clay coatings in order to identify an argillic horizon (*sensu stricto*) produced varied results between operators (see also McKeague, 1983; Murphy et al., 1985). This is why Bullock et al. (1985) wisely chose to keep broad groupings in their Frequency and Abundance scales. Additionally, although coarse mineral grains, void space, major microfabric and faunal excrement types can be accurately estimated, small inclusions such as rare fragments of bone can best be recorded on the Abundance scale of Bullock et al. (1985). Point counting at normal intervals (e.g., 1,000 points per standard geological slide) may well miss very small and rare inclusions. That is why in archaeological studies, where microscopic inclusions may be crucial to an interpretation, area estimation/counting is generally preferred.

In fully funded research projects, thin sections can be counted at practical intervals of 0.5 cm. The Wareham Experimental Earthwork study involves image analysis (by Tim Acott, University of Greenwich), which is being employed to count the amount and shape of voids, mineral grains, organic fragments, and the organic matrix, whereas manual counting (Macphail et al., in preparation) is being used for the numerical analysis of faunal droppings and the different types of plant fragments and their distribution. Traditional descriptive soil micromorphology is also being used to check the accuracy of digitized images, which can then be more accurately and more confidently quantified. The combined soil study also involves chemical analysis of samples from 1 to 2 cm spits taken from the same locations (Macphail et al., in preparation).

Since the early 1990s many archaeological deposits were first described, and then counted, so that the stratigraphical distribution of selected materials and features could be more fully appreciated. Reasonable results have been achieved at the 1 cm scale. Here, a thin section (7.5–13.5 cm) takes some 3 to 5 hours to count. Data may be more rapidly obtained by area counting each archaeological context. These data can be extracted largely from the initial soil micromorphological description and do not require large amounts of extra time. Whichever scale is selected, however, it is essential that the micromorphologist should first examine the slide and gain a general understanding of the soil prior to counting. For example, the soil micromorphologist must be able to differentiate between a natural soil, a washed sediment, and a trampled floor deposit before counting is undertaken. Otherwise counting is a waste of time. Although this basic understanding of the slide may require learned skills and/or advice, it is an absolutely vital step. In fact, numerical data (for its own sake) in soil micromorphology can produce nonsense (Stoops, personal communication, 1997). What is advocated here is the thoughtful gathering of numerical data from thin sections that are already well understood. After counting, slides and counted features can again be analyzed as the understanding of the soil micromorphology deepens. Further benefits arise from the fact that counted data are useful when more than one soil micromorphologist is involved in a single project, because findings can be compared rapidly. Additionally in our experience, during a long-term project, it takes less time to refamiliarize ourselves with our thin section when we have counts than when we have only long descriptions to read.

2.3.1. Presentation of Soil Micromorphological Data (Courty et al., 1989; Romans and Robertson, 1983; Simpson and Barrett 1996)

In 1994, one of the authors (Macphail) presented a seminar paper to the Archaeological Soil Micromorphology Working Group at Rennes University, France. The object was to demonstrate and discuss the many ways in which soil micromorphological data can be presented and to note the views of the members of the working group. For example, full-page descriptions as per Bullock et al. (1985) were compared with tables summarizing data and their interpretation and schematic diagrams to express numbers of features present (per thin section/horizon). Bullet points were employed in the last example. This simple idea came from a paper by Simpson and Barrett (1996) and has been used by other authors (R. Kemp, Royal Holloway University of London, personal communication 1995; A. Gebhardt, Rennes University, personal communication, 1997). At more recent meetings of the working group (Cambridge, London, and Pisa) some soil micromorphological data were expressed as percentages (e.g., Matthews et al., 1997, Fig. 3a-b), with counted data from experiments illustrated as bar graphs, bullet points (Crowther et al., 1996; Macphail, 1998) and on Frequency and Abundance scales.

Nonsoil micromorphologists may examine data from seed, bone, and palynological studies because these are presented graphically, but soil micromorphological findings have generally been obscured by its presentation either in jargon or as interpretation. The present authors have therefore been endeavoring to make soil micromorphology more user friendly to other scientists. This does not mean, however, that they will fully understand the nitty-gritty of soil micromorphology any more than they would the intricacies of pollen taphonomy and mineralized seed identification, but they can at least see how interpretations are constructed on the logical registration of data as expressed graphically.

The present authors and their colleagues continue to produce soil micromorphological descriptions as the basis of "counted" microfabrics and components. Professor Stoops (University of Gent), although acknowledging the need to summarize data for publication, has also suggested that sufficient data should still be available to enable the reader to judge the scientific merit of the work (Stoops, personal communication, 1997). In papers produced for our peers, this is certainly crucial, but in archaeology we also have to deal with a lay audience. The same must be true for soil micromorphologists reporting to agronomists and to Quaternary scientists. It is therefore up to us to both produce and present data that are both acceptable to our peers and understood by our audience (e.g., archaeologists, paleoenvironmentalists, and field Quaternary scientists).

3. Research Base

Soil micromorphologists working in archaeology need to break new ground because most publications on soil micromorphology have dealt only with natural soils. To achieve this, workers have developed their own specific reference collections, analyzed specific archaeological materials, studied ethnologically interesting sites, and carried out experiments (Courty et al., 1989, 1994; Crowther et al., 1996; Gebhardt, 1992; Goldberg and Whitbread, 1993; Wattez and Courty, 1987; Wattez et al., 1990).

In our case, this approach to archaeological soil micromorphology has been supported by two major strategies, as follows:

First, "counting" has been applied to thin-section studies of deposits formed by ethnoarchaeological experiments, in order to try and identify key semi-numerical microfabric signatures, that may be of significance in the archaeological record.

A second approach has been to identify from our experience some specific components and microscopic inclusions that regularly occur in archaeological deposits and to analyze examples of these intensively. This is a way to identify the archaeological significance of these, especially when recorded semi-numerically, just as counted pollen or seed types may be given anthropogenic weighting according to, for example, established floras and ecological groupings. Where possible, soil micromorphological findings have been combined with chemical data, macrofossil, and palynological studies of the same horizons and components.

3.1. Experimental Findings

The Ancient (Iron Age) Farm at Butser, Hampshire, U. K. is situated on the chalk of southern England, and is well known in Europe for being a focus of experimental studies in agriculture, arable soils, architectural structures, and their floors (Gebhardt, 1990, 1992; Macphail and Goldberg, 1995; Macphail et al., 1990; Reynolds, 1979). To be consistent with the approach to the study of soils at the Experimental Earthwork at Overton Down (Crowther et al., 1996; Macphail and Cruise, 1996) and at numerous current archaeological sites, it was decided to restudy the floors from the Moel-y-gar House (animal stabling) and the Pimperne House (domestic occupation) at Butser, using counted soil micromorphological data. At the same time, bulk samples were run for chemical and palynological analyses. This approach would then provide an experimental example of multidisciplinary microstratigraphic studies as the preferred approach of the authors. Our work at the Moel-y-gar and Pimperne House floors are examples of soil micromorphology counting and how resulting data can be linked to complementary data from chemistry and palynological studies.

At the Moel-y-gar stable house, three distinct layers were identified (Tables 9.1a and 9.2a): an uppermost cemented crust of layered, long monocotyledonous plant fragments, a "stable soil" of phosphate stained chalk and soil, and a phosphate-contaminated buried subsoil (Macphail and Goldberg, 1995, Fig. 2, Plates 3 and 4). The uppermost layer was further characterized by microprobe and X-ray diffraction analyses to confirm the view that this plant-rich layer that is autofluorescent under ultraviolet light, is cemented by calcium phosphate in the form of hydroxyapatite. Key microstratigraphic features were counted (Tables 9.1a and 9.2a).

At the Pimperne House at Butser, a very different kind of microstratigraphy had developed, with an uppermost trampled/beaten floor layer overlying a buried soil (Tables 9.1b and 9.2b). In addition to the soil microfabric differences, complementary studies found, in comparison with the Moel-y-gar "crust," a more strongly enhanced magnetic susceptibility, but less organic matter (LOI 18%) and phosphate (2400 ppm P), the last being dominantly in an organic form (P ratio 2.2–3.4). Furthermore, pollen concentrations were considerably lower but contained a far more diverse herbaceous and weed pollen assemblage.

How do these findings compare with archaeological data? At the Italian Neolithic cave of Arene Candide, Liguria, phosphate-stained stabling layers composed of layered and compacted oak twig wood (leaf hay foddering) can be differentiated from sublamina, massive structured mineralogenic domestic floors (Macphail et al., 1997). At the Roman London site of 23, Bishopsgate, two counted samples from a red charred floor context were composed of semi-layered plant fragments/cattle dung-like material with total phosphate averaging 9,000 ppm, thus indicating that a likely stable layer had been found (an hypothesis now supported by macrobotanical findings; Macphail et al., in press). Sites ranging from prehistoric to recent from Scotland through Switzerland, Italy, and southern France to north Africa have yielded further comparative examples of floors with covered (roofed) stable and domestic areas having microstratigraphic signatures consistent with the experimental findings from Butser (e.g., Boschian, 1997; Cammas, 1994; Cammas et al., 1996; Davidson et al., 1992; Del Lucchese and Ottomano, 1996; Guélat et al., 1998). At the London Guildhall site two types of Anglo–Danish (1060–1120 A.D.) floors were differentiated on the basis of soil micromorphology, chemistry, and palynology. One floor type has a poorly preserved but diverse pollen assemblage in a heterogeneous mineralogenic (LOI 9%) soil with an enhanced magnetic susceptibility (assumed domestic structure).

Table 9.1a. Counted Microstratigraphy of Floor and Buried Soil of the Moel-y-gar Stabling House 1990 (Center of the House)

a	B	AI 1	AI 2	AI 3	AI 4	AI 6	AI 7	AI 8	NI 2	NI 3	NI 5	NI 6	NI 9	S1	S2	S3	S4	S5	S7	EX 2	EX 3
1	B1(B2)		aa		a	ffff f		aa	f					f	ffff						
2	B1		a	aaa		ffff		aaaa						f	ffff						
3	B1					ffff f		a aaaa						f	ffff						
4	B1/B2	fff				ff		a		a				ff	ffff						
5	B2	f				ff	a	a	f	aaaa	a		a	ff	ffff					a	
6	B2(B1/3)	fff f					aaaa			a	a	a	aa	ff		ffff	ffff	ffff	ffff f	a	
7	B2(B1)	fff f				f	a	a	f	a	a	a	a	ffff	ff	ffff	f	ffff	ffff f	a	
8	B1/B2					ff	a	aaaa	f	a	a	a	a	ff	ffff	ffff	f	ffff f	ffff	a	aa
9	B2/B3 (B1)					f	a	a	fff	a	a	aaa	aa	ff		ffff f	ffff	ffff	ffff	aaa	
10	B2(B3)								ff	aa	a	aa	aa	ff		ffff f	ffff	f	f	aa	
11	B2(B3)						a		fff	aa	aa	aa	aa	fff		ffff f	ffff f	f	f	aaa	
12	B2(B3)								ffff	aa	a	aa	aaa	ff		ffff f	ffff f	f	f	aa	
13	B2(B3)								ffff	aa	aa	aa	aaa	fff		ffff	ffff	ffff		aaa	

Table 9.1b. Counted Microstratigraphy of Floor and Buried Soil of the Pimperne House 1990 (Sample from Near Oven Location)

a	B	AI 4	AI 5	AI 6	AI 7	AI 8	AI 9	NI 2	NI 4	NI 5	NI 6	NI 8	S1	S3	S5	S6	S7	S8	FF 1	EX 1	EX 2
1	B4	f	a	f	aa	a	aaa	ff	a	a	a	aa	ff				ffff	f	aa	a	aa
2	B4	f	a	f	aaa	a	aaa	fff	a	a	aa	aa	fff				ffff	f	a		a
3	B4/B5	f	a		aa	a	a	fff		aa	a	aaa	fff	ffff	fff		ff	fff		a	aaa
4	B5				aa			fff		aaa	aaa	aaa	fff	ffff	ffff	f		f		aaaa	aaa
5	B5				a			fff		aa	aaa	aaa	fff	fff	ffff	ffff		f		aaaa	aaa
																f					
6	B5							fff		aaa	aa	a	fff	f	ffff	ffff		ffff		aaa	aaa

Key to Tables 9.1a–9.1b
a = Depth below surface (cm).
B = Microfabric type.
B1 = Dominantly organic; very dominant pale to dark brown speckled and dotted (PPL), isotic to very low interference colors (weak to strong crystalline b-fabric) (XPL), dill brown (OIL) and high autofluorescence (UVL); dominant amorphous organic matter and tissue fragments, with occasional to abundant calcite crystals, occasional to many phytoliths, rare to occasional calcium oxalates (druses) and rare to very abundant patches of <20 μm size spherulites; single-spaced porphyric; C:F 70:30 (limit as 10 μm); Coarse mineral—very dominant, well-sorted silt-size quartz and very few medium sand-size chalk inclusions; Coarse organic—dominant longitudnal sections of *Poaceae* cuticles (tissue remains) and frequent stem cuticle and parenchyma cells (organ remains); much cellulose is poorly birefringent and browned (humified).
B2 = Dominantly mineral very dominantly blackish brown (PPL), very low interference colors (weak crystalline micritic b-fabric) (XPL), pale whitish yellow (OIL) and moderately autofluorescent (UVL); many to abundant amorphous organic matter fragments; single-spaced porphyric; C:F 40:60; Coarse organic—few amorphous fragments; Coarse mineral—common stained chalk clasts (rims autofluorescent under UVL), common silt, very few biogenic calcite (probable earthworm granules) and dark-stained mollusc shell fragments.

Soil Micromorphologist as Team Player

B3 = Dominantly mineral—very dominant finely speckled brown (PPL), medium/high interference colors (XPL), pale yellow (OIL) and non-autofluorescent (UVL); occasional to many amorphous organic matter fragments. As B2.

B4 = Dominantly mineral; very dominant speckled and dotted yellowish brown (PPL), low to high-interference colors (weak to strong crystallitic micritic b-fabric); pale yellow to yellowish brown with frequent black specks (OIL), non-autofluorescent (UVL); abundant amorphous and charred organic matter fragments; rare calcium oxalates (druses), <20 μm size spherulites and wood ash crystals; single-spaced porphyric; C:F 60:40; Coarse mineral—dominant silt, common chalk and flint, few mollusc shell and biogenic calcite, very few burned soil and argillic (Bt) sandy soil inclusions; Coarse organic—few charcoal and very few browned (humified) tissue fragments and organ remains.

B5 = Dominantly mineral—speckled, darkish reddish brown (PPL), medium interference colors (XPL), yellowish orange brown (OIL), non-autofluorescent (UVL); abundant amorphous organic matter; single-spaced porphyric; C:F 55:45; Coarse mineral—dominant silt, common chalk and flint, few mollusk shell and biogenic calicite; Coarse organic—very few root traces (organs).

NB: PPL—plane polarized light; XPL—crossed polarized light; OIL—oblique incident light; UVL—ultraviolet light.

Major anthropogenic inclusions

AI1 = stained chalk (with dark UVL autofluorescent rims); AI2 = micrite; AI3 = silty micrite (with well-sorted silt-size quartz); AI4 = burned soil (rubified under OIL); AI5 = soil inclusions (e.g., argillic subsoil clay; daub mixture); AI6 = partially layered plant fragments (>1 mm); AI7 = coarse charcoal; AI8 = amorphous organic matter; AI9 = ash and fine charred fragments.

Major natural inclusions

NI1 = chalk; NI2 = flint; NI3 = sandy soil; NI4 = biogenic calcite; NI5 = mollusk shells; NI6 = plant fragments (<1 mm); NI7 = roots; NI8 = root traces/fragments.

Structure

S1 = voids; S2 = laminar structure; S3 = prismatic structure; S4 = spongy structure; S5 = subangular blocky structure; S6 = crumb structure; S7 = massive structure; S8 = burrows.

Fabric pedofeatures

FF1 = horizontal slickenside/clay smear?

Excrements

EX1 = organo-mineral excrements (>500 μm); EX2 = organo-mineral excrements (<500 μm); EX3 = organo-excrements.

Frequency

ffff >70%, very dominant; fff 50–70%, dominant; ff 30–50%, common; f 15–30%, frequent; f 0–15%, very few and few.

Abundance

aaaaa >20%, very abundant; aaaa 10–20%, abundant; aaa 5–10%, many; aa 2–5%, occasional; a <2%, rare.

Table 9.2a. Key Microstratigraphic Features at the Moel-y-gar House

Depth and layer	Soil micromorphology	Some complementary studies
1–3 cm ("crust")	Very dominant partially layered plant fragments >1 mm in length (grass stems, AI6) and abundant amorphous organic matter (humified dung, AI8) set in a specific, chiefly autofluorescent (under UVL) microfabric (B1) with little evidence of small animal mixing (EX2 and EX3).	Dominant grass stem fragments (W. Cazrruthers, personal communication, 1995), high amounts of organic matter (LOI 41%) and inorganic phosphate (6000 ppm P; citric acid P ratio <1.0; 0.5–1.3% elemental P), and unexpectedly high concentration of pollen 822 (grains × 1000) per cm^3 dominated by grass pollen, compared to only 77 (grains × 1000) per cm^3 in the subsoil.
3–9 cm ("stable soil")	Dominant dark stained chalk clasts (AI1) and microfabric type B2, with chalk clast rims and soil ped edges autofluorescent under UVL; dominant subangular blocky structures and burrows within an overall prismatic structure (S3, S5, and S7) (Fig. 14.1).	Organic (LOI 32%) and phosphatic (2840 ppm P; citric acid P ratio <1.0).
9–13 cm ("buried subsoil")	Only rare anthropogenic inclusion (AI) occur in an unstained chalk (NI1) dominated, increasingly by calcareous soil (B3), with a subangular blocky and spongy structure associated with many very thin to thin organo-mineral excrements (EX2).	Comparatively less organic (23% LOI) and phosphatic (1460 ppm P; citric acid P ratio >1.0).

Table 9.2b. Key Microstratigraphic Features at the Pimperne House

Depth and layer	Soil micromorphology	Complementary studies
1–3 cm ("trampled floor")	1–3 cm ("trampled floor"): few burned soil fragments (AI4), with many coarse charcoal (AI7), ash and fine charred materials (AI9) and rare amorphous organic matter (dung, AI8) fragments, set in a highly heterogeneous mineralogenic soil (microfabric type B4) with a dominant (but now cracked) massive structure (S7) (Fig. 14.2).	Moderately humic (LOI 20%) and phosphatic (2430 ppm P; citric acid P ratio >1.0). Highest magnetic susceptibility (47 Si units SiKg10-8, 4.7% MS conversion at 550 °C) compared with Moel-y-gar floor (16–27) and surropunding fields (mean 22).
3–6 cm (buried soil")	3–6 cm ("buried soil"): only rare to occasional anthropogenic anthropogenic inclusions (AI) in calcareous soil (microfabric type B5) containing many natural inclusions (NI), and featuring first prisms (S3) and increasing amounts of subangular blocky (S5) and crumb (S6) structures, associated with many to abundant organo-mineral excrements (EX1 and EX2).	Similarly humic (LOI 20%) and phosphatic (2310 ppm P; citric acid P ratio >1.0), with magnetic susceptibility at 28 Si units (SiKg10-8).

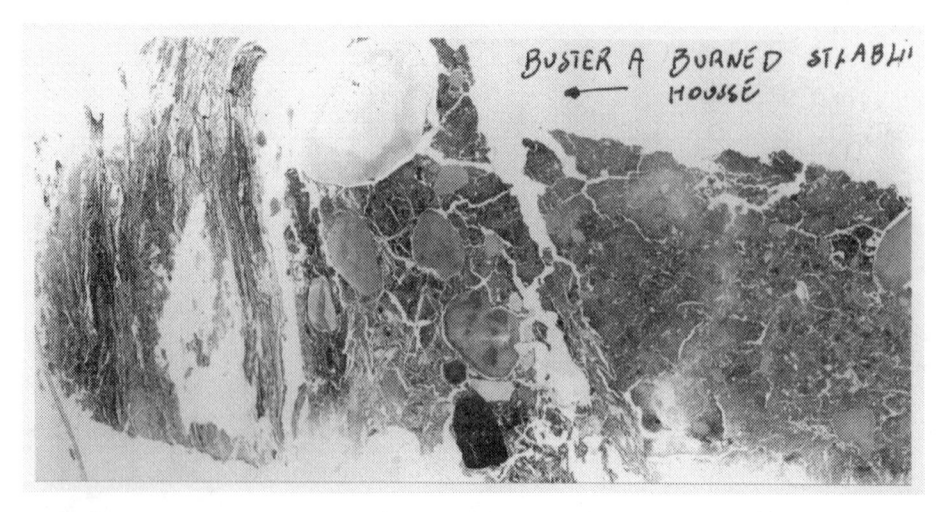

Figure 9.1. Butser Ancient Farm, Moel-y-gar stabling floor layers; center of floor; uppermost 3 cm—crust of calcium phosphate cemented layered monocotyledonous plant fragments and dung; middle 4 cm—stable floor of phosphate stained chalk and soil; lowermost 6 cm—phosphate stained buried colluvial rendzina soil (for detailed soil micromorphology see Table 9.1a). Computer-enhanced image of 13 × 6.5 cm thin section.

In contrast, other *in situ* floors and floor deposits have well-preserved grass- and cereal-dominated pollen assemblages in highly organic (LOI 30%) and phosphatic (e.g., 6,000 ppm P) deposits characterized by layered plant fragments or probable cattle dung (Cruise and Macphail, in press; Macphail and Cruise, 1995). The latter contexts are interpreted as stable floors and deposits. In order to begin the process of interpreting soils, individual microfeatures need to be characterized.

Some of these are better understood than others. For example, at Overton Down much was made of the apparent transformation of earthworm-worked soils (1–5 mm wide mammilated excrements) into soils featuring 100 to 500 μm thin Enchytraeid-like excrements as a result of changed soil conditions induced by burial (Crowther et al., 1996). The relationship between the excrements of soil fauna and soil conditions is well understood in general (Babel, 1975; Bal, 1982). Equally, the presence of vivianite and related features are seen as indicative of the presence of phosphate, and such features occur in bog ores, occupation deposits, and floors (Landyudt, 1990; Macphail, 1983, 1994). In the following sections, we give two examples of how two important anthropogenic soil components, "dark clay coatings" and "phosphatic nodules" were characterized in order to determine their composition and their implied archaeological significance.

3.1.1. Dark Clay Coatings

A feature common to archaeological sites, but which is poorly understood, is dark-reddish brown clay coatings. For many years it was suspected that these were

Soil Micromorphologist as Team Player 255

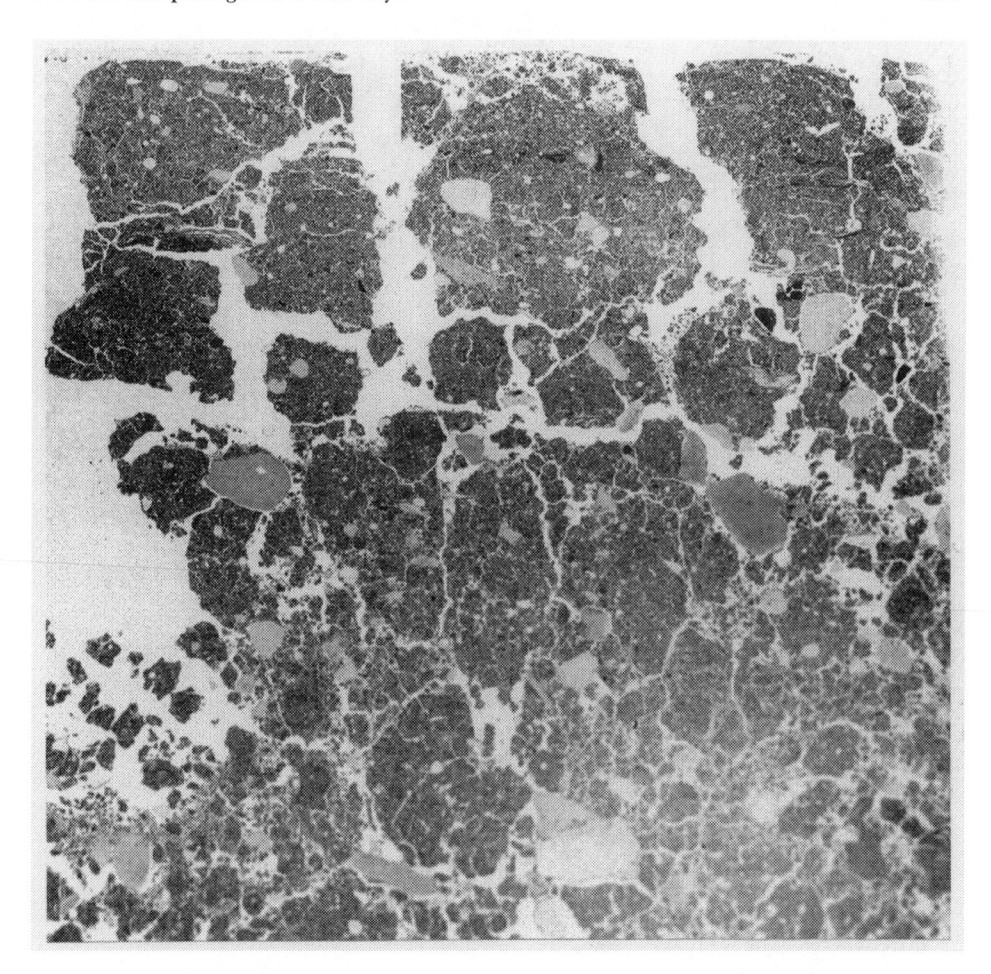

Figure 9.2. Butser Ancient Farm, Pimperne House floor; center, juxtaposed to hearth; uppermost 3 cm—massive, dry trampled crust containing various anthropogenic inclusions; lowermost 3 cm—prismatic structures of soil underlying the trampled crust, give way to subangular blocky and crumb structures, relic of the natural buried colluvial rendzina soil. Computer-enhanced image of 6.5 cm long thin section.

of an organic/phosphatic character and were associated with animal concentrations, (Nörnberg and Courty, 1985, plate 5). The present authors have also found them while counting urban soils in London and at rural sites in Northamptonshire and Bedfordshire (see Figs. 9.5 and 9.7). The long-term study of the important site of Raunds, Northamptonshire—where large numbers of these dark clay coatings are present in the counted microfabrics—permitted a detailed study of these features (Courty et al., 1994, photo 6; Windell et al., 1990). They were counted alongside coarse textural features such as micropans and impure clay coatings and infills in soils that featured one barrow burial of a hundred cattle skulls and another buried soil that contained dung beetles.

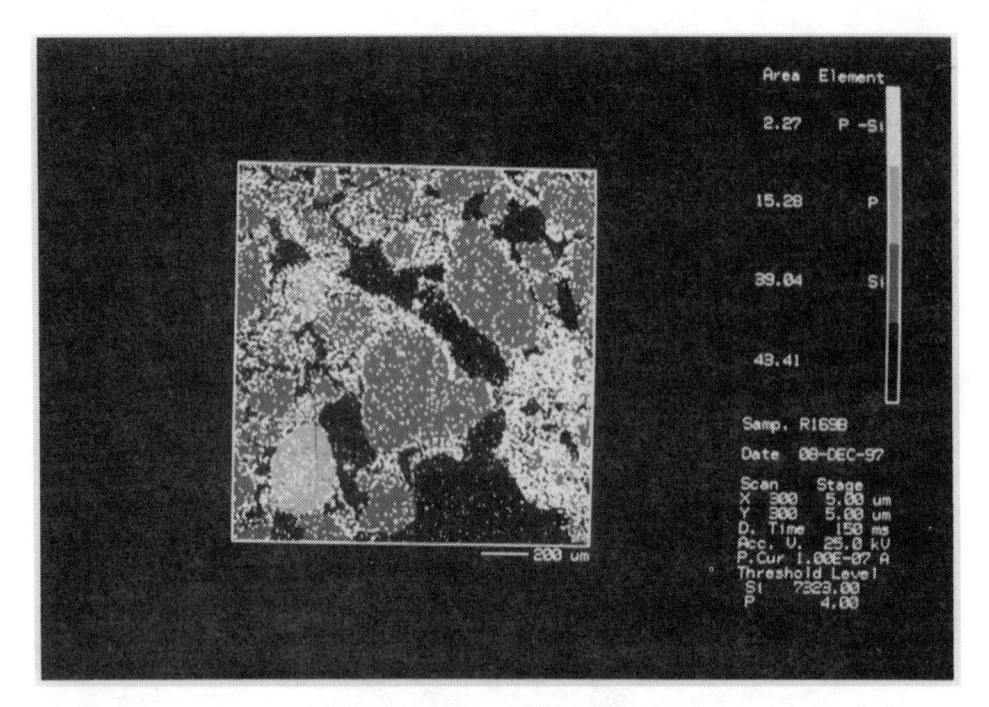

Figure 9.3. Raunds prehistoric barrow cemetery (example from West Cotton, Barrow 5, sample 9); Bronze Age barrow-buried subsoil; example of elemental mapping of dark clay coatings by microprobe; dark clay coatings, 50–100 microns thick, coat sand grains of diagonal void; elemental P averages 0.2%; such features are coincident with enhanced levels of organic phosphate and may be indicative of animal concentrations (see text). (Analysis by Kevin Reeves, University College London). Computer-enhanced image.

We tested the hypothesis of a link between these dark clay coatings and animal management at Raunds. First, total phosphate analyses of bulk soil samples revealed an association between these features and phosphate, a simple finding consistent with the literature on phosphate and animal activity (Proudfoot, 1976; Quine, 1995). Moreover, the dark clay coatings are most abundant in layers dominated by concentrations of organic matter and organic phosphate (P ratio 2.3–3.5; Engelmark and Linderholm, 1996; Macphail et al., in press). The dark reddish colors of the clay coatings already implied that they were humic in character. Finally, the concentration of phosphorus within the dark clay coatings themselves, rather than the soil matrix surrounding them, was confirmed by microprobe studies of numerous examples from two uncovered thin sections. Data from two examples are presented in Table 9.3, with Fig. 9.3 illustrating the mapped presence of P in Sample 9. We can therefore conclude that organic matter and organic P are apparently concentrated in these dark clay coatings.

Dark clay coatings in natural Bt horizons of Alfisols have long been known to contain organic matter and phosphorus, which are related to natural clay

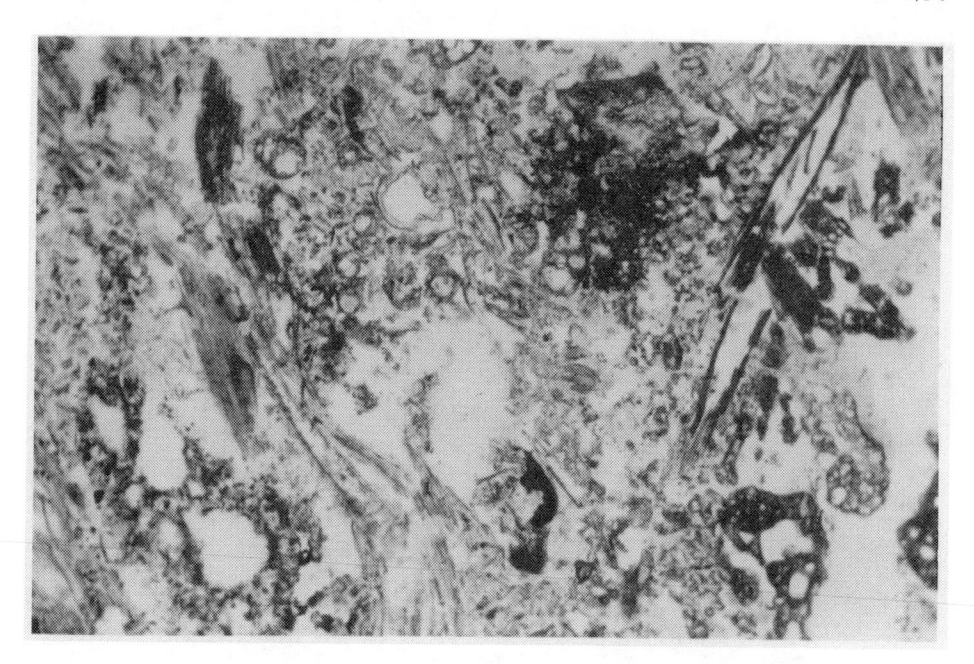

Figure 9.4. Potterne Early Iron Age "Deposit"; Sample 21b, reference fused amorphous material; "fused ash," composed of a colorless to dark gray non-birefringent cement, that is autofluorescent under ultraviolet light and has a siliceous and calcium phosphate character; inclusions include vesicular silica (from melted phytoliths) slag (left-hand corner), residual phytoliths (center), blackish (rubified under oblique incident light) burned soil and charred cereal awn fragments (right); plane polarized light (PPL), frame length is ~0.33 mm. (see Table 9.4).

translocation with fulvic acid under conifer woodland (e.g., Gray forest soils of Duchaufour, 1982:301; e.g., boreal paleosols of Fedoroff and Goldberg, 1982). Thus any link between dark clay coatings and animal management has to be argued carefully. At Raunds, humic topsoils of Spodosols and acidic Alfisols were present in prehistory, and liquid animal waste passing through these may have mobilized fulvic acid to produce these dark reddish brown clay coatings, which occur alongside other textural features indicative of animal trampling (M. A. Courty, CNRS, Paris, personal communication, 1992). Obviously, this hypothesis of a process active at the microscale is worthy of further testing. But, as fieldwork, bulk chemistry, and soil micromorphology studies have yielded comparable interpretations from nine barrows dating from the Neolithic to the Bronze Age, proxy soil landscapes and their land use can be reconstructed on the scale of kilometers for this part of the Nene river valley. As similar paleosols have been analyzed in the nearby Ouse valley, such findings have implications for regional proxy soil landscape and land-use reconstruction. Past soils of the chalk downlands of southern England have already been modeled in this way (Allen 1992; Evans, 1972; Whittle et al., 1993).

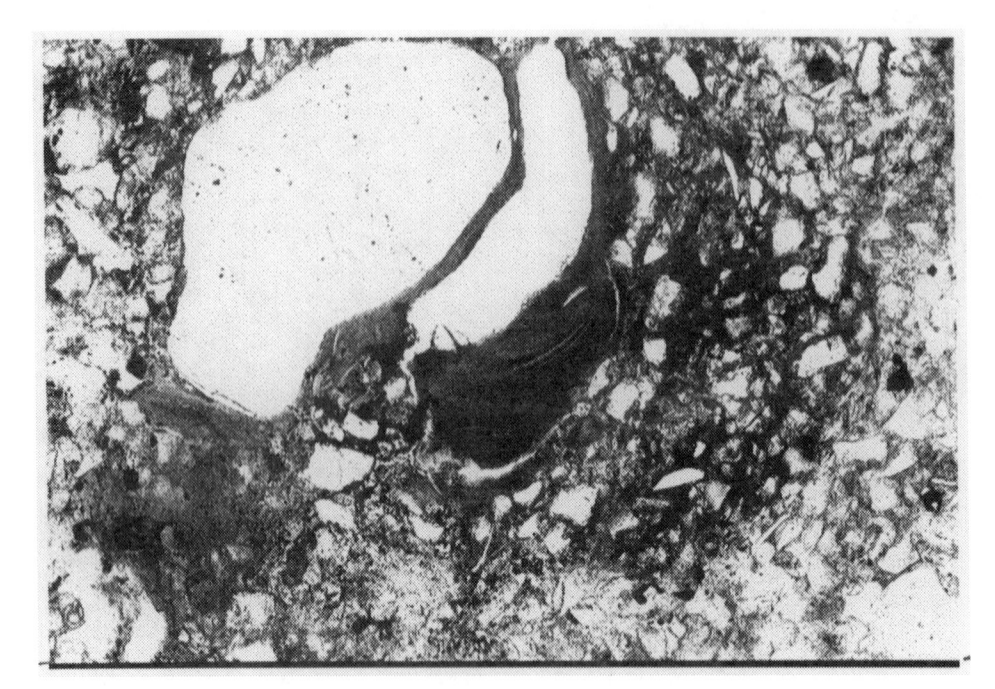

Figure 9.5. Colchester House, London; microlaminated dark reddish clay void coatings in brickearth subsoil beneath 3rd Century truncated, 1st–2nd Century Roman occupation; dark coatings are coincident with enhanced levels of organic phosphate; the authors have argued for a possible phase of animal activity on site predating 3rd century constructions (see text). PPL, frame length is ∼1.3 mm. Computer-enhanced image.

3.1.2. Phosphatic Nodules

Three enigmatic materials with specific features under PPL, XPL, OIL, and UVL, were identified in thin sections of an occupation deposit at Potterne (Late Bronze Age/Early Iron Age, Wiltshire; Lawson, 1994). Individual fragments were made into thin sections, with residues being studied under microprobe, through bulk chemistry, and through macroplant remains and pollen analysis. These materials, termed for convenience as "pale nodules" (possible cess-pit nodules), "fused ash" (burned and fused cereal processing waste), and "burned and cemented soil" (often burned, possible stable soil floor deposit), were all autofluorescent under ultraviolet light and contained around 12%, 7%, and 1% P, respectively. Table 9.4 shows an example of how one of these components was defined and then interpreted to become an established microscopic indicator of the presence of domestic cereal processing waste at a site. This description and characterization is a crucial step before such counted components can be given any significance in site reconstruction.

Subsequent to this work, fused ash, cess-pit nodules, and dark clay coatings were found at a number of midden and occupation sites, their semi-quantified presence added to the collage of information available for the interpretation of sites with complicated site formation processes.

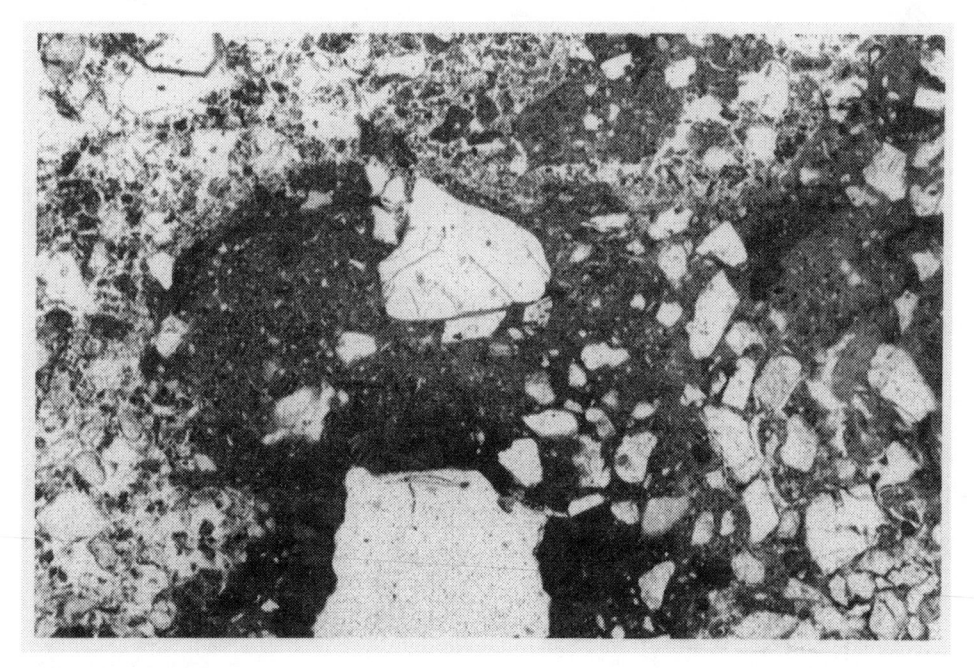

Figure 9.6. Haynes Park, Bedfordshire; Roman to Norman rural activity on a catena; "water hollow" location at base of slope, upper fill; humic sandy soils with very abundant phytoliths impregnated with probable poorly crystalline iron compounds, forming a nodule around a void that is coated by fibrous crystalline material (goethite or Fe/P compound), all indicative of dominant waterlogging; layer is coincident with enhanced amounts of inorganic phosphate; other micromorphological features and palynology indicate the presence of animals and inputs of dung (see text). PPL, frame length is ~3.4 mm. Computer-enhanced image.

4. Discussion

How successful has this fully integrated microstratigraphical approach been? We have already cited our study at Folly Lane where soil micromorphology and microprobe studies were combined with the identification and semi-quantitative analysis of diatoms in thin sections, as one example of a multi-disciplinary investigation of rural Romano–British soils (Macphail et al., 1998b). Such an approach allowed us to go further with our interpretations than would otherwise have been the case if only single or non-integrated techniques had been applied. A consensus understanding of what happens to soils when buried at the Overton Down Experimental Earthwork drew on palynological, microbiological, chemical, soil micromorphological, and archaeological excavation data, and again this led to confident extrapolations when discussing archaeologically buried soils such as at nearby Easton Down (Bell et al., 1996; Crowther et al., 1996; Cruise and Macphail in Whittle et al., 1993). Many other cases have yet to be published, but they can be briefly cited here. As examples, we summarize relevant findings from the Roman site of Colchester House, London, and the Roman to Norman site of Haynes Park, Bedfordshire. At Haynes Park we show how we have graphically

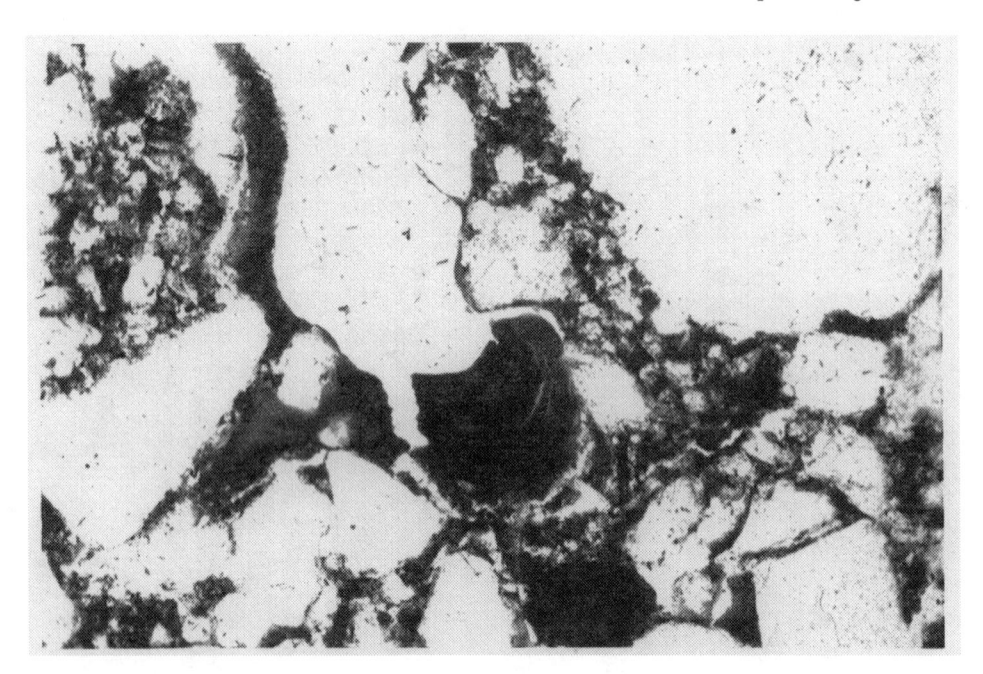

Figure 9.7. Haynes Park, Bedfordshire; Roman to Norman rural activity on a catena; colluvial Roman soil below lynchet; major soil micromorphological features are burrowing and mixing by soil fauna and abundant dark, microlaminated dusty void clay coatings (illustrated) and likely amorphous phosphate infills; layer is also coincident with 1,000 ppm P (2% citric acid extract)— a maxima in these slope soils; this layer is interpreted as an animal trampled soil, leading down to the waterhole. PPL, frame length is ~1.4 mm. Computer-enhanced image.

presented summarized soil micromorphogical, chemical, and pollen data to illustrate and support our arguments as reported to the archaeologists working on the site.

A number of mechanisms were identified that accelerate weathering of Roman to medieval urban stratigraphy and the formation of a cumulative anthrosol termed "dark earth" (Macphail, 1994). One atypical urban land use is the stocking of animals, the trampling and rooting-up of soils that could homogenize earth-based (timber and clay) buildings. At Colchester House, London, the coincidence of organic phosphate in subsoils with counted dark clay coatings (Fig. 9.5) allowed the hypothesis that a phase of animal activity could have contributed to the reworking of clay and timber buildings believed to have been on the site before construction of a stone-founded structure in the third century A.D. (Macphail and Cruise, 1997b).

At Haynes Park, Bedfordshire, a catenary sequence contains wet hollows at the bottom of the slope (Macphail and Cruise, 1997a). Fieldwork, excavation, and macrofossil studies suggested that the Roman to Norman deposits were likely the result of dominant arable activity, as indicated by the presence of a Roman corn dryer, charred cereal grains, and substantial lynchet. Soil micromorphology was

Table 9.3. Mean Values of P in Dark Clay Coatings (Microprobe Line Analysis) and Background Bulk Chemistry at Raunds (see Fig. 14.3)

Sample	Context	% LOI	Total P ppm nitric acid	Po ppm citric acid	Ptot ppm citric acid after ignition at 550 °C	P ratio	Coating P ppm (probe)
11	buried soil upper	3.8	1410	n.d.	n.d.	n.d.	570 (17 points)
9	buried soil lower	3.3	n.d.	130	350	2.6	2380 (9 points)

linked to chemical studies of the dry soils, whereas in the wet hollows, pollen cores were first evaluated before sampling for thin sections and chemistry. Although cultivated soils were broadly identified, the preserved presence of dung fragments and anthropogenic inclusions such as chalk, ashes, and igneous rock (grindstone), along with the magnetic susceptibility and phosphate chemistry additionally implied that manuring had taken place. Furthermore, the palynological study indicated inputs of fresh manure in a landscape where animals grazed on herb-rich grasslands, acid heath, and wet valley bottoms. Microscopic crust and pan fragments alongside phytolith and diatom-rich microfabrics that featured amorphous organic inputs (dung) and concentrations vivianite and poorly crystallized iron phosphate (Fig. 9.6), further implied the on-site presence of animals (Fig. 9.8). Drier soils up slope also contained dark clay coatings (Fig. 9.7) and other features of trampling. When reconstructing the site's past land use and proxy vegetation history, it became clear from modern studies of the same soil type at nearby Woburn that a probable mixed farming regime had been practiced at Haynes Park to offset the susceptibility of the soils to erosion (Catt, 1992; Macphail and Cruise, 1997b).

It may be considered that the wetter the site, the better pollen may be preserved, but the less potential there is for soil micromorphology and chemical analysis, and few peat bogs have been studied using our preferred combined approach. Nevertheless, at Bargone, Liguria, Italy, the colluvial peat bog edge of mountain peat bog was studied in this way in 1994 (Cruise et al., 1996). The site had already been cored several times in its center and fully analyzed for pollen in the late 1980s. Here again, the palynological evaluation of the new cores from the trenched excavation of the bog edge guided the multidisciplinary investigation. Layers of interest within the cores were subsampled for chemistry and soil micromorphology, as well as being chosen for radiocarbon dating. Of particular relevance to this chapter is the discovery that changes in vegetation as recorded by palynological analysis are coincident, for example, with different chemistry and soil micromorphological indications of the peat bog drying out or animal trampling or colluviation. We also have archaeological and diatom data to add to the debate. Such a multidisciplinary approach is a great advance on traditional palynological investigations.

Table 9.4. "Fused Ash": An Example of Potterne's Components and Microfabrics

Material	Sample number examples	Soil micormorphology (SM), bulk data (BD) SEM/EDXRA (XR), microprobe (probe) and elemental map (EM)	Interpretation and comments
Fushed ash	SM20, SM21b (2673/2227)	SM: (a) colorless to pale yellow to dark grayish brown with fine black dots (PPL), non-birefringent (XPL) (rarely crystallitic), very pale gray, with few fine red material (OIL), with extremely abundant (siliceous) plant fragments up to 800 μm and phytoliths in general, occasional plant pseudomorphs and occasional long charred fragments; common vesicular areas; occasional charred and amorphous organic matter; frequent quartz silt and very fine sand; contains pale yellow, cloudy (PPL), non-birefringent patches, except for fine blackened (XPL) and whitish grey (OIL) areas; all matrix is generally highly UVL autofluorescent; includes patches of silt and fine sand, whereas areas of plant remains contains only a few fine silt-size quartz. fine sand, whereas areas of plant remains contains only a few fine silt-size quartz. (Macroremains: awn fragments of *Triticum* sp.; Carruthers, personal communication) (Pollen preparation: microscopic fragments of uncharred organic matter, charcoal and unidentifiable nebulous "ghosted bodies"; Wiltshire, personal communication) (b) associated material is dark grayish with black specks (PPL), nonbirefrinent (XPL) and gray (OIL), with coarse and fine vesicles; material is non-UVL autofluorescent). XR and Probe: Si (10.5%), Al (0.73%), Fe (1.4%), Ca (14.8%), P (6.8), K (0.4%), Mg (0.15%) Ti (0.007%), S (0.19%), and Mn (0.007%)	(a) probable product of moderately low high-temperature burning (500–600°C) of cereal-rich Poaceae material, with loss of calcitic ash, partial melting of the phytoliths (high Si) in places and their (yellow) staining probably with phosphate (hence high Ca and P and UVL autofluorescent appearance — calcium phosphate); blackish dots are probably composed of carbon; (b) totally melted (above 650°C) silica (Poaceae phytoliths) to produce a vesicular glasslike material with staining by pure carbon; high temperature has probably driven off K found in lower temperature ashes (Courty, personal communication); Straker (personal communication) suggests such fused ashes probably formed under reducing conditions.

4.1. A Final Cautionary Tale

Working within a team can have its own complications. For example, while working on the 500,000 year old site of Boxgrove, West Sussex, UK, findings from the widest imaginable environmental team were debated openly (Roberts and Parfitt, 1999; Roberts et al., 1997; Stringer et al., 1998). Soil sediments with cold formation signatures were associated with cold faunas, and marls contained pond-living mollusks and alluvial deposits had amphibian and fish faunas. On the other hand, Unit 4c, which had all the micromorphological hallmarks of a sediment, was the focus of human activity and full of mammal bone remains and was considered to be a land surface. The described soil microfabric, including that from several thin sections through "chipping floors," initially led to an interpretation of Unit 4c as a sediment. It was only after repeated study that some small residual pedological features were identified, and this together with reference to analogues from drowned coastal sites in the UK and ripened polders in Holland allowed the overturning of the original strictly sedimentary hypothesis. Thus, Unit 4c could safely be identified as a bona fide ripened soil (Macphail, 1996). This was not a compromise interpretation to meet the other specialists halfway, but a soil micromorphological contribution from an equal. Counting of soil micromorphological features for its own sake will not yield interpretations, and at Boxgrove because of postburial transformation, less than 5% of the microfabric contained clues to Unit 4c's pedological history. There is therefore always the danger that the counting of "identifiable" features, components, and the like may become a mechanical substitute for accurate, thoughtful analysis of a thin section and its interpretation.

5. Conclusions

1. Soil micromorphology can produce extremely accurate semi-quantitative data that is most convincing to non-specialist soil micromorphologists when expressed graphically.
2. Experimental soils when characterized through counted soil micromorphology have specific signatures that are replicated in the archaeological record.
3. The specific analysis of individual microscopic components and pedofeatures that are counted can lead to the identification of features of archaeological significance.
4. The multidisciplinary approach has shown that specific microstratigraphies can have coincident and related chemical and fossil signatures that immensely aid the task of arriving at convincing interpretations of archaeological sites.

ACKNOWLEDGMENTS. It should be noted here that the chemical component of our studies has been provided by Cyril Bloomfield (retired from Rothamsted Experimental station), John Crowther (University of Wales, Lampeter), and

Jöhan Linderholm (University of Umeå, Sweden), whom we gratefully acknowledge. The authors thank the many people and organizations who have supported our work (e.g., Bedfordshire County Archaeological Service, English Heritage, Museum of London Archaeological Service, Soprintendenza Archeologica della Liguria, Wessex Archaeology), with special thanks to Roger Engelmark (University of Umeå) and Peter Reynolds (Butser Ancient Farm) for their collaboration. The authors thank the editors and Plenum for this opportunity to present their approach in *Earth Science and Archaeology*. Finally, we specifically thank Paul Goldberg and Vance T. Holliday for their comments on this chapter.

6. References

Acott, T. G., Cruise, G. M., and Macphail, R. I., 1997, Soil Micromorphology and High Resolution Images. In *Soil Micromorphology: Diversity, Diagnostics and Dynamics*, edited by S. Shoba, M. Gerasimova, and R. Miedema, pp. 372–378. International Soil Science Society, Moscow-Wageningen.

Allen, M. J., 1992, Products of Erosion and Prehistoric Landuse of the Wessex Chalk. In *Past and Present Soil Erosion*, edited by M. Bell and J. Boardman, pp. 37–52. Oxbow Monograph 22, Oxbow Books, Oxford.

Arpin, T., Macphail, R. I., and Boschian, G., 1998, Summary of the Spring 1998 Meeting of the Working Group on Archaeological Soil Micromorphology, February 27–March 1, 1998. *Geoarchaeology* 13 (6):645–647.

Avery, B. W., 1964, *The Soils and Land-Use of the District Around Aylesbury and Hemel Hempstead*, Her Majesty's Stationary Office, London.

Babel, U., 1975, Micromorphology of Soil Organic Matter. In *Soil Components*, Volume 1: *Organic Components*, edited by J. F. Giesking, pp. 369–473. Springer Verlag, New York.

Bal, L., 1982, *Zoological Ripening of Soils*, Agricultural Research Reports No. 850, Pudoc, Wageningen.

Barham, A. J., and Macphail, R. I. (eds.), 1995, *Archaeological Sediments and Soils: Analysis, Interpretation and Management*, Institute of Archaeology, London.

Bell, M. G., Fowler, P. J., and Hillson, S.W., 1996, *The Experimental Earthwork Project, 1960–1992*. Council for British Archaeology Research Report 100, York, UK.

Boschian, G., 1997, Sedimentology and Soil Micromorphology of the Late Pleistocene and Early Holocene Deposits of Grotta dell'Edera (Trieste Karst, NE Italy), *Geoarchaeology* 12:227–250.

Bullock, P., Fedoroff, N., Jongerius, A., Stoops, G. J., and Tursina, T., 1985, *Handbook for Soil Thin Section Description*. Waine Research Publishers, Wolverhampton, UK.

Cammas, C., 1994, Approche Micromorphologique de la Stratigraphie Urbaine à Lattes: Premiers Résultats. In *Lattara* 7:181–202. A.R.A.L.O., Lattes.

Cammas, C., Wattez, J., and Courty, M. A., 1996, L'enregistrement Sédimentaire des Modes d'Occupation de l'Espace. In *XIII International Congress of Prehistoric and Protohistoric Sciences Forlì`-Italia-8/14 September 1996*, edited by L. Castelletti and M. Cremaschi, pp. 81–86. A.B.A.C.O., Forlì`.

Castelletti, L. and Cremaschi, M. (eds.), 1996, Paleoecology. In *XIII International Congress of Prehistoric and Protohistoric Sciences Forli-Italia-8/14 September 1996*, Volume 3. A.B.A.C.O., Forli.

Catt, J. A., 1992, Soil Erosion on the Lower Greensand at Woburn Experimental Farm, Bedfordshire—Evidence, History and Causes. In *Past and Present Soil Erosion*, edited by M. Bell and J. Boardman, pp. 67–76. Oxbow Monograph 22, Oxbow Books, Oxford.

Clark, A. J., 1990, *Seeing Beneath the Soil*, Batsford, London.

Courty, M. A., Goldberg, P., and Macphail, R. I., 1989, *Soils and Micromorphology in Archaeology*, Cambridge University Press, Cambridge, UK.

Courty, M. A., Goldberg, P., and Macphail, R. I., 1994, Ancient People—Lifestyles and Cultural Patterns. In *Proceedings of International Soil Science Society*, Volume 6a, edited by L. Wilding, pp. 250–269. International Soil Science Society, Acapulco.

Cornwall, I. W., 1958, *Soils for the Archaeologist*, Phoenix House, London.
Cremaschi, M., 1985, Geoarchaeology: Earth Sciences in Archaeological Research. In *Homo: Journey to the Origins of Man's History*, pp. 183–191. Cataloghi Marsilio, Venice.
Crowther, J., and Barker, P., 1995, Magnetic Susceptibility: Distinguishing Anthropogenic Effects from the Natural, *Archaeological Prospection* 2:207–215.
Crowther, J., Macphail, R. I., and Cruise, G. M., 1996, Short-Term Post-Burial Change in a Humic Rendzina, Overton Down Experimental Earthwork, England, *Geoarchaeology* 11:95–117.
Cruise, G. M. and Macphail, R. I., in press, Microstratigraphical Signatures of Experimental Rural Occupation Deposits and Archaeological Sites. *Interpreting Stratigraphy* (Bedford, July 1996). York University, York.
Cruise, G. M., Macphail, R. I., Maggi, R., Haggart, B. A., Linderholm, J., and Moreno, D., 1996, New Approaches to Old Problems: Neolithic to Medieval Land-Use at Bargone, Eastern Liguria. In *XIII International Congress of Prehistoric and Protohistoric Sciences Forli-Italia-8/14 September 1996*, Volume 1, edited by L. Castelletti and M. Cremaschi, pp 401–413 A.B.A.C.O., Forlì.
Davidson, D. A., Carter, S. P., and Quine, T. A., 1992, An Evaluation of Micromorphology as an Aid to Archaeological Interpretation, *Geoarchaeology* 7(1):55–65.
Del Lucchese, A., and Ottomano, K., 1996, Micromorphology of the Neolithic Sequence of "Pian del Ciliego" Shelter (Savona-Italy). In *XIII International Congress of Prehistoric and Protohistoric Sciences Forli-Italia-8/14 September 1996*, edited by L. Castelletti, and M. Cremaschi, M., pp. 151–160. A.B.A.C.O., Forlì.
Dimbleby, G. W., 1962, *The Development of British Heathlands and Their Soils*, Clarendon Press, Oxford, UK.
Dockrill, S. J., Bond, J. M., Milles, A., Simpson, I., and Ambers, J., 1994, Tofts Ness, Sandy, Orkney. An Integrated Study of a Buried Orcadian Landscape. In *Whither Environmental Archaeology*, edited by R. Luff and P. Rowley-Conwy, pp. 115–132. Oxbow Monograph 38, Oxbow Books, Oxford, UK.
Duchaufour, P., 1982, *Pedology*, Allen and Unwin, London.
Engelmark, R., and Linderholm, J., 1996, Prehistoric Land Management and Cultivation. A Soil Chemical Study. In *6th Nordic Conference on the Application of Scientific Methods in Archaeology, Esjberg 1993*, pp. 315–322 P.A.C.T., Brussels, Belgium.
Evans, J. G., 1972, *Landsnails in Archaeology*, Seminar Press, London.
Fedoroff, N., and Goldberg, P., 1982, Comparative Micromorphology of Two Late Pleistocene Palaeosols (in the Paris Basin), *Catena* 9:227–251.
Gebhardt, A., 1990, *Evolution du Paléopaysage Agricole dans le Nord-Ouest de la France. Aport de la Micromorphologie*. Thèse de l'Université de Rennes I.
Gebhardt, A., 1992, Micromorphological Analysis of Soil Structural Modification Caused by Cultivation Implements. In *Prehistoire de la Agriculture: Nouvelles Approches Experimentales et Ethnographiques*, edited by P. Anderson, pp. 373–392. Monographie du CRA No. 6. Editions du CNRS., Centre Nationale de la Recherche Scientifique, Paris.
Goldberg, P., 1983, Applications of Soil Micromorphology in Archaeology. In *Soil Micromorphology*. Volume 1: *Techniques and Applications*, edited by P. Bullock and C. P. Murphy, pp. 139–150. A. B. Academic Publishers, Berkhamsted.
Goldberg, P., and Whitbread, I., 1993, Micromorphological Studies of Bedouin Tent Floors. In *Formation Processes in Archaeological Context*, edited by P. Goldberg, D.T. Nash, and M. D. Petraglia, pp. 165–188. Monographs in World Archaeology No. 17, Prehistory Press, Madison, WI.
Guélat, M., Paccolat, O., and Rentzell, P., 1998, Une Etable Gallo-Romaine à Brigue-Glis, Waldmatte; Evidence Archéologiques et Micromorphologiques, *Annuaire de la Société Suisse de Préhistoire et d'Archéologie* 81:171–182.
Heathcote, J. L., in preparation, Recognition of the Effects of Intense Vertebrate Activity — *Structural, Geochemical and Mineralogical Signatures*. Ph.D. diss, Institute of Archaeology, London.
Iversen, J., 1964, Retrogressive Vegetational Succession in the Post-Glacial, *Journal of Ecology* 52:59–70.
Kemp, R. A., 1985, The Decalcified Lower Loam at Swanscombe, Kent: A Buried Quaternary Soil, *Proceedings of the Geologists Association* 96:343–355.
Kubiena, W. L., 1938, *Micropedology*, Collegiate Press, Ames, IA.
Landuydt, C. J.,1990, Micromorphology of Iron Minerals from Bog Ores of the Belgian Campine Area. In *Soil Micromorphology: A Basic and Applied Science*, edited by L. A. Douglas, pp. 289–294. Elsevier, Amsterdam.

Lawson, A. J., 1994, Potterne. In *The Iron Age of Wessex: Recent Work*, edited by A. P. Fitzpatrick and E. L. Morris, pp. 42–46. Association Française D'Etude de l'Age du Fer, Paris.

Macphail, R. I., 1983, The Micromorphology of Dark Earth from Gloucester, London and Norfolk: An Analysis of Urban Anthropogenic Deposits from the Late Roman to Early Medieval Periods. In *Soil Micromorphology*, edited by P. Bullock and C. P. Murphy, pp. 367–388. A. B. Academic Publishers, Berkhamsted.

Macphail, R. I., 1987, A Review of Soil Science in Archaeology in England. In *Environmental Archaeology: A Regional Review*, Volume 2, edited by H. C. M. Keeley, pp. 332–379. English Heritage Occasional Paper No. 1, London.

Macphail, R. I., 1992, Soil Micromorphological Evidence of Ancient Soil Erosion. In *Past and Present Soil Erosion*, edited by M. Bell and John Boardman, pp. 197–216. Oxbow Monograph 22, Oxbow Books, Oxford.

Macphail, R. I., 1994, The Re-Working of Urban Stratigraphy by Human and Natural Processes. In *The Archaeology of Town and Country: Economic Connections and Environmental Contrasts*, edited by A. Hall and H. Kenward, pp. 13–44. Oxbow Books, Oxford.

Macphail, R. I., 1996, The Soil Micromorphological Reconstruction of the 500,000 Year Old Hominid Environment at Boxgrove, West Sussex, UK. In *XIII International Congress of Prehistoric and Protohistoric Sciences Forlı`-Italia-8/14 September 1996*, edited by L. Castelletti and M. Cremaschi, pp. 133–142. A.B.A.C.O., Forlı` .

Macphail, R. I., 1998, A Reply to Carter and Davidson's "An Evaluation of the Contribution of Soil Micromorphology to the Study of Ancient Agriculture," *Geoarchaeology* 13(6):549–564.

Macphail, R. I., and Cruise, G. M., 1995, *Guildhall Yard East (GYE92): Brief Assessment of Microstratigraphy (Soil Micromorphology and Pollen)*. Unpublished report to Museum of London Archaeological Service, London.

Macphail, R. I., and Cruise, G. M.,1996, Soil Micromorphology. In *The Experimental Earthwork Project 1960–1992*, edited by M. Bell, P. J. Fowler, and S. W. Hillson, pp. 95–106. CBA Research Report 100, Council for British Archaeology, York, UK.

Macphail, R. I., and Cruise, G. M., 1997a, *Report on the Soil Micromorphology, Chemistry and Palynology of Haynes Park, Bedfordshire*. Unpublished report to Bedfordshire County Archaeology Service, Bedford.

Macphail, R. I., and Cruise, G. M., 1997b, *7–11, Bishopsgate and Colchester House (PEP89), London: Preliminary Report on Soil Microstratigraphy and Chemistry*. Unpublished report to Museum of London Archaeological Service, London.

Macphail, R. I., and Goldberg, P., 1995, Recent Advances in Micromorphological Interpretations of Soils and Sediments from Archaeological Sites. In *Archaeological Sediments and Soils: Analysis, Interpretation and Management*, edited by A. J. Barham and R. I. Macphail, pp. 1–24e. Institute of Archaeology, London.

Macphail, R. I., Courty, M. A., and Gebhardt, A., 1990, Soil Micromorphological Evidence of Early Agriculture in North-West Europe, *World Archaeology* 22(1):53–69.

Macphail, R. I., Courty, M. A., Wattez, J., and Hather, J., 1997, The Soil Micromorphological Evidence of Domestic Occupation and Stabling Activities. In *Arene Candide: A Functional and Environmental Assessment of the Holocene Sequences Excavated by L. Bernabo' Brea (1940–1950)*, edited by R. Maggi, pp. 53–88. Istituto Italiano di Paleontologia Umana, Rome.

Macphail, R. I., Cammas, C., Gebhardt, A., Langohr, R., and Linderholm, J., 1998a, Anthropogenic Influences on Soils in the Late Quaternary. In *Proceedings World Congress of Soil Science August 1998*, edited by N. Fedoroff and J. Catt. International Soil Science Society, Montpellier. CD-ROM, Symposium 16, paper 855.

Macphail, R. I., Cruise, G. M., Mellalieu, S., and Nisbet, R. 1998b, Micromorphological Interpretation of a Turf-Filled Funerary Shaft at Folly Lane, St. Albans, *Geoarchaeology* 13(6):617–644.

Macphail, R. I., Cruise, G. M., and Linderholm, J., in press, Integrating Soil Micromorphology and Rapid Chemical Survey Methods: New Developments in Reconstructing Past Rural Settlement and Landscape Organization. In *Interpreting Stratigraphy*, edited by S. Roskams. University of York, York, UK.

Macphail, R.I., Acott, A.G., Crowther, J., and Cruise, G.M., in preparation, The Experimental Earthwork at Wareham, Dorset after 33 Years: Changes to the Buried Soil. *Journal of Archaeological Science*.

Maggi, R., 1997, *Arene Candide: A Functional and Environmental Assessment of the Holocene Sequences Excavated by L. Bernabo' Brea (1940–1950)*. Istituto Italiano di Paleontologia Umana, Rome.

Matthews, W., and Postgate, J. N., 1994, The Imprint of Living in an Early Mesopotamian City: Questions and Answers. In *Whither Environmental Archaeology*, edited by R. Luff and P. Rowley-Conwy, pp. 171–212. Oxbow Monograph 38, Oxbow Books, Oxford.

Matthews, W., French, C. A. I., Lawrence, T., Cutler, D. F., and Jones, M. K., 1997, Microstratigraphic Traces of Site Formation Processes and Human Activities. *World Archaeology* 29(2):281-308.

McKeague, J. A., 1983, Clay Skins and Argillic Horizons. In *Soil Micromorphology*, edited by P. Bullock and C. P. Murphy, pp. 367–388. A. B. Academic Publishers, Berkhamsted.

Moore, P. D., Webb, J. A., and Collinson, M. E., 1991, *Pollen Analysis*, Blackwell Scientific Publications, Oxford.

Murphy, C. P., McKeague, J. A., Bresson, L. M., Bullock, P., Kooistra, M. J., Miedema, R., and Stoops, G., 1985, Description of Soil Thin Sections: An International Comparison, *Geoderma* 35:15–37.

Nörnberg, P., and Courty, M.A., 1985, Standard Geological Methods Used on Archaeological Problems. In *Proceedings 3rd Nordic Conference on the Application of Scientific Methods to Archaeology*, edited by T. Edgren and H. Junger, pp. 107–117. ISKOS, Finnish Antiquarian Society, Helsinki.

Oldfield, F., Krawiecki, A., Maher, A., Taylor, J. J., and Twigger, S., 1985, The Role of Mineral Magnetic Measurements in Archaeology. In *Palaeoenvironmental Techniques, Design and Interpretation*, edited by N. G. R. Fieller, D. D. Gilbertson, and N. G. A. Ralph, pp. 29–43. British Archaeological Reports, International Series 258, Oxford.

Preece, R. C., 1992, Episodes of Erosion and Stability Since the Late-Glacial: The Evidence from Dry Valleys in Kent. In *Past and Present Soil Erosion*, edited by M. Bell and J. Boardman, pp. 175–184. Oxbow Monograph 22, Oxbow Books, Oxford.

Preece, R. C., Kemp, R. A., and Hutchinson, J. N., 1995, A Late-Glacial Colluvial Sequence at Watcombe Bottom, Ventnor, Isle of Wight, England. *Journal of Quaternary Science* 10(2):107–121.

Proudfoot, V. B., 1976, The Analysis and Interpretation of Soil Phosphorus in Archaeological Contexts. In *Geo-archaeology: Earth Science and the Past*, edited by D. A. Davidson and M. L. Shackley, pp. 93–113. Duckworth, London.

Quine, T. A., 1995, Soil Analysis and Archaeological Site Formation Studies. In *Archaeological Sediments and Soils: Analysis, Interpretation and Management*, edited by A. J. Barham and R. I. Macphail, pp. 77–98. Institute of Archaeology, London.

Reynolds, P. J., 1979, *Iron Age Farm. The Butser Experiment*, British Museum Publications Ltd., London.

Roberts, M. B., and Parfitt, S. A. (eds.), 1999, *The Middle Pleistocene Site at A. R.C. Eartham Quarry, Boxgrove, West Sussex, UK*, English Heritage Archaeological Report 17, London.

Roberts, M. B., Parfitt, S. A., Pope, M. I., and Wenban-Smith, F. F., 1997, Boxgrove, West Sussex: Rescue Excavations of a Lower Palaeolithic Landsurface (Boxgrove Project B, 1989–91), *Proceedings of the Prehistoric Society* 63:303–358.

Romans, J. C. C., and Robertson, L., 1983, The General Effects of Early Agriculture on the Soil. In *The Impact of Aerial Reconnaissance on Archaeology*, edited by G. S. Maxwell, pp. 136–41. CBA Research Report No. 49, Council for British Archaeology, London.

Simpson, I. A., and Barrett, J. H., 1996, Interpretation of Midden Formation Processes at Robert's Haven, Caithness, Scotland Using Thin Section Micromorphology. *Journal of Archaeology Science* 23:543–556.

Stringer, C. B., Trinkhaus, E., Roberts, M. B., Parfitt, S. A., and Macphail, R. I., 1998, The Middle Pleistocene Human Tibia from Boxgrove, *Journal of Human Evolution* 34:509–547.

Taylor, J. J., Innes, J. B., and Jones, M. D. H., 1994, Locating Prehistoric Wetland Sites by an Integrated Palaeoenvironmental/Geophysical Survey Strategy at Little Hawes Water, Lancashire. In *Whither Environmental Archaeology*, edited by R. Luff and P. Rowley-Conwy, pp. 13–24. Oxbow Monograph 38, Oxbow Books, Oxford.

Wattez, J. and Courty, M. A., 1987, Morphology of Some Plant Materials. In *Soil Micromorphology*, edited by N. Fedoroff, L. M. Bresson, and M. A. Courty, pp. 677–683. AFES, Plaisir, France.

Wattez, J., Courty, M. A., and Macphail, R. I., 1990, Burnt Organo-Mineral Deposits Related to Animal and Human Activities in Prehistoric Caves. In *Soil Micromorphology: A Basic and Applied Science*, edited by L. A. Douglas, pp. 431–440. Elsevier, Amsterdam.

Windell, D., Chapman, A., and Woodiwiss, J., 1990, *From Barrows to Bypass: Excavations at West Cotton, Raunds, Northamptonshire 1985–1989.* Northamptonshire County Council, Northampton, UK.

Whittle, A., Rouse, A. J., and Evans, J. G., 1993, A Neolithic Downland Monument in its Environment: Excavations at the Easton Down Long Barrow, Bishops Cannings, North Wiltshire, *Proceedings of the Prehistoric Society* 59:197–239.

10

Buried Artifacts in Sandy Soils

Techniques for Evaluating Pedoturbation versus Sedimentation

DAVID S. LEIGH

1. Introduction

Since the 1960s, it has become increasingly clear that once an artifact is deposited on a surface it is subject to many different types of displacement and burial processes that come under the general heading of pedoturbation (Hole, 1961; Johnson, 1990; Wood and Johnson, 1978). Pedoturbation refers to many different soil mixing processes that can result in the displacement, movement, and burial of artifacts (Table 10.1). Bioturbation, which includes faunalturbation and floralturbation, is the common mixing agent at most sites. Equifinality of artifact burial by pedoturbation or by various forms of sedimentation is a common problem that confronts archeological survey and site investigations, particularly in loose sandy soils and sediments. The goal of this chapter is to identify and discuss techniques suited to evaluate the relative importance of pedoturbation versus sedimentation processes at sandy sites with buried components.

DAVID S. LEIGH • Department of Geography, University of Georgia, Athens, Georgia 30602-2502.

Earth Sciences and Archaeology, edited by Paul Goldberg, Vance T. Holliday, and C. Reid Ferring. Kluwer Academic/Plenum Publishers, New York, 2001.

Table 10.1. Ten Categories of Pedoturbation

Process category	Soil-mixing agents
Aeroturbation	Gas, air, wind
Aquaturbation	Water
Argilliturbation	Swelling and shrinking of clays
Cryoturbation	Freezing and thawing
Crystalturbation	Growth and wasting of salts
Faunalturbation	Animals (especially burrowers)
Floralturbation	Plants (treefall, root growth, etc.)
Graviturbation	Mass wasting (creep, solifluction, etc.)
Impacturbation	Comets, meteoroids, artillery shells
Seismiturbation	Earthquakes

Source: Adapted from Johnson and Stegner, 1978, 1990; and Hole, 1961.

Many archeological sites contain buried components, but in some settings (particularly sands) the process of burial can be attributed entirely to bioturbation (Mitchie, 1990). Although pedoturbation operates in virtually any sort of soil texture, it appears to be particularly vigorous in sands (>85 % sand by weight). The consistence of sand is inherently susceptible to bioturbation because of its loose and noncohesive nature, which leads to the loss of the integrity of cultural stratigraphy. The Coastal Plain of the southeastern United States is a region where problems of site burial in sandy soils are common, and some examples familiar to the author are drawn from this region.

The problem is significant because pedoturbation destroys the *in-situ* nature of archeological contexts, and correct paleoenvironmental interpretations about site formation hinge on accurate determination of pedoturbation, sedimentation, or a combination of both processes. For example, if eolian sediment is recognized at a site, then a windy period of blowing sand is indicated that possibly invokes aridity and lack of vegetation. However, misinterpretation of the process of artifact burial can lead to serious errors in this interpretation. If pedoturbation is the main site burial process, then paleoclimatic inferences are difficult (perhaps impossible) to conclude.

Hole (1961) cited nine categories of pedoturbation that relate to particular types of disturbance processes, and Wood and Johnson (1978) later amended and refined the list to ten processes with the addition of impacturbation (Table 10.1). Buol et al. (1989) cited seven kinds of pedoturbation. Excellent reviews of pedoturbation as an agent of artifact burial are provided by Wood and Johnson (1978), Schiffer (1987), Johnson (1990), Johnson and Watson-Stegner (1990), and Waters (1992).

Perhaps the most important group of pedoturbation processes from an archaeological perspective is bioturbation (Butler, 1995; Hole, 1981; Johnson and Watson-Stegner,1990; Mitchell, 1988; Schiffer, 1987; Waters, 1992; Wood and Johnson, 1978), which includes faunalturbation and floralturbation. *Faunalturbation* refers to processes where animals (especially burrowers) are the agents of disturbance, whereas *floralturbation* refers to processes stimulated by the activity of plants. Common examples of faunalturbation include burrowing by animals such as badgers, gophers, earthworms, ants, termites, tortoises, and humans. Examples of floralturbation include treefall and uprooting, root growth, and root decay. Burrowing activities of ants, earthworms, termites, and other small animals have been cited to favor the downward movement of coarse particles (i.e., artifacts) whereas fine particles (sands) are moved up to the surface (Johnson and Watson-Stegner,1990), and this may result in the formation of subsurface stonelines and artifact zones. Fine examples of stonelines resulting from faunalturbation and floralturbation are provided by Johnson (1990). However, note that other processes have been cited as the cause of stonelines in stratigraphic sections, particularly erosion and the formation of a lag gravel such as in the pedimentation process (Ruhe, 1959). Faunalturbation, particularly by burrowing animals such as the gopher tortoise or the badger, may bring buried artifacts to the surface as well as move artifacts downward as roof material collapses down into the burrow. Thus, the result of bioturbation may depend largely on the types of organisms involved.

Although a comprehensive discussion of pedoturbation processes is beyond the scope of this chapter, it is important to recognize that pedoturbation may be involved in both progressive (proanisotropic) as well as regressive (proisotropic) pedogenesis as noted by Johnson and Watson-Stegner (1990). Progressive pedogenesis involves the development of distinct soil horizons, soil upbuilding, and deepening. In contrast, regressive pedogenesis operates to homogenize soil profiles and curtail horizonation, retard soil development, and favor soil erosion. The soil profile itself is not necessarily an indicator of the past vigor of pedoturbation in archeological contexts. For example, progressive pedogenesis may result in a distinct A-E-Bt-C horizon sequence (Johnson and Watson-Stegner, 1990), where the A and E horizons experience active bioturbation that may even accentuate the distinction of the horizons in the profile. Later in the weathering history, bioturbation and regressive pedogenesis may cause the destruction of a Bt horizon and the simplification of the profile to an a A-Bw-C horizon sequence.

1.1. Equifinality of Pedoturbation and Sedimentation

In addition to pedoturbation, one typically needs to consider eolian, fluvial, and colluvial sedimentation in an analysis of site formation and burial processes. In many cases one or a combination of sedimentary processes may have led to burial of artifacts independent of, or in conjunction with, pedoturbation. Upon cursory field inspection it may be impossible to discern the dominant process of site burial. Even with close field and laboratory analysis, in some complex cases it may be impossible to finally distinguish what sort of burial processes dominated. For example, the land surface of eolian sands may have had artifacts that were undergoing burial by slow bioturbation in conjunction with limited eolian reworking and sedimentation. However, in many cases, particularly on upland interstream divides, artifact burial may result solely from pedoturbation (Leigh, 1998), thus restricting any paleoenvironmental implications about eolian, fluvial, or colluvial sedimentation. Thus, one focus of this chapter is on identifying whether or not certain types of sedimentation processes could have operated as a burial mechanism, or whether the entire process of burial should be explained by pedoturbation. Of course, the role of time is critical in the analysis of artifact burial by pedoturbative processes, and there is typically a positive relationship between depth of pedoturbative artifact burial and time, at least down to a certain depth representing the biotically active zone of soil. However, both sedimentation and pedoturbation can operate at the same time, and thus their discrimination may be difficult.

2. Techniques

Several techniques are available to the geoarchaeologist to evaluate the relative importance of pedoturbation versus sedimentation. Unfortunately, many of these techniques are not incorporated into standard excavation practices. Elucidation of these methods should encourage their use and help to resolve uncertainties about site formation and artifact burial processes in sandy settings.

2.1. Geomorphic Setting

The geomorphic setting of a site is obviously the first variable that should be examined in the analysis of artifact burial in sandy soils. Upland interstream divides and flat high-level Pleistocene terraces are about the only places where one can virtually eliminate colluvial and alluvial sedimentation processes. In such settings the research focus can be directed toward evaluating pedoturbation versus eolian sedimentation as site burial processes as illustrated by Leigh (1998). Convex shoulder slopes and flat (<2% slopes) upland divides are generally considered nondepositional parts of a landscape from the standpoint of colluvial and alluvial hillslope sedimentation. In contrast, concave lower midslopes, footslopes, and toeslopes should be viewed as settings that are subject to sedimentation (Fig. 10.1). In fact, even the lower slopes and bottoms of zero-order and first-order tributaries should be considered as potential loci for sedimentation. In some cases, undulations and irregularities on interstream divides may be subject to minor colluvial and alluvial sedimentation. A widely referenced hillslope classification is that of Dalrymple et al. (1968), which is the basis for the nomenclature used in Fig. 10.1. Dalrymple et al. (1968) indicated that the main zone of sediment deposition is on the colluvial footslope and the alluvial toeslope.

The potential for relatively rapid sedimentation on sandy slopes can be ascertained by observing present-day sedimentation patterns following intense rainfall. In some cases (i.e., well-sorted sand) erosion and sedimentation may not occur, even during extreme rainfall. County soil survey reports published by the United States Department of Agriculture (USDA Government Printing Office, are valuable resources for evaluating the erosional/depositional potential. These county soil survey reports are available at county U.S.D.A. offices and federal repository libraries. The soil survey reports include information about the slope

Buried Artifacts in Sandy Soils

Hillslope Components

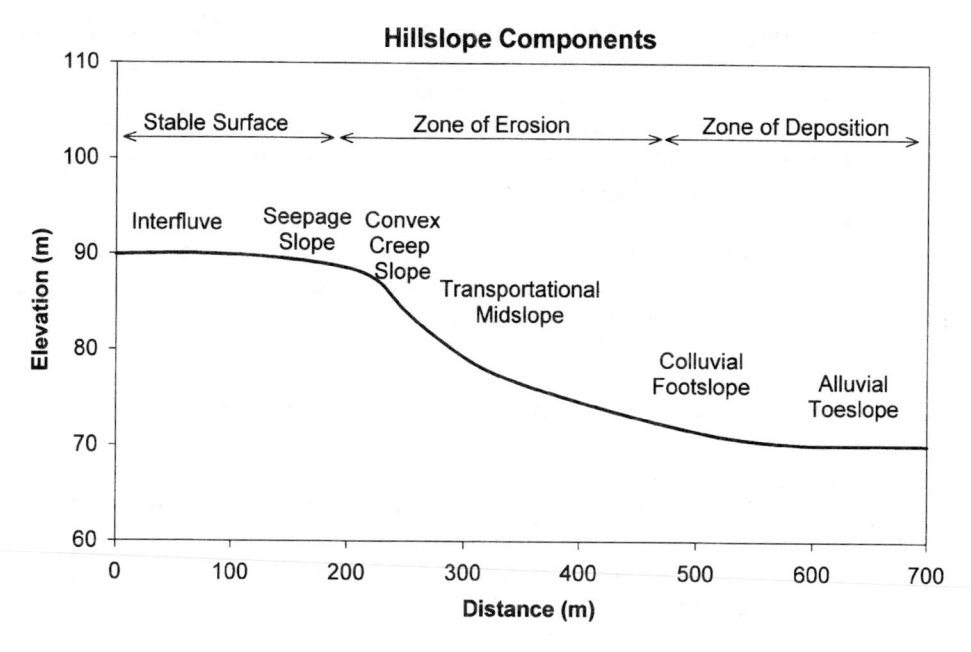

Figure 10.1. Cross-section illustrating typical elements of hillslopes using the terminology of Dalrymple et al. (1968). The data used for this plot were from a first-order tributary (blue line method) in the Sandhills of North Carolina at the Fort Bragg Military Reservation, taken from U.S.G.S. a 7.5-minute quadrangle with 10 ft contour intervals. A spline interplolation was used to smooth the line.

class of the mapped soil series, and typically will contain quantitative tables of erosion factors (K) and permeability rates that can be related to the potential for erosion and sedimentation. The erosion factor (K) is a relative index of susceptibility of bare cultivated soil to particle detachment and transport by rainfall simulations adjusted to a 9 percent slope that is used in the Revised Universal Soil Loss Equation (Renard et al., 1991). Permeability is equivalent to saturated hydraulic conductivity, which can be used to estimate infiltration rates that must be exceeded by rainfall rates in order to produce overland flow and sediment transport. However, caution must be used in evaluating the soil survey reports based on the present-day landscape, because agricultural land use often represents accelerated erosion and sedimentation patterns relative to prehistoric conditions that might have been radically different. In other words, bare agricultural ground will be more susceptible to erosion and sedimentation than natural forest cover.

In alluvial valleys the potential for flood sedimentation must be considered for surfaces that were within the floodplain during the period of human occupation. Beware that it is possible that the floodplain during prehistoric time is now a terrace, so careful consideration of the alluvial stratigraphy must be made to fully realize sites that were subject to burial by fluvial processes. In North America, terminal Pleistocene surfaces (usually terraces) are generally the highest surfaces that could have experienced sedimentation because artifacts were left on the surface. However, variation in neotectonic and climatic histories can greatly influence terrace chronologies and the relative elevations of terraces.

Eolian dunes present the greatest difficulty in discerning burial by sedimentation versus bioturbation processes, because limited resedimentation of dune sand may deposit thin strata of sands that are relatively quickly bioturbated. Furthermore, dune sand is already presorted, and reworking may not significantly change the grain size distribution. In dunes, recognition of sedimentary structures and strata that are indicative of site burial, such as buried A horizons, are key elements toward discerning burial by sedimentation versus bioturbation.

2.2. Sedimentary Structures, Stratigraphy, and Pedology

Sedimentary structures exposed in excavations can be extremely useful in evaluating the relative importance of pedoturbation versus sedimentation. Thoroughly bioturbated sediments are typically characterized by an unbedded, massive, and homogenous appearance, possibly with fresh evidence of bioturbation. However, collapse burrows and other types of evidence of mixing may not be readily apparent in very homogenous sands, as noted by Waters (1992, p. 314). Root and burrow holes that are infilled with darker or lighter colored sediment than the surrounding matrix (krotovinas) are also typical in bioturbated soils. Less extensive bioturbation will sometimes show primary sedimentary structures preserved between individual burrows (Ahlbrandt, 1979). Nonbioturbated sediments are characterized by primary sedimentary structures, such as thinly laminated and stratified sands and cross-bedding, that extend up to the ground surface (Fig. 10.2a). Bioturbation inherently destroys primary sedimentary structures and may erase all evidence of primary sedimentation. In dunes, the zone of bioturbation (solum) typically is recognized down to the level where primary sedimentary structures become apparent (Fig. 10.2b). Ahlbrandt (1979) noted that certain types of faunal and floral turbators (i.e., ants and trees) can completely destroy sedimentary structures in modern dunes. In fact, they noted that the sedimentary structures of entire dunes may be destroyed where ant colonies are relatively numerous.

A problematic scenario occurs when primary sediments are deposited as massive, unbedded units consisting of texturally and mineralogically uniform sands. Such sands may show no trace of their original sedimentation. Examples may include eolian sediments that were blown into standing vegetation or overbank sands that fell out of suspension. In such cases the discrimination of bioturbation versus sedimentation may be quite difficult, based on the sedimentary structures. However, stratigraphic discontinuities may exist between the two massive units, which can provide critical information about possible sedimentation.

A stratigraphic discontinuity is another line of evidence that is indicative of sedimentation rather than bioturbation. An abrupt erosional contact or bounding surface overlain by a different layer of sand is clear evidence of sedimentation (Fig. 10.2a). In many cases particle size analysis (and other types of physical and

Buried Artifacts in Sandy Soils

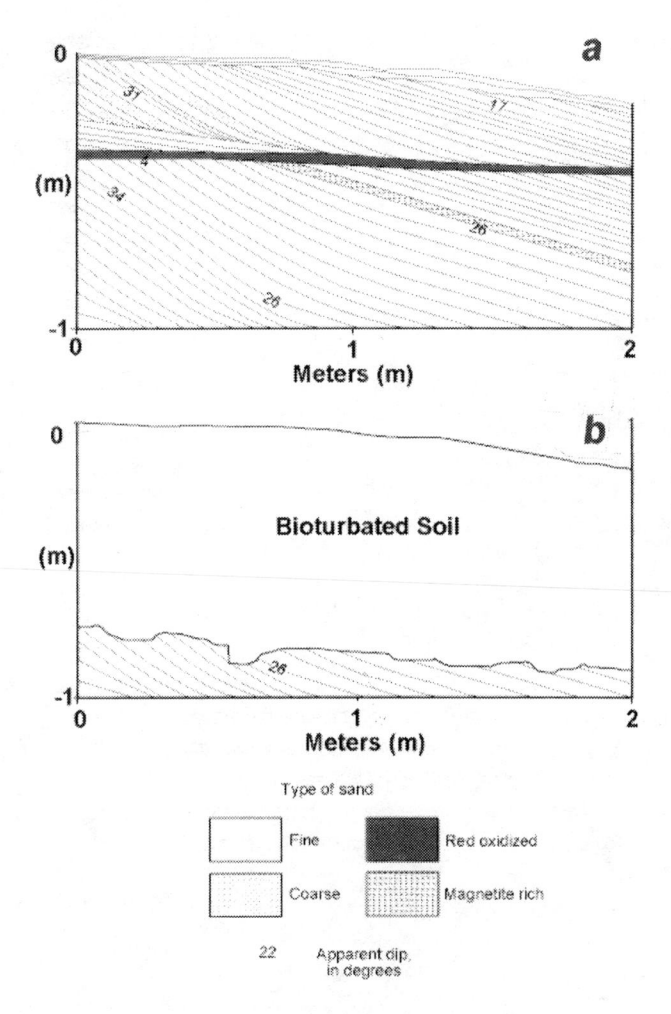

Figure 10.2. Diagram of typical cross-bedded sedimentary structures found in eolian sand dunes (a) and the obliteration of those sedimentary structures due to bioturbation (b). This diagram is adapted from an example on the lee side of a barchanoid dune (wind flow from left to right) in the Killpecker dune field of Wyoming, U.S.A., which is illustrated by McKee (1979, p. 92).

chemical measurements) is useful in defining lithologic discontinuities, which is fully discussed later in this chapter. Also, a buried A or B horizon may be overlain by younger sands, clearly marking a stratigraphic boundary (Fig. 10.3). Buried A horizons are some of the best indicators of stratification and episodic sedimentation in sands. Buried A horizons are typified by a darker color (brown or dark gray), greater carbon content, greater content of charcoal flecks, and greater evidence of biotic activity (i.e., burrows or worm casts) than the bounding sediments. Down-profile measurements of carbon concentrations can be used to document the presence of a buried A horizon, and many different sorts of carbon analyzers are available for measuring the carbon content of soil. In well-drained

Figure 10.3. Photograph illustrating a buried soil with an Ab-Eb-Bwb horizon sequence from source-bordering riverine dunes near Ludowici, Georgia. The frame height is 1.2 m, and the buried A horizon is at about 75–82 cm depth. The sediment that buries the soil is probably historical windblown sand deflated from a nearby railroad cut through a dune field. The photograph was provided by Andrew H. Ivester.

sands the carbon in buried A horizons may become oxidized after thousands of years, making the buried soil difficult to recognize. However, buried soils are also indicated by an abrupt increase in opal phytoliths (biogenic silica; Wilding et al., 1977:534), which are more apt to remain in the stratigraphic section through time than carbon. Thus, down-profile measurements of phytolith concentrations also can be used to document a buried A horizon. Phytoliths (<2.4 g/cm^3) have a lower density than crystalline quartz (2.65 g/cm^3), which allows them to be isolated and quantified by separation in heavy liquid. Alternately, Jones (1969) indicated that variance in phytolith abundance can be estimated by the amount of silica recovered from a 20-minute digest in a boiling solution of 0.5N NaOH.

2.3. Particle Size Analysis

Particle size analysis is useful for the documentation of stratified sedimentary deposits and for the confirmation of specific types of sedimentary environments, particularly eolian sediments. Particle size analysis may be used to confirm subtle thin stratification in sands that is not readily apparent on visual inspection. Particle size analysis for sandy soils is typically measured by fractionation of the sands according to the Wentworth phi (ϕ) scale ($\phi = -\log_2 D$, where D is the sediment diameter in millimeters) at whole phi intervals (Table 10.2) using nested sieves (Ingram, 1971). In some cases researchers may prefer to use half or quarter phi intervals. Whole phi intervals are sufficient to determine if a sediment sample can reasonably be classified as eolian, but quarter phi intervals are desirable to achieve statistical significance of problems such as short-distance eolian transport. If additional fractionation of silt and clay is desired, it can be accomplished by using the pipette or hydrometer technique (Gee and Bauder, 1986) or by using an automated laser, X-ray, or electronic particle size analyzer.

In many cases separate depositional units of sandy sediments can be discriminated based on subtle changes in particle size composition. Brooks and Sassaman (1990) indicated that separate phases of Holocene fluvial deposition could be discerned with close-interval (5 cm) sampling and phi size analysis of sands. In addition, Brooks et al. (1996) indicated that separate periods of eolian sedimentation were represented by subtle changes in the statistical properties of particle size distributions. Knox (1987) showed that individual historical floods could be discerned based on down-core variations in percentages of individual sand fractions within fine-grained overbank sediments. In order to discriminate individual flood events, Knox (1987) relied on 1 to 3 cm interval samples. At larger sample intervals the resolution of individual flood strata would be lost. In

Table 10.2. Particle Size Classes According to the Phi Scale

Phi size interval	Millimeters	Size class
<-1	>2.0	gravel
-1 to 0	2.0 to 1.0	very coarse sand
0 to 1	1.0 to 0.5	coarse sand
1 to 2	0.5 to 0.25	medium sand
2 to 3	0.25 to 0.125	fine sand
3 to 4	0.125 to 0.0625	very fine sand
-1 to 4	2.0 to 0.0625	sand
4 to 9	0.0625 to 0.002	silt
<9	<0.002	clay

cases where there is a question about the process of artifact burial, particle size analysis may be used to document subtle changes in the vertical columns that are related to the depositional history. In the case of bioturbated sands, abrupt and significant changes in close-interval particle size analyses is not expected because of mixing and homogenization of the initially stratified sediment.

Particle size analysis also can be used to confirm or deny the presence of eolian sands in situations where other types of sedimentation can be ruled out, such as on upland interstream divides as illustrated by Leigh (1998). Methods for distinguishing eolian sediment have been studied extensively (e.g., Ahlbrandt, 1979; Blatt et al., 1980; Boggs, 1987; Friedman, 1961, 1979). The process of eolian sand transport is a natural winnowing and sorting process that tends to eliminate larger than 2 mm and smaller than 0.1 mm particles (Ahlbrandt, 1979), producing a relatively well-sorted particle size distribution in the fine through coarse sand fraction. Although some sedimentologists have concluded that particle size distributions and moment statistics are not completely diagnostic of sedimentary origin and may involve considerable errors in environmental classification (Tucker and Vacher, 1980), eolian deposits are nonetheless some of the most distinctive types of terrestrial sediments. Indeed, it is possible that fluvial sediment can appear quite like eolian sediment, especially when it is composed of fluvially retransported eolian sediment, and vice versa along rivers with dune belts. Also, interdune sediments tend to lack the distinctive sedimentary textures of dune deposits (Ahlbrandt, 1979).

Despite the problems of diagnosing eolian sediment with 100% accuracy, the distinctive lack of coarse and fine tails in eolian particle size distributions allows a clear assessment of the likelihood of past eolian transport and deposition. That is, if the particle size distribution of a sample contains measurable amounts of >2 mm particles ($>0.02\%$) and a considerable proportion ($>10\%$) of silt plus clay (<0.063 mm), then a fluvial, colluvial, or ancient marine origin should be suspected and the sample is not a good candidate for having had eolian transport. The most diagnostic elements of eolian sediment that have been presented in the literature are presented in Table 10.3. If a soil or sediment does not meet the criteria outlined in Table 10.3, then a fluvial or colluvial origin should be considered more probable than an eolian origin. In summary, particle size analysis at least can be used as an initial screening test to determine whether or not a sample can reasonably be interpreted as having an eolian origin.

A graphical example of this screening process is provided by samples obtained from upland sands at the Fort Bragg Military Reservation (Fig. 10.4), where a sample from modern eolian sand in a clear-cut area (DZ3) is compared to a sample of the soil from which the sand was blown (Lakeland series). The eolian sand is then compared to a sample of sediment that buries archeological materials at a nearby site (31HT285). The eolian sample is much better sorted, more peaked, and contains less silt and clay (phi >4) than the other samples. The sample from 31HT285 appears more like the Lakeland soil than the eolian sand. The comparison indicates that an eolian process of site burial is not supported by the particle size data.

Particle size analysis of the heavy mineral fraction is a variant of particle size analysis to confirm the presence (or absence) of eolian sediment. Heavy minerals

Buried Artifacts in Sandy Soils

Table 10.3. Typical Grain Size Attributes that are Characteristic of Eolian Sand

Eolian sand attribute	References
Lack of >2 mm particles	Numerous citations; see Ahlbrandt, 1979
>2 mm particles comprise <0.02% of the total sample weight	Leigh, 1998; Leigh and Ivester, 1998
1–2 mm particles comprise <1% of the total sample weight	Friedman, 1979, Fig. 18, $n = 191$ inland dunes
1–2 mm particles comprise <2% of the <2 mm sample weight	Leigh and Ivester, 1998; source bordering riverine dunes in south Georgia
>90% sand or <0.063 mm particles (silt + clay) as <10% of the <2 mm sample weight	Daniels and Hammer, 1992, p. 35
phi coefficient of variance (CV) of 0.063–2.0 mm fraction is typically <55%	Apparent in the data of Friedman, 1979, Fig. 17; Thames, 1982; Leigh, 1998; Leigh and Ivester, 1998

that were transported in air typically will have a different particle size distribution relative to quartz, based on the differential settling velocities of light and heavy minerals (Friedman, 1961; Hand, 1967; Leigh, 1998; Lowright, 1973; Steidtmann and Haywood, 1982; Watson, 1969). That is, the relative sizes of hydraulically equivalent quartz and heavy minerals deposited in water are different from

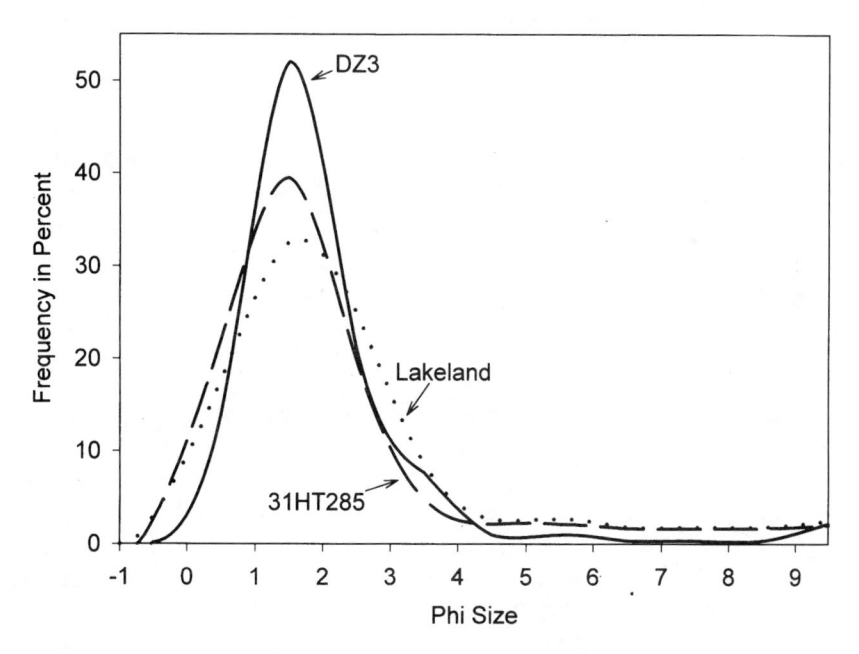

Figure 10.4. Frequency curves of recent eolian sand (DZ3), Lakeland sand soil series, and sands that bury and upland archeological site (31HT285) at the Fort Bragg Military Reservation. The eolian sand was deflated from an area mapped as Lakeland sand. Comparison of the samples indicates that the sample from 31HT285 is not similar to the eolian sample (DZ3).

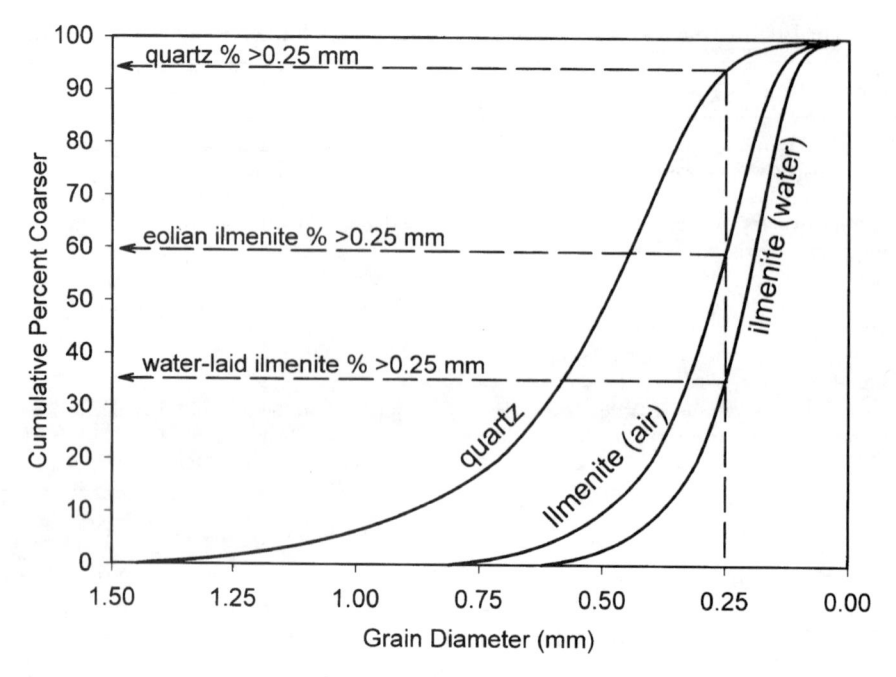

Figure 10.5. Plot of the comparative grain size distributions for quartz, ilmenite settling in air, and ilmenite settling in water. The quartz sample is from a dune at Fort Stewart, Georgia. The quartz-equivalent ratios used were 0.56 for air and 0.43 for water as specified by Watson (1969). Note that a large difference in cumulative percentage is predicted between ilmenite in air versus water, particularly at about 0.25 mm size.

those deposited in air. The size distribution of heavy minerals deposited in air is theoretically closer to the distribution of corresponding quartz grains than heavy minerals that were deposited in water (Fig. 10.5). Based on this principle Leigh (1998) was able to clearly discriminate fluvial from eolian samples using the ratio of the percent of light minerals larger than 0.25 mm versus the percent of heavy minerals larger than 0.25 mm (Fig. 10.6).

2.4. Distribution and Integrity of Cultural Materials and Features

The distribution of cultural artifacts and features is a key aspect of evaluating the relative roles of pedoturbation versus sedimentation at a site. Michie (1987, 1990), who based his work in sandy soils of the South Carolina Coastal Plain, indicated that several aspects of the archaeological record are indicative of bioturbation, including the following:

1. A correlation between maximum artifact depth and the visibly apparent depth of bioturbation
2. Artifacts tilted at $>0°$ to $90°$ to the horizontal

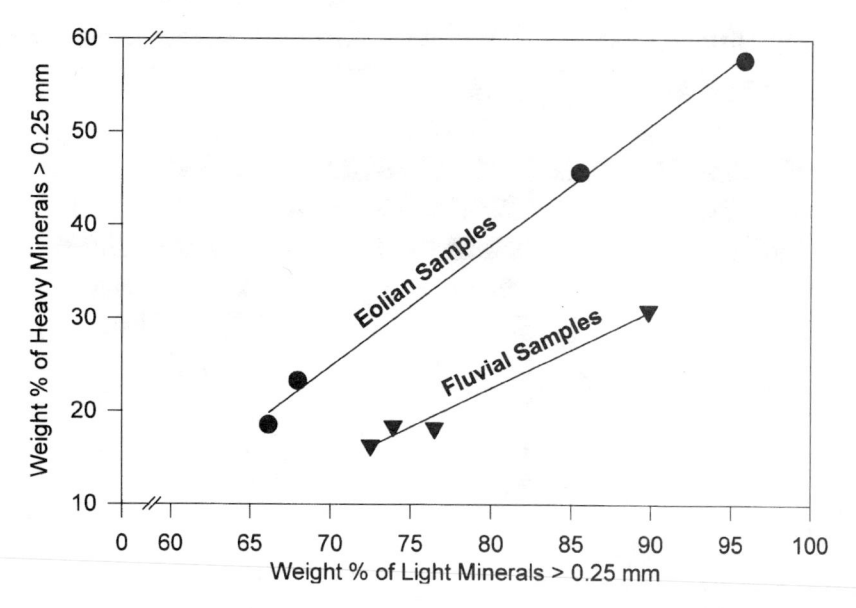

Figure 10.6. Plot illustrating the differences between eolian and fluvial control samples, based on the percentages of >0.25 mm particles in light mineral versus heavy mineral separates. These same samples were used in a comparative analysis by Leigh (1998). The heavy minerals were separated in sodium polytungstate liquid at a density of 2.89 g/cm^3 and are dominated by ilmentite but contain a mixture of many other heavy minerals. The light minerals are dominantly quartz.

3. Absence of intact features such as hearths
4. Vertical and horizontal displacement of clustered artifacts that would otherwise be found together
5. Evidence of a single behavioral activity (i.e., core reduction) or artifact (broken pot) mixed throughout the soil
6. Stratification of cultural horizons.

Items 1 through 5 are good indicators of bioturbation. However, Michie's (1987, 1990) model is somewhat unique, because Item 6, stratification of cultural horizons, contrasts with other models (e.g., Bocek, 1986) that may favor total disruption of cultural stratigraphy. These discrepancies may be related to differences in soil texture and in bioturbating organisms. Michie argued that older artifacts have more time to descend in the profile and that vertical separation of cultural stratigraphy will remain relatively intact. He hypothesized the following cultural association versus depth relationships: Paleoindian and early Archaic at about 0.4 to 0.7 m, middle Archaic at about 0.3 to 0.5 m, late Archaic at 0.2 to 0.4 m, Woodland at about 0.1 to 0.3 m, and Mississippian at 0.0 to 0.1 m (Michie, 1990:44). In fact, the degree of vertical separation of cultural material probably depends largely on the types of organisms that have operated at the site and the grain size distribution of the sand. Ants and worms are not capable of moving large artifacts up to the ground surface, which may help preserve vertical separation of cultural strata, whereas larger animals (i.e., gopher, tortoise) are

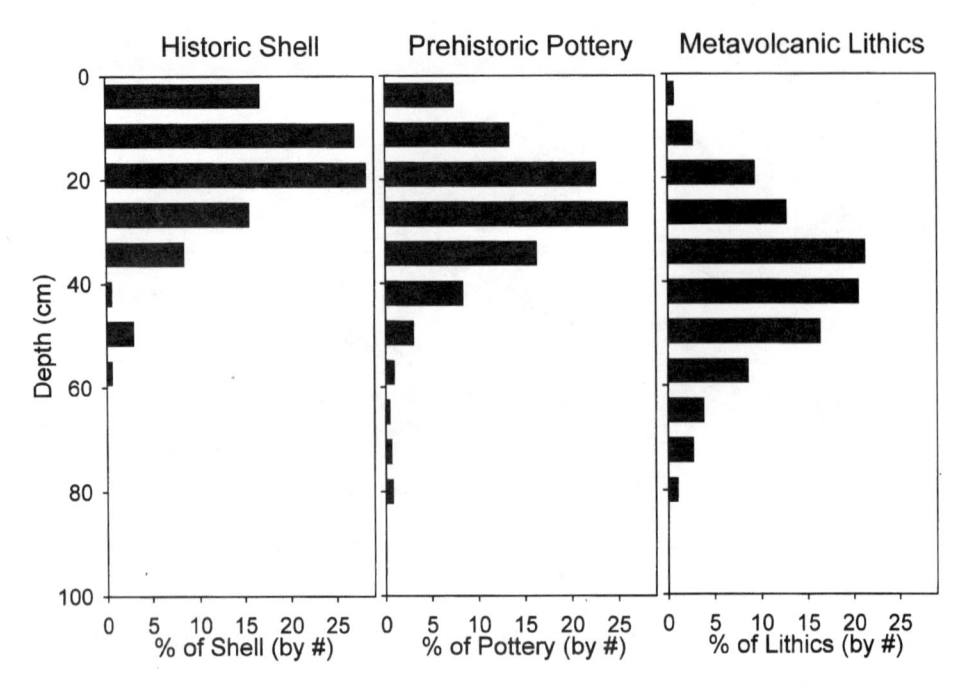

Figure 10.7 Histogram plots of data presented by Michie (1990) illustrating the frequency versus depth of various types of artifacts. The historic shells were deposited in the plow zone (top 15 cm) during the 1820s. The pottery and lithics are from site 38GE261 and probably represent large time spans of deposition on the surface, but the mean age of lithics probably predates that of the pottery.

capable of bringing artifacts larger than 3 cm up to the surface. Thus, while the overall integrity of cultural stratigraphy may remain intact, some degree of mixing is to be expected.

Mitchie (1990) stated that some of the most convincing evidence for bioturbative burial of artifacts comes from the observation of buried historical shell fragments at an antebellum plantation in Georgetown County, South Carolina. He noted that shell fragments that were deposited in the early 1800s (as agricultural lime) were distributed from 0 to 61 cm below the ground surface on sandy upland soils that were not subject sedimentation processes (Fig. 10.7). The known surface of shell deposition was the plow zone in an old cultivated field that became inactive and was replaced by forest in about 1870. At least 35 percent of the crushed shell had moved below the plow zone in a period of slightly more than 100 years. The only known mode of burial for the shell fragments was by bioturbation. Michie (1990) noted frequency distributions of prehistoric artifacts from a nearby site that were strikingly similar to that of the shell fragment distribution (Fig. 10.7). That is, the number of artifacts in each successive level was proportionate to the previous level and appears to indicate a wave of artifacts moving down from the surface. Note also in Fig. 10.7 that the modal depths of shell, pottery, and lithics are successively deeper, which logically follows the relative time of introduction of each type of material and more time for burial by

Buried Artifacts in Sandy Soils 283

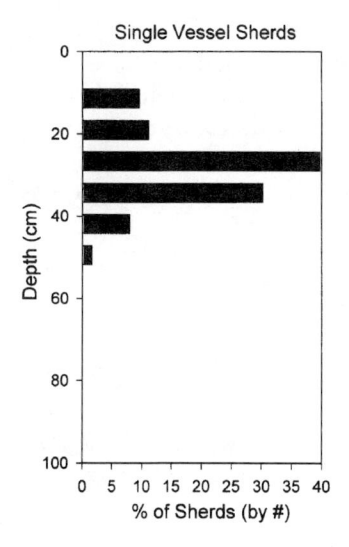

Figure 10.8. Histogram plot data presented by Michie (1990) illustrating the vertical distribution of pot sherd fragments from a single vessel found in an excavation unit at site 38GE261.

bioturbation. The prehistoric pottery and lithics are from a multicomponent site where the lithics may have been deposited on the surface as early as Paleoindian or early Archaic time, whereas the introduction of pottery probably began during the late Archaic at the earliest.

Mitchie (1990) also noted the vertical displacement of pottery from a single vessel that was distributed throughout a vertical distance of 46 cm (Fig. 10.8). Again, bioturbation is the only mechanism that reasonably could explain the distribution of the fragments of the single vessel. Additional examples of the vertical movement of artifacts in soil profiles is provided by Matthews (1965), Moeyersons (1978), and Rowlett and Robbins (1982) and Bocek (1986).

Michie (1990) and Gunn and Foss (1992, 1997) have suggested that the degree of mixing and downward displacement of artifacts is a function of artifact size. In sandy soils of the South Carolina Sandhills, Gunn and Foss indicated that artifacts smaller than 5 cm appeared to move rapidly downward at rates ranging from 3 to 7 cm per 1000 years, whereas artifacts larger than 5 cm were more stable and tended to be better indicators of the spatial and temporal attributes of the cultural assemblage. In addition, Gunn and Foss (1997) suggested that the rate of vertical displacement of artifacts is a function of the particle size composition of sandy soils. They indicated that the greatest rates of downward movement of artifacts were apparent in fine-textured sands.

In summary, if sedimentation by alluvial, colluvial, or eolian processes has buried occupation levels at a site, then clear stratigraphic separation of cultural horizons by culturally sterile zones (or at least low-density artifacts zones) is possible. However, the degree of stratigraphic separation is totally contingent on the periodicity of occupation and the rate of sedimentation. If periodic (but rapid) sedimentation of relatively thick deposits occurs, then it is likely that burial will favor a multimodal distribution of artifacts with depth. In contrast, if bioturbation has disrupted and buried a surface site, then the cultural material is likely to be spread throughout a wide vertical zone as a unimodal distribution (as in Fig. 10.7) and mixtures of artifacts from different cultural periods within the same level will be common. Bioturbation can ultimately bury an entire site, provided that some of the sediment is progressively brought to the surface through time.

2.5. Micromorphology

Micromorphology has many applications in archeology (Brewer, 1976; Bullock et al., 1985; Courty et al., 1989; Fitzpatrick, 1980) that typically are not utilized in routine survey and data recovery excavations but could be valuable in analyzing the role of pedoturbation versus sedimentation. Sedimentary structures that are not apparent macroscopically may be readily apparent microscopically as microlaminae that are segregated on the basis of subtle grain size or mineralogical differences. Living surfaces or floors that have not been bioturbated may show a greater degree of compaction and lower porosity than surrounding soils, which can be quantitatively measured in thin sections (Goldberg and Whitbread, 1993). In addition, undisturbed floors of structures may show preferred orientations of the long axis of grains that are not apparent outside of the structure. Evidence of bioturbation can also be documented by noting the disruption of microsedimentary structures and by documenting features such as animal burrows and worm fecal pellets.

Gé et al. (1993) reviewed the micromorphological study of various archeological contexts and concluded that micromorphology aided in the identification of sedimentary microstructures caused by human activity. They concluded that archaeologists typically have not considered mechanisms of human-induced modifications at sedimentary scales, which explains why occupation surfaces are rarely noted. Examples provided by Gé et al. (1993) are discussed in the context of disaggregation, compaction, accumulation, and redistribution mechanisms at a sedimentary scale. These mechanisms were then related to human activities and occupation surfaces. They suggested that more systematic studies are needed at all scales, and that more ethnoarchaeological studies (i.e., Goldberg and Whitbread, 1993) are needed to build a better understanding of micromorphological attributes of living surfaces.

2.6. Dating Techniques

Luminescence dating techniques for sediment include thermoluminescence (TL), optically stimulated luminescence (OSL), and infrared stimulated luminescence (IRSL), (see Rink, Chapter 14, this volume; Prescott and Robertson, 1998; Wintle, 1998), which are potentially applicable to resolve the relative importance of pedoturbation versus sedimentation in sandy soils. Thermoluminescence (TL) dating is best applied to eolian sediments, but OSL is suited to dating eolian, fluvial, and colluvial sediments (Prescott and Robertson, 1998) because of the shorter zeroing time required to erase the OSL signal. Thermoluminescence dating has the advantage of being applied to heat-treated flints and pottery and thus could be used to document the mixing of similar age artifacts throughout the soil profile or to show that very different age artifacts are mixed together in the same level. Both patterns of mixing would favor bioturbation. Luminescence dating also can be used to discern temporally separate cultural horizons and sediments, which could provide support for other types of stratigraphic differentiation (i.e., particle size analysis). A major drawback of luminescence dating is a relatively high cost and a continuing experimental nature of the dating technique. However, many new luminescence labs are being developed around the world and ease of access to luminescence labs is bound to improve through 2010.

Luminescence dates provide an estimate of the last time that sand grains (or silt grains) were exposed to sunlight, and the dates can be used to refine the stratigraphy and geochronology at a site. Most geochronologists who use luminescence dating tend to avoid the upper meter of a soil profile because that is the active zone in terms of bioturbation. Bioturbation introduces surface material, which has been exposed to light, into subsurface settings and negates significance of the date in terms of primary sedimentation. The result can be a date that is younger than the time of sedimentation. However, such a date could actually be used to support the idea of bioturbation. At present, the author is not aware of any studies that have used luminescence dating to specifically target the problem of bioturbation, but manufacturers of luminescence dating equipment recently have made devices that can scan continuously down a core section and identify major time-stratigraphic breaks. Such a continuous scan has great potential for evaluating the zone of bioturbation. One would expect a bioturbated zone to show a wide range of variability in luminescence properties and dates, whereas a stratigraphic unit that was deposited as a single event should show less variance.

The luminescence technique is in some ways superior to radiocarbon dating because it averts the problem of dating intrusive plant materials (e.g., charred roots) that are not representative of the time of sedimentation. However, it suffers from the need to date *in situ* sediment beneath the zone of bioturbation in order to produce an accurate date of the time of sedimentation. Charred roots of individual trees can extend more than a meter beneath the surface, potentially producing the same age radiocarbon material throughout the soil profile. In addition, conventional radiocarbon dates from sandy soils typically require obtaining a composite sample of charcoal from a particular level or strata that may contain charcoal originating from several different plants growing on the surface over a period of thousands of years. Therefore, although radiocarbon dating may provide a useful date pertaining to the time of death of intrusive plant material, it can be of limited use for estimating the time of sedimentation in sandy bioturbated soils. In fact, Moeyersons (1978) noted that radiocarbon samples may have no clear relation to artifacts found a the same depth for Kalihari sands.

An advantage of luminescence dating (over radiocarbon) is that it could be used to document age inversions, which would be consistent with bioturbation, or to determine whether or not the age of the sediment is compatible with the age of the cultural materials in the sediment matrix. If artifacts are buried by sedimentation, then the artifacts presumably will be older than the overlying sediment and the law of superposition will apply to dates from the stratigraphic sequence. In contrast, with radiocarbon dates in sands there are problems related to dating of intrusive material (such as burned roots) that can make the interpretation of the dates questionable.

Problems associated with "composite" ages from samples that may contain more than one bioturbation event may be overcome by the single-grain OSL techniques (Murray and Roberts, 1997). Dating single grains from an individual strata or level that has not been significantly bioturbated should produce a cluster of similar ages, whereas bioturbated sediment would be expected to produce wider range of ages from the same level. This principle can be applied to down-profile measurements to help discriminate the zone of bioturbation. For example, results of 30 single-grain measurements from a strata excavated in a bioturbated zone should produce much greater variance than 30 single-grain measurements from an underlying undisturbed strata of primary eolian sand.

2.7. Other Techniques

Many methods of measuring physical and chemical properties of soil are available and possibly could be applied to solve the relative importance of pedoturbation versus sedimentation at a site. Common methods that could be used to help define buried occupation surfaces include magnetic susceptibility (see chapter 13, this volume), phytolith analysis, and phosphorus chemistry (Herz and Garrison, 1998). In general, any technique that can be used to confirm separate depositional strata and paleosols can be applied to help resolve the question of burial processes by pedoturbation versus sedimentation. The reader is referred to Page (1982), Singer and Janitzky (1986), Klute (1986), Gale and Hoare (1991), and Sparks (1996) for excellent compilations of methods for the analysis of sediment and soil.

3. Case Example

A case study, which exemplifies many of the techniques discussed previously, comes from the sand hills of Fort Bragg, North Carolina. Here, a column of sediment samples associated with buried artifacts at an upland site, 31HT285 (Chinaberry Site), were analyzed to evaluate bioturbation versus eolian sedimentation as a site burial process. In addition, the sedimentology of nearby upland soils that obviously had been reworked by eolian processes were analyzed for comparative purposes to determine the sedimentology of reworked upland soils.

Site 31HT285 is situated in a wooded area on the western side of an upland interfluve where slopes are less than 1 percent (Fig. 10.9). The mapped soil series at the site is Candor sand (Spangler, 1994), which is assigned the lowest possible erosion factor ($K = 0.1$) and is characterized by an A-E-Bt-E'-B't horizon se-

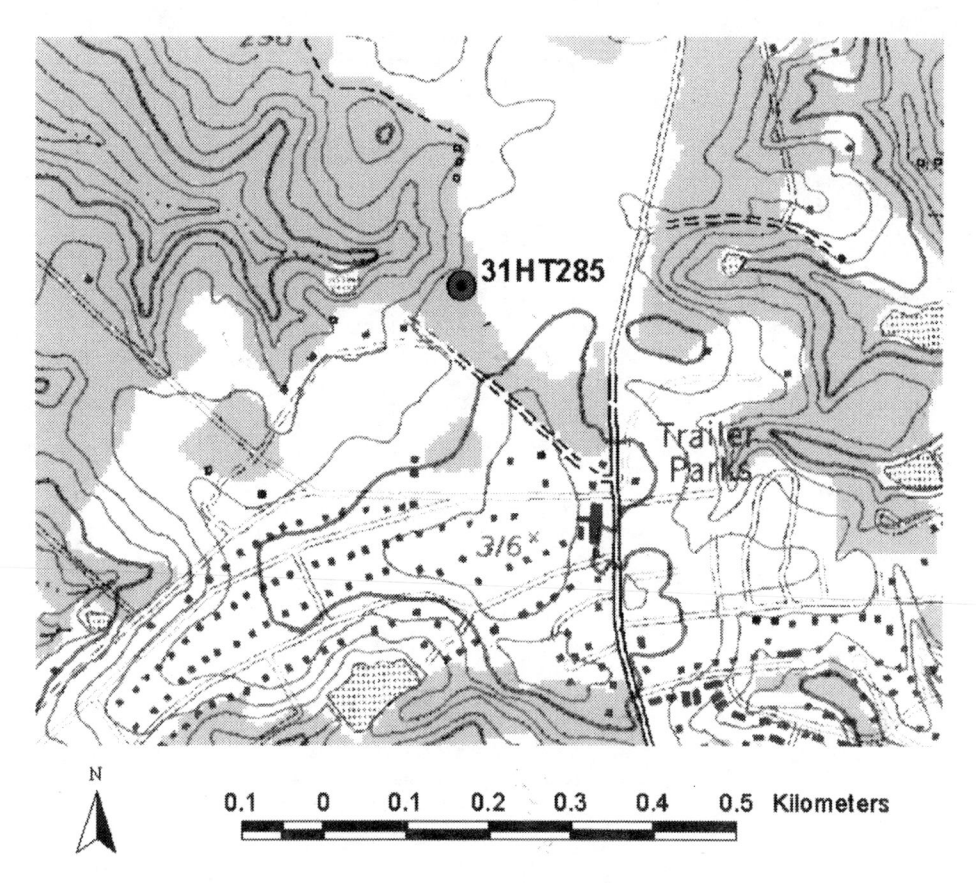

Figure 10.9. Map showing the location of site 31HT285 at Fort Bragg, North Carolina. The map is part of the Manchester, North Carolina 7.5-minute U.S.G.S. topographic quadrangle, illustrating 10 ft contour intervals.

quence. There are no sedimentary structures visible in the massive sands, yet lithic artifacts are distributed throughout the E horizon, with the greatest concentration at 30 to 40 cm depth (Fig. 10.10). No pottery was recovered from the excavation block, and therefore, it is possible that the entire lithic scatter is middle Archaic or older. The vertical distribution of artifacts is a right-skewed, bell-shaped histogram that matches Michie's (1987, 1990) hypothetical depth of middle to late Archaic artifacts with a mode at 0.3 to 0.5 m depth. Comparison of the artifact distribution to particle size, total carbon, and phytolith data fails to support the presence of a buried A horizon that corresponds to the cluster of artifacts (Fig. 10.10). Furthermore, the particle size data are characterized by a very uniform down-profile pattern that fails to indicate any lithologic discontinuity corresponding to possible buried strata containing the artifacts (Fig. 10.10, Tables 10.4 and 10.5).

In order to determine the sedimentology of reworked upland soils, five samples from rippled sands were collected from the ground surface at the

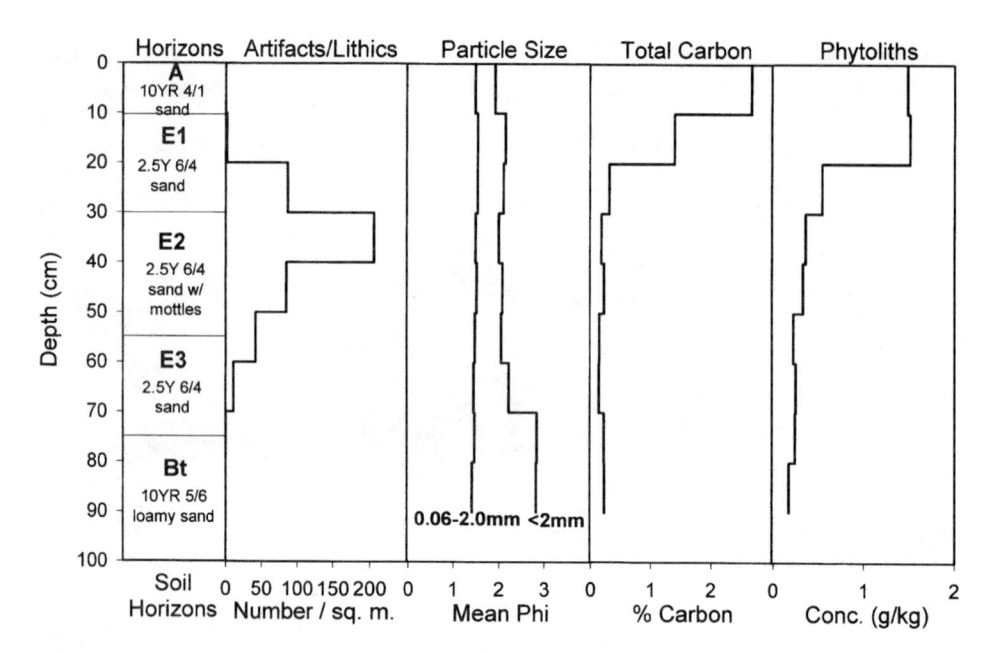

Figure 10.10. Plots of data from a 2 × 4 m excavation block at site 31HT285 at Fort Bragg, North Carolina. No pottery was found in this excavation unit. The arithmetic mean phi is given for the <2 mm fraction, which shows the finer texture of the Bt horizon, and for the sand (0.063–2.0 mm) fraction, which eliminates the pedogenic bias of translocated clay in the Bt horizon. Total carbon was measured on the <0.25 mm fraction to eliminate the bias of macro-organics and to concentrate humus. Phytoliths were measured from the silt fraction (0.002–0.063 mm), but phytolith concentration is reported on the basis of the whole sample weight.

Normandy Drop Zone at Fort Bragg. This drop zone is a parachute training range where a 1 km by 3 km rectangular upland area has been stripped of vegetation and active eolian sedimentation clearly occurs during dry and windy periods. The drop zone was built on Lakeland sand, which had very similar particle size properties to site 31HT285 prior to eolian reworking (Fig. 10.4). Comparison of the particle size distribution of the eolian sand from the Normandy Drop Zone to the upper 30 cm of sediment at site 31HT285 fails to suggest an eolian origin for the sand that buries lithics at the site (Tables 10.4 and 10.5). The eolian sands from the Normandy Drop Zone are much better sorted (lower SD and CV), contain less silt, and contain fewer coarse particles (>1 mm) than the sands at site 31HT285. Furthermore, the sands burying the artifacts at 31HT285 (top 30 cm) fail to meet most of the criteria that are typical of eolian sand (Tables 10.3, 10.4, and 10.5). The sands from the Normandy Drop Zone represent minimal transport by eolian processes, with most of the samples having moved only about 10 to 50 m from their source. However, four out of five of these samples still meet the criteria presented in Table 10.3. Sample DZ1 is the exception (Tables 10.4 and 10.5), which is unusual because it contains a large portion of coarse sand (30.6% of 0.5–1.0 mm sand). Sample DZ1 was collected only about 10 m from a deflation zone, which may explain its coarse texture.

Table 10.4. Particle Size Fractions from Eolian Normandy Drop Zone Samples (DZ) Compared to 31HT285 Samples

Site I.D.	Depth (cm)	% gravel >2 mm	% sand	% silt	% clay	% phi −1–0	% phi 0–1	% phi 1–2	% phi 2–3	% phi 3–4	% phi 4–6	% phi 6–9
DZ 1	0–5	0.04	95.7	2.5	1.8	2.6	30.6	37.4	18.5	6.6	1.8	0.7
DZ 2	0–5	0.00	96.2	1.6	2.2	0.3	14.1	51.6	23.7	6.5	0.8	0.8
DZ 3	0–5	0.00	94.9	2.9	2.2	0.2	13.9	52.0	21.1	7.7	1.9	1.0
DZ 4	0–5	0.00	95.2	2.8	2.0	0.6	8.5	47.2	31.7	7.2	1.5	1.3
DZ 5	0–5	0.00	95.7	2.6	1.8	0.2	7.7	54.2	27.6	6.0	1.4	1.2
31HT285	0–10	0.02	91.2	7.8	1.0	2.2	23.5	44.2	17.6	3.7	4.8	3.0
31HT285	10–20	0.05	88.8	9.2	2.0	2.7	22.0	39.5	19.8	4.9	4.2	5.0
31HT285	20–30	0.05	89.7	8.3	2.0	2.5	21.9	41.0	19.6	4.6	3.9	4.5
31HT285	30–40	0.05	91.0	7.5	1.6	2.3	23.7	42.7	17.9	4.4	3.1	4.4
31HT285	40–50	0.03	89.6	8.5	2.0	2.8	23.1	39.2	19.7	4.7	4.5	4.0
31HT285	50–60	0.07	89.7	8.5	1.8	2.9	24.3	39.9	17.8	4.8	3.9	4.7
31HT285	60–70	0.13	87.3	9.0	3.8	3.1	23.9	38.9	17.8	3.7	3.8	5.2
31HT285	70–80	0.11	80.3	8.8	10.9	2.9	21.4	35.6	16.2	4.1	3.9	5.0
31HT285	80–90	0.13	80.9	6.8	12.4	3.9	22.9	33.9	16.5	3.6	2.8	4.0

Note. Percent gravel is calculated on the basis of the whole sample, whereas all other percentages are of the <2 mm fraction.

Table 10.5. Arithmetic Moment Statistics for Eolian Normandy Drop Zone Samples (DZ) Compared to 31HT285 Samples

Arithmetic moment statistics for the <2 mm fraction

Site	Depth (cm)	Mean	Standard deviation	Skewness	Kurtosis	CV
DZ 1	0–5	1.72	1.61	2.84	14.43	0.94
DZ 2	0–5	1.98	1.54	3.40	17.53	0.78
DZ 3	0–5	2.04	1.60	3.15	15.30	0.78
DZ 4	0–5	2.16	1.53	3.15	15.76	0.71
DZ 5	0–5	2.09	1.47	3.45	17.86	0.70
31HT285	0–10	1.90	1.70	2.30	9.52	0.89
31HT285	10–20	2.14	2.02	2.07	7.47	0.94
31HT285	20–30	2.09	1.97	2.18	8.11	0.94
31HT285	30–40	1.99	1.88	2.30	8.81	0.95
31HT285	40–50	2.07	1.95	2.16	8.18	0.94
31HT285	50–60	2.04	1.97	2.14	7.90	0.97
31HT285	60–70	2.21	2.29	2.04	6.78	1.03
31HT285	70–80	2.83	3.00	1.51	3.97	1.06
31HT285	80–90	2.82	3.11	1.50	3.87	1.10

Arithmetic moment statistics for the <2 mm–0.063 mm fraction

Site	Depth (cm)	Mean	Standard deviation	Skewness	Kurtosis	CV
DZ 1	0–5	1.46	0.95	0.38	2.61	0.65
DZ 2	0–5	1.73	0.79	0.36	3.00	0.46
DZ 3	0–5	1.73	0.80	0.47	3.00	0.46
DZ 4	0–5	1.88	0.78	0.11	3.03	0.41
DZ 5	0–5	1.83	0.72	0.38	3.21	0.39
31HT285	0–10	1.47	0.84	0.22	2.99	0.57
31HT285	10–20	1.53	0.90	0.16	2.77	0.59
31HT285	20–30	1.52	0.88	0.17	2.83	0.58
31HT285	30–40	1.48	0.87	0.25	2.91	0.58
31HT285	40–50	1.50	0.90	0.17	2.75	0.60
31HT285	50–60	1.47	0.90	0.23	2.82	0.61
31HT285	60–70	1.44	0.88	0.17	2.81	0.61
31HT285	70–80	1.46	0.90	0.19	2.82	0.62
31HT285	80–90	1.41	0.92	0.14	2.72	0.65

However, it still contains less silt and clay, and is better sorted, than the 0 to 30 cm samples from 31HT285.

If eolian sedimentation had occurred at 31HT285, then one would expect the upper part (top 30 cm) of the soil profile to be lithologically different from the underlying sand and possibly contain a buried A horizon. However, this is not the case. Furthermore, the particle size characteristics of sand at 31HT285 do not resemble eolian sediment, even that which was blown only a short distance in similar landscape and soil conditions at the Normandy Drop Zone. In summary, it would be irrational to conclude that eolian sediment buries 31HT285. Instead, bioturbation is the most logical site burial process.

4. Conclusions

Determining the relative importance of pedoturbation (mainly bioturbation) versus sedimentation as burial agents and site formation processes in sandy soils admittedly can be a difficult task. However, solving such a mystery can have important implications about the paleoenvironmental conditions since the time of artifact burial that make the detailed analysis and effort worthwhile. For earth scientists the techniques outlined here can be applied to other stratigraphic settings, besides just archeological sites, to help resolve questions about the environment of sedimentation or postdepositional changes. For the archaeologist, application of the techniques could significantly enhance the conclusions about site formation processes at sites in sandy soils.

Careful consideration and measurement of the geomorphic setting, stratigraphy, pedology, sedimentology, micromorphology, and geochronology can help significantly to resolve the processes of site burial. In some cases (e.g., upland divides) the research question may simply involve the discrimination of burial by pedoturbation versus eolian sedimentation. In such a case, if eolian processes can be ruled out, then pedoturbation will remain as the probable cause of burial. In other cases, such as footslopes and sand dunes, the interplay of pedoturbation and sedimentation may be difficult (if not impossible) to resolve. Despite the difficulties posed in uniform sandy soils, a combination of the methods discussed here should provide guidance in evaluating the potential for pedoturbation, sedimentation, or both processes as viable site formation processes. Without application of such tests the interpretations about site burial processes will remain speculative.

5. References

Ahlbrandt, T. S.,1979, Textural Parameters of Eolian Sands. In *A Study of Global Sand Seas*, edited by E. D. McKee, pp. 21–54. U.S. Geological Survey Professional Paper 1052, U.S. Government Printing Office, Washington, DC.

Blatt, H., Middleton, G., and Murray, R., 1980, *Origin of Sedimentary Rocks*, Prentice-Hall, Englewood Cliffs, NJ.

Bocek, B., 1986, Rodent Ecology and Burrowing Behavior: Predicted Effects on Archaeological Site Formation, *American Antiquity* 51:589–603.

Boggs, S., 1987, *Principles of Sedimentology and Stratigraphy*, Merrill, Columbus, OH.

Brewer, R.,1976, *Fabric and Mineral Analysis of Soil*. Huntington, Krieger, New York.

Brooks, M. J., and Sassaman, K. E., 1990, Point Bar Geoarchaeology in the Upper Coastal Plain of the Savannah River Valley, South Carolina. In *Archeological Geology of North America. Geological Society of America Centennial Special Volume 4*, edited by N. P. Lasca and J. Donahue, pp. 183–197. Geological Society of America, Boulder, CO.

Brooks, M. J., Taylor, B. E., and Grant, J. A.,1996, Carolina Bay Geoarchaeology and Holocene Landscape Evolution on the Upper Coastal Plain of South Carolina. *Geoarchaeology* 11: 481–504.

Bullock, P., Fedoroff, N., Jongerius, A., Stoops, G., and Tursina, T., 1985, *Handbook for Soil Thin Section Description*. Waine Research, Mount Pleasant, Wolverhampton, UK.

Buol, S. W., Hole, F. D., and McCracken, R. J., 1989, *Soil Genesis and Classification*. Iowa State University Press, Ames.

Butler, D. R.,1995, *Zoogeomorphology*, Cambridge University Press, New York.

Courty, M. A., Goldberg, P., and Macphail, R. I., 1989, *Soils and Micromorphology in Archaeology*, Cambridge University Press, New York.
Daniels, R. B., and Hammer, R. D., 1992 *Soil Geomorphology*. Wiley, New York.
Dalrymple, J. B., Blong, R. J., and Conacher, A. J., 1968, An Hypothetical Nine-Unit Landsurface Model. *Zeitschrift Fur Geomorphology* 12:60–76.
Fitzpatrick, E. A., 1980, *The Micromorphology of Soils*. Department of Soil Science, University of Aberdeen, Aberdeen, Scotland.
Friedman, G. M.,1961, Distinction Between Dune, Beach, and River Sands from their Textural Characteristics. *Journal of Sedimentary Petrology* 31:514–529.
Friedman, G. M.,1979, Address of the Retiring President of the International Association of Sedimentologists: Differences in Size Distributions of Populations of Particles Among Sands of Various Origins, *Sedimentology* 26:3–32.
Gale, S. J., and Hoare, P. G.,1991, *Quaternary Sediments: Petrographic Methods for the Study of Unlithified Rocks*, Wiley, New York.
Gé, T., Courty, M. A., Matthews, W., and Wattez, J., 1993, Sedimentary Formation Processes of Occupation Surfaces. In *Formation Processes in Context*. Monographs in World Archaeology No. 17, edited by P. Goldberg, D. Nash, and M. Petraglia, pp.149–163, Prehistory Press, Madison, WI.
Gee, G. W., and Bauder, J. W., 1986, Particle Size Analysis. In *Methods of Soil Analysis: Part 1, Physical and Mineralogical Methods*, 2nd ed, edited by A. Klute, pp. 383–411. Soil Science Society of America, Madison, WI.
Goldberg, P. G., and Whitbread, I.,1993, Micromorphological Study of a Bedouin Tent Floor. In *Formation Processes in Context*. Monographs in World Archaeology No. 17, edited by P. Goldberg, D. Nash, and M. Petraglia, pp. 165–188, Prehistory Press.
Gunn, J. D., and Foss, J.,1992, Copperhead Hollow 38CT58: Middle Holocene Upland Conditions on the Piedmont-Coastal Plain Margin, *South Carolina Antiquities* 24:1–17.
Gunn, J. D., and Foss, J. E., 1997, Variable Artifact Displacement and Replacement in a Holocene Eolian Feature. In *Proceedings of the Second International Conference on PedoArchaeology*, edited by A. C. Goodyear and K. E. Sassaman, pp. 53–74. South Carolina Institute of Archaeology and Anthropology, Columbia.
Hand, B. M., 1967, Differentiation of Beach and Dune Sand Using Settling Velocities of Light and Heavy Minerals, *Journal of Sedimentary Petrology* 37:514–520.
Herz, N., and Garrison, E.G., 1998, *Geological Methods for Archaeology*, Oxford University Press, New York.
Hole, F. D., 1961, A Classification of Pedoturbations and Some other Processes and Factors of Soil Formation in Relation to Isotropism and Anisotropism, *Soil Science* 91:375–377.
Hole, F. D., 1981, Effects of Animals on Soil, *Geoderma* 25:75–112.
Ingram, R. L., 1971, Sieve Analysis. In *Procedures in Sedimentary Petrology*, edited by R. E. Carver, pp. 49–67. Wiley, New York.
Johnson, D. L., 1990, Biomantle Evolution and the Redistribution of Earth Materials and Artifacts, *Soil Science* 149:84–102.
Johnson, D. L., and Watson-Stegner, 1990, The Soil-Evolution Model as a Framework for Evaluating Pedoturbation in Archeological Site Formation. In *Archeological Geology of North America. Geological Society of America Centennial Special Volume 4*, edited by N. P. Lasca and J. Donahue, pp. 541–558. Geological Society of America, Boulder, CO.
Jones, R. L., 1969, Determination of Opal in Soil by Alkali Dissolution Analysis, *Soil Science Society American Proceedings* 33:976-978.
Klute, A., 1986, *Methods of Soil Analysis: Part 1, Physical and Mineralogical Methods*, 2nd ed. Soil Science Society of America, Madison, WI.
Knox, J. C., 1987, Stratigraphic Evidence of Large Floods in the Upper Mississippi Valley. In *Catastrophic Flooding*, edited by L. Mayer and D. Nash, pp. 155–179. Allen and Unwin, Boston.
Leigh, D. S., 1998, Evaluating Artifact Burial by Eolian Versus Bioturbation Processes, South Carolina Sandhills, USA, *Geoarchaeology* 13:309–330.
Lowright, R. H., 1973, Environmental Determination Using Hydraulic Equivalence Studies. *Journal of Sedimentary Petrology* 43:1143–1147.
Matthews, J. M., 1965, Stratigraphic Disturbanced: The Human Element. *Antiquity* 39:295–298.

McKee, E. D., 1979, Sedimentary Structures in Dunes. In *A Study of Global Sand Seas*, edited by E. D. McKee, pp. 83–134. U.S. Geological Survey Professional Paper 1052, U.S. Government Printing Office, Washington, DC.

Michie, J., 1987, *Bioturbation and Gravity as a Potential Site Formation Process: The Open Area Site, 38GE261, Georgetown County, South Carolina*. South Carolina Institute of Anthropology and Archaeology, University of South Carolina, Columbia.

Michie, J., 1990, Bioturbation and Gravity as a Potential Site Formation Process: The Open Area Site, 38GE261, Georgetown County, South Carolina. *South Carolina Antiquities* 22:27–46.

Mitchell, P. B., 1988, The Influences of Vegetation, Animals, and Micro-Organisms on Soil Processes. In *Biogeomorphology*, edited by H. A. Viles, pp. 43–82. Blackwell, New York.

Moeyersons, J., 1978, The Behavior of Stones and Stone Implements Buried in Consolidating and Creeping Kalahari Sand, *Earth Surface Processes* 3:115–128.

Murray, A. S., and Roberts, R. G., 1997, Determining the Burial Time of a Single Grain of Quartz Using Optically Stimulated Luminescence, *Earth and Planetary Science Letters* 152:163–180.

Page, A. L., *Methods of Soil Analysis*, American Society of Agronomy, Madison, WI.

Prescott, J. R. and Robertson, G. B., 1998, Sediment Dating by Luminescence: A Review, *Radiation Measurements* 27:893–922.

Renard, K. G., Foster, G. R., Weesies, G. A., and Porter, J. P., 1991, RUSLE: Revised Universal Soil Loss Equation, *Journal of Soil Water Conservation* 48:30–33.

Rowlett, R. M., and Robbins, M. C., 1982, Estimating Original Assemblage Content to Adjust the Post-Depositional Vertical Artifact Movement, *World Archaeology* 14:73–83.

Ruhe, R. V., 1959, Stone Lines in Soils, *Soil Science* 87:223–231.

Schiffer, M. B., 1987, Formation Processes of the Archaeological Record. University of New Mexico Press, Albuquerque.

Singer, M. J., and Janitzky, P., 1986, *Field and Laboratory Methods Used in a Soil Chronosequence Study*. U.S. Geological Survey Bulletin 1648, U.S. Government Printing Office, Washington DC.

Spangler, D. G., 1994, *Soil Survey of Harnett County, North Carolina*. United States Department of Agriculture, Soil Conservation Service, U.S. Government Printing Office, Washington DC.

Sparks, D. L.,1996, *Methods of Soil Analysis: Part 3, Chemical Methods*. Soil Science Society of America, Madison, WI.

Steidtmann, J. R., and Haywood, H. C., 1982, Settling Velocities of Quartz and Tourmaline in Eolian Sandstone Strata, *Journal of Sedimentary Petrology* 52:95–399.

Thames, H. R., 1982, *Origin of Sand Ridges Along Streams in Southeastern Georgia*. Unpublished Masters Thesis, Emory University, Atlanta. GA.

Tucker, R. W., and Vacher, H. L., 1980, Effectiveness of Discriminating Beach, Dune, and River Sands by Moments and the Cumulative Weight Percentages, *Journal of Sedimentary Petrology* 50:165–172.

Waters, M. R., 1992, *Principles of Geoarchaeology: A North American Perspective*. The University of Arizona Press, Tucson.

Watson, R. L., 1969, Modified Rubey's Law Accurately Predicts Sediment Settling Velocities, *Water Resources Research* 5:1147–1150.

Wintle, A. G., 1998, Luminescence Dating: Laboratory Procedures and Protocols, *Radiation Measurements* 27:769–817.

Wood, R. W., and Johnson, D. L., 1978, A Survey of Disturbance Processes in Archaeological Site Formation. In *Advances in Archaeological Method and Theory*, edited by M. B. Schiffer, pp. 315–381. Academic Press, New York.

IV

Specific Techniques

The Role of Petrography in the Study of Archaeological Ceramics

11

JAMES B. STOLTMAN

1. Introduction

Petrographic microscopy is a venerable geological technique that has been used in the service of archaeology at least since the 1930s (e.g., Shepard, 1936, 1939). Compared to newer, sophisticated, "high-tech" approaches to the study of the compositional analysis of ceramics (e.g., neutron activation or acid extraction), petrography surely rates the appellation "old fashioned." The goal of this chapter is to describe and evaluate critically the current status and potential of ceramic petrography as an approach to the compositional analysis of archaeological ceramics, especially in light of the increasingly widespread and successful application of newer technologies for determining the elemental composition of ceramics that might be seen as rendering petrography obsolete (useful earlier reviews of ceramic petrography may be found in Freestone, 1991 and 1995; Peacock, 1970; Williams, 1983).

The basic thesis of this chapter is that petrography, far from being obsolete, is a valuable and underutilized approach that offers unique and important

---

JAMES B. STOLTMAN • Department of Anthropology, University of Wisconsin–Madison, Wisconsin 53706.

Earth Sciences and Archaeology, edited by Paul Goldberg, Vance T. Holliday, and C. Reid Ferring. Kluwer Academic/Plenum Publishers, New York, 2001.

insights into the physical composition of archaeological ceramics. In discussing some of the virtues and limitations of ceramic petrography, lest my exuberance be misunderstood, I should like to stress at the outset that petrography and elemental analyses are complementary rather than competing approaches and that together they provide deeper insights into ceramic composition than either can alone. It is sometimes difficult to appreciate this fact from the literature because most applications of the two approaches (thus the published results) are produced by scholars situated in laboratories that are devoted to only one of them (due mainly to funding and staffing limitations). Using examples derived primarily from recent research into North American prehistory, this chapter attempts to demonstrate the value of petrography to archaeology by discussing four basic issues pertaining to archaeological ceramics that petrography can help to illuminate: classification, engineering, production, and exchange.

The successful implementation of petrographic analysis of ceramics requires, minimally, that three conditions be met: (1) the availability of properly prepared ceramic thin sections; (2) the availability of a suitable petrographic microscope; and (3) the requisite training in geology and experience in the use of the petrographic microscope, or the aid of a qualified petrographer.

To meet the first condition it is important to realize that each specimen must be cut with a saw to create a flat surface that will be attached to a glass slide. Because of the friable nature of most pottery, a sherd is normally impregnated with epoxy to stabilize it before it is cut. The ensuing cut may cross the sherd at any suitable position but is normally transverse to the plane of the sherd within a few millimeters of an edge so as to do minimal damage. One advantage of transverse cuts, which expose both sherd surfaces on the thin sections, is that certain surface finishes like slipping and burnishing may often be reliability identified. Alternately, sherds may be cut within the plane of the sherd, which has the advantage of producing a thin section with a greater area but has the disadvantages of sacrificing larger portions of the sherds and providing no views of sherd surfaces. The newly cut surface is then ground with fine abrasives until it is as flat as possible and then is mounted with a special epoxy onto a standard 46 mm × 27 mm glass slide (larger slides are available but are rarely used for ceramic thin sections). The slide dimensions dictate the maximum amount of a sherd that can be mounted as a thin section. Once the epoxy has cured, the mounted specimen is sawed a second time parallel to the first cut, leaving a 1 to 2 mm thick slice of the sherd on the slide; the remainder of the sherd can be returned to the collection. The "thick section" on the glass slide is then ground again with fine abrasives until it is as close as possible to a standard thickness of 30 microns (0.03 mm). A glass cover slip is then usually mounted over the finished thin section, although this step is sometimes omitted when the same slide is to be subjected to scanning electron microscopy (SEM) or electron microprobe analysis.

The net result of this process is normally three items: a "voucher" fragment, typically epoxy impregnated, that was detached by the first cut; the thin section mounted on its glass slide; and the surviving sherd after having been cut twice, all of which subsequently remain as permanent records for future reference. For unusually small specimens (<3 cm), however, there is some risk that only the thin section and the edge fragment would survive the mounting process. In such cases, especially when dealing with a unique specimen, the benefits of having a thin section made may be considered too costly.

The second condition, the availability of a petrographic microscope, is usually not a major problem in university settings because these are standard pieces of apparatus in most geology departments. Unlike standard biology microscopes, which illuminate specimens with light from above (i.e., incident light), petrographic microscopes employ "transmitted light," that is light that is passed through the thin sections from below. By observing how the transmitted light behaves as it passes through the thin sections, especially by making use of a standard feature of petrographic microscopes that allows one to employ either plane-polarized or cross-polarized light, one can identify minerals and rocks with great accuracy. For quantitative as opposed to qualitative analyses of thin section contents, however, a special stage attachment is required, which is not always so readily available as the microscopes themselves.

Meeting the third condition, learning how to use a petrographic microscope, is perhaps the biggest deterrent because the significant outlay of time required to learn the basics not only of petrography, but also mineralogy and petrology (large portions of at least 2 years) is a period lengthy enough to discourage most archaeology students. Enlisting the aid of a skilled petrographer, of course, is a viable alternative to learning these skills oneself, but that is unlikely to be more than a short-term or a one-time solution because such people are likely to have research interests of their own.

2. Basic Principles of Ceramic Petrography

Petrography—the microscopic analysis of rocks and minerals in thin section—is applicable to archaeological ceramics but only with important constraints. From a petrographic perspective, a ceramic thin section is composed of two basic ingredients: clays (which are plastic when moist) and coarser grained inclusions, usually mineral (often referred to as aplastics), of silt, sand, and gravel sizes (See Fig. 11.1). The term *clay* as used by earth scientists pertains both to a size grade—grains less than 0.002 mm in diameter no matter what the chemical composition—and to a set of specific minerals. In this chapter the term clay is used in the former sense. In a similar way the terms *silt* (0.002–0.0625 mm), *sand* (0.0625–2.00 mm), and *gravel* (>2.00 mm) are used to pertain solely to size grades (see Griffiths, 1967:76–77). The major constraint to petrographic analysis of ceramics is that the clays that normally comprise well over 50% by volume of any ceramic artifact appear through the petrographic microscope as an amorphous groundmass or matrix. Clays are simply too fine grained at ca. 1 micron in thickness to be identified beyond the generic level in a thin section that is 30 microns thick (but see Whitbread, 1989, for various observations on groundmass character and structure that are possible petrographically).

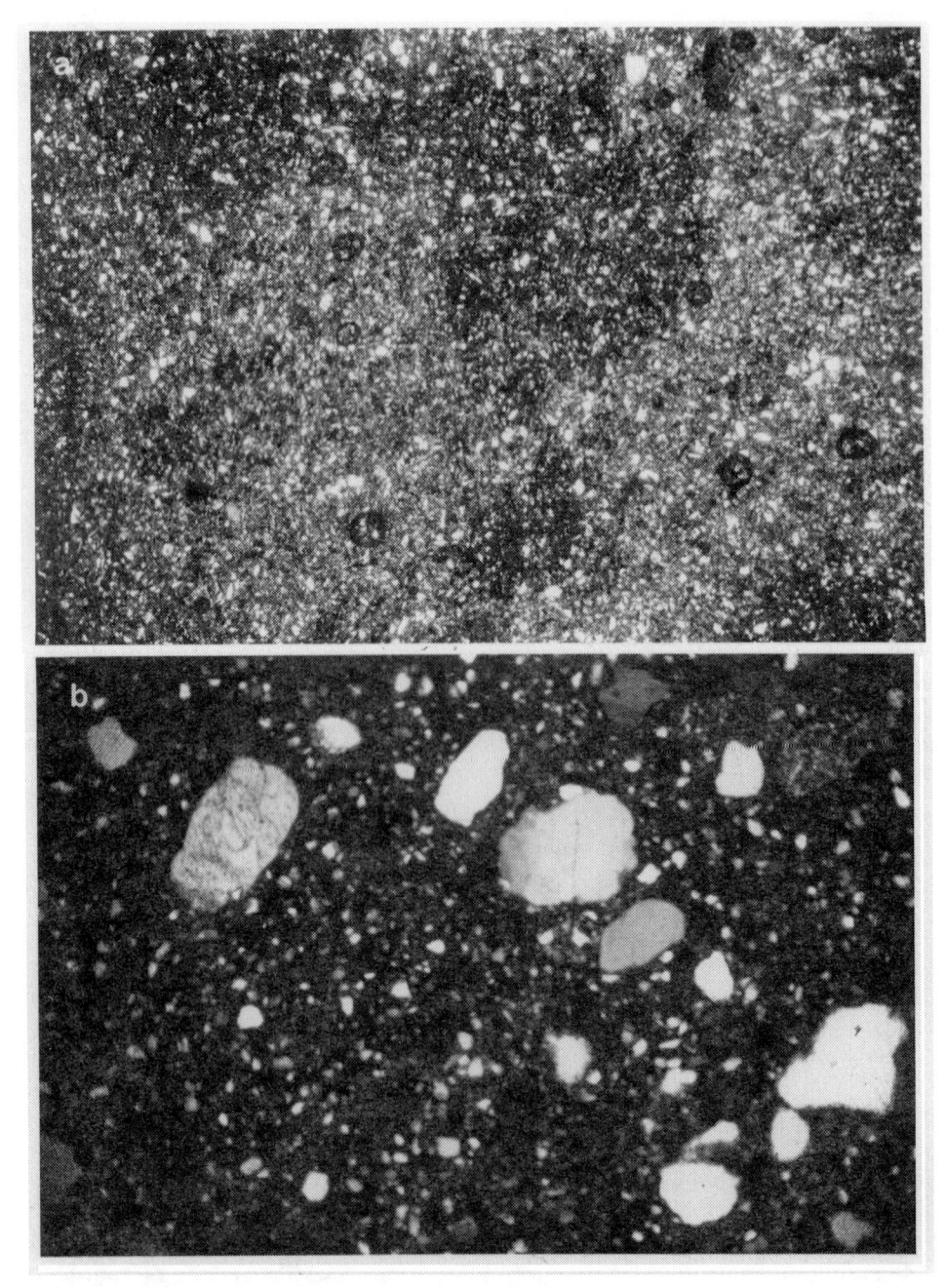

Figure 11.1. Photomicrographs of untempered clays; both viewed under crossed polars at 10 × magnification. (a) Subsoil sediment from the Fred Edwards site, Wisconsin. Scale: quartz (clear) grain, upper right = .15mm. (b) Subsoil sediment from Hartley Fort, Iowa. Scale: chert grain, largest speckled grain in upper right corner = .5mm.

2.1. Qualitative Observations

The main application of petrography in the analysis of archaeological ceramics is in the qualitative identification of mineral inclusions in the silt–sand–gravel size ranges—many other materials such as grog, bone, shell, fossils, and plants are also readily identifiable when present (see Fig. 11.2). In this domain petrography is unexcelled in both its effectiveness and reliability, although there are some techniques, such as X-radiography, that have certain advantages, such as being cheaper and less destructive (see Carr and Komorowski, 1995). Besides their mineral composition, the precise dimensions and shapes of individual grains in the silt–sand–gravel size ranges may be easily and reliably determined from thin sections with a petrographic microscope.

Although the composition, size, and shape of aplastic inclusions in the silt–sand–gravel size ranges can be reliably determined in ceramic thin sections through petrography, the issue is more complex because such grains are typically derived from two separate sources that should be discriminated. On the one hand, virtually all clay-rich sediments used in ceramic manufacture contain coarser grained inclusions whose character and/or relative abundance may be source-specific. On the other hand, it is commonplace for potters intentionally to add aplastic materials—temper—in order to enhance the workability of the clays or to ensure that the intended products survive the drying and firing stages of production. Distinguishing the temper from natural inclusions is important because it allows one to make a number of inferences concerning production and exchange that would otherwise be impossible.

As discussed in an earlier paper (Stoltman, 1991:111), distinguishing temper from natural inclusions can frequently be done objectively through petrography (see also Shepard, 1956:156–168). This is especially so when such tempers as grog (any fired clay product used as temper; see Cuomo Di Caprio and Vaughn, 1993; Porter, 1964; Whitbread, 1986), shell, limestone, and volcanic ash are present. It is more difficult when the temper is a quartz-rich rock—"grit" (e.g., granite or quartzite)—because smaller fragments of the temper may be difficult to distinguish from naturally occurring sand grains, which are most commonly quartz (various common temper types are shown in Fig. 11.2). In such cases a careful inventory of the mineral composition of the clearly identifiable temper grains must be one of the first steps in the analysis and, if at all possible, local soil and sediment samples should be collected, fired, and processed into thin sections in order to provide direct evidence of what minerals occur naturally in the local soils and sediments.

There is at least one situation in which the distinction between temper and natural inclusions may be impossible to make objectively: when "sand" is used as temper (cf. Figs. 11.1b and 11.2a). The addition to clays of sand-size rock and mineral fragments derived from unconsolidated sediments—that is, sand temper—essentially nullifies the criteria of greater size, greater angularity, and polymineralic makeup that normally serve to distinguish grit (crushed rock) temper from natural inclusions. In such cases the mineral species and size of each "sand" grain may still be determined and bulk compositional indices (expressed

Figure 11.2. Photomicrographs of various common tempers, all at 10× magnification. (a) *Sand temper* viewed under crossed polars. Note well-rounded character of inclusions. Scale: longest quartz (clear) grain, upper left = .65mm. (b) *Grit (granite) temper* viewed under crossed polars. Note large size, high angularity, and polymineralic (mainly twinned feldspar and quartz) character of larger grains. Scale: largest grain = 1.98mm. (c) *Grog temper* viewed in plane-polarized light. Note greater optical density and sandiness of the large grog grain (Scale: = 1.325mm) compared to the surrounding paste.

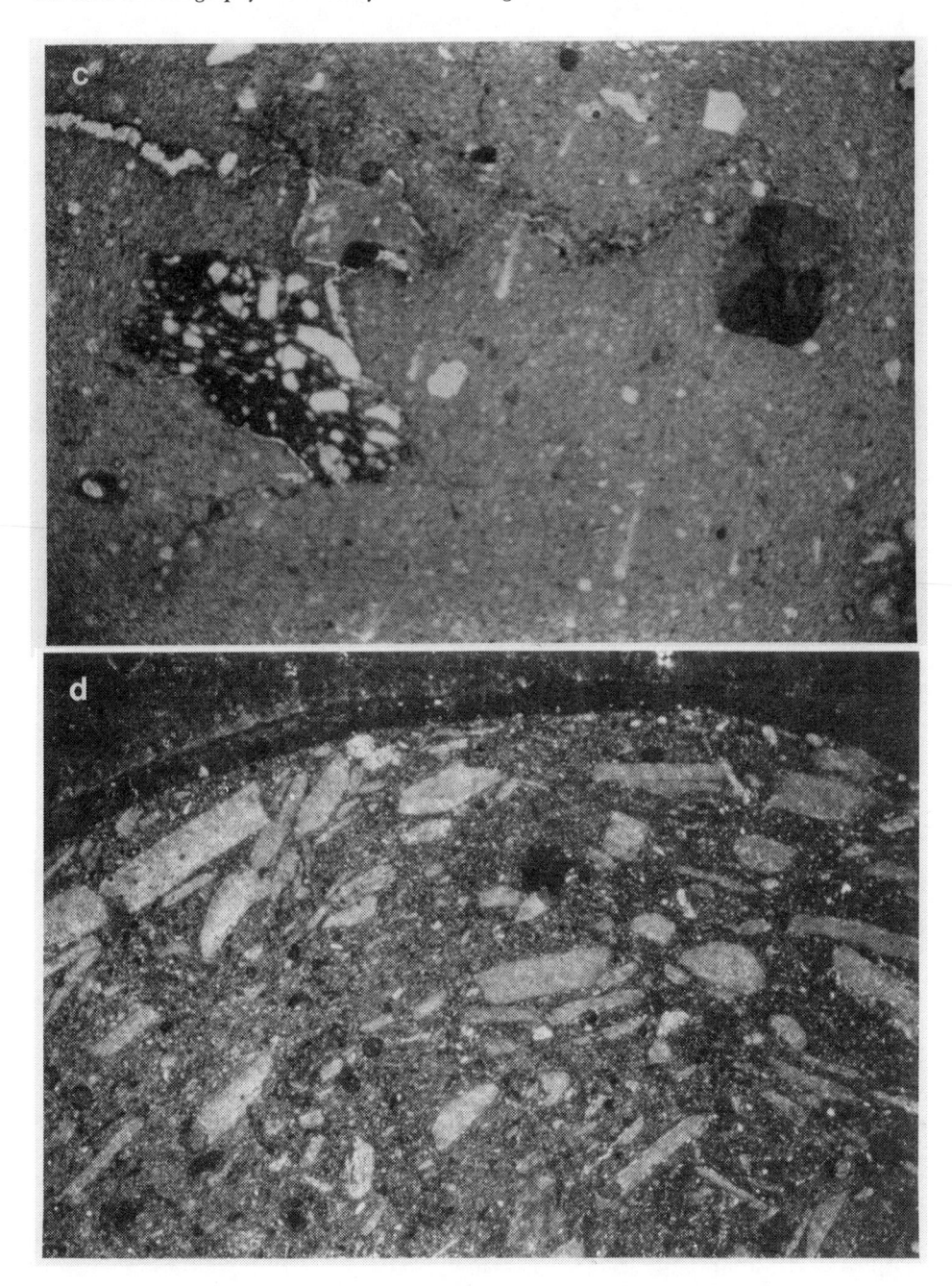

Figure 11.2. Continued. (d) *Shell temper* viewed under crossed polars. This is an imported jar of the type Ramey Incised from the Fred Edwards site similar to lower right vessel in Fig. 11.3. Note the slipped surface visible in upper field. Scale: longest shell grain, upper left = .6mm. (e) *Leached shell temper* viewed in plane-polarized light. This is a locally made vessel from the Fred Edwards site similar to lower left vessel in Fig. 11.3 — note similarities in low-sand, high-silt properties of this paste and the sediment shown in Fig. 11.1a. Scale: longest diagonal shell void = 1.3mm.

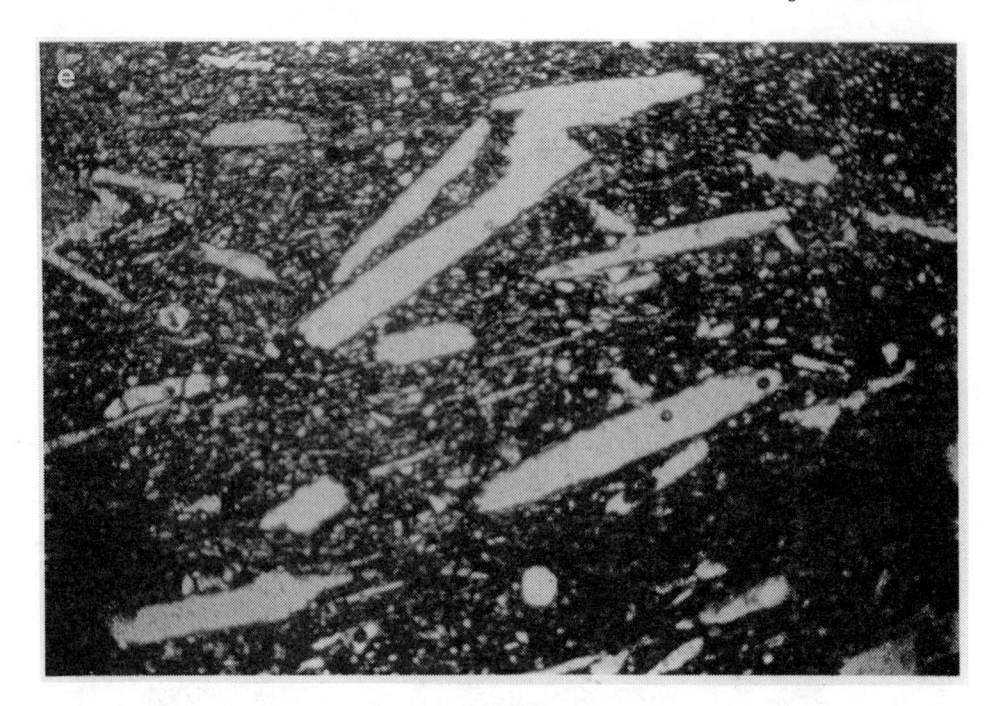

Figure 11.2. Continued.

quantitatively either as volume or percent of grains) can be tabulated to characterize individual thin sections (e.g., Dickinson and Shutler, 1971, 1974, 1979; Lombard, 1987; Miksa and Heidke, 1995), but only a comprehensive comparison of the local clay-rich sediments might reveal whether temper was added or the potters used naturally sandy sediments (see Barnett, 1991).

Besides the use of sand temper, there are two ceramic production processes that can render source determination of pastes an impossible task for petrography: the mixing of clays from various sources and levigation (artificial removal of coarse particles by grinding and/or water separation). Both have been reported among traditional and commercial potters (Rye, 1981:31; Shepard, 1956:52, 182). In each case, but for opposite reasons, that is, enrichment and depletion, the thin sections will reflect mixtures of materials that cannot be matched through direct analysis of sediment samples. Happily, these practices do not appear to have been widely adopted in prehistory.

In the following discussion the value of making the distinction between natural and intentionally added inclusions is illustrated through a number of examples. In order to facilitate this discussion two important concepts, *body* and *paste*, must be introduced. The former pertains to the bulk composition of a vessel, including clays, all coarser natural inclusions, plus any human additives (temper). By contrast, paste pertains only to the mix of natural materials, clays, and coarser inclusions, found in the raw sediments collected by the potters before tempers are added (Fig. 11.1). In a sense paste is everything else after the temper has been excluded, and it is only paste, not body, that can be compared productively to raw sediment samples with any realistic hope of ever being able to match ceramic vessel composition with raw material sources. This is the main problem faced by both of the major elemental analysis techniques, neutron activation (e.g., Arnold et al., 1991:75) and weak acid extraction (Burton and Simon, 1993). For, as techniques that analyze bulk samples of powdered sherds, the results are always primarily a measure of the elemental composition of body, that is, paste + temper, which is a humanly produced mixture that has no precise natural equivalents.

2.2. Quantitative Observations

Although traditionally considered a qualitative technique (mainly for identifying kinds of minerals and rocks), petrography also may be effectively employed to provide a number of quantitative measures of aplastic inclusions observed in ceramic thin sections, including mean grain sizes, the percentage of grains of each mineral species, and the percentage of artifact volume comprised by each mineral species (e.g., see Middleton et al., 1985; Schubert, 1986; Wandibba, 1982). It is important to realize that such quantitative measures may be applied both to paste and to body (e.g., see Stoltman, 1989 and 1991), although far too often no distinction is made between temper and natural inclusions so that one is left to wonder exactly what mineral grains were observed in a petrographic analysis.

There are two methods commonly used to describe the contents of thin sections quantitatively. The simplest but less precise method involves making visual comparisons between thin sections and specially prepared test tiles containing measured amounts of mineral inclusions. For example, Matson (1955:44) illustrated six test tiles containing from 5% to 60% shell and stated that by using such standards one can usually estimate the amount of shell temper in a thin section to within 10% (See also Mathews, et al., 1991; Matson, 1970; Rye, 1981:50–52).

The second method for deriving quantitative data pertaining to the composition of thin sections—referred to as modal analysis by geologists (e.g., Chayes, 1956)—is point counting (Daniels et al., 1968; Galehouse, 1971; Griffiths, 1967). Point counting is a more time-consuming and tedious sampling technique than that using visual estimates and may be done in a number of ways (Galehouse, 1971). The two most common point-counting approaches, the line and the Glagolev–Chayes methods, require a special attachment for the microscope stage that allow one to advance the thin sections beneath the crosshairs at specified intervals (e.g., 0.5 mm or 1 mm). The line method involves advancing the thin section beneath the crosshairs along equally spaced, parallel lines and recording all grains encountered along each line until a preset total is attained, usually 200–400 grains (see Barnett, 1991; Dickinson and Shutler, 1971, 1974, 1979; Ferring and Pertulla, 1987). For most archaeological applications, the grains counted are in the sand size range and may or may not include tempers other than sand (cf., contrast Dickinson and Shutler, 1979, vs. Ferring and Pertulla,

1987). An important limitation of the line method is that it produces number frequencies (of grains) that cannot be precisely correlated with area, volume, weight, or even percentage (see Galehouse, 1971:391–392).

The more common and robust form of point counting, sometimes called the Glagolev–Chayes method (Galehouse, 1971:389), involves, in effect, superimposing a grid over a thin section and recording observations (not just of grains, but of matrix, voids, or whatever appears beneath the crosshairs) at every grid intersection point. In actuality, rather than superimposing a grid onto the thin section, the stage attachment allows one to advance the thin section back and forth beneath the eyepiece crosshair at specified increments at which the desired observations and measurements are made (e.g., Miksa and Heidke, 1995; Schubert, 1986; Stoltman, 1989).

Because of the distinctive character of ceramic thin sections as contrasted with most geologic specimens—they have identifiable mineral grains widely scattered within an amorphous groundmass (see Fig. 11.2)—certain compromises must be made. For example, it is common for geologists to count 1000 or more points on a single mineral/rock thin section (e.g., Hutchison, 1974:58; Nesbitt, 1964). Although such high numbers of counts have also sometimes been done on archaeological ceramics (e.g., Schubert, 1986), in most cases the coarseness of the mineral inclusions mitigates against such large numbers of point counts on ceramic thin sections.

The efficacy of the Glagolev–Chayes method (henceforth referred to simply as point counting) is founded upon the so-called Delesse, or area–volume, relation originally reported by A. Delesse in 1848 and amply confirmed by subsequent research. As noted by Chayes (1956), "...provided certain simple rules of sampling are observed, areal measurements made in thin section are theoretically sound and unimpeachable estimates of volumetric proportions (37–38)."

One of the most important rules of sampling concerns selecting an appropriate sampling interval. Obviously, the smaller the sampling interval, the greater the number of observation points that can be made on a thin section, thus the greater the precision of the analysis. However, a counterconsideration to ensure the randomness of each observation requires that "the point distance chosen should be larger than the largest grain fraction that is to be included in the analysis" (Plas and Tobi, 1965:89). This requirement presents a problem for the ceramic analyst because mineral inclusions in the coarse sand size range (>1.0 mm) and larger are commonplace in archaeological ceramics. For most ceramic thin sections, however, a point-counting interval larger than 1.0 mm would yield so few total points (certainly fewer than 100) that the reliability of the analysis would be questionable.

As a compromise to these competing considerations, I have found that the use of a 1-mm sampling interval can be effectively and reliably utilized to point count archaeological ceramics (Stoltman, 1989). This interval normally ensures a total count of from 100 to 300 points per thin section, which yields results that are reliable within a range of ±3.5% (Stoltman, 1989:150–151). Another advantage of the use of a 1-mm point-counting interval is that it allows one simultaneously to record individual grain counts as well as areal counts. The important consideration here is that, even with a 1-mm interval, coarser grains will often receive multiple counts. This is no problem for volumetric computations—according to the Delesse relation, volume is proportional to area; however, for recording properties of individual grains, such as mineral species, size, and shape, which would then be used to calculate grain class rather than volumetric percentages, each grain must be counted only once. With a 1-mm counting interval, it is relatively easy, even for the coarsest ceramics, to maintain records of all multicount grains and thus to count each such grain as one for grain-class percentages and to use the raw, areal counts for volumes. If a smaller counting interval is used, however, the number of multicount grains and the number of counts per grain for larger grains become so great that it becomes virtually impossible to keep track of individual grains (see Middleton et al., 1985, for a discussion of this issue as it pertains to grain-size distributions in thin sections).

Because of its inherent strengths and limitations, petrographic analysis of archaeological ceramics has understandably focused on the qualitative characterization of temper, that is, intentionally added aplastic inclusions. As the foregoing discussion has emphasized, and as the ensuing discussion documents, distinguishing natural from artificial inclusions and applying quantitative measures to them both can significantly expand the power and scope of petrography in the service of archaeology. Unlike elemental analyses, petrography has the capacity to be more than a bulk-compositional technique. Petrography has the capacity to transcend temper identification, its traditional role in archaeology, and thereby to provide unique and robust data that are directly relevant to a number of archaeological issues, that is, issues concerning past human behavior.

3. Archaeological Problems Amenable to Petrographic Analysis

Basically, petrography is a technique that can reliably identify the kinds and amounts of mineral species of grains in the silt and larger size ranges visible in ceramic thin sections. Presented as a list or an inventory, such data are typically relegated to little-used appendices. Placed within the context of a problem-oriented analysis, however, such data can be of enormous value in addressing at least four ceramic issues of interest to archaeologists—classification, function, production, and exchange. Through the use of selected examples, the relevance of petrography in the service of archaeology is documented for each of the four problem areas in turn.

3.1. Ceramic Classification

Temper is a major, often the dominant, criterion used by archaeologists to distinguish ceramic classes like wares, series, and types, especially in North America (e.g., Griffin, 1950; Hawley, 1950). Thus, the reliable identification of temper types and the objective characterization of ceramic body are minimal requisites for ceramic classes to be useful and reliable tools for archaeological systematics. Yet, despite the importance of temper to ceramic classification and the unexcelled power and precision of petrography to identify reliably aplastic inclusions in pottery, rarely does one encounter the use of petrography to characterize temper types in the process of defining ceramic classes.

Shepard (e.g., 1939, 1942) set a realistic and economical standard for how petrography could be effectively utilized in the domain of ceramic classification in her early studies of Southwestern ceramics. In these pioneering studies she initially identified temper type from petrographic observations of selected sherds and, based on these findings, then used a binocular microscope to assign the remainder of the sherds to the appropriate classes. This approach is still current in the Southwest (e.g., Habicht-Mauche, 1993; Miksa and Heidke, 1995), but is neither as widely nor commonly utilized as is warranted.

James Porter (1962) provided a valuable cautionary tale concerning the relative unreliability of traditional, megascopic (as opposed to petrographic) identification of temper. Based on a megascopic examination of 35 sherds traditionally considered to be grit tempered from four Mississippi Valley sites in Illinois, he concluded that 24 were grit tempered and 9 were grog tempered. Thin sections of the same sherds were then analyzed petrographically with the result being a dramatic reversal of tempers identified: 24 were grog tempered and only 9 were grit tempered. This study dramatizes a real problem that seems to be underappreciated by archaeologists: megascopic temper identifications, unless informed by prior petrographic observations, are apt to be highly subjective, imprecise, and even inaccurate.

As an example of the use of petrography to assist in resolving a taxonomic issue, an analysis of Upper Mississippi Valley ceramics once assigned to a single ware may be cited (Stoltman, 1989). In the 1980s while involved in archaeological research in southwestern Wisconsin we encountered distributional evidence that one ceramic type, Spring Hollow Incised (Logan, 1976), was not associated with the other four ceramic types—Spring Hollow Cordmarked, Spring Hollow Plain, Levsen Stamped, and Levsen Punctuated—that had all been assigned to Linn Ware, a Middle Woodland taxon (Logan, 1976:109). As an independent test of the status of Spring Hollow Incised as a member of this Middle Woodland ware, selected sherds of this type were thin sectioned and the results evaluated under the null hypothesis that their compositions should be closely similar to those of the other Linn Ware types.

Initially, a petrographic analysis was conducted on thin sections derived from a sample of five Spring Hollow Incised vessels and nine undeniable Linn Ware vessels from a single locality. The results of this analysis revealed the former to be sand tempered, with only occasional traces of grit, whereas the latter was grit tempered (Stoltman, 1989). Confirmation of this difference was sought by adding 11 additional vessels, six of which were of types formerly unclassified but now suspected (based on new contextual evidence) as being companion types of Spring Hollow Incised. The combined data set, consisting of 11 classic Linn Ware vessels versus 14 Spring Hollow Incised and Spring Hollow Incised-associated vessels revealed conclusively that two distinct ceramic bodies—one grit tempered and the other sand tempered—were represented (cf. Figs. 11.2a and 11.2b).

The possibility that the observed compositional differences could be attributed to functional considerations on the part of a single group of potters is an alternate explanation that could not be discounted on the basis of the petrographic evidence alone. However, the results of the petrographic analysis in combination with the contextual evidence that originally inspired this inquiry, and has subsequently become even more compelling (e.g., Stoltman, 1986), leave no reasonable doubt that Spring Hollow Incised and its newly recognized companion types represent an entirely different grouping from Linn Ware. As a result, Prairie Ware, a new Early Woodland ceramic series was defined, whereas the type Spring Hollow Incised was redesignated Prairie Incised in order to reflect its simultaneous deletion from Linn Ware (whose Middle Woodland affiliation remains in tact) and its inclusion within a Early Woodland taxon. In this case ceramic petrography provided evidence that was pivotal not only to the reclassification of a known ceramic type, but to the refinement of a local cultural sequence resulting in a more complete and accurate conceptual framework for interpreting the prehistory of the Upper Mississippi Valley. This example should demonstrate that classification in archaeology is far more important than simply name calling.

3.2. Ceramic Engineering/Functional Considerations

A second problem area to which petrography can make significant contributions concerns the identification of the different ways that potters may design or engineer their products to perform different tasks: ceramic body is the index of primary consideration here. For the most part, the traditional archaeological literature prior to the 1970s treated ceramic variability as if it were primarily stylistic—a reflection of shared cultural values—and used it primarily to address issues of time–space systematics. Under this paradigm, what Rye (1976, p. 106) has referred to as "cultural determinism," the role of "Pots as Tools" (Braun, 1983) was generally overlooked.

In the 1970s and 1980s archaeologists began to investigate seriously the relationship between vessel composition and performance under "the functional hypothesis" that potters sometimes engineered their products in order to improve performance characteristics. Drawing inspiration from the material sciences, these studies have attempted to understand how the performance characteristics of ceramic vessels are affected by the kinds, amount, and sizes of temper (e.g., Bronitsky, 1986; Bronitsky and Hamer, 1986; Ericson et al., 1972; Rye, 1976; Steponaitis, 1984). For the most part, this research has involved the experimental manipulation of different materials, as well as their amounts and sizes, in the laboratory and then the testing of the various byproducts to evaluate potentially significant performance characteristics such as thermal shock resistance, impact resistance, strength, and porosity (see also Budak, 1991; Feathers, 1989; Hoard et al., 1995; Neupert, 1994). This valuable research provides both theoretical and empirical bases for interpreting characteristics of ceramic materials in functional terms. Based on this foundation, petrographic analysis is well suited to continue to expand our insights into prehistoric ceramic technologies.

3.2.1. Temper Type as a Functional Variable

In a pioneering study of La Plata Valley ceramics of southwestern Colorado and adjacent New Mexico, Shepard (1939) employed petrography to investigate temper differences between two classes of pottery that were distinguished on functional grounds—culinary versus nonculinary wares. The functional underpinnings of her analysis, however, went generally unnoticed as attention focused on her controversial view, which eventually expanded to include Chaco Canyon farther south in New Mexico, that the temper differences she observed—grog in the nonculinary wares and igneous rock in the culinary wares (e.g., Figs. 11.2b and 11.2c) in some Pueblo II and III contexts—were best attributed to exchange from specialized production centers (Shepard, 1939:277–281, 1954; see also Stoltman, 1999).

What Shepard (1939) observed in her petrographic analyses of La Plata and Chaco ceramics—that functionally distinct ceramic types may be tempered with different materials within a single community—has been duplicated in a number of other Southwestern settings since that time (e.g., Rugge, 1976; Stoltman, et al., 1992; Stoltman, 1996). However, still unconfirmed is the functional alternative that underlies these observations; namely, that no matter where they were produced, culinary and nonculinary wares may have been made from different materials because they had been specially engineered to perform different tasks.

3.2.2. Temper Sizes and Amounts as Functional Variables

Rather than relying on different tempering materials as a criterion for identifying different functional classes of ceramic vessels, different sizes and amounts of temper might provide a more explicit test of the functional hypothesis. One of the few applications of petrography to the issue of temper size and amount as related to functional variation within a ceramic assemblage is Steponaitis' (1984) study of shell-tempered ceramics from Moundville, Alabama. Whereas most of this study involved megascopic measurement of grain sizes and physical testing of sherds in the laboratory, the percentage of shell by volume was determined for four jars and six bowls/bottles by analyzing ten thin sections. By comparing the temper percentages obtained from the microscopic analysis with the megascopic grain size determinations, he found that, with some overlap, jars generally had more abundant and coarser temper than did bowls and bottles. Employing basic materials sciences principles, Steponaitis (1984:115) interpreted these differences as "fundamentally technological, rather than stylistic, in nature." In particular he suggested that the cooking vessels (i.e., the jars) had more abundant and coarser shell as an effective way to maximize resistance to crack propagation induced by thermal shock (Steponaitis, 1984:106).

Later experiments with shell-tempered test tiles by Bronitsky and Hamer (1986:96–97) supported this conclusion, which makes it tempting to generalize these findings farther afield. Although both the theoretical and experimental data seem to warrant the conclusion that the use of higher percentages and coarser grades of shell temper will enhance the thermal shock resistance of a ceramic body, it must be remembered that a third variable, temper type, is also involved in this relationship.

Shell, unlike most other tempering materials, has a number of unique properties that may override the variables of size and amount in resisting thermal shock-induced crack propagation. As Steponaitis (1984:112–113) pointed out, not only does shell possess claylike thermal expansion properties, but it also fractures readily into thin, broad plates with large surface areas per unit of volume (See Figs. 11.2d and 11.2e). These two properties combine to provide greater resistance to crack propagation than is true for such other common tempering materials as grit, grog, or sand, all of which normally occur as, or fracture into, equiaxial grains having surface areas that are relatively small compared to volume (see also Shepard, 1956:27). Until carefully controlled experimental tests are conducted on various grits and grog—most of the experiments on the effects of tempers on fired ceramic bodies so far pertain to shell and sand—it would be premature to assume that larger, coarser temper other than shell can be expected to enhance the thermal shock resistance of fired clay bodies.

Although there is little experimental evidence available pertaining to performance characteristics of ceramic vessels with nonshell tempers, there is some empirical evidence to suggest that different sizes and amounts of such tempers may be associated with functionally distinct ceramic classes. For example, Garrett's (1986) petrographic analysis of Black Mesa ceramics in northern Arizona shows that the main utilitarian grayware type of Pueblo II times had both coarser and more abundant sand temper than the three main black-on-white painted types of the same period. Similarly, the utilitarian brownwares of Wind Mountain in southwestern New Mexico had more and coarser grit temper on average than did the painted vessels during the Mimbres occupation of the site (Stoltman, 1996:370).

Because so many variables besides temper size and amount are involved in each of these cases, it is difficult to draw firm conclusions about the functional hypothesis (see also Rice, 1996a:142). On the one hand it is important to remember Rye's admonition (1976:109) that there are multiple ceramic properties that potters typically attempt to control and that the temper employed, including its size and amount, may reflect a concern, say, for enhancing workability rather than firing behavior. On the other hand, the "cultural hypothesis"— that the distinction between coarser and finer bodies is attributable to effort minimization on the part of the potters who expended less time and effort in temper preparation for coarse wares (e.g., Steponaitis, 1984:86)—cannot be categorically dismissed. Until better experimental evidence is available concerning the technological characteristics of grit and grog tempers in particular, it is perhaps premature to accept too readily the functional hypothesis for the seemingly widespread association between coarse and abundant temper and "utilitarian" pottery containers. However, no matter the outcome, it should be evident that ceramic petrography has a pivotal role to play in generating solid empirical data relevant to the issue of ceramic composition as an expression of functional considerations on the part of potters.

3.3. Ceramic Production

The third problem toward which petrography has an important contribution concerns ceramic production, particularly location and specialization. The first step on recognition of a new ceramic type should be to demonstrate that it was locally produced. Petrography can add considerable weight to the inference of local production (which is usually just assumed on stylistic grounds) by documenting whether the vessels' paste and temper were locally available materials. The identification of specialized ceramic production is a trickier issue, but (in the absence of direct evidence of production facilities like kilns) it is often founded on the debated supposition that product uniformity covaries with the numbers of producers within a community (Rice, 1996b). The objective measurement of paste and body properties of vessels from a suspected production locus is well within the capacity of petrography and, when presented quantitatively and used comparatively, can provide robust data concerning the relative uniformity/variability of ceramic products. Each of these two aspects of ceramic production are addressed in turn in order to make explicit the relevance of petrography to them.

3.3.1. Location of Production

To demonstrate that a ceramic vessel or type was locally produced, any of four independent petrographically based approaches can be employed:

1. Compare mineral inclusions in vessel bodies to locally available sands, minerals, and rocks
2. Compare vessel pastes to local sediments
3. Compare tempers, bodies, and/or pastes of vessels being investigated to those of a sample of accepted locally produced ceramics
4. Compare between-site variation in tempers, bodies, or pastes of a single class of vessels.

The first two tests are analogous to what is sometimes referred to as "the provenience postulate" under which artifact sources are identified by matching artifact materials with local raw materials (e.g., Barnett, 1991; Bishop, 1980; Lombard, 1987; Miksa and Heidke, 1995; Weigand et al., 1977:24). By contrast, the third test may be referred to as the "local-products-match postulate" because the items being compared include no raw materials but only manufactured products (i.e., ceramic vessels vs. other ceramic vessels). The fourth test, the "spatial pattern postulate," also involves vessel-to-vessel comparisons across space, but only within a single ceramic class. In each of the first three tests the null hypothesis is that if locally produced, the vessels under investigation will differ in no significant way either qualitatively or quantitatively from local materials or products. Under the spatial pattern postulate, local production of a ceramic class is accepted if vessel compositions display a pattern of within-site homogeneity along with between-site heterogeneity.

3.3.1.a. The Provenience Postulate and Petrography. In order to illustrate the use of petrography and the provenience postulate to verify the local production of ceramics the case of the Fred Edwards site, a hypothesized site-unit intrusion into southwestern Wisconsin is considered (Finney and Stoltman, 1991). The ceramic assemblage at this site contains an unusual diversity of ceramic types (See Fig. 11.3), the preponderance of which consists of grit-tempered (with hematite), cord-marked, and often cord-decorated types (eventually assigned to a new series, Grant) that were previously unknown in Wisconsin. Stylistically, these vessels closely resemble the Late Woodland Canton Ware of northwestern Illinois (Fowler, 1955), thus raising doubts concerning their local production. Intermixed in the assemblage were lesser numbers of smooth-surfaced, shell-tempered, Middle Mississippian vessels of the types Powell Plain and Ramey Incised that are associated with the American Bottom culture of southern Illinois (Griffin, 1949; Bareis and Porter, 1984) and were suspected imports. In addition to these two broad groups there were a number of vessels that had smooth surfaces, hematite temper, and distinctive Powell/Ramey sharp-shouldered jar forms that obviously reflected a blending of Late Woodland and Mississippian practices (these, too, were eventually assigned to a new series, Potosi).

Figure 11.3. Various ceramic types from the Fred Edwards site. (*top left and top center*) Locally made Grant Cord-Impressed, a Late Woodland type with grit temper; (*top right*) Potosi Plain; grit-tempered, plain Powell/Ramey-like jar made of local materials; (*lower left*) locally made Powell Plain jar; unslipped, shell tempered, and made from local sediments; (*lower center*) imported Powell Plain jar; slipped, shell tempered, and with a nonlocal paste; (*lower right*) imported Ramey Incised jar; slipped, shell tempered, and with a nonlocal paste.

Table 11.1. Mean Body and Paste Values for Ceramic Vessels from Fred Edwards by Series

Type	N	% Matrix	% Sand	% Temper	Temper type	Temper size index*
			BODY			
Grant Series	12	80.0 ± 5.2	2.0 ± 1.2	18,0 ± 5.0	Hematite	3.22 ± .50
Potosi Plain	4	80.0 ± 2.9	1.0 ± 8	19.0 ± 3.2	Hematite	3.26 ± .50
Local Powell Plain	6	87.7 ± 4.4	1.5 ± 1.4	10.8 ± 5.0	Shell	3.52 ± .53
Exotic Powell/Ramey	5	71.8 ± 4.8	0	28.2 ± 4.8	Shell	2.17 ± .36
Local subsoil clay	1	98	2	0	—	—

Type	N	% Matrix	% Silt	% Sand	Sand Size Index*
			PASTE		
Grant Series	12	78.5 ± 3.6	19.2 ± 3.2	2.3 ± 1.4	1.27 ± .40
Potosi Plain	4	82.5 ± 2.4	16.25 ± 2.1	1.25 ± 1.0	1.00 ± 0
Local Powell Plain	6	83.0 ± 4.4	15.3 ± 5.2	1.7 ± 1.4	1.08 ± .65
Exotic Powell/Ramey	5	96.8 ± 1.3	3.2 ± 1.3	0	—
Local subsoil clay	1	80	18	2	1.25

*Size index is a mean value for all measured grains of sand and gravel sizes using an ordinal scale of 1–5.

A petrographic analysis of these three ceramic classes was conducted and the results were compared to local materials to ascertain which of them was locally produced (Stoltman, 1991). The quantitative data derived from this analysis were obtained by point counting using a 1 mm interval. These data are presented in two parts in Table 11.1, one for body and one for paste. Body in Table 11.1 (and in Table 11.2 that follows) is a volumetric measure of bulk vessel composition that is composed of three parts: percentage matrix (clay and silt fractions), percentage sand (natural mineral inclusions in sand and gravel sizes), and percentage temper (all humanly added inclusions, whatever the sizes). Paste, similarly, is a three-part volumetric index but with temper excluded; it is expressed as percentage matrix (the clay fraction only), percentage silt, and percentage sand. Size indices are also recorded separately in Tables 11.1 and 11.2 for temper (under body) and for natural sand inclusions (under paste). This particular size index is an ordinal scale ranging in value from 1 to 5. It was computed for each thin section by assigning all grains recorded in the point counts to size classes based on maximum diameters as follows: (1) 0.0625 to 0.249 mm; (2) 0.25 to 0.499 mm; (3) 0.50 to 0.99 mm; (4) 1.00 to 1.99 mm; (5) greater than 2.00 mm. The index itself is a single number that reflects the mean grain size in this ordinal scale for temper and for sand for each thin section or ceramic class.

Because both hematite and shell were recovered in context on the site, there can be no doubt that the main materials used as temper at Fred Edwards had been locally procured. Besides the tempers, local sediments were collected on and around the site for comparison with the vessel pastes.

As can be seen from Table 11.1, a subsoil sample from the site (Fig. 11.1a) has a paste index—80% matrix, 18% silt, and 2% sand—that is virtually identical

The Role of Petrography in the Study of Archaeological Ceramics

Table 11.2. Mean Body and Paste Values for Powell Plain/Ramey Incised Vellels from Cahokia Tract 15A and Four Northern Sites

Type	N	% Matrix	% Sand	% Temper	Variability index	Temper type	Temper size index*
			BODY				
Cahokia Powell/Ramey							
Tract 15A	16	75.6 ± 7.7	0.8 ± 0.7	23.6 ± 8.0	5.47	Shell	2.23 ± .35
Exotic ? Powell/Ramey							
Rench, IL	5	73.0 ± 4.8	0.4 ± 0.6	26.6 ± 5.2	3.53	Shell	2.73 ± .26
Hartley Fort, IA	1	71.0	1.0	28.0	—	Shell	2.46
Fred Edwards, WI	5	71.8 ± 4.8	0 ± 0	28.2 ± 4.8	3.20	Shell	2.17 ± .36
Aztalan, WI	5	71.2 ± 2.3	0.6 ± 0.6	28.2 ± 2.4	1.77	Shell	2.05 ± .26
Mean, Exotic Vessels	16	71.9 ± 3.8	0.4 ± 0.5	27.7 ± 3.9	2.73		2.32 ± .40

Site	N	% Matrix	% Silt	% Sand	Variability index	Sand Size index
			PASTE			
Cahokia Powell/Ramey						
Tract 15A	16	95.5 ± 3.5	3.5 ± 3.1	1.0 ± .8	2.53	1.09 ± .30
Exotic ? Powell /Ramey						
Rench, IL	5	95.2 ± 3.1	4.4 ± 3.0	0.4 ± 0.6	2.23	1.00 ± .0
Hartley Fort, IA	1	94.0	5.0	1.0	—	1.00
Fred Edwards, WI	5	96.8 ± 1.3	3.2 ± 1.3	0 ± 0	0.87	—
Aztalan, WI	5	96.8 ± 1.3	2.6 ± 1.5	0.6 ± 0.6	1.13	1.00 ± 0
Mean, Exotic Vessels	16	96.1 ± 2.1	3.5 ± 2.0	0.4 ± 0.5	1.53	1.00 ± 0
Local Late Woodland						
Rench, IL.	5	71.8 ± 3.0	7.8 ± 2.6	20.4 ± 3.0	2.87	1.49 ± .19
Hartley Fort, IA	14	76.4 ± 5.6	12.2 ± 4.1	11.4 ± 3.5	4.40	2.01 ± .29
Fred Edwards, WI*	16	79.5 ± 3.7	18.4 ± 3.2	2.1 ± 1.3	2.73	1.14 ± .47
Aztalan, WI	12	85.8 ± 7.3	6.2 ± 3.7	8.0 ± 6.6	5.87	1.41 ± .34

*Local Late Woodland Vessels from the Four Northern Sites also included. Includes both Grant Series and Potosi Plain vessels recorded in Table 11.1

to the mean for the twelve Grant Series vessels selected for thin sectioning. In terms of both temper and paste the Grant Series can be seen to have been manufactured from local materials. Figure 11.4 provides visual support for this conclusion, showing the paste values of the twelve Grant Series vessels arrayed in a ring around the site's subsoil sample. Similarly, the paste values for four Potosi Plain vessels, along with their hematite temper, suggest that they were also manufactured locally (Table 11.1; Fig. 11.4).

An unexpected outcome of the comparative analysis of the shell-tempered vessels with the local ceramics and sediments was the discovery that some were apparently made of local materials whereas others were clearly different. Based on the paste indices, it was concluded that six of the eleven shell-tempered vessels were actually made from Fred Edwards soils (Table 11.1; Fig. 11.4); whereas five of these vessels, which have no sand and only about 3% silt (as contrasted with

50% SILT

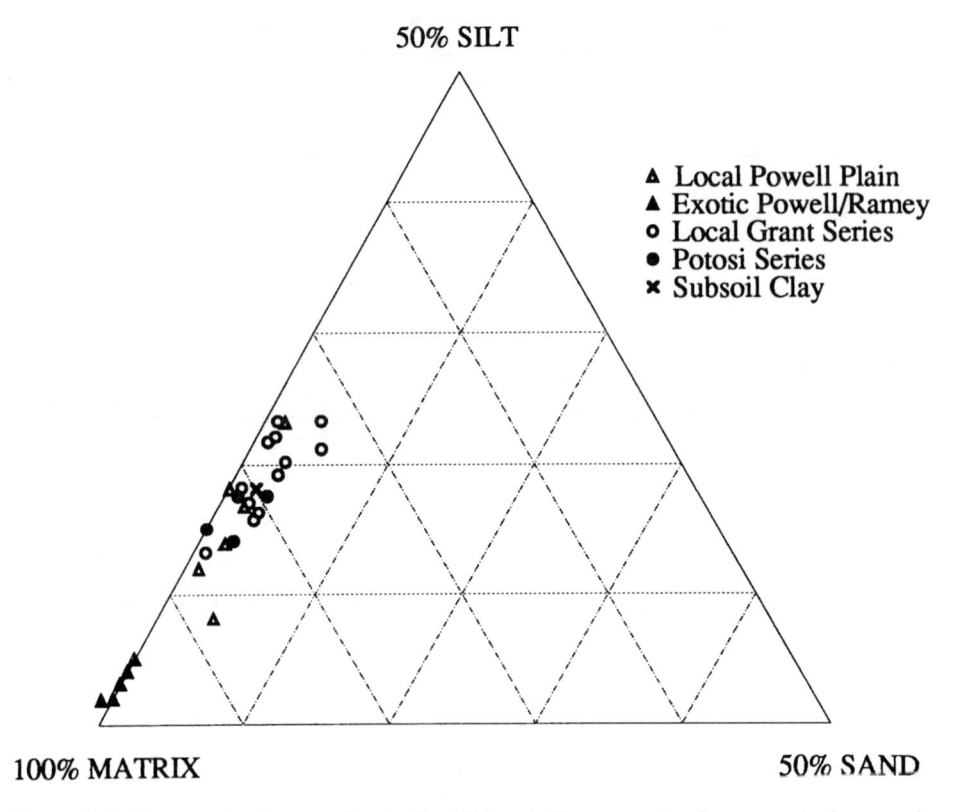

▲ Local Powell Plain
▲ Exotic Powell/Ramey
○ Local Grant Series
● Potosi Series
✕ Subsoil Clay

100% MATRIX   **50% SAND**

Figure 11.4. Ternary plot for paste for the Fred Edwards Site comparing four ceramic classes and a local subsurface clay.

15–20% silt for the local products) most likely were imported. Further support for this inference of separate origins for the two classes of shell-tempered vessels can be seen in Fig. 11.5, which shows a mutually exclusive distribution of the body values: the "local" vessels have less than 20% temper whereas all "exotic" vessels have more than 20%. Moreover, the temper size indices (Table 11.1) are also mutually exclusive ($3.52 \pm .53$ vs. $2.17 \pm .36$), consistent with the inference that the two classes of shell-tempered pottery had been manufactured by different potters using different recipes (see Stoltman, 1991:108–109, for a more complete discussion of this particular size index).

3.3.1.b. The Local-Products-Match Postulate and Petrography. When only sherds, but no local raw materials, are available for petrographic analysis (as is common with long-curated collections), the provenience postulate cannot be applied. In such cases the local-products-match postulate can still be employed. Under this postulate, a local derivation for a problematic vessel or type is accepted if the paste, temper, and body values of the specimens under investigation match those of a sample of demonstrably local vessels (see Stoltman, 1992, Stoltman and Snow, 1998). In the case of Fred Edwards, even without the local soil sample, the inference that the Potosi Plain and "local" Powell Plain vessels were made locally is consistent with the expectations of the local-products-match

50% TEMPER

▲ Local Powell Plain
▲ Exotic Powell/Ramey

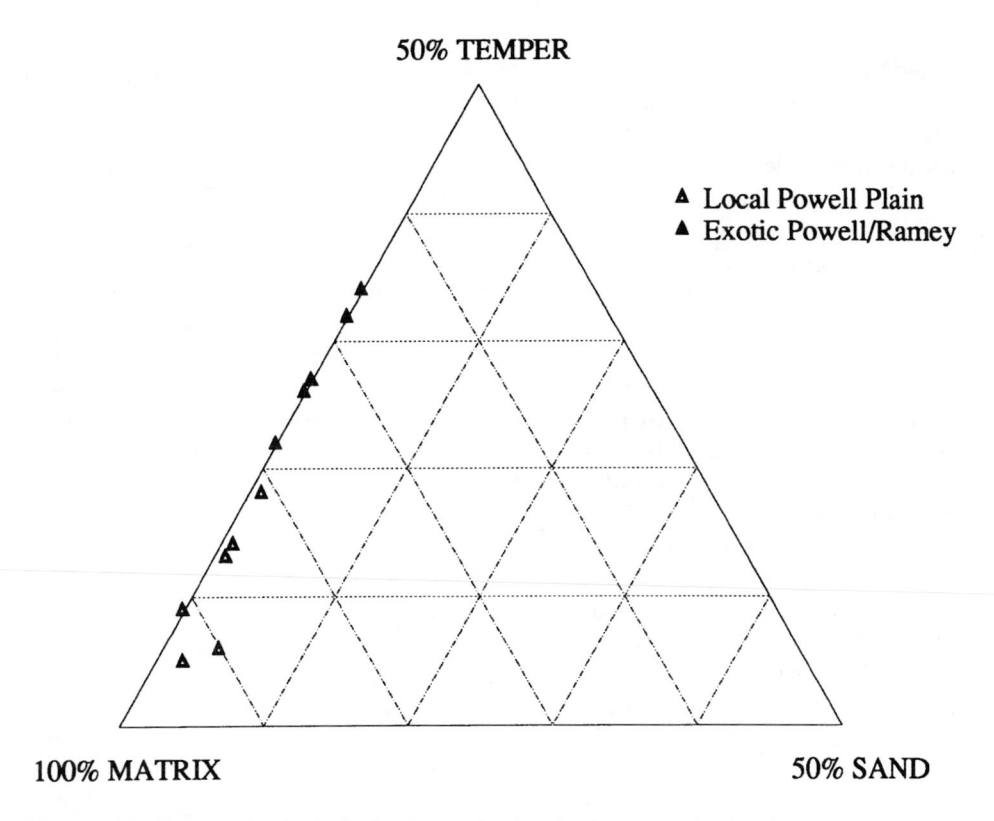

100% MATRIX   50% SAND

Figure 11.5. Ternary plot for body for the Fred Edwards Site comparing local versus exotic shell-tempered jars.

postulate because their basic properties so closely resemble those of the twelve Grant Series vessels whose local derivation seems assured (see Fig. 11.4).

3.3.1.c. The Spatial Patterning Postulate and Petrography. In a unique study, Ferring and Pertulla (1987) employed petrography to evaluate the compositional variability of a single ceramic type—Sanders Plain—in Texas and in Oklahoma. Using a sample of 28 vessels from 14 sites distributed across six drainage basins, they observed that the sands in these vessels had compositions that were similar within each drainage and dissimilar between drainages. On this basis—using what is here termed the spatial patterning postulate—they concluded that local production better accounted for the observed patterning than limited production and exchange. This study differs from applications of the provenience and local-products-match postulates in that no other materials, whether sediments, tempers, or other vessels, were analyzed petrographically for comparative purposes.

3.3.2. Specialization of Ceramic Production

The uniformity of ceramic products is a much-debated indicator of specialized ceramic production (Arnold and Nieves 1992; Rice, 1981, 1989, 1996b). As Rice

(1996b:178) noted, "Most of the analyses that have been carried out to date have relied on dimensional attributes [i.e., measurements of various attributes of vessel form] as indicators of standardized production." Without attempting to resolve this debate here, it is important to point out that the relative rarity of relevant compositional data is a notable shortcoming in the debate and that petrography can play an important role in rectifying the situation.

As an example of how petrography can provide data pertinent to the issue of compositional uniformity of ceramics, an ongoing study of possible trade vessels from the American Bottom of southern Illinois and four sites to the north in Illinois, Iowa, and Wisconsin are cited (Table 11.2). At present the sample contains a total of 32 vessels of the related types Powell Plain and Ramey Incised, 16 from Tract 15A at Cahokia and 16 from the Rench (Illinois Valley), Hartley Fort (northeast Iowa), Fred Edwards (southwest Wisconsin), and Aztalan (south-central Wisconsin) sites. In addition, 47 locally produced, grit-tempered, Late Woodland vessels from each of the four northern sites are included for comparative purposes.

Table 11.2 records not only the mean body and paste values for each of the five sites but also a variability index for each site by ceramic type. This index was created to contrast the relative amount of variability in the body and paste indices of a ceramic class from a site with those of comparable ceramics from other sites. This index is computed by summing the standard deviation values for each of the three components of the body (matrix/sand/temper) and paste (matrix/silt/sand) indices and then dividing by three to provide a mean standard deviation value for each site sample. This index suffers the same limitation pointed out by Arnold and Nieves (1992) for coefficients of variability, namely that no significance levels can be determined to aid in assessing between-site differences. Nonetheless, because it is based on standard deviations, it should be a realistic and explicit gauge of relative sample variability, which in these cases is always constrained by the use of percentage data.

Examining the variability indices presented in Table 11.2, a common pattern is evident for the four northern sites; namely, both the bodies and the pastes of the grit-tempered Late Woodland vessels have higher indices (i.e., are more variable) than those of the shell-tempered Mississippian vessels from the same site. More interesting, however, are the comparative indices of the shell-tempered vessels from the four northern sites, where these vessels are presumably exotic, versus Tract 15A at Cahokia, where these vessels were certainly local products. As can be seen from Table 11.2, variability indices for the Powell/Ramey vessels from Tract 15A are higher for both body and paste than are the comparable indices for Powell/Ramey vessels from all four northern sites combined—5.47 vs. 2.73 for body and 2.53 vs. 1.53 for paste. Accepting that the Powell/Ramey vessels were not locally made at the four northern sites (in all four cases they are minority types in assemblages dominated by grit-tempered ceramics), the relative homogeneity of the Mississippian ceramics of the Powell Plain/Ramey Incised types at the four northern sites (whose linear distances apart range from 60 km to over 275 km) strongly suggests a common origin. These data—greater within-site materials variability in Powell/Ramey ceramics at Tract 15A than exists among the four northern sites—raises the interesting implication that a single, not-yet-recognized production center may exist somewhere within the American Bottom. Although these data are only suggestive, hopefully they illustrate the potential for petrography to provide compositional data relevant to the issue of limited/specialized ceramic production.

3.4. Ceramic Exchange

Shepard's classic 1936 analysis of the ceramics of Pecos pueblo, which focused on the issue of exchange, first brought ceramic petrography to the attention of archaeologists generally (see also Shepard, 1942). Her study, like a supernova, burst on the scene, and nothing quite like it has appeared since. Based on the petrographic identification of rock tempers, which she correctly recognized as locally unavailable, she concluded that during Glaze I times, "Pecos potters had apparently ceased to make decorated ware," instead importing these wares from the Galisteo and Santa Fe regions to the west (Shepard, 1936:579). A veritable bombshell, this conclusion was "difficult to accept in view of the generally prevalent concept of the pueblos as independent economic units. The findings suggest a degree of specialization and industrialization which had not been anticipated" (Shepard, 1936:581).

The issue of exchange is really the "flip side of the coin" of local ceramic production and can be addressed similarly, that is, by applying any or all of the provenience, local-products-match, and spatial patterning postulates. The provenience postulate is rejected when the temper and/or the paste of the suspected exotic vessel or type fail to match locally available materials, whereas the local-products-match postulate is rejected when there is no match between the paste and body properties of the suspected exotic vessel or type and a sample of demonstrably local vessels. The spatial patterning postulate is rejected when the observed variation in composition fails to show marked between-site heterogeneity within a ceramic class expected under conditions of local production, as is the case for the Powell/Ramey ceramic vessels from the four sites in the northern hinterlands of Cahokia discussed previously (see Table 11.2).

In Shepard's (1936) case only that variant of the provenience postulate pertaining to tempers was applied (see also Peacock, 1968, 1969, for similar studies of ceramic exchange in England). A much stronger case for exchange can be made if pastes are also considered and if the local-products-match and spatial patterning postulates are also tested. However, no matter how many of these hypotheses are tested and rejected, a nagging question normally persists in such studies; namely, was the full range of locally available materials sampled? As with any tested hypotheses in science, we accept positive tests only tentatively, for "We never reach certainty, but our theories become more and more probable all the time" (Kemeny, 1959:120).

Assuming that testing on the basis of the provenience, local-products-match, and/or spatial patterning postulates leads one to reject the hypothesis of local production in a specific case, it is both possible and desirable to evaluate this hypothesis further by seeking a compositional match with an external source. In

Figure 11.6. Hartley Cross-Hatched vessels. (*above*) Two from the Hartley Fort site; (*below*) vessel recovered at the Fred Edwards site.

this way one can move from the purely negative posture of asserting that a vessel or type was not locally produced to a more positive suggestion concerning where the suspected exotic vessel or type was produced.

To accomplish this objective either or both of the provenience and local-products-match postulates can be applied, this time to local materials and/or vessels from sites suspected to have produced the postulated exotic items in what can be termed two-sided tests. To exemplify how this can be done let us return to the case of the Fred Edwards site.

Besides the shell-tempered, Mississippian vessels, a number of grit-tempered, Woodland vessels were also recovered at Fred Edwards that on stylistic grounds appeared to be nonlocal products. On reviewing the literature and then the actual collections, a close stylistic match for some of the suspected exotic vessels from Fred Edwards was observed for the Hartley Fort site located ca. 100 km away in northeastern Iowa (McKusick, 1964; Tiffany, 1982; Fig. 11.6). A sample of nine Woodland vessels from Hartley Fort was obtained from the University of Iowa for comparative thin section analysis. Employing the local-products-match postulate in two-sided fashion, thin sections from 5 of the suspected exotic vessels from Fred Edwards were compared with those from the 16 local Late Woodland vessels from Fred Edwards and the 9 local Late Woodland vessels from Hartley Fort. The results of this analysis revealed mutually exclusive paste and body indices for the two sites and a close match for the

The Role of Petrography in the Study of Archaeological Ceramics

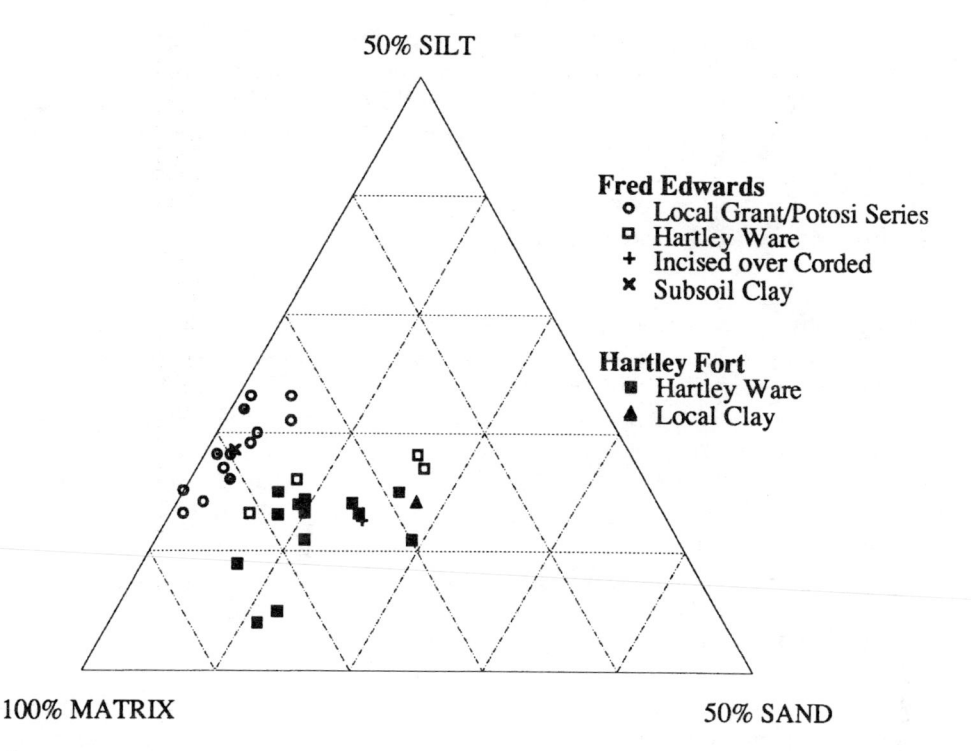

Figure 11.7. Ternary plot for paste comparing local ceramics and soils from the Fred Edwards and Hartley Fort sites along with five suspected trade vessels from Hartley Fort that were recovered at Fred Edwards (cf. Figs. 11.1a and 11.1b, noting the greater siltiness and lesser sandiness of the Fred Edwards soil).

suspected exotic vessels from Fred Edwards only with Hartley Fort, suggesting strongly that these five vessels were, indeed, imports from Hartley Fort (Stoltman, 1991:112–113; Fig. 11.7).

One loose end in this study involved the unavailability of local raw materials from Hartley Fort. Thus, the provenience postulate could not be applied as had been done for Fred Edwards where the local subsoil clays had proved to be virtually identical to the paste of the locally produced ceramics (Stoltman, 1991:113). This situation was rectified a few years later when subsoil sediment samples were collected from Hartley Fort (Fig. 11.1b). Figure 11.7 shows the results of the two-sided comparative analyses of all Late Woodland vessels and local soils from both Fred Edward and Hartley Fort (four additional thin sections from Hartley Fort brought its local-vessel sample to 13). As can be seen from the mutually exclusive clusters of vessels and soils by site in Fig. 11.7, the expectations of a two-sided application of the provenience postulate are met unambiguously. In addition, the five stylistically "exotic" vessels recovered at Fred Edwards are shown convincingly to conform to the expectations of both the provenience and local-products-match postulates that they were produced at Hartley Fort. Some unexpected qualitative evidence provides further support for this view: chert grains were observed to be present in the pastes of 12 of the 13 Hartley Fort vessels, were present in the Hartley Fort subsoil clay, and were also observed in all five of the exotic vessels recovered at Fred Edwards, whereas none of the 16 local Fred Edwards vessels had a single grain of chert in their pastes.

4. Summary and Conclusions

Petrography is a venerable geological technique well known to archaeologists for its unrivaled precision in identifying the kinds of tempers employed by ancient potters. However, the qualitative identification of ceramic tempers hardly exhausts the archaeological potential of petrographic analysis. Indeed, temper identification, rather than being a goal, is better regarded as the initial step in the compositional analysis of archaeological ceramics that can lead to valuable insights into the broad sociocultural context in which the pottery was produced.

In this chapter the contributions of ceramic petrography to four major topics of interest to archaeologists—classification, function, production, and exchange—were outlined and documented with examples. Although in most cases the first analytical step involves the qualitative identification of temper, it should be clear that further analytical steps—which can provide reliable estimates of the sizes and amounts not only of tempers, but also of the natural mineral inclusions in the pastes of ceramic vessels—are both feasible and uniquely informative. By distinguishing paste from body and employing quantitative measures of each, it is possible for petrographers to generate data that are simultaneously readily intelligible to archaeologists and sensitive indicators of past human behavior.

The advent of sophisticated new technologies that provide a comprehensive inventory of the elemental composition of ceramic vessels is a relatively recent innovation in ceramic compositional analysis (e.g., Bishop et al., 1982; Neff, 1992) that has perhaps changed the popular perception of the continued usefulness of such an "old-fashioned" technique as petrography. A major goal of this chapter is to demonstrate that it is too soon to consider the petrographic microscope to be an outmoded instrument in the service of archaeology.

Of primary importance in appreciating the differences in the contributions that the two techniques can make is the distinction between body and paste: elemental analyses, whether neutron activation or weak acid extraction, provide signatures derived from the former, whereas petrography has the potential to characterize the two parameters independently. More important, temper is not the sole source of "noise" (e.g., Bishop, 1980) in elemental analyses, for these techniques are so powerful that they can also be expected to record elements that were added post depositionally. Cooking residues can, and normally are, readily removed during sample preparation, but that is not the case for soluble elements contained in groundwater. These latter, diagenetic elements are difficult if not impossible to identify as such, but they are extremely important to the issue of ceramic exchange because they have the capacity to distort the chemical composition to the point of unrecognition for vessels produced in one setting but traded and deposited in another. For example, in most neutron activation analyses a significant number of vessels inevitably are "ungrouped," "outliers," or "unassigned" (e.g., Mainfort et al., 1997; Steponaitis et al., 1996), and it seems reasonable to suspect that some of these are actually trade vessels whose chemical signatures have been altered by diagenesis.

Diagenesis, for the most part, has no adverse effect on petrography because it normally occurs at the elemental level. However, there are contexts, especially in hot, dry climates, within which minerals such as calcite (e.g., soil carbonates) can be observed as surface coatings or fissure fillings in thin section. These are clearly postdepositional and can be readily recognized as such in thin section. Sometimes these carbonate deposits are visible only in internal fissures, which makes real the fear that diagenesis may not always be recognized from visual inspection of hand specimens nor compensated for by careful surface cleaning prior to elemental analysis. Diagenesis is one of those sticky issues in need of much further investigation. Perhaps here, as in many other areas, elemental analysis and petrography can find common ground on which not merely to coexist but to cooperate actively in order to provide a richer understanding of ceramic production and distribution in antiquity.

5. References

Arnold, D., and Nieves, A., 1992, Factors Affecting Ceramic Standardization. In *Ceramic Production and Distribution*, edited by G. Bey, III and C. Pool, pp. 93–113. Westview Press, Boulder, CO.

Arnold, D., Neff, H., and Bishop, R., 1991, Compositional Analysis and "Sources" of Pottery: An Ethnoarchaeological Approach, *American Anthropologist* 93:70–90.

Bareis, C., and Porter, J., 1984, *American Bottom Archaeology*, University of Illinois Press, Urbana and Chicago.

Barnett, W., 1991, The Identification of Clay Collection and Modification in Prehistoric Potting at the Early Neolithic Site of Balma Margineda, Andorra. In *Recent Developments in Ceramic Petrology*, edited by A. Middleton and I. Freestone, pp. 17–37. British Museum Occasional Paper No. 81. British Museum Research Laboratory, London.

Bishop, R., 1980, Aspects of Ceramic Compositional Modeling. In Models and Methods in Regional Exchange, edited by R. E. Fry. *SAA Papers* 1:47–65.

Bishop, R., Rands, R., and Holley, G., 1982, Ceramic Compositional Analysis in Archaeological Perspective. In *Advances in Archaeological Method and Theory*, Volume 5, edited by M. Schiffer, pp. 275–330. Academic Press, New York.

Braun, D., 1983, Pots as Tools. In *Archaeological Hammers and Theories*, edited by A. Keene and J. Moore, pp. 107–134. Academic Press, New York.

Bronitsky, G., 1986, The Use of Materials Science Techniques in the Study of Pottery Construction and Use. In *Advances in Archaeological Method and Theory*, Volume 9, edited by M. Schiffer, pp. 209–276. Academic Press, New York.

Bronitsky, G., and Hamer, R., 1986, Experiments in Ceramic Technology: The Effects of Various Tempering Materials on Impact and Thermal-Shock Resistance, *American Antiquity* 51:89–101.

Budak, M., 1991, The Function of Shell Temper in Pottery, *Minnesota Archaeologist* 50:53–59.

Burton, J., and Simon, A., 1993, Acid Extraction as a Simple and Inexpensive Method for Compositional Characterization of Archaeological Ceramics, *American Antiquity* 58:45–59.

Carr, C., and Komorowski, J., 1995, Identifying the Mineralogy of Rock Temper in Ceramics Using X-Radiography, *American Antiquity* 60:723–749.

Chayes, F., 1956, *Petrographic Modal Analysis*, Wiley, New York.

Cuomo Di Caprio, N., and Vaughn, S., 1993, An Experimental Study in Distinguishing Grog (Chamotte) from Argillaceous Inclusions in Ceramic Thin Sections, *Archeomaterials* 7:21–40.

Daniels, R., Gamble, E., Bartelli, L., and Nelson, L., 1968, Application of the Point-Count Method to Problems of Soil Morphology, *Soil Science* 106:149–152.

Dickinson, W., and Shutler, R., Jr., 1971, Temper Sands in Prehistoric Pottery of the Pacific Islands, *Archaeology and Physical Anthropology of Oceania* 6:191–203.

Dickinson, W., and Shutler, R., Jr., 1974, Probable Fijian Origin of Quartzose Temper Sands in Prehistoric Pottery from Tonga and the Marquesas, *Science* 185:454–457.

Dickinson, W., and Shutler, R., Jr., 1979, Petrography of Sand Tempers in Pacific Island Potsherds, *Geological Society of America Bulletin* 90:1644–1701.

Ericson, J., Read, D., and Burke, C., 1972, Research Design: The Relationships Between the Primary Functions and the Physical Properties of Ceramic Vessels and Their Implications for Ceramic Distributions on an Archaeological Site, *Anthropology UCLA* 3:84–95.

Feathers, J., 1989, Effects of Temper on Strength of Ceramics: Response to Bronitsky and Hamer, *American Antiquity* 54:579–588.

Ferring, C. R., and Pertulla, T., 1987, Defining the Provenance of Red Slipped Pottery from Texas and Oklahoma by Petrographic Methods, *Journal of Archaeological Science* 14:437–456.

Finney, F., and Stoltman, J., 1991, The Fred Edwards Site: A Case of Stirling Phase Culture Contact in Southwestern Wisconsin. In *New Perspectives on Cahokia: Views from the Peripheries*, edited by J. Stoltman, pp. 229–252. Monographs in World Archaeology, No. 2. Prehistory Press, Madison, WI.

Fowler, M., 1955, Ware Groupings and Decorations of Woodland Ceramics in Illinois, *American Antiquity* 20:213–235.

Freestone, I., 1991, Extending Ceramic Petrography. In *Recent Developments in Ceramic Petrology*, edited by A. Middleton and I. Freestone, pp. 399–410. British Museum Occasional Paper No. 81. British Museum Research Laboratory, London.

Freestone, I., 1995, Ceramic Petrography, *American Journal of Archaeology* 99:111–115.

Galehouse, J., 1971, Point Counting. In *Procedures in Sedimentary Petrology*, edited by R. Carver, pp. 385–408. Wiley Interscience, New York.

Garrett, E., 1986, A Petrographic Analysis of Black Mesa Ceramics. In *Spatial Organization and Exchange*, edited by S. Plog, pp. 114–142. Southern Illinois University Press, Carbondale.

Griffin, J., 1949, The Cahokia Ceramic Complexes. In Proceedings of the Fifth Plains Conference for Archaeology, compiled by John L. Champe. *University of Nebraska Laboratory of Anthropology Notebooks* 1:44–58.

Griffin, J. (ed.), 1950, *Prehistoric Pottery of the Eastern United States*, Museum of Anthropology, University of Michigan, Ann Arbor.

Griffiths, J., 1967, Scientific Method in Analysis of Sediments. MacGraw-Hill, New York.

Habicht-Mauche, J., 1993, The Pottery from Arroyo Hondo Pueblo, New Mexico: Tribalization and Trade in the Northern Rio Grande. *Arroyo Hondo Archaeological Series*, 8:1–163. School of American Research Press, Santa Fe, NM.

Hawley, F., 1950, *Field Manual of Prehistoric Southwestern Pottery Types*, rev. ed. University of New Mexico Bulletin, Anthropological Series, Vol. 1, No. 4. Albuquerque.

Hoard, R., O'Brien, M., Khorasgany, M., and Goplaratnam, V., 1995, A Materials-Science Approach to Understanding Limestone-tempered Pottery from the Midwestern United States. *Journal of Archaeological Science* 22:823–832.

Hutchison, C., 1974, *Laboratory Handbook of Petrographic Techniques*. Wiley, New York.

Kemeny, J., 1959, *A Philosopher Looks at Science*. Van Nostrand Reinhold, New York.

Logan, W., 1976, *Woodland Complexes in Northeastern Iowa*. Publications in Archaeology No. 15, National Park Service, Washington, DC.

Lombard, J., 1987, Provenance of Sand Temper in Hohokam Ceramics, Arizona. *Geoarchaeology* 2(2):91–119.

Mainfort, R., Jr., Cogswell, J., O'Brien, M., Neff, H., and Glascock, M., 1997, Neutron Activation Analysis of Pottery from Pinson Mounds and Nearby Sites in Western Tennessee: Local Production vs. Long-Distance Importation. *Midcontinental Journal of Archaeology* 22:43–68.

Mathews, A., Woods, A., and Oliver, C., 1991, Spots Before the Eyes: New Comparison Charts for Visual Percentage Estimation in Archaeological Materials. In *Recent Developments in Ceramic*

Petrology, edited by A. Middleton and I. Freestone, pp. 211–263. British Museum Occasional Paper No. 81. British Museum, London.
Matson, F., 1955, Ceramic Archaeology. *The American Ceramic Society Bulletin* 34:33–44.
Matson, F., 1970, Some Aspects of Ceramic Technology. In *Science in Archaeology*, edited by D. Brothwell and E. Higgs, pp. 592–602. Praeger Publishers, New York.
McKusick, M., 1964, Discovering the Hartley Fort. *The Palimpsest* 45:487–494.
Middleton, A., Freestone, I., and Leese, M., 1985, Textural Analysis of Ceramic Thin Sections: Evaluation of Grain Sampling Procedures. *Archaeometry* 27:64–74.
Miksa, E. and Heidke, J., 1995, Drawing a Line in the Sands: Models of Ceramic Temper Provenance. In *Roosevelt Community Development Study*, Volume 2: *Ceramics, Chronology, Technology, and Economics*, edited by J. Heidke and M. Stark, pp. 133–205. Anthropological Papers No. 14. Center for Desert Archaeology, Tucson, AZ.
Neff, H. (ed.), 1992, Chemical Characterization of Ceramic Pastes in Archaeology. *Monographs in World Archaeology*, No. 7. Prehistory Press, Madison, WI.
Nesbitt, R., 1964, Combined Rock and Thin Section Modal Analysis. *The American Mineralogist* 49:1131–1136.
Neupert, M., 1994, Strength Testing Archaeological Ceramics: A New Perspective. *American Antiquity* 59:709–723.
Peacock, D., 1968, A Petrological Study of Certain Iron Age Pottery from Western England. *Proceedings of the Prehistoric Society* 34:414–427.
Peacock, D., 1969, Neolithic Pottery Production in Cornwall. *Antiquity* 43:145–149.
Peacock, D., 1970, The Scientific Analysis of Ancient Ceramics: A Review. *World Archaeology* 1:375–389.
van der Plas, L. and Tobi, A., 1965, A Chart for Judging the Reliability of Point Counting Results. *American Journal of Science* 263:87–90.
Porter, J., 1962, Temper in Bluff Pottery from the Cahokia Area. *Southern Illinois University Museum Lithic Laboratory, Research Report* 1.
Porter, J., 1964, Comment on Weaver's "Technological Analysis of Lower Mississippi Ceramic Materials." *American Antiquity* 29:520–1.
Rice, P., 1981, Evolution of Specialized Pottery Production: A Trial Model. *Current Anthropology* 22:219–240.
Rice, P., 1989, Ceramic Diversity, Production, and Use. In *Quantifying Diversity in Archaeology*, edited by R. Leonard and G. Jones, pp. 109–117. Cambridge University Press, Cambridge.
Rice, P., 1996a, Recent Ceramic Analysis: 1. Function, Style, and Origins. *Journal of Archaeological Research* 4:133–163.
Rice, P., 1996b, Recent Ceramic Analysis: 2. Composition, Production, and Theory. *Journal of Archaeological Research* 4:165–202.
Rugge, D., 1976, *A Petrographic Study of Ceramics from the Mimbres River Valley, New Mexico*. Master's thesis, Department of Anthropology, University of New Mexico, Albuquerque.
Rye, O., 1976, Keeping Your Temper Under Control: Materials and Manufacture of Papuan Pottery. *Archaeology and Physical Anthropology in Oceania* 11:106–137.
Rye, O., 1981, *Pottery Technology*. Taraxacum, Inc., Washington, DC.
Schubert, P., 1986, Petrographic Modal Analysis—A Necessary Complement to Chemical Analysis of Ceramic Coarse Ware. *Archaeometry* 28:163–178.
Shepard, A., 1936, The Technology of Pecos Pottery. In *The Pottery of Pecos* Volume II: *The Glaze-Paint, Culinary, and Other Wares*, edited by A. Kidder, pp. 389–588. Yale University Press, New Haven.
Shepard, A., 1939, Appendix A—Technology of La Plata Pottery. In *Archaeological Studies in the La Plata District*, by E. Morris, pp. 249–287. Carnegie Institution of Washington Publication 519. Washington, DC.
Shepard, A., 1942, Rio Grande Glaze Paint Ware. A Study Illustrating the Place of Ceramic Technological Analysis in Archaeological Research. *Carnegie Institution of Washington Publication* 528:129–262.
Shepard, A., 1954, Letter excerpt. In *The Material Culture of Pueblo Bonito*, by N. M. Judd, pp. 236–238. Smithsonian Miscellaneous Collections Vol. 124. Washington, DC.
Shepard, A., 1956, *Ceramics for the Archaeologist*. Carnegie Institution of Washington Publication, 609. Washington, DC.

Steponaitis, V., 1984, Technological Studies of Prehistoric Pottery from Alabama: Physical Properties and Vessel Function. In *The Many Dimensions of Pottery*, edited by S. Van der Leeuw and A. C. Pritchard, pp. 79–122. Universiteit van Amsterdam, Amsterdam.

Steponaitis, V., Blackman, M. J., and Neff, H., 1996, Large-Scale Patterns in the Chemical Composition of Mississippian Pottery. *American Antiquity* 61:555–572.

Stoltman, J., 1986, The Prairie Phase: An Early Woodland Manifestation in the Upper Mississippi Valley. In *Early Woodland Archaeology*, edited by K. Farnsworth and T. Emerson, pp. 121–136. Center for American Archeology Press, Kampsville, IL.

Stoltman, J., 1989, A Quantitative Approach to the Petrographic Analysis of Ceramic Thin Sections. *American Antiquity* 54:147–160.

Stoltman, J., 1991, Ceramic Petrography as Technique for Documenting Cultural Interaction: An Example from the Upper Mississippi Valley. *American Antiquity* 56:103–120.

Stoltman, J., 1992, Petrographic Observations on Six Ceramic Vessels Bearing on the Issue of External Influences at Obion. In The Obion Site, An Early Mississippian Center in Western Tennessee. *Report of Investigations* 7:208–213. Cobb Institute of Archaeology, Mississippi State University.

Stoltman, J., 1996, Petrographic Observations of Selected Sherds from Wind Mountain. In *Mimbres Mogollon Archaeology*, edited by A. Woosley and A. McIntyre, pp. 367–371. University of New Mexico Press, Albuquerque.

Stoltman, J., 1999, The Chaco–Chuska Connection: In Defense of Anna Shepard. In *Pottery and People: Dynamic Interactions*, edited by J. Skibo and G. Feinman, pp. 9–24. University of Utah Press, Salt Lake City.

Stoltman, J., and Snow, F., 1998, Cultural Interaction Within Swift Creek Society: People, Pots, and Paddles. In *A World Engraved: Archaeology of the Swift Creek Culture*, edited by M. Williams and D. Elliott, pp. 130–153. The University of Alabama Press, Tuscaloosa.

Stoltman, J., Burton, J., and Haas, J., 1992, Chemical and Petrographic Characterizations of Ceramic Pastes: Two Perspectives on a Single Data Set. In *Chemical Characterization of Ceramic Pastes in Archaeology*, edited by H. Neff, pp. 85–92. Monographs in World Archaeology, No. 7. Prehistory Press, Madison, WI.

Tiffany, J., 1982, Hartley Fort Ceramics, *Proceedings of the Iowa Academy of Science* 89:133–150.

Wandibba, S., 1982, Experiments in Textural Analysis. *Archaeometry* 24:71–75.

Weigand, P., Harbottle, G., and Sayre, E., 1977, Turquoise Sources and Source Analysis: Mesoamerica and the Southwestern United States. In *Exchange Systems in Prehistory*, edited by T. Earle and J. Ericson, pp. 15–34. Academic Press, New York.

Whitbread, I., 1986, The Characterization of Argillaceous Inclusions in Ceramic Thin Sections. *Archaeometry* 28:79–88.

Whitbread, I., 1989, A Proposal for the Systematic Description of Thin Sections Towards the Study of Ancient Ceramic Technology. In *Archaeometry, Proceedings of the 25th International Symposium*, edited by Y. Maniatis, pp. 127–138. Elsevier, Amsterdam.

Williams, D., 1983, Petrology of Ceramics. In *Petrology of Archaeological Artifacts*, edited by D. Kemp and A. Harvey, pp. 301–329. Clarendon Press, Oxford.

Microartifacts

SARAH C. SHERWOOD

1. Introduction

Although size is a basic empirical observation in most earth science research, such as sedimentology, pedology, and stratigraphy, attention to this basic physical parameter is relatively undeveloped in archaeology and in the study of artifacts. Artifacts are particles generated by human activity or by anything displaying one or more attributes resulting from human activity (Spaulding, 1960). Archaeological analysis of artifacts has conventionally concentrated on attributes other than size, such as composition and morphology, which are used to make inferences about technology, chronology, and style. Such attributes are observed in macro- or in large artifacts. Therefore, archaeologists have assumed that as an artifact's size decreased so did its ability to provide useful information (Dunnell and Stein, 1989). As archaeologists' questions evolved, however, beyond a narrow technological and chronological focus to include the formation of the archaeological record (Binford, 1978, 1981; Schiffer, 1987), the potential has been recognized for the size distributions of materials in an archaeological deposit to provide information on site locations and their complex depositional histories.

Microartifacts, small artifacts that generally require magnification for identification, are the most abundant kind of artifact in the archaeological record. They bypass standard techniques of field collection (manual surface collection or screening using 0.25 in mesh or 0.5 cm mesh) and are part of the matrix, or

---

SARAH C. SHERWOOD • Department of Sociology and Anthropology, Middle Tennessee State University, Murfreesboro, Tennessee 37132.

Earth Sciences and Archaeology, edited by Paul Goldberg, Vance T. Holliday, and C. Reid Ferring. Kluwer Academic/Plenum Publishers, New York, 2001.

sediment, surrounding the larger artifacts. Larger traditional counterparts, macroartifacts, are typically far fewer in number and can be identified without magnification. Archaeological studies by Hassan (1978), Fladmark (1982), and Rosen (1986, 1989) brought initial recognition to the interpretive potential of microartifacts. These studies and others pointed to the value of these small artifacts in helping archaeologists identify and interpret activity areas, sort out site formation processes, and examine reduction stages in stone tool technology. Dunnell and Stein (1989) placed microartifacts firmly within the wider theoretical context of artifact studies. They suggested that the information archaeologists derive from larger artifacts can be supplemented and complemented by microartifacts (Dunnell and Stein, 1989:31).

Microartifact analysis is probably the most simple of all geoarchaeological techniques currently in use. The technique is uncomplicated, requiring minimal specialized training and equipment. Nevertheless, it is still infrequently applied. Time required for the recovery and identification of microartifacts and the resulting cost are often cited by researchers as reasons not to undertake microartifact analysis. Recent studies, however, have demonstrate that these prohibitive factors can be overcome and that microartifact analysis is an important part of the archaeological record.

This chapter has two primary goals: to define the types of research questions that can be studied using microartifact data and to discuss microartifact analysis protocol, emphasizing cost effectiveness and compatibility among analyses. The chapter begins with an examination of the principles underlying microartifacts, including their size threshold, identification, and theoretical framework. The methods required to design and implement microartifact analyses follows with a section on recovery, size divisions, sample quantification, and graphic representation. Microartifact studies are finally examined at two scales—small-scale site research and large-scale regional research. The kinds of questions that can benefit from these small-size data are discussed within these two scales and demonstrate the potential of microartifact analysis.

2. Defining Microartifacts

The size threshold between macro- and microartifacts is not universally defined. Hassan (1978), following the early microarchaeological analyses of middens (e.g., Cook and Treganza, 1947; Ambrose, 1963) suggested 1.0 mm as the threshold for separating micro- from macroartifacts. Fladmark (1982), working with microdebitage suggested that the cutoff be less than 1.0 mm. Microartifacts were most recently defined as artifacts less than 2.0 mm (Dunnell and Stein, 1989; Rosen, 1989), the sedimentological threshold between gravel and sand. These points of departure, however, ignore those artifacts that fall between this 2.0 mm designation and the traditional field recovery size of greater than 6.35 mm (>0.25 inch).

The relative proportion of artifacts between 2.0 mm and 6.35 mm might be assumed small and insignificant. Based on experimental lithic reduction and resulting size distribution frequencies, however, microdebitage is in fact most abundant between 6.35 mm and 1.0 mm (Shott, 1994). The use of a 6.35 mm threshold to define the largest microartifacts, therefore, is advantageous in that it does not exclude this significant part of the size distribution. This expanded definition allows for a full-scale artifact analysis linking the macroartifact and microartifact size range. Stein and Teltser (1989) have recommended using a geometric progression to look at the spectrum or all grain sizes of artifacts.

The smallest size threshold within a microartifact analysis is restricted by the size at which the material can be properly identified under a binocular microscope at magnifications from $10 \times$ to $100 \times$. As size decreases, specific microartifacts can break down into their constituent components, making the identification of their original form no longer possible. For example, fired clay material such as daub and ceramics can only be identified with confidence down to 0.5 mm. In sizes smaller than 0.5 mm the particles begin to break down into separate fired clay fragments and temper media such as shell fragments and quartz grains, and they can no longer be properly identified (Madsen and Dunnell, 1989; Stein and Teltser, 1989; Sherwood, 1991). Studies that analyze only microdebitage are often able to confidently identify lithic microartifacts at a much smaller size, down to 0.063 mm, the sand/silt boundary (Fladmark, 1982). Based on the success of past studies and the potential for interobserver error, the lower threshold for more than one material class should rarely be less than 0.5 mm, and should never be below 0.25 mm. Below these size thresholds, depending on the material type, human activity should be identified by chemical signatures within the matrix (e.g., Eidt, 1985; Middleton and Price, 1996), or by micromorphological examination of microstratigraphy (e.g., Courty et al., 1989; Gé et al., 1993; Matthews et al., 1997).

2.1. Theoretical Framework

One archaeological deposit may contain particles (artifacts and nonartifacts) of numerous materials and size classes derived from various sources. The particles and their arrangement typically reflect a complex depositional history of natural and cultural processes. The size distribution of these various materials can offer insights into their complex depositional histories, including their sources and transport agents and postdepositional processes (Stein, 1987). The first step in reconstructing a depositional history is to distinguish which particles are natural and which ones are cultural.

The particles derived from natural processes are interpreted in a sedimentary framework. Natural particles constitute anything in the deposit that is not an artifact. Geological studies involving sedimentary rock or unconsolidated material rely on empirical sedimentological laws for grain distributions. The size distribution of particles within a geological deposit is a measure of the sizes available in the source, the energy of the transport medium, the configuration of the locus of deposition, and postdepositional processes such as bioturbation and diagenesis. The distribution of these sizes have modal peaks when graphed by frequency, and analysis of the distributions can be used to identify subpopulations (Reineck and Singh, 1986).

Cultural particles constitute anything in the deposit from cultural processes and are interpreted both in archaeological and sedimentological frameworks. These particles can include artifacts of all sizes, as well as particles, which although not directly identified as artifacts, are present due to human activity. This includes materials such as plant remains that do not show evidence of human modification, but their presence in a specific context could not have been dictated by natural processes. This distinction between cultural and natural particles requires a comprehensive understanding of the environment of deposition and the postdepositional processes that might be active in a specific environment. Microartifacts, due to their abundance and incorporation into the "natural matrix" of a deposit constitute a significant part of the cultural particles present.

Microartifacts represent data collected and observed at a small size, or at a fine scale of resolution. These data can be used to address questions representing different spatial resolutions of interpretation. Microartifact studies can be divided into two categories of questions based on scale: small-scale site research and large-scale regional research. This distinction is important as each scale incorporates a different interpretive framework within the "human scale" (*sensu* Stein, 1993) and focuses on a different spatial resolution of interpretation.

Site-scale research includes recovery of microartifacts and the interpretation of them within a high spatial resolution paradigm, focusing on the spatial context of site structure and the organization of activity within the limits of a site. The relationships between microartifacts and macroartifacts at this scale are critical to the inferences drawn from the artifacts and their patterning. At the site scale, microartifacts cannot be used in place of their larger counterparts but must be considered within the broader artifact spectrum (Dunnell and Stein, 1989). Different sizes within different material classes of varying durability and representing variable activities offer clues to different aspects of the history of a deposit.

Regional-scale research concentrates on paleoenvironmental reconstruction and human land use in the context of landscape evolution (Butzer, 1982). At the regional scale, microartifacts are used as an indication of human activity on the landscape. At this writing, there are few applications of microartifact analysis in regional investigations. The reason so few studies have been done may be the interdisciplinary nature of this application. Much of the analytical framework for microartifacts was originally developed in archaeology for site-scale research questions. Protocol development has been slow to evolve with the changing and expanding recovery needs at the regional scale. Most of the initial work in microartifacts was developed in strict adherence to sedimentological grain-size analysis, using standardized procedures applied to site scale questions and a limited number of samples (e.g. , Stein and Teltser, 1989; Vance, 1989). These procedures often involve processing steps not necessarily required in regional scale questions. The following discussion of methods reports on recent developments that encourage microartifact analysis in both site and regional contexts.

3. Methods

Designing a protocol for microartifact analysis embodies simple procedures based on several steps. These steps include isolation of the material classes or composition types present and determining the distinct attributes used to differentiate between the various classes or types. Such procedures assure accuracy by determining the size distribution and size classes to be sampled and quantified, the appropriate cost-effective quantification procedure, and, the clearest representation of the resulting data.

3.1. Microartifact Identification

Microartifact identification relies on the isolation of specific attributes, such as color, shape, structure, and surface texture, that are unique to each material class. Material classes are established based on the macroartifact assemblage (if present) and on observations made on comparative collections. In order to reduce observer error and define the attributes used to identify microartifacts at specific magnifications, a comparative collection should be organized. For any given study, a comparative collection consists of macroartifacts (large artifacts from the site or context in question) that are reduced in size to match the microsize distributions to be analyzed. This collection allows the observation and establishment of signature attributes of known material under magnification. Standardized comparative collections representative of many sites containing materials not necessarily present at the macro scale are also necessary to assure material recognition. Depending on the region, these materials might include small charred seeds, fish scales, and rocks and minerals such as ocher and mica.

The attributes used to define each material class in a microartifact analysis must be stated clearly. For example, attributes for microdebitage might be the color of a specific resource type (e.g., chert, obsidian), thinness (resulting in translucent particles), morphological attributes including remnants of conchoidal fracture, and angularity (Hull, 1987; Sherwood et al., 1995). Attributes of bone might be based on the unique color, structure, and shape of cortical and cancellous bone. Ceramic microattributes for shell-tempered ceramics might be the dark red to black color of the fired clay, the white subrounded lenticular shell inclusions, or the lenticular, plate-shaped voids where the shell has weathered. The important point is to define clearly the attributes used for each material type.

Off-site control samples are essential (Stein, 1985). These samples determine the particles in the deposits derived from natural processes and thus distinguish particles derived from cultural processes, insuring that those attributes composing material classes are truly unique to the microartifact assemblage. Without control samples, particles such as sand-sized quartz grains can be mistaken for microdebitage, because they are similar in size and shape. A comparative collection, however, can reveal the attributes of natural sand grains. These natural grains often have a frosted surface texture and are rounded (implying extensive transport) compared to angular, glassy surface textures of microdebitage. The latter physical features imply that these rock or mineral particles are close to their source.

The combination of comparative collections and control samples from natural sediments allows one to distinguish between cultural and natural particles at the micro-scale, and the unique attributes that define the material classes can be designated. The result is a lower degree of observer error and comparable micromaterial classes between assemblages or sites.

3.2. Microartifact Recovery

Field and laboratory sampling protocol are best designed on the basis of the types of research questions asked and the nature of the deposits involved. Analyses that seek to define and interpret activity areas at the site scale (where horizontal variability over a distance of meters is key) requires a sampling strategy comparable between artifact scales. Microartifact samples must be comparable to macroartifact samples in order to identify fine-resolution spatial variability within the same overall artifact assemblage.

Several different collection strategies for microartifact samples are suggested. Scrape samples are recommended in cases where specific spatial questions are important (Sherwood et al., 1995). The procedure is to scrape sediment from the same area as the macroartifact collection area (typically a 1.0 m or 0.50 m square unit) to collect a 200 g or greater sample (Sherwood, 1991). A scrape sample results in a homogeneous bulk or matrix sample from across the entire unit, representing the same excavation unit and the same depth (or stratum) as the macroartifact assemblage. Often bulk samples are collected from a designated corner of a unit and do not represent the entire unit area.

Another technique is to collect sediment bulk samples from a vertical face. Core samples are also acceptable, although this pertains more to regional scale survey where open excavations and vertical exposures are rare. Typically, vertical face samples are collected at regular intervals down an exposed profile. The disadvantage of such vertical sampling is that it cannot be used to represent the horizontal spatial distribution of microartifacts. Their advantage is that they are easy to collect and do not have to be incorporated into the excavation protocol. This vertical sampling scheme is best used for the consideration of postdepositional processes within a site (often encompassing several lithostratigraphic units or soil horizons) and their effect on the vertical translocation of particles by size (e.g., Hofman, 1992; Madsen, 1992).

In the laboratory, micro-sized particles are separated from the finer matrix to facilitate identification and quantification. The procedures employed vary greatly, depending on the texture of the material, the nature of the material classes, and the scale of the research questions. The basic procedure entails soaking a set volume or weight of air-dried sediment in a dispersing solution (typically sodium hexametaphosphate). The saturated samples are gently agitated and washed through a standardized screen to separate the desired fraction. The contents of each screen are washed into a labeled container and oven dried. In specific instances sampling a range of sediment sizes is desired (e.g., in vertical sequence sampling). In such an instance the finest materials (silt and clay) are separated from the sand fraction, retained, and processed using pipette or hydrometer analysis.

Certain conditions can make laboratory recovery of microartifacts difficult. In deposits containing large amounts of clay (approximately 50% or greater), disaggregation can be arduous, resulting in extended processing times in order to acquire clean sand fractions. Deposits rich in organic material can cause aggregation of coarse particles, creating larger coarse fractions and skewing the size distribution toward the coarse range. In cases where material classes such as bone and shell are present, pretreatment to remove high concentrations of organic matter, soluble salts, or carbonates is discouraged, as pretreatment will damage these materials. Typically, a slight revision of the washing and drying procedures prevents aggregate formation (see Sherwood, 1991), but such treatments should be carefully assessed for their research potential versus their cost because they can add processing time and destroy artifacts.

In the event that the finer sediment (silt and clay) is not of interest or where sample volumes or numbers are large, it is best to use alternate procedures. These procedures are expedient and therefore cost effective. One recommended procedure for site-scale research provides compatibility between macro- and microartifact analysis through the use of flotation samples. Sampling for recovery of botanical remains using flotation is becoming routine in archaeology. Typically, a large-volume sample (> 10 liters) is collected from specific deposits, feature fills, or vertical control columns in archaeological excavations in order to retrieve fragile, charred plant remains (Pearsall, 1989; Watson, 1976). The heavy fraction, clasts in the sample that do not float, are collected in a sieve (typically >1 mm) at the bottom of the flotation tank. Charred materials float and are collected as runoff when running water or compressed air agitates the solution in the tank. When the lower retrieval units in these tanks are mounted with a durable ASTM standardized mesh, the heavy fraction can be used as a microartifact sample. The heavy fraction, collected with the standardized mesh, can provide a representative grain size distribution of the sample, thus eliminating the need for independent field sampling and moving the lab analysis straight to the separation of designated size fractions. A less than 1.0 mm mesh in the bottom of a float tank is generally undesirable, as the finer mesh restricts water flow and can result in an ineffective float process. If microartifacts that are smaller than 1.0 mm are targeted in the research, then this technique will not work.

A second alternate procedure for recovering microartifacts targets both site-scale and regional-scale questions of site location and landscape analysis. This technique, developed by Stafford (1993, 1995), is appropriate when large numbers of samples (often hundreds) are needed. Briefly, this procedure includes the submergence of samples in 0.2 mm mesh nylon bags in a tank of dispersant solution (up to 20 at a time depending on the tank size). The bags are then individually washed and dried and the contents are size sorted. Processing numerous samples in the initial stages of the protocol, and eliminating steps of oven drying and transferring samples from glassware to sieves and back again, greatly decreases the time involved, whereas the 0.2 mm mesh provides smaller grain sizes. This protocol is currently in use at the University of Tennessee, Knoxville, where it has produced positive, expedient results. The nylon bags allow for more thorough cleaning of the samples, permitting quicker size sorting.

3.3. Size Distribution

Once separated, the coarse sand and small-gravel fraction can be divided into the size grades desired for analysis. As in any empirical observation or comparison of

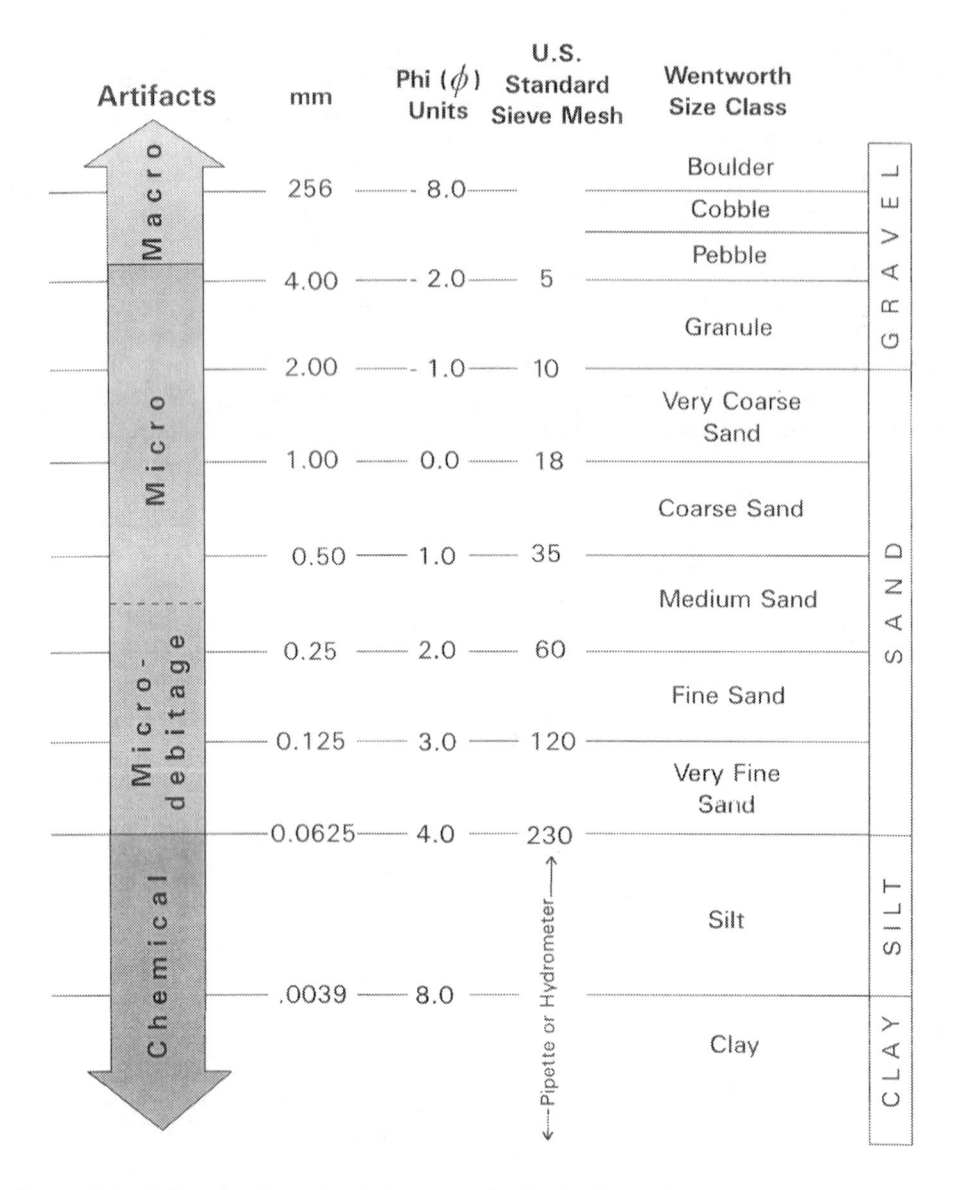

Artifacts	mm	Phi (φ) Units	U.S. Standard Sieve Mesh	Wentworth Size Class	
Macro	256	8.0		Boulder	GRAVEL
				Cobble	
				Pebble	
	4.00	2.0	5		
				Granule	
Micro	2.00	1.0	10		SAND
				Very Coarse Sand	
	1.00	0.0	18		
				Coarse Sand	
	0.50	1.0	35		
				Medium Sand	
Micro-debitage	0.25	2.0	60		
				Fine Sand	
	0.125	3.0	120		
				Very Fine Sand	
	0.0625	4.0	230		SILT
Chemical				Silt	
	.0039	8.0			CLAY
				Clay	

Figure 12.1. Artifact size classes in relation to standardized sedimentological units.

data, standardization into ordinal data is crucial for systematically characterizing the nature of a population. This standardization permits comparison between populations. Size distributions of particles can range from microns to meters. Several scales were derived in sedimentology and soil science to divide a spectrum, both logarithmically and geometrically, into discrete, meaningful units with upper and lower size limits (Fig. 12.1). The scale commonly applied by sedimentologists is the Wentworth Scale, where a fixed ratio exists between successive elements of the series (Folk, 1980). The result is a scale divided into four major categories, encompassing clay, silt, sand, and gravel. The Wentworth Scale was later modified into a logarithmic scale expressed as phi (ϕ), units of equal value for ease in statistical calculations and graphs (Krumbien and Sloss, 1963).

Determining which size grade(s) to employ for microartifact study is based on physical principles (derived from sedimentology) and experimentation based on archaeological actualistic and experimental studies (see the discussion that follows under "Research Questions") (Stein and Teltser, 1989). The divisions used depend on the kinds of questions being addressed and the compatibility of the results. Microartifacts are best studied using the phi scale, which enables comparison between cultural and physical processes. If possible, the macroartifact and microartifact distributions should be organized within the same logarithmic system (e.g., Stein and Teltser, 1989; Buck, 1990).

Macroartifact analyses related to size also attempt to divide the spectra into comparable and meaningful intervals. Studies focusing on lithic reduction represent the majority of the macroartifact size studies and have served as templates for size interval studies in other artifact classes. Lithic reduction studies concerned with size have relied on logarithmic divisions of the standardized (and commercially available) mesh sizes. Several of these studies are extensions of Ahler's (1975, 1989) mass analysis technique. In this technique, Ahler organizes debitage by standardized size grades (e.g., size grade $3 = 0.50 - 0.25$ inch, size grade $4 = 0.25 - 0.125$ inch). These size grades are used in a discriminant analysis and the resulting functions are applied to distributions observed in archaeological assemblages.

Ceramic analyses by size are few but are typically based on the same size divisions, though they tend to focus only on size grades larger than 0.25 inch (6.35 mm) (Bradley and Fulford, 1980; Chase, 1985; Hally, 1983). The processes responsible for the ceramic size distributions have not been as systematically explored experimentally as those of lithics, and they remain poorly understood. Ultimately, the size distributions one uses in both macroartifact and microartifact analyses must be explicitly stated in order for others to assess the research and compare results.

3.4 Microartifact Quantification

Due to their size and abundance, microartifacts are best sampled and quantified using standard sedimentological techniques. The size of the sample used to represent the percentage of microartifacts in an archaeological population requires one of several possible variables: a designated weight range, a single phi fraction, or a designated number of grains per sample (see Sherwood and Ousley, 1995). There are limitations and disadvantages inherent in each of these variables. Standardized quantification therefore remains an unresolved issue (Shott, 1994), but specific techniques have consistently produced more reliable, cost-effective results.

Point counting (Galehouse, 1971) offers the most reliable process of microscopically quantifying microartifacts. In general, when applied to unconsolidated sediment the method involves the systematic counting by each material class within a size grade. Point counting was operationalized in several microartifact studies by identifying a portion of each size fraction, usually 1000 grains (Stafford, 1995; Stein, 1987; Stein and Teltser, 1989). The proportion of each material class (e.g. lithics, ceramics) represented among those 1000 grains is converted to a percentage of weight. This counting procedure requires that the analyst first separate the grains into equal size classes, so that the number of grains counted represent the same volume. The analyst must also determine not only the number of material classes present but also the rarity of a given material class. Sample sizes in general are usually determined by the derived accuracy of estimates for materials with low ($<10\%$) frequencies. Therefore, in order to acquire a statistically representative sample, extensive preliminary analyses must be carried out to determine a standardized weight or number. This process can add significant time to the quantification process.

Point counting can be expedited in two ways. The sediment should be spread evenly across a gridded dish (a 10 cm glass or plastic square dish is recommended with a 0.5 cm grid etched in the bottom). The dish is carefully moved by hand, following the designated counting system within the grid. First, only a representative number of grains within a material class should be counted. Second, the counting and sampling process can be streamlined by counting directly into a computer, programmed to estimate proportions based on the standard error for each proportion. Any number of computer programs can be developed to perform this calculation, and Sherwood and Ousley (1995) have employed the relational database program Paradox (Borland International, Inc. 1992; available on the Internet, see Sherwood and Ousley, 1995 for address). The technique assumes a binomial distribution of discrete variables, but because there are usually more than two material classes, comparisons are actually multinomial (Sherwood and Ousley, 1995:426). The result is a statistically representative sample requiring the minimal counting time. With the availability of such simple programs to expedite the counting process, quantification of microartifacts will no longer be considered a deterrent to acquiring these data.

Other less efficient techniques have been implemented, such as frequency estimates using visual percentage charts (Rosen, 1986, 1993). This method involves observing a sample through a microscope and estimating the representation of each material class by comparison to standardized percentage charts. Though expedient, the method of visual percentage charts lacks precision and reliability between analysts and is not recommended (Stein and Teltser, 1989).

A technique requiring the sorting of material types within classes in order to count or weigh the results has also been implemented (Metcalfe and Heath, 1990; Simms and Heath, 1990). In order to use weight for representation, the material types must be physically sorted to determine each material class. This sorting substantially increases processing time, and for this reason weight is not recommended. In many instances, counts by material type within a size class provide the most expedient and representative frequency of microartifacts present within a sample. Where small, consistent volume samples produced by subsurface testing techniques (e.g., coring) are implemented, the entire sample of sand size and greater particles are usually quantified. But when sample sizes are large (as is typically the case with abundant small-size particles), counting the entire sample is an onerous task (cf. Simms and Heath, 1990). Point counting, expedited through statistically representative samples, is the most cost-effective and reliable technique of quantifying microartifacts.

3.5. Data Representation

In sedimentology, granulometric data are typically depicted graphically and numerically. The graphical methods include a histogram or frequency curve of particle size and cumulative curves where the size divisions are plotted against individual or cumulative weight percents (these data are plotted on either an arithmetic or probability scale). In sedimentology, mathematical measures describe the grain-size distribution with basic descriptive statistics (e.g., mean, standard deviation, skewness, and kurtosis) that ultimately are used to describe the depositional environment or to search for environmentally diagnostic patterns (Folk, 1980; Visher, 1969).

Because microartifact material classes undergo different formation processes in their initial production and subsequent postdepositional alteration, summary statistics such as skewness and kurtosis lack archaeological significance. Graphic illustrations remain the most informative representation method. In research involving high-resolution horizontal distributions, density maps and variations in histograms are most effective. Density contour maps illustrate frequency variability across horizontal space. The accuracy of these density maps relies on the presence of an adequate number of samples and their systematic distribution to represent an area. The sampling scheme (interval and size) is dependant on the research or survey design.

Microartifact data are difficult to portray in regional studies due to the high number of samples involved as well as the large scale of the study area. Histograms or area diagrams are useful tools to compare the frequencies of microartifacts among core samples, but such an illustration cannot present more than a few samples at a time. Where samples are collected systematically or represent specific landforms, a tool such as a Geographic Information System (GIS) could be used to incorporate microartifact distributions as a layer or data theme in a broader relational database. GIS analysis (Kvamme, 1989) could integrate microartifacts into a regional context and evaluate them with other forms of data such as soils, vegetation, paleotopography, and elevation. Microartifact analysis performed and combined within GIS analysis could ultimately augment settlement pattern or land-use models.

Vertical distributions of microartifacts in both site- and regional scale research are plotted as frequency distributions, where sample depth is correlated against frequency. An advantage of this presentation technique is that more than one sample can be illustrated at a time while additional information such as soil horizons can also be incorporated into the graph.

4. Research Questions

The two categories of microartifact research, small-scale site research and large-scale regional research, are separated based on their different data distributions, sampling procedures, and interpretative frameworks. Microartifact analyses in site-scale research emphasize the identification of formation processes, the characterization of assemblage composition, and the delineation of activity areas within the confines of a site. At the regional scale, microartifact analyses focus on detection of sites across a landscape, specifically on sites that are difficult to detect, such as buried sites. The following discussion addresses each of these interpretive frameworks beginning with the site scale. Examples to illustrate microartifact research at the particular scale follow each section.

4.1. Site-Scale Research

A large part of the initial interest in microartifacts lays in their abundance and their potential to provide information on spatial patterning and intrasite structure. The movement of macroartifacts, either due to cultural or natural processes, may reduce our ability to reveal the use of space using these traditional artifacts. Results of ethnoarchaeological (Hayden and Cannon, 1983; O'Connell, 1987, Simms, 1988) and ethnographic (Murray, 1980) observation of site structure prompted some researchers to posit that small-scale artifacts generated during human activity are better indicators of activity location than are larger artifacts. This assumption is based on the notion that activity areas are maintained by removing larger materials to a separate location (referred to as cultural sorting by Gallagher, 1977), and the observation that sand-sized artifacts are incorporated into the floor by trampling (Butzer, 1982; Clark, 1986). A considerable number of activity area studies using microartifacts have rested on this assumption (Hull, 1987; Metcalfe and Heath, 1990; Reese, 1986; Rosen, 1993; Vance, 1986).

This assumption, that microartifact distributions represent primary context or activity areas, may be problematic, oversimplified, and in need of experimental verification and subsequent refining (Nelson, 1987; Schick, 1986; Sherwood et al., 1995; Villa, 1982). The distribution of microartifacts is affected by a number of sources, transport mechanisms, and postdepositional processes at the site scale. Unlike most natural processes, humans do not transport a consistent size mode; we transport objects of varying sizes (Binford, 1981; Stein, 1987). Different activities, however, incorporate different materials that may enter the record at a variety of sizes. So although we may not be able to relate size distributions directly to human depositional signatures, we can address specific activities, such as stone tool production and use, based on the nature of the material itself. Knowledge of the physical properties of a material type provides information about the potential for size reduction and sorting. For example, the physical properties of lithic material as well as the size distributions relating to various lithic reduction strategies are relatively well understood (Ahler, 1975;

Shott, 1994; Stahle and Dunn, 1982). In addition, the durable physical structure of the material results in limited postdepositional size reduction from mechanical and physical weathering.

Different material classes may reflect a different series of depositional or post-depositional processes depending on their durability and original size. Two basic types of durability have been defined: stable (e.g., lithics and some metal) and unstable (e.g., ceramics, shell, bone, charcoal, daub) (Sherwood et al., 1995). The stable artifacts are relatively constant with respect to size once they are introduced into the archaeological record. The unstable materials, affected by chemical and mechanical weathering processes, can break down systematically, thus generating microartifacts after deposition.

Actualistic studies have demonstrated that processes, such as trampling and sweeping, disassociate both micro- and macroartifact distributions from their original activity area. For example, Nielson (1991b) swept a lithic reduction area following pickup cleaning (cultural sorting). From the remaining material an interesting pattern emerged: Flakes less than 1.0 mm were greatly displaced as a result of the sweeping process, whereas flakes falling within the range of 1.0 to 8.0 mm remained in their original locations (Nielsen, 1991b:5). Trampling studies in archaeology have also addressed the movement and breakage of artifacts due to foot traffic in relation to site structure and taphonomy (Nielsen, 1991a; Rosen, 1989; Shea and Klenck, 1993). Nielsen (1991a, 1991b) conducted another actualistic study that applied foot traffic to macroceramics to assess microartifact production rates. The production of microsherds due to the trampling of ceramic macrosherds resulted in far more particles than the amount produced by initial vessel breakage. This indicated that high concentrations of microceramics may be the result of postdepositional trampling rather than prehistoric domestic activities involving ceramic vessel use (Nielsen, 1991b:7).

Natural processes such as sheet wash and some forms of bioturbation (e.g., termites) are shown to disassociate micro- and macroartifacts postdepositionally (cf. Baumler, 1985; McBrearty, 1990; Petraglia and Nash, 1987). The vertical translocation of microartifacts in specific depositional environments can also affect artifact size distributions. This phenomena should be carefully considered in porous deposits such as shell middens (Madsen, 1992) and where soil formation may have significantly altered archaeological deposits. These kinds of postdepositional processes affect all material classes, archaeological and non-archaeological, and can usually be identified through patterns in overall grain size distributions and the identification of microstratigraphic structures and features (see Courty et al., 1989, for a discussion of the micromorphological analysis of postdepositional processes). Systematic vertical sampling schemes, looking at particle size variability down a profile, can be used to identify the postdepositional downward movement of smaller artifacts in a profile.

Plowing, affecting a large percentage of open-air sites, can reduce macroartifacts after they have entered the archaeological record (Stein and Teltser, 1989). Research on the effects of tillage on archaeological sites located in plow zones indicates that there are numerous factors that govern how, and to what extent, artifacts are broken (Dunnell and Simek, 1995; Lyman and O'Brien, 1987; O'Brien and Lewarch, 1981). The primary factor affecting artifact breakage due to plowing appears to be material durability (affected by composition and morphology) rather than time under cultivation (Dunnell and Simek, 1995). This phenomenon should have a direct bearing on our interpretation of microartifact assemblages in plowed contexts. Although microartifacts can be generated as a result of tilling, they can still offer coarse-resolution spatial data that might not be available from their larger counterparts. Macroartifacts, for example, often have low densities on the surface of tilled sites, making it difficult to detect patterned distributions, or even to identify a site at all (Madsen and Dunnell, 1989). Microartifacts can be good predictors of sites in these situations. Small artifacts (artifacts < 1 cm) also typically undergo limited displacement due to plowing compared to larger artifacts (Lewarch and O'Brien, 1981), further suggesting that microartifacts can also reveal spatial patterning, as well as site locations in tilled contexts.

Microartifacts alone, it appears, cannot be used to identify site structure, just as macroartifacts should not be used alone. These small artifacts must be collected and analyzed in conjunction with their larger counterparts, and only then can the primary context or activity areas be discovered and interpreted within the complex combination of site formation processes. As the following Loy site example illustrates, various material types at distinct scales reveal different processes within a depositional history.

4.1.1. The Loy Site

Research at the Loy site demonstrates the utility of spatial distributions within material size classes through the examination of concordance relationships between micro- versus macroartifact scales. Microartifact analysis was carried out on a Dallas phase (Late Mississippian Period, ca. 1400) domestic house floor deposit at the Loy site in East Tennessee (Sherwood, 1991; Sherwood et al., 1995). The primary goal of the study was to distinguish primary activity areas versus areas of discard. The confined floor deposit offered an excellent setting for this analysis, because the structure was abandoned, subsequently burned, and then collapsed, sealing the living floor.

The spatial structure of Dallas phase houses is based on a model grounded in observations of macroartifact distributions from similar archaeological structures in conjunction with ethnographic descriptions. These provided details designating the interior division of the houses into private versus public space (Polhemus, 1985; Fig. 12.2). This model of house organization and the assumptions surrounding microartifacts in the context of activity areas suggested two patterns of artifact distributions. The public areas would primarily contain microartifacts resulting from "cultural sorting" of public activity areas and subsequent trampling. The private areas, outside the main traffic areas (beneath benches and in the corners), were reserved for storage and nonartifact-producing activities such as sleeping and therefore should contain predominantly macroartifacts. Six material classes were counted: lithic, bone, shell, charcoal, daub, and ceramic. Only the lithics (stable material) and the ceramics and bone (unstable material) are discussed here.

Microartifacts

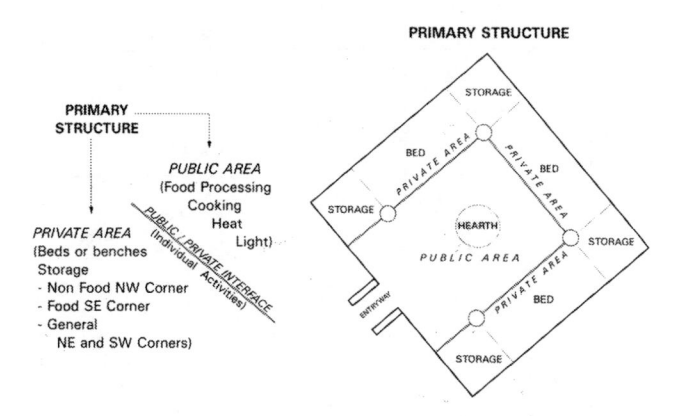

Figure 12.2. Model for the interpretation of the use of space in the minimal settlement unit within the Dallas phase, Mississippian period. (after Polhemus, 1985, Figure 3.6)

Eighty samples were collected in 76 cm (2.5 ft) square excavation units located within the site provenience system (Fig. 12.3). The microartifact size distribution combined the 1 and 0 ϕ fractions (2.0–0.5 mm), whereas macroartifacts consisted of materials systematically collected within a 6.35 mm mesh sieve. Macroartifact distributions are illustrated as total counts per unit, and microartifact distributions are shown as volume percentages based on proportions derived from standard errors (Sherwood and Ousley, 1995).

Lithic macroartifacts were concentrated around the central hearth with a cluster toward the northern limit of the central public area. The distribution of microlithics clustered in the southern corner to the right of the entrance partition (Fig. 12.4). Each size class has a concentrated distribution, but concentrations are not concordant. As lithics are considered stable material, these distributions are interpreted as representing different activities. The corner containing microdebitage may have provided a safe flint-knapping area from which finished products were carried away to be utilized elsewhere.

The initial model proposed for public versus private space does not provide for this kind of specialized work area within the structure, and it clearly needs modification. In accordance with the assumptions stated previously, a lithic reduction area is identified by the concentration of microdebitage. Perhaps due to cultural sorting, at least some of the larger debitage was removed from this work location and was disposed of elsewhere, most likely outside the structure. Lithic artifacts at the site consisted of small and extensively reworked pieces of poor-quality Knox Black Chert that crops out locally as small nodules. As a result, debitage dumps should be limited and extremely difficult to identify as only a few small pieces; much of the macrodebitage may have been saved for use at a later time. The macrolithic (including tools and debitage) distribution was concentrated around the hearth (Fig. 12.4). The lack of microlithic in the proximity of the hearth indicates an area where stone tools were probably used but not made.

The ceramic macroartifacts were found in greatest concentrations around the hearth, whereas the microsherds were scattered along the walls (Fig. 12.5).

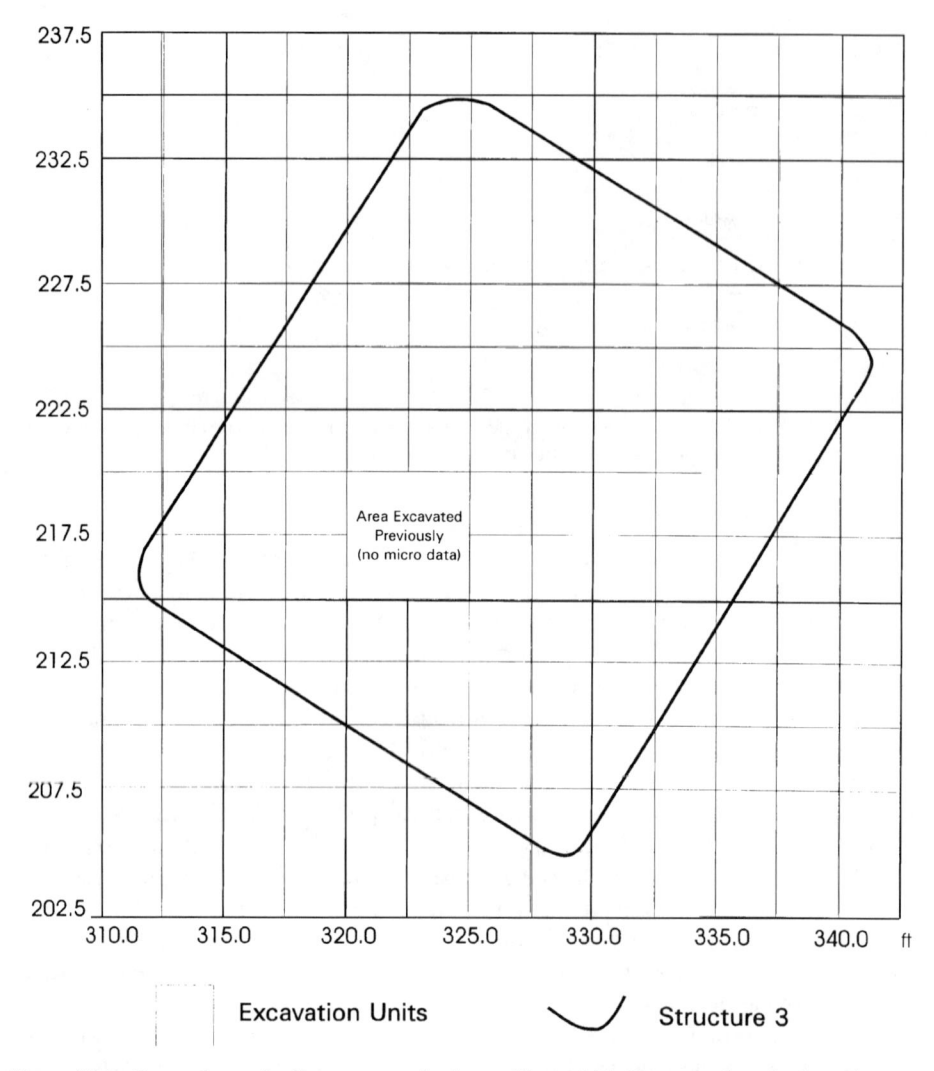

Figure 12.3. Excavation units for macro- and microartifacts within Structure 3 at the Loy Site.

These same patterns were also observed for bone (see Sherwood et al., 1995). Similar distributions of microartifacts along the interior walls of a structure have been observed in other studies (see Metcalfe and Heath, 1990). Ceramics and bone are materials affiliated with food preparation and cooking and their association around the hearth indicate domestic activity. The distributions of microceramics and bone along the walls were interpreted as the result of sweeping, clearing the public space of small refuse. The microsherd clustering in the entryway could indicate trampling. Trampling generates more microsherds and increases their likelihood of being missed during sweeping as these sand-sized particles are embedded in the finer matrix of the floor in the high-traffic

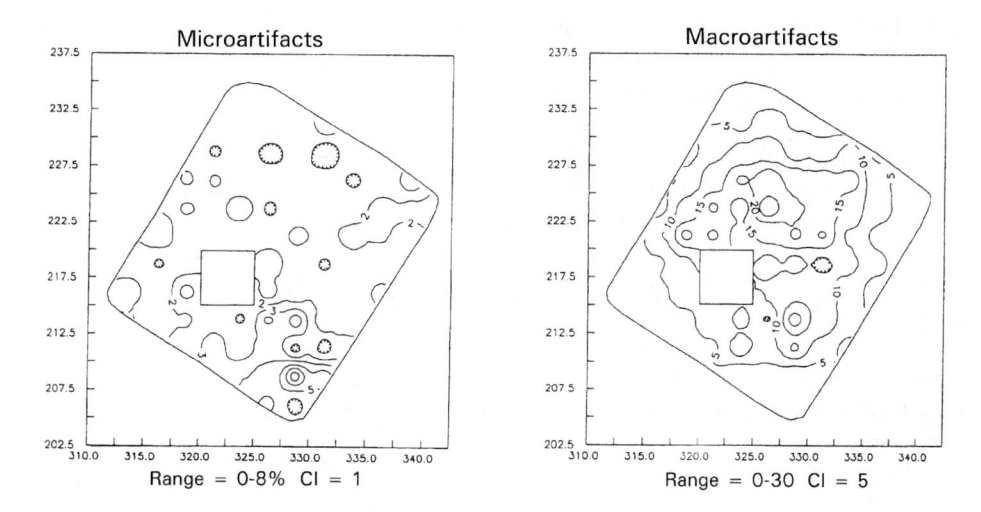

Figure 12.4. Lithic distribution in Structure 3 at the Loy site. (The range for the microartifacts is reported in volume percentages; the range for macroartifacts is reported in total counts. The CI indicates the contour interval.)

areas. This study of a Mississippian Dallas phase house floor illustrates the need for both micro- and macroartifact analysis in order to define activity areas. Distributions of only a single size class would have offered a limited if not misleading picture.

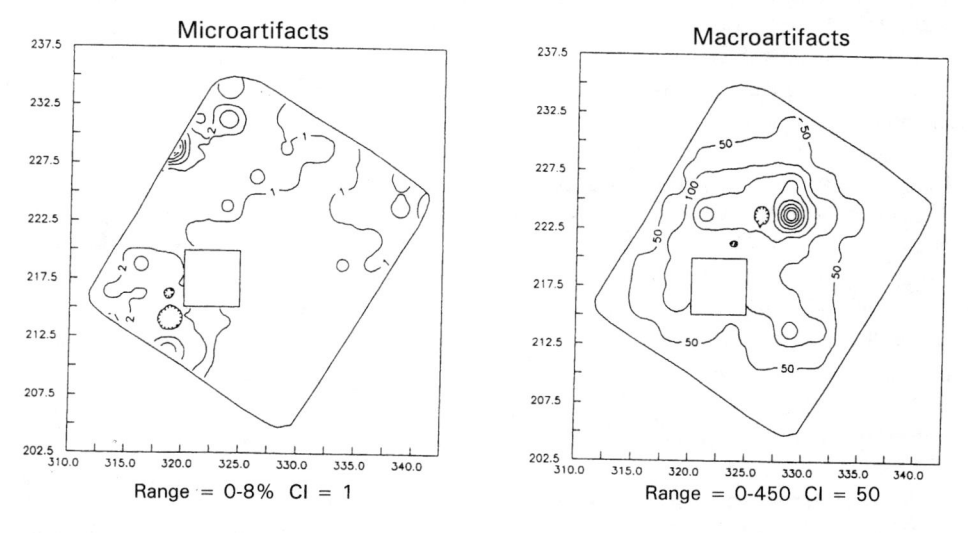

Figure 12.5. Ceramic distribution in Structure 3 at the Loy Site. (The range for the microartifacts is reported in volume percentages; the range for macroartifacts is reported in total counts. The CI indicates the contour interval.)

4.2. Landscape-Scale Research

The cultural depositional history for a region includes the detection of surficial and buried artifact accumulations (Ebert, 1992; Fish and Kowalewski, 1989; Linse, 1993). The identification of buried sites can distinguish a depositional hiatus and signify the presence of buried soils. To the landscape ecologist, buried surfaces attest to the presence of extinct or extant ecosystems. To the archaeologist, buried sites are a clue about potential variation in regional diachronic settlement patterns and changing environments. Since the 1980s, the research potential of buried archaeological sites and their prospective National Register status has become more important in the assessment of the presence of cultural resources in an area. Federal and state cultural resource management guidelines stipulate that the potential for buried sites be investigated and their National Register status considered.

Regional landscape studies have enlisted geoarchaeologists, geomorphologists, and soil scientists to identify subsurface deposits, often relying on limited testing techniques (deep trenching and coring; e.g., Brakenridge, 1984; Delcourt, 1980; Hajic, 1990). Sampling in increments across the landscape, if restricted to macroartifacts with their potentially lower densities at this scale of observation, may fail to identify buried sites. Microartifacts, given their relative abundance and higher probability of recovery in a small volume sample (Madson and Dunnell, 1989; Stafford, 1995), offer the necessary quantitative data to identify artifact densities across a landscape.

Buried surfaces can usually be detected through vertical lithologic changes and the presence of paleosols. Yet sometimes surfaces agrade at a rate, or in an environment, where soil formation is slow or the new sedimentary deposit is not appreciably different from the underlying sediments. The result is an ephemeral or absent buried A horizon, or a poorly developed B horizon, that is difficult to identify due to its similarity with the surrounding units (i.e., entisols or inceptisols). On such rapidly buried surfaces the only record of stability may be surviving evidence of human occupation—artifacts. The regional scale of questioning applied to microartifacts negates unique interpretations of the small-sized artifacts and treats them simply as artifacts. They are merely sand-sized particles produced by human actions, representing artifact accumulations on the landscape.

Because numerous processes affect differently sized particles, this approach is not without limitations. The use of microartifacts at the regional scale must therefore be considered carefully if applied in aeolian or coarse-grained depositional environments, where natural processes may winnow out smaller debris (Nicholson, 1983). Better suited are environments (especially alluvial) composed of finer grained loam, silt, and clay deposits. These environments, characterized by low-energy transport agents and deposition of fine sand, silt, and clay particles, generally produce reliable correlations between site location and microartifacts (Stafford, 1993).

In order to illustrate the application of microartifact analysis in the identification of relic surfaces, two brief examples are offered; Stright's (1986) work on the Continental Shelf and a pilot study reported from Stafford's (1993, 1995) ongoing research in the Ohio River Valley.

4.2.1. Sabine River Valley

In an effort to identify archaeological sites on submerged relic landforms along the continental shelf, Stright (1986) employed both high-resolution seismic profiles and sedimentological analyses. The seismic profiles were used to locate and identify Late Wisconsin relic fluvial landforms along the ancient Sabine River Valley in the Gulf of Mexico. Five of the landforms were then cored (vibracores) and the cores were subjected to physical analyses to confirm the presence or absence of an archaeological site. These analyses used criteria established by base-line studies (grain size, point count, and geochemical analyses) performed on different analogous site-types in the Gulf of Mexico coastal region.

The baseline study considered several types of landforms and included both on- and off-site samples. The results from the grain size and point count analyses indicated that the particle size distribution in general was not a useful indicator for archaeological deposits. Discriminant function analysis of the sand-sized point-count data, however, identified the presence of bone and charred material as significant predictors for on-site versus off-site deposits. Flakes (or angular lithic particles with the same attributes as microdebitage) were suprisingly found in both on- and off-site samples and were therefore insignificant site predictors. Microdebitage (where chipped stone technology was in use), because of its durability and distinctive attributes, is usually considered a reliable site indicator and should be carefully considered when designing a microartifact analysis.

Stright's (1986:363) preliminary findings require refinement through the addition of other marine environments in the baseline study (specifically deposits resulting from natural processes). Still, this study illustrates the utility of sand-sized artifact analysis where sampling opportunities are restricted as well as an emphasis on the research toward determining where sites are located.

4.2.2. Ohio River Floodplain

Using augered core samples in conjunction with 2×2 m excavation units on the Ohio River floodplain, Stafford (1993, 1995) illustrated the usefulness of microartifacts as indicators of buried sites. The core samples were collected at 10 cm vertical intervals and divided into fractions greater than -2.0ϕ (>4 mm), -1.0ϕ and -2.0ϕ (2–4 mm), and -1.0ϕ and 0.0ϕ (1–2 mm), in which debitage, ceramic, oxidized sediment, bone, wood, and nutshell were counted. The corresponding 2×2 m excavation units, following the same levels as the cores, collected macroartifacts only (>6.35 mm; 0.25 inch). The -1.0 and 0.0ϕ (1.0–2.0 mm) core fraction produced the greatest percentage of artifacts even though the volume of the sediment in the excavation units measured 200 times the volume of the auger samples (Stafford, 1993:8). The comparison between the core samples and the adjacent excavation units show a basic correlation, as both methods identified artifact concentrations at approximately the same depths (Fig. 12.6). There is some variability, in that the core results identified microartifact concentrations that did not appear in macroartifact distributions.

This exercise illustrates the use of microartifacts to identify artifact concentrations at a regional scale while also showing that different sizes have variable

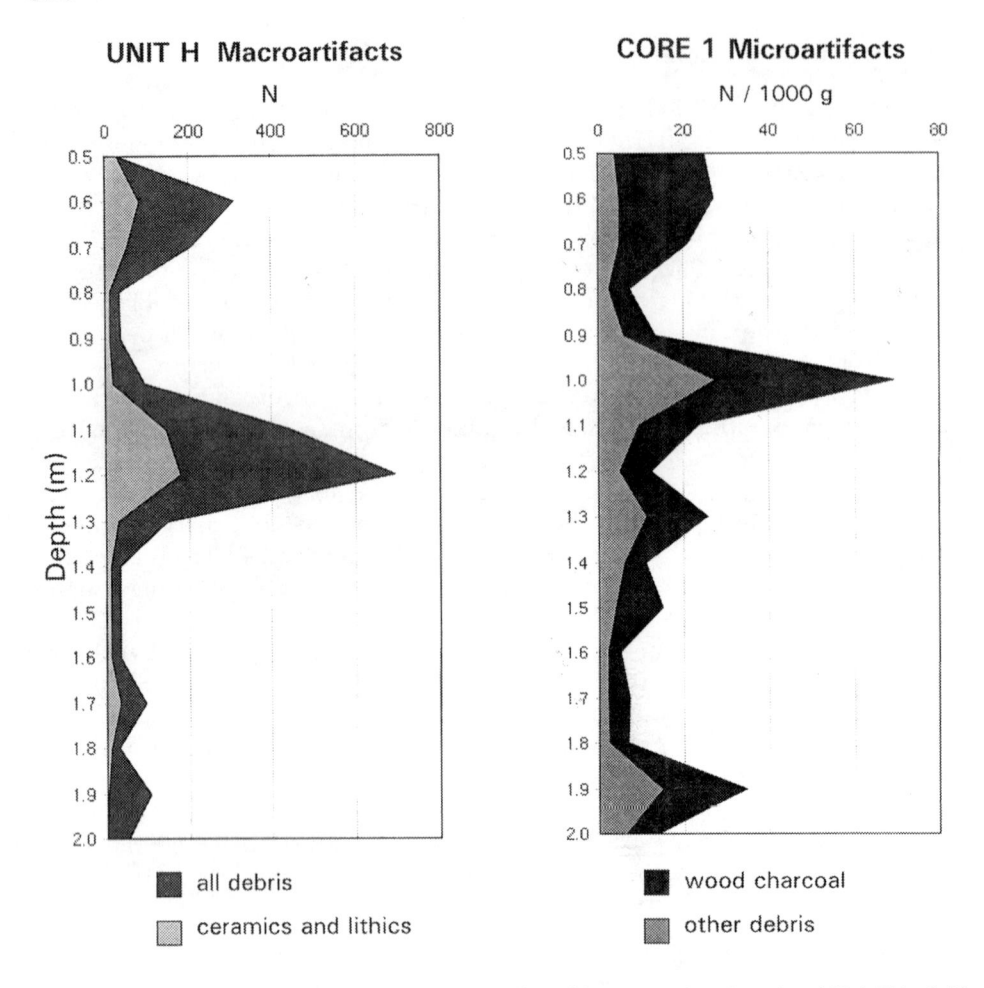

Figure 12.6. Frequency distributions of materials collected from Core 1 (>1 mm) and Unit H (>6.35 mm). (after Stafford, 1995, Figure 5)

concordance relationships. Overall the microartifacts distinguished the buried surfaces identified during the more time-consuming excavation of the 2 × 2 m unit. Stafford is continuing the coring technique with microartifact analyses in the Ohio River Valley. The results of this study will be one of the first full regional scale microartifact studies.

5. Conclusions

Microartifacts are an abundant and informative aspect of artifact size distributions. This chapter has sought to familiarize the reader with the principles underlying the study of microartifacts and the potential research applications of this largely untapped resource. The suggestions made in regard to streamlining the protocol make the process of recovering and quantifying microartifacts more efficient and therefore easier to implement. Using either the heavy fraction from a flotation sample or the nylon bags of Stafford's (1993, 1995) protocol can significantly reduce the process of retrieving microartifacts from the matrix. Employing computer programs to quantify microartifact size fractions expedites the counting process and enters the data in a proportional database ready for analysis.

Microartifact analysis cannot be applied in all depositional environments in a reliable and efficient way. In porous deposits, such as shell middens and burned rock middens, systematic vertical samples should first be analyzed to determine if microartifacts may have moved downward, falsely skewing artifact size distributions (e.g., Madsen, 1992). Deposits with high clay contents may prolong the processing time required to separate microartifacts from the rest of the matrix. These conditions are relatively rare among archaeological deposits in general, and most contexts—including plowed sites and buried alluvial environments—can benefit from microartifact analysis.

The Mississippian period structure in the Loy site example illustrates the necessity of combining micro- and macroartifact analyses to provide the clearest view of activities and formation processes in a site-scale context. The behavioral interpretations generated from the Loy site distributions are admittedly tenuous. Improved interpretive frameworks for site-scale questions involving site structure, assemblage composition, and natural site formation processes are needed to reliably interpret fine-scale horizontal distributions. These frameworks must be generated through experimental studies structured to create a theoretical framework for separate material types in different environments. Ethnoarchaeological research in site structure has consistently observed the potential for artifact size to define activity areas (Hayden and Cannon, 1983; O'Connell, 1987; Simms, 1988). Further, Gamble (1991) emphasized the potential for artifact size to provide cross-cultural comparisons relating to behavioral patterns of spatial organization and of artifact production, use, and discard. In spite of the recognized potential, the dynamics of causation behind the variability in archaeological microartifact material classes are still largely unknown. We know the most about size distribution in lithic material due to experiments in replicating reduction sequences and their attention to size distributions. More experimental research, such as Nielson's work (1991a, 1991b) and ethnoarchaeological research firmly linked to archaeological practice and interpretation (Simms, 1992), is essential before the behavioral processes generating artifact size distributions are understood well enough to make reliable behavioral interpretations.

Microartifact analysis in regional scale research has been underutilized. The success of Stafford's (1993, 1995) research with identifying buried sites in alluvial contexts using limited-volume coring should inspire further applications. Making use of expedient processing protocols and quantification techniques in general should encourage the implementation of microartifact analysis in the identification of buried sites and regional scale research.

Extending the threshold of microartifacts up to 6.35 mm to include the gap between the existing definition of 2 mm and the limits of typical field sieving is essential to the collection, sampling, and interpretation of artifact size distributions. Presently, very few studies have spanned this gap, overlooking a significant portion of the archaeological record. In the future, taking microartifact distributions into account, earth scientists in general will be better equipped to address questions of natural and cultural depositional histories at both site and regional scales.

ACKNOWLEDGMENTS. My sincere thanks to J. Lev-Tov, H. N. Qirko, J. F. Simek, J. K Stein, and the editors of this volume for their valuable comments on various drafts of this chapter. I am also grateful to C. R. Stafford for sharing both his research and his nylon bag idea for microartifact processing.

6. References

Ahler, S. A., 1975, *Pattern and Variability in Extended Coalescent Lithic Technology.* Ph.D. diss., University of Missouri, Columbia. University Microfilms International, Ann Arbor, MI.
Ahler, S. A., 1989, Mass Analysis of Flaking Debris: Studying the Forest Rather than the Trees. In *Alternative Approaches to Lithic Analysis,* edited by D. O. Henry and G. H. Odell, pp. 85–118. Archaeological Papers of the American Anthropological Association Number 1.
Ambrose, W., 1963, Shell Dump Sampling, *New Zealand Archaeological Association Newsletter* 6(3):155–159.
Baumler, M. F., 1985, On the Interpretation of Chipping Debris Concentrations in the Archaeological Record, *Lithic Technology* 14:120–125.
Binford, L. R., 1978, Dimensional Analysis of Behavior and Site Structure: Learning From an Eskimo Hunting Stand, *American Antiquity* 43:330–361.
Binford, L. R., 1981, *Bones,* Academic Press, New York.
Borland International, 1992, *Paradox User's Guide, Version 4.0* Borland International, Scotts Valley, California..
Bradley, R., and Fulford, M., 1980, Sherd Size in the Analysis of Occupation Debris, *Bulletin of the Institute of Archaeology* 17:85–94.
Brakenridge, G. R., 1984, Alluvial Stratigraphy and Radiocarbon Dating along the Duck River, Central Tennessee: Implications Regarding Flood-plain Origin, *Geological Society of America Bulletin* 95:9–25.
Buck, P. E., 1990, Deposits in the Western Nile Delta, Egypt: A Geoarchaeological Example from Kom EL-Hisn. Ph.D. diss. University of Washington.
Butzer, K. W., 1982, *Archaeology as Human Ecology,* Cambridge University Press, New York.
Chase, P., 1985, Whole Vessels and Sherds: An Experimental Investigation of their Quantitative Relationships, *Journal of Field Archaeology* 12:213–218.
Clark, J. E., 1986, Another Look at Small Debitage and Microdebitage, *Lithic Technology* 15:21–33.
Cook, S. F., and Treganza, A. E., 1947, The Quantitative Investigations of Aboriginal Sites: Comparative Physical and Chemical Analysis of Two California Indian Mounds, *American Antiquity* 13:135–142.
Courty, M.-A., Goldberg, P., and Macphail, R., 1989, *Soils and Micromorphology in Archaeology,* Cambridge University Press, London.
Delcourt, P. A., 1980, Quaternary Alluvial Terraces of the Little Tennessee River Valley, East Tennessee. In *The 1979 Archaeological and Geological Investigations in the Tellico Reservoir Area,* edited by J. Chapman, *Report of Investigations* 29. Department of Anthropology, University of Tennessee.
Dunnell, R. C., and Stein, J. K., 1989, Theoretical Issues in the Interpretation of Microartifacts, *Geoarchaeology* 4(1):31–42.
Dunnell, R. C., and Simek, J. F., 1995, Artifact Size and Plowzone Processes, *Journal of Field Archaeology* 22(3):305–319.

Ebert, J. I., 1992, *Distributional Archaeology*, University of New Mexico Press, Albuquerque.
Eidt, R. C., 1985, Theoretical and Practical Considerations in the Analysis of Anthrosols. In *Archaeological Geology*, edited by G. R. Rapp, pp. 155-190. Academic Press, San Diego, CA.
Fish, S. K., and Kowalewski, S. A., 1989, *The Archaeology of Regions: A Case for Full-Coverage Survey*, Smithsonian Institution Press, Washington, DC.
Fladmark, K. R., 1982, Microdebitage Analysis: Initial Considerations, *Journal of Archaeological Science* 9:205-220.
Folk, R. L., 1980, *Petrology of Sedimentary Rocks*, Hemphill Publishing Co., Austin, TX.
Galehouse, J. S., 1971, Point Counting, In *Procedures in Sedimentary Petrology*, edited by R. Carver, pp. 385-407. Wiley, New York.
Gallagher, J. P., 1977, Contemporary Stone Tools in Ethiopia: Implications for Archaeology, *Journal of Field Archaeology* 4:407-414.
Gamble, C., 1991, An Introduction to the Living Spaces of Mobile Peoples. In *Ethnoarchaeological Approaches to Mobile Campsites: Hunter-Gatherer and Pastoralist Case Studies*, edited by C. Gamble and W. A. Boismier, pp. 1-24, Prehistory Press, International Monographs in Prehistory, Madison, WI.
Ge, T., Courty, M.-A., Matthews, W., and Wattez, J., 1993, Sedimentary Formation Processes of Occupation Surfaces. In *Formation Processes in Archaeological Context*, edited by P. Goldberg, D. T. Nash, and M. D. Petraglia, pp. 149-164. Prehistory Press, Madison, WI.
Hajic, E. R., 1990, Late Pleistocene and Holocene Landscape Evolution, Depositional Subsystems, and Stratigraphy in the Lower Illinois River Valley and Adjacent Central Mississippi River Valley diss, University of Illinois.
Hally, D. J., 1983, The Interpretive Potential of Pottery from Domestic Contexts, *Midcontintental Journal of Archaeology* 8:163-196.
Hassan, F. A., 1978, Sediments in Archaeology: Methods and Implications for Paleoenvironmental and Cultural Analysis, *Journal of Field Archaeology* 5:197-213.
Hayden, B., and Cannon A., 1983, Where the Garbage Goes: Refuse Disposal in the Maya Highlands, *Journal of Anthropological Archaeology* 2(2)117-163.
Hofman, J. L., 1992, Defining Buried Occupation Surfaces in Terrace Sediments. In *Piecing Together the Past: Applications of Refitting Studies in Archaeology*, edited by J. L. Hofman and J. G. Enloe, pp. 128-150 BAR International Series, Volume 578. British Archaelogical Reports, Oxford.
Hull, K. L., 1987, Identification of Cultural Site Formation Processes through Microdebitage Analysis, *American Antiquity* 52:772-783.
Krumbein, W. C., and S. S., Sloss, 1963, *Stratigraphy and Sedimentation*, W.H. Freeman, San Francisco.
Kvamme, K. L., 1989, Geographic Information Systems in Regional Archaeological Research and Data Management. In *Archaeological Method and Theory* Volume 1, edited by M. B. Schiffer, pp. 139-204, University of Arizona Press, Tucson.
Lewarch, D. E., and O'Brien, M. J., 1981, The Expanding Role of Surface Assemblages in Archaeological Research. In *Advances in Archaeological Method and Theory* Volume 4. edited by M. B. Schiffer, pp. 297-342, Academic Press, New York.
Linse, A. R., 1993, Geoarchaeological Scale and Archaeological Interpretation: Examples from the Central Jornada Mogollon. In *Effects of Scale on Archaeological Geoscientific Perspectives*, edited by J. K. Stein and A. R. Linse, pp. 11-28. Geological Society of America, Boulder, CO.
Lyman, R. L., and O'Brien, M. J., 1987, Plow-Zone Zooarchaeology: Fragmentation and Identifiability, *Journal of Field Archaeology* 14:3-498.
Madsen, M. E., 1992, Lithic Manufacturing at British Camp: Evidence from Size Distributions and Microartifacts. In *Deciphering A Shell Midden*, edited by J. K. Stein, pp. 193-210 Academic Press, San Diego, CA.
Madsen, M. E., and Dunnell, R. C., 1989, *The Role of Microartifacts in Deducing Land Use from Low Density Records in Plowed Surfaces*. Paper presented at the 54th annual meeting of the Society for American Archaeology, Atlanta, GA.
Matthews, W., French, C. A. I., Lawrence, T., Cutler, D. F., and Jones, M. K., 1997, Microstratigraphic Traces of Site Formation Processes and Human Activities, *World Archaeology* 29(2):281-308.
McBrearty, S., 1990, Consider the Humble Termite; Termites as Agents of Post-Depositional Disturbance at African Archaeological Sites, *Journal of Archaeological Science* 17(2):111-143.

Metcalfe, D., and Heath, K. M., 1990, Microrefuse and Site Structure: The Hearths and Floors of the Heartbreak Hotel, *American Antiquity* 55(4):781-796.

Middleton, W. D., and Price, T. D., 1996, Identification of Activity Areas by Multi-Element Characterization of Sediments from Modern and Archaeological House Floors Using Inductively Coupled Plasma-Atomic Emission Spectroscopy, *Journal of Archaeological Science* 23:673-687.

Murray, P., 1980, Discard Location: The Ethnographic Data, *American Antiquity* 45:490-502.

Nelson, M., 1987, Site Contents and Structure: Quarries and Workshops in the Maya Highlands. In *Lithic Studies Among the Contemporary Highland Maya*, edited by B. Hayden, pp. 120-147. University of Arizona Press, Tucson.

Nicholson, B. A., 1983, A Comparative Evaluation of Four Sampling Techniques and of the Reliability of Microdebitage as a Cultural Indicator in Regional Surveys, *Plains Anthropologist* 28:273-281.

Nielsen, A. E., 1991a, Trampling the Archaeological Record: An Experimental Study, *American Antiquity* 56:483-503.

Nielsen, A. E., 1991b. *Where Do microartifacts Come From.* Paper presented at the 56th annual meeting of the Society for American Archaeology, New Orleans, Louisiana.

O'Brien, M. J., and Lewarch, D. E. (eds.), 1981, *Plowzone Archaeology: Contributions to Theory and Technique*, Vanderbilt University Press, Nashville, TN.

O'Connell, J. F., 1987, Alyawara Site Structure and Its Archaeological Implications, *American Antiquity* 52:74-108.

Pearsall, D. M., 1989, *Paleoethnobotany: A Handbook of Procedures*, Academic Press, San Diego, CA.

Petraglia, M. D., and Nash, D. T., 1987, The Impact of Fluvial Processes on Experimental Sites. In *Natural Formation Processes and the Archaeological Record*, edited by D. T. Nash and M. D. Petraglia, pp. 108-130. BAR International Series, Volume 352. British Archaeological Reports, Oxford.

Polhemus, R., 1985, Mississippian Architecture: Temporal, Technological, and Spatial Patterning of Structures at the Toqua Site (40MR6). M.A. thesis, University of Tennessee, Knoxville.

Reese, J. A., 1986, *Microarchaeological Analysis of the Chinese Worker's Area at the Warrendale Cannery Site, Oregon.* Paper presented at the 51st annual meeting of the Society for American Archaeology, New Orleans, Louisiana.

Reineck, H. E., and Singh, I., 1986, *Depositional Sedimentary Environments*, Springer-Verlag, New York.

Rosen, A. M., 1986, *Cities of Clay: The Geoarchaeology of Tells*, University of Chicago Press, Chicago.

Rosen, A. M., 1989, Ancient Town and City Sites: A View from the Microscope, *American Antiquity* 54(3):564-578.

Rosen, A. M., 1993, Microartifacts as a Reflection of Cultural Factors in Site Formation. In *Formation Processes in Archaeological Context*, edited by P. Goldberg, D. T. Nash, and M. D. Petraglia, pp. 141-148. Prehistory Press, Madison, WI.

Schick, K. A., 1986, *Stone Age Sites in the Making: Experiments in the Formation and Transformation of Archaeological Occurrences*, BAR International Series, Volume 163. British Archaeological Reports, Oxford.

Schiffer, M. B., 1987, *Formation Processes of the Archaeological Record*, University of New Mexico Press, Albuquerque.

Shea, J. J., and Klenck, J. D., 1993, An Experimental Investigation of the Effects of Trampling on the Results of Lithic Microwear Analysis, *Journal of Archaeological Science* 20:175-194.

Sherwood, S. C., 1991, Microartifact Analysis of a Dallas Phase House Floor. M.A. thesis, University of Tennessee, Knoxville.

Sherwood, S. C., and Ousley, S. D., 1995, Quantifying Microartifacts Using a Personal Computer, *Geoarchaeology* 10(6):423-428.

Sherwood, S. C., Simek, J. F., and Polhemus, R. R., 1995, Artifact Size and Spatial Process: Macro- and Microartifacts in a Mississippian House, *Geoarchaeology* 10(6):429-455.

Shott, M. J., 1994, Size and Form in the Analysis of Flake Debris: Review and Recent Approaches, *Journal of Archaeological Method and Theory* 1:69-110.

Simms, S. R., 1988, The Archaeological Structure of a Bedouin Camp, *Journal of Archaeological Sciene* 15:197-211.

Simms, S. R., 1992, Ethnoarchaeology: Obnoxious Spectator, Trivial Pursuit, or the Keys to a Time Machine. In *Quandaries and Quests: Visions of Archaeology's Future*, edited by L. Wandsnider, pp. 186-198. Center for Archaeological Investigations, Carbondale, IL.

Simms, S. R., and Heath, K. M., 1990, Site Structure of the Orbit Inn: An Application of Ethnoarchaeology, *American Antiquity* 55:797-813.

Spaulding, A. C., 1960, The Dimensions of Archaeology. In *Essays in the Science of Culture*, edited by G. E. Dole., and R. L. Carniero, pp. 437-456 T.Y. Crowell, New York.

Stafford, C. R., 1993, *Applying Distributional Archaeology to the Subsurface: Some Initial Observations*, Paper presented at the 58th Annual Meeting of the Society for American Archaeology, St. Louis, MO.

Stafford, C. R., 1995, Geoarchaeological Perspectives on Paleolandscapes and Regional Subsurface Archaeology, *Journal of Archaeological Method and Theory* 2(1):69-104.

Stahle, D. W., and Dunn, J. E., 1982, An Analysis and Application of the Size Distribution of Waste Flakes from the Manufacture of Bifacial Stone Tools, *World Archaeology* 14:84-97.

Stein, J. K., 1985, Interpreting Sediments in Cultural Settings. In *Archaeological Sediments in Context*, edited by J. K. Stein and W. R. Farrand, pp. 5-20 Center for the Study of Early Man, Orono, ME.

Stein, J. K., 1987. Deposits for Archaeologists. In *Advances in Archaeological Method and Theory* Volume 11, edited by M. B. Schiffer, pp. 337-392. Academic Press, Orlando, FL.

Stein, J. K., 1993, Scale in Archaeology, Geosciences, and Geoarchaeology. In *Effects of Scale on Archaeological Geoscientific Perspectives*, edited by J. K. Stein and A. R. Linse pp. 1-10. Geological Society of America, Boulder, CO.

Stein, J. K., and Teltser, P. A., 1989, Size Distributions of Artifact Classes: Combining Macro- and Micro-fractions, *Geoarchaeology* 4(1):1-30.

Stright, M. J., 1986, Human Occupation of the Continental Shelf During the Late Pleistocene/Early Holocene: Methods for Site Location, *Geoarchaeology* 1(4):347-364.

Vance, D. E., 1986, Microdebitage Analysis in Activity Analysis: An Application, *Northwest Anthropological Research Notes* 20:179-189.

Vance, D. E., 1989, The Role of Microartifact in Spatial Analysis. Ph.D. diss, University of Washington, Seattle.

Visher, G. S., 1969, Grain Size Distributions and Depositional Processes, *Journal of Sedimentary Petrology* 39:1074-1106.

Villa, P., 1982, Conjoinable Pieces and Site Formation Processes, *American Antiquity* 47:276-290.

Watson, P. J., 1976, In Pursuit of Prehistoric Subsistence: A Comparative Account of Some Contemporary Flotation Techniques, *Midcontinental Journal of Archaeology* 1:77-100.

13

Current Practices in Archaeogeophysics

Magnetics, Resistivity, Conductivity, and Ground-Penetrating Radar

KENNETH L. KVAMME

1. Introduction

As the cost of conducting archaeological research continues to rise, so has the need to define site structure while minimizing excavation efforts. At the same time there is a growing ethic for site conservation and the use of noninvasive procedures for site exploration. Near surface geophysical methods are increasingly seen as useful and cost-effective tools for archaeological exploration because they provide a means to remotely sense what lies beneath the earth. By raising the probability of encountering sought-for features, data recovery efficiency through excavation is increased and costs are reduced. Yet, the very success of geophysical methods has led to a misconception that they are suited only for discovery purposes. A growing perspective is that they are useful, in themselves, for acquiring primary data of relevance to a variety of research questions (Dalan, 1993; Summers et al., 1996). In some contexts, for example, and depending on

---

KENNETH L. KVAMME • Department of Anthropology and Center for Advanced Spatial Technologies, University of Arkansas, Fayetteville, Arkansas 72701.

Earth Sciences and Archaeology, edited by Paul Goldberg, Vance T. Holliday, and C. Reid Ferring. Kluwer Academic/Plenum Publishers, New York, 2001.

research goals, the need for excavation may be precluded as when geophysical surveys provide accurate plans of buried architectural remains that offer sufficient information for cultural resource management purposes. Indeed, the simple mapping of architectural and other anthropogenic features (e.g., trenches, ditches, pits, middens, pathways) through geophysics now constitutes a principal means of site inventory in the United Kingdom and elsewhere (David, 1995; Payne, 1996).

Whereas near-surface geophysical methods and instrumentation have been available for some time in archaeology (see Clark [1990] for an historical overview and Conyers and Goodman [1997] for a history of ground-penetrating radar), since the late 1980s, a number of significant changes have occurred owing to advances in technology. One change is the greater sensitivity of the instruments through better electronics. More subtle signals or responses can be measured with greater spatial resolution and accuracy. A second change is a 50- to 100-fold increase in the speed of some of the instruments. For example, in the mid-1980s a proton precession magnetometer might yield a measurement in 5 to 7 seconds; in the year 2000, magnetic gradiometers output as many as ten readings per second. Third, computer memory devices like data loggers (essentially large arrays of computer memory chips) are routinely employed to store measurements, precluding the need for manual recording in the field, and allowing a means for rapid linkage with computers for processing. Taken together, these changes have profoundly altered the practice, conduct, and, indeed, some of the goals of archaeological geophysics.

The greater speed of many of the instruments means that survey of very large areas may be contemplated in relatively short amounts of time (e.g., Summers et al., 1996). Surveys using contemporary magnetic gradiometers, electrical resistivity, and electromagnetic conductivity meters routinely approach coverage of from one half to one hectare a day or more. At the same time, the intensity of survey in a given area may be increased by sampling measurements more densely, yielding greater spatial detail. A consequence of these practices is that it is not unusual to collect tens of thousands of measurements per instrument, per day (Scollar et al., 1990:491)!

The only way to handle this volume of data is through computerization, and this circumstance alone has ushered in the greatest change in contemporary archaeogeophysics. Each instrument yields a matrix of measurements that can be treated as imagery by the computer. Various image processing procedures therefore become important in the treatment of these data to remove noise, improve brightness, adjust contrast, enhance edges, define linear features (common cultural expressions), and find subtle details that may not be readily apparent in the raw data (Eder-Hinterleitner et al., 1996; Music, 1995; Neubauer et al., 1996; Scollar et al., 1990). In many contexts it is possible to visualize culturally generated patterns in geophysical data with sufficient clarity that even the nonspecialist can understand and interpret the output, as when the floor plan of a prehistoric house or the layout of an entire village is clearly expressed (e.g., Becker, 1995; Dawson and Gaffney, 1996; Scollar et al., 1990:509–511; Summers et al., 1996). This outcome is probably the single factor most responsible for the growing acceptance of geophysical survey results in the larger archaeological community.

These changes in the practice of archaeological geophysics have not come without a price. Competency now requires extensive skills in the computer processing and image analysis of the data. Current practitioners of archaeological geophysics tend to be more the computer scientist than were their colleagues of 1990, and there is a fundamental change in goals. No longer is the geophysicist out in the field studiously hand-contouring sparse matrices of measurements, attempting to ascertain the nature of subsurface content through deductive reasoning of physical science principles. Rather, the typical goal today is to capture as much data over as large an area as possible and to process the data to clarify culturally formed patterns in the deposits. The latter arises from a basic tenet recognized in air photo interpretation (Avery and Berlin, 1985): patterned geometries in the landscape like circles, ellipses, squares, rectangles, and lines generally are of cultural origin (they occur much less frequently as a result of natural processes). Survey of large areas increases the probability that features with regular, interpretable geometries will be encountered (e.g., complete houses or house clusters, walls, fortifications, etc.). Moreover, with tens of thousands of geophysical measurements distributed over a large area it is no longer practical to apply deductive reasoning to more than a few of the more interesting contexts. The practice of contemporary archaeogeophysics is therefore much more inductive, with a focus on the end-image and patterns in it, rather than deductive.

The computerization of geophysical surveys has also caused a scale change. In past decades, with relatively few measurements, only small areas could be examined. In the year 2000, surveys now map entire settlements and examine interrelationships between individual houses, lanes between them, and other features like middens and fortifications (e.g., Keay et al., 1991; Payne, 1996; Summers et al., 1996; see Weymouth and Nickel, 1977, and Scollar, 1971, for early large-area surveys).

The contemporary archaeogeophysicist, then, must combine skills in a rather eclectic mix of physics, geology, archaeology, and computer science, a daunting task to the newcomer (Schurr, 1997). In the space of this chapter only selected aspects of these pursuits can be presented. I review basic geophysical principles and technologies, contemporary instrumentation, field methods, and outline general data processing methods. Finally, several case studies are examined that illustrate appropriate uses of geophysics in a variety of archaeological contexts.

2. Geophysical Prospection Principles

Archaeological geophysical prospection involves a variety of techniques designed to record the physical properties of near-surface deposits. In the context of archaeology, "near-surface" generally refers to the uppermost 1 to 2 m, although deeper prospection is also practiced. The various sensing technologies each possess depth limitations that vary according to particular earth properties that occur at a site. Active and passive sensing methods are utilized. Active methods might pass an electrical current or radar energy through the earth, for example, and record how they respond to subsurface characteristics. Passive methods, on the other hand, measure inherent or natural properties detectable at the surface. Although a bewildering array of techniques and technologies have been investigated in each domain (Clark, 1990; Gaffney et al., 1991; Weymouth, 1986; Wynn, 1986), four principal methods are focused on here: magnetometry, electrical resistivity, electromagnetic conductivity, and ground-penetrating radar (GPR). Except for magnetometry, a passive technique, all are active prospection methods.

Regardless of the method employed, useful geophysical findings are the result of contrasts between archaeological deposits and the natural background geology. In other words, if archaeological deposits or features possess physical properties different from the surrounding matrix, then they may offer a contrast against the natural background in terms of magnetic characteristics, resistance to an electrical current, or their ability to reflect radar energy. Such contrasts are commonly referred to as "anomalies" until they can be identified. A buried limestone block foundation, for example, might be somewhat less magnetic, more resistant to an electrical current (low conductivity), and might better reflect radar energy than the surrounding earth. Frequently, contrasts may not be large enough to be identified or they may not illustrate sufficient pattern for clear interpretation. Consequently, successful applied work is something of an art that combines scientific knowledge with a good amount of judgement based on experience and context. Archaeological testing of identified anomalies remains the principal means to confirm their identity, but it cannot always be undertaken, especially when geophysical surveys are performed over large areas and numerous anomalies are encountered.

The field task of the archaeogeophysicist is to collect and assemble a matrix of geophysical measurements taken systematically across a surface in order that significant contrasts may be identified and mapped for purposes of archaeological interpretation.

3. Field Survey Methods

Area-focused geophysical surveys in archaeology typically are conducted in a series of grids that control the placement of the instruments over the landscape. Grid sizes of 10×10 m to 50×50 m are commonly employed. Each grid is established by staking a series of ropes parallel to each other on the ground. The length of each rope equals the grid size and each possesses meter marks through its length. The ropes might be placed 1 m apart to form the grid. The geophysical instruments are then moved along each rope where measurements are recorded in meter or submeter increments. The meter marks insure that each measurement is correctly located spatially. Geophysical surveys thus record data in a series of equally spaced parallel lines, with data recorded at regular intervals along each line, forming a matrix of measurements (see Scollar et al., 1990:478–488). Depending on survey needs and instrumentation, data might be sampled along a line every 0.5 to 1 m (resistivity, conductivity), 0.1 to 0.5 m (magnetic gradiometry), or through thousands of pulses per meter (GPR). High-precision surveys employ line separations of 0.5 m or even 0.25 m. After the survey of a grid, another is established, often adjacent to the previous, where survey commences again.

Level fields with short mowed grass are best suited for geophysical survey as dense vegetation and other impediments can hinder efficient movement of the instruments. Steep slopes can make movement of heavy instruments difficult (e.g., large GPR antennas). Additionally, some instruments are based on a uniform rate of movement in a given time interval, which is made difficult by steep slopes. Field methods unique to specific instruments are discussed in the following sections.

4. Geophysical Methods and Instruments

4.1. Magnetic Methods

Magnetic surveys measure minute variations in earth magnetic properties across an area. These fluctuations can be extremely subtle owing to low levels of iron in earth deposits that stem from iron compounds like hematite, magnetite, and maghaemite (Weymouth, 1986:342). A prehistoric ditch that was subsequently filled with sediments, for example, might yield a magnetic contrast with the surrounding matrix because the fill is slightly more (or less) magnetic. The degree of soil magnetism is referred to as its *magnetic susceptibility*. Fired materials, such as baked clays around hearths or a burned house floor, tend to possess elevated magnetic properties, referred to as *thermoremanent magnetism*. Magnetic survey methods are therefore ideal for locating burned areas, hearths, kilns, fired bricks, and the like, owing to the strong thermoremanent response. Historic iron artifacts possess large induced magnetic fields that are readily detected, which can be a blessing if one is after a few iron-bearing artifacts, but can be a nightmare on certain historic period sites (or sites with modern surface trash) where such items as a rain of nails, steel cans, or wire can obscure subtle magnetic details beneath. Strongly magnetic artifacts or features tend to yield a dipole field, expressed as paired positive and negative extremes (much like the north and south poles of a magnet), that are frequently aligned on a north–south axis unless the source has a principal axis aligned in a different direction (see Clark, 1990; Scollar et al., 1990; Weymouth, 1986, for other details).

Magnetic field strength is measured in nanoteslas (nT; 10^{-9} Tesla). In North America and much of Europe the background magnetic field strength ranges from about 40,000 to 60,000 nT (Weymouth, 1986:341). This is noteworthy because magnetic anomalies of potential archaeological interest often lie well within ± 5 nT, and soil unit differences can be as subtle as 0.1 nT and less (recent work by Becker, 1995, shows anomalies in the picotesla [.001 nT] range). Magnetic survey instrumentation therefore is incredibly sensitive, capable of detecting less than one part in a half million. The instrument operator must be free of ferrous material. Steel fences, passing automobiles, and magnetic fields generated by power lines all pose survey difficulties. An additional problem is the diurnal variation of the Earth's magnetic field, which changes continuously. On a typical day it might vary over 40 to 100 nT (Weymouth and Lessand, 1986), but occasional magnetic "storms" can produce changes in the field that range over hundreds of nT in the course of hours. Survey methods and instrument design have been established specifically to allow for this phenomenon.

The proton precession magnetometer is one of the earliest magnetic sensing technologies used in archaeology (Clark, 1990). These instruments, capable of 0.1 nT resolution, measure the total magnetic field strength, typically requiring 5 to 7 seconds for a reading. The diurnal variation problem is usually handled through use of a second instrument—located at a fixed base station—that records magnetic variation simultaneously with the roving field unit. The roving unit records magnetic variations across space and time, while the base unit records changes in time only. Simply by differencing the two data sets one derives magnetic measurements that vary across space only, obtained by the roving field unit. With the advent of fast magnetic gradiometers, proton magnetometers are now used less frequently in applied work.

Fluxgate and cesium-vapor gradiometers are popular instruments in archaeology, capable of yielding 8 to 10 measurements per second at less than 0.1 nT resolution (Scollar et al., 1990). As gradiometers they do not measure total magnetic field strength. Rather, they employ two sensors vertically separated by about 0.5 m. The difference between the two readings yields a measure of the vertical gradient of the magnetic field while the simultaneous readings eliminate diurnal change effects.

The FM-36 fluxgate gradiometer, by Geoscan Research (1995), is specifically designed for archaeological application. The sensors are separated by 0.5 m, its resolution is less than 0.1 nT, it is fully computerized, and capable of storing 16,000 measurements for later downloading and processing (Fig. 13.1a). In an automatic recording mode the FM-36 is moved at a uniform pace along each transect line with great care to insure that it is aligned with meter marks as it gives an audible signal, the speed of which may be varied according to the user's pace. The matching of the audible signal with the meter marks ensures that the data, up to 8 measurements per interval, are properly located spatially.

Depth of penetration in magnetic sensing depends on the magnetic susceptibility of the materials being sensed. It is usually confined to 1 to 2 m in archaeological sites, with a limit of about three meters (Clark, 1990:78–80; see Table 13.1 for a summary of advantages and disadvantages of various sensing techniques).

4.2. Electrical Resistivity

Earth resistance to an electrical current depends on a number of factors including moisture, dissolved ion content, and the structure of soil particles and components. A resistivity survey utilizes two probes that establish a current through conductive earth, which is measured. Two other probes measure voltage, and the ratio of voltage to current yields resistance, according to Ohm's Law. In the

Current Practices in Aracheogeophysics

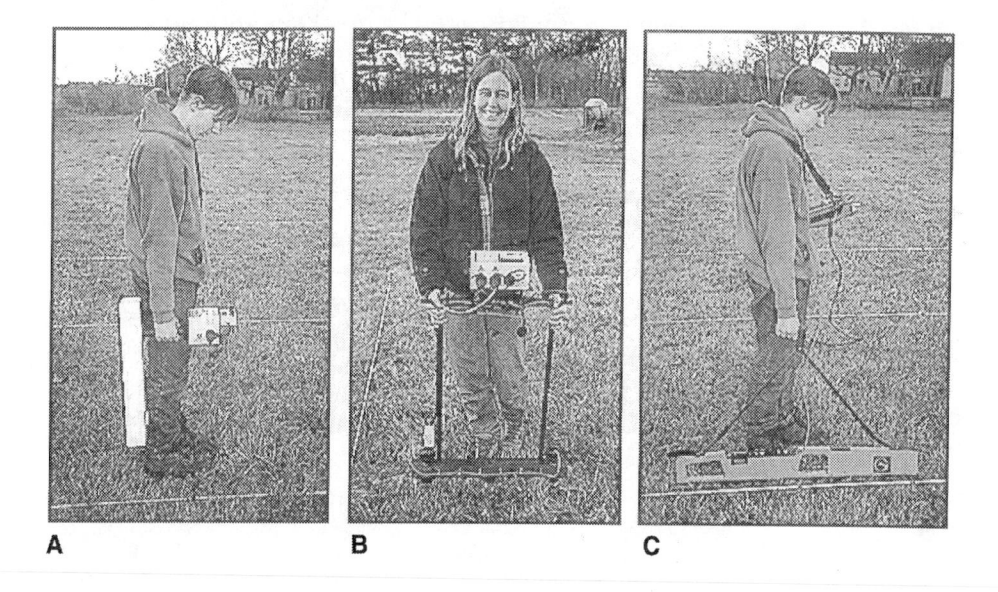

Figure 13.1. Geophysical survey instruments: (a) the FM-36 fluxgate gradiometer by Geoscan Research, (b) the RM-15 electrical resistance meter by Geoscan Research, (c) the EM-38 electromagnetic conductivity meter by Geonics Limited, (d) a 120 MHz GPR transducer (1), power source (2; a 12 v car battery), and control unit (3) of the SIR-System 3, by GSSI.

traditional Wenner configuration the four probes are each separated along a line by an equal distance, with the current probes on either end. Probe separation controls depth of prospection. In a uniform matrix (e.g., an alluvial fan of sand) voltage varies with distance from the current probes in regular hemispheres. If voltage is measured on the surface 1 m from a current probe, the value recorded is equivalent to the voltage at 1 m below the surface, allowing a means to control depth of prospection (Clark, 1990; Scollar et al., 1990; Weymouth, 1986).

In practice, the earth is not a uniform matrix, nor would we expect it to be in an archaeological deposit. Moist fill in a buried ditch feature might provide an

Table 13.1. Characteristics of Four Principal Subsurface Prospection Methods

	Magnetic	Resistivity	EM	GPR
Units	nT	ohm/m	mS/m	nS
Common depth	<1.5 m	0.25–2 m	0.75–6 m	*500 MHz: .5–3 m 300 MHz: 1–9 m
Typical				
Low sampling	1/m	1/m	1/m	>1 m transect spacing
High density	16/m	4/m	4/m	.5–1 m transect spacing
Survey time (20 m grid of 20 lines)	20–30 min	45 min	20 min	60 min
Area/day	.5–1 ha	.5 ha	.5–1 ha	.25–.4 ha
Sensitivity to metals	ferrous only	no	any	any
Situations to avoid	Metallic debris, igneous areas	Surface very dry, saturated earth, shallow bedrock	High resistance areas, very dry or saturated earth, metallic debris	Highly conductive clays, salts, rocky glacial deposits (e.g., moraines)
Tree effects	Impede survey, invisible in data	Impede survey, positive anomaly	Impede survey, negative anomaly	Impede survey, roots yield anomalies
Advantages	Speed, hearths, burned area detectable	Good feature definition, specific depth settings	Speed, ease of use	Vertical profiles, stratigraphy, results in real time
Disadvantages	Restricted depth, need open parkland for speed, iron clutter detrimental, sensor facing critical, constant pace of movement, high cost, must process data for results	Probe contacts slow, must deal with cables, must process data for results	Less spatial detail, metal clutter detrimental, must maintain constant ground angle, need open parkland for speed, must process data for results	Equipment bulky, difficult data processing, interpretations difficult, constant speed of movement, high cost
Daily data volume	High	Low	Low	High
Data processing complexity	Moderate	Low	Moderate–Low	High
Costs (USD)	$5k–25k	$600–12k	$6k–18k	$25k–70k

*Depends on soil properties.

easy pathway for the current, but it must flow around highly resistant features like a stone foundation, which alters the potential (voltage) and therefore resistance, creating anomalies. One consequence is that the probe separation or depth criterion becomes only an approximation in complex deposits. As the recorded resistance in ohms is dependent on probe spacing and configuration, these values are normally converted to earth resistivity, a measure of the bulk properties of the ground, not of a particular probe arrangement. For the Wenner array, resistivity in ohms per meter is given by $ohm/m = 2\pi Rd$, where R is resistance in ohms and d is the interprobe distance in meters (Weymouth, 1986:323).

A twin electrode configuration is typically employed for large-area surveys in archaeology. It offers the advantage of greater speed because only two probes are moved and they may be fixed in a frame to facilitate their simultaneous placement. It also yields a clearer response than the Wenner array, and one that is less confusing for interpretation purposes. For reasons too complex to elaborate here, the Wenner array will yield two resistivity peaks over a single feature (like a buried wall), whereas the twin array returns only one (Clark, 1990:37–53). The twin array is simply a modified Wenner configuration where pairs of current and voltage probes are separated by a large distance. Only one pair is moved to record resistivity, however, while the other pair remains fixed in the ground.

The RM-15 electrical resistance meter, a twin probe array by Geoscan Research (1996), is specifically designed for archaeological application, fully computerized with an integrated data logger (capable of storing 30,000 measurements), and optimized for speed (Fig. 13.1b). It consists of a rigid frame holding one current and one voltage probe, and two remote probes linked by a cable. The probe spacing in the frame may be set between 0.25 to 2 m, which allows one to vary prospection depth. A new multiplexer attachment, the MPX-15, allows data from multiple probes or depths to be acquired simultaneously. Resistance measurements are automatically sensed and recorded as fast as the frame can be lifted and moved to the next recording station (typically 0.5 to 1 m away), allowing 10 to 20 measurements per minute.

Under very dry conditions it may not be possible to promote flow of a current through the earth in resistivity surveys (solutions are to wet the ground prior to the survey or to insert the probes to a greater depth in order that the current may find a pathway to moister earth below). Conversely, significant contrasts between features may not occur if the ground is completely saturated. Obviously, ground moisture plays a major role in resistivity survey and markedly different results can be obtained at the same site depending on ground moisture conditions (Al Chalabi and Rees, 1962).

Resistivity surveys of the same piece of ground at different sensing depths yield profiles of data along any transect. When multiple adjacent profiles are considered over an area one achieves a resistivity volume that may be sliced along any axis to investigate vertical relationships, stratigraphy, and plan views at various depths. This method is known as *resistivity tomography*, and it represents the current state of the art in resistivity surveys (Aspinall and Crummett, 1997; Griffiths and Barker, 1994; Noel, 1992; Szymanski and Tsourlos, 1993). Resistivity tomography therefore is one geophysical method that potentially allows examination of vertical relationships between sediments, features, and stratigraphy (the principal method is GPR, discussed in a following section).

4.3. Electromagnetic Conductivity

Soil conductivity is the theoretical inverse of resistivity, making much of the previous discussion relevant here. Fundamentally different non-contact instruments are used to record earth conductivity, however. Electromagnetic (EM) conductivity meters employ widely separated transmitting and receiving coils. An EM signal sent out by the transmitter induces a current in the soil that creates a secondary magnetic field sensed and measured by the receiving coil. Characteristics of this second component, out of phase with the transmitted signal, are determined by various electrical and magnetic properties of the earth. In general, highly resistant features (e.g., stone walls, foundations, dry sands) possess low conductivity whereas low resistivity features (like the moist fill of a buried ditch or house pit) are highly conductive. Conductivity is measured in millisiemens (mS) per meter (10^{-3} siemens), and the theoretical relationship with resistivity is given by $mS/m = 1000/ohm/m$ (Bevan, 1983; McNeill, 1980).

The EM-38 electromagnetic conductivity meter, by Geonics Limited (1992), operates at a frequency of 14.6 kHz (thousand Hertz or cycles per second) and houses a 1 m coil separation (Fig. 13.1c). This instrument is capable of delivering two measurements per second to a nonintegrated data logger as it is skidded across the ground surface. Soil conductivity is measured as a weighted average through an earth volume of about 1.5 m in what is termed a vertical dipole mode (with peak sensitivity at about 0.4 m depth). Placed horizontally (in a horizontal dipole mode), the EM-38 records the average conductivity through a 0.75 m depth (with sensitivity decreasing from the surface). As there are no probes or cables associated with this instrument, with two readings per second it is possible to conduct very rapid surveys over sizable areas.

Unlike resistivity surveys, EM instruments are sensitive to buried metals, ferrous and nonferrous, that show up as extreme values (metals are highly conductive). As in magnetic surveys, this may be an advantage or a disadvantage depending on the character of the site and the research goals (whether one wants to locate metallic artifacts or avoid metallic litter). Moreover, in desert landscapes where it may not be possible to promote current flow through upper dry sands using the contact probes of a resistivity meter, EM signals can sometimes penetrate to lower, moist layers (Frohlich and Lancaster, 1986). Electromagnetic conductivity meters are also able to measure magnetic susceptibility by focusing on the in-phase component of the received signal (Clark, 1990:105), but this feature will not be examined here.

Geonics Limited also manufactures the EM-31, which operates at a frequency of 9.8 kHz and maintains a 4-m coil separation that allows greater depth of penetration (up to about 6 m in a vertical dipole mode; 3 m in a horizontal dipole mode), but much less spatial resolution (McNeill, 1980). Dalan (1991) has used this instrument with great success to prospect for buried mounds and large geomorphic features like sand ridges and associated swales at the Cahokia Mounds State Park, in Illinois.

Although theoretically equivalent, EM and resistivity instruments employ fundamentally different sensing technologies that evaluate unequal earth volumes and react or do not react to metallic artifacts (metals are nearly invisible to resistivity meters), so results are not as redundant as might be expected (see below for an actual comparison). Despite their slowness and bulky frame with encumbering cables, resistivity meters can more readily explore a variety of depths and probably yield better spatial resolution owing to the greater placement precision allowed by the contact probes and the smaller earth volume evaluated. With the recent introduction of the GEM-300, however, a variable-frequency (330 Hz to 20 kHz) EM profiler by Geophysical Survey Systems, Inc. (GSSI), simultaneous prospection of multiple depths will now be very easy using EM methods.

4.4. Ground-Penetrating Radar (GPR)

Ground-penetrating radar is very different from the other sensing technologies. Most GPR used in archaeology send continuous pulses of radar energy vertically into the ground along the full length of a survey transect. These pulses reflect off buried features like stratigraphic contacts, walls, house floors, pits, rubble, or middens. The return times of echoes from these pulses give information on depth, and their magnitudes indicate something of the nature of the subsurface reflectors. The outcome mimics a section or profile along the length of the survey line. Thus, GPR data in their native form are ideally suited for gaining information in the vertical plane, including stratigraphic relationships (Conyers and Goodman, 1997).

GPR employs a transducer, an antenna–receiver combination commonly in the 100 to 1,000 MHz range, with 300 to 500 MHz most popularly used in archaeology. As radar pulses are transmitted into the earth their velocity changes as a function of the electrical properties of the various materials through which they are traveling. Relative dielectric permitivity measures how easily a material polarizes when subjected to electromagnetic radiation. Metals possess a nearly infinite dielectric coefficient, which means they will give very strong reflections. The dielectric properties of other materials vary considerably; in general, the higher the coefficient the slower the velocity of radar waves passing through them. The velocity of electromagnetic energy in air is about 30 cm/nS (nS = nanoseconds = 10^{-9} seconds), but much less in soils. Bevan and Kenyon (1975) reported a velocity of 19 cm/nS in dry sand, 6 cm/nS in wet sand, and about 7.4 cm/nS in wet clay, for example, at a frequency of 100 MHz. GPR measures travel times and velocity can be estimated through knowledge of soil properties, yielding one means to estimate depth to anomalies. The GPR signal attenuates with depth, however, and with increases in soil conductivity. Penetration therefore is less in wet soils than in dry ones, and ion-laden deposits with high clay, salt, or even organic content can limit penetration (Weymouth, 1986). Signal attenuation is also a function of frequency, but so is the resolving power of GPR. The reflection returned by a subsurface feature depends on its size relative to the wavelength or frequency employed. Features smaller than the mean wavelength will yield weaker reflections. With a low-frequency antenna, subsurface features must be very large to even have the potential of being "seen" (Heimmer and De Vore, 1995) but, as with EM methods, low frequencies allow greater depth of prospection than higher frequencies.

The actual shape of a transmitted radar signal is a cone (with a 60–90° angle). This wide arc means that the transducer senses echoes from a reflector for some distance before, while on top, and for some distance after moving across it. The distance, and therefore time, the transmitted and reflected energy must travel is at a minimum when the transducer is vertically above the reflector, and it is larger before or after the approach. The consequence is a hyperbolic return from reflectors of small size. The multiple pulses yield a series of parallel reflections or bands, another feature of GPR output. Finally, there is a dead zone of interference from the ground surface where no propagation occurs, usually expressed as a series of meaningless reflection bands near the top of each profile (Fletcher and Spicer, 1995).

In a GPR record, the two-way travel time scale (in nS) of the vertical axis must be calibrated to establish an approximate depth to anomalies. The most accurate way to accomplish this is simply to excavate to the indicated feature and calibrate actual depths with travel times. Alternately, theoretical means may be employed to estimate depth by conducting velocity analyses based on known dielectric properties of various soil types (Conyers and Goodman, 1997).

GSSI manufactures a number of GPR systems that are popular among archaeologists. A complete system is typically composed of a control unit, a display unit, a transducer, a number of cables, and a power source (Fig. 13.1d). Transducer sizes vary with frequency. The archaeologically common 300 MHz transducer is about 1 m in length and weighs 30 kg whereas a 1,000 MHz unit will almost fit in a shoebox. The control unit houses mass data storage devices (hard disk, tape drive), and the display unit contains a color computer screen. One advantage of GPR is that initial results may be seen in real time as the survey is being undertaken (in contrast with the other technologies), making instant field interpretations possible. In a GPR survey the transducer is pulled on the ground surface along a transect line. Because pulses are continuous, the unit must be moved at a uniform pace (a pace of 0.15 to 0.5 m/sec is commonly employed). As the transducer is pulled, the operator manually presses a button that sends a marker signal to the output that indicates the locations of meter marks as they are passed (Conyers and Goodman, 1997).

When GPR data are acquired in a series of closely spaced parallel transects, significant reflections in adjacent profiles may be cross-correlated to gain a three-dimensional understanding of the subsurface. Recent state-of-the-art advances in the computer processing of GPR data allow interpolation between the profiles, providing a means to generate horizontal plans of significant subsurface anomalies. Goodman, Nishimura, and Rogers (1995) have illustrated what they term "horizontal time slices" (horizontal slices that vary according to time or depth) at a number of sites. Milligan and Atkin (1993) actually reconstructed the topography of a paleosurface through GPR reflection data. Bradley and Fletcher

(1996) have pursued an analogous approach, but they focus on the variance or activity of the reflected radar signal as it varies across a site, showing that significant cultural features may be horizontally defined through this statistic alone. In general, these interpolated views greatly aid in the interpretation of GPR data by making horizontally expressed cultural patterns more easily recognizable, a factor that has caused much recent interest in this technology. GPR is clearly one of the technically most demanding sensing technologies in archaeology; it may also yield the greatest analytical potential.

A comparison of characteristics, costs, and the relative advantages and disadvantages of the four subsurface sensing methods discussed, magnetic gradiometry, electrical resistivity, EM conductivity, and GPR, is given in Table 13.1.

5. Computer Methods

The computer processing of archaeogeophysical data deserves a serious paper of its own. For non-GPR data, Scollar et al. (1990) introduced a number of key issues and concepts, and several papers describing specific operations exist (e.g., Music, 1995; Neubauer et al., 1996). One of the best summary guides is by Walker and Somers (1994), in the form of the GEOPLOT software manual (commonly used geophysical software by Geoscan Research), where numerous processing strategies, alternatives, and case studies are presented. The very specialized and extensive processing steps required of GPR data are not well presented, although the recent book by Conyers and Goodman (1997) is a good place to start (see also Malagodi et al., 1996).

For non-GPR data, computer processing requires a series of common steps:

1. *Concatenation* of the data from individual survey grids into a single composite matrix;
2. *Clipping and despiking* of extreme values (that may result, for example, from introduced pieces of iron in magnetic data);
3. *Edge matching* of data values in adjacent grids through balancing of brightness and contrast (i.e., means and standard deviations);
4. *Filtering* to emphasize high-frequency changes and smooth statistical noise in the data;
5. *Contrast enhancement* through saturation of high and low values or histogram modification; and
6. *Interpolation* to improve image continuity and interpretation.

Several other more specialized processing steps may also be required as cases warrant (e.g., detrending, edge enhancement, see the following section). One benefit of the ongoing computer revolution is that initial data processing may now be pursued on-site with portable field computers. This gives an important and nearly immediate link with the data that can influence survey decisions by guiding work to areas of greater archaeological potential.

6. Case Studies I: Field Methods and Results

6.1. Whistling Elk Village, South Dakota

Whistling Elk is a large prehistoric earthlodge village (the earthlodge was an earth-covered house overlying a frame of wood) surrounded by a bastioned fortification ditch that is partially revealed by subtle vegetation markings in aerial photographs. The fortified perimeter of Whistling Elk (determined by geophysical survey) measures about 170 × 110 m. The village is located on the Missouri River, near Pierre, South Dakota, in an area that was historically farmed, although more than 90 percent of it has existed within a federally protected wildlife refuge since the 1960s. Excavations at similar sites show that the interiors of fortification ditches and bastions were lined with a palisade of timbers. In the late 1970s two earthlodges were excavated in the eroding embankment. The form of these houses (square with long entranceways to the east or southeast), the artifact assemblages, and radiocarbon dates clustering around A.D. 1300, assigned this village to the Initial Coalescent variant of the Plains Village pattern (Steinacher and Toom, 1984). Both of the excavated structures were destroyed by fire and contained a large number of functional artifacts and concentrations of foodstuffs (maize, beans), pointing to a hasty abandonment. This evidence suggests the village was vacated under duress, probably in response to an enemy's attack. Aside from artifacts and archaeological deposits eroding from the cutbank, no indications of cultural features are expressed on the surface of this site.

Whistling Elk is encased in conductive silt-loam. At the embankment the depth to cultural features varied between 0.8 to 1.5 m. A plow zone about 0.3 m thick exists near the surface. The conductive earth and surface-apparent salts from irrigation practices did not recommend GPR. The known depth to cultural features approach the limits of magnetic methods, unless magnetic susceptibility is very high. Consequently, greatest emphasis was placed on resistivity survey, with contrasts expected from house and fortification ditch fills. Probe separation, and therefore approximate prospection depth, was set at 1 m using the twin probe array of the RM-15. To investigate similarities, differences, and relative advantages between the technologies, a conductivity survey was also carried out utilizing the EM-38. Finally, a magnetic gradiometry survey was performed with the FM-36. About 34,000 resistivity, 33,200 conductivity, and 64,800 magnetic gradiometry measurements were acquired over a 2-week period in the approximate 1.7 ha of the village. These data, after substantial computer processing, are illustrated in Fig. 13.2 (see Toom and Kvamme, 2001, for further details).

The electrical resistivity data, sampled at two readings per meter, reveal remarkable detail about this site (Fig. 13.2a). The complete fortification ditch, with five bastion loops, a berm line near the Missouri River (reflecting the government's efforts to protect this site from bank erosion with a fiberglass wrap and rock facing), and numerous indications of houses ("blob"-like black features), are clearly portrayed. Of some importance is an indication of an inner fortification ditch, previously unknown, with perhaps four bastion loops encircling an increased density of houses. With clear evidence of an attack on this village shown

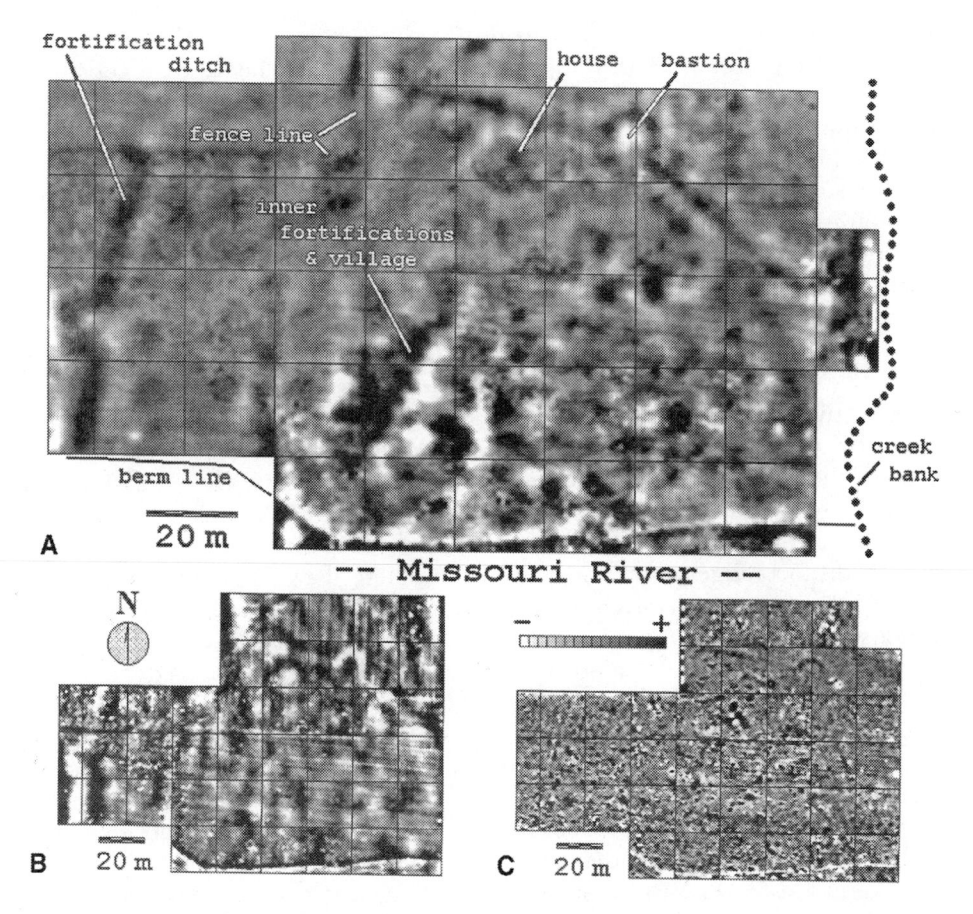

Figure 13.2. Geophysical results at Whistling Elk, South Dakota: (a) electrical resistivity, (b) electromagnetic conductivity, and (c) magnetic gradiometry.

by the 1970s excavations, it is quite possible that Whistling Elk was reoccupied by the survivors and consolidated into a tighter, more defensible settlement, a circumstance not unknown in other village sites along the Missouri River (Johnson, 1998). The data also show linear suggestions of historic plow marks, intensive speckling in regions of great rodent activity, and the locus of a contemporary steel wire fence marking the boundary between government and private property (the fence line exhibits a resistivity contrast because the dense vegetation beneath and near it somewhat reduces soil moisture). The weaker resistivity response to cultural features on private property north of the fence line may be due to altered moisture content in this agricultural field or perhaps the effects of an additional 30 years of plowing. An unusual aspect of the resistivity data is that cultural features exhibit higher measurements, indicating that the site's ditches and house pits are filled with more resistant materials (typically, ditch and house pit features reveal low resistivity due to enhanced moisture levels).

The EM survey, also sampled at two readings per meter and conducted in the vertical dipole mode, was approximately 50 percent faster than resistivity. The presence of the wire fence and steel fence posts precluded survey in a portion of the northwest quadrant of the village (Fig. 13.2b; this metallic feature is expressed in some of the adjacent survey grids as unusually high or "washed-out" readings). Besides the obvious fortification ditch, bastions, the riverside berm, and house indications, the most obvious feature in the EM data is the historic plow marks. Two fields are indicated with a clear boundary between them. The north field has furrows running north–south, whereas the southern field has furrows oriented east–west, paralleling the river.

The heightened response of the plow marks in the EM data is undoubtedly due to the peak sensitivity of the EM-38, which lies at a depth of about 0.4 m, close to the bottom of the plow zone, even though this instrument assesses conductivity through a bulk volume of 1.5 m. Whatever the case, it is obvious that some of the cultural features are less distinct (e.g., some of the houses), or are indistinct (e.g., the inner fortification ditch), compared to the resistivity data where prospection was set to the meter depth of the archaeological deposits. The lack of distinct edges, or "fuzziness" of the EM data, is better shown in a comparison of feature details in Figs. 13.3a and 13.3b where a clear, square house with a southeastern facing entranceway is revealed by resistivity, but the entranceway or even the house's shape are not well indicated by conductivity.

That electromagnetic conductivity is the theoretical inverse of resistivity is revealed by the reversal of the gray scale in Fig. 13.2b, but what exactly are the relationships between the measurements derived from two very different sensing technologies? A 20×40 m portion of the site, centered over the western fortification ditch and bastion, is illustrated in Figs. 13.3d–g. A plot of the 1,600 measurements acquired at identical locations in this area by the RM-15 (Fig. 13.3d) and EM-38 (Fig. 13.3e) shows considerable variation about the theoretical expectation, indicating that different information is obtained by these instruments (Fig. 13.3h). A principal components analysis run on these data yields a first component—accounting for over 95 percent of the variance—that summarizes the common information derived from these instruments (Fig. 13.3f). The second component, explaining only 5 percent of the variance, represents information *not* in common, which therefore emphasizes differences obtained between the instruments (Fig. 13.3g). It is interesting that the mapped Component 2 reveals clear indications of linear features not apparent in the raw data that parallel the fortification ditch. These features may represent berms associated with ditch or palisade construction, with the right, interior berm the approximate locus of the palisade.

The magnetic gradiometry data, obtained by the FM-36, appear very noisy at a global glance (Fig. 13.2c). Strongly apparent are the northern fortification ditch, a bastion, the riverside berm, and some historic plow markings. The effects of the steel fence are revealed by the zebra stripes in the upper center of the figure—a series of dipoles each associated with a steel fence post—and by the strong negative measurements that wash out the image along the northwest fence line. The grainy appearance is due to numerous magnetic data spikes, or extreme

Current Practices in Aracheogeophysics

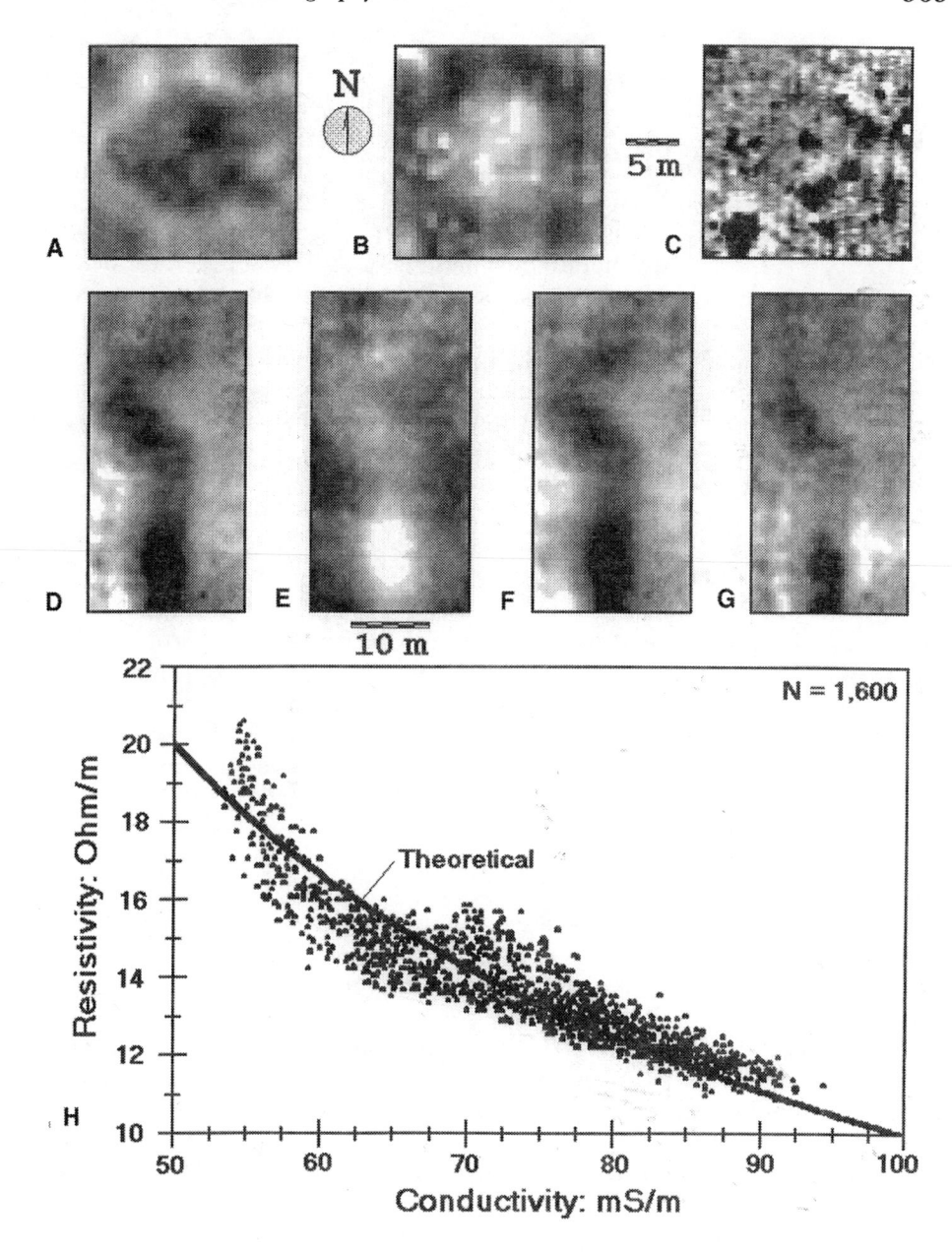

Figure 13.3. Details from Whistling Elk: (a) house feature resistivity data, (b) house feature conductivity data, (c) house feature magnetic data, (d) fortification ditch resistivity data, (e) fortification ditch conductivity data, (f) principal component 1 of resistivity–conductivity, (g) principal component 2 of resistivity–conductivity, (h) relationship between conductivity and resistivity measurements.

values, that stem from historic iron debris (e.g., broken plow parts) and prehistoric burned features. On close inspection, this data set is rich in large-scale detail. For example, the enlargement in Fig. 13.3c reveals interior details of a house feature. That the house was burned is revealed by the large magnetic values that define its perimeter. The central hearth is also strongly apparent. Initial Coalescent houses are known to contain four stout roof support posts placed in quadrants about the central hearth, and these too are apparent in the magnetic data. A 2 × 6 m trench excavated to this house floor at 0.95 m depth verified the locus of the house wall, the hearth, one support post, and the fact of its burning (Toom and Kvamme, 2001).

The geophysical data from Whistling Elk provide a clear plan of the village with outer and inner fortifications and house distributions. At a larger scale, interior features and details of individual houses are indicated. Whistling Elk therefore demonstrates the use of geophysical survey results as primary information for the investigation of settlement pattern.

6.2. Menoken Village, North Dakota

Menoken Village, a National Historic Landmark near Bismarck, North Dakota, is another fortified village site in the Great Plains, but very different in character from Whistling Elk. The site has never been plowed, the cultural features are very shallow, about 0.4 to 0.5 m in depth, and the surface expresses ample evidence of the villages' structure. A fortification ditch, four bastion loops, and as many as 16 house depressions are visible in this 1 ha site, indicated by subtle and large relief changes as great as 0.5 m.

Poorly reported excavations in the 1930s revealed a palisade with shallow post holes along the interior of the ditch. Several houses were also excavated or tested but the reporting does not allow a confident reconstruction of their form. Ahler (1993) has argued the importance of Menoken, which dates to approximately 1100 A.D., because it is a fortified village manifestation thought to immediately predate the introduction of maize horticulture. The only excavated house from a similar site of this period is at the Flaming Arrow site, about 70 km away. This house was oval in shape, measured 8.2 × 9.8 m, and possessed an unusual interior entranceway consisting of a sloping ramp (this feature is relevant to findings that that follow).

The success of the resistivity survey at the culturally similar Whistling Elk site suggested its use at Menoken, and the possibility of burned features and shallow depth recommended magnetic gradiometry. Both surveys were undertaken over the approximate 1 ha area of the village, with sampling densities for resistivity of four measurements per meter and magnetic gradiometry of 16 measurements per meter. With relatively little depth to features, the magnetic gradiometry data at Menoken offered robust contrasts (Fig. 13.4a). The data clearly indicated the fortification ditch, four bastion loops, trails, multiple large magnetic features that suggest burned houses (later verified through excavation), the large circular open hole left by the 1930s excavations, and the effects of a steel power pole in the lower left corner (Fig. 13.4a). A number of the point anomalies indicate histori-

Current Practices in Aracheogeophysics 371

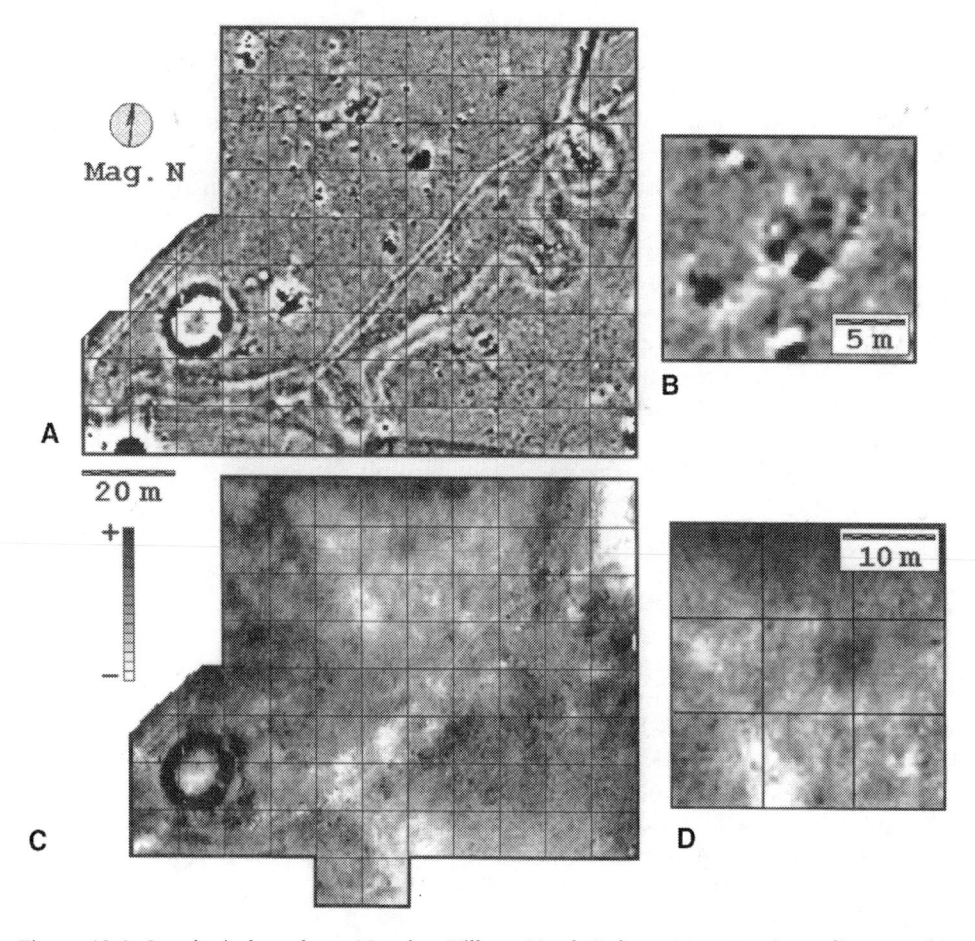

Figure 13.4. Geophysical results at Menoken Village, North Dakota: (a) magnetic gradiometry, (b) magnetic details of house features, (c) electrical resistivity, (d) resistivity detail of possible trail winding around the large house feature in (b).

cally introduced iron and steel artifacts, verified through a metal detector survey and excavation. It is noteworthy that the interior of the fortification ditch and three of the four bastion loops possess large magnetic anomalies that quite likely indicate burned features, possibly of the palisade that once stood in these areas, and suggestive of an attack on the village (excavation has not yet confirmed this inference). Of particular interest are several large burned features, most likely houses, near the central portion of the village. Although some are associated with clear surface depressions, and were therefore suspected house locations prior to the survey, others are not expressed on the surface and represent new findings (one of which has been verified as a burned house by excavation). One house gives an unusually clear magnetic signature. It is oval in form, measures about 6.5 × 9 m, and reveals interior details that suggest an entranceway ramp and possible hearth features (Figure 13.4b). The size and form of this house is almost identical to the excavated house at Flaming Arrow. The magnetic data alone therefore provide evidence that helps link Menoken to this site and the early cultural complex it represents (Kvamme, 1999).

The resistivity data obtained at Menoken are much less clear in their definition of features seen in the magnetic data, with only the 1930s excavation hole plainly visible, and the many trails, the fortification ditch, and bastion loops only faintly revealed (Fig. 13.4c). Although numerous small anomalies of high and low resistivity that measure 2 to 4 m in diameter exist across the data set, archaeological testing has not yet been performed on them and their source is uncertain. One feature that close inspection of the resistivity data does reveal are numerous linear and curvilinear features, not seen magnetically, that are interpreted as trails (such resistivity contrasts can arise from increased soil compaction or from a different fill if trails are incised). One trail is shown in Fig. 13.4d that, when match against the corresponding magnetic data, suggests a trail winding between several of the houses revealed magnetically or as surface depressions.

6.3. Sluss Cabin, Kansas

The Sluss Cabin site represents a pioneering homestead on the Kansas frontier of the late 1870s. The site is located on a small parcel measuring about 25 m wide, located in a contemporary farmyard. Overgrown in dense vegetation, the tops of two stone foundation blocks are visible at the surface. The surrounding farmyard contains an abundance of abandoned iron and steel farm machinery, water tanks, a barbed wire fence with an iron gate surrounds the parcel, and strands of wire, nails, and other recent debris litter the ground. These circumstances posed a challenge to geophysical survey. The profusion of iron-bearing and other metallic artifacts precluded magnetic and EM surveys. The surface expression of parts of the foundation indicated a depth too shallow for conventional GPR unless an unusually high frequency antenna was employed (i.e., about 1,000 MHz). Consequently, electrical resistivity methods were selected using a 0.5 m target depth and a dense sampling interval of 0.5 m.

The results of this survey illustrate quite well some of the positive benefits of resistivity methods (Fig. 13.5). The stone foundation, in the form of a rectangle and defining the perimeter of the structure, yields high resistivity values and therefore a good contrast against the surrounding earth of much lower resistivity. The data reveal that the area of the cabin is bisected with an interior foundation line (labeled "2" in Fig. 13.5). The south half possesses a floor of high resistivity similar to the foundation, surmised as also being of stone (and later verified with a small test excavation). The north half of the cabin evidently possessed only an earthen floor, as resistivity values are identical with regions outside of the cabin. These features are clearly evident in plots of the data and in a resistivity profile (Figs. 13.5a and b).

Outside of the cabin two additional features occur. Subtly revealed linear alignments about 1 m wide extend outward from the north and west cabin walls (Fig. 13.5a). They are expressed by slightly elevated measurements that may

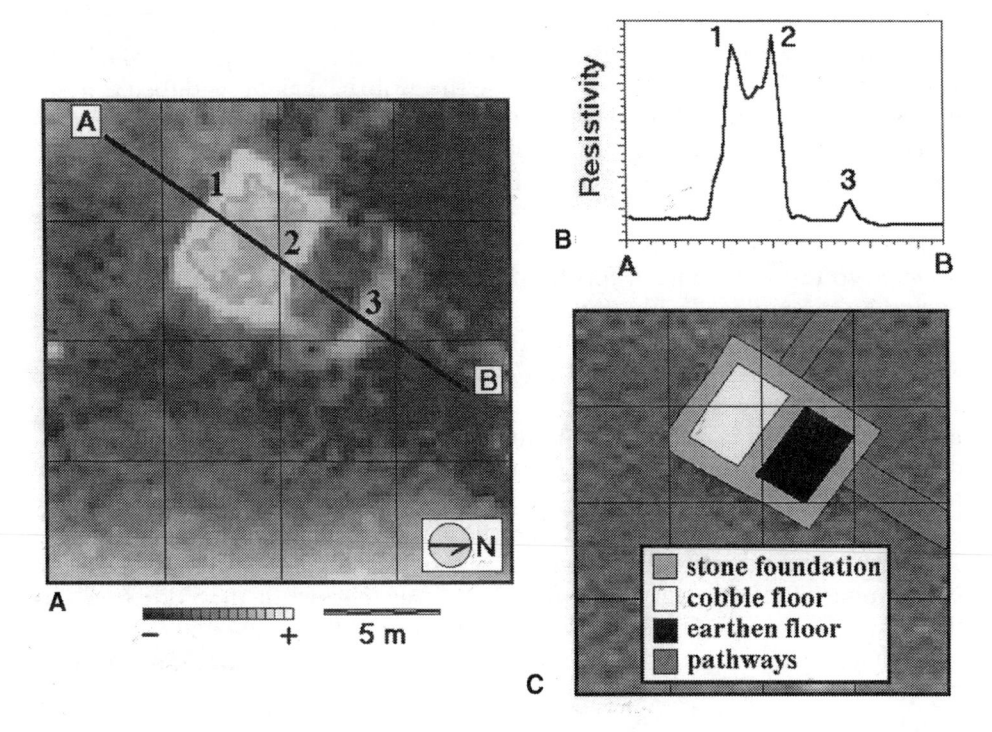

Figure 13.5. (a) Electrical resistivity results at Sluss Cabin, Kansas, with transect line and three foundation walls numbered, (b) resistivity profile across transect, and (c) interpretations.

represent walkways where more compacted earth has increased ground resistivity. If they are walkways then the loci of entrances to the cabin may be inferred (Fig. 13.5c).

6.4. Breed's Hill, Massachusetts

Breed's Hill, now part of the Bunker Hill National Monument, is an historic battlefield site of the American Revolutionary War, located near Charlestown, Massachusetts (Ketchum, 1962). During the night and early morning of June 17, 1775, American patriots created a redoubt and breastworks on the crest of the hill. It was promptly and successfully attacked with great loss by the British, who were occupying nearby Boston. An accurate plan of the redoubt was never made, and its form remains largely a matter of conjecture today.

Breed's Hill is actually one of New England's many drumlins, characterized by well-drained sands and little soil formation at the surface. This geological setting suggested the suitability of GPR; a resistivity survey was also undertaken. As this site now lies in an urban setting, with numerous parked and moving cars in close proximity, large iron fences, sign posts, park lights, statues, an underground sprinkler system, and power conduits, magnetic and conductivity surveys were not considered. The goal of this project was to locate some of the famous battleground's fortifications. A complicating factor is that subsequent to the battle a fort was established at the same site by the British, making it difficult, if not impossible, to distinguish between American and British features. Further complexity to interpretation arises from the fact that the modern surface of the park has been heavily landscaped for over 150 years.

In one area of the site 27 parallel GPR transects were recorded. These were generally of 20 to 25 m in length, and separated by one meter with direction oriented to cross anticipated fortification features perpendicularly. A GSSI SIR System 3 was employed with 300 and 500 MHz transducers that revealed strong anomalies in each transect. Two roughly parallel features, one about 3 to 5 m wide and the other 1 to 2 m wide, were revealed that span the length of the area surveyed, with varying distances between them (Fig. 13.6). The reflections are most likely derived from the bottom surfaces of historically excavated ditches that could represent aspects of the hilltop's fortifications. These features are estimated to lie between 1 and 1.5 m beneath the surface through a soil velocity analysis.

A resistivity survey, conducted with the RM-15 using a prospection depth of 1 m, reveals the extent of these features in the horizontal dimension through an overlapping 20 × 20 m survey block (Fig. 13.7a; the loci of the GPR transects illustrated in Fig. 13.6 are also indicated). These linear anomalies, possible fortification features, exhibit higher resistivities than the surrounding matrix,

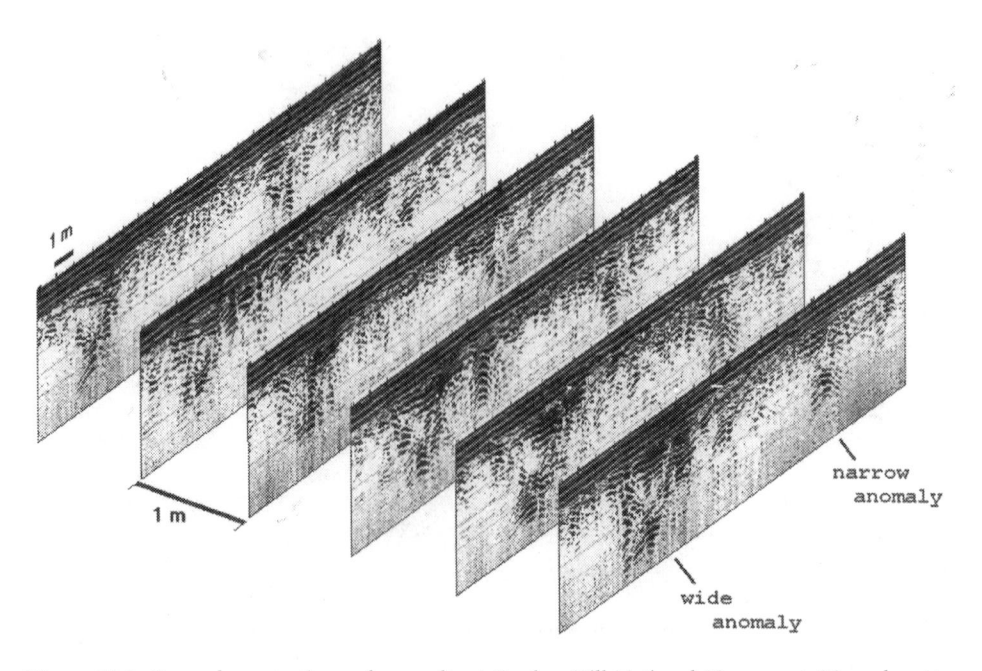

Figure 13.6. Ground-penetrating radar results at Bunker Hill National Monument, Massachusetts. Each transect is 20 m long. Note the meter tick marks at the surface, the near-surface interference zone, the subsurface hyperbolic reflections, and the wide and narrow anomalies in contiguous transects suggesting continuity of these features.

Current Practices in Aracheogeophysics 375

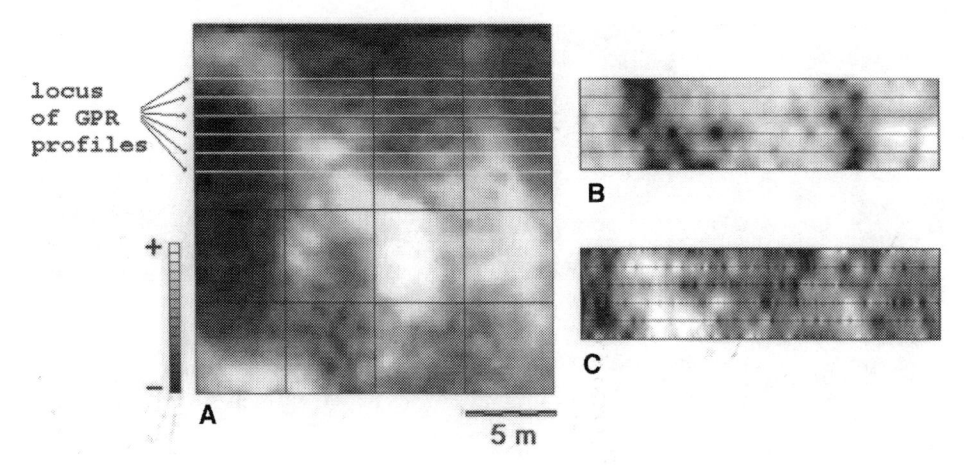

Figure 13.7. Geophysical results at Bunker Hill National Monument, Massachusetts: (a) resistivity results, (b) interpolated GPR data at a 20 nS two-way travel time "depth" (taken from data in Fig. 13.6), (c) variance of GPR activity interpolated across the region shown in Fig. 13.6. The horizontal lines indicate the locations of the GPR transects given in Fig. 13.6.

suggesting a fill with different dielectric properties that caused the pronounced radar reflections. Although these indications show great promise, future work is needed to clarify these findings. Geophysical survey of a larger area, for example, could help to define the nature of these anomalies if their geometries coincide with the anticipated form of fortifications from that battle and period, for example. Test excavations can help to clarify depth, the nature of the fill, and sources of the radar echoes.

7. Case Studies II: Advanced Geophysical Data Processing

In addition to appropriate geophysical methods and field strategies for the context at hand, correct data processing is essential in order to realize the full potential of the data. Otherwise, significant and meaningful cultural patterns that exist in a data set might not be realized. This is especially true of subtle patterns that must be "teased" out of the data in order to bring them to light. Frequently, the application of these methods occurs months after fieldwork (in some cases researchers have gone back to decades-old data sets, e.g., Weymouth, 1997), yielding significant new discoveries in the laboratory. Several case studies briefly illustrate the nature of these procedures and findings.

7.1. Navan Fort, Northern Ireland

This Iron Age site, located on a drumlin with components that date to about 100 B.C., plays a prominent role in the pagan mythology of Northern Ireland. A total

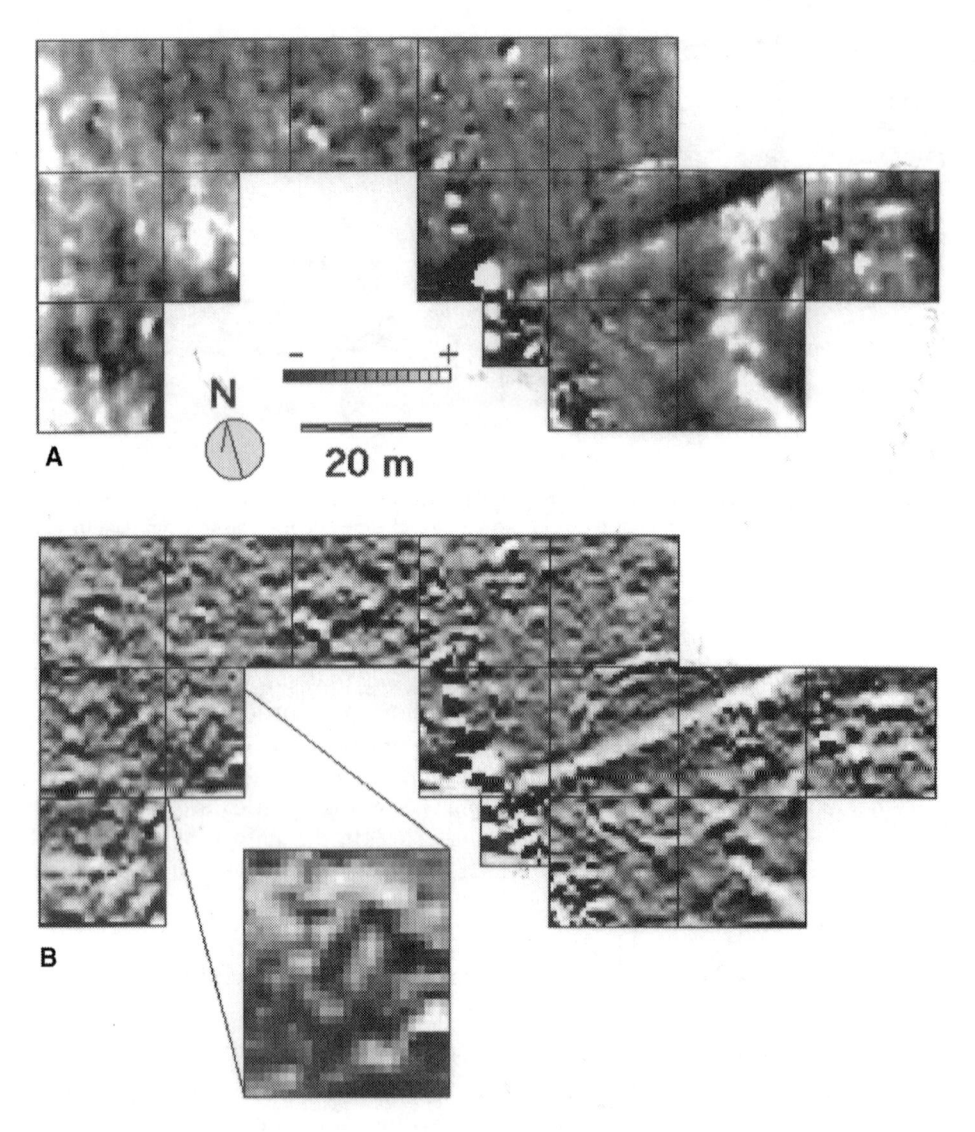

Figure 13. 8. Proton precession magnetometry results from Navan Fort, Northern Ireland: (a) total magnetic field, (b) shadow image.

field magnetic survey with a proton precession magnetometer revealed a number of obvious subsurface features, circular in form, that are consistent with other Iron Age findings (Fig. 13.8a). Analytical surface shading, a technique that illuminates a surface to cast shadows, was also employed. By placing the light source at a low angle to the north, subtle details in the magnetic surface were enhanced, revealing a square feature about 8 m wide in an otherwise unremarkable portion of the raw magnetic surface (Fig. 13.8b). This finding excited the local archaeologists because it possibly indicates the locus of a medieval hall, reputedly erected

Current Practices in Aracheogeophysics   377

by Niall O'Neill in 1387, "for the entertainment of the learned men of Ireland," according to the *Annals of Ulster* and *Annals of the Four Masters* (Kvamme, 1996).

7.2. 3D Ranch, Kansas

This historic pioneer period settlement of the 1860–1870s frontier was a dairy farm according to oral history. Located close to a creek bed (now ponded), and undoubtedly under some depth of colluvial wash from a nearby slope, initial resistivity results revealed little pattern aside from a strong trend of changing resistivity with proximity to the pond area. In order to remove this trend and ascertain if other patterns might exist hidden within this data set, a least-squares plane was fit to these data and was then subtracted from the raw measurements. This treatment removed the trend and the residuals expressed a clear pattern, suggestive of a building foundation, possibly a barn (Fig. 13.9).

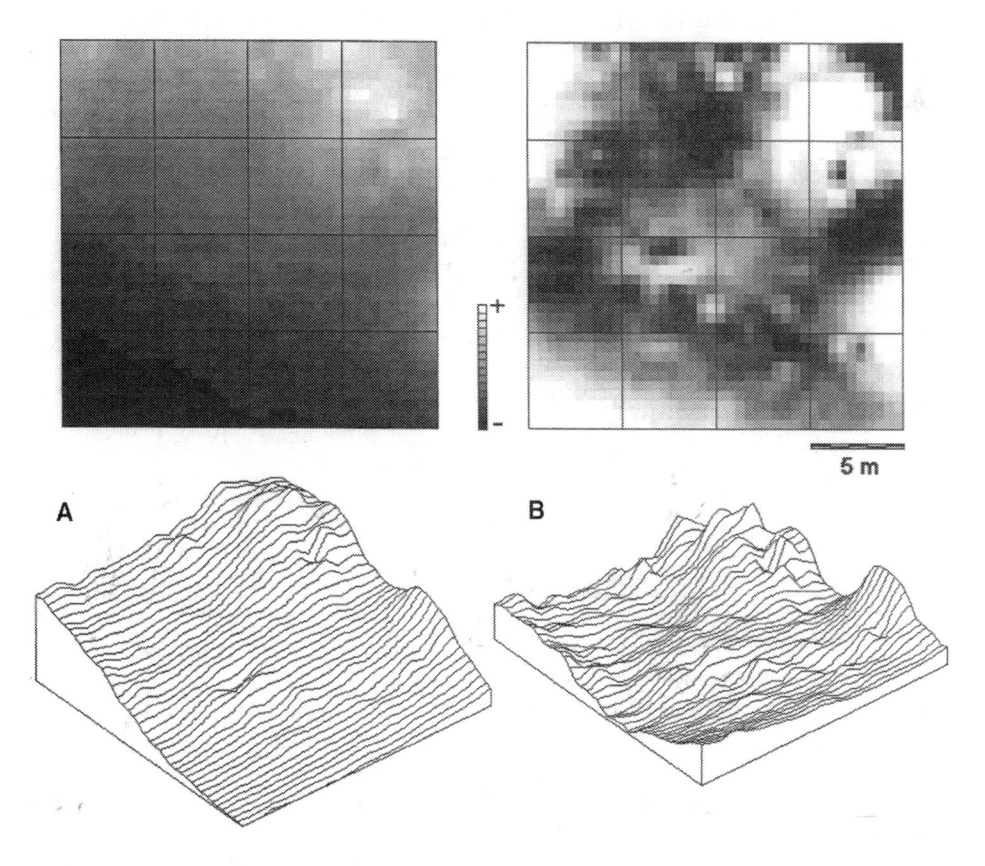

Figure 13.9. Electrical resistivity results at 3D Ranch, Kansas: (a) raw data illustrating a strong regional trend, (b) detrended data after subtraction of a least-squares plane.

7.3. Whistling Elk Village, South Dakota

The resistivity data from this site, discussed earlier, behaved much like the 3D Ranch data, with a profound regional trend that varied with distance from the nearby Missouri River. Although linear or quadratic least-squares solutions are able to remove broad regional trends, they typically leave smaller regions of extreme values that can obscure features within them. The high-pass filter offers an alternative by contrasting, or differencing, the value of a measurement with neighboring measurements. By differencing measurements in a narrow window — the adjacent eight neighbors in a geophysical data raster — high-frequency or spatial resolution changes are emphasized. Using a large window (e.g., a radius of 10 measurement loci) tends to remove trends of larger size. At Whistling Elk a filter of the latter form successfully revealed a number of features, probably prehistoric houses, within pockets of high overall resistance (Fig. 13.2a).

7.4. Breed's Hill, Massachusetts

The parallel GPR transects illustrated in Fig. 13.6 were subjected to advanced processing in order to express the continuity of the pronounced anomalies in plan view. Two methods were investigated. In the first, a time slice was extracted from each transect corresponding to approximately a 20 nS two-way travel time below the surface (soil velocity analysis suggests an equivalent depth of about 1 m). The slices were subjected to computer interpolation methods to yield an approximate plan view of the reflector data (Fig. 13.7b). The second method merely computed the amplitude variance in 0.2 m intervals along the length of each transect, an indicator of signal "activity" (Bradley and Fletcher, 1996). These data were then interpolated to yield a horizontal surface indicating the overall variance of radar activity (Fig. 13.7c). Both plan views give a sense of the layout of these features in the horizontal dimension that compare favorably with resistivity findings (Fig. 13.7a).

8. Conclusions

This overview of geophysical survey practice in archaeology has attempted to convey a sense of the principal methods, theory, instruments, survey considerations, data processing needs, and the nature of contemporary findings. Of necessity, there are many details that could not be discussed, and a suite of lesser used geophysical methods have been ignored (e.g., self-potential [Drahor et al., 1996] or seismic exploration [Overdon, 1994]). Nevertheless, it should be evident from the foregoing that archaeogeophysics is a rapidly evolving and increasingly specialized discipline, and one that is of growing utility to the larger archaeological community.

Weymouth (1986:374) emphasized the complementary nature of magnetic, resistivity–conductivity, and GPR data, each of which generally derives from different earth properties. Numerous studies, including those shown here, illustrate the positive benefits of employing several geophysical methods simultaneously to the same region (see also Corney et al., 1994; Piro, 1996). Geophysical data should be interpreted in concert with each other and with other archaeological information. The nature of prior archaeological discoveries at a site or at nearby sites must be reviewed. Additionally, microtopographic surface expressions of buried features occasionally exist (Newman, 1993), subtle vegetation patterns that may correlate with the subsurface can be visible in aerial photography (Scollar et al., 1990), and surface artifact distributions visible in plowed fields or in arid lands contexts can provide important information about overall site structure (Dunnell and Dancey, 1983). A comprehensive remote sensing study should consider all sources of information in order to assemble as complete a picture as possible about a site (see Keay et al., 1991). Of most importance, archaeogeophysicists should have good knowledge of, and experience with, the kinds of archaeological features that might be encountered and the typical responses of those features to the various sensing technologies. Whenever possible, test excavations should be conducted at the locus of significant anomalies to ascertain or confirm their identity.

With all the benefits, possibilities, and successes of contemporary geophysics, the archaeological community must nevertheless make a commitment to support training in this technically demanding field. A present shortcoming is a lack of instructors in the academic community, particularly in North America. At the same time, archaeological students and professionals must be made aware of the benefits of geophysical prospection as well as its limits. After all, the success of geophysical methods in archaeology will depend on whether there is sufficient demand for it.

ACKNOWLEDGMENTS. I wish to thank Lew Somers, Bruce Bevan, and John Weymouth for sharing their knowledge and experience with me over the past several years. Jo Ann Christein and Bob Burgess have consistently assisted with fieldwork. Surveys in the Dakotas were facilitated by a grant from the National Center for Preservation Technology and Training, U.S. National Park Service. Work at Whistling Elk was conducted under a federal ARPA permit from the U.S. Army Corps of Engineers and was greatly facilitated by Dennis Toom and the University of North Dakota field school. The Menoken project was conducted with the assistance of Fern Swenson and Paul Picha of the State Historical Society of North Dakota, Stan Ahler of the PaleoCultural Research Group, Flagstaff, Arizona, and the University of Missouri field school, directed by W. Raymond Wood. Work at Bunker Hill National Monument was undertaken with the encouragement of the U.S. National Park Service.

9. Glossary

Active prospection methods Geophysical techniques that inject electrical, radio, or radar energy into the earth. The nature of the response gives indications of subsurface properties.

Anomaly A term used in geophysics referring to measurements that stand out as being different from typical background measurements, indicating the presence of a local geological, biological, or archaeological feature with different physical properties.

Archaeogeophysics The application of geophysical methods, principles, and theory to archaeological contexts.

Attenuation The loss of power in a transmitted signal that occurs with all active geophysical methods, particularly in reference to GPR.

Clipping An image-processing technique that removes a percentage (usually 5–10%) of the most extreme values in a data set to allow focus on the more characteristic central measurements, usually resulting in a contrast enhancement.

Concatenation An image-processing technique used in geophysics to merge data from multiple survey grids into a single data set and image.

Contrast enhancement An image-processing technique that accentuates image detail by redistributing the range of color or gray display values, often by assignment to a narrower range of prevalent data values (see *Clipping*).

Data logger A computer memory device used to automatically store large volumes of information.

Despiking An image-processing technique for the removal of isolated extreme measurements in a scene. These measurements are often regarded as noise that may result from modern litter, such as metallic debris, in geophysical surveys.

Edge matching An image-processing technique for the normalization of multiple grids of geophysical survey data to a common level. Differences in instrument calibration or soil conditions between surveys can cause arbitrary changes in mean measured values. Mean values of one edge of a matrix of measurements are matched with the mean value of an adjacent edge in another matrix through the addition or subtraction of a constant.

Electrical resistance Resistance to the flow of an electrical current as determined by Ohm's law. The measurement of soil resistance is partially dependent on probe configuration. Resistance is used to estimate soil resistivity, an intrinsic soil property.

Electrical resistivity An active geophysical survey technique that injects a current into the earth to assess resistivity to that current, an intrinsic soil property measured in ohm-meters (see *Ohm-meter*).

Electromagnetic conductivity (EM) An active geophysical survey technique that transmits radio frequency energy into the earth, which induces a current and secondary magnetic field proportional to soil conductivity sensed by a receiver. Measured in millisiemens/meter (see *Millisiemens/meter*).

EM See *Electromagnetic conductivity*.

Ground-penetrating radar (GPR) An active geophysical remote sensing device that transmits radar signals into the earth that are reflected by discontinuities caused by changes in soil dielectric properties.

GPR transducer A GPR antenna for transmitting radar energy into the earth, usually coupled with a receiving antenna for sensing signals reflected from subsurface discontinuities.

Hertz (Hz) Cycles per second, used to express frequency, typically in thousands (kHz) or millions (mHz) of cycles per second.

Horizontal dipole mode See vertical and horizontal dipole mode.

Image processing Manipulation of image data or other similarly structured matrices by computers.

kHz See *Hertz*.

Magnetic dipole Magnetic poles of opposite sign located in close proximity and resulting from a single, highly magnetic source that mimics a bar magnet.

Magnetic gradiometer A magnetometer with twin, vertically separated sensors, that

simultaneously measure the difference, or gradient, in the magnetic field, eliminating the effects of diurnal variation.
Magnetic susceptibility The ability of a substance, usually a soil, to become magnetized.
mHz See *Hertz*.
Millisiemens/meter (mS/m) The unit of measurement of electromagnetic conductivity, an intrinsic bulk property of a soil that depends on type, particle size, compaction, dissolved ion content, moisture, and other factors. 0.001 siemens/meter; the siemen/meter = 1.0/ohm-meter.
Nanosecond (nS) 10^{-9} second.
Nanotesla (nT) 10^{-9} Tesla, a measure of magnetic flux density per unit area (formerly the gamma).
Ohm's Law Allows determination of earth resistance as the ratio of voltage to current: $R = V/I$, where R is resistance, V is volts, and I is current.
Ohm-meter The unit of resistivity measurement, an intrinsic property of a soil that depends on type, particle size, compaction, dissolved ion content, moisture, and other factors. One approximation gives: ohm-m = $2\pi aR$, where a is the interprobe spacing (in a Wenner array) and R is the measured resistance in ohms.
Passive prospection methods Geophysical techniques that record native earth properties like magnetism or gravitation over the earth's surface.
Pixel A picture element in a computer display representing a single measured value in the underlying data.
Proton precession magnetometer A magnetic sensor containing a coil in a proton-rich liquid. Application of a current polarizes the protons, which then precess to the magnetic field of the earth when the current ceases. The frequency of the precession is proportional to the strength of the field.
Relative dielectric permitivity The ability of a substance to propagate radar energy, calculated as the ratio of its electrical permitivity to that of a vacuum.
Resistivity See *Electrical resistivity*.
Resistivity tomography The slicing of a data cube of resistivity measurements horizontally, to generate plan views at various depths, and vertically, to create sections revealing stratigraphic relationships. The data cube is obtained through horizontal survey over an area with multiple depth readings taken at each station.
Vertical and horizontal dipole modes Two modes of use for EM instruments that align transmitting and receiving coils perpendicular to or parallel with the ground surface, affecting depth of prospection.
Thermoremanent magnetism A permanent magnetism derived from heating a material beyond its Curie point. The orientation of the magnetism reflects its relative position in the earth's magnetic field at the time of cooling.
Total field magnetic survey A survey over an area that systematically records variations in the total magnetic field stemming cumulatively from all sources, including underlying geology, properties of soils, and metallic objects. Measured in nanoteslas (nT; see *Nanotesla*).
Twin electrode array A modification of the Wenner array where the probes are separated into two pairs, each consisting of current and voltage probes. Only one pair is moved during a survey, facilitating data gathering and generating information with some advantages for interpretation. The other pair remains fixed at a remote location.
Wenner array A linear, equally spaced, four-probe configuration for measuring soil resistivity, where the probe interval width is approximately equal to the depth of prospection. The outer probes inject a current while the inner probes measure voltage, allowing computation of resistance in ohms and resistivity in ohm-meters.

10. References

Ahler, S. A., 1993, Plains Village Cultural Taxonomy for the Upper Knife-Heart Region. In *The Phase I Archaeological Research Program for the Knife River Indian Villages National Historic Site, Part IV: Interpretation of the Archaeological Record*, edited by T.D. Thiessen, pp. 57–108. Midwest Archaeological Center Occasional Studies in Anthropology No. 27, National Park Service, Midwest Archaeological Center, Lincoln, NE.

Al Chalabi, M. M., and Rees, A. I., 1962, An Experiment on the Effect of Rainfall on Electrical Resistivity Anomalies in the Near Surface, *Bonner Jahrbücher 162:226*-271.

Aspinall, A., and Crummett, J. G, 1997, The Electrical Pseudo-Section. *Archaeological Prospection* 4:37–47.

Avery, T. E., and Berlin, G. L., 1985, *Fundamentals of Remote Sensing and Airphoto Interpretation*, 5th ed. Macmillan, New York.

Becker, H., 1995, From Nanotesla to Picotesla—A New Window for Magnetic Prospecting in Archaeology. *Archaeological Prospection* 2:217–228.

Bevan, B.W., 1983, Electromagnetics for Mapping Earth Features. *Journal of Field Archaeology* 10:47–54.

Bevan, B. W., and Kenyon, J., 1975, Ground-Penetrating Radar for Historical Archaeology. *MASCA Newsletter* 11:2–7.

Bradley, J., and Fletcher, M., 1996, Extraction and Visualization of Information from Ground Penetrating Radar Surveys. In *Interfacing the Past: Computer Applications and Quantitative Methods in Archaeology, CAA95*, Volume 1, edited by H. Kamermans and K. Fennema, pp.103–110. Analecta Praehistorica Leidensia, No. 28, University of Leiden, The Netherlands.

Clark, A., 1990, *Seeing Beneath the Soil: Prospection Methods in Archaeology*. B. T. Batsford, London.

Conyers, L. B., and Goodman, D., 1997, *Ground-Penetrating Radar: An Introduction for Archaeologists*. Alta Mira Press, Walnut Creek, CA.

Corney, M., Gaffney, C. F., and Gater, J. A., 1994, Geophysical Investigations at the Charlton Villa, Wiltshire (England). *Archaeological Prospection* 1:121–128.

Dalan, R. A., 1991, Defining Archaeological Features with Electromagnetic Surveys at the Cahokia Mounds State Historic Site. *Geophysics* 56:1280–1286.

Dalan, R.A., 1993, Issues of Scale in Archaeological Research. In *Effects of Scale on Archaeological and Geoscientific Perspectives*, edited by J. K. Stein and A. R. Linse, pp. 67–78. Geological Society of America Special Paper 283, Boulder, CO.

David, A., 1995, *Geophysical Survey in Archaeological Field Evaluations*. Ancient Monuments Laboratory, English Heritage Society, London.

Dawson, M., and Gaffney, C. F., 1996, Correction to "The Application of Geophysical Techniques With a Planning Application to Norse Road, Bedfordshire (England)". *Archaeological Prospection* 2:237.

Drahor, M. G., Akyol, A. L., and Dilaver, N., 1996, An Application of the Self-Potential (SP) Method in Archaeogeophysical Prospection. *Archaeological Prospection* 3:141–158.

Dunnell, R. C., and Dancey, W. S., 1983, The Siteless Survey: A Regional Scale Data Collection Strategy. In *Advances in Archaeological Method and Theory*, Volume 6, edited by M.B. Schiffer, pp. 267–287. Academic Press, New York.

Eder-Hinterleitner, A., Neubauer, W., and Melichar, P., 1996, Restoring Magnetic Anomalies. *Archaeological Prospection* 3:185–197.

Fletcher, M., and Spicer, D., 1995, Simulation of Ground Penetrating Radar. In *Computer Applications and Quantitative Methods in Archaeology 1993*, edited by J. Wilcock and K. Lockyear, BAR International Series No. 598, pp. 45–49. Tempus Reparatum, Oxford.

Frohlich, B., and Lancaster, W. J., 1986, Electromagnetic Surveying in Current Middle East Archaeology: Application and Evaluation. *Geophysics* 51:1414–1425.

Gaffney, C., Gater, J., and Ovendon, S., 1991, *The Use of Geophysical Techniques in Archaeological Evaluations*. Institute of Field Archaeologists Technical Paper 9, University of Birmingham, UK.

Geonics, Limited, 1992, *EM-38 Operating Manual*. Geonics Limited, Mississaugua, Ontario.

Geoscan Research, 1995, *Instructional Manual: Fluxgate Gradiometer FM-9, FM-18, FM-36*. Geoscan Research, Bradford, UK.

Geoscan Research, 1996, *Instructional Manual: Resistance Meter RM-15*, Geoscan Research, Bradford, UK.
Goodman, D., Nishimura, Y., and Rogers, J. D., 1995, GPR Time-Slices in Archaeological Prospection. *Archaeological Prospection* 2:85–89.
Griffiths, G. H., and Barker, R. D., 1994, Electrical Imaging in Archaeology. *Journal of Archaeological Science* 21:153–158.
Heimmer, D. H., and De Vore, S. L., 1995, *Near-Surface High Resolution Geophysical Methods for Cultural Resource Management and Archaeological Investigations*. U.S. Department of the Interior National Park Service, Denver, CO.
Johnson, C., 1998, The Coalescent Tradition. In *Archaeology on the Great Plains*, edited by W. R. Wood, pp. 308–344. University Press of Kansas, Lawrence.
Keay, S., Creighton, J., and Jordan, D., 1991, Sampling Ancient Towns. *Oxford Journal of Archaeology* 10:371–383.
Ketchum, R. M., 1962, *The Battle for Bunker Hill*. Doubleday, Garden City, NY.
Kvamme, K. L., 1996, A Proton Magnetometry Survey at Navan Fort, *Emania* 14:83–88.
Kvamme, K. L., 1999, Remote Sensing at Menoken Village. In *Interim Report and Work Plan for Continuing Archaeological Studies at Menoken Village State Historic Site, 32BL2 Burleigh County, North Dakota*, edted by S. A. Ahler. Report prepared for the State Historical Society of North Dakota, Bismarck.
Malagodi, S., Orlando, L., Piro, S., and Rosso, F., 1996, Location of Archaeological Structures Using GPR Method: Three-Dimensional Data Acquisition, *Archaeological Prospection* 3:15–23.
McNeill, J. D., 1980, *Electromagnetic Terrain Conductivity Measurements at Low Induction Numbers*. Technical Note TN-6, Geonics Limited, Mississaugua, Ontario.
Milligan, R., and Atkin, M., 1993, The Uses of Ground-Probing Radar Within a Digital Environment on Archaeological Sites. In *Computing the Past: Computer Applications and Quantitative Methods in Archaeology*, edited by J, Andresen, T. Madsen, and I. Scollar, pp. 21–23. Aarhus University Press, Aarhus, Denmark.
Music, B.,1995, On-Site Prospection in Slovenia: The Case of Rodik. *Archaeological Computing Newsletter* 43:6–15.
Neubauer, W., Melichar, P., and Eder-Hinterleitner, A., 1996, Collection, Visualization, and Simulation of Magnetic Prospection Data. In *Interfacing the Past: Computer Applications and Quantitative Methods in Archaeology, CAA95*, Volume 1, edited by H. Kamermans and K. Fennema, pp.121–129. Analecta Praehistorica Leidensia, No. 28, University of Leiden, The Netherlands.
Newman, C., 1993, The Tara Survey: Interim Report. In *Discovery Programme Reports 1: Project Results 1992*, pp. 70–93. Royal Irish Academy, Dublin.
Noel, M., 1992, Multielectrode Resistivity Tomography for Imaging Archaeology. In *Geoprospection in the Archaeological Landscape*, edited by P. Spoerry, pp. 89–99. Oxbow Monograph 18, Oxbow Books, Oxford.
Overdon, S. M., 1994, Application of Seismic Refraction to Archaeological Prospecting, *Archaeological Prospection* 1:53–64.
Payne, A., 1996, The Use of Magnetic Prospection in the Exploration of Iron Age Hillfort Interiors in Southern England. *Archaeological Prospection* 3:163–184.
Piro, S., 1996, Integrated Geophysical Prospecting at Ripa Tetta Neolithic Site (Lucera, Foggia—Italy). *Archaeological Prospection* 3:81–99.
Schurr, M. R., 1997, Using the Concept of the Learning Curve to Increase the Productivity of Geophysical Surveys. *Archaeological Prospection* 4:69–83.
Scollar, I., 1971, A Magnetometer Survey of the Colonia Ulpin Trajana Near Xanten, West Germany. *Prospezioni Archeologiche* 6:83–92.
Scollar, I., Tabbagh, A., Hesse, A., and Herzog, I., 1990, *Archaeological Prospection and Remote Sensing*. Cambridge University Press, Cambridge, UK.
Steinacher, T.L., and Toom, D.L., 1984, Archaeological Investigations at the Whistling Elk Site (39HU242), 1978–1979. In *Archaeological Investigations Within Federal Lands Located on the East Bank of the Lake Sharpe Project Area, South Dakota: Final Report*, Volume II, Appendix 1, edited by C. Falk. Technical Report No. 83-04, prepared for the U.S. Army, Omaha District Corps of Engineers, NE.
Summers, G. D., Summers, M. E. F., Baturayoglu, N., Harmansah, Ö., and McIntosh, E., 1996, The Kerkenes Dag Survey: An Interim Report. *Anatolian Studies* 46:201–234.

Szmanski, J. E., and Tsourlos, P., 1993, The Resistance Tomography Technique for Archaeology. *Archeologia Polona* 31:5–32.

Toom, D. L., and Kvamme, K. L., 2001, The "Big House" at Whistling Elk Village (39HU242): Geophysical Findings and Archaeological Truths. *Plains Anthropologist*, in press.

Walker, R., and Somers, L., 1994, *Geoplot 2.01, Instruction Manual*. Geoscan Research, Bradford, UK.

Weymouth, J .W., 1986, Geophysical Methods of Archaeological Site Surveying. In *Advances in Archaeological Method and Theory*, Volume 9, edited by M.B. Schiffer, pp. 311–395. Academic Press, New York.

Weymouth, J. W., 1997, *Knife River Revisited or Old Data in New Clothes*. paper Presented at the 55th Plains Anthropological Conference, Boulder, CO.

Weymouth, J.W., and Lessard, Y.A., 1986, Simulation Studies of Diurnal Corrections for Magnetic Prospection. *Prospezioni Archeologiche* 10:37–47.

Weymouth, J. W., and Nickel, R. K., 1977, A Magnetometer Survey of the Knife River Indian Villages. *Plains Anthropologist Memoir* 22(13):104–118.

Wynn, J. C., 1986, A Review of Geophysical Methods Used in Archaeology. *Geoarchaeology* 1:245–252.

Beyond ¹⁴C Dating

A User's Guide to Long-Range Dating Methods in Archaeology

W. JACK RINK

1. Introduction

In the 1980s and 1990s a range of new dating methodologies has emerged that are based on radiation exposure effects in host materials. These methodologies include electron spin resonance (ESR) and various forms of luminescence dating (OSL, IRSL, and TL). These techniques require careful sampling strategies that are best accomplished by archaeologists and geochronologists working together. This chapter aims to provide archaeologists and geochronologists with an introduction to the sampling approaches and range of problems that can be solved in archaeology, with an accompanying description of the physical basis of the methods. These radiation exposure techniques, along with fission-track methods, are compared with recently improved conventional radiogenic isotopic dating approaches such as mass spectrometric uranium series and $^{40}Ar/^{39}Ar$ dating. Emphasis is placed on considerations of how to identify datable materials

---

W. JACK RINK • School of Geography and Geology, McMaster University, Hamilton, Ontario, L8S 4M1, Canada.

Earth Sciences and Archaeology, edited by Paul Goldberg, Vance T. Holliday, and C. Reid Ferring. Kluwer Academic/Plenum Publishers, New York, 2001.

and selection of the appropriate technique for dating in a site, with due consideration to various problems and advantages in the various approaches.

1.1. Scope and Current Issues

The range of methods available to earth scientists and archaeologists that can date events older than the range of radiocarbon dating expanded rapidly during the 1980s. The scope of this chapter includes many of the modern dating techniques in archaeological science that are useful beyond the range of ^{14}C dating (about 40 ka), with only brief mention of parallel applications in the earth sciences. I have chosen to discuss only those dating methodologies that fall into two contrasting categories: those based on radiogenic isotopes versus those based on radiation exposure. In the former, radioactive decay serves as the clock, whereas the latter is based on the accumulation of radiation exposure effects in a host material. The discovery and application of these radiation exposure dating techniques has rapidly filled the dating gap which existed for the time range of about 40,000 to about 200,000 years ago. Additions to the toolkit for dating Pleistocene age deposits include new methods such as electron spin resonance, optical luminescence and thermoluminescence.

The expansion in dating technologies has brought forward a dilemma for archaeologists because application of many of these new methods requires certain shifts in excavation strategy in order to optimize the potential for dating results. The needs of the dating specialist share certain similarities to that of other technical specialists such as pollen analysts and archaeozoologists. The timing and nature of sample collection among the specialists at the same site are beginning to merge and in some cases this produces potential conflicts in sample demands. However, joint interpretive activity with a team focus can enhance and contribute to the significance of the findings. I focus in some detail on the planning and collaboration that are needed for success in these approaches and discuss the implications of the results.

Aside from the sampling demands of the new archaeochronologists, a number of controversial dating results since the mid-1980s provided evidence that has stretched the views of the most liberally inclined thinkers in paleoanthropology and archaeology. For example, ESR dating of tooth enamel and luminescence dating provided dramatic evidence that modern humans overlapped in time and space with Neanderthals and *Homo erectus*, and that humans reached Australia at least 50,000 to 60,000 years ago. These findings have caused a change in the nature of teaching the subject of human evolution, which now includes a more critical appraisal of the new dating technologies.

2. How to Choose the Right Dating Methods

Five questions should be asked by the archaeologist when attempting to develop a secure chronology for an archaeological site:

1. What datable materials are present?
2. Are these materials useful in the expected time range for the deposit?
3. What is actually dated by using the chosen material?
4. Is the dating method(s) chosen likely to provide sufficient precision to answer the questions of interest?
5. Are the datable materials younger, older, or nearly coeval with the archaeological finds?

These questions are sometimes difficult to answer. Tables 14.1 and 14.2 summarize a number of relevant characteristics of the dating methods discussed in this chapter. ^{14}C is a technique that requires calibration using an independent dating method, which must be considered for Question 4. The fifth question is often the most difficult to answer with confidence and should always color the interpretation of dating results. Problems such as bioturbation and other postdepositional processes, as well as complex modes of deposition, require experienced skills of interpretation to decipher the sequence of events represented in a deposit. The best way to increase confidence in dating results is to use several methods on

Table 14.1. Isotopic Dating Methods Useful in Archaeometry

Method	Materials	Half-life (years)	Dating range (ka)	Typical age uncertainty ($\pm\%$)	Reference
^{14}C	Wood, seeds shell, charcoal, bone, painting	0.780×10^3	0.2–40	1–2	1
^{40}Ar/^{39}Ar	Feldspar, pumice	1.250×10^9	10–>1 × 10^7	1–2	2, 3
Uranium series (Closed system) ^{234}U + ^{230}Th	Calcite	2.48×10^5 (U) 7.52×10^4 (Th)	0.1–350	1–2	4, 5, 6
^{234}U + ^{230}Th (Open system)	Teeth, bone	2.48×10^5 (U) 7.52×10^4 (Th)	See text	See text	4, 5, 7
^{235}U + ^{231}Pa (closed system)	Calcite	7.04×10^8 (Pa)	0.1–200	1–5	4, 5
^{235}U + ^{231}Pa (open system)	Teeth, bone	7.04×10^8 (Pa)	See text	See text	4, 5, 7

Sources: 1. Harris, 1987; 2. McDougall, 1995; 3. Swisher et al., 1994; 4. Schwarcz and Blackwell, 1991; 5. Blackwell and Schwarcz, 1993; 6. Dickin, 1995; 7. Simpson and Grün, 1998.

Table 14.2. Radiation Exposure Dating Methods Useful in Archaeometry

Method	Materials	Dating range (ka)	Typical age uncertainty (±%)	Reference
Electron Spin Resonance (ESR)	Shell	5–200	10–20	1, 2
	Calcite speleothem*	10–500	20–30	1, 3
	Burned quartz Sediment*	10–500	10–20	1, 4
	Burned quartz rock*	10–500	20–30	1, 5
	Windblown quartz	20–1000	10–20	1, 6
	Waterlain quartz*	20–1000	10–20	1, 7
	Tooth enamel	10–2000	20–50	1, 8
Coupled ESR/U-series	Tooth enamel dentine/cementum	10–2000	10–50	1, 9, 10, 13
Thermoluminescence (TL)	Burned silex	1–500	5–10	14, 15
	Flint/chert	4–400	10–20	16, 17, 18
	Burned quartz rock	4–400	10–20	16, 17, 18
	Burned quartz sediment	4–400	10–20	18
	Tephra	1–800	10–20	18
	Quartz	1–100	10–20	18
	Beach and dune sands feldspar	1–800	10–20	18
	Loess (windblown) Eolian and Waterlain sediments	1–200	10–20	18
Optically Stimulated Luminescence (OSL)	Windblown/Waterlain quartz	0.5–200	10–20	16, 19, 20
	Quartz in wasp nest	0.5–200	10–20	16, 19, 21
Infrared Stimulated Luminescence (IRSL)	Windblown/Waterlain feldspar	0.5–250	5–15	18, 19
	Windblown/Waterlain quartz	0.5–500	10–20	18, 19, 22
Fission Track Dating	Zircon			23
	Natural glass			24

*Method has a short track record.
Source: 1. Rink, 1997; 2. Blackwell, 1995; 3. Bahain et al., 1994; 4. Falguères et al., 1994; 5. Monnier et al., 1994; 6. Yoshida, 1996; 7. Laurent et al., 1994; 8. Schwarcz et al., 1994; 9. Rink et al., 1998; 10. Rink et al., 1995; 11. McDermott et al., 1993; 12. Blackwell and Schwarcz, 1993; 13. Chen et al., 1997; 14. Mercier et al., 1995a; 15. Valladas, 1992; 16. Roberts, 1997; 17. Valladas and Valladas, 1987; 18. Berger, 1995; 19. Aitken, 1998; 20. Roberts et al., 1994; 21. Roberts et al., 1997; 22. Huntley et al., 1985; 23. Gleadow, 1980; 24. Westgate and Naeser, 1995.

samples spanning the stratigraphic range of the deposit, preferably through a coordinated team approach involving several dating specialists.

What if more than one dating method seems appropriate for a single type of sample? An example of this would be dating of windblown or waterlain sediment (Table 14.2) for which you could use three different methods: OSL, TL,

Table 14.3. Selected Applications in Geochronology Related to Paleoclimate or Paleoenvironmental Studies

Application	Datable material	Method	Reference
Age of glacial epoch	Quartz or feldspar grains in loess or glaciofluvial silt	OSL, IRSL, TL	1, 2
Age of interglacial, sea level	Speleothem calcite, coral	U-series, ESR	3, 4
Age of coastal deposit	Quartz or feldspar grains dune or raised beach	OSL, IRSL	5, 6
Age of arid phase	Quartz from desert sand feature	OSL, IRSL, TL	7
Age of limnic or humid phase	Quartz from lake sediment	OSL, IRSL, TL	5
Age of fluvial deposit	Quartz or feldspar from sandy unit	OSL, IRSL	1
	Feldspar grains or pumice fragment	$^{40}Ar/^{39}Ar$	8
Age of terrace development	Quartz or feldspar from river terrace	OSL, IRSL	1

Sources: 1. Prescott and Robertson, 1997; 2. Forman, 1989; 3. Dickin, 1995; 4. Li et al., 1989; 5. Stokes, 1992; 6. Ollerhead et al., 1994; 7. Rendell et al., 1994; 8. Swisher et al., 1994.

or ESR. In general, the best choice would be to use the one that assures more complete zeroing (OSL), but if the sample is thought to be older than about 500 ka, then it would be appropriate to try ESR. Whenever possible, it is best to choose at least one method that has a relatively high resolution (lower age uncertainty). An example of this would be to choose mass spectrometric U-series dating of calcite instead of ESR dating of calcite, unless it is already known that the practical limit of U-series dating has been exceeded (>350 ka).

Beyond direct dating of archaeological finds, the applications in earth science can provide relevant chronologies for paleoenvironmental or paleoclimatic events. (Table 14.3). The melding of geo- and archaeo-based chronologies provides links among spatial and temporal dimensions of the human environment, and the response by humans reflected in evolutionary and behavioral patterning. Moreover, these chronologies form an important source of data needed for modeling ancient human migration patterns.

3. Radiogenic Isotopes for Dating: Physical Basis

A chemical element is defined on the basis of the number of protons in the nucleus of the atom. Isotopes of a chemical element have the same number of protons but a different number of neutrons. Radioactive decay, in its various forms, is a random process of transformation of one isotope into another chemical element or into another isotope of the original (or parent) isotope. The radioactive decay process (law of radioactive decay) is characterized by a time constant, such that

$$N = N_0 \exp(-t) \qquad (1)$$

where N is the remaining number of parent atoms after decay time t, and N_0 is the initial number of parent atoms. The "half-life" of the process is most easily understood by considering the events that occur at the outset. In this case, the half-life ($T1/2$) of the parent is the amount of time required for $N = 0.5N_0$, that is, when the process reduces the initial quantity to one half of its initial size (Fig. 14.1). Each decay event produces one daughter isotope, therefore after $T_{1/2}$, the amount of radioactive daughter atoms equals N (the number of remaining parent atoms). As the process continues beyond $T_{1/2}$, the decay constant requires that this halving occur again after $T_{1/2}$ has passed again. The relationship between the half-life (units = years) and the decay constant (units = per year) is given by

$$T_{1/2} = (\ln 2)/\lambda \qquad (2)$$

Figure 14.1 shows the parent decay and daughter growth curves governed by Equation 1. It is clear that the daughter growth curve is complementary to that of the decay curve. The y-axis units can either be the fraction of isotope present (N/N_0) or the number of disintegrations of isotope per unit time, also termed the activity, which is simply N_0. The average number of decay events per second is

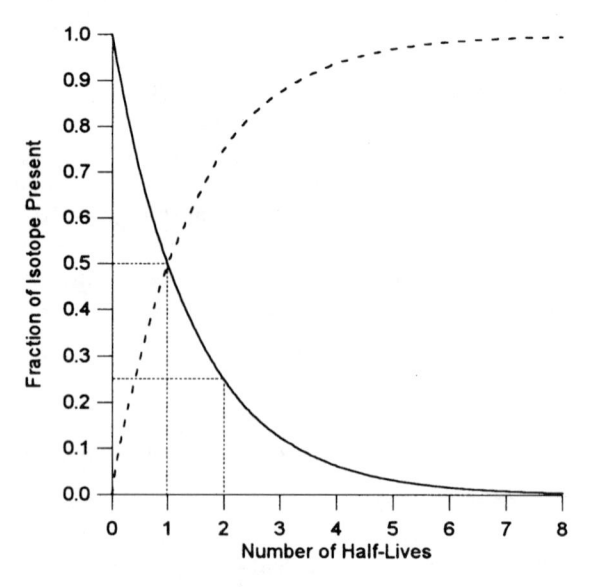

Figure 14.1. Behavior of radioactive parent isotope (solid line) and daughter (dashed line) in a closed nuclear decay system, showing the proportion of isotope present after a given half-life of the parent.

thus proportional to the number of parent atoms present, and therefore the activity has a definite average value (over a longer time period) even though this number should vary from second to second within certain limits predictable according to probability theory. This latter characteristic of radioactive decay allows its use as a chronometer provided that certain assumptions can be reasonably made (Faure, 1974).

For our purposes, the main concern is whether the chemical system of the mineral used for dating is "closed" or "open," that is, whether or not exchange of parent or daughter with the surrounding environment has occurred during burial. For a closed system, the ratio of parent to daughter at any time t can be measured and used to estimate the time since radioactive decay commenced, provided that we assume that no daughter was present at the outset. The age equation for a given isotopic dating method relates the decay constant to the parent and daughter concentrations (or ratios) in the sample. For an open system, accommodation is made for gain or loss of parent or daughter through a series of simplifying assumptions and can involve models for exchange between the environment and the sample. Open-system dating results are often of limited utility because they only provide a constraint on the sample age rather than an absolute age. However, such constraints are of vital importance in certain circumstances. A special use of open-system U-series dating is to constrain the age of burial of tooth enamel through coupled use with electron spin resonance dating (see section 4.6.2). In some circumstances, discovery of "open-system" behavior in a material which normally exhibits "closed-system" behavior places direct constraints upon the scope of the conclusions that can be drawn from the study. This is discussed further in section 5.

3.1. Applications of Radiogenic Isotope Dating

Examples of isotopic dating methods useful in the Quaternary period are ^{14}C, U-series, and $^{40}Ar/^{39}Ar$ dating (Table 14.1). The latter two methods can be used to determine the time of mineral formation that corresponds to the age of crystallization of a closed geochemical system, whereas radiocarbon dating (^{14}C) is used to measure the time since death of an organism. Table 14.1 lists important characteristics of selected radiogenic isotope dating methods. In uranium series (U-series) and ^{14}C dating, the half-lives are short, making these isotopic systems useful over relatively short periods of geological time. Accelerator mass spectrometry (AMS) has greatly reduced the sample size needed for ^{14}C dates, and has improved its resolution. In the case of the $^{40}Ar/^{39}Ar$ method, the much longer half life makes its application much more difficult for samples younger than about 100,000 to 200,000 years (100–200 ka), although samples as young as about 12,000 years have been dated under favorable circumstances (Hu et al., 1994). More problematic is the availability of materials of volcanic origin in most archaeological sites. Older samples in the 1 to 5 million year range are much easier to date, which enabled high-resolution dating of early hominids in Kenya (Walter et al., 1991), in Asia (Swisher et al., 1994) and in Ethiopia (Walter, 1994; Kimbel et al., 1994).

3.2. ^{40}Ar/^{39}Ar Dating

Radioactive potassium (^{40}K) is the parent of radiogenic ^{40}Ar. The parent material, ^{40}K, occurs abundantly as a major element in volcanic minerals such as feldspar and in volcanic glasses such as pumice. Because the ratio of ^{39}K to ^{40}K is a constant in nature, ^{40}K is determined by proxy in ^{40}Ar/^{39}Ar dating by measuring ^{39}Ar that is produced artificially from ^{39}K in a nuclear reactor by neutron bombardment. This is done before determining isotope ratios in the sample. After irradiation, a mass spectrometer is used to measure the ^{40}Ar/^{39}Ar ratio of the gas evolved during incremental heating or total fusion melting of the sample by a laser beam (SCLF; single crystal laser fusion) or within a resistance furnace. The laser beam approach pioneered by York et al. (1981) allows use of individual crystals from volcanic rocks (individual grains as small as a few mg in size for Pleistocene age samples, and a small as 1 mg for older samples), which can be analyzed with very high precision. Grains of differing age introduced by sedimentary processes into reworked volcanic units of a single crystallization age can be identified and ignored (e.g. Walter et al., 1991).

3.3. Closed-System Uranium-Series Dating

This method is mainly applied to calcite in archaeological applications (Blackwell and Schwarcz, 1995; Schwarcz and Blackwell, 1991). The ^{234}U/^{230}Th dating method is possible due to the preferential solubility of uranium in fresh groundwaters relative to its daughter product thorium (or protactinium for the ^{235}U/^{231}Pa method). Thus, just after crystallization, calcite precipitated from aqueous solution contains uranium but without any significant quantity of its natural radiogenic daughter thorium. This state of disequilibrium in the uranium decay chain opens a dating range window of about 1 to 500 ka in which the daughter of ^{234}U (^{230}Th) will grow into isotopic equilibrium with its parent. The state of isotopic equilibrium is characterized by equal activity levels at each stage in a decay-chain sequence of isotopes that transform from one to another by nuclear decay at different rates. Because both the ^{234}U and the ^{230}Th are radioactive (the ^{230}Th decays into ^{226}Ra), the half-lives of both isotopes are important in the age equation. The window of opportunity for the ^{235}U/^{231}Pa system is somewhat shorter because of the relatively faster decay rate of the ^{231}Pa. In both systems, the long-lived isotopes of ^{238}U (for the ^{234}U/^{230}U system) and ^{235}U (for the ^{235}U/^{231}Pa system) are essentially invariant.

Thermal ionization mass spectrometry (TIMS) produced a major advance for all of the U-series dating methods (Edwards et al., 1987). This technique allows precise measurement of isotope ratios of heavy elements like uranium and thorium. The older technology of alpha counting required much longer analysis time and larger samples, but chemical separation of the uranium and thorium (or protactinium) is still needed. For samples with high uranium concentrations (tens of parts per million [ppm]) only tens of milligrams of sample are needed for TIMS U-series, but tens of grams are still required when uranium concentrations are only a few parts per billion (ppb; as is often true for cave speleothems).

3.4. Open-System Uranium-Series Dating

This method is used to constrain the length of burial time for objects that absorb uranium into their structure. The most typical materials are bone, enamel, dentine, and cementum; ostrich egg shell also falls into this group (Schwarcz and Blackwell, 1991). Open-system U-series dating is possible because uranium in groundwater may be incorporated into buried material, which then sets up a situation similar to that of closed-system U-series dating in that the same clock is used, but different in that the clock is partially reset whenever more uranium is incorporated. Each time this occurs, the state of disequilibrium is disturbed further. Open-system dating often relies on the "early uranium uptake assumption," which states that the time required to absorb the uranium is small relative to the time period of burial, in which case the age represents a good approximation of the burial age. However, this may or may not be a valid assumption, and in many cases the open-system age probably represents the minimum age, which can be an important age constraint in an archaeological deposit (see discussion on dating of bone in section 6). Of significance is the use of this method to constrain possible uranium uptake histories in ESR dating of teeth.

3.5. Sampling Requirements for Uranium-Series and ^{40}Ar/^{39}Ar Dating

For both of these methods the mineral of interest contains all of the dating information. The sample size is important, and it depends on the abundance of the parent and daughter isotopes in the sample. The amount of daughter product is a function of the initial abundance of parent and the time elapsed since crystallization.

For U-series dating of calcite it is best to collect 10 to 20 gm of clean, dense, hard material, which will provide enough sample in most cases, even if the isotopic concentrations are very low and the age is quite young. For calcite that is softer and more friable ("dirty calcite"), even larger amounts of sample should be obtained (Schwarcz and Latham, 1989). For U-series dating of bone, it is desirable to collect both the dense cortical portions and the more porous elements. If the bone is important from an archaeological point of view, a small sample can be first studied to determine the minimum amount needed, but where use of abundant faunal elements is possible, 5 to 10 gm is generally sufficient. In U-series dating of teeth, only a portion of the dentine, cementum, or enamel in a tooth is generally needed when mass-spectrometric methods are used (e.g. Swisher et al., 1996), but alpha-spectrometric U-series dating of enamel often requires complete destruction of the tooth (McKinney, 1998). Gradients in U-series age within teeth have recently been discussed by Pike and Hedges (in press). Nondestructive gamma-spectrometric methods for U-series dating of bone and teeth are also under development, which employ the radioactivity emitted by the whole sample (Schwarcz et al., in press; Simpson and Grün, 1998). These are particularly useful for analysis of human bone fragments.

For ^{40}Ar/^{39}Ar dating of volcanic rocks or transported sediment containing volcanogenic mineral grains, there is always a variety of minerals present that are undesirable. With modern $^{40}Ar/^{39}Ar$ methods, single grains to be dated can be isolated by hand picking under a binocular microscope. The amount of sample to be collected depends on the approximate age of the sample (the younger the sample the less radiogenic ^{40}Ar available for measurement) and the potassium content of the minerals to be dated (potassium-rich sanidine feldspars as opposed to potassium-poor plagioclase or hornblende). Usually the coarsest grain-size fraction of the tephra is collected and rarely is the collection of more than several kilograms of sample necessary to obtain sufficient abundance of datable grains. In cases where pumice fragments can be identified in the sediment, these should be taken as individual samples.

4. Radiation Exposure Dating

There are a large number of materials and situations that can be dated using these methods, which were rapidly developed throughout the 1980s and 1990s (Table 14.2). In contrast to more traditional approaches based on radiogenic isotopes, radiation exposure dating is based on the interaction between the sample and radioactivity. Regardless of whether the mineral "system" is chemically open or closed, minerals are continually exposed to the effects of internal and external radiation exposure. The effects of such exposure can be detected through a variety of means depending on the nature of the radiation passage through the mineral lattice structure. The effects may be subtle and on an atomic scale or conspicuous and easily detected by simple optical microscopy. These rather newer methods of chronometry are possible by quantitative measurement of the accumulated effect that has increased during the time period of interest. The age equations relate the quantified radiation effect to the rate of accumulation. The accumulated quantities are usually expressed in terms of radiation dose, which is a measure of the energy deposited per unit mass in the material, but in some cases is in terms of efficiency of effect per radiation event. For dose-based methods, the rates are dose per unit time, whereas some are based on the events per unit time. The age equation for dose-based methods are all based on an equation of the form:

$$\text{Age} = \text{Dose Units/Dose Rate Units per year} \tag{3}$$

4.1. Physical Basis of Fission-Track Dating

Fission-track dating is based on the accumulation of structural damage in the crystal due to natural fission decay of internal uranium and thorium impurities. The structural damage can be easily measured because each decay event produces an extended zone of damage along the path of an emitted the fission fragment, which breaks chemical bonds throughout much of the zone due to its very large kinetic energy. These zones are called *fission tracks* and remain intact for hundreds of thousands of years after the event as long as the mineral temperature is not raised above a certain threshold, which causes relaxation of the damaged zone back into the general state of the undamaged crystal. Minerals such as sphene (a calcium–titanium–silicate), apatite (a calcium phosphate) and zircon (zirconium silicate) are useful because uranium is present in significant abundance. Natural glass is often the only suitable component of volcanic tephras, which are commonly archaeologically relevant. In glass, the fission tracks anneal at ambient temperature, but major improvements in the technique have led to preheating procedures that overcome this difficulty (Westgate, 1988, 1989).

The age equations relate the observed number of fission tracks per unit volume to the number of decays events expected per uranium atom per unit time in unit volume of the material. The fission tracks (usually about 0.02 mm long) are made visible in the optical microscope by preferential chemical dissolution of these zones on a polished surface of the mineral and must be counted one by one.

4.2. Applications of Fission-Track Dating

The applications of fission track dating in archaeology are summarized by Wagner and Van den Haute (1992). Examples of very young objects dated by fission tracks include pottery and heated obsidian. In the time range beyond that of ^{14}C, fission-track dating is useful for natural volcanic glasses and detrital mineral grains of volcanic origin. When using detrital minerals to establish the age of a unit that encloses them, it is essential to ensure the sample is fresh and not reworked from older volcanic materials. A fission-track age of 1.87 ± 0.04 million years was obtained by Gleadow (1980) in good agreement with an age of 1.88 ± 0.02 million years by conventional ^{40}Ar/^{39}Ar for the KBS tuff at Lake Turkana in northern Kenya. Using the labor-intensive isothermal plateau fission-track (ITPFT) method, recently reviewed by Wintle (1996), Chesner et al. (1991) were able to date glass from the Toba eruption to 68 ± 7 ka in good agreement with a SCLF ^{40}Ar/^{39}Ar date of 73 ± 3 ka. This was in good agreement with its location at the oxygen isotope stage 5/4 boundary in nearby marine cores.

4.3. Sampling Requirements for Fission-Track Dating

A tiny portion of the sample of interest is used for the dating analysis, but it must be sufficiently large so that a statistically valid number of fission tracks can be counted. The number of tracks in a given mineral grain depends on the age of the sample and the amount of uranium per unit volume in the sample. For samples older than about 1 million years with high uranium content, ages can be obtained on single grains. When tracks must be counted in multiple grains, the area of sample containing the grains is highly variable depending on the sample. A section must be sawed and ground thin prior to acid etching and the final microscopic analysis.

4.4. Physical Basis of ESR and Luminescence Dating

ESR and luminescence dating are based on the accumulation of electronic charge in atomic lattice sites that are generally distributed throughout the crystal volume rather than ones that are grouped together along zones of damage as in the case of fission tracks. They are widely distributed because they are associated either with impurities in the lattice that are incorporated diffusely during crystallization or they are found at widely dispersed atomic-scale disturbances of the normal latticework of the crystal structure. There is some evidence that clusters of defects may be present in some systems. Because these defect sites are only slightly different from the normal lattice, they are only slightly different from the electronic states (charge states) that are characteristic of electronic states in chemical bonds that bind the lattice together. Figure 14.2 shows that their energy

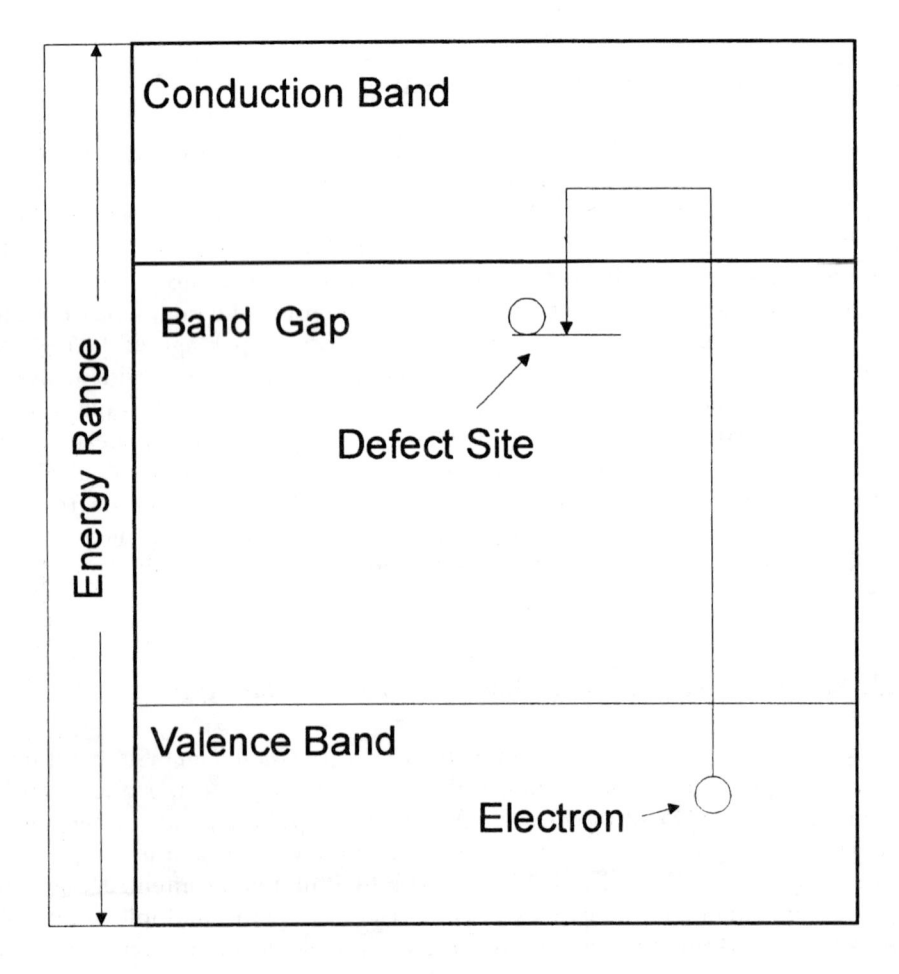

Figure 14.2. Effect of ionizing radiation on electrons in insulating materials that leads to accumulation of electrons in the energy region of the band gap at defect sites.

lies in a broad band that is of higher energy than that of the valence (bonding) electrons. These states are often referred to as "traps" in luminescence dating and "paramagnetic centers" in ESR dating. During extended time periods, the crystal is "charged" at these sites because radiation causes excitation of the bonding electrons up to the higher energy levels. This charging occurs because during passage through the crystal the radiation transfers its energy into ionization of atoms, which ejects electrons into the defects with higher energies, hence the idea of "trapping" them at sites with energies above their usual energies.

Charge in defect sites may be detected using ESR only when the electrons are unpaired, which makes their magnetic quality "visible" to microwaves. Only some defects in a crystal have charge in an unpaired state that can be studied by ESR, whereas a generally larger number can be studied with luminescence detection. In ESR detection, the microwaves cause the electron energy to be changed only slightly, but enough for detection as microwave absorption between two magnetic states in the defect. In luminescence, the charge is completely ejected from the defect and moves to a lower energy as it gives off visible or ultraviolet light (luminescence). ESR is less sensitive than luminescence because the number of electrons that can be forced between magnetic states is limited by the population difference that exists during resonance between two states of very similar energy. Therefore samples younger than about 10,000 years cannot be dated with ESR, but samples as young as about 100 years can be dated by luminescence, provided that sufficient radiation levels exist in the vicinity of the sample. In either detection technique, the aim is to quantify the amount of dose (paleodose) that was needed to establish the population of charge found in the as-collected sample and to couple that with an estimate of radiation dose rate emanating from the local natural radioactive sources within and outside of the dated material. When the paleodose is estimated in the laboratory it is called the *equivalent dose* (ED or DE).

4.5. Dosimetry Requirements

The external radiation environment is a concern for all of the types of ESR and luminescence dating. This environment delivers gamma and cosmic radiation doses to all of the types of datable samples and thus its analysis is common to all of these forms of dating. Assessment of the environmental radiation levels is often accomplished using thermoluminescence dosimeters implanted into the datable layers of the site (Fig. 14.3). The dosimeters are usually heated at the site just before emplacement so that effects from previous radiation doses during travel to the site are erased. The dosimeters contain highly radiation-sensitive mineral grains (e.g., CaF_2 or $CaSO_4$) that allow determination of the combined gamma and cosmic radiation dose after a period of only 6 to 12 months. The actual number of days of burial must be recorded in order to calculate the annual radiation dose (dose rate to the sample).

Gamma radiation to a datable sample originates at distances up to 30 cm away from the dated sample, hence the implanted dosimeters must be surrounded by at least 30 cm of sediment in all directions in order to sense all of

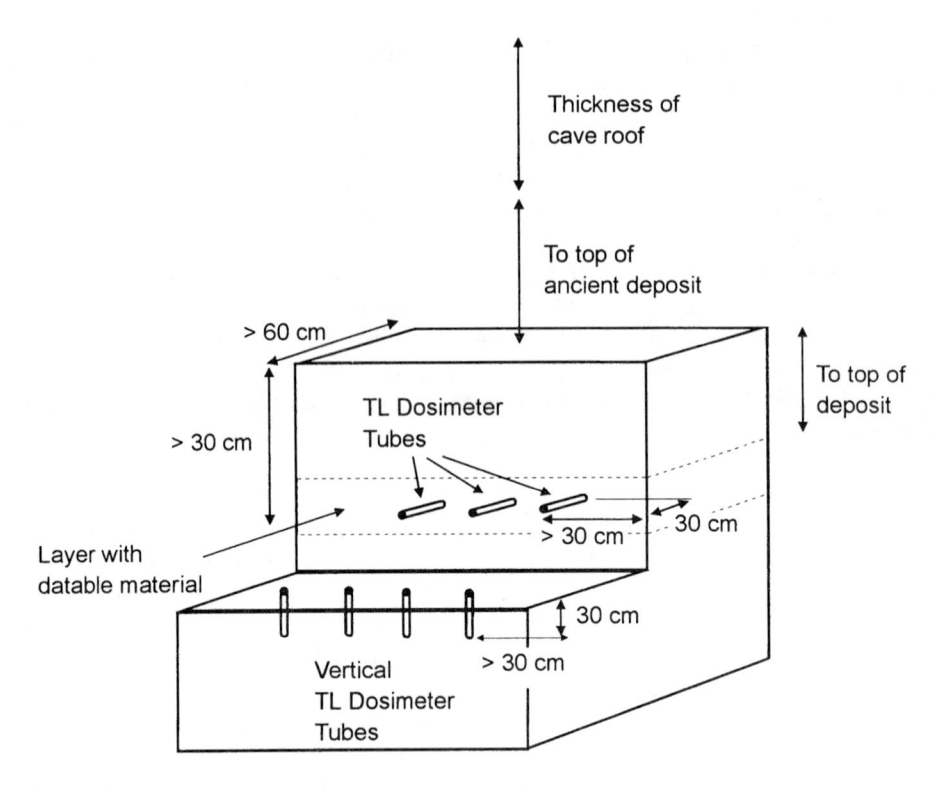

Figure 14.3. Scheme for emplacement of thermoluminescence (TL) dosimeters into a profile or excavation floor, showing minimum numerical dimensions important for emplacement and the correct method to measure other quantities needed (top of datable layer to top of present deposit, distance to original deposit height, and thickness of cave roof where appropriate). Vertical emplacements in excavation floor must be done to 50 cm depths when a cave roof is not present.

the important gamma radiation. Because the cosmic radiation is detected simultaneously and its intensity is dependent on the overlying thickness of the deposit, certain measurements of the overlying deposit (see Fig. 14.3) are important for appropriate assessment of the cosmic radiation dose rate during the burial history. Horizontal dosimeter emplacement in profiles is often used to approximate the gamma plus cosmic dose rate in nearby excavated layers that contained datable material. Vertical dosimeter emplacement in the excavation floor is a convenient way to determine the gamma plus cosmic dose rate of materials that will be excavated in later years, and is the optimal strategy for these measurements (see discussion that follows).

For deposits with low densities of archaeological materials, an electronic gamma spectrometer can be used in the same geometry shown in Fig. 14.3. This device allows a measurement of the radiation field in less than one hour but has the disadvantage that it only takes a snapshot of the dose rate. The TL dosimeters average the effects of moisture variation throughout the year (if burial is for 12 full months). If TL dosimeters are inadvertently or deliberately left buried for several years, they will still have properly recorded the annual radiation dose provided that a record of the total burial time is kept. The electronic instrument requires 30-cm deep holes that must be about 8 cm in diameter rather than the 1 cm that is needed for TL dosimeter capsules.

The excavation team and geochronologist usually work together for at least two consecutive seasons and develop a strategy for optimal timing of site visits, but the geochronologist usually visits at the very end of the season. Details of sample collection methods should be discussed in advance of the excavation season, so that finds can be properly collected and can become valuable datable samples. Dosimeter emplacement strategy evolves according to the distribution of sample finds in a given year, with horizontal emplacements chosen according to the highest density of useful finds in close proximity to witness sections. A key element in excavation planning is the need to retain witness sections within a few meters of datable samples that can remain undisturbed for about one year. These surveys are needed for extended surveys of the environmental radioactivity of each dated archaeological level using implanted radiation dosimeters. However, environmental radioactivity surveys are most ideally carried out in the floor of the active excavation rather than in witness sections. The geochronologist and excavators may determine at the end of the field season where to emplace vertical positions (dosimeter tubes or holes for the gamma spectrometer that are 30 cm deep) that should intersect with zones of tooth finds in the following year. If datable samples are found in the subsequently excavated layers, further horizontal dosimetry is not needed later in the following year. Furthermore, this approach provides the actual dosimetry in the layers containing datable samples, rather than an approximation in the laterally adjacent areas sampled by horizontal dosimetry. In extended excavations, it also allows for the development of a three-dimensional model of the radiation fields.

Figure 14.4 shows the sampling method for determining the porosity of the sediment in a layer. This method is used to estimate the maximum moisture content of the sediment and is important for appropriate estimates of uncertainty for the age of the samples. The sampling tube should be made of thin steel so that compaction during sampling is kept to a minimum. The ends must be carefully closed with spacers (when necessary) so that the sample does not shift during transport. It is also useful to determine the moisture content of this sample, so it should be sealed moisture-tight. Moisture content analyses of sediments that are collected adjacent to dated samples are also carried out.

4.6. Applications of ESR Dating

Teeth are prevalent in the archaeological record. Therefore, tooth enamel is the most widely used material for ESR dating in archaeological context. Consisting of pure, hard calcium phosphate, its large, tightly packed crystals have been found highly superior to bone calcium phosphate for ESR dating purposes because of their tendency to resist diagenetic alteration. Although highly arid or acidic environments can accelerate weathering of enamel, well-preserved enamels suitable for ESR dating are common in karst-hosted deposits throughout the

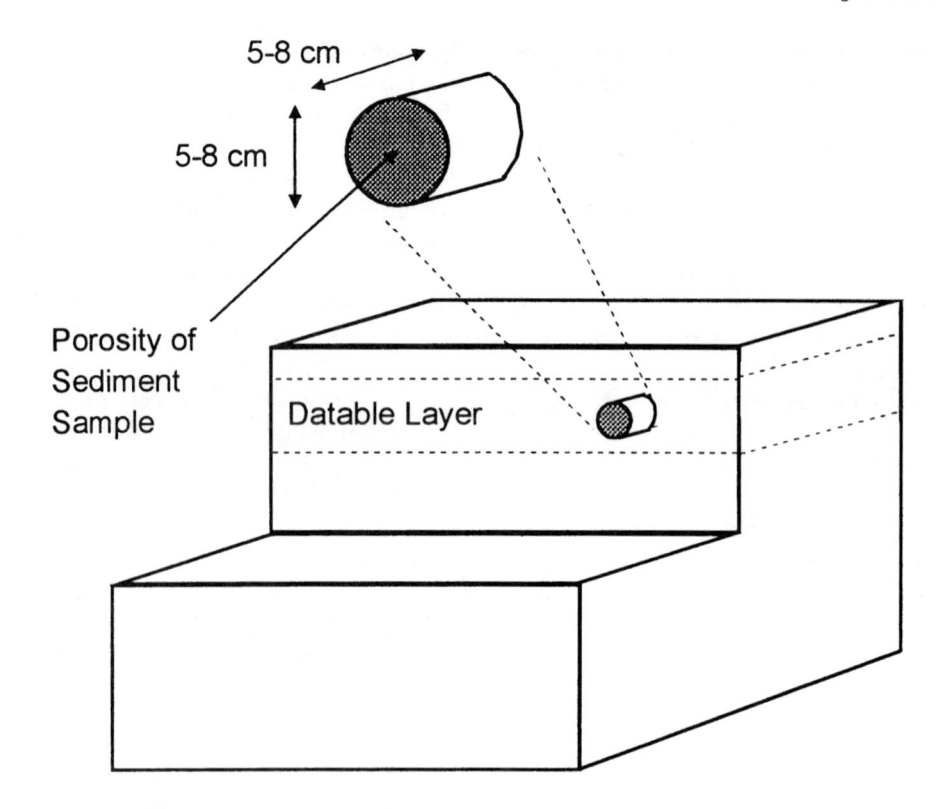

Figure 14.4. Method to collect samples of sediment needed to measure the porosity in a datable layer.

Pleistocene. The greatest advantage of using enamel is that it does not carry a relict geological irradiation dose at the time of death of its host, whereas luminescence dating materials must experience sufficient light or heat exposure at burial in order to zero or reset the geological dose.

Tooth enamel has become an important tool in dating evolution of humans and their tool assemblages, as reviewed recently by Rink (1997). It was widely applied to Neanderthals in the Levant (Schwarcz et al., 1988, 1989) and in Europe (e.g., Rink et al., 1995), and to modern humans in Africa (Grün et al., 1990). More recently it was used to date early humans in Asia (Chen et al., 1997; Huang et al., 1995), early modern humans in China (Chen et al., 1994), and very late *H. erectus* (Swisher et al., 1996).

4.6.1. Sampling Requirements for ESR Dating of Tooth Enamel

R. Grün was the first to publish ESR dating results on tooth enamel (Grün, 1985). Geologists Grün, Schwarcz, and Zymela (1987) were the first to test and publish a method for ESR dating of enamel, which remains in use but with recent modifications (Brennan et al., 1997, Yang et al., 1998). Methodological aspects are discussed by Grün (1989), Blackwell (1995), and Rink (1997). From an

Beyond ^{14}C Dating

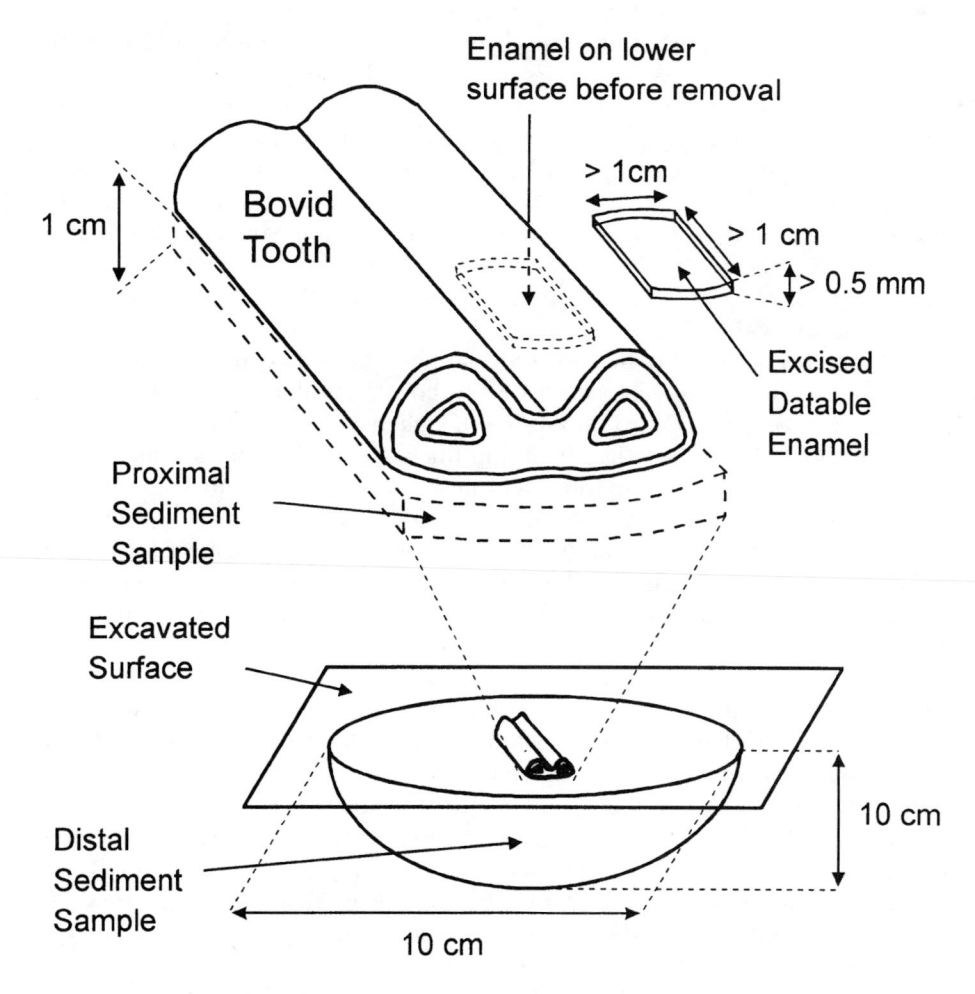

Figure 14.5. Requirements for collecting tooth samples for ESR dating of tooth enamel, showing locations and approximate dimensions of two different sediment samples that are needed to assess different aspects of the nearby radiation fields, and showing a portion of enamel that would be excised in the laboratory, preferably from the underside of the tooth in contact with the proximal sediment sample.

archaeologist's vantage point, there are some specific sampling requirements that are important to successful ESR dating (Fig. 14.5). The emphasis should be on careful extraction of the tooth with the adjacent few grams of sediment (<1 cm away), and with an additional larger sample of sediment coming from the zone within 10 cm of the tooth. The latter sample is used for a moisture content measurement, and both sediment samples are needed to assess different parts of the environmental radiation dose to the tooth enamel. The thin sample nearest the tooth is needed for the beta radiation dose rate, whereas the thicker sample is used to cross-check the results of the gamma radiation dose derived from the TL dosimetry or gamma spectrometry (Fig. 14.3).

From the viewpoint of the geochronologist, close consultation with the archaeological team is essential. Negotiations with the faunal specialist are a special concern because mutual agreement about selection of teeth is important to both parties. Special precautions for removal of sediment are also discussed so that faunal studies and casting can be done in the interval between the tooth removal date and shipment to the ESR laboratory. The amount of damage to specimens is limited to an area of about 1.5 cm^2 of the enamel and underlying dentine layers, and the occlusal surfaces should never sustain damage. Multiple photographs are made before enamel is excised, and they become part of the permanent archival record.

What kinds of teeth are best for ESR dating? The most commonly used teeth in the middle Pleistocene age sites are from bovid, equid, cervid, and ovid taxa. The best rule of thumb is to choose teeth that have reasonably thick enamel. Ideally, it should be greater than 0.75 mm thick, although 0.5 mm is an absolute minimum (Fig. 14.3). The adjacent dentine just inside the enamel should be at least 1.5 mm thick. The best teeth are from rhinoceros and hippopotamus because multiple large areas can be removed for a special variant of ESR dating called the isochron method. This technique allows the geochronologist to measure internal variations in radiation dose to enamel portions that have received uniform external doses. Isochrons provide an independent measure of the gamma dose rates that removes the uncertainty associated with moisture content variations in the surroundings (Blackwell and Schwarz, 1993). Mammoth and elephant teeth have proven problematic for ESR dating (Rink et al., 1996a, 1996b), and bear teeth are very difficult to process in the laboratory because of their thin enamel. The mammoth and elephant teeth are very large and therefore displace a large volume of sediment, making assessment of their self-gamma dose necessary. This situation is problematic because of the internal variability of uranium distribution, and it is impossible for incomplete fragments (that still have plenty of enamel for dating) because they have experienced a change in their self-gamma dose at the time of breakup.

The sampling strategy for ESR (and luminescence) dating is much more demanding than that for radiocarbon dating, and the laboratory work is almost always more time consuming. On average, a single ESR date costs about twice as much as an AMS radiocarbon date. Why do it? There are many reasons, the main one being that you can determine ages in the problematic range for radiocarbon (>30–40 ka) and well beyond. Although most ESR dates on enamel are in the range of 50–200 ka (Rink, 1997), the method was shown to be valid back to about 2 million years (Schwarcz et al., 1994). The age uncertainty in ESR results depends strongly on the circumstances of burial. Although shown to be accurate (Rink et al., 1996c), the precision suffers dramatically in circumstances where uranium uptake has caused significant radioactivity to build up in the enamel and other dental tissues. In some cases, with very large uptake, the uncertainty in the age can be as large as 80%, but in less drastic conditions of burial, where uranium uptake is buffered by carbonate rich source waters, typical uncertainty in ESR ages is 10 to 20%.

4.6.2. Coupled ESR/Uranium-Series Dating of Teeth

The relatively large uncertainties in ESR model ages (early and linear model ages) arise from simplifying assumptions about uranium uptake into the dental tissues. These early and linear uptake model ages are invoked to constrain the age to a range based on reasonable geological assumptions about behavior of uranium in groundwaters. The early uptake model assumes that all of the uranium in the tooth at the time of collection was absorbed within a short time after burial, whereas the linear uptake model is based on the idea that the uranium was accumulated gradually in linear fashion throughout the burial time. As proposed by Grün et al. (1988), U-series and ESR can be done on the same teeth in order to attempt to correct for the problems of uranium uptake on ESR ages. Coupled ESR/U-series dating of teeth (McDermott et al., 1993) suggested that rapid or gradual uptake of uranium can be modeled sufficiently well to reduce the large uncertainties associated with assumed uptake models. Cases of recent uptake were proven using U-series dating of the enamel and dentine in the same teeth that the enamel was dated by ESR (Schwarcz and Grün, 1993, Monigal et al., 1997), but early and linear uptake have also been proven (Rink et al., 1998). The same approach allows detection of complex uptake histories in teeth, such as uranium loss (Swisher et al., 1996). Improved age determinations through constraining the U-uptake model using the coupled approach are not yet possible whenever uranium loss is detected.

4.6.3. Optical ESR Dating and ESR Dating of Burned Materials

There are several other uses of ESR dating in archaeological contexts. ESR dating has also been done on heated sediment (Miallier et al., 1994), heated rocks from hearths (Monnier et al., 1994), and heated silex (a French term encompassing chert and flint in English; Porat et al., 1994; Walther, 1995; Walther and Zilles, 1994), but all need further study to become routine methods. The heating event removes the effects of prior geological irradiation in the sample, a process called "zeroing" or "resetting" discussed further in the section on luminescence dating. Another method under development is optical ESR dating of sediment (Yokoyama et al., 1985; Yoshida, 1996; Laurent et al., 1998). The samples for optical ESR dating are collected in the same way as those for optical luminescence dating (see Fig. 14.7). ESR dating of heated silex requires the same approach as that for TL dating of burned silex (Fig. 14.6), whereas those for ESR dating of burned sediment are similar to those for optical luminescence dating (Fig. 14.7) except that protection from light during collection is not required. Sample collection for ESR (and TL) dating of burned hearth rocks is quite complex and beyond the scope of this chapter.

4.6.4. ESR Dating of Shell and Calcite Speleothem

Shells and calcitic deposits are often associated with archaeological contexts. For coastal sites, unaltered aragonitic mollusk shell can often be used for dating of

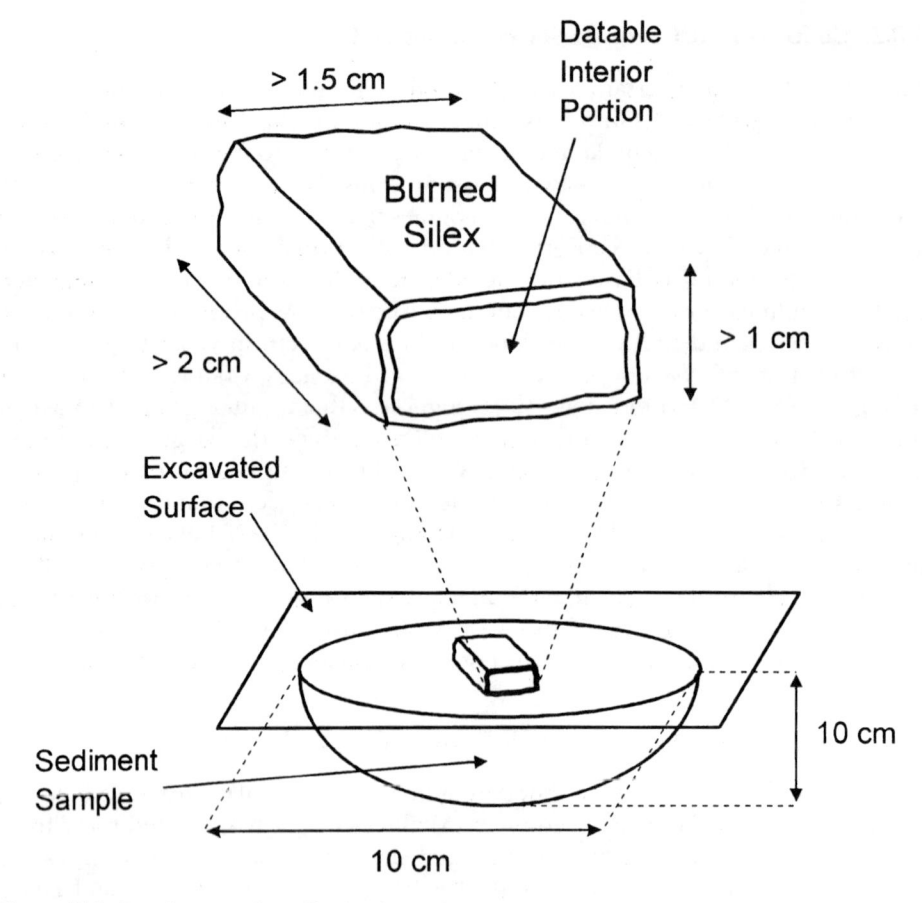

Figure 14.6. Requirements for collecting burned silex (flint/chert) for luminescence dating, showing the minimum sample dimensions and sediment samples needed. The datable interior portion is excised in the laboratory by removing the outermost layer (about 2 mm thick).

shellfish harvesting activity or associated cultural remains, but with caution to avoid recrystallized mollusk shell (Bahain et al., 1995). Recrystallization is characterized by the transition of aragonite to calcite. ESR of shell can also be used to rapidly and inexpensively determine whether the site is beyond the range of radiocarbon dating (Radtke, 1988). The sampling requirements are very similar to those of dating tooth enamel (Fig. 14.3 and 14.5).

Calcite speleothems are often dated by U-series, but they can be older than the 350 to 500 ka limits of this technique. Recent work on ESR dating of speleothems has shown it can be used beyond the 350 ka limit (Bahain et al., 1994; Engin et al., 1999). For speleothems, the sampling requirements are much like those used for collecting sediment for OSL (Figs. 14.3 and 14.7), except that 10 to 20 gm of hard dense calcite should be chipped out of the layer of solid calcite. Whenever the speleothem layer is thinner than about 15 cm, great care must be used to sample both the calcite layer and the sediment above and below.

4.7. Applications of Luminescence Dating

4.7.1. Thermoluminescence Dating of Burned Silex (Chert and Flint)

Luminescence dating of silex is also a very widely applied method in prehistoric archaeology. Silex is a siliceous (SiO_2-rich) rock composed of different chemical compounds but dominated by two different mineral phases composed of SiO_2: chalcedony, which is often is a fibrous crystalline quartz and standard crystalline quartz. The grain sizes of these two different phases can range from cryptocrystalline (submicroscopic; approx. <1 μm) to microcrystalline (grain sizes of about 1-20 μm). Cherts may also contain varying amounts of relict biogenic opal (noncrystalline SiO_2) from the skeletons of marine organisms. Other minerals in chert or flint include calcite, mineralogical clays, and iron oxides.

Following the pioneering work of Göksu et al. (1974), Valladas (1992) and Mercier et al. (1995a) have refined the method into a reliable chronometer. Although burned silex is not as ubiquitous as teeth in the Middle Paleolithic, it has a number of very favorable properties for radiation exposure dating. It has been used to date a large number of sites important to human evolution in the Levant (Valladas et al.1987, 1988; Mercier et al., 1995b) and in Europe (Mercier et al. 1991).

4.7.2. Sampling Requirements for Dating Burned Silex

The concerns of the excavation team are much the same as in the case of ESR dating of enamel. One important difference is the sensitivity of burned silex to daylight. Samples that appear burned should be removed from the excavation floor without delay and placed in aluminum foil or black plastic wrapping. Most burned silex is opaque enough to protect the innermost portions from light exposure, but it is safer to protect the material from light by wrapping the sample in an opaque material very soon after excavation. Figure 14.6 shows that sediment within 10 cm of the sample should be collected for assessment of the proximal gamma radiation dose rate and for moisture content analyses, but sediment exclusively from in the nearest 1 cm is not needed as is the case for ESR dating. *In situ* dosimetry in witness sections or in the excavation floor is carried out in precisely the same approach as described for ESR dating (Fig. 14.3).

In contrast to the potential acquisition of internal radioactivity in teeth during the burial period, the internal radioactivity of silex is constant, providing a distinct advantage for the age calculations. The internal dose forms an invariant component of the total dose, which can often be high enough to constitute 50% or more of the total radiation dose rate. This offsets one of the main sources of age uncertainty shared by silex and enamel, which is the range of potential variability in the moisture content of the surroundings. In other words, if the internal radioactivity of the flint is fortuitously high, the proportion of the external dose rate becomes smaller, leading to a reduction in the estimated uncertainty in the total dose rate and the age. Conversely, high internal radioactivity of teeth leads to a greater uncertainty in the age than in teeth with lower radioactivity, unless U-series dating of the teeth yields conclusive results about the question of uranium uptake.

4.7.3. Dating the Last Exposure to Light—Thermoluminescence and Optical Luminescence Dating

Sediments may contain two main types of light-sensitive mineral grains. Quartz and feldspar are framework-structure silicate minerals, built from a strong network of pyramidal-shaped tetrahedrons that contain one silicon and four oxygen atoms. Quartz is more resistant to weathering due to the fact that all of the oxygen atoms are common to more than one structural unit and the silicon–oxygen bond is very strong. On the other hand, feldspars have a certain proportion of tetrahedrons containing aluminum instead of silicon and have other atoms such as sodium, calcium, and potassium that introduce weaker bonding among the strong building blocks. Thus the feldspars are more easily weathered but do occur in many kinds of unconsolidated sediments. Natural grain mixtures containing quartz and feldspar are often used in thermoluminescence dating of fine-grained sediments (e.g., loess and fluvio-glacial silt) because they are easily prepared and the feldspar's stronger luminescence dominates the light emission of the mixture. Zircon is another mineral found in sediments that is luminescent (Aitken, 1985) but has not become a mineral of choice for luminescence dating of optically bleached sediments.

Quartz and feldspar grains can become part of the sedimentary context of a site through processes of transport in air, water, or by mass-wasting (gravity-driven motion of sediments and rocks). Colluvium is sediment transported by overland or sheet-flow of water. Colluvial transport of grains can involve movement of particles en masse in dense form, whereas transport in air and water disperses the grains and allows them to receive light exposure. Light exposure during transport to the burial location is often the event that is dated using luminescence, but the date of heating of grains is also possible provided that the temperature reached about 4500°C for more than about 15 minutes. Luminescence dating provides the age of the last of the two possible events. If the grain was exposed to light during transport, then buried, but heated some 5,000 years later, then the age determined by luminescence will be that of the heating event. The light exposure event or heating event is generally referred to as zeroing or resetting because it reduces the previous levels of accumulated radiation exposure inside the grains.

The degree of resetting by light exposure can be estimated in the laboratory, but in the end it cannot be established precisely in certain circumstances. Reliance on geological interpretations of the mode of transport is essential to proper application of these methods. If windblown (eolian) deposition is a clear geological interpretation of the burial context, then complete resetting is a safe assumption. For cases where water or colluvial transport are inferred from geological circumstances, doubt arises regarding the level of resetting. Geochronologists can test for partial resetting and provide an estimate of the true burial age in fortuitous circumstances. Berger (1995) provides an excellent overview of problems and solutions related to partial resetting of signals in luminescence dating.

4.7.4. Applications of Luminescence Dating of Sediment

Great advances in radiation exposure dating were made with the discovery of optically stimulated luminescence (OSL) in quartz (Huntley et al., 1985) and in feldspar (Hütt et al., 1988). OSL has also been referred to as photo-stimulated luminescence dating (PSL) or optical dating. A recent book on the topic of optical luminescence dating was written by Aitken (1998). Technically, feldspar luminescence is conveniently stimulated using infrared excitation, and is often referred to is IRSL. It is widely applied in geological contexts. OSL and IRSL offer two significant advantages over the more conventional thermoluminescence (TL) method that is used on quartz and feldspar separates or mixtures containing both minerals. The first is that effectively complete resetting to a true zero level is possible with OSL and IRSL, and the second is that the time required to reset the signal to near zero is only a few minutes in full sunlight for both quartz and feldspar. In contrast, TL is never completely reset by daylight exposure and requires 10 to 20 hours of exposure to be reduced to the minimum possible level.

OSL is an emerging tool in archaeology that is extensively applied in arid parts of Australia and Africa. Roberts et al. (1990, 1994) applied the method to obtain surprisingly old ages of greater than 50 ka for human activities in Australia, and to cave paintings covered by mud-wasp nests containing quartz (Roberts et al., 1997). More recently, TL was used to obtain very old ages of about 170 ka for human occupation in Siberia (Waters et al., 1997) and greater than 100 ka in Australia (Fullagar et al., 1996), raising fears that partial resetting of TL may be a problem in these sites (Gibbons, 1997). Using OSL, the TL results at Jinmium were refuted, suggesting that incomplete resetting of deposited grains had occurred (Roberts et al., 1998). Applications of luminescence dating in archaeology (Roberts, 1997), and luminescence dating of sediments in earth sciences (Prescott and Robertson, 1997) are two relevant recent review papers.

4.7.5. Sampling Strategy for Optical Luminescence Dating of Sediments

The various approaches aim at preventing natural light exposure to the sediment during collection (Fig. 14.7). Tubes may be inserted and packed tightly on the ends to prevent shifting of material during transport. Alternately, an exposure on the excavation floor or in the profile may be scraped clean and sampled into a light-tight container using digging tools either at night or while situated under a black tarp. If the sediments appear to be relatively homogeneous (uniform color and lie within a sand/silt/clay particle size range), the radiation dosimetry is usually done by laboratory chemical analysis of the same sediment sampled for dating. However, additional samples for dosimetry are often taken at a few points 10 to 15 cm away from the dated sample without precautions for light protection and are later analyzed for their radioactivity in the laboratory as a test for homogeneity of the radioactivity within the range of gamma radiation affecting the sample. Gamma plus cosmic dosimetry (Figure 14.3) should also be carried out wherever possible, particularly where the sediment layers are not homogeneous in color or lithology.

5. Dating Intercomparisons and General Problems with Interpretation of Dating Results

The statistical analysis of dating results can be used as a benchmark for consideration of their validity, but it involves a number of different concepts that cannot always be applied in the same way across the spectrum of dating methodologies. Moreover, a large body of statistical terminology appears in the dating literature, which is sometimes rather loosely applied. Accuracy is the difference between the true answer and the obtained answer but is rarely a measurable quantity because the true answer is not usually known. Precision is the quantifiable repeatability of a result through multiple analysis of portions of a sample that has had the same geological history. But the use of this term is often extended to sets of samples that are presumed to have had the same history, leading to statements such as "seven samples from layer X yielded a precise date of Y."

The term "analytical uncertainty" is a widely used term applied to a single dating result, which combines analytical and geological sources of uncertainty in various ways, often not clearly defined by the author. It might involve unstated ways of treating sources of random and systematic uncertainty by combining the squares (generally referred to as the quadrature method) of the error contributions from each source of error, or it may simply combine equally the sources of error in quadrature. In the former case, the statistical calculation is generally called "standard error." Each of these approaches are ways of estimating the accuracy of a result. Sources of random error are those that are associated with measurement error and other errors likely to be different from sample to sample. An example of this is the error associated with counting the number of ^{14}C atoms in a known mass of sample. Systematic uncertainties arise from errors that would affect all the samples in a series buried in the same contemporaneous unit, such as the calibration correction in ^{14}C dating that is associated with the changing budget of atmospheric ^{14}C over geological time.

Methods of quantifying the precision of a result always require a series of samples that can be assumed to be contemporaneous. Calculations of the mean and standard deviation of a series of results provide a measure of the variance (standard deviation) of the individual results in relation to the average (mean) result. Its value is increasingly meaningful as the number of samples tested rises ($n = 2$ to 3 is not particularly useful, but $n = 9$ to 10 is considerably more meaningful). This statistical measure is often wrongly thought of as a measure of the accuracy of the set of results.

In addition to measurement errors, all dating results have an inherent level of uncertainty that arise from geological sources of uncertainty, which cannot be quantified. An example of a source of uncertainty arising from geological conditions is contamination of a ^{14}C dating sample with a quantity of parent or daughter atoms that were not present at the time of burial in the sample. In

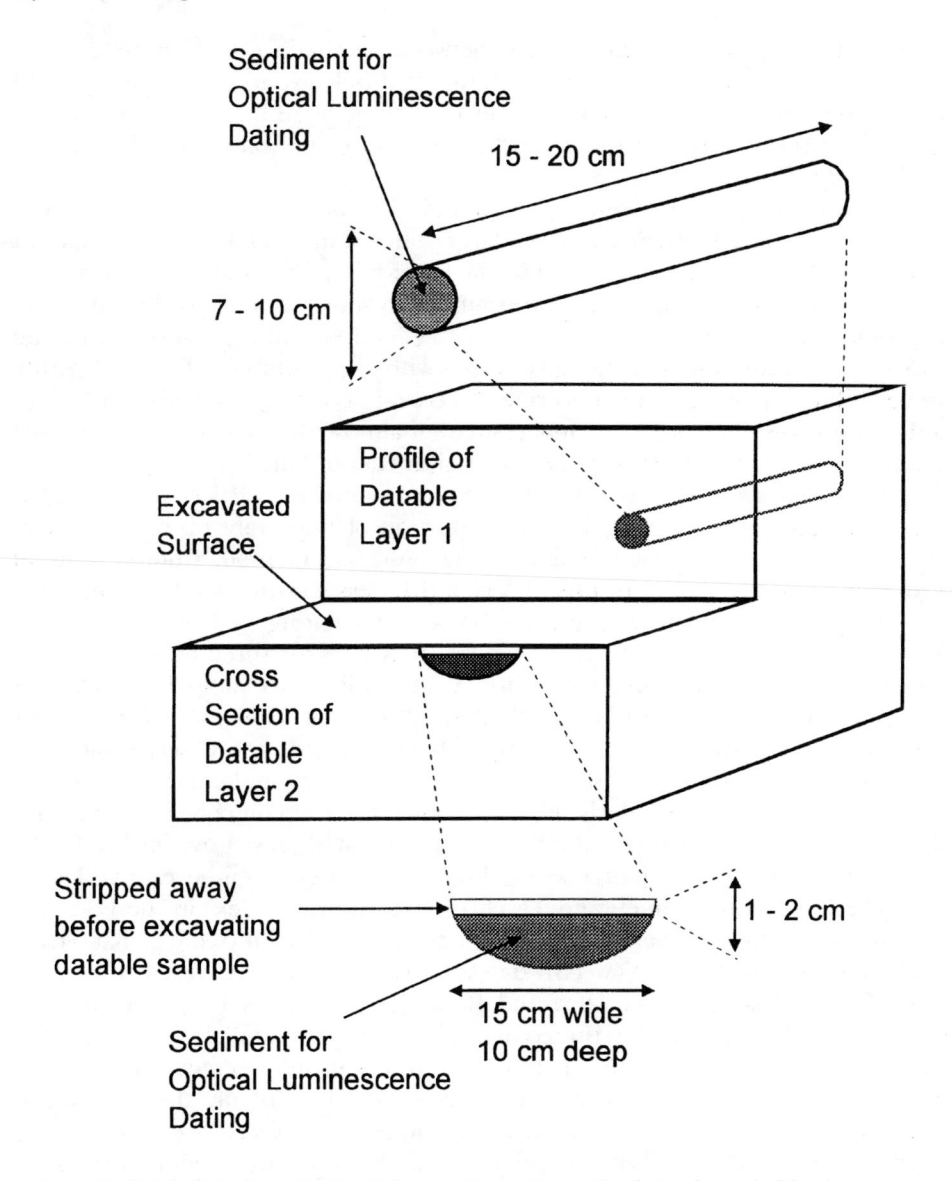

Figure 14.7. Methods for collecting light-sensitive sediment for dating by optical luminescence.

radiation exposure dating, an example would be the moisture content of the surroundings. This may be a systematic or random error depending on the conditions of burial, because it may be different in various parts of the same unit. The use of multiple dating methods can help us to ascertain when geological sources of uncertainty have affected the results. It is best to have multiple methods that are based on different physical principles, which means that the sources of uncertainty and the assumptions in the methods are quite independent. For example, ^{14}C and ESR dating are considerably independent of each other. The degree of independence between methods can help to exclude unknown sources of geological error. However, if only two methods are used and they don't agree, the problematic one cannot be identified, but with three, agreement between two of them can be taken as an indicator that the third is problematic.

Isotopic methods are generally considered to be more accurate than radiation exposure methods. Some of the larger quoted uncertainties ("$\pm$" values of >50%) in individual dates obtained by the latter methods have led to this general view. However, these larger age uncertainties in some sites can overshadow the capability of the same method for lower age uncertainty in other sites. The radiation exposure methods can give results with uncertainties of less than $\pm 10\%$ in favorable situations. These uncertainties contain the combined effects of about a dozen sources of error in the final result, but almost always include estimates of many of the sources of error due to geological circumstances.

The multimethod approach has been used routinely in Paleolithic archaeology, although dates for a significant proportion of important events in human history rest on single age estimates using only one method. Comparisons of radiation exposure dating results to ^{14}C and U-series results for large numbers of sites have been made by Rink (1997) for ESR dating and by Prescott and Robertson (1997) for luminescence dating. It is rare for three different dating methods to be compared in a single site. At El Castillo Cave in Spain (Bischoff et al. 1992; Cabrera-Valdes and Bischoff, 1989; Rink et al., 1997), it was shown that ESR dating of tooth enamel, ^{14}C dating of bone collagen, and U-series dating of stalagmitic floors are in good stratigraphic agreement throughout the time range between 40 and 90 ka. Similarly, ESR on shell, ^{14}C dating on charcoal, OSL, and TL showed good agreement at 15 to 20 ka at Batadomba Cave in Sri Lanka (Abeyratne et al., 1997), though some discrepancies were seen beyond 20 ka.

The ESR (tooth enamel) and TL chronology (burned silex) in the Levant is in remarkably good agreement except for the site of Tabun (see e.g. Bar-Yosef, 1998, for a compilation). New ESR dates on tooth enamel recovered *in situ* and using *in situ* dosimetry (Schwarcz and Rink, 1998) give results in much closer agreement with TL ages for the youngest part of the Tabun E lithic sequence (Valladas et al., 1998). Though it can be shown that disagreement between the radiation exposure methods and other methods does occur, there is ample evidence that the radiation exposure techniques are basically reliable despite their relatively larger margins of cited error. The most problematic method of the radiation exposure techniques has been dating of optically zeroed sediment by TL, which is now rapidly being replaced in the older time ranges by optical luminescence techniques (OSL and IRSL), particularly in environments where complete zeroing of the signal is not expected to have occurred.

6. Potential Problems with Various Dating Methods

Consideration must be given to any potential differences between the event dated and the object dated in the burial context, and to be aware of potential errors in the particular dating method. ^{14}C ages always provide the age of death of the parent organism of the dated material, and do not necessarily represent the burial age of the dated material. Coupled ESR/U-series ages on tooth enamel provide an estimate of the burial age for a single burial event but cannot yield the true age of the burial context if the teeth are redeposited from older burial contexts.

TL dates on burned silex or on burned hearth materials yield the age of only the last heating of sufficient duration to about 450°C. Optical luminescence ages (or optical ESR ages) yield the age of the last light exposure before burial assuming that the transport event produced complete zeroing, otherwise the age will be too old. U-series dating of calcite will yield the age of crystallization unless the calcite accepted uranium into its structure later or was partially redissolved and reprecipitated (with younger uranium impurity) in areas of the analyzed sample, which will produce an age that is too old. In general, all results that are obtained at or near the limits of the datable time range (or dose range, in radiation exposure dating) should be viewed cautiously. This is also true when a new technique has been developed within the technology of that dating method and is applied for the first time. Similarly, general acceptance of dating results without independent verification by another laboratory working on the same problem has become commonplace.

Whereas some dating results are clearly spurious and difficult to interpret, some obvious aspects of dating limitations are often overlooked. One of the most common is that associated with dating of materials such as bone and teeth using systems that rely exclusively on the open-system behavior of the materials. Whenever teeth and bone are dated exclusively by U-series the result might be the true age, but it can also be too young or too old due to the open-system chemical behavior during burial. If uranium was not accumulated to its present-day value during the earliest part of the burial history ("early uptake assumption") then the age result will probably be too young because of the additional episodes of uranium accumulation during any later stages of the burial history. These later episodes reset the isotopic clock. Conversely, if uranium was leached from the bone or tooth in significant quantity after an initial period of uptake, then the age might be significantly too old. Therefore, U-series dates on bone cannot stand alone as absolute ages, and U-series dates on teeth can only be useful to clarify the uranium uptake history into teeth used for ESR dating. Unfortunately, U-series dates on both bone and teeth are often viewed as absolute dates, but they rely on a geologically precarious assumption of early uptake.

For the radiation exposure dating methods, one must generally assume that the moisture content and sediment geochemistry of the surrounding deposit is similar to that of present-day conditions. It is now well known that ash diagenesis can be associated with dramatic changes in soil chemistry that affect the radioactivity levels in the deposit, but these can be easily identified and avoided (Shiegl et. al., 1996). The use of isochron methods in TL dating of burned silex (Aitken and Valladas, 1992) and in ESR dating of tooth enamel (Blackwell and Schwarcz, 1993) can be applied under certain fortunate circumstances, and this allows one to test the validity of these assumptions and to make corrections for them. In general, desirable samples have high internal radioactivity (potassium feldspar) or are geochemically resistant (silex and quartz) because this reduces the potential effects of external variations in sediment geochemistry and moisture content on the external radiation dose rate. In all cases, the dating results should include a properly weighted estimate of the potential effects of moisture content variation on the quoted age, rather than using the standard deviation of the present-day moisture content in a unit. This is a systematic error that cannot be quantified because long periods with moisture content different from that of the present-day value may have occurred. The true range of variability can be linked to the estimated maximum moisture content of the sediment on the basis of porosity analysis (Aitken, 1985). Assessments of the assumptions involved in ESR dating of tooth enamel (which are in many cases general assumptions for radiation exposure dating) are given by Rink (1997).

7. Summary

In recent years the advent of mass spectrometry in U-series and ^{14}C dating has reduced the sample size requirements and increased the precision of these techniques. Introduction of refined methods of radiation exposure dating has opened a new toolkit for dating in archaeological sites, which includes ESR and luminescence dating of a wide range of archaeological materials. Though relatively new, and more challenging from a sampling point of view, they have provided some now widely accepted dating evidence that has challenged our perceptions about the appearance of modern humans in the middle Pleistocene. As a result, dating results and methodology now receive more intense scrutiny than ever before and therefore are becoming a more important theme in the curriculum of archaeological science. Archaeologists need to be more familiar with the demanding sampling requirements in order to make appropriate plans for funding and excavation. It is hoped that this work will be a useful guide in these areas and in the capabilities and problems of the newer dating methods.

ACKNOWLEDGMENTS. This work would not have been possible without financial support to the author from the Natural Sciences and Engineering Council of Canada, the L.S.B. Leakey Foundation (USA), NATO (Brussels), and support from the National Science Foundation (USA) to H. P. Schwarcz. I thank A. P. Dickin (McMaster University) and C. C. Swisher III (Berkeley Geochronology Center) for useful discussions that helped to improve this manuscript.

8. References

Abeyratne, M., Spooner, N. A., Grün, R., and Head, J., 1997, Multidating Studies of Batadomba Cave, Sri Lanka, *Quaternary Science Reviews (Quaternary Geochronology)* 16:243–255.
Aitken, M. J., 1985, *Thermoluminescence Dating*, Academic Press, London.
Aitken, M. J., 1998, *An Introduction to Optical Dating. The Dating of Quaternary Sediments by the Use of Photon*–stimulated Luminescence, Oxford University Press, New York.

Aitken, M. J. and Valladas, H., 1992, Luminescence Dating Relevant to Human Origins, *Philosophical Transactions of the Royal Society of London. Series B* 337:139–144.

Bahain, J.-J., Yokoyama, Y., Falguères, C., and Bibron, R., 1994, Choix du Signal à Utiliser Lors de la Datation par Résonance Paramagnétique Electronique (RPE) de Calcites Stalagmitiques Quaternaires. *Compt. Ren. Acad. Sci. Paris, Ser. II*, 318:375–379.

Bahain, J.-J., Yokoyama, Y., Falguères, C., and Bibron R., 1995, Datation par Resonance de Spin Electronique (ESR) de Carbonates Marins Quaternaires, *Quaternaire* 6:13–19.

Bar-Yosef, O., 1998, The Chronology of the Middle Paleolithic of the Levant. In *Neanderthals and Modern Humans in Western Asia*, edited by T. Akazawa, K. Aoki, and O. Bar-Yosef, pp. 39–56. Plenum Press, New York.

Berger, G. W., 1995, Progress in Luminescence Dating Methods for Quaternary Sediments. In *Dating Methods for Quaternary Deposits*, edited by N. W. Rutter and N. R. Catto, pp. 81–104. Geotext 2, Geological Association of Canada. C/o Department of Earth Sciences Memorial University of Newfoundland, St. John's, Newfoundland, A1B 3X5, Canada.

Bischoff, J. L., Garcia, J. F., and Straus, L. G., 1992, Uranium–series Dating at El Castillo Cave (Cantabria, Spain): The "Acheulian"/"Mousterian" Question. *Journal of Archaeological Science* 19:1949–1962.

Blackwell, B. A., 1995, Electron Spin Resonance Dating. In: *Dating Methods for Quaternary Deposits*, edited by N. W. Rutter and N. R. Catto, pp. 209–268. Geotext 2, Geological Association of Canada. C/o Department of Earth Sciences Memorial University of Newfoundland, St. John's, Newfoundland, A1B 3X5, Canada.

Blackwell, B. A., and Schwarcz, H. P., 1993, ESR Isochron Dating for Teeth: A Brief Demonstration in Solving the External Dose Calculation Problem. *Applied Radiation and Isotopes* 43:243–252.

Blackwell, B. A., and Schwarcz, H. P., 1995, The Uranium Series Disequilibrium Dating Methods. In *Dating Methods for Quaternary Deposits*, edited by N. W. Rutter and N. R. Catto, pp. 167–208. Geotext 2, Geological Association of Canada. C/o Department of Earth Sciences Memorial University of Newfoundland, St. John's, Newfoundland, A1B 3X5, Canada.

Brennan, B. J., Rink W. J., McGuirl, E. L., Schwarcz, H. P., and Prestwich, W. V., 1997, Beta Doses in Tooth Enamel by "One–Group" Theory and the ROSY ESR Dating Software, *Radiation Measurement.* 27:307–314.

Cabrera-Valdés, V., and Bischoff, J. L., 1989, Accelerator ^{14}C Dates for Early Upper Paleolithic at El Castillo Cave. *Journal of Archaeological Science* 16:577–584.

Chen T., Quan Y., and En, W., 1994, Antiquity of *Homo sapiens* in China. *Nature* 368:55–56.

Chen, T.-M., Yang, Q., Hu, Y.-Q., Bao, W.-B., and Li, T.-Y., 1997, ESR Dating of Tooth Enamel from Yunxian *Homo erectus* Site, China, *Quaternary Science Reviews (Quaternary Geochronology)* 16:455–458.

Chesner, C. A., Rose, W. I., Deino, A., Drake, R., and Westgate, J. A., 1991, Eruptive History of Earth's Largest Quaternary Caldera (Toba, Indonesia) Clarified, *Geology* 19:200–203.

Dickin, A. P., 1995, *Radiogenic Isotope Geology*, Cambridge University Press, Cambridge, UK.

Edwards, R. L., Chen J. J., and Wasserburg, G. J., 1987, ^{238}U–^{234}U–^{230}Th–^{232}Th sytematics and the Precise Measurement of Time Over the Past 500,000 Years, *Earth and Planetary Science Letters* 90:371–381.

Engin, B., Güven, O. and Köksal, F., 1999, Electron spin resonance age determination of a travertine sample from the Southwestern part of Turkey. *Applied Radiation and Isotopes* 51:689–699.

Falguères, C., Miallier, D., and Sanzelle, S., 1994, Potential Use of the E' Center as an Indicator of Initial Resetting in TL/ESR of Volcanic Materials, *Quaternary Science Reviews (Quaternary Geochronology)* 13:619–623.

Faure, G., 1986, *Principles of Isotope Geology*, 2nd ed. Wiley, New York.

Forman, S. L., 1989, Applications and Limitations of Thermoluminescence to Date Quaternary Sediments, *Quaternary International* 1:47–59.

Fullagar, R. L. K., Price, D. M., and Head, L. M., 1996, Early Human Occupation of Northern Australia: Archaeology and Thermoluminescence Dating of Jinmium Rock-Shelter, Northern Territory, *Antiquity* 70:751–773.

Gibbons, A., 1997, Doubts Over Spectacular Dates, *Science* 278:220–222.

Gleadow, A. J. W., 1980, Fission-Track Age of the KBS Tuff and Associated Hominid Remains in Northern Kenya, *Nature* 284:225–230.

Göksu, H. Y., Fremlin, J. H., Irwin, H. T., and Fryxell, R., 1974, Age Determination of Burned Flint by a Thermoluminescent Method, *Science* 183:651–654.

Grün, R., 1985, Beiträge Zur ESR-Datierung, *Geologisches Institut der Universität zu Köln Sonderveröffentlichungen 59.*

Grün, R., 1989, Electron Spin Resonance (ESR) Dating, *Quaternary International* 1:65–109.

Grün, R., Schwarcz, H. P., and Zymela, S., 1987, ESR Dating of Tooth Enamel, *Canadian Journal of Earth Sciences* 24:1022–1037.

Grün, R., Schwarcz, H. P., and Chadam, J. M., 1988, ESR Dating of Tooth Enamel: Coupled Correction for U—uptake and U-series Disequilibrium, *Nuclear Tracks and Radiatiation Measurements* 14:237–241.

Grün, R., Beaumont, P. B., and Stringer, C. B., 1990, ESR Dating Evidence for Early Modern Humans at Border Cave in South Africa, *Nature* 344:537–539.

Harris, D. R., 1987, The Impact on Archaeology of Radiocarbon Dating by Accelerator Mass Spectrometry, *Philosophical Transactions of the Royal Society London* A323:23–43.

Hu, Q., Smith, P. E., Evensen, N. M., and York, D., 1994. Lasing in the Holocene: Extending the ^{40}Ar/^{39}Ar Laser Probe Method into the ^{14}C Age Range, *Earth and Planetary Science Letters* 123:331–336.

Huang, W., Ciochon, R., Gu, Y., Larick, R., Fang, Q., Yonge, C., de Vos J., Schwarcz, H. P., and Rink, W. J, 1995, Earliest Hominids and Artifacts from Asia: Longgupo Cave, Central China, *Nature* 378:275–278.

Huntley, D. J., Godfrey-Smith, D. I., and Thewalt, M. L. W., 1985, Optical Dating of Sediments, *Nature* 313:105–107.

Huntley, D. J., Hutton, J. T., and Prescott, J. R., 1993, Optical Dating Using Inclusions within Quartz Grains, *Geology* 21:1087–1090.

Hütt, G., Jaek, I., and Tehonka, J., 1988, Optical Dating: K-feldspars Optical Response Stimuation Spectra, *Quaternary Science Reviews* 7:381–385.

Hutt, G., Jack, I. and Tehonka, J. (1988) Optical dating: K-feldspars Optical Response Stimulation Spectra. *Quaternary Science Reviews* 7:381–385.

Kimbel, W. H., Johanson, D. C., and Ray, Y., 1994, The First Skull and Other New Discoveries of *Australopithecus afarensis* at Hadar, Ethiopia, *Nature* 368:449–451.

Laurent, M., Falguères, C., Bahain, J.-J., and Yokoyama, Y., 1994, Géochronologie du Système de Terrasses Fluviatiles Quaternaires du Bassin de la Somme par Datation RPE sur Quartz, Déséquilibres des Familles de l' Uranium et Magnétostratigrapie. *Comp. Rend. Acad. Sci. Paris Ser. II* 318:521–526.

Laurant, M., Falguères, C., Bahir, J. J. Rousseau, L. and van Vliet-Lanoë, B (1998), ESR Dating of Quartz Extracted from Quaternary and Neogene Sediments: Methods, Potential and Actual Limits, Quaternary Science Reviews. *Quaternary Geochronology* 17:1057–1062.

Li, W. X., Lundberg, J., Dickin, A. P., Ford, D. C., Schwarcz, H. P., McNutt, R. H., and Williams, D., 1989, High-Precision Mass-Spectrometric Uranium-Series Dating of Cave Deposits and Implications for Paleoclimate Studies, *Nature* 339:534–536.

McDermott, F., Grün, R., Stringer, C. B., and Hawkesworth, C. J., 1993, Mass-Spectrometric U-Series Dates for Israeli Neanderthal/Early Modern Hominid Sites, *Nature* 363:252–255.

McDougall, I., 1995, Potassium–Argon Dating in the Pleistocene. In *Dating Methods for Quaternary Deposits*, edited by N. W. Rutter and N. R. Catto, pp. 1–14. Geotext 2, Geological Association of Canada. C/o Department of Earth Sciences Memorial University of Newfoundland, St. John's, Newfoundland, A1B 3X5, Canada.

McKinney, C. R., 1998, Uranium Series Dating of Enamel, Dentine and Bone from Kabazi II, Starosele, Kabazi V and Gabo. In *Palaeolithic of the Crimea Series*, Volume 1, edited by A. E. Marks and V. P. Chabai, pp. 341–354. ERAUL, Liege.

Mercier, N., Valladas, H., Joron, J.-L., Reyss, J.-L., Lévêque, F., and Vandermeersch, B., 1991, Thermoluminescence Dating of the Late Neanderthal Remains from Saint-Césaire, *Nature* 351:737–739.

Mercier, N., Valladas, H., and Valladas, G., 1995a, Flint Thermoluminescence Dates from the CFR Laboratory at GIF: Contributions to the Study of the Chronology of the Middle Palaeolithic, *Quaternary Science Reviews (Quaternary Geochronology)* 14:351–364.

Mercier, N., Valladas, H., Valladas, G., Reyss, J.-L., Jelinek, A., Meignen, L., and Joron, J.-L., 1995b,

TL Dates of Burned Flints from Jelinek's Excavations at Tabun and Their Implications. *Journal of Archaeological Science* 22:495–509.

Miallier, D., Sanzelle, S., Fain, J., Montret, M., Pilleyre, T., Soumana, A., and Falguères, C., 1994, Intercomparisons of Red TL and ESR Signals from Heated Quartz Grains, *Radiatiation Measurements* 23:143–153.

Monigal, K., Marks, A. E., Demidenko, Yu. E., Rink, W. J., Schwarcz, H. P., Ferring, C. R., and McKinney, C., 1997, Nouvelles Decouvertes de Restes Humains au Site Paleolithique Moyen de Starosele, Crimee (Ukraine), Préhistoire Européene 11.

Monnier, J.-L., Hallégouet, B., Hinguant, S., Laurent, M., Auguste, P., Bahain, J. J., Falguères, C., Beghardt, A., Marguerie, D., Molines, N., Morazdec, H., and Yokoyama, Y., 1994, A New Regional Group of the Lower Paleolithic in Brittany (France), Recently Dated by ESR, *Comp Rend. Acad. Sci. Paris Ser. II* 319:155–160.

Ollerhead, J., Huntley, D. J., and Berger, G. W., 1994, The Evolution of Buctouche Spit, New Brunswick, *Canadian Journal of Earth Sciences* 31:523–531.

Pike, A. W. G., and Hedges, R. E. M., in press. Sample Geometry and U-Uptake in Archaeological Teeth: Implications for U-Series and ESR Dating.

Porat, N., Schwarcz, H. P., Valladas, H., Bar-Yosef, O., and Vandermeersch, B., 1994, Electron Spin Resonance Dating of Burned Flint from Kebara Cave, Israel, *Geoarchaeology* 9:393–407.

Prescott, J. R., and Robertson, G. B., 1997, Sediment Dating by Luminescence: A Review, *Radiation Measurements* 27:893–922.

Radtke, U., 1988, How to Avoid "Useless" Radiocarbon Dating. *Nature* 333:304–308.

Rendell, H. M., Lancaster, N., and Tchakerian, V. P., 1994, Luminescence Dating of Late Quaternary Aeolian Deposits at Dale Lake and Cronese Mountains, Mojave Desert, California, *Quaternary Science Reviews (Quaternary Geochronology)* 13:417–422.

Rink, W. J., 1997, Electron Spin Resonance (ESR) Dating and ESR Applications in Quaternary Science and Archaeometry, *Radiation Measurements* 27:975–1025.

Rink, W. J., Schwarcz, H. P., Smith, F. H., and Radovcic, J., 1995, ESR Dating of Krapina Hominids, *Nature* 378:24.

Rink, W. J., Schwarcz, H. P., Valoch, K., Seitl, L., and Stringer, C. B., 1996a, Dating of the Micoquian Industry and Neanderthal Remains at Kulna, Czech Republic, *Journal of Archaeological Science* 23:889–901.

Rink, W. J., Schwarcz, H. P., Stuart, A. J., Lister, A. M., Marseglia, E., and Brennan, B. J., 1996b, ESR dating of the Type Cromerian Fresh Water Bed at West Runton, UK, *Quaternary Science Reviews (Quaternary Geochronology)* 15:727–738.

Rink, W. J., Schwarcz, H. P., Lee, H. K., Cabrera Valdés, V., Bernaldo de Quiros, F., and Hoyos, M., 1996c, ESR Dating of Tooth Enamel: Comparison with AMS ^{14}C at El Castillo Cave, Spain, *Journal of Archaeological Science* 23:945–951.

Rink, W. J., Schwarcz, H. P., Lee, H. K., Cabrera Valdés, V., Bernaldo de Quiros, F., and Hoyos, M., 1997, ESR Dating of Mousterian Levels at el Castillo Cave, Cantabria, Spain, *Journal of Archaeological Science* 24:593–600.

Rink, W. J., Lee H. K., Rees-Jones, J., and Goodger, K. A., 1998, Electron Spin Resonance (ESR) and Mass-Spectrometric U-Series Dating of Teeth in Crimean Middle Palaeolithic Sites: Starosele, Kabazi II and Kabazi V. In *Palaeolithic of the Crimea Series*, Volume 1, edited by A. E. Marks and V. P. Chabai, pp. 323–340. ERAUL, Liege.

Roberts, R. G., 1997, Luminescence Dating in Archaeology: From Origins to Optical, *Radiation Measurements* 27:819–892.

Roberts, R. G., Jones, R., and Smith, M. A., 1990, Thermoluminescence Dating of a 50,000 Year-Old Human Occupation Site in Northern Australia, *Nature* 345:153–156.

Roberts, R. G., Jones, R., Spooner, N. A., Head, M. J., Murray, A. S., and Smith, M. A., 1994, The Human Colonization of Australia: Optical Dates of 53,000 and 60,000 Years Bracket Human Arrival at Deaf Adder Gorge, Northern Territory, *Quaternary Science Reviews (Quaternary Geochronology)* 13:575–583.

Roberts, R. G., Walsh, G., Murray, A., Olley, J., Jones, R., Morwood, M., Tuniz, C., Lawson, E., Macphail, M., Bowdery, D., and Naumann, I., 1997, Luminescence Dating of Rock Art and Past Environments Using Mud-Wasp Nests in Northern Australia, *Nature* 387:696–699.

Roberts, R., Bird, M., Olley, J., Galbraith, R., Lawson, E., Laslett, G., Yoshida, H., Jones, R., Fullagar, R., Jacobsen, G., and Hua, Q., 1998, Optical and Radiocarbon Dating at Jinmium Rock Shelter in Northern Australia, *Nature* 393:358–362.

Schwarcz, H. P., 1985, ESR Studies on Tooth Enamel, *Nuclear Tracks and Radiation Measurements* 10:865–867.
Schwarcz, H. P., and Blackwell, B., 1991, Archaeological Applications. In *Uranium Series Disequilibrium Applications to Environmental Problems*, 2nd ed., edited by M. Ivanovich and R. S. Harmon, pp. 513–552, Oxford University Press, Oxford, UK.
Schwarcz, H. P., and Grün, R., 1993, Electron Spin Resonance (ESR) Dating of the Lower Industry at Hoxne. In *The Lower Palaeolithic Site at Hoxne*, edited by B. G. Gladfelter and J. J. Wymer, pp. 210–211. University of Chicago Press, London.
Schwarcz, H. P., and Latham, A. G., 1989, Dirty Calcite 1: Uranium Series Dating of Contaminated Calcites Using Leachate Alone, *Chemical Geology* 80:35–43.
Schwarcz, H.P., and Rink, W.J., 1998, Progress in ESR and U-Series Chronology of the Levantine Paleolithic. In *Neanderthals and Modern Humans in Western Asia*, edited by T. Akazawa, K. Aoki, and O. Bar-Yosef, pp. 57–68. Plenum Press, New York.
Schwarcz, H. P., Grün, R., Vandermeersch, B., Bar-Yosef, O., Valladas H., and Tchernov, E., 1988, ESR Dates for the Hominid Burial Site of Qafzeh in Israel, *Journal of Human Evolution* 17:733–737.
Schwarcz, H. P., Buhay, W. M., Grün, R., Valladas, H., Tchernov, E., Bar-Yosef, O., and Vandermeersch, B., 1989, ESR Dating of the Neanderthal Site of Kebara Cave, Israel, *Journal of Archaeological Science* 16:653–659.
Schwarcz, H. P., Grün, R., and Tobias, P. V., 1994, ESR Dating Studies of the Australopithecine Site of Sterkfontein, South Africa, *Journal of Human Evolution* 26:175–181.
Schwarcz, H. P., Simpson, J. J., and Stringer, C. B., 1998, Neanderthal Skeleton from Tabun: U-Series Date by Gamma Ray Spectrometry, *Journal of Human Evolution* 35:635–645.
Shiegl, S., Goldberg, P., Bar-Yosef, O., and Weiner, S., 1996, Ash Deposits in Hayonim and Kebara Caves, Israel: Macroscopic, Microscopic and Mineralogical Observations, and their Archaeological Implications, *Journal of Archaeological Science* 23:763–781.
Simpson, J. J., and Grün, R., 1998, Non-Destructive Gamma Spectrometric U-Series Dating, *Quaternary Science Reviews (Quaternary Geochronology)* 17:1009–1022.
Stokes, S., 1992, Optical Dating of Young (Modern) Sediments Using Quartz: Results from a Selection of Depositional Environments, *Quaternary Science Reviews* 11:153–159.
Swisher, C. C., III, Curtis, G. H., Jacob, T., Getty, A. G., Supriji, A., and Widiasmoro, 1994, Age of the Earliest Known Hominids in Java, Indonesia, *Science* 263:1118–1121.
Swisher, C. C., III, Rink, W. J., Antón, S C., Schwarcz, H. P., Curtis, G. H., Suprijo, A., and Widiasmoro, 1996, Latest *Homo erectus* of Java: Potential Contemporaneity with *Homo sapiens* in Southeast Asia, *Science* 274:1870–1874.
Valladas, H., 1992, Thermoluminescence Dating of Flint, *Quaternary Science Reviews* 11:1–5.
Valladas, H., and Valladas, G., 1987, Thermoluminescence Dating of Burned Flint and Quartz: Comparative Results, *Archaeometry* 29:214–220.
Valladas, H., Joron, J. L., Valladas, G., Arensburg, B., Bar-Yosef, O., Belfer-Cohen, A., Goldberg, P., Laville, H., Meignen, L., Rak, Y., Tchernov, E., Tillier, A. M., and Vandermeersch, B., 1987, Thermoluminescence Dates for the Neanderthal Burial Site at Kebara Cave in Israel, *Nature* 331:614–616.
Valladas, H., Reyss, J. L., Joron, J. L., Valladas, G., Bar-Yosef, O., and Vandermeersch, B., 1988, Thermoluminescence Dating of Mousterian "Proto-Cro-Magnon" Remains from Israel and the Origin of Modern Man, *Nature* 331:614–616.
Valladas, H., Mercier, N., Joron, J.-L., and Reyss, J.-L., 1998, GIF Laboratory Dates for the Middle Paleolithic Levant. In *Neanderthals and Modern Humans in Western Asia*, edited by T. Akazawa, K. Aoki, and O. Bar-Yosef, pp. 69–77. Plenum Press, New York.
Wagner, G. A., and Van den Haute, P., 1992, *Fission-Track Dating*. Kluwer Academic, Dordrecht.
Walter, R. C., 1994, Age of Lucy and the First Family, *Geology* 22:6–10.
Walter, R. C., Manega, P. C., Hay, R. L., Drake, R. E., and Curtis, G. H., 1991, Lase-fusion $^{40}Ar/^{39}Ar$ Dating of Bed I, Olduvai Gorge, Tanzania, *Nature* 354:145–149.
Walther, R., 1995, Elektronen-Spin-Resonanz-Datierung an Silikaten. Ph.D. diss., University of Heidelberg, Germany.
Walther, R., and Zilles, D., 1994, ESR studies on Flint with a Difference-Spectrum Method, *Quaternary Science Reviews (Quaternary Geochronology)* 13:635–639.

Waters, M. R., Forman, S. L., and Pierson, J. M., 1997, Diring Yuriakh: A Lower Paleolithic Site in Central Siberia, *Science* 275:1281–1284.

Westgate, J. A., 1988, Isothermal Plateau Fission Track Age of the Late Pleistocene Old Crow Tephra, Alaska, *Geophysical Research Letters* 15:376–379.

Westgate, J. A., 1989, Isothermal Plateau Fission-Track Ages of Hydrated Glass Shards from Silicic Tephra Beds, *Earth and Planetary Science Letters* 95:226–234.

Westgate, J. A., and Naeser, N. D., 1995, Tephrochronology and Fission-Track Dating. In *Dating Methods for Quaternary Deposits*, edited by N. W. Rutter and N. R. Catto, pp. 15–28. . Geotext 2, Geological Association of Canada. C/o Department of Earth Sciences Memorial University of Newfoundland, St. John's, Newfoundland, A1B 3X5, Canada.

Wintle, A. G., 1996, Archaeologically Relevant Dating Techniques for the Next Century: Small Hot and Identified by Acronyms, *Journal of Archaeological Science* 23:123–138.

Yang, Q., Rink, W. J., and Brennan, B. J., 1998, Experimental Determinations of Beta Attenuation in Planar Dose Geometry and Application to ESR Dating of Tooth Enamel, *Radiation Measurements* 29:663–671.

Yokoyama, Y., Falguères, C., and Quaegebeur, J. P., 1985, ESR Dating of Quartz from Quaternary Sediments: First Attempt, *Nuclear Tracks and Radiation Measurements* 10:921–928.

York, D., Hall C. M., Yanase, Y., Hanes, J. A., and Kenyon, W. J., 1981, ^{40}Ar/^{39}Ar Dating of Terrestrial Minerals with a Continuous Laser, *Geophysical Research Letters* 8:1136–1138.

Yoshida, H., 1996, *Quaternary Dating Studies Using ESR Signals, with Emphasis on Shell, Coral, Tooth Enamel and Quartz*. Ph.D. diss. Australian National University, Canberra, Australia.

Stable Carbon and Oxygen Isotopes in Soils

Applications for Archaeological Research

LEE C. NORDT

1. Introduction

The most common use of stable carbon (C) and oxygen (O) isotopes in archaeology is to infer paleodiet, artifact provenance, and paleoenvironment (Herz, 1990; Herz and Garrison, 1998).

Paleodiet studies often apply stable C isotope analysis to human bone to estimate the timing of corn domestication (Hard et al., 1996; van der Merwe, 1982; van der Merwe and Vogel, 1977). Provenance investigations may use stable O isotopes to infer the source area of marble procured as construction material (H. Craig and Craig, 1972; Herz and Dean, 1986). Further, stable O isotopes of mollusk shells associated with human occupation reflect seasonal temperature changes (Shackleton, 1973), whereas stable C isotopes of animal bone can be a proxy for reconstructing prehistoric human habitats (Ambrose and DeNiro, 1989).

The distribution of stable C and O isotopes in soils and in paleosols is also a powerful indicator for many kinds of pedologic and paleoenvironmental interpretations. Stable C isotope analysis of soil organic matter is commonly

---

LEE C. NORDT • Department of Geology, Baylor University, Waco, Texas 76798.
Earth Sciences and Archaeology, edited by Paul Goldberg, Vance T. Holliday, and C. Reid Ferring. Kluwer Academic/Plenum Publishers, New York, 2001.

employed to quantify organic matter turnover rates (Balesdent and Mariotti, 1996; Bernoux et al., 1998) and to reconstruct past vegetation communities (Boutton et al.; 1998b; Boutton, 1996; Cerling, 1992; Cerling et al., 1993; Fredlund and Tieszen, 1997; Kelly et al., 1991b, 1993; Nordt et al., 1994; Wang et al., 1996b). Stable C isotopes of pedogenic carbonate are used for quantifying rates of pedogenesis, estimating pathways of soil development, and inferring landscape age (Amundson et al., 1988, 1989; Kelly et al., 1991a; Marion et al., 1991; Nordt et al., 1996, 1998). As with soil organic matter, stable C and O isotopes of pedogenic carbonate can be interpreted as a proxy for paleoclimate reconstruction (Cerling and Hay, 1986; Cerling and Quade, 1993; Cole and Monger, 1994; Humphrey and Ferring, 1994; Pendall and Amundson, 1990; Quade et al., 1989).

The primary application of stable C and O isotopes in soils to archaeological research is for paleoenvironmental reconstruction. This method is particularly effective in semiarid to subhumid climates where plant communities have experienced shifts in the ratio of C_3 to C_4 species. These areas typically include tropical/temperate grasslands, semidesert and dry steppes, and tropical shrub/woodlands (Fig. 15.1). Humid or tropical climates that have been continuously forested, polar regions with cool season grasses, and deserts are less likely to have supported C_4 plant communities (Fig. 15.1). The purpose of this chapter is to (1) review the theory of stable C and O isotopes as applied to pedology, (2) review laboratory procedures employed to determine stable C and O isotope values, and

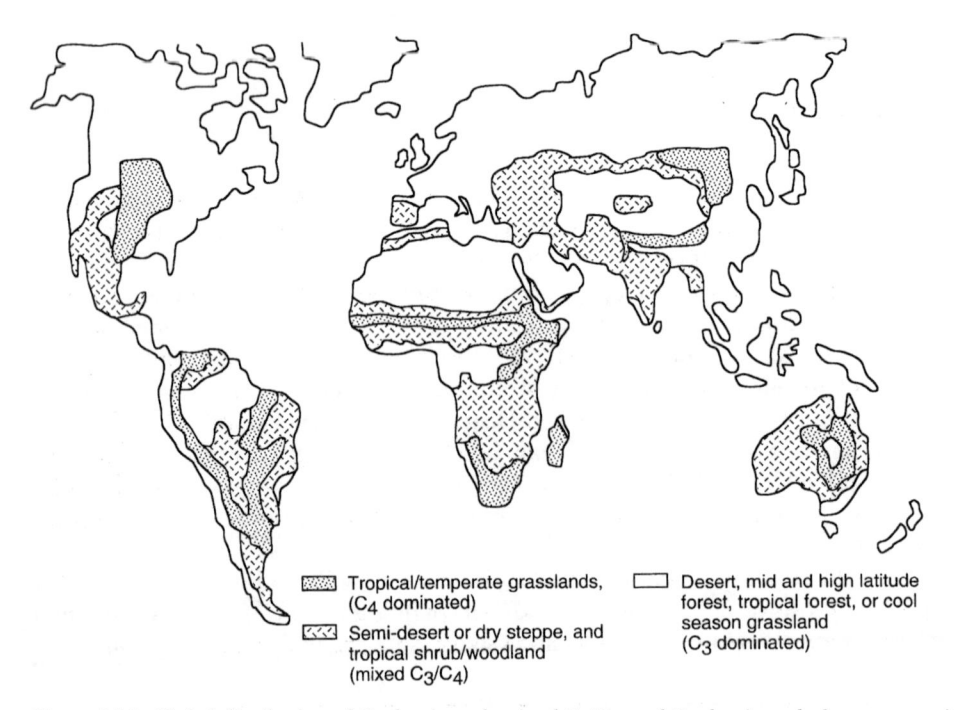

Figure 15.1. Global distribution of C_4 dominated, mixed C_3/C_4, and C_3 dominated plant communities. Marginal areas may have experienced shifts in C_3/C_4 plant ratios as well (modified from Cerling and Quade, 1993).

the potential errors involved, (3) and present case studies where stable C and O isotopes in soils are used to enhance archaeological problem solving, particularly with respect to paleoclimates.

2. Theory of Isotope Pedology

2.1. Soil Genesis

Surface soils and buried soils store important paleoenvironmental information applicable to solving archaeological problems because pedogenesis is strongly influenced by climate and vegetation (Jenny, 1941). Soils contain two components that can be analyzed isotopically for paleoenvironmental reconstruction: organic matter and pedogenic carbonate. Both of these materials contain carbon that is derived from decaying plant matter, which is the link between soil and vegetation/climate.

   Organic matter consists mainly of decomposed plant litter in surface and near-surface layers (A or O horizons) that imparts a dark color to the soil. Organic matter accumulates in the early stages of soil development until inputs and outputs become equal (steady state), typically within 1000 years (Birkeland, 1984; Buol et al., 1997). The amount of soil organic matter decreases with depth as biological activity decreases, whereas the whole-soil amount of soil organic matter decreases in drier climates where plant biomass production is lower. In contrast, most pedogenic carbonate typically accumulates in lower soil layers (Bk horizons) at the terminus of a wetting front, from CO_2 degassing, and from dewatering from evapotranspiration (Wilding et al., 1990). These processes enrich the soil solution in calcium and bicarbonate ions to the point that a carbonate precipitate forms. Pedogenic carbonate typically accumulates as nodules, filaments, or indurated masses that are morphologically visible in the field, and in a disseminated form that is only visible at the microscopic level (Wilding et al., 1990). In nonleaching environments, and with an abundant source of calcium and carbonate ions, the accumulation of soil carbonate increases with time (Birkeland, 1984; Buol et al., 1997).

2.2. Stable C Isotopes of Soil Organic Matter

There are three carbon isotopes in nature: ^{14}C, ^{13}C, and ^{12}C (Hoefs, 1987). The ^{14}C isotope is unstable and undergoes radioactive decay and is commonly used for radiocarbon dating late Quaternary materials. The ^{12}C and ^{13}C isotopes are stable, and the ratio of these two isotopes can be influenced by biological, physical, and chemical reactions. The amount of fractionation that occurs between ^{13}C and ^{12}C is measured as the relative deviation between a sample and standard, and is expressed in $\delta^{13}C$ notation (see section 4.2). Three plant groups occur isotopically based on their degree of discrimination against atmospheric

```
                    ┌─────────────────────┐
                    │  Atmospheric CO₂    │
                    │  δ¹³C_PDB = -7 ‰    │
                    └─────────┬───────────┘
                         photosynthesis
         ┌────────────────────┴────────────────────┐
┌────────────────────┐                  ┌──────────────────────────┐
│   C₄ plants        │                  │   C₃ plants              │
│ warm season grasses│                  │ cool season grasses,     │
│ δ¹³C_PDB = -13 ‰   │                  │ trees, shrubs, forbs     │
│                    │                  │ δ¹³C_PDB = -27 ‰         │
└────────────────────┘                  └──────────────────────────┘
```

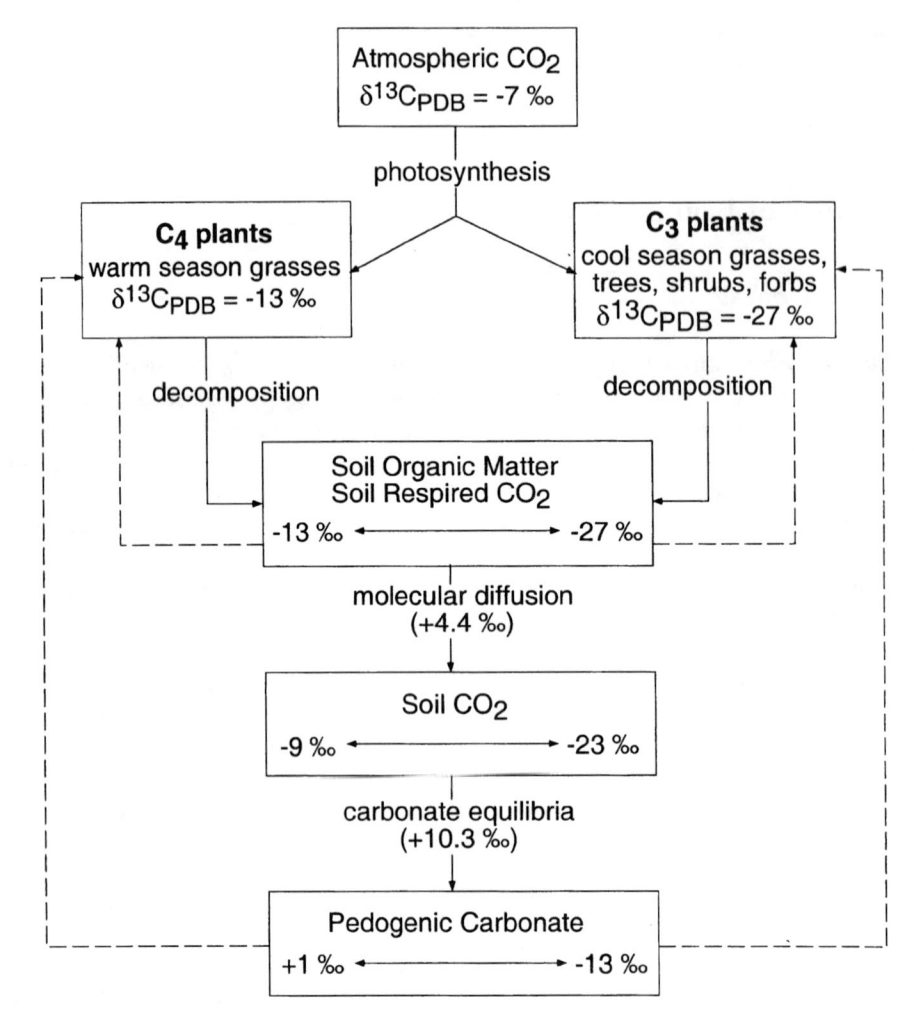

Figure 15.2. Photosynthetic pathway for C_4 and C_3 plants and associated $\delta^{13}C_{PDB}$ values for soil organic matter, respired soil CO_2, and pedogenic carbonate. Isotopic values are rounded to the nearest whole number. Dashed lines represent ways by which the proportion of C_4 and C_3 plants can be estimated knowing soil organic matter or pedogenic carbonate $\delta^{13}C_{PDB}$ values.

$^{13}CO_2$ during photosynthesis (Boutton, 1991b; Deines, 1980; Smith and Epstein, 1971). The C_3 plants discriminate most against atmospheric $^{13}CO_2$, and in normal conditions produce $\delta^{13}C$ values around $-27‰$ (Fig. 15.2). Plants that discriminate much less against atmospheric $^{13}CO_2$ are C_4, which produce $\delta^{13}C$ values near $-13‰$ (Fig. 15.2). The C_3 plants consist mainly of trees, shrubs, forbs, and cool-season grasses, whereas C_4 plants consist mainly of warm-season grasses. Crassulacean acid metabolism (CAM) plants typically have a photosynthetic pathway similar to C_4 plants. However, facultative CAM plants can generate values that span the entire spectrum of C_3 to C_4 photosynthesis.

Virtually no isotopic fractionation occurs during decomposition of plant litter and the subsequent incorporation of organic by-products into the soil organic matter pool (Dzurec et al., 1985; Melillo et al., 1989; Nadelhoffer and Fry, 1988). Thus, the $\delta^{13}C$ value of soil organic matter reflects the proportion of C_3 and C_4 plants contributing organic material to the soil (Fig. 15.2). For example, a pure C_4 plant community would produce soil organic matter with a $\delta^{13}C$ value of approximately $-13‰$. Some authors have discovered that the isotopic composition of plants has fluctuated during the late Quaternary by as much as 2‰ in response to changes to atmospheric CO_2 levels (Marino et al., 1992; Penuelas and Azcon-Bieto, 1992). This variation still does not obscure major shifts in biomass production as reflected in the $\delta^{13}C$ values.

Plants with C_3 photosynthesis may grow in all environmental settings throughout the world, making them difficult to use as a paleoclimatic indicator. Facultative CAM plants are also problematic because they may yield both C_3 and C_4 isotopic signatures. The contribution of C_4 plants to the soil organic matter pool, however, is a powerful proxy for paleoclimatic reconstruction because C_4 species abundance (Teeri and Stowe, 1976) and biomass production (Boutton et al., 1980; Tieszen et al., 1979) are strongly and positively related to environmental temperature. Therefore, the $\delta^{13}C$ of soil organic matter can be used to interpret past vegetation communities and from that, general interpretations about past environmental temperatures (Fig. 15.2).

The following mass balance equation illustrates how the proportion of C_4 and C_3 contributions to the soil organic matter pool can be estimated knowing a bulk soil $\delta^{13}C$ value:

$$\delta^{13}C = (\delta^{13}C_{C4})(x) + (\delta^{13}C_{C3})(1 - x) \tag{1}$$

where $\delta^{13}C$ is the value of the whole sample, $\delta^{13}C$ C_4 is the average $\delta^{13}C$ value of the C_4 components of the sample, x is the proportion of carbon from C_4 plant sources, $\delta^{13}C$ C_3 is the average $\delta^{13}C$ value of the C_3 components of the sample, and $1 - x$ is the proportion of carbon derived from C_3 plant sources. Figure 15.3a displays a theoretical mixing line calculated by rearranging Equation 1 and solving for x. Some natural variation exists in the end-member $\delta^{13}C$ values for C_3 and C_4 plants (generally $\pm 1‰$). Consequently, it may be necessary to construct an isotopic mixing line unique to the area of investigation.

2.3. Stable C and O Isotopes of Pedogenic Carbonate

Soil organic matter is converted to respired CO_2 by microbial respiration and does not isotopically fractionate when fluxing from the ground surface (Cerling et al., 1991b; Hesterberg and Siegenthaler, 1991; Fig. 15.2). During the instantaneous flux of respired CO_2 the lighter $^{12}CO_2$ isotope migrates more rapidly to the surface by molecular diffusion, thus theoretically enriching the remaining soil CO_2 by 4.4‰ (Amundson et al., 1998; Cerling, 1984; Cerling et al., 1991b; Dorr and Munnich, 1980; Fig. 15.2). This soil CO_2 component takes part in carbonate equilibria reactions, which leads to an additional 10.3‰ enrichment in $\delta^{13}C$ (at

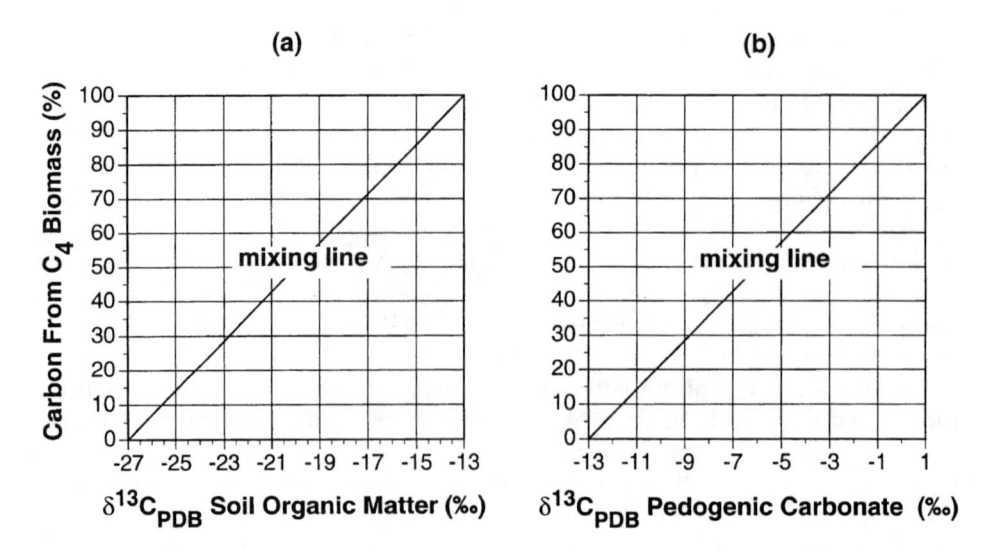

Figure 15.3. Theoretical isotopic mixing lines illustrating the relationship between C$_4$ biomass production and the $\delta^{13}C_{PDB}$ of (a) soil organic matter and (b) pedogenic carbonate. Isotopic end members may vary slightly depending on local environmental conditions.

20°C) as soil CO$_2$ is transferred to the solid pedogenic carbonate phase (Cerling, 1984; Cerling et al., 1989; Deines, 1980, (Fig. 15.2)). Although this reaction is temperature dependent, the influence on isotopic values is minor within the temperature range of most soils (1 to 1.5‰). Thus, the pedogenic carbonate component of most soils can in theory be isotopically estimated by adding approximately 14 to 15‰ to soil organic matter or respired CO$_2$ as shown by the following equation (Nordt et al., 1998):

$$\delta^{13}C \text{ pedogenic carbonate} = \alpha_{CaCO_3 - CO_2}(\delta^{13}C_{SOM} + 1004.4) - 1000 \quad (2)$$

where ^{a}CaCO$_3$ − CO$_2$ is the fractionation factor between pedogenic carbonate and soil CO$_2$ (1.0103, which calculates to 10.3‰), and SOM is soil organic matter. Cerling and Quade (1993) and Cerling (1984) both showed, however, that ^{13}C-enriched atmospheric CO$_2$ can mix with biologically produced soil CO$_2$ in arid regions where biological activity is low. Generally, this is not a factor at the depth at which pedogenic carbonate precipitates in subhumid to humid environments, nor below a depth of 50 cm in semiarid or arid climates. Computer modeling provides the best means of estimating the $\delta^{13}C$ of pedogenic carbonate in areas where atmospheric mixing is significant (Cerling, 1984; Nordt et al., 1998; Quade et al., 1989).

Pedogenic carbonate accumulates in disseminated forms, nodular forms, and as indurated layers (Buol et al., 1997; Wilding et al., 1990). The $\delta^{13}C$ value of these forms, if not contaminated with lithogenic or detrital pedogenic carbonate, can be used as a paleoclimatic proxy (Cerling, 1984; Cerling and Quade, 1993).

Such estimates are achieved by subtracting approximately 14 to 15‰ (at 20°C) from the $\delta^{13}C$ value of pedogenic carbonate to estimate the proportion of C_3 and C_4 plants growing at the time of carbonate precipitation (Fig. 15.2). For example, a $\delta^{13}C$ value of $-20‰$ obtained from a pedogenic carbonate nodule would indicate equal contributions from C_3 and C_4 plants to soil organic matter production during the time of carbonate formation (Fig. 15.3b).

There are also three stable oxygen isotopes in nature: ^{18}O, ^{17}O, and ^{16}O. Ratios of the two most abundant isotopes, ^{18}O and ^{16}O, are used to estimate relative and absolute shifts in environmental temperature (Cerling, 1984; Cerling and Quade, 1993; Hays and Grossman, 1991; Quade et al., 1989). The basis of the relationship between ^{18}O content of pedogenic carbonate and environmental conditions is that pedogenic carbonate obtains its ^{18}O signature from meteoric water. Dansgaard (1964) demonstrated that at latitudes poleward from 40°N and 40°S, $\delta^{18}O$ values of meteoric water are strongly and inversely related to temperature. This relationship is caused by differences in vapor pressure between $H_2^{18}O$ and $H_2^{16}O$ described as a Rayleigh Distillation process (Hays and Grossman, 1991). Thus, each rainfall event that originates in a cloud mass in the tropics or subtropics preferentially depletes the heavier ^{18}O isotope, concentrating the lighter ^{16}O isotope in the remaining cloud vapor in its migration poleward.

By assessing a world database, Rozinski et al. (1993) showed more complex relationships between $\delta^{18}O$ values of meteoric water and temperature in latitudes between 20°N and 20°S. Among the many factors affecting this relationship, rainfall amount and origin of air mass were the most important. The latter is a particular problem in the North America plains because of the potential convergence of air masses originating from the Gulf of Mexico, the Atlantic Ocean, and the Pacific Ocean. These air masses are typically subjected to different degrees of Rayleigh Distillation during their evolution, making it difficult to correlate isotopic signatures of meteoric water with local climatic conditions. It follows that the $\delta^{18}O$ of pedogenic carbonate may produce similar results that are difficult to interpret.

Cerling and Quade (1993) observed a strong positive correlation in many climates between $\delta^{13}C$ and $\delta^{18}O$ values of pedogenic carbonate (Fig. 15.4). This occurs because C_3 plants are more dominant in cooler environments where meteoric $\delta^{18}O$ water values are more depleted. The correlation in Mediterranean climates is not as strong because precipitation falls mainly in the winter resulting in C_3 plant dominance (Fig. 15.4). In addition, evaporative effects can enrich soil water in $\delta^{18}O$, creating an unpredictable deviation between meteoric water and pedogenic carbonate (Amundson and Lund, 1987; Cerling, 1984; Cerling and Quade, 1993).

3. Field Application of Stable C and O Isotopes in Soils

3.1. Radiocarbon Dating and $\delta^{13}C$ Depth Distributions

Radiocarbon ages of soil organic matter increase with depth because surface horizons are replenished with modern carbon daily, as more resistant and older

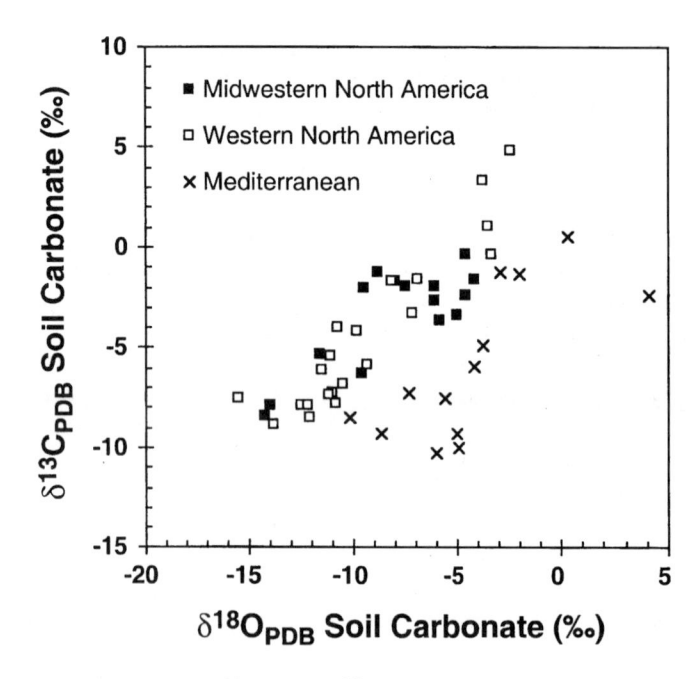

Figure 15.4. Relationship between d^{13}C$_{PDB}$ and d^{18}O$_{PDB}$ of pedogenic carbonate for selected areas in North American and Mediterranean environments (reproduced from Cerling and Quade, 1993; copyright by the American Geophysical Union).

organic components accumulate in subsoils (Sharpenseel and Neue, 1984; Trumbore, 1996; Wang et al., 1996a). This depth function, coupled with contamination from soil mixing and differential rooting depths, presents problems when comparing bulk humate ^{14}C ages to the timing of ^{13}C inputs for the same depth. The following is an example demonstrating the difficulties in making this comparison.

Figure 15.5a illustrates a steady-state condition for soil organic matter inputs with a uniform distribution of δ^{13}C values and increasing ^{14}C ages with depth. If there is a rapid shift in vegetation isotopically (e.g., from C$_4$ to C$_3$), but no shift in root distribution such that steady state is maintained at all depths, the ^{14}C age with depth will not change (Fig. 15.5b). However, the δ^{13}C values will reflect the vegetation shift to the maximum depth of rooting (Fig. 15.5b). If the same isotopic shift in vegetation occurs, but the rooting depth increases appreciably, all ^{14}C ages below this depth will decrease as modern carbon is added (Fig. 15.5c). The corresponding δ^{13}C values will also shift, and to greater depths, in the direction of the vegetation change that occurred. If, on the other hand, the vegetation shift occurs in association with a decrease in rooting depth, the ^{14}C age at a 1 m depth becomes older as modern organic inputs are reduced (Fig. 15.5d). The surface horizon in all cases best reflects the most recent change in vegetation, although at steady state its ^{14}C age never changes.

Stable Carbon and Oxygen Isotopes in Soils 427

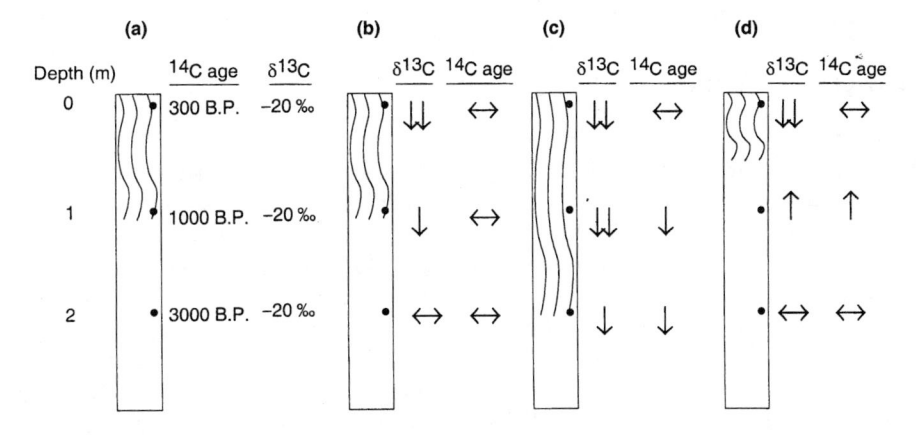

Figure 15.5. An example of the complex interactions that a rapid change in vegetation can have on the depth distribution of soil $d^{13}C_{PDB}$ values of soil organic matter and bulk humate ^{14}C ages. Lines within columns represent rooting depths, arrows show direction of change in either $d^{13}C_{PDB}$ values or ^{14}C ages, and double arrows show greater magnitudes of change than single arrows. (a) Steady-state condition for soil organic matter; (b) change in vegetation isotopically, but with constant rooting depth; (c) change in vegetation isotopically, but with an increase in rooting depth; and (d) change in vegetation isotopically, but with rooting depth decreasing. All vegetation changes illustrate increasing C_3 plant biomass production.

With these scenarios it would be difficult to estimate the timing of vegetation shifts and corresponding $\delta^{13}C$ values based on ^{14}C dating of soil humates. Establishing a numerical chronology of the geologic and soil stratigraphic framework using radiocarbon-datable material such as charcoal or wood provides the best method for interpreting the timing of $\delta^{13}C$ inputs. Interpreting vegetation shifts from soils developed in ancient uplands is also problematic because of possible mixing of differing $\delta^{13}C$ inputs through time and because of the difficulty of dating soils or deposits in these settings. Needless to say, modeling efforts are necessary to resolve the problems presented in Fig. 15.5.

Soil texture plays an important role in turnover rates of organic matter, which in turn may also affect the relation between organic matter $\delta^{13}C$ values and humate ^{14}C ages with depth (Balesdent and Mariotti, 1996; Boutton et al., 1998). Boutton et al. (1998) found that during historic changes in vegetation in south Texas, the stable C isotopes of sandy soils reflected the vegetation shift best and to relatively great depths. In contrast, fine-grained soils recorded the same vegetation shift only in the surface horizons where the greatest inputs of modern plant litter occurred. It appears that organic carbon and preexisting isotopic signatures are protected within the fine pores of clayey soils. It follows that ^{14}C ages and $\delta^{13}C$ values may track each other fairly closely with depth in sandy soils, such that the timing of the vegetation shift could be approximated with a bulk humate ^{14}C age. In fine-grained soils where organic matter turnover rates are much slower, this probably will not be the case. Here, there will be greater mixing of organic matter from two sources that include the existing organic matter and that reflecting the recent change in vegetation.

3.2. Sources of Soil Carbon

3.2.1. Organic Matter

There are two sources of soil organic matter that commonly occur in late Quaternary environments (Nordt et al., 1994). The first is detrital, which is the organic fraction transported with mineral particles in alluvium, colluvium, windblown deposits, or even glacial deposits. For example, if the drainage basin of a stream or the source area of eolian sediment does not cross major climatic or vegetation boundaries, a detrital organic carbon source is desirable because it represents local conditions. On the other hand, detrital isotopic signatures from large drainage basins or from distant eolian sources probably represent an average of several climatic and vegetation zones and cannot be interpreted as one environmental condition. Generally, transported organic carbon originates from the most easily erodible materials in the drainage basin or area in question. This tends to be from upland topsoils or older alluvial topsoils that carry a modern organic carbon isotopic signal. Again, this organic component is desirable for interpreting paleoenvironmental conditions. Some organic carbon may be derived from bedrock, but this component is much more likely to contain low quantities of organic carbon and consequently have its signal diminished by modern carbon sources. Detrital isotopic values are more likely to be encountered in subsoils (B horizons) where pedogenic organic inputs are lower.

The second source of soil organic matter is pedogenic. This source is clearly desirable for paleoenvironmental reconstruction because the isotopic values reflect the vegetation growing directly in the soil. Pedogenic isotopic signatures of organic matter will be more evident in surface horizons where they are superimposed on detrital signatures. Mixing of detrital and pedogenic sources should also provide favorable isotopic results if the detrital component was transported a relatively short distance.

The stable C isotope composition of organic matter in upper horizons of buried paleosols may provide a discrete view of the isotopic plant community present during soil formation. This is provided that appreciable erosion of the surface horizon did not occur during soil burial or that significant biological mixing not occur after burial. If the soil developed in uplands from sediments that have never been buried, it becomes more problematic interpreting either surface or subsurface isotopic values because long-term surface exposure may average isotopic conditions from two or more climatic intervals. Surface soils no more than a few thousand years old may also exhibit problems similar to older upland soils, depending on the frequency of climatic shifts in a particular area. In this case, data is needed from younger surface soils that can help partition differing vegetation and climatic intervals preserved in older soils.

An example demonstrating the various conditions that can be encountered when collecting samples for stable isotope analysis of organic matter in buried soils is presented in Fig. 15.6. If the paleosol is undisturbed, and pedogenic organic matter contents high, "good" conditions for sampling are generally encountered. Under these conditions, the source of organic material is local and derived directly from the soil being sampled. If organic matter contents are low,

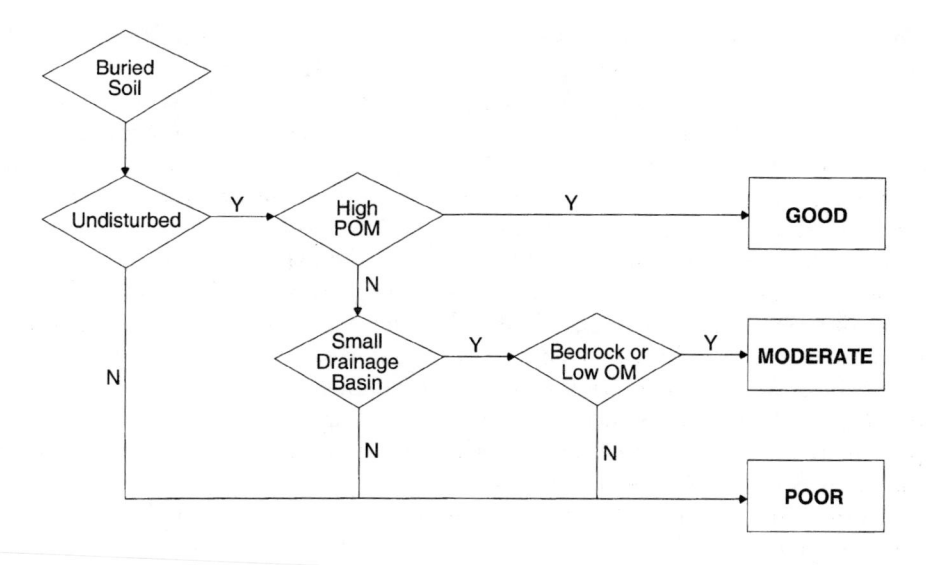

Figure 15.6. Probability of obtaining reliable paleoenvironmental information from stable C isotope samples of paleosols in different alluvial settings. POM = pedogenic organic matter; Y = yes; N = no.

however, then a greater proportion of the organic material will be of detrital origin. Here, "moderate" sampling results can be obtained, but only if the drainage basin is small and does not have easily erodible and organic rich bedrock. Low soil organic matter contents within large drainage basins are the most likely to yield "poor" sampling results in that the organic material is mainly detrital and is derived from unknown climatic or vegetation regions.

These scenarios generally apply to sampling either A or B horizons. Parent material, or C horizons, are more problematic because virtually none of the organic matter is pedogenic. A further caveat is to always sample within the same depositional facies through time, regardless of soil horizon. This strategy avoids comparing isotopic results from riparian and well-drained floodbasin settings, for example.

3.2.2. Pedogenic Carbonate

Samples for stable C and O isotope analysis of pedogenic carbonate must also be interpreted with caution. The most common problem occurs when carbonate nodules appear to be pedogenic, but instead are: (1) lithogenic; (2) pedogenic, but engulfing lithogenic components; (3) pedogenic, but detrital; or (4) produced in association with ground water. Fortunately, most limestone $\delta^{13}C$ values ($\approx 0‰$) are greater than pedogenic carbonate values that form in equilibrium with organic matter produced from C_3 or mixed C_3/C_4 plant communities (Fig. 15.3b). Unfortunately, pedogenic carbonate produced in association with organic matter from a pure C_4 plant community will yield $\delta^{13}C$ values similar to limestone

(Fig. 15.3b). This problem can sometimes be resolved by insuring that the $\delta^{13}C$ of co-existing soil organic matter and pedogenic carbonate isotopically differ by 14 to 15‰ as theoretically predicted for most soils (see Equation 2). If, for example, the difference in $\delta^{13}C$ values between pedogenic carbonate and soil organic matter is less than 14 to 15‰, it is likely that a vegetation shift to increasing C_4 plant abundance occurred during carbonate genesis that was reflected in the organic matter, but because of slower kinetics, was never recorded in the pedogenic carbonate. Another technique would be to use petrographic thin section analysis because in many instances pedogenic and lithogenic carbonates have different micromorphic attributes (West et al., 1988). One example of this situation is that carbonate formed as root casts normally indicate *in situ* pedogenesis. Pedogenic carbonate that is eroded and redeposited is also difficult to detect, but the depth of nodules should occur systematically with climate. For example, pedogenic carbonate nodules accumulate at greater depths in subhumid climates than arid climates, all else being equal.

Some carbonate nodules precipitate in equilibrium with groundwaters supersaturated in calcium and carbonate ions. This would yield isotopic values that are not necessarily in equilibrium with biologically produced soil CO_2. The presence of iron depletion zones or oxidized root channels observed in the field or in petrographic thin section (Vepraskas, 1994) may indicate whether groundwater was associated with formation of the carbonate. Carbonates precipitating within a capillary fringe of a groundwater table, however, may not pose a problem if a significant amount of biologically produced CO_2 takes part in the reaction.

Small carbonate nodules are ideal for paleoenvironmental reconstruction because they generally form during a relatively short interval of pedogenesis in association with one climatic interval. Large nodules may represent successive layers of carbonate that accumulated during several climatic intervals, making climatic interpretations difficult. If analysis of large nodules is needed it may be necessary to microsample concentric increments within the nodule (Pendall et al., 1994).

The analysis of $\delta^{18}O$ can be performed on pedogenic carbonates, but not on soil organic matter, for paleoenvironmental reconstruction. Before being utilized for oxygen isotopes, the carbonate nodules must be assessed for pedogenic purity as discussed previously for the $\delta^{13}C$ analysis. In addition, if carbonate nodules formed from groundwater, the signatures may not be meteoric.

3.2.3. Soil-Stratigraphic Example

Depending on the presence or absence of paleosols, the interval and pathway of paleosol pedogenesis, and the presence or absence of erosional unconformities, estimating the timing of organic and inorganic carbon inputs may be difficult in soil-stratigraphic sections. As an example of these potential problems, Fig. 15.7 illustrates an alluvial soil-stratigraphic section with four units, two buried paleosols, and six ^{14}C ages obtained from charcoal; and the proportion of the isotopic record allocated to pedogenic organic inputs, basinwide organic inputs, and erosion.

Stable Carbon and Oxygen Isotopes in Soils 431

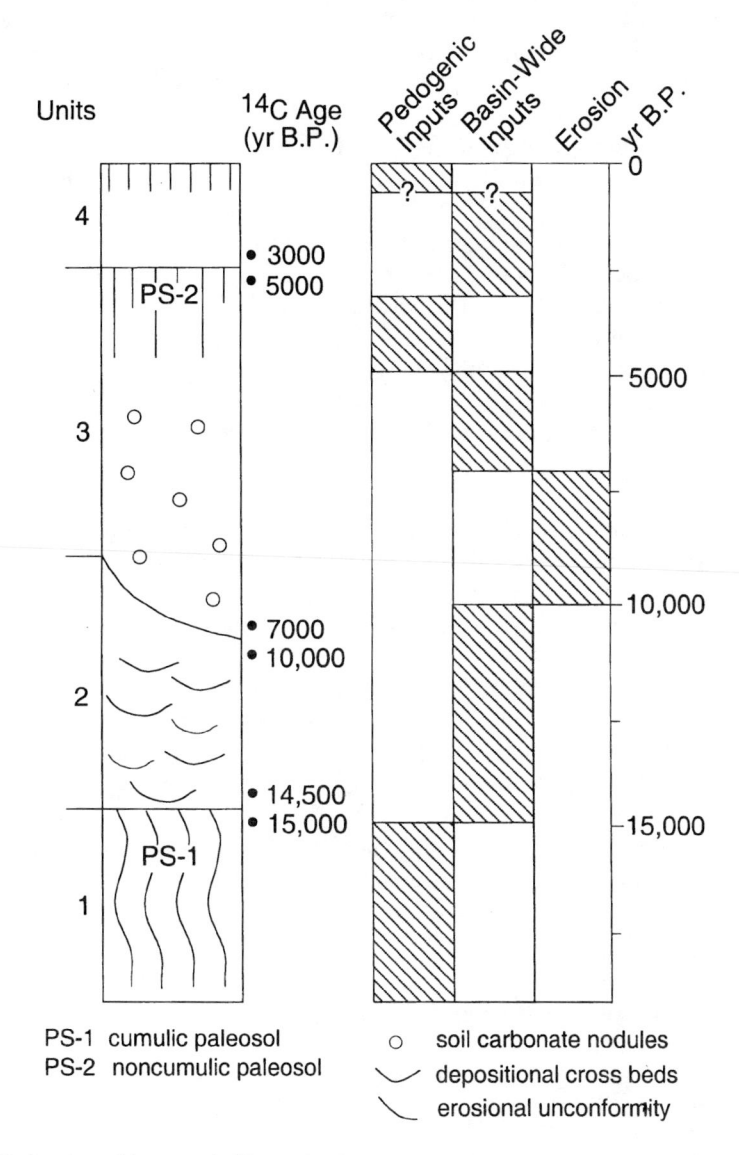

Figure 15.7. Stratigraphic example illustrating how periods of erosion, deposition, and soil formation influence the depth distribution and chronological distribution of organic matter inputs and associated stable C isotope values.

With cumulic formation of the PS-1 soil in the upper part of Unit 1 (e.g., pedogenesis keeping pace with deposition), a combination of detrital and pedogenic organic carbon isotopic signatures will be observed. However, organic inputs from pedogenic sources should dominate the isotopic record and reflect local vegetation conditions at the time of soil formation (Fig. 15.7). Aggradation of Unit 2 is marked by an increase in depositional rate that terminated formation of PS-1 (Fig. 15.7). Because little time elapsed during this depositional transition, all isotopic values in PS-1 of Unit 1 were recorded at, or prior to, 15,000 yr BP. Stable isotope values in Unit 2 will reflect basinwide organic inputs deposited between 14,500 and 10,000 yr BP. The boundary between Unit 2 and Unit 3 was created by an episode of erosion that removed 3000 years of sediment (Fig. 15.7). Thus, there is no isotopic record between 10,000 and 7000 yr BP.

The PS-2 soil in the upper part of Unit 3 formed between 5000 and 3000 yr BP during a period of landscape stability. Hence, organic carbon in the buried A horizon of PS-2 also accumulated and imparted its isotopic signature during this time. The same rationale would apply to the carbonate nodules in the B horizon because they would have formed in association with CO_2 produced from organic matter decomposition. The isotopic record between 7000 and 5000 yr BP is contained in the detrital organic matter component of the B horizon of PS-2, where postdepositional pedogenic organic inputs were lower. Deposition began again at 3000 yr BP, bringing detrital organic matter back into the area. Although deposition continued for an unknown time, isotopic values in the uppermost profile are modern.

With this example, approximately 16% of the stratigraphic record is missing from erosion, whereas 52% of the isotopic record is from basinwide detrital sources and 32% from local pedogenic inputs. For this location, it would be important to insure that the drainage basin is small and not crossing major climatic and vegetation zones because over one half of the paleoenvironmental record would be based on stable C isotopes derived from detrital organic carbon.

This soil-stratigraphic example demonstrates that a stable C isotope record should be interpreted within context of at least several pedological and geological factors (1) erosion, which removes part of the isotopic record; 2) rapid deposition, which creates a record strongly influenced by basinwide vegetation conditions and isotopic signatures; and (3) minimal deposition, leading to pedogenesis and local vegetation conditions and isotopic signatures.

4. Stable Isotope Laboratory Procedures

4.1. Sample Collection

Bulk soil samples can be collected for stable C isotope analysis on a horizon-by-horizon basis or in predetermined depth increments. Minimum sample size depends on laboratory preparation procedures and mass spectrometer characteristics (Boutton, 1991a; Midwood and Boutton, 1998). Most laboratories use 0.5 to 1.0 mg of carbon for a single isotopic analysis. To meet this requirement, a minimum of 500 mg of bulk soil with 1 to 3% organic matter is typically needed. If the bulk sample contains carbonate, a larger sample will be necessary because removal of the carbonate during pretreatment may significantly reduce the sample size. If a radiocarbon age is desired from the same zone, the sample will be small enough that an accelerator mass spectrometer (AMS) assay will probably be required. It is important to avoid large roots and layers that have noticeable mixing of materials from different horizons. This could occur from erosion, biological mixing, or from shrink–swell activity in the soil.

Pedogenic carbonate comes in many sizes and shapes. If the soil developed from noncalcareous parent material, then the carbonate in the bulk sample is probably pedogenic (disseminated) and suitable for stable C and O analysis for paleoenvironmental reconstruction. Many situations, however, warrant collecting carbonate nodules for isotopic analysis to reduce potential contamination from lithogenic sources. In either case, a minimum of 10 mg of pure carbonate is needed to meet the minimum requirement of 0.5 to 1.0 mg of carbon for isotopic analysis. In most cases more will be needed to account for the impurities that occur in most nodules (Boutton, 1991a). Ideally, it would be better to analyze several nodules independently within a soil horizon to get a sense of the isotopic variability likely to be encountered in a particular study area. As discussed in the previous section, avoid sampling large carbonate nodules for isotopic analysis, or else analyze discrete zones within large nodules.

4.2. Procedures for Soil Organic Matter

In preparation for $\delta^{13}C$ analysis of soil organic matter, laboratories remove carbonate carbon with HCl (Boutton, 1991a; Midwood and Boutton, 1998). The residue is then converted to CO_2 during dry combustion, typically with CuO. The CO_2 is analyzed for variations in ^{13}C and ^{12}C content with an isotope ratio mass spectrometer. The difference in abundance between ^{13}C and ^{12}C in a particular substance is most commonly reported in units of $\delta^{13}C$, which is the relative deviation of the $^{13}C/^{12}C$ ratio of the sample from a standard. The standard for soil organic matter and pedogenic carbonate is Pee Dee belemnite (PDB) limestone that has an assigned $\delta^{13}C$ value of 0‰ (Craig, 1957). The equation for $\delta^{13}C$ determination is

$$\delta^{13}C_{PDB}‰ = [(R_{sample}/R_{standard}) - 1] \times 10^3 \qquad (3)$$

where $\delta^{13}C$ has units of parts per thousand (permil, or ‰) and R is the $mass_{45}/mass_{44}$ of sample or standard CO_2 (Craig, 1957). Negative $\delta^{13}C$ values represent a depletion in ^{13}C relative to the standard and positive $\delta^{13}C$ values represent an enrichment relative to the standard.

4.3. Procedures for Pedogenic Carbonate

For $\delta^{13}C$ and $\delta^{18}O$ analysis, pedogenic carbonate is typically decomposed to CO_2 by reaction with H_3PO_4 (Barrie and Prosser, 1996; Boutton, 1991a). As with soil organic matter, isotope ratio mass spectrometers are used to determine the isotopic ratios of carbonate. The standard for $\delta^{18}O$ and $\delta^{13}C$ of pedogenic carbonate is also PDB, with results reported in units of ‰ or permil. As with carbon isotopes, Equation (3) is used to calculate the $^{18}O/^{16}O$ deviation of the sample from the standard. Standard mean ocean water (SMOW) is used for measuring $\delta^{18}O$ values of meteoric water. The conversion from SMOW in permil to $\delta^{18}O_{PDB}$ in permil is as follows (Hoefs, 1987):

$$d_{SMOW} = 1.03086 \; d_{PDB} + 30.86‰ \qquad (4)$$

4.4. Laboratory Comparisons

The most reliable means of determining $\delta^{13}C$ values for paleoenvironmental reconstruction is by stable isotope geochemistry laboratories designed specifically for this task. $\delta^{13}C$ values are also produced in association with ^{14}C ages generated in radiocarbon laboratories. This is done to correct the $^{14}C/^{12}C$ ratio of the radiocarbon sample for variations in ^{13}C content produced by living organisms during photosynthesis (Lowe and Walker, 1984; Trumbore, 1996). If a subsample of the radiocarbon sample is analyzed using standards and procedures of a stable isotope geochemistry laboratory, the $\delta^{13}C$ results should be suitable for paleoenvironmental reconstruction. However, the user must be aware that high-level precision and accuracy for the $\delta^{13}C$ correction is not necessary for correcting ^{14}C ages. For example, each per mil difference between the $\delta^{13}C$ value generated by ^{14}C dating and the stable isotope standard for ^{14}C corrections (generally wood with a $\delta^{13}C$ value of $-25‰$ relative to PDB), accounts for only about 16 years. Therefore, a several per mil fluctuation in the accuracy or precision of the $\delta^{13}C$ value generated by a radiocarbon laboratory is still within the 1 sigma standard deviation commonly reported for ^{14}C ages (typically 50 to 100 years). In contrast, a several per mil shift in $\delta^{13}C$ from a soil sample indicates about a 21% shift in the ratio of C_4 to C_3 plants! That is not to say that the $\delta^{13}C$ values produced by ^{14}C laboratories are necessarily incorrect, but rather it's not their primary goal to insure the level of accuracy and precision for paleoenvironmental reconstruction. If the user wishes to obtain $\delta^{13}C$ results for paleoenvironmental analysis, it is important to discuss the matter with the ^{14}C laboratory to get assurances as to the desired results.

5. Geoarchaeology Case Studies

5.1. Arid Southwest

5.1.1. New Mexico, West Texas

As part of geoarchaeological investigations in the Fort Bliss Military Reservation of southern New Mexico and West Texas, Monger (1995) and Cole and Monger (1994) reconstructed paleoenvironments for the last 10,000 years of human occupation in alluvial fan settings. Three proxies for paleoclimate were used: erosion–sedimentation history, palynology, and stable C and O isotope analysis of pedogenic carbonate from buried soils. The carbon isotope analysis of paleosols showed that the $\delta^{13}C$ of pedogenic carbonate decreased from between -1 and $-4‰$ to between -10 and $-6‰$ at approximately 8000 yr BP. This change was interpreted as a response to a major increase in C_3 biomass production that coincided with a significant increase in Cheno-am pollen. Together these data suggested a decline in grasslands and increase in shrublands in the early Holocene. Rates of upland erosion were also increasing at this time, probably in response to reduced vegetative cover from a spreading C_3 shrub plant community. The $\delta^{18}O$ values of pedogenic carbonate varied little during this time, indicating that temperature or precipitation was not impacting vegetation changes. Consequently, it was inferred that increasing atmospheric CO_2 concentrations beginning 8000 yr BP contributed to the decline in grasslands in fragile ecosystems of alluvial fan settings.

Monger (1995) noted that at the time of declining grasslands, there was a regionwide shift from Paleoindian to Archaic cultures in this region that may have affected settlement patterns and preservation potentials. It was inferred that many upland archaeological sites were destroyed from erosion and that others were preserved in stratified contexts in areas with increased deposition. In sum, stable C and O isotopes were used to determine the timing of vegetation deterioration in the American Southwest, which may have lead to a shift in prehistoric subsistence strategies, widespread erosional and depositional events, and differential preservation of the prehistoric archaeological record.

5.1.2. South Texas

In south Texas in the Rio Grande alluvial valley, Nordt (1998) used stable C isotope from organic matter of surface soils to infer changes in climate and prehistoric resource distribution during the Holocene. Based on a combination of radiocarbon dating and surface diagnostic artifacts, a middle Holocene stratigraphic unit was identified that was deposited between about 7500 and 4000 yr BP in a low order tributary (Fig. 15.8a). At depths below 175 cm (lower B horizon), $\delta^{13}C$ values of -23 to $-25\permil$ indicated minor contributions from C_4 plants to the soil organic matter pool (Fig. 15.8a). According to a ^{14}C age, $\delta^{13}C$ values within this depth range were probably emplaced prior to 7000 yr BP by detrital alluvial deposition. Between depths of 175 and 30 cm (upper B horizons), there is a significant increase in C_4 contributions (-23 to $-15\permil$) to soil biomass production that was recorded by a mix of detrital alluvial deposition and postdepositional pedogenesis (Fig. 15.8a). However, because of the clayey subsoil, most organic carbon present at this depth is probably detrital, not having been affected as much by postdepositional inputs. Based on the timing of alluvial deposition, this C_4 plant community was present in both the larger drainage basin and in the local floodplain, and it began flourishing shortly after 7000 yr BP and persisted until at least 4000 yr BP. The more negative isotopic values in the upper 30 cm were probably emplaced by pedogenic inputs in the late Holocene after termination of alluvial deposition. This shift reflects an increase in C_3 plant production.

A second isotopic profile was constructed from a late Holocene alluvial soil that formed cumulicly (Fig. 15.8b). These isotopic data show that during the period between 2200 and 1200 yr BP only minor vegetation shifts occurred and that the $\delta^{13}C$ values were similar to those in the upper horizon of the middle

(a)

δ¹³C$_{PDB}$(‰)
Soil Organic Matter

(b)

δ¹³C$_{PDB}$(‰)
Soil Organic Matter

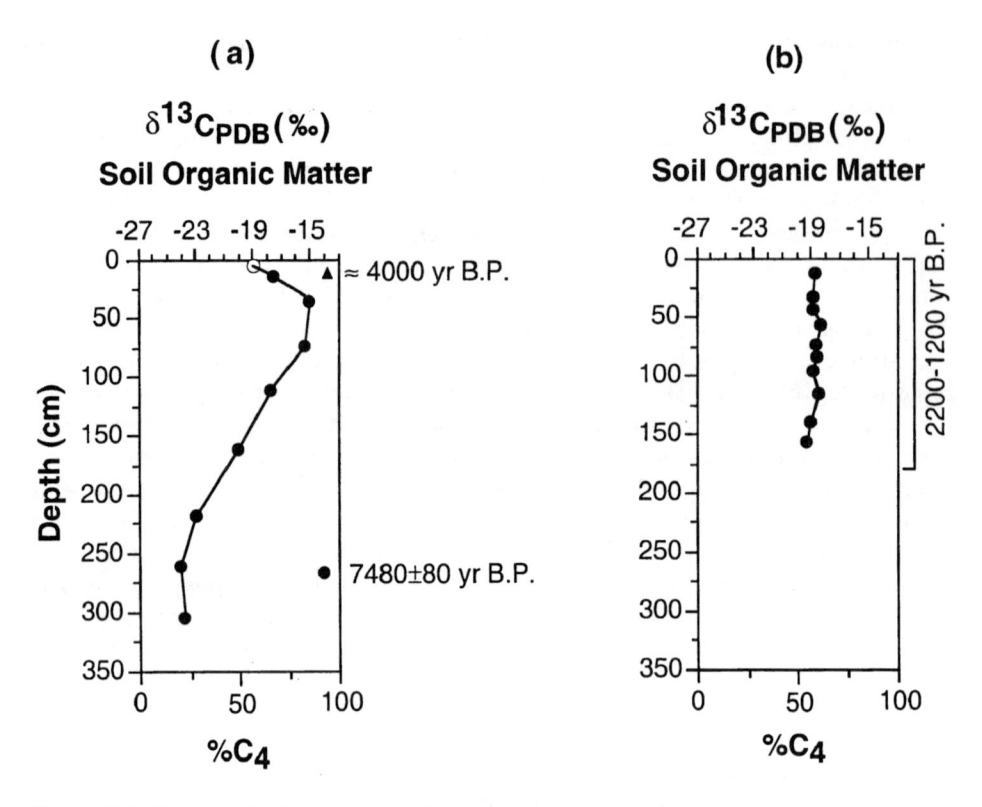

Figure 15.8. Isotopic depth distribution for organic matter in two Rio Grande alluvial soils in semi-arid south Texas. (a) Middle Holocene alluvium with the age of deposition based on a deep bulk humate ^{14}C sample (solid circle) and on a surface age estimated from time-diagnostic prehistoric artifacts (solid triangle); (b) late Holocene alluvial soil with the age of deposition based on ^{14}C dating of charcoal. The open circle near the surface of the middle Holocene soil illustrates that the direction of vegetation change corresponds with that occurring in the late Holocene soil (Nordt, 1998).

Holocene alluvial soil (Fig. 15.8a, open circle). Consequently, it was inferred that the shift from a dominance of C$_4$ grasses as recorded in the B horizons of the middle Holocene soil, to a mixed C$_4$/C$_3$ grass or shrub community as recorded in both in the surface horizon of the middle Holocene soil and in all horizons of the late Holocene soil, occurred sometime between 4000 and 2200 yr BP. The relatively dark and thick surface horizon of the late Holocene soil further indicates that grasses were an important part of the plant community. Thus, the shift to increasing C$_3$ species in the late Holocene suggests cooler conditions and the presence of cool-season grasses or trees, and not hot/dry conditions associated with an influx of C$_3$ desert shrubs.

These isotopic results demonstrate that in the middle Holocene, this region of south Texas did not support a desert shrub community. Consequently, prehistoric subsistence was responding to resources provided by a grassland during a warm middle Holocene, and to possibly a mixed grassland/savanna with cooler conditions in the late Holocene. In fact, in the study area, there was more evidence of human occupation in buried contexts in late Holocene soils than in middle Holocene soils. This difference may reflect a climate yielding greater resource availability in the late Holocene.

5.2. Southern Great Plains

5.2.1. North-Central Texas

Humphrey and Ferring (1994) and Ferring (1995) conducted paleoclimatological and geoarchaeological investigations along the Trinity River in the Southern Plains of North Texas using stable C and O isotopes. After establishing the Holocene alluvial chronology, stable C and O isotopes of pedogenic carbonate from buried soils were used to estimate vegetation and temperature changes. The objective was to determine whether there was a relationship between alluvial history, climate, and prehistoric subsistence strategies.

Humphrey and Ferring (1994) observed relatively low C_4 biomass production between 10,000 and 8000 yr BP based on $\delta^{13}C$ values between -7.5 and $-8‰$ from pedogenic carbonate (Fig. 15.9a). There was a significant increase in C4 biomass production in buried soils formed in the middle Holocene between 8000 and 4000 yr BP (-5 to $-6‰$), and again in the late Holocene around 1500 yr BP (-3.5 to $-4‰$) (Fig. 15.9a).

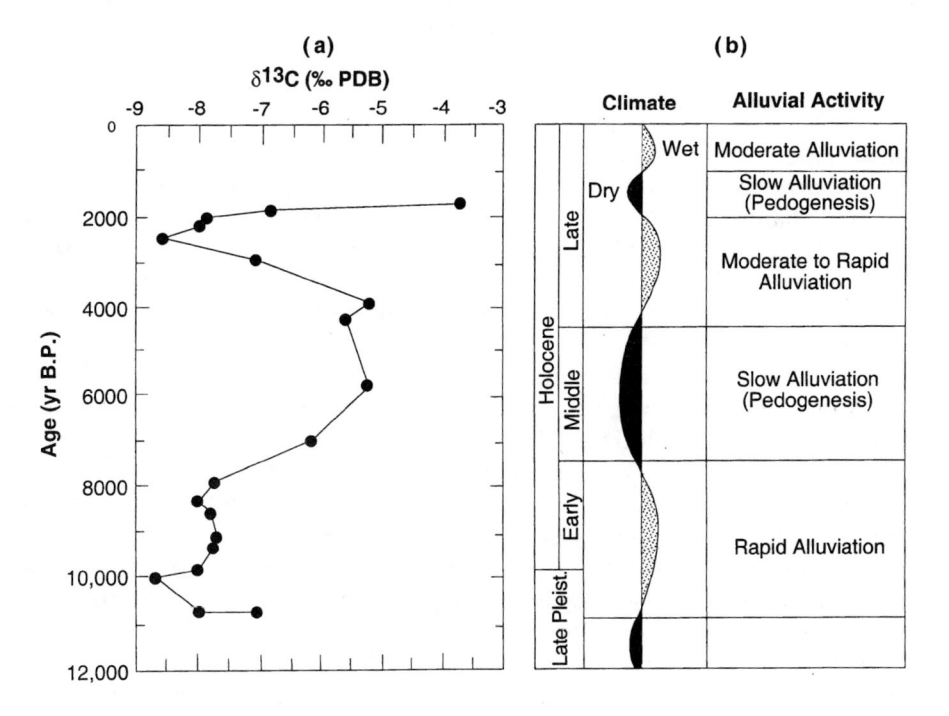

Figure 15.9. Isotopic record (a) and inferred climatic and alluvial depositional changes (b) for the Holocene based on $\delta^{13}C_{PDB}$ values of paleosol pedogenic carbonate in the upper Trinity River floodplain, Texas (reproduced from Humphrey and Ferring, 1994; copyright by Academic Press, Inc.).

Depleted $\delta^{18}O$ values in late Pleistocene lacustrine deposits indicated that temperatures were cooler than at present between 10,000 to 12,000 yr BP (not shown). However, $\delta^{18}O$ values from Holocene paleosols were stable, suggesting that C_4/C_3 vegetation shifts were responding to changes in seasonal precipitation and not average precipitation or temperature.

Humphrey and Ferring (1994) were then able to correlate periods of floodplain stability and soil formation to dry climates and periods of floodplain aggradation to wet climates (Fig. 15.9b). In the absence of isotopic data, these relationships were also used to extend the climatic record to the last 2000 years (Fig. 15.9b). Several archaeological conclusions were drawn from the paleoenvironmental isotopic record along the Trinity River (Ferring, 1990, 1995):

1. Increased rates of alluvial aggradation during wet intervals led to the preservation of stratified prehistoric sites.
2. Decreased rates of aggradation during dry intervals correlated with periods of landscape stability, soil formation, and vertical compression of buried sites.
3. Renewed late Holocene floodplain aggradation in response to wetter conditions deeply buried most of the Early and Middle Archaic record.
4. The middle Holocene dry interval led to a possible decrease in plant biomass production and the reduction of bison as a resource for human subsistence.
5. Prehistoric population densities may have decreased in the middle Holocene due to a drier or warmer climate.

5.2.2. High Plains, Texas

Holliday (1995) conducted a stable isotope study on buried soils in dry valleys on the Southern High Plains of Northwest Texas. The purpose was to assess vegetation changes for the last 12,000 years and to make paleoclimatic correlations to other proxy established in the project area (e.g., pollen, palaeontological, environments of deposition). The focus of the isotopic investigation was on organic matter because of its ubiquity in the study area compared to pedogenic or lacustrine carbonates.

Holliday (1995) calculated the following contributions of C_4 grasses to soil biomass production from the isotopic record of buried soils: 28% in the latest Pleistocene to early Holocene, 79% in the middle Holocene, and nearly 0% during the last 500 years. Two caveats were provided in association with this data set. First, it was discovered that lacustrine and valley axis deposits supported plant communities with low C_4 production relative to valley margins, regardless of time period. The importance of this observation is that isotopes from similar environments of deposition should always be compared through time for paleoclimatic reconstruction. Second, the cause of the dramatic increase in $\delta^{13}C$ values for the last 500 years was unknown.

Geoarchaeologically, the major shift to higher C_4 plant production and inferred warmer and drier conditions in the middle Holocene may have contributed to lowering of water tables on the Southern High Plains, the reduction of available resources for bison, a decrease in prehistoric population densities, and the concentration of prehistoric occupations near natural springs (Holliday, 1989).

5.2.3. Central Texas

In the Southern Plains of Central Texas, Nordt et al. (1994) conducted an isotopic study on organic matter from surface and buried alluvial soils to assess vegetation and climatic changes during the last 15,000 years. The investigation was part of a geoarchaeological assessment of the prehistoric archaeological record in the Fort Hood Military Reservation (Nordt, 1992).

The isotopic data shown in Fig. 15.10 summarize most of the late Quaternary vegetation and climatic interpretations for the study area. These isotopic values were determined to a depth of over 3 m in a soil developed in alluvium deposited in the late Pleistocene (15,000 yr BP). Three notable shifts in the depth distribution of $\delta^{13}C$ values for soil organic matter are evident. It was inferred that at depths greater than 150 cm, Pleistocene plant signatures were emplaced mainly by detrital alluvial deposition. Thus, based on the lower profile having the most depleted $\delta^{13}C$ values ($-21‰$), greater contributions from cool-season grasses or trees occurred in the late Pleistocene. Early Holocene values from buried soils in other alluvial localities in the area consistently revealed $\delta^{13}C$ values

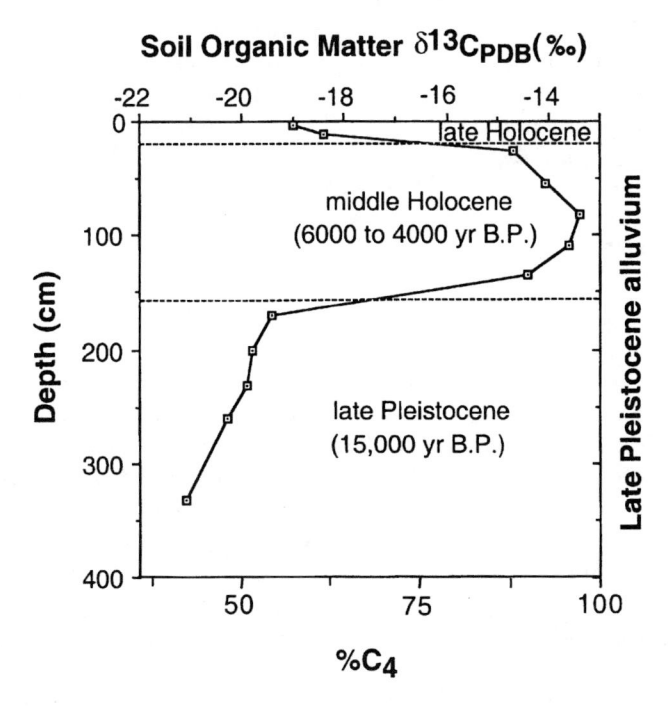

Figure 15.10. Stable C isotope depth distribution in a late Pleistocene alluvial soil from central Texas. Ages within the soil column represent estimated times of isotopic emplacement during soil formation, superimposed on sediment deposited 15,000 yr BP (modified from Nordt et al., 1994).

of around -18 to $-19\permil$, suggesting a gradual increase in C_4 biomass production during this time. The dramatic increase in $\delta^{13}C$ values in the middle profile (-13 to $-14\permil$) probably occurred from the pedogenic superimposition of organic matter from C_4 plants onto the late Pleistocene detrital isotopic values (Fig. 15.10). The increase in C_4 biomass production was interpreted as the onset of a middle Holocene warm interval because of the absence of isotopic values in any late Holocene surface or subsurface contexts that were greater than $-18\permil$. The shift to depleted $\delta^{13}C$ values and less C_4 biomass production in the upper horizons indicated a return to cooler conditions in the late Holocene (Fig. 15.10). The shallow depth marking the middle/late Holocene isotopic boundary probably resulted from high clay content in the subsoil, which protected isotopic values that were emplaced from deep-rooted C_4 grasses during the middle Holocene.

The climatic record established with stable C isotopes in the Fort Hood Military Reservation permitted general interpretations regarding the archaeological record. Widespread valley filling occurred during the middle Holocene, which coincides with increasing C_4 biomass production, warmer temperatures, and perhaps reduction in upland vegetative cover associated with erosion. This event resulted in the deep burial of early Archaic and Paleo-Indian sites in floodplain settings and probably the destruction of sites dating to this same time period in the uplands. Reduced sedimentation rates in the late Holocene, possibly in response to an increase in C_3 biomass production, cooler conditions, and landscape stability led to vertical compression of middle and late Archaic sites in a veneer of alluvium that buried the middle Holocene alluvium. It was also concluded that more resources were available for human subsistence in the early and late Holocene than in the middle Holocene. Further, most buried archaeological sites were discovered in late Holocene alluvium, possibly suggesting greater prehistoric population densities during a cooler and more biologically productive climatic interval.

5.3. East Africa

5.3.1. Kenya Rift Valley

Stable C isotopes from surface soils were used in the highlands of the Kenya Rift Valley to document fluctuations in the position of the forest–savanna boundary during the Holocene (Ambrose and Sikes, 1991). The primary objective was to assess the influence of shifts in this important ecotonal boundary on hunter–gatherer and agricultural settlement patterns. Samples were collected for isotopic analysis within the upper 50 cm of surface soils along an altitudinal transect. Numerous forest soils up to 300 m in elevation above the modern ecotonal line displayed a significant increase in $\delta^{13}C$ values with depth (e.g., from -24 to $-15\permil$), suggesting the presence of a C_4 grass community sometime in the past. By a combination of radiocarbon dating and correlation to lake and pollen records in adjacent valley floors, it was concluded that the $\delta^{13}C$-enriched values in the subsoil were registered sometime between 3000 and 6000 yr BP. Thus, the forest-savanna boundary was considerably higher in elevation during the middle

Holocene. The boundary migrated downslope to near its modern position sometime after 3000 yr BP.

Several conclusions were drawn from the isotopic data set that helped explain prehistoric settlement patterns in the highland study area of the Kenya Rift Valley:

1. There was diminished agricultural activity during the middle Holocene warm interval when the forest–savanna boundary was at least 300 m higher in elevation.
2. There was intensification of Neolithic agriculture after 3000 yr BP when conditions were cooler and the forest–savanna boundary had migrated below the elevation of the study area.
3. Early Holocene sites known to have formed during wet climatic intervals were located well below the modern forest–savanna line, indicating that occupation had occurred during cooler conditions.

This investigation clearly showed that shifting ecotonal boundaries associated with changing climates can have a profound affect on prehistoric settlement patterns.

5.3.2. Kenya and Tanzania

East Africa is thought to be the place where early hominids evolved from hominoids during the middle to late Miocene (Cerling et al., 1991a; Kingston et al., 1994). Although many theories have been proposed, numerous researchers believe that the development of hominid bipedality was an adaptive response to the flourishing of open grasslands (see Cerling et al., 1991a; Kingston et al., 1994). Unfortunately, varying interpretations about environmental conditions during the late Miocene in East Africa have been provided, and they range from closed forests inferred from faunal assemblages (Kappelman, 1991) to open grasslands based on paleosol morphology (Retallack et al., 1990).

To help resolve this issue, Cerling (1992) set out to determine the proportion of C_3 and C_4 plants growing in east Africa using the isotopic composition of pedogenic carbonate in a series of paleosols spanning the last 16 million years. The intent was to assess whether closed-canopy forested conditions, mixed-forest grasslands, or grasslands existed at the time of early hominid development. Based on low paleosol carbonate $\delta^{13}C$ values (-11 to $-15‰$), it was concluded that between 10 to 16 million yr BP a closed canopy dry woodland existed in the region (Cerling, 1992; Cerling et al., 1991a; Fig. 15.11a). The appearance of C_4 plants began shortly after 10 million yr BP ($> -11‰$), but still represented less than one half of the total biomass production (Cerling, 1992, Fig. 15.11a). Thus, the plant community at this time was interpreted as a grassy woodland. The $\delta^{18}O$ values of paleosol carbonate also revealed a shift to higher values shortly after 10 million yr BP (Fig. 15.11b). This change indicated somewhat drier or warmer conditions, which corresponded to the development of more drought-resistant C_4 grasses as determined by the $\delta^{13}C$ record. Pure C_4 grass communities did not develop until the Quaternary (Fig. 15.10b), leading Cerling to conclude that

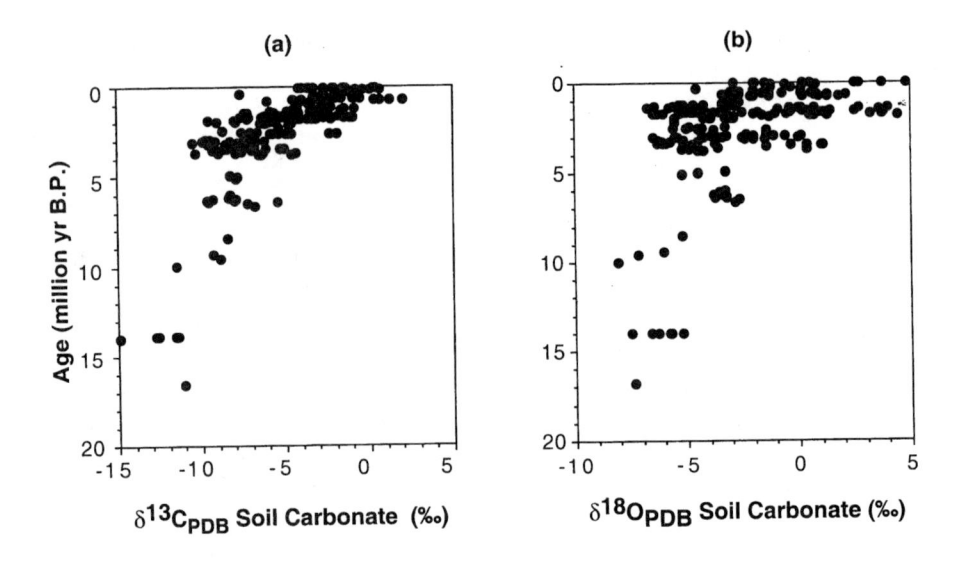

Figure 15.11. Stable C (a) and O (b) isotope values from pedogenic carbonate in paleosols in East Africa from localities spanning the last 16 million years (reproduced from Cerling, 1992; copyright by Academic Press, Inc.).

although hominids may have co-evolved with the spread of grasslands, a mixed C_3/C_4 ecosystem was present during this important period.

Kingston et al. (1994) provided a slightly different isotopic interpretation of paleosols spanning the Miocene in East Africa. They did not find significant shifts in the proportion of C_3 and C_4 plant biomass production during the last 14 million yr BP based on the analysis of paleosols from a wide variety of environmental settings. From this, they concluded that a heterogeneous ecosystem existed continuously during the last 14 million years, and that changes in vegetation probably did not play an important role in the development of hominids in East Africa during the Miocene.

Regardless of the controversy surrounding human evolution and climate, stable C and O isotopes of pedogenic carbonate still proved to be a powerful means of reconstructing past vegetation communities in East Africa. Because of this work, it is now known that mixed C_3 and C_4 plant communities have existed in the area at least since the Pliocene. This is especially important given the paucity of available pollen and palaeontological data and given the uncertainties in using soil morphology as a climatic indicator.

5.4. China

Homo erectus, thought to have developed in tropical/subtropical climates in east Africa, was occupying central China by about 1.15 million yr B.P. (Wang et al., 1997). Conflicting climatic theories prevail for China for this time period and include the presence of a glacial steppe (sedimentological data), warm temperate woodland (fossil pollen), and subtropical forest (faunal evidence). Using a combination of stable C and O isotopes of carbonates from paleosols, Wang et al. (1997) proposed to determine which of these climatic theories best described conditions at the time of occupation in the study area. The relationship between stable C and O isotopes in paleosols from the last glacial/interglacial episode was used as a modern analogue because climates are well known from this time period. These relationships were then compared with the paleosol isotopic record associated with early Pleistocene occupation in the area.

Isotopic ratios revealed low $\delta^{18}O$ values (pedogenic carbonate) in association with low soil organic matter $\delta^{13}C$ values 1.15 million yr BP. According to the modern analogue (last glacial/interglacial), this isotopic combination suggested a dominantly C_3 nonglacial environment influenced by monsoonal precipitation emanating from the Indian Ocean. From this it was inferred that a cold/cool dry winter and warm/mild semihumid summer and fall existed at this time. This interpretation clearly placed the *Homo erectus* occupation in a temperate climate regime 1.15 million years ago in China, which is 150,000 years earlier than the first *Homo* occupation in temperate climates of Europe.

5.5. Summary of Case Studies

As shown in this section the use of stable C isotopes from soil organic matter or pedogenic carbonate for paleoenvironmental research and archaeological application is site specific and depends on the C source being analyzed. The method is only applicable to those areas that have experienced significant shifts in the ratio of C_4 to C_3 plants. For example, in many semiarid to subhumid regions of the Great Plains in the United States, C_3 or mixed C_3/C_4 plant communities transformed into nearly pure C_4 communities during the middle Holocene. In contrast, C_4 grasses were replaced by C_3 desert shrubs in parts of the Chihuahuan Desert of New Mexico during the middle Holocene. In either case, changes in the ratio of C_3 to C_4 grasses were effectively able to portray a major climatic shift during the middle Holocene.

Most investigations reviewed in this chapter applied paleosol stable C isotopes to organic matter (Ambrose and Sikes, 1991; Holliday, 1995) or stable C and O isotopes to pedogenic carbonate (Cerling, 1992; Humphrey and Ferring, 1994; Monger, 1995; Wang, et al., 1997) for solving archaeological problems. In contrast, some investigators used organic matter from modern surface soils for isotopic analysis as applied to archaeological research (Nordt, 1998; Nordt et al., 1994). With this application, the surface soils must be young enough to have formed under only one climatic condition, or if not, a series of surface soils of different ages must be analyzed in order to partition vegetation and climatic events adequately. In all cases, resource availability and periods of erosion and deposition were used to predict preservation potentials and settlement patterns of the prehistoric archaeological record.

The $\delta^{18}O$ of pedogenic carbonate was utilized in four investigations presented in this chapter. The application to late Cenozoic paleosols in East Africa revealed an increase in $\delta^{18}O$ of paleosol carbonate beginning 10 million yr BP that may have tracked the spreading of C_4 grasslands (Cerling, 1992). In China, when combined with $\delta^{13}C$ values of soil organic matter, the $\delta^{18}O$ of pedogenic carbonate was used to demonstrate that a temperate climate existed at the time of *Homo erectus* occupation of central China (Wang et al., 1997). In two studies of Holocene paleosols, however, no shifts in $\delta^{18}O$ were observed even though changes in the $\delta^{13}C$ values of organic matter was evident (Cole and Monger, 1994; Humphrey and Ferring, 1994). This indicates that $\delta^{18}O$ of pedogenic carbonate may not be a viable proxy for climate change in the Holocene in some areas.

6. Conclusions

The use of stable C and O isotopes in soil organic matter and of pedogenic carbonate for archaeological research is in its infancy. The primary application thus far has been as a tool for paleoenvironmental reconstruction. Examples include inferring changes in vegetation communities through time and estimating paleotemperatures and paleoprecipitation levels. This information has in turn been used to estimate rates of alluvial aggradation and upland erosion and associated prehistoric preservation potentials, to estimate the availability of prehistoric resources such as plants and water, and to assess the possibility of human evolution in association with vegetation changes.

Further work is needed in the field of stable C and O isotopes in soils and in paleosols for archaeological research. Most important, a larger database is needed from regions other than those discussed in this chapter. Regardless, isotopic data may be combined with other indicators to develop a robust paleoclimatic record that can be used to enhance archaeological interpretations with respect to settlement patterns, subsistence strategies, and preservation potentials.

ACKNOWLEDGMENTS. I thank Darden Hood, Austin Long, and Sue Trumbore for insightful conversations regarding radiocarbon dating techniques, Chris Caran for theoretical discussions on ^{13}C and ^{14}C laboratory procedures, and Garry Running for generously discussing his stable isotope data from North Dakota paleosols. I also thank Tom Boutton, Reid Ferring, Paul Goldberg, and Vance Holliday for reviewing an earlier draft of this manuscript. Also thanks to Thure Cerling for kindly donating isotopic figures.

7. References

Ambrose, S. H., and DeNiro, M. J., 1989, Climate and Habitat Reconstruction Using Stable Carbon and Nitrogen Isotope Ratios of Collagen in Prehistoric Herbivore Teeth from Kenya, *Quaternary Research* 31:407–422.

Ambrose, S. H., and Sikes, N. E., 1991, Soil Carbon Isotope Evidence for Holocene Habitat Change in the Kenya Rift Valley, *Science* 253:1402–1405.

Amundson, R. G., and Lund, L. J., 1987, The Stable Isotope Chemistry of a Native and Irrigated Typic Natrargid in the San Joaquin Valley of California, *Soil Science Society of America Journal* 51:761–767.

Amundson, R., Stern., L., Baisden, T., and Wang, Y., 1998, The Isotopic Composition of Soil and Soil-Respired CO_2, *Geoderma* 82:83–114.

Amundson, R. G., Chadwick, O. A., Sowers, J. M., and Doner, H. E., 1988, The Relationship Between Modern Climate and Vegetation and the Stable Isotope Chemistry of Mojave Desert Soils, *Quaternary Research* 29:245–254.

Amundson, R. G., Chadwick, O. A., Sowers, J. M., and Doner, H. E., 1989, The Stable Isotope Chemistry of Pedogenic Carbonate at Kyle Canyon, Nevada, *Soil Science Society of America Journal* 53:201–210.

Balesdent, J., and Mariotti, A., 1996, Measurement of Soil Organic Matter Turnover Using ^{13}C Abundance. In *Mass Spectrometry of Soils*, edited by T.W. Boutton and S. Yamasaki, pp. 83–112. Marcel Dekker, New York.

Barrie, A., and Prosser, S. J., 1996, Automated Analysis of Light-Element Stable Isotopes by Isotope Ratio Mass Spectrometry. In *Mass Spectrometry of Soils*, edited by T. W. Boutton and S. Yamasaki, pp. 1–46. Marcel Dekker, New York.

Bernoux, M., Cerri, C. C., Neill, C., and de Moraes, J. F. L., 1998, The Use of Stable Carbon Isotopes for Estimating Soil Organic Matter Turnover Rates, *Geoderma* 82:43–58.

Birkeland, P. W., 1984, *Soils and Geomorphology*, Oxford University Press, New York.

Boutton, T. W., 1991b, Stable Carbon Isotope Ratios of Natural Materials (II). Atmospheric, Terrestrial, Marine, and Freshwater Environments. In *Carbon Isotope Techniques*, edited by D. C. Coleman and B. Fry, pp. 173–185. Academic Press, San Diego, CA.

Boutton, T. W.,1991a, Stable Carbon Isotope Ratios of Natural Materials (I). Sample Preparation and Mass Spectrometric Analysis. In *Carbon Isotope Techniques*, edited by D. C. Coleman and B. Fry, pp. 155–171. Academic Press, San Diego, CA.

Boutton, T.W., 1996, Stable Carbon Isotope Ratios of Soil Organic Matter and their Use as Indicators of Vegetation and Climate Change. In *Mass Spectrometry of Soils*, edited by T. W. Boutton and S. Yamasaki, pp. 47–82. Marcel Dekker, New York.

Boutton, T. W., Harrison, A. T., and Smith, B. N., 1980, Distribution of Biomass of Species Differing in Photosynthetic Pathway Along an Altitudinal Transect in Southeastern Wyoming Grassland, *Oecologia* 45:287–298.

Boutton, T. W., Archer, S. R., Midwood, A. J., Zitzer, S. F., and Bol, R., 1998, $\delta^{13}C$ Values of Soil Organic Carbon and their Use in Documenting Vegetation Change in a Subtropical Savanna Ecosystem, *Geoderma* 82:5–42.

Buol, S. W., Hole, F. D., McCracken, R. J., and Southard, R. J., 1997, *Soil Genesis and Classification*, 4th ed. Iowa State University Press, Ames.

Cerling, T. E., 1984, The Stable Isotopic Composition of Modern Soil Carbonate and its Relationship to Climate, *Earth and Planetary Science Letters* 71:229–240.

Cerling, T. E., 1992, Development of Grasslands and Savannas in East Africa during the Neogene, *Palacogesgraphy, Palaeoclimatogogy, Palaeoecology* 97:241–247.

Cerling, T. E., and Hay, R. L., 1986, An Isotopic Study of Paleosol Carbonate from Olduvai Gorge, *Quaternary Research* 25:63–78.

Cerling, T. E. and Quade, J., 1993, Stable Carbon and Oxygen Isotopes in Soil Carbonates. In *Climate Change in Continental Isotopic Records*, edited by P. K. Swart, K. C. Lohmann, J. McKenzie, and S. Sarin, pp. 217–232, Geophysical Monograph 78. American Geophysical Union, Washington, DC.

Cerling, T. E., Quade, J., Wang, Y., and Bowman, J. R., 1989, Carbon Isotopes in Soils and Palaeosols as Ecology and Palaeoecology Indicators, *Nature* 341:138–139.

Cerling, T. E., Quade, J., Ambrose, S. H., and Sikes, N., 1991a, Fossil Soils, Grasses, and Carbon Isotopes from Fort Ternan: Grassland or Woodland? *Journal of Human Evolution* 21:295–306.

Cerling, T. E., Solomon, D. K., Quade, J., and Bowman, J. R., 1991b, On the Isotopic Composition of Carbon in Soil Carbon Dioxide, *Geochimica Cosmochimica et Acta* 55:3403–3405.

Cerling, T. E., Wang, Y., and Quade, J., 1993, Expansion of C_4 Ecosystems as an Indicator of Global Ecological Change in the Late Miocene, *Nature* 361:344–345.

Cole, D. R., and Monger, H. C., 1994, Influence of Atmospheric CO_2 on the Decline of C_4 Plants During the Last Deglaciation, *Nature* 368:533–536.

Craig, H., 1957, Isotopic Standards for Carbon and Oxygen and Correction Factors for Mass Spectrometric Analysis of Carbon Dioxide, *Geochimica Cosmochimica et Acta* 12:133–149.

Craig, H., and Craig, V., 1972, Greek Marbles: Determination of Provenance by Isotopic Analysis, *Science* 176:401-403.

Dansgaard, W., 1964, Stable Isotopes in Precipitation, *Tellus* 16:436–468.

Deines, P., 1980, The Isotopic Composition of Reduced Organic Carbon. In *Handbook of Environmental Geochemistry (1) The Terrestrial Environment*, edited by P. Fritz and J. C. Fontes, pp. 329–406. Elsevier, Amsterdam.

Dorr, H., and Munnich, K., 1980, Carbon-14 and C-13 in soil CO_2, *Radiocarbon* 22:909–918.

Dzurec, R. S., Boutton, T. W., Caldwell, M. M., and Smith, B. N., 1985, Carbon Isotope Ratios of Soil Organic Matter and Their Use in Assessing Community Composition Changes in Carlew Valley, Utah, *Oecologia* 66:17–24.

Ferring, C. R., 1990, Archaeological Geology of the Southern Plains. In *Archaeological Geology of North America*, edited by N. P. Lasca and J. Donahue, pp. 253–266. Geological Society of America, Centennial Special Volume 4, Boulder, CO.

Ferring, C. R., 1995, Middle Holocene Environments, Geology, and Archaeology in the Southern Plains. In *Archaeological Geology of the Archaic Period in North America*, edited by E. A. Bettis, pp. 21–36. Geological Society of America Special Paper No. 297, Boulder, CO

Fredlund, G. G., and Tieszen, L. L., 1997, Phytolith and Carbon Isotope Evidence for Late Quaternary Vegetation and Climate Change in the Southern Black Hills, South Dakota, *Quaternary Research* 47: 206–217.

Hard, R. J., Mauldin, R. P., and Raymond, G. R., 1996, Mano Size, Stable Carbon Isotope Ratios, and Macrobotanical Remains as Multiple Lines of Evidence of Maize Dependence in the American Southwest, *Journal of Archaeological Method and Theory* 3:253–318.

Hays, P. D., and Grossman, E. L., 1991, Oxygen Isotopes in Meteoric Calcite Cements as Indicators of Continental Paleoclimate, *Geology* 19:441–444.

Herz, N., 1990, Stable Isotope Geochemistry Applied to Archaeology. In *Archaeological Geology of North America*, edited by N. P. Lasca and J. Donahue, pp. 585–596. Geological Society of America Centennial Special Volume 4, Boulder, CO.

Herz, N., and Dean, N. E., 1986, Stable Isotopes and Archaeological Geology, the Carrara Marble, Northern Italy, *Applied Geochemistry* 1:139–151.

Herz, N., and Garrison, E. G., 1998, *Geological Methods for Archaeology*, Oxford University Press, New York.

Hesterberg, R., and Siegenthaler, U., 1991, Production and Stable Isotopic Composition of CO_2 in a Soil Near Bern, Switzerland, *Tellus* 43B:197–205.

Hoefs, J., 1987, *Stable Isotope Geochemistry*, 3rd ed. Springer-Verlag, New York.

Holliday, V. T., 1989, Middle Holocene Drought on the Southern High Plains, *Quaternary Research* 31:74–82.

Holliday, V. T., 1995, Stratigraphy and Paleoenvironments of Late Quaternary Valley Fills on the Southern High Plains, *Geological Society of America Memoir 186, Boulder, CO*.

Humphrey, J. D. and Ferring, C. R., 1994, Stable Isotopic Evidence for Latest Pleistocene and Holocene Climatic Change in North-Central Texas, *Quaternary Research* 41:200–213.

Jenny, H., 1941, *Factors of Soil Formation*. McGraw-Hill, New York.

Kappelman, J., 1991, The Paleoenvironments of *Kenyapithecus* at Fort Ternan, *Journal of Human Evolution* 20:95–129.

Kelly, E. F., Amundson, R. G., Marino, B. D., and DeNiro, M. J., 1991a, Stable Carbon Isotopic Composition of Carbonate in Holocene Grassland Soils, *Soil Science Society of America Journal* 55:1651–1658.

Kelly, E. F., Amundson, R. G., Marino, B. D., and DeNiro, M. J., 1991b, The Stable isotope ratios of carbon in phytoliths as a quantitative method of monitoring vegetation and climatic change, *Quaternary Research* 35:222–233.

Kelly, E. F., Yonker, C., and Marino, B., 1993, Stable Carbon Isotope Composition of Paleosols: Application to Holocene. In *Climate Change in Continental Isotopic Records*, edited by P.K. Swart,

K.C. Lohmann, J. McKenzie, and S. Savin, pp. 233–239. Geophysical Monograph 78, American Geophysical Union, Washington, DC.

Kingston, J. D., Marino, B. D., and Hill, A.,1994, Isotopic evidence for Neogene Hominid Paleoenvironments in the Kenya Rift Valley, *Science* 264:955–959.

Lowe, J. J., and Walker, M. J. C., 1984, *Reconstructing Quaternary Environments*, Longman, New York.

Marino, B. D., McElroy, M. B., Salawitch, R. J., and Spaulding, W. G, 1992, Glacial-to-Interglacial Variations in the Carbon Isotopic Composition of Atmospheric CO_2, *Nature* 357:461–466.

Marion, G. M., Introne, D. S., and Van Cleve, K., 1991, The Stable Isotope Geochemistry of $CaCO_3$ on the Tanana River Floodplain of Interior Alaska, U.S.A.: Composition and Mechanisms of Formation, *Chemical Geology* 86:97–110.

Melillo, J. M., Aber, J. D., Linkins, A. E., Ricca, A., Fry, B., and Nadelhoffer, K. F., 1989, Carbon and Nitrogen Dynamics Along a Decay Continuum: Plant Litter to Soil Organic Matter. In *Ecology of Arable Land: Perspectives and Challenges*, edited by M. Clarholm and L. M. Bergstrom, pp. 53–62. Kluwer Academic, Dordrecht, Netherlands.

Midwood, A. J., and Boutton, T. W., 1998, Soil Carbonate Decomposition by Acid has Little Effect on $\delta^{13}C$ of Organic Matter, *Soil Biology and Biochemistry* 30:1301–1307.

Monger, H. C., 1995, Pedology in Arid Lands Archaeological Research: An Example from Southern New Mexico-Western Texas. In *Pedological Perspectives in Archaeological Research*, edited by M.E. Collins, B. J. Carter, B. G. Gladfelter, and R. J. Southard, pp. 35–50. Soil Science Society of America Special Publication Number 44, Madison, WI.

Nadelhoffer, K. F., and Fry, B., 1988, Controls on Natural Nitrogen-15 and Carbon-13 Abundances in Forest Soil Organic Matter, *Soil Science Society of America Journal* 52:1633–1640.

Nordt, L. C., 1992, *Archaeological Geology of the Fort Hood Military Reservation, Fort Hood, Texas*. U.S. Army Fort Hood, Archaeological Resource Management Series, Research Report 25. Texas A&M University, College Station.

Nordt, L. C. 1998, Geoarchaeology of the Rio Grande and Elm Creek in the Vicinity of Site 41MV120. In *41MV20: A Stratified Late Archaic Site in Maverick, County, Texas*, edited by B. J. Vierra, pp. 43–77. Archaeological Survey Report No. 251, Center for Archaeological Research, The University of Texas at San Antonio.

Nordt, L. C., Boutton, T. W., Hallmark, C. T., and Waters, M. R., 1994, Late Quaternary Vegetation and Climate Changes in Central Texas Based on the Isotopic Composition of Organic Carbon, *Quaternary Research* 41:109–120.

Nordt, L. C., Wilding, L. P., Hallmark, C. T., and Jacob, J. S., 1996, Stable Carbon Isotope Composition of Pedogenic Carbonate and their Use in Studying Pedogenesis. In *Mass Spectrometry of Soils*, edited by T. W. Boutton and S. Yamasaki, pp. 133–154. Marcel Dekker, New York.

Nordt, L. C., Hallmark, T. C., Wilding, L. P., and Boutton, T. W., 1998, Quantifying Pedogenic Carbonate Accumulations Using Stable Carbon Isotopes, *Geoderma* 82:115–136.

Pendall, E., and Amundson, R., 1990, The Stable Isotope Chemistry of Pedogenic Carbonate in an Alluvial Soil from the Punjab, Pakistan, *Soil Science* 149:199–211.

Pendall, E. G., Harden, J. W., Trumbore, S. E., and Chadwick, O. A., 1994, Isotopic Approach to Soil Carbonate Dynamics and Implications for Paleoclimatic Interpretations, *Quaternary Research* 42:61–70.

Penuelas, J., and Azcon-Bieto, J., 1992, Changes in Leaf $\delta^{13}C$ of Herbarium Plant Specimens During the Past 3 Centuries of CO_2 Increase, *Plant Cell Environments* 15:485–489.

Quade, J., Cerling, T. E., and Bowman, J. R., 1989, Systematic Variations in the Carbon and Oxygen Isotopic Composition of Pedogenic Carbonate along Elevation Transects in the Southern Great Basin, United States, *Geological Society of America Bulletin* 101:464–475.

Retallack, G. J., Dugas, K. P., and Bestland, E. A., 1990, Fossil Soils and Grasses of a Middle Miocene East African Grassland, *Science* 247:1325–1328.

Rozinski, K., Araquas-Araguas, L., and Gonfiantini, R.,1993, Isotopic Patterns in Modern Global Precipitation. In *Climate Change in Continental Isotopic Records*, edited by P.K. Swart, K. C. Lohmann, J. McKenzie, and S. Savin, pp. 1–36. Geophysical Monograph 78, American Geophysical Union, Washington, DC.

Shackleton, N. J., 1973, Oxygen Isotope Analysis as a Means of Determining Season of Occupation of Prehistoric Midden Sites, *Archaeometry* 15:133–157.

Sharpenseel, H., and Neue, H., 1984, Use of Isotopes in Studying the Dynamics of Organic Matter in Soils. In *Organic Matter and Rice*, pp. 273–310. International Rice Research Institute, Manila, the Philippines,

Smith, B. N., and Epstein, S., 1971, Two Categories of $^{13}C/^{12}C$ ratios for Higher Plants, *Plant Physiology* 47:380–384.

Teeri, J. A., and Stowe, L. G., 1976, Climatic Patterns and the Distribution of C_4 Grasses in North America, *Oecologia* 23:1–12.

Tieszen, L. L., Senyimba, M., Imbamba, S., and Troughton, J., 1979, The Distribution of C_3 and C_4 Grasses and Carbon Isotope Discrimination along an Altitudinal and Moisture Gradient in Kenya, *Oecologia* 37:337–350.

Trumbore, S. E., 1996, Applications of Accelerator Mass Spectrometry to Soil Science. In *Mass Spectrometry of Soils*, edited by T. W. Boutton and S. Yamasaki, pp. 311–340. Marcel Dekker, New York.

van der Merwe, N. J., 1982, Carbon Isotopes, Photosynthesis, and Archaeology, *American Scientist* 70:595–606.

van der Merwe, N. J., and Vogel, J., 1977, Isotopic Evidence for Early Maize Cultivation in New York State, *American Antiquity* 42:238–242.

Vepraskas, M. J., 1994, *Redoximorphic Features for Identifying Aquic Conditions*. North Carolina Agricultural Research Service, North Carolina State University, Technical Bulletin 301. Raleigh, NC.

Wang, Y., Amundson, R., and Trumbore, S., 1996a, Radiocarbon Dating of Soil Organic Matter. *Quaternary Research* 45:282–288.

Wang, Y., Cerling, T. E., and Effland, W. R., 1996b, Stable Isotope Ratios of Soil Carbonate and Soil Organic Matter as Indicators of Forest Invasion of Prairie Near Ames, Iowa, *Oecologia* 95:365-369.

Wang, H., Ambrose, S. H., Liu, C. L. J., and Follmer, L. R., 1997, Paleosol Stable Isotope Evidence for Early Hominid Occupation of East Asian Temperate Environments, *Quaternary Research* 48:228–238.

West, L. T., Drees, L. R., Wilding, L. P., and Rabenhorst, M. C., 1988, Differentiation of Pedogenic and Lithogenic Carbonate Forms in Texas, *Geoderma* 43:271–287.

Wilding, L. P., West, L. T., and Drees, L. R., 1990, Field and Laboratory Identification of Calcic and Petrocalcic Horizons. In *Proceedings of the Fourth International Soil Correlation Meeting (ISCOM IV) Characterization, Classification, and Utilization of Aridisols. Part A: Papers*, edited by J. M. Kimble and W. D. Nettleton, pp. 79–92. U.S. Department of Agriculture, Soil Conservation Service, Lincoln, NE.

Sourcing Lithic Artifacts by Instrumental Analysis

NORMAN HERZ

1. Introduction

The best preserved artifacts of the oldest hominid sites are lithics found at the earliest Oldowan sites of Africa, securely dated by K/Ar dating and by magnetic polarity stratigraphy at 2.5 my. The first stone tool industries of the sites used local volcanics, most of which were trachyte (Semaw et al., 1997) but also some rhyolite and basalt found in nearby stream conglomerates.

Lithic artifacts remain identifiable long after alluvial burial, fire, and even prolonged weathering, and they can yield much information on a culture. For many preliterate societies, lithics may be the only material artifacts available. The ability to source lithics will yield important information on the technology and trade patterns of the time.

More often in the past, but occasionally even now, lithic artifacts are assigned a provenance based on subjective esthetic judgments—best known as a "gut feeling"—and are used to "prove" a preconceived theory. This practice has led to unresolvable controversies. An example of this practice is the head of Pan limestone sculpture in the Cleveland Museum (Herz et al., 1989). It was allegedly found on the north slope of the Athenian Acropolis, near the site where the monument commemorating the victory over the Persians in 490 B.C. had been erected. Many classical archaeologists considered it among the foremost artistic

---

NORMAN HERZ • Department of Geology, University of Georgia, Athens, Georgia 30602.

Earth Sciences and Archaeology, edited by Paul Goldberg, Vance T. Holliday, and C. Reid Ferring. Kluwer Academic/Plenum Publishers, New York, 2001.

memorials to the victory, and as such it figured in several art historical reference works.

The sculpture was studied with petrographic thin sections and was analyzed by neutron activation analysis (NAA), scanning electron microscope analysis (SEM), and stable isotope analysis (SIA). The results were compared to documented fragments of the monument itself, carved from Acropolis limestone, now in the Parthenon Museum. The Cleveland head was found to differ unequivocally from the Acropolis limestone in petrology, paleontology, and trace elements; variations in isotopic values were too great to draw any conclusions about exact sources. The head was carved from a lower Eocene biosparite limestone deposited in deep marine water; the Acropolis pieces were post-lower Eocene, shallow brackish water oolitic microsparite. The moral to draw from the study is that, in addition to field evidence, questions of provenance and authenticity can often be resolved by routine laboratory analyses.

The first step in any provenance study of lithic artifacts is to accumulate databanks of the physical characteristics of potential raw materials sources. Samples are systematically collected from quarries or worked outcrops and are analyzed by methods that are appropriate for the rock type (Shackley, 1998). Hopefully, such a study will find distinctive geochemical, petrographic, or isotopic fingerprints that can identify each potential source. The artifacts can then be analyzed by the same methodology and compared to the databases for a matching source. A collection of lithic artifacts correctly identified petrographically, geochemically, and to source can often yield information on the following:

1. An approximate time of fabrication of an object. In places in the eastern Mediterranean, obsidian totally replaced chert in the transition from Mesolithic to the Neolithic period. In Franchthi Cave, Greece for example, the obsidian to flint ratio is 10% in the upper Mesolithic (7250 to 6000 B.C.) but reaches 95% by the final Neolithic (4000 to 3000 B.C., Jacobsen, 1976). Chert sources were invariably local but in most places obsidian had to be imported from outside sources. Later, in Classical Greek and Roman times, only the purest white marbles were selectively quarried for important statuary and monuments. Because the time of operation of most quarries is well documented, the monuments can also be dated.

2. Information on trading patterns. In Franchthi Cave in Mesolithic Greece, obsidian artifacts suddenly appear, obtained from the island of Milos, some 400 km away in the Aegean (Renfrew and Aspinall, 1990). At the same time, bones of large deep-sea fish are also found on the site showing a change in food sources and the development of a seafaring trade (Jacobsen, 1976).

3. Insight into changing esthetic tastes. The earliest societies used locally available stone for strictly utilitarian purposes. With time, ornamental stone and gemstones became highly prized for jewelry, grave objects, and as personal and civic decoration. Widespread exchange of soapstone (Truncer et al., 1998) and copper in the New World and Baltic amber in Neolithic–Bronze Age Greece, Egypt, and Mesopotamia in the second millennium B.C. (Beck, 1986) evince changing tastes and the early development of sophistication in societies.

4. Improving technology. Extracting, working, processing, and shipping

stone requires technology. The simplest quarrying technique, widespread in Neolithic Europe, was to select blocks of stone that could easily be worked out from an outcrop using antlers, harder stone, or wooden wedges, taking advantage of the natural foliation and jointing of the blocks (Waelkens, 1990). Tools of stone and metal were developed with time so that new varieties of material, including harder or more compact stone, could be extracted. Means of transportation over both land and sea were improved, leading to such remarkable achievements as the movement and emplacement of huge stone blocks overland to Stonehenge and over hundreds of sea miles to Rome.

5. Detect modern forgeries or ancient copies of original works. A continuing problem for museums and archeologists is the authenticity of an artifact. If a lithic object is allegedly from a certain culture and time period, and if databases exist of well-documented artifacts and their sources, then a comparative analysis of the object in question should be the first test. Obsidian from Central Hungary should not be found in a Neolithic Aegean island site nor should Roman marble be used for an Archaic Greek sculpture.

6. Assembling broken fragments of artifacts. The correct assembly of the fragments of an artifact is a common vexing problem for museums and archaeologists. Important artifacts including sculpture and inscriptions may have been broken in the course of burial, during transport by natural processes, or by humans. Each broken piece should have identical signatures in order to be correctly associated.

2. Instrumental Analysis

A great variety of analytical systems are now used for determining the provenance of artifacts and their source materials (e.g., Pollard and Heron, 1996, chap. 2). The first necessary step should be a field description by hand lens followed by thin section study with the petrographic microscope. This "eyeball" study, unfortunately frequently overlooked by archaeologists not trained in mineralogy and in optical microscopy, affords the first useful data for a secure provenance assignment. The importance of a correct petrographic description, (e.g., mineralogy, rock texture, and structures), cannot be overestimated. If this is not done correctly and carefully, it will be impossible to correlate artifacts from one site to another. Kempe and Harvey (1983) is a good general reference for petrography (rock description) and petrology (rock genesis) of lithic artifacts.

Petrographic study is a bridge between field/hand lens study and microanalysis. It is essential before the most appropriate type of geochemical analysis can be selected. Most analytical techniques utilize different parts of the electromagnetic spectrum. The sample is irradiated and the response of individual elements to specific wavelengths is recorded. Detailed explanations of the instrumental geochemical techniques cited can be found in Herz and Garrison (1998). Depending on the system, different information can be obtained, such as the following: (1) the elements present and their amounts, by instrumental NAA; (2) the minerals comprising the sample, by X-ray diffraction analysis (XRD); or the fabric and chemical composition of the sample itself, by SEM and the electron microprobe. Analytical systems using the visible or near-visible part of the electromagnetic spectrum include optical emission spectroscopy (OES), atomic absorption spectroscopy (AAS), and inductively coupled plasma emission spectroscopy (ICP). More sensitive varieties of ICP are ICP-MS, whereby a mass spectrometer is coupled to an ICP, and ICP-AES, whereby an atomic emission spectrometer is linked to an ICP in which temperatures >8,000–10,000°C are reached (Tykot and Young, 1996).

Cathodoluminescence (CL) is based on photon emission of in the visible range of the electromagnetic spectra after excitation by high-energy electrons in an electron microscope. The irradiated mineral produces luminescence colors that originate from various impurities, such as manganese within the carbonate crystal lattice (Barbin et al., 1992). The luminescence has been shown to be distinctive for many different Mediterranean marble quarries and is used as another criterion for sourcing artifacts.

X rays are used in X-ray diffraction (XRD) and in X-ray fluorescence (XRF) with its applications in the scanning electron microscope (SEM) and the electron microprobe ("probe"). In proton-induced X-ray emission (PIXE) analysis, compared to conventional XRF, the sample is bombarded with ionized hydrogen nuclei or protons instead of electrons as the incident particles. PIGME (proton-induced gamma-ray emission) is a variation of PIXE. Each crystalline material diffracts X rays differently, as determined by its unique unit cell. In XRD analysis diffraction angles are measured allowing the minerals to be identified. Glasses such as the matrix of obsidian will not diffract but included phenocrysts present will.

XRF determines the elemental composition of a powdered sample, as does SEM and the probe. The latter two however, generally use intact pieces of the sample allowing a visual observation, as well as energy-dispersive (ED) or wavelength-dispersive (WD) detectors, which can carry out detailed elemental analysis at the same time. WD has higher detection limits, (0.05 to 0.26% of the element present), than ED. Although neither is as sensitive as normal XRF analysis, they have the added advantage of selectability of the area to be analyzed using optical or electron microscopy.

Other important analytical systems include NAA, electron spin resonance spectroscopy (ESR), nuclear magnetic resonance (NMR), and stable isotope mass spectrometry. ESR and NMR are resonance techniques in which the sample is subjected to two magnetic fields, one stationary and the other varying at a known frequency. The interaction of the subatomic particles with the magnetic fields produces a characteristic spectrum. ESR has been used as a dating technique (Schwarcz and Grün, 1992), as well as in lithic thermal (Maniatis and Mandi, 1992) and provenance studies (Armiento et al., 1997). NMR is a popular technique for studies of organic material, such as amber (Lambert et al., 1996)

NAA involves (1) irradiation of a sample with neutrons to produce radioactive isotopes; (2) measuring the radioactive emanations from the sample; and (3) determining the radioactive isotopes present through the energy, type, and half-life of the emanations produced by radiation. Because each element has a distinctive radioactive decay pattern, the amount and kind of isotopes produced will identify the presence of an element and its amount.

The method was first used in the late 1950s in Mediterranean ceramic studies (Sayre and Dodson, 1957) and quickly became a major technique for analysis of artifacts. In recent years, the research reactor at the University of Missouri, with support from the National Research Council, has been one of the most productive laboratories in the analysis of lithic archaeological artifacts (Glascock, 1992).

Mass spectrometric techniques have been used extensively in archaeology for age determination, paleodiet studies, and sourcing lithics using stable isotopic ratios of carbon, oxygen, sulfur, strontium and lead (Stos-Gale, 1995; see also Nordt, Chapter 15, this volume). Measurements of stable isotopic ratios are carried out with a mass spectrometer that, in the newer, state-of-the-art machines, uses less than 5 mg of sample for an analysis, an amount readily acquired without causing harm to artifacts. Typically, the precise measurement of the isotopic ratios $^{18}O/^{16}O$ and $^{13}C/^{12}C$ is expressed as a deviation from an international standard, either the PDB or SMOW. This deviation, called δ, expressed as $\delta^{13}C$ or $\delta^{18}O$ is measured in parts per thousand (or per mil, ‰). Stable isotopes have been used especially to determine the provenance of (a) carbonates by O, C, and Sr (Herz, 1992), (b) chert by O (Shaffer and Tankersley, 1989), (c) gypsum by Sr and S (Gale et al., 1988), and (d) obsidian by Sr (Gale, 1981).

Sourcing of artifacts by magnetic properties is a promising technique that has many advantages, including low cost—0.1% that of NAA—and being nondestructive (Tarling, 1990). Magnetic fingerprinting is based on three simply measured magnetic properties—initial remanance, low-field susceptibility, and high-field remanance—that have already been shown to be characteristic for most geologic sources.

3. Determining Provenance of Lithic Materials

3.1. Obsidian

Obsidian is a glassy product of a highly viscous acidic magma, that is, one with high SiO_2 and alkalies. It is typically dacitic, rhyodacitic, or rhyolitic in composition, a result of rapid quenching of a magma that either extruded on to the surface of the earth or intruded at shallow depths of only a few kilometers, at high temperatures of the order of 750° to 950°C, and contains less than 1% water by weight (Williams et al., 1982). Its index of refraction, which has sometimes been used as a field method to differentiate source flows, is low and varies according to the chemical composition of the flow. Although phenocrysts may be present, high viscosity inhibits crystal growth so aphyric (lacking phenocrysts) varieties are common. With abundant crystal growth, the rocks grade into rhyolite. If the flow was high in gasses, a highly vesiculated pumice may result.

Obsidian is normally shiny and dark, black, or gray, although compositional varieties may be colorless, red, or green. It possesses a conchoidal fracture that makes it easy to work into sharp flakes and blades as well as into decorative goods, including jewelry, statuettes and small vessels. Obsidian artifacts are found from Paleolithic to Bronze Age sites in the Eastern Mediterranean where it was traded for more than 10,000 years (Williams-Thorpe, 1995). The Aegean island of Melos was the most popular source but others included Asia Minor, the Balkans, and Hungary.

Major element analysis cannot generally be used as a tool for sourcing obsidian because of a relatively narrow variation in major elements. At Franchthi Cave in Greece obsidian was imported, starting during the late Mesolithic, because no obsidian sources exist anywhere near the site. Renfrew and Aspinall (1990) reviewed the studies carried out on the artifacts and the analytical methods used. These methods included OES, NAA, fission-track dating, observations of color in transmitted light, translucency, and other physical characteristics, such as luster and vesicles.

Assessing the different methods used to source the artifacts, they concluded the following:

1. Appearance alone is not a reliable guide. Detailed study of material from a single source showed a variety of appearances.
2. Simple physical parameters, such as specific gravity and refractive index, were also too variable to be useful. The restrictive range in major element chemical compositions from one source to another makes such analysis of limited use.
3. Trace element analysis by any system—OES, NAA, and XRF in this case—proved to be the most useful. Great care had to be exercised in comparing data because the results were not consistent from one analytical system to another. Each system, however, was internally consistent, so interpretations based on any one system alone are valid. Using the raw data of the analyses showed that no single element or pair of elements could distinguish among all possible sources, although they could be separated using discriminant function analysis.
4. Fission-track (FT) analysis proved useful. Most obsidian older than 1 my with high enough uranium can generate enough fission tracks needed for a good statistical count. Obsidian from potential sources had wide enough age ranges, (e.g., Hungary, about 3.6 my, and Melos, about 8.7 my) to make FT analysis useful in this study.

Many other analytical methods (reviewed in Williams-Thorpe, 1995) have been used to source obsidian, including magnetic susceptibility (McDougall et al., 1983), Sr isotope analysis (Gale, 1981), Mossbauer spectra (Tenorio et al., 1998), and obsidian hydration dating (Bigazzi et al., 1990). Williams-Thorpe concluded that NAA is probably the best proven method to source obsidian (e.g., Beardsley et al., 1996, on Easter Island; Brooks et al., 1997, in the Andes).

Two other increasingly popular systems are ICP-MS (Tykot and Young, 1996) and the probe (Tykot, 1997). In addition to chemical analysis of the glass matrix, phenocrysts, if present, can be selected visually and analyzed, giving additional discriminators.

3.2. Basalt

Basalt and andesite are by far the most abundant of all volcanic rocks and their total surface exposure is greater than all other igneous rocks combined (Williams et al., 1982, p. 95). They cover extensive areas in the oceans, in the northwestern United States, in southern Brazil, in western India, along the African–Jordan rift system, and elsewhere where they erupted as flows from fissures, or, as in Iceland and Hawaii, from shield volcanoes. Texturally, basalts are fine grained as flows; as shallow intrusives they are medium-grained diabases. Compared to rhyolite, basalts are much less glassy and more crystalline. They have lower viscosity as a magma because of low alkali and silica and high iron and magnesium content that allows a greater amount of crystallization.

The Olmec culture of central Mexico (500–100 B.C.) produced some of the most dramatic basalt sculptures of the Western Hemisphere. Human heads of basalt over 1.8 m high and 4.5 to 5.5 m in diameter (Clewlow, 1967) have been found. Williams-Thorpe (1988) was able to source 113 of 146 Roman millstones, largely crafted from basalt, but some also of ignimbrite and leucite–phyric lava. She characterized each by petrography and by chemical analysis using wavelength-dispersive X-ray fluorescence analysis (WDXRF) to determine major and trace elements. Most Mediterranean basic lavas are well known, so the artifacts could be assigned to a general source region. Discriminant function analysis of trace elements permitted sourcing to more specific areas. Millstones were found to have been transported as much as 1500 km from their lithic sources.

3.3. Granitic and Other Felsic Igneous Rocks

Granite has always been admired and used as a construction and decorative stone wherever it is found. Local and imported granite was used on megalithic and Roman sites throughout western Europe, in Russia, on the acropolis of Zimbabwe in Africa, and in Egypt since the first dynasty (Kempe and Harvey, 1983). Because granites have distinctive mineralogy — including variations in kinds and amounts of feldspars, micas, amphiboles, and accessory minerals — as well as their textures and structures, many granite artifacts can be sourced on the basis of petrography alone (Galetti et al., 1992). Major and trace element chemistry is often needed for ancillary information to resolve questionable provenance assignments.

Egypt was a source of granites and granitic-appearing igneous rocks throughout its history and remained a source for the Romans until the end of the fourth century A.D. Detailed studies of petrography and chemistry carried out on known quarries in the Eastern Desert of Egypt have provided databases for determining provenance (Brown and Harrell, 1995). Other important granite sources found in Sardinia and in other islands of Italy and Turkey have also been described petrographically and by major and trace element chemistry (Galetti et al., 1992).

Magnetic susceptibility variation, a nondestructive analysis, has been used to source Roman columns to specific quarries and even to parts of quarries

(Williams-Thorpe et al., 1996). A study was carried out at Mons Claudianus, an important source for Roman columns, where magnetic susceptibility was measured at 91 of the 130 quarries. Systematic variations found in the quarries could be traced in actual columns in Rome. Artifacts were provenanced to within an area of about 700 × 700 m allowing a chronological development of granodiorite extraction and yielding evidence for reuse of columns in Rome.

3.4. Serpentine and Related Rocks

Serpentine was used for over 4,000 years along the Labrador coast where Eskimo and Indian cultures fashioned serpentine plummets, lamps, and cooking pots (Allen et al., 1984). Although serpentine is severely depleted in trace elements, sourcing by NAA was found feasible in a test at the Fleur de Lys quarry in Newfoundland. The absolute concentrations of the trace elements varied, but chondrite-normalized distribution patterns of the rare earth elements (REE) remained parallel and provided a fingerprint for each quarry source.

A study of steatite sources in eastern North America, from New York to Virginia, found that seven transition metals, Co, Cr, Fe, Sc, Zn, Mn, and V, provided better signatures than the REE (Truncer et al., 1998). The study was the most intensive intrasource sampling of prehistoric steatite quarry material done in North America, and it showed that NAA had the potential to assign provenance to steatite artifacts, at least to a regional level.

3.5. Marble

Because of its importance in the history and prehistory of European civilization going back to the Neolithic period, more work has been done to provenance marble than with all other lithics combined. The International Association for the Study of Marble and Other Stones in Antiquity (ASMOSIA) organized in 1988, has held meetings every two and half years and has published the transactions for each (Herz and Waelkens, 1988; Maniatis et al., 1995; Waelkens et al., 1992).

In the 19th century and for most of the first three quarters of the 20th, archaeologists determined the provenance of classical Greek and Roman marble artifacts by comparing their physical characteristics to the quarry descriptions of Lepsius, published in 1890. Because databanks for determining provenance were lacking until quite recently, Lepsius' descriptions remained archaeological gospel.

In 1964, trace elements Na and Mn were determined by NAA in an attempt to characterize Greek marbles (Rybach and Nissen, 1965). The variation was found to be too great, with factors of over 100 within the same quarry, so the method was judged not satisfactory for sourcing individual artifacts. Trace element studies require many samples and a statistical handling of the data to overcome the inherent variability in marble. Today, large databases of NAA are available (Matthews, 1997) that, when used together with other types of analysis,

Sourcing Lithic Artifacts

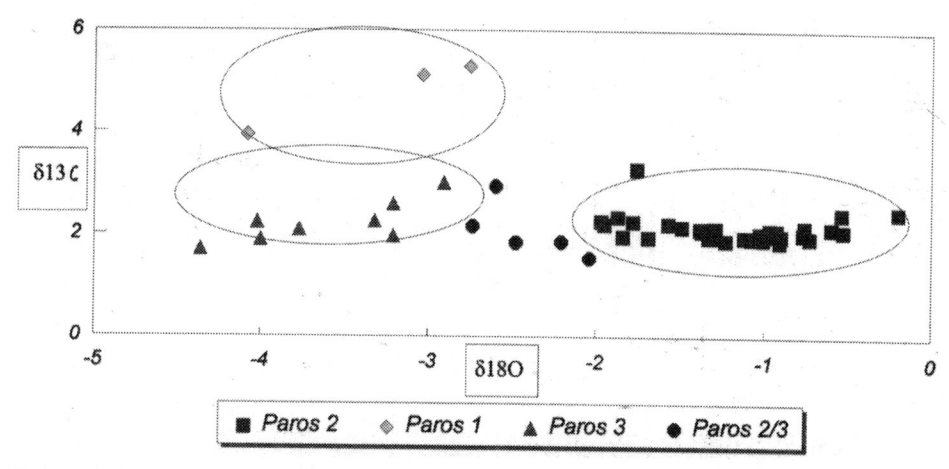

Figure 16.1. Stable isotopic assignment of ancient Greek marble statuary to specific quarries on the island of Paros. Shown are the fields for each quarry and the results of analyses of each statue. Paros 2/3 are either from quarries 2 or 3 (after Moltesen, et al., 1992).

have vastly improved the discriminating powers of NAA used alone (Matthews et al., 1995).

ESR spectra of Mn^{2+} have been used with some success (Lloyd et al., 1988). Preliminary work suggests that some quarries can be distinguished but that detailed work to establish inter- and intraquarry variation is needed. An adequate database of quarry samples is now being compiled to make the method viable.

CL of white marble has proven to be a successful discriminator (Barbin et al., 1992). More than 500 white marble samples from classical quarries have been analyzed and a database of "cathodomicrofacies" has been compiled. Another database using CL ultraviolet (UV) spectra is also being compiled (Blanc, 1995).

Stable isotopic signatures, first suggested by Craig and Craig (1972), are the most widely used system today for determining marble provenance. Isotopic plots on a $\delta^{18}O - \delta^{13}C$ diagram show that many individual quarries can be distinguished (Herz, 1992). Most classical quarries have relatively homogeneous isotopic variations that can serve as valid geochemical signatures (Fig. 16.1). In addition, other isotopic data bases now exist for marble of the Mediterranean Basin, from Neolithic Aegean to 18th century Italian and French marble.

In addition to provenance determination, marble analytical systems, including NAA, SIA, and SEM, have been used to expose fakes (Herz et al., 1989). SIA and mineralogy were used to show that a statue was a composite assembly of unrelated marble fragments (Herz, 1990). A disputed marble portrait, allegedly of Livia, the wife of Augustus Caesar, was found to be a composite of three different marbles: her skullcap was marble from the ancient Roman quarry of Ephesos, her head was from the Greek island of Paros, and her nose was Italian, from Carrara (Table 16.1). Removing her head and nose, she is now relabeled as Agrippina (Moltesen et al., 1992).

Table 16.1. Isotopic Analysis of the "Livia" Head, Copenhagen, Ny Carlsberg Glyptotek, cat. no. 614. Results in per mil, Relative ppb.

Piece	$\delta^{13}C$	$\delta^{18}O$	Composition	Source
Skull cap	+5.00	−2.99	calcite + dolomite	Ephesos
Head	+5.38	−3.85	calcite	Paros
Nose	+2.09	−2.64	calcite	Carrara

3.6. Sandstone and Quartzite

Sandstone and other clastic rocks are common sources of artifactual material. They are widespread and relatively resistant to weathering, especially if free of carbonate cement. Their mineral content, fabric, and structures alone can frequently be used for sourcing. Sandstone consists largely of sand-size particles, most abundantly quartz, that form a framework. In argillaceous sandstone, abundant clay- and silt-sized particles are packed between the quartz grains (Williams et al., 1982). Most sandstone also contains a large amount of other minerals including feldspar, micas, and resistates, as well as lithic grains. With increased amounts and kinds of other components, rocks grade into graywackes and arkoses.

Graywackes typically consist of angular sand and silt grains bound together in a lithified dark matrix. Many are quartz-poor and some consist almost entirely of lithic debris. Arkoses are characterized by quartz as well as abundant feldspar.

Chemical analysis, if necessary, is done after petrographic characterization. In a study of sandstone provenance in the Apennines in Italy, the first grouping used mineral and clastic content—either feldspathic/volcanolithic or feldspathic/quartzose (Fornelli and Piccarreta, 1997). This study was carried out with a hand lens, followed by thin section analysis. Although this study was primarily geological, petrography obviated the need for numerous chemical analyses. Chemistry was also not needed in a study of provenance of molasse sandstone from Switzerland where heavy mineral and carbonate content allowed sourcing to an exact quarry (de Quervain, 1972).

NAA was successfully used to trace the source of quartzite used for two huge ancient statues at Thebes, "the Colossi of Memnon" (Heizer et al., 1973). The Colossi represent King Amenhotep III, (14th century B.C.) with smaller figures of his family members. Carving had been done on a 720-ton monolithic block composed of a medium- to coarse-grained ferruginous quartzite. The statues were partially destroyed in an earthquake in 27 B.C. and then were repaired by the Roman emperor Septimus Severus at the end of the 2nd century A.D. using a lower quality local quartzite. Six quartzite quarries were known to be worked in ancient times. In addition to local quarries, the closest of which was only 60 km upriver from the site; an important quarry was one near Cairo 1125 km north of

Thebes. Most archaeologists had argued that the local quarry, being closer, must have supplied the stone. Petrographic study, however, showed greater similarities between the artifacts and the Cairo quartzite than with the local quarries. To resolve the problem definitively, NAA was conducted.

Over 110 samples taken from seven quarries and nine sculptures were analyzed by NAA for major, minor, and trace elements. The results showed first that europium (Eu) versus iron (Fe) could distinguish the Cairo quarry from the local sources (Figure 16.2). On a Eu versus Fe plot, samples from the original statuary clearly fell into the Cairo field. Finally, local stone was used by Severus for repair. Unfortunately it is of much lower quality than the Cairo original, weathers more rapidly, and poses great problems for the conservation of the site.

3.7. Chert and Other Siliceous Sediments

Confusion reigns supreme in the archaeological literature on the terminology of the siliceous chemical sediments. A student archaeologist working on a southeastern United States site asked if some artifacts were flint or chert. They were actually neither, but metaquartzite. However, the form he was required to fill out only gave him those two choices, so he rephrased his question: "Well then, which is it more like, flint or chert?'

In England, the term *chert* is used for all siliceous deposits of a sedimentary nature that are lithified by redistributed silica, excluding, of course, the detrital rocks. Flint came into use to describe siliceous nodules found in the Upper Cretaceous White Chalk of Western Europe (Hatch and Rastall, 1965). Then in the United States, flint was used to describe varieties of gray to black chert that possessed a conchoidal fracture, useful in the manufacture of tools. Some other terms that have been used include (1) novaculite for certain white chert found in Arkansas and Texas; (2) jasper, a red or brown chert made up of microgranular quartz and colored with ferric oxides; and (3) porcellanite, not as tough or vitreous as typical chert, composed largely of cryptocrystalline opal, and with a dull luster resembling unglazed porcelain (Williams et al., 1982). Although historically flint is an older term than chert, all standard petrographic references prefer the general term chert for chemical siliceous sediments: a dense cryptocrystalline rock composed of chalcedony (microcrystalline fibrous silica and microfibrous amorphous silica or opal) and cryptocrystalline quartz.

In a study of chert artifacts, the first tool should be petrography. Prothero and Lavin (1990) studied chert sources and artifacts from nine sites in the Delaware River Valley using thin sections. They could source virtually all the artifacts and concluded that with a larger database, eventually all chert artifacts could easily be identified to source. It should be noted that this could be done with a total expenditure of less than $10 per artifact, the cost of a thin section.

Although many prehistoric sites yielded chert artifacts, Finland possesses no natural chert resources (Mataiskainen et al., 1989). The earliest artifacts are found in the Subneolithic Combed Ware period (5500–800 yr BP) made of chert from central and northeastern Russia. Later, during the Early Metal and Bronze ages (4000–2500 yr BP), cultural connections suggest that the new flint sources were

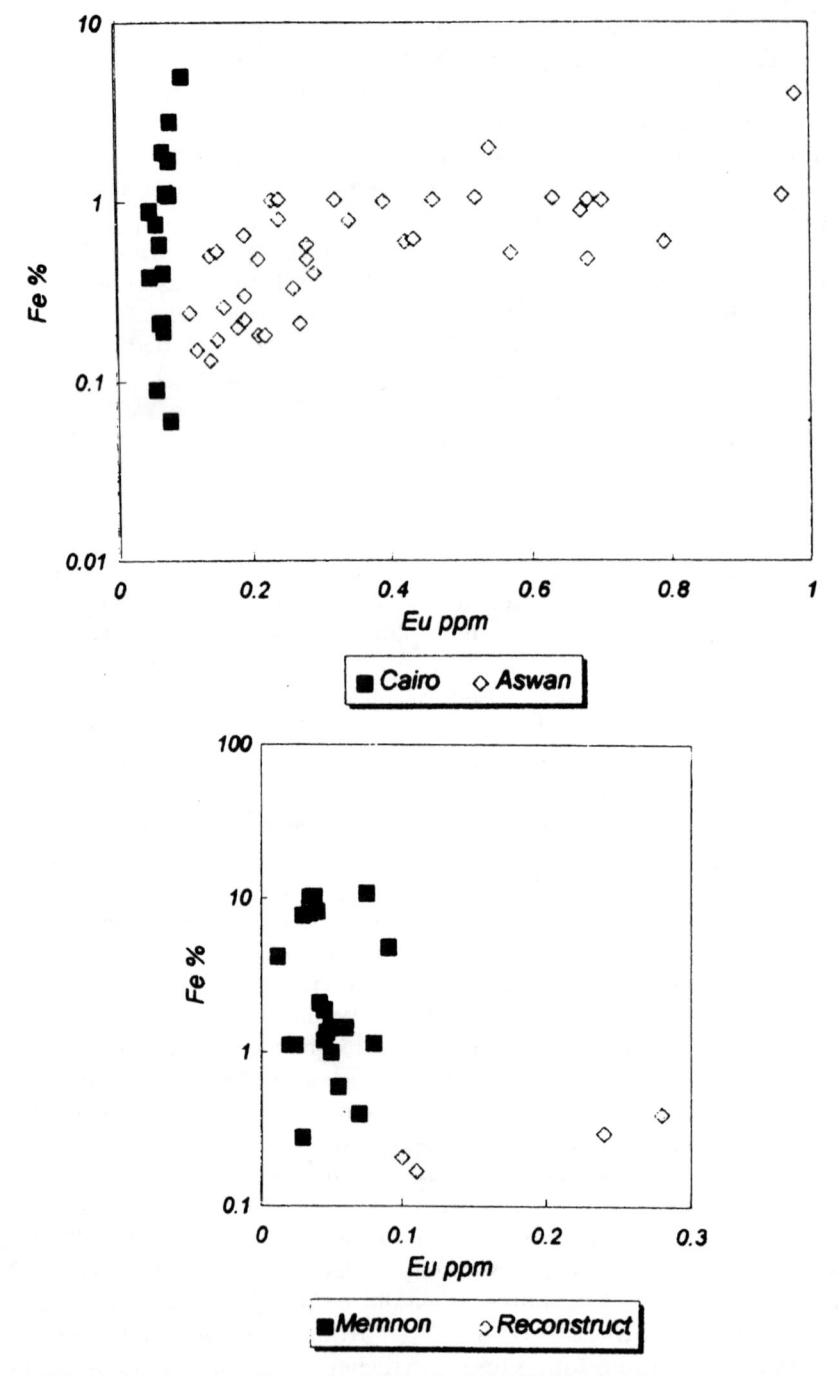

Figure 16.2. NAA assignment of quartzite samples by Fe vs. Eu. Results of analysis from the quartzite quarry in Cairo vs. the quarries near Aswan shown above. Results of analysis of the giant Memnon statue and its later Roman repair shown below (after Heizer et al., 1973).

from southern Scandinavia. AAS was used to analyze for 20 elements in artifacts and from potential sources. Color and textures were also noted in the study.

It was found that each chert source has its own distinctive geochemistry. The earlier and later chert could easily be distinguished from each other by trace elements but only to a general geographic locale. Because of the great variation in amount of each element, specific sourcing to a quarry was not possible.

In a study of Columbia Plateau chert associated with basalt flows in Oregon, thin sections revealed one group containing microfossils, but NAA was needed to discriminate among the various possible sources (Hess, 1996). Nine trace elements comprised the best base for discriminant analysis. The colors of the fresh interior and patinated surface of the cherts were determined visually by comparison to a rock color chart. No exact correlation using color was found although many older cherts were often shades of gray and the younger cherts were generally light greenish or yellowish.

Jasper, a popular form of chert with prehistoric people, was used extensively along the northeastern seaboard of the United States for tool production. Eighty jasper artifacts from 52 sites from Maine to Virginia were analyzed by XRF (King et al., 1997). Previous to this study, all artifacts were believed to have come from the Reading Prong, a geological unit in Pennsylvania-New Jersey–New York, the so-called Pennsylvania jasper. Although the results are still preliminary, the conclusions were that Pennsylvania jasper was used only locally, that is within 160 km of the source, and artifacts from elsewhere were made from other sources.

ESR spectroscopy has been used both to source cherts and also to date time of settlement. Heating chert will reset the ESR signal, so ESR can date only the last reheating. In any event, correct dates depend on a complete resetting of the previous radiation damage by heating above a required minimum temperature (Skinner and Rudolph, 1996). Dates obtained will be older than the actual occupancy if that temperature was not reached. Maniatis et al. (1989) found that variability in peaks made sourcing by ESR difficult, though temperatures attained during firing could be determined.

Because chert consists almost entirely of SiO_2, oxygen isotopic composition has been used to determine provenance (e.g., Shaffer and Tankersley, 1989). Determination of these ratios in fine-grained quartz extracted from ochre was found to discriminate between red ochre sources in central Australia (Smith and Pell, 1997).

3.8. Carbonates

The carbonate rocks include limestone, composed principally of calcite and aragonite, and dolomite. Dolomite is either a mineral or a rock term though dolostone is also commonly used today. A variety of descriptive and taxonomic terms exist, based on accessories such as clay and quartz, and on depositional textures. Micrite is composed of microcrystalline calcite; included grains are allochems (see Williams et al., 1982, for an ample terminology). Most limestones, especially those used for the manufacture of artifacts, are quite pure, with clay minerals and quartz under 5%.

Trace elements and stable isotopic ratios in carbonate rocks are controlled by (1) the aluminosilicate minerals, largely clay; (2) the carbonate minerals calcite and dolomite; and (3) diagenetic sulfides and iron oxides. Thus samples will vary greatly depending on their amounts of clay minerals, carbonate, or diagenetic cement. Kelepertsis (1983) found that the clay minerals controlled the greatest part of the trace elements and the carbonates controlled the smallest.

Using petrography and NAA, limestone sources in northern France could easily be distinguished (Holmes et al., 1994). Petrography alone could identify the source geological formation. Limestone species, that is, oolitic, crinoidal, micritic, or pisolitic, and accessories, largely quartz, clay minerals, and fossil content, were noted and used as variables in the linear discriminant analysis.

Twenty-three elements were determined by NAA. Correlation coefficients were calculated for all possible pairs of elements. Coefficients exceeded 0.5 for nearly half the element pairs and exceeded 0.7 for most of the rare earths. These coefficients are statistically significant and were used in the discriminant analysis. Only Na, Sr, Ba, Mn, and U showed smaller coefficients and Ca showed negative coefficients, as expected. This paper also concluded that the clay minerals controlled the greatest part of the trace element content.

The conclusions of the study were that stone from any one quarry face was compositionally homogeneous and could be distinguished from stone from other localities. Multi-element analysis provided enough variation to allow a geographic resolution of samples.

Stable isotope analysis and petrography were used in a study of two ancient Greek limestone quarries, Corinth and Neapolis (Wenner and Herz, 1992). Samples were carefully selected for homogeneity and calcrete coatings were removed. Isotopically, the quarries delineated restricted ranges on a $\delta^{18}O$ versus $\delta^{13}C$ plot and so could be used for provenance determination.

Sampling of opposite sides in several hand specimens differed by 1 per mil in oxygen and slightly less in carbon. This variation is far outside the normal sample reproducibility of 0.1 to 0.2 per mil found in marble and clearly indicates microscale isotopic heterogeneity in limestone. Limestone metamorphism to marble results in homogenization of both the isotopic and chemical compositions (Wenner et al., 1988).

Analysis of calcrete coatings compared to the fresh limestone showed great differences in isotopic ratios. The calcrete could be either depleted or enriched, which was unexpected. In one case the calcrete was enriched 6 per mil in $\delta^{13}C$ and 3 per mil in $\delta^{18}O$ compared to the fresh rock. Clearly for provenance determination, samples must be completely clear of surface coatings, and portions selected for analysis should be truly representative of the rock itself.

Hayward (1996) attempted to source the oolitic limestone used for construction in ancient Corinth to specific quarries. Because the location of the quarries was well known, the problem was to resolve which quarries were used at what times and for which monuments. At least 1.5×10^6 m^3 of oolitic limestone was taken from the area from the 7th century B.C. to the 2nd century A.D., mainly for construction. Because of the uniform macroscopic appearance of the oolites, it was not possible to assign any stone used in a particular ancient construction to a quarry. However, by thin section study of the shapes and inclusions of the oolites, the cement, and quartz and chert grains it was possible to distinguish limestones from as close stratigraphically and laterally as 2 and 40 m, respectively. Hayward called his work a *high-resolution provenance study* and showed that petrography could resolve problems at a very fine scale.

3.9. Amber

Amber is a fossil resin and being organic has widely varying physical and chemical properties. Amber has been given many names, depending mostly on its place of origin and chemical composition. Although there is no generally agreed on classification, ambers have been divided into three groups: (1) the succinites, containing high succinic acid and no sulfur, the true "Baltic amber"; (2) a high-sulfur group; and (3) the retinites, which lack succinic acid and sulfur (Hey, 1962). Baltic amber is of highest quality of all, with artifacts widely distributed in many old-world archaeological sites. Geologically it is thought to have originated in southern Scandinavia in the Oligocene, derived from a now-extinct species of pine. The amber was widely disbursed throughout northern Europe and Russia by glaciation and by stream transport during the Pleistocene

The most significant fraction of amber is an insoluble polymer, which can be characterized by infrared spectrophotometry analysis (Beck, 1986). These data have provided a large database of amber worldwide that included over 1200 entries in 1986, 300 of which are from the Baltic. Other methods of analysis now include carbon isotopic ratios, X-ray crystallography, and ^{13}C nuclear magnetic resonance spectroscopy (Lambert et al., 1996).

One of the earliest efforts to source archaeological artifacts was on amber by Otto Helm, an apothecary from Gdansk, Poland toward the end of the 19th century (Beck, 1986). His work was also one of the first systematic applications of the natural sciences to archaeology. The problem was to find the source for over 2,000 amber beads uncovered by Schlieman at Mycenae. Schlieman had written in his excavation monograph that "It will forever remain a secret whether this amber is derived from the Baltic or from Italy..." (Beck, 1986). Helm might also have been motivated by local pride because an Italian mineralogist, Capellini, stated that at that early time in Europe, trade routes were not highly developed, so the source must have been Italy which was, after all, much closer to Greece than the Baltic. Helm measured the succinic acid content of Baltic amber, using a technique first described by Georg Bauer (Agricola) over 400 years earlier. Although Helm analyzed only Baltic amber, which has the highest succinic acid content of any amber in the world, it matched some of Schlieman's artifacts. His conclusion, now backed up by many new analyses including Beck's (1986) database, proved to be correct: amber found in Early Bronze Age Mycenae did indeed come from the Baltic. We know now that trading in both amber and marble had started in the Paleolithic and was well established by the Neolithic Period, and that the vast majority of amber found in prehistoric Europe and the Near East was indeed from the Baltic.

4. Summary and Conclusions

Many types of analysis are used for lithic artifacts (Table 16.2). Omitted from Table 16.2 are analytical systems appropriate for organic materials, such as amber, which are mentioned in the section on amber.

The first step in an analysis of lithic material should be petrographic, preferably with thin sections. For greater detail of the stone fabric that can be seen in the petrographic microscope, the SEM, or the probe can be used. Individual minerals, such as phenocrysts in obsidian or in basaltic lavas, can be identified by thin sections or by XRD. The SEM and probe will give a major element chemical analysis of minerals and sometimes their morphology, but in many cases, they will not resolve the exact mineral species. Thus pyroxenes and amphibole species may be commonly confused, calcite versus aragonite cannot be resolved; these minerals are easily identified by XRD. Given appropriate databases, a correct mineral identification can often identify the provenance of the artifact.

Chemical analysis of major and and/or trace elements is necessary in many cases. The system used must be carefully chosen. If other similar artifacts have been analyzed with NAA, or with whatever system, the study in question should make every effort to use the same system, and preferably the same lab. Because the response of an element in any system of analysis depends on several factors, including the methodology of the lab, different systems and labs may yield different results. No analytical method in use today measures the amounts of an element directly but measures them by the response of the element to excitation by electromagnetic radiation. Gravimetric methods, that is, determining the actual amount by direct weighing, the "wet chemistry" systems of the past, are no longer followed.

Serious problems that arise in analytical data can be caused by the matrix effect and an inherent interlab variation. Some elements present in a sample can enhance or diminish the response of the element in question, depending on the system used. The equipment and setup in each lab can also produce a variation of amounts reported of an element. Renfrew and Aspinall's (1990) study of Aegean obsidian stress the need for caution. The amounts determined by XRF and NAA for Ba and Zr did not agree with previously published results using OES: "The absolute and relative amounts of Ba and Zr are not in good accord with those obtained by optical emission... . [The] previous work was self consistent and the conclusions drawn from such work are not invalidated by subsequent reassessments of the absolute concentrations" (p. 261).

Thus caution must be exercised in combining data derived from independent laboratories, even when the same analytical standards are used. This is especially important when using a database based on data obtained by unknown methods of analysis.

Some of the newer techniques that rely on physical rather than chemical parameters must also be used with caution. CL data, for example, has been indexed by colors produced under CL (Barbin et al., 1992) and also by intensity and wavelength of the spectra (Blanc, 1995). ESR databases for marble have been

Table 16.2. Analytical Methods Useful for Provenance Determination of Archaeological Artifacts

Method[a]	Petrography	XRD	INAA	SEM, Probe	ICP	XRF, etc	SIA	CL	ESR	Mag	Fiss Track
Sample	thin section	powd	powd	ts or slice	powd	powd	powd	slice	powd	slice or entire piece	thin sec
Size	large	small	small	medium	small	small	small	med	small	varies	med
Cost[b]	low	low	high	moderate	low	low	med-high	mod	low	low	high
Rock type[c]											
Obsidian	high	high	high	high	high	high	Sr, O	—	—	poten	poten
Basalt	high	high	high	high	high	high	Sr, O	—	—	poten	oldest rocks[d]
Granite	high	high	high	high	high	high	Sr, O	—	—	poten	—
Serpentine	high	high	high	high	high	high	Sr	—	—	—	—
Marble	high	high	high	high	high	high	C, Sr, O	—	high	—	—
Sandstone	high	high	high	high	high	high	poten Sr	—	—	—	—
Chert	high	high	high	high	high	high	O	poten	—	—	—
Carbonate	high	high	high	high	high	high	C, O, Sr	high	high	—	—

[a] The type of analysis for major and trace elements should be determined by: (a) what is the usual method used with this rock type; (b) what kinds of labs are available to the researcher; (c) how much funding is available.

[b] Cost—costs for any type of analysis will depend on factors such as the laboratory—is it non-profit university or commercial; the grant—many universities have a sliding scale, graduate student research vs. faculty Federal grant, etc.

[c] The types of analysis done with each rock type are evaluated as (a) high—highest priority, a commonly used system; (b) poten—a potentially useful method not commonly used or not used to date; and (c)—lowest priority, not useful or not proven to be useful.

[d] For fission track measurements to be meaningful, the rock should have a minimum content of U as well as a minimum age. Felsic rocks, such as obsidian generally meet the first requirement; mafic rocks such as basalt rarely do.

compiled by different standards and methods of analysis (cf. Armiento et al., 1997; Mandi et al., 1992). Although each new method may be internally consistent, as with chemical analysis, comparing data from a site to the literature must be done with extreme caution.

Keeping in mind these caveats, extensive bibliographies can be consulted on sourcing different kinds of lithic materials. The questions answered by sourcing artifacts from a site can contribute mightily to its understanding: culture, technology, trade, and sophistication are among the most important. Whatever analysis is done and databases utilized, if the potential sources have been identified, then the search for provenance should be by comparison of the analytical data of artifacts to a database. Under ideal conditions any type of lithic materials can been sourced, if not to a quarry, as with limestone and marble, then at least to a general area, as with chert and amber.

5. References

Allen, R., Hamroush, H., Nagle, C., and Fitzhugh, W., 1984, Use of Rare Earth Element Analysis to Study the Utilization of Soapstone along the Labrador Coast. In *Archaeological Chemistry III*, edited by J. B. Lambert, pp. 3–18. American Chemical Society Symposium Series 205, Washington, DC.

Armiento, G., Attanasio, D., and Platania, R., 1997, Electron Spin Resonance Study of White Marbles from Tharros (Sardinia): A Reappraisal of the Technique, Possibilities and Limitations. *Archaeometry* 39:309–320.

Barbin, V., Ramseyer, K., Decrouez, D., Burns, S. J., Chamay, J., and Maier, J. L., 1992, Cathodoluminescence of White Marbles: An Overview, *Archaeometry* 34:175–184.

Beardsley, F. R., Goles, G. G., and Ayres, W. S., 1996, Provenance Studies on Easter Island Obsidian: An Archaeological Application. In *Archaeological Chemistry*, edited by M. V. Orna, pp. 47–63. American Chemical Society Symposium Series 625, Washington, DC.

Beck, C. W., 1986, Spectroscopic Studies of Amber, *Applied Spectroscopy Reviews* 22:57–110.

Bigazzi, G., Marton, P., Norelli, P., and Rozloznik, L., 1990, Fission Track Dating of Carpathian Obsidian and Provenance Determination, *Nuclear Tracks and Radiation Measurements* 17:391–396.

Blanc, P., 1995, A Cathodoluminescence Spectrometer Built for an SEM JSM840A: First Results on Minerals and White Marbles. In *The Study of Marble and Other Stones Used in Antiquity*, edited by Y. Maniatis, N. Herz, and Y. Basiakos, pp. 137–142. Archetype Press, London.

Brooks, S. G., Glascock, M. D., and Giesso, M., 1997, Source of Volcanic Glass for Ancient Andean Tools, *Nature* 386:449–450.

Brown, V. M., and Harrell, J. A., 1995, Topographical and Petrological Survey of Ancient Roman Quarries in the Eastern Desert of Egypt. In *The Study of Marble and Other Stones Used in Antiquity*, edited by Y. Maniatis, N. Herz, and Y. Basiakos, pp. 221–234. Archetype. Press, London.

Clewlow, C. W., 1967, *Colossal Heads of the Olmec Culture*, University of California, Archeological Research Facility, Berkeley.

Craig, H., and Craig, V., 1972, Greek Marbles: Determination of Provenance by Isotopic Analysis, *Science* 176:401–403.

Fornelli, A., and Piccarreta, G., 1997, Mineral and Chemical Provenance Indicators in some Early Miocene Sandstones of the Southern Apennines, *European Journal of Mineralogy* 9:433–447.

Gale, N. H., 1981, Mediterranean Obsidian Source Characterization by Strontium Isotope Analysis, *Archaeometry* 23:41–51.

Gale, N. H., Einfalt, H. C., Hubberton, H. W., and Jones, R. E., 1988, The Sources of Mycenean Gypsum. *Journal of Archaeological Science* 15:57–72.

Galetti, G., Lazzarini, L., and Maggetti, M., 1992, A First Characterization of the Most Important Granites Used in Antiquity. In *Ancient Stones: Quarrying, Trade and Provenance*, edited by M.

Waelkens, N. Herz, and L. Moens, pp. 167–177. Acta Archaeol. Lovaniensia Monographiae 4, Leuven University Press, Leuven, Belgium.

Glascock, M. D., 1992, Neutron Activation Analysis. In *Chemical Characterization of Ceramic Pastes in Archaeology*, edited by H. Neff, pp. 11–26. Monographs in World Archaeology 7, Prehistory Press, Madison, WI.

Hatch, F. H., and Rastall, R. H., 1965, *Petrology of the Sedimentary Rocks* (revised by R. H. Greensmith). Allen and Unwin, London.

Hayward, C. L., 1996, High-Resolution Provenance Determination of Construction-Stone: A Preliminary Study of Corinthian Oolitic Limestone Quarries at Examilia, *Geoarchaeology* 11:215–234.

Heizer, R. F., Stross, F., Hester, T. R., Albee, A., Perlman, I., Asaro, F., and Bowman, H., 1973, The Colossi of Memnon Revisited, *Science* 182:1219–1225.

Herz, N., 1990, Stable Isotope Analysis of Greek and Roman Marble: Provenance, Association, and Authenticity. In *Art Historical and Scientific Perspectives on Ancient Sculpture*, J. Paul Getty Museum, Malibu, pp. 101–110.

Herz, N., 1992, Provenance Determination of Neolithic to Classical Mediterranean Marbles by Stable Isotopes, *Archaeometry* 34:185–194.G23

Herz, N., and Garrison, E. G., 1998, *Geological Methods for Archaeology*, Oxford University Press, New York.

Herz, N., and Waelkens, M. (eds.), 1988, *Classical Marble: Geochemistry, Technology, Trade*. NATO ASI Series E., Volume 153, pp. 369–377. Kluwer, Dordrecht.

Herz, N., Grimanis, A. P., Robinson, H. S., Wenner, D. B., and Vassilaki-Grimani, M., 1989, Science versus Art History: The Cleveland Museum Head of Pan and the Miltiades Marathon Victory Monument, *Archaeometry* 31:161–168.

Hess, S. C., 1996, Chert Provenance Analysis at the Mach Canyon Site, Sherman County, Oregon: An Evaluative Study, *Geoarchaeology* 11:51–81.

Hey, M. H., 1962, *An Index of Mineral Species and Varieties Arranged Chemically*, British Museum, London.

Holmes, L. L., Harbottle, G., and Blanc, A., 1994, Compositional Characterization of French Limestone: A New Tool for Art Historians, *Archaeometry* 36:25–38.

Jacobsen, T. W., 1976, 17,000 Years of Greek Prehistory, *Scientific American* 234(6):76–87.

Kelepertsis, A. E., 1983, Major and Trace Element Association and Distribution through a Lower Carboniferous Sequence from Anglesey Island (Great Britain), *Chemie der Erde* 42:205–219.

Kempe, D. R. C. and Harvey, A. P., 1983, *The Petrology of Archaeological Artefacts*, Clarendon Press, Oxford.

King, A., Hatch, J. W., and Scheetz, B. E., 1997, The Chemical Composition of Jasper Artifacts from New England and the Middle Atlantic: Implications for the Prehistoric Exchange of "Pennsylvania Jasper," *Journal of Archaeological Science* 24:793–812.

Lambert, J. B., Johnson, S. C., and Poinar, G. O., Jr., 1996, Nuclear Magnetic Resonance Characterization of Cretaceous Amber, *Archaeometry* 38:325–335.

Lepsius, R., 1890, *Griechische Marmorstudien*, Konigl. Akademie der Wissenschaften, Berlin.

Lloyd, R. V., Tranh, A., Pearce, S., Cheeseman, M., and Lumsden, D. N., 1988, ESR Spectroscopy and X-Ray Powder Diffractometry for Marble Provenance Determination. In Herz, N., and Waelkens, M., editors, *Classical Marble: Geochemistry, Technology, Trade*. NATO ASI Series E., vol. 153: 369–377, Dordrecht: Kluwer.

Mandi, V., Maniatis, Y., Bassiakos, Y., and Kilikoglou, V., 1992, Provenance Investigation of Marbles from Delphi with ESR Spectroscopy: Further Developments. In *Ancient Stones: Quarrying, Trade and Provenance*, edited by M. Waelkens, N. Herz, and L. Moens, pp. 443–452. Acta Archaeol. Lovaniensia Monographiae 4, Leuven University Press, Leuven, Belgium.

Maniatis, Y., and Mandi, V., 1992, Electron-Paramagnetic-Resonance Signals and Effects in Marble Induced by Working, *Journal of Applied Physics* 71:4859–4867.

Maniatis, Y., Aloupi, H., and Hourmouziadi, A., 1989, An Attempt to Identify Flint Origin and Heat Treatment by ESR Spectroscopy. In *Archaeometry: Proceedings of the 25th International Symposium*, edited by, Y. Maniatis, pp. 645–659. Elsevier, Amsterdam.

Maniatis, Y., Herz, N., and Basiakos,, Y. (eds.), 1995, *The Study of Marble and Other Stones Used in Antiquity*, Archetype Press, London.

Matiskainen, H., Vuorinen, A., and Burman, O., 1989, The Provenance of Prehistoric Flint in Finland. In *Archaeometry: Proceedings of the 25th International Symposium*, edited by Y. Maniatis, pp. 625–643. Elsevier, Amsterdam.

Matthews, K. J., 1997, The Establishment of a Data Base of Neutron Activation Analyses of White Marble, *Archaeometry* 39:321–332.

Matthews, K. J., Leese, M. N., Hughes, M. J., Herz, N., and Bowwman, S. G. E., 1995, Establishing the Provenance of Marble Using Statistical Combinations of Stable Isotope and Neutron Activation Analysis Data. In *The Study of Marble and Other Stones Used in Antiquity*, edited by Y. Maniatis, N. Herz, and Y. Basiakos, pp.171–180. Archetype Press, London.

McDougall, J. M., Tarling, D. H., and Warren, S. E., 1983, The Magnetic Sourcing of Obsidian Samples from Mediterranean and Near Eastern Sources, *Journal of Archaeological Science* 10:441–452.

Moltesen, M., Herz, N., and Moon, J., 1992. The Lepsius Marbles. In *Ancient Stones: Quarrying, Trade, and Provenance*, edited by M. Waelkens, N. Herz, and L Moens, pp. 277–281. Acta Archaeologica Lovaniensia Monographt 4, Leuven University Press, Leuven, Belgium.

Pollard, A. M., and Heron, C., 1996, *Archaeological Chemistry*, Royal Society of Chemistry, Cambridge.

Prothero, D. R., and Lavin, L., 1990, Chert Petrography and its Potential as an Analytical Tool in Archaeology. In *Archaeological Geology of North America*, edited by N. P. Lasca and J. Donahue, pp. 561–584. Geological Society of America Centennial Special Volume 4, Boulder, Colorado.

de Quervain, F., 1972, Herkunft und Beschaffenheit des Steinernen Werkstoffes Kulturhistorisch Bedeutsamer Bau- und Bildwerke in Graubünden. *Rätischen Museums Chur. heft* 13: 3–40.

Renfrew, C., and Aspinall, A., 1990, Aegean Obsidian and Franchthi Cave. In *Les Industries Lithiques de Franchthi (Argolide, Grèce), Tome II: Les Industries de Mésolithique et du Néolithique Initial*, Fasc. 5, edited by C. Perlès, pp. 258–270. Indiana University Press, Bloomington.

Rybach, L., and Nissen, H.-U., 1965, Neutron Activation of Mn and Na Traces in Marbles Worked by the Ancient Greeks. In *Radiochemical Methods of Analysis*, Volume 1, pp. 105–117. International Atomic Energy Agency, Vienna.

Sayre, E. V., and Dodson, R. W., 1957, Neutron Activation Study of Mediterranean Potsherds, *American Journal of Archaeology* 61:35–41.

Schwarcz, H. P., and Grün, R., 1992, Electron Spin Resonance (ESR) Dating of the Origin of Modern Man, *Philosophical Transactions of the Royal Society of London B* 337:145–148.

Semaw, S., Renne, P., Harris, J. W. K., Felbel, C. S., Bernor, R. L., Fesseha, N., and Mowbray, K., 1997, 2.5 Million-Year-Old Stone Tools from Gora, Ethopia. *Nature* 385:333–336.

Shackley, M. S., 1998, Gamma Rays, X-rays and Stone Tools: Some Recent Advances in Archaeological Geochemistry, *Journal of Archaeological Science* 25:259–270

Shaffer, N. R., and Tankersley, K. B., 1989, Oxygen Isotopes as a Method of Determining the Provenience of Silica-Rich Artifacts, *Current Research in the Pleistocene* 6:47–50.

Skinner, A. F., and Rudolph, M. N., 1996, Dating Flint Artifacts with Electron Spin Resonance: Problems and Prospects. In *Archaeological Chemistry*, edited by M. V. Orna, pp. 37–46. American Chemical Society Symposium Series 625, Washington, D.C.

Smith, M.A., and Pell, S., 1997, Oxygen-Isotope Ratios in Quartz as Indicators of the Provenance of Archaeological Ochres, *Journal of Archaeological Science* 24: 773–778.

Stos-Gale, Z., 1995, Isotope Archaeology—A Review. In *Science and Site*, edited by J. Beavis and K. Barker, pp. 12–28. Bournemouth University School of Conservation Studies, Bournemouth, UK.

Tarling, D. H., 1990, Some Uses for Magnetic Properties of Materials in Archaeological Sites. In *Archaeological Geology of North America*, edited by N. P. Lasca and J. Donahue, pp. 597–602. Geological Society of America Centennial Special Volume 4, Boulder, CO.

Tenorio, D., Cabral, A., Bosch, B., Jimènez-Reyes, M., and Bulbulian, S., 1998, Differences in Colored Obsidians from Sierra de Pachuca, Mexico, *Journal of Archaeological Sciences* 25:229–234.

Truncer, J., Glascock, M. D., and Neff, H., 1998, Steatite Source Characterization in Eastern North America: New Results Using Instrumental Neutron Activation Analysis, *Archaeometry* 40: 23–44.

Tykot, R. H., 1997, Characterization of the Monte Arci (Sardinia) Obsidian Sources, *Journal of Archaeological Sciences* 24:467–479.

Tykot, R. H., and Young, S. M. M., 1996, Archaeological Applications of Inductively Coupled Plasma—Mass Spectrometry. In *Archaeological Chemistry*, edited by M. V. Orna, pp. 116–130. American Chemical Society Symposium Series 625, Washington, DC.

Waelkens, M. (ed.), 1990, *Pierre Éternelle du Nil au Rhin: Carrieres et Prèfabrication*, Crèdit Communal, Brussels.
Waelkens, M., Herz, N., and Moens, L. (eds.), 1992, *Ancient Stones: Quarrying, Trade and Provenance*. Acta Archaeologica Lovaniensia Monographiae 4, Leuven University Press, Leuven, Belgium.
Wenner, D. B., and Herz, N., 1992, Provenance Signatures for Classical Limestone. In *Ancient Stones: Quarrying, Trade, and Provenance*, edited by M. Waelkens, N. Herz, and L. Moens, pp. 199–202. Acta Archaeologica Lovaniensia Monograph 4, Leuven University Press, Leuven, Belgium.
Wenner, D. B., Havert, S., and Clark, A., 1988, Variations in Stable Isotopic Compositions of Marble: An Assessment of Causes. In *Classical Marble: Geochemistry, Technology, Trade*, edited by N. Herz and M. Waelkens, pp. 325–338. NATO ASI Series E., Volume 153, Kluwer, Dordrecht.
Williams, H., Turner, F. J., and Gilbert, C. M., 1982, *Petrography: An Introduction to the Study of Rocks in Thin Sections*, W. H. Freeman, San Francisco.
Williams-Thorpe, O., 1988, Provenancing and Archaeology of Roman Millstones from the Mediterranean Area, *Journal of Archaeological Science* 15:253–305.
Williams-Thorpe, O., 1995, Obsidian in the Mediterranean and the Near East: A Provenancing Success Story, *Archaeometry* 37:217–248.
Williams-Thorpe, O., Jones, M. C., Tindle, A. G. and Thorpe, R. S., 1996, Magnetic Susceptibility Variations at Mons Claudianus and in Roman Columns: A Method of Provenancing to Within a Single Quarry, *Archaeometry* 38:15–41.

VI

A Prehistorian's Perspective

17

A Personal View of Earth Sciences' Contributions to Archaeology

OFER BAR-YOSEF

1. Opening Remarks

Archaeologists have been working closely with geologists and paleontologists since the beginning of the 19th century, and the basic concepts employed for the time ordering of prehistoric sites were borrowed by archaeologists from geology. The most common example is the subdivision of the prehistoric sequence into Lower, Middle, and Upper Paleolithic, following the tripartite division of the Pleistocene. Geologists participated in prehistoric excavations and in many regions became responsible for describing the site stratigraphy and the paleo climatic interpretation of the deposits. As the approach of regional archaeology evolved, mainly after the 1950s, geoscientists studied the changes of past landscapes with the archaeologists. The rapid evolution of dating techniques opened a new avenue for the involvement of earth scientists in archaeology, and in particular in the dating of prehistoric periods. Finally, the advancements in

---

OFER BAR-YOSEF • Department of Anthropology, Peabody Museum, Harvard University, Cambridge, Massachusetts 02138.

Earth Sciences and Archaeology, edited by Paul Goldberg, Vance T. Holliday, and C. Reid Ferring. Kluwer Academic/Plenum Publishers, New York, 2001.

fields such as mineralogy, magnetic susceptibility, remote sensing, and others increased the potential interactions between archaeologists and earth scientists. At the end of the 20th century, there are probably thousands of archaeologists who have had at least some experience working closely with a variety of earth scientists. It would therefore be presumptuous to assume that my personal view can represent such a large community. I accepted the invitation of the editors of this volume to write this chapter because I felt that over my 40-year career I have been involved with numerous practitioners of science in archaeology. I was originally trained as both an archaeologist and a geomorphologist, but as one cannot do both full-time, I chose to be responsible only for archaeological research. In this domain, I was lucky to be engaged in the study of the entire sequence of the Paleolithic and Neolithic in Western Asia and in some adjacent regions, and I am well aware of the importance of successful cooperation with geoscientists.

I have chosen three categories in which the close cooperation between earth scientists and archaeologists is crucial, including the sites and their environments, site formation processes, and geochronology. Finally, as, let us say, an "experienced" teacher, I thought it would be useful to propose, within various sections of this chapter a few practical suggestions.

Before delving into the various issues, it seems appropriate to remind ourselves of the original motivation for practising earth sciences in archaeology. Sometimes, while reading papers or site reports, I gain the impression that the impetus behind publishing a series of dates, a reconstructed past landscape, or a sequence of paleoenvironmental fluctuations lies in the recipe for "how to do and succeed in modern-day archaeology". In numerous cases, one is surprised by the brevity or almost total lack of explanation of how the information is related to an archaeological or an evolutionary query and whether the current results corroborate other sources of information. In such published instances, one can find the name of one or more archaeologists who either were involved in the excavation of the site or were solely the submitters of the samples. It is my contention that in a large number of joint publications the interdisciplinary integration of information is rarely well worked out. I detail these criticisms in the sections that follow. Here, I assume that most scientific research is carried out because earth scientists are interested in the archaeological implications of their results. Of course, not every scientist can be involved in the intricacies of the archaeological information, which in recent decades, like other fields of investigation, has enjoyed an explosion of knowledge. In addition, when most of the Paleolithic sequence is dealt with, interpretations of the evolutionary trends, the classification of fossils, and the validity of nuclear and molecular studies are constantly debated, by the majority of archaeologists. For example, in most cases when archaeologists are presented with the various dating techniques, even those with which they consider themselves familiar, in a new series or volume (e.g., Aitken et al., 1993; Taylor, 1997; Wagner, 1998), they are only qualified to evaluate the results in relation to the archaeological information and generally are not able to decipher the assumptions, measurements, and calculations behind the dates, whether they consider them correct or not.

2. Open-Air Sites and Their Environments—Are We Doing What Is Needed?

The study of sites within their settings began in earnest in the 1960s. The continental scale of paleoenvironmental reconstruction that characterized Quaternary geology provided a sound basis for the more limited, regional approach. The "site catchment analysis" was developed by C. Vita-Finzi and E. Higgs and their students and was first practised in the Mediterranean (Vita-Finzi and Higgs, 1970 Jarman et al., 1982). Mapping potential resources within the radius of a one-hour walk or 5 km combined the known instances of daily forays by hunter-gatherers with knowledge of Holocene environments. The naive assumptions at the time were that use of a site's environs is radial and that recording the features of the current landscape would somehow recall past situations. Although the surveyors took into account the determinants of the local topography, and potential easy access by natural routes, it was almost impossible to reconstruct the type of the prevailing vegetation, especially during the Terminal Pleistocene. In addition, studies of recent hunter-gatherers indicate that linearly aligned foraging trajectories are perhaps more common than previously assumed (e.g., Laden, 1992). The gradually acquired recognition that natural geomorphic changes occurred even during the Holocene led to increasing concentration on more recent, historical environments where the spatial distribution of natural resources can be more accurately plotted (e.g., Sagalassos in Turkey, Waelkens et al., 1999; or in southern Sweden, Berglund 1991; Larsson et al., 1992).

The Holocene period has seen the transition to agriculture, the emergence of urban societies and states, and later, the Industrial Revolution. Studying the fluctuations in the vegetational cover—which determine the carrying capacity for all bio-elements—in conjunction with the alluvial-colluvial degradation and deposition around sites is arguably an achievable goal, These data sets reflect not only the paleoclimatic effects, but also the anthropogenic impact, which has received increasing attention since the early 1970s. This aspect, which interests archaeologists and anthropologists, echoes the growing awareness among practitioners of applied sciences and social researchers of the human-induced environmental degradation through gas emissions into the atmosphere and pollution of terrestrial and marine environments. As correctly noted by K. Butzer (1997), archaeologists can contribute the time depth to these types of studies that is needed for calibrating the rate of environmental change (e.g., Roberts et al., 1997) and that should assist policy makers.

However, this optimistic scenario does not apply to most Pleistocene situations. In the majority, if not all Lower, Middle, and Upper Paleolithic localities, the reconstruction of past environments remains an intellectual exercise in which quantitative studies of animal bones (including mammals, reptiles, birds, and insects), as well as pollen and charcoal, are employed. Climatic fluctuations are seen as reflected by the types of sediments (if not shown to be essentially of anthropogenic/biogenic origin (see below), magnetic susceptibility, and pollen (e.g., Aitken, 1990; Wagner, 1998).

It is not surprising that the achievements in temperate European and Mediterranean paleoenvironmental investigations motivated archaeologists researching early hominid sites in Africa. A new term "landscape archaeology," was coined, in order to distinguish this new type of study from the more general geological study of Pliocene and Pleistocene formations. Early hominid sites, especially in East Africa, are embedded within lacustrine, fluvial, and limnic sedimentary cycles and are often exposed by erosion. The ongoing degradation of the deposits provides a continuous series outcrop that follows the curves and breaks in the various gullies (or korongos). This sort of "meandering" exposure of the same formation, member, or bed serves as a basis for the overall reconstruction of regional paleoenvironments (e.g., Hay, 1990; Feibel, 1997) but is also taken to represent the particular, more localised environments (e.g., Stern, 1993; Blumenschine and Peters, 1998, and references therein). The supposedly random sampling provided by natural agencies is assumed to represent the two-dimensional geographic space, the landscape through which early hominids roamed during their lifetime, scavenging, hunting, food sharing, mating, manufacturing stone tools, carrying raw materials over some distance, and so forth. The third dimension, the time during which the remains of the scatters and the patches accumulated, unfortunately, cannot be easily resolved and often the time slice is somewhere in the order of 100 ka (e.g., Stern 1993). However, in some instances, the resolution of $^{40}Ar/^{39}Ar$ can reduce the time dimension to 10 ka (Walter et al., 1991). The good intentions of all field workers who practise archaeology, often under harsh conditions, are acknowledged, however, their simplistic approach to the potential degree of environmental resolution is somewhat surprising. Commonly, the basic studies are geological in nature, as the stratigraphic position and the age of the formations are the essential requisites for placing the new fossils and the artifact clusters in the overall human evolutionary trajectory. The details, especially of the immediate environments, as disclosed by micromorphology (e.g., Courty et al., 1989), have hardly ever been practised in the African sites.

3. What Do We Expect to Learn from Site Formation Processes in Caves and Rockshelters?

The study of site formation processes is one of the foremost crucial aspects of any excavation project, and despite the major contributions made since the early 1970s, I feel that our cumulative knowledge is still in its infancy. We definitely need to know more about the processes responsible for the accumulation of what we excavate. Whether prehistoric caves or rockshelters, the traditional approach has been to view the Paleolithic deposits as accumulating by natural agencies with minimal contribution by humans (e.g., Laville et al., 1980). Later sites, from the Neolithic through the Historical period, are viewed as the results of human actions (e.g., Butzer, 1982; Rosen, 1986; Schiffer, 1987; Matthews et al., 1997).

We only need to remind ourselves that the investigations of cave sediments began in temperate Europe, where historically, the paleontological–geological approach dominated prehistoric research (e.g., Sackett, 1989). In general, within the temperate belt, there is a correlation between regional climatic fluctuation and the igniting of physicochemical processes within the sites, such as frost shattering, or the incorporation of wind and water-laid sediments. However, human occupations create their own effects, besides leaving behind broken bones and stone artifacts, and until the introduction of micromorphology, in the early 1970s, these were poorly known (e.g., Courty et al., 1989).

Sites situated in the lower latitudes: in the Mediterranean basin, and in particular in its southern margins, demonstrate that a major portion of the observable volume accumulates due to a mixture of anthropogenic and biogenic activities (e.g., Schiegl et al., 1994, 1996; Goldberg and Bar-Yosef, 1998, and references therein). Research aimed at elucidating these complex processes is currently underway in the Levant and in Greece but one only needs to examine excavation reports from sites in Italy or Spain in order to gain the overall impression that similar phenomena can be observed in other localities. Often, the deposits bear no clear paleoclimatological signals. Efforts to correlate them with oxygen isotope stages or marine regressions and transgressions have been met with criticism (e.g., Farrand, 1979; Bar-Yosef, 1989). On the other hand, the study of the intricate relationship between geogenic and anthropogenic processes opened new avenues of paleoanthropological research (e.g., Goldberg and Bar-Yosef, 1998; Goldberg and Laville, 1991; Weiner et al. 1993; Schiegl et al. 1994, 1996). The following comments are intended purely for the illustration of these studies.

In numerous cave and rockshelter sites excavated by modern recording techniques, archaeologists are seeking to decipher the distributions of animal bones (in addition to the evidence for intentional breakage, cut and gnaw marks, etc.). In one example, at Kebara cave, the overall distribution of animal bones in certain middle Palaeolithic layers was linked to diagenetic processes (Bar-Yosef et al., 1992; Weiner et al., 1993). However, in that portion of the site that escaped the destructive effects, as revealed by the mineralogical analyses, the spatial distribution of hearths, ashes, bones, and lithics indicates that the inhabitants of some 60 to 55 ka BP maintained a certain separation between their various activities. Part of their actions included the removal and dumping of ashes near the cave wall and discarding a considerable percentage of the large debitage products (unused blanks and exhausted cores) in the same area. In the central space, animal bones and small debitage products accumulated adjacent to the hearths but were restricted in their distribution. In the main ash lenses (perhaps intentionally spread for sleeping), only dispersed artifacts were recovered.

The mineralogical studies in Kebara and Hayonim caves disclosed the importance of ashes as a major contributor to the volume of preserved sediments, reflecting the importance of firewood, branches, and leaves and grasses for bedding, as well as plant food. Together with other brought-in components such as animal tissues and lithics, the anthropogenically derived sediments become the major depositional component (e.g., Bar-Yosef, 1993; Schiegl et al., 1994, 1996;

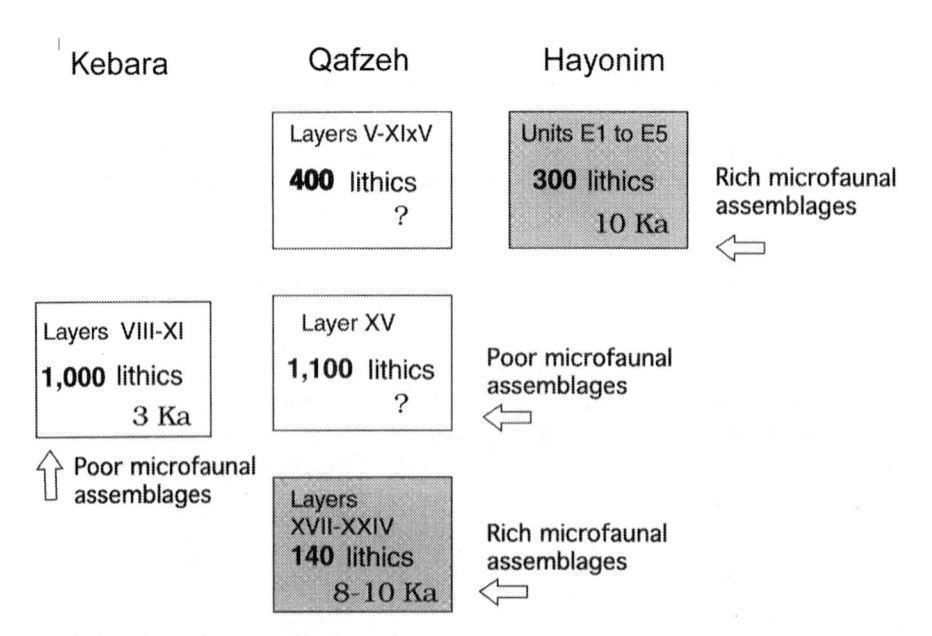

Kebara	Qafzeh	Hayonim
	Layers V-XIxV **400** lithics ?	Units E1 to E5 **300** lithics 10 Ka	Rich microfaunal assemblages ⇐
Layers VIII-XI **1,000** lithics 3 Ka	Layer XV **1,100** lithics ?	Poor microfaunal assemblages ⇐
⇑ Poor microfaunal assemblages	Layers XVII-XXIV **140** lithics 8-10 Ka	Rich microfaunal assemblages ⇐

Figure 17.1. Relative densities of lithics and microvertebrates in Kebara, Qafzeh and Hayonim Caves. Each box represents one cubic meter. Approximate dates based on TL are represented where available.

Bar-Yosef, 1998; Goldberg and Bar-Yosef, 1998). The effects of ashes resulting from in-cave campfires were explored in both sites (Schiegl et al., 1994, 1996; Weiner et al., 1995).

The history of the hearths in Kebara, which has turned out to be the best preserved Middle Paleolithic site in the Levant to date, is telling. Hearth building, for providing warmth and light as well as a means for parching and roasting, began by digging or scraping an often rounded or oval pit, some 40 to 60 cm in diameter (Courty et al., 1989; Meignen et al., 1989). The main combustible was wood in most cases. The burned wood produced white calcitic ashes, and where the original hearth structure is well-preserved, it overlies a reddish sediment. The charred remains of plant tissues and burnt bone fragments were recovered during the excavation and are also observed in thin sections. Hearths, and especially the ashes were later altered to a series of phosphate minerals as part and parcel of postdepositional processes. As they are some of the most vulnerable features of a campsite, their disappearance from the visible record occurs frequently (as areas that served for sleeping). This explains why, in many sites, burned flints are available for TL dating, but the hearths or ashes are no longer visible to the naked eye. In many of these situations, thin sections reveal the presence of microscopic charcoal remains (Courty et al., 1989; Goldberg, 1992). Therefore, the original volume of deposits in most sites is reduced, and in some cases, perhaps what is left for the archaeologists is barely 10% of the original accumulations. This volumetric reduction, in conjunction with the complexity of human behavior, is exemplified from our own work in the Levant accumulations.

In Levantine caves, most of the accumulations are due to human activities and therefore one does not recognize sterile layers, as in the northern Mediterranean basin or in temperate Eurasia. The amount of human activities can be surmised from taking into account the frequencies of microvertebrates, the number of artifacts per cubic meter, and the suggested rae of accumulation based on available TL and ESR dates. In a few cave sites in the Levant such calculation is now feasible (Fig. 17.1). Thus in Kebara Cave each cubic meter contains about 1000 stone artifacts (larger than 2 cm), with meager quantities of microfauna, accumulated over about 3,,, TL years (Bar-Yosef, 1998). This can be compared to Hayonim Cave where each cubic meter contains about 300 lithics (larger than 2 cm), with a relatively abundant collection of microfauna, that accumulated approximately over 10/15,000 TL years. The information from the lower layers at Qafzeh Cave, known for their middle Paleolithic burials, is similar to Hayonim Cave, but the upper layers are different (Fig. 17.1).

Therefore, as most of the sediments are of biogenic origins, we might compare them by "decompressing" the stratigraphies. Using this approach, the number of artifacts at Hayonim would be reduced to about 100 per cubic meter. The microfauna indicates repeated habitations by barn owls and other small predators. In terms of settlement pattern, this means that the human occupations at Hayonim Cave, some 200 to 100 ka ago, were ephemeral and the social unit moved often between different sites. Thus, the relative population size was small in comparison to later Mousterian times, which is corroborated by the faunal analysis (Stiner et al., 1999).

In Qafzeh Cave, we notice a shift from the lower part of the stratified deposits where a low degree of intensity of occupation is expressed by the richness in microvertebrates and relative paucity of artifacts. The TL dates may indicate a range of 5 to 10 ka per cubic meter. The upper layers of the Mousterian sequence are rich in artifacts and poor in microfaunal remains, but due to the lack of TL dates it is as yet difficult to estimate the intensity of occupation.

When both sites are compared to Kebara, which dates to the later part of the Levantine Mousterian, the richness in artifacts and rarity of microvertebrates, as well as archaeobotanical and faunal analyses, indicate that Kebara cave was used for "central place foraging" (Isaac, 1984) with clear indications for seasonal mobility (Lev and Kislev, 1993; Meignen et al., 1998).

In sum, as archaeologists we expect to gather from the converging lines of evidence some clear indications of human behaviour that may lead us to ask further questions. One of the obvious conclusions from the previously discussed studies is that the length of time during which identifiable sequences of several *chaînes opératoires* (Meignen, 1995) — also called "core reduction strategies" — were practised may have been in the order of 60 to 100 ka. This observation raises the possible interpretation of a long-held mental template in a population that continuously maintained the way in which they made their stone tools, despite climatic changes and shifts in resource availability and accessibility.

4. Geochronology — Is it Simply a Game of Numbers?

It is easily understood that without chronology there is little point in archaeology, especially if one views this discipline as the tool for producing an anthropologically interpretable culture history. Even when one studies a region or a culture over a limited amount of time, dates are a prerequisite. In a major portion of Southwest Asia, the use of the Egyptian calendrical chronology is achieved by correlating assemblages to the historically known Egyptian finds and thus building a timetable of the last five millennia. However, this practice (which is not devoid of unresolved problems) is not applicable to neighboring regions such as Iran, central Asia, or Southeast Europe. Radiocarbon dating emerged as a response to the need for accurately placing the pre-fifth millennium BP sites in the Near East, as well as for providing a chronometric scale for the rest of the world. This scientific step motivated the development of additional radiometric techniques, such as uranium-series dating (e.g., Aitken, 1990; Wagner, 1998).

4.1. Radiocarbon Chronology

We are all aware that the use of radiocarbon dating has revolutionized the archaeological studies of the last 40 to 30,000 years. Recently, the calibration curve and its improved version (e.g., Stuiver and Braziunas, 1993; Stuiver et al., 1998), based on tree rings, has allowed us to translate radiocarbon years into calendrical years. This provides a tool with which we can measure the duration of cultural processes and the continuity and discontinuity of prehistoric entities and correlate archaeological phenomena with climatic fluctuation over the last 40,000 years (e.g., van der Plicht, 1999; contra van Andel, 1998). The calibration curve has its "problematic" portions, namely, the "plateaux" in which radiocarbon readings could have various calendrical spans (depending on the SD of the dated sample) sometimes of as much as 1000 years.

In this discussion, I also leave aside the problems that emerge from the actual measurement of the ^{14}C in the laboratories (as demonstrated a few years ago in a study of interlaboratory calibration that uncovered an unsatisfactory situation), and I concentrate solely on the archaeological aspects.

In dating an archaeological unicultural site or a layer, it is undoubtedly better to have more than one analysis. In some cases, the spread of the calibrated dates indicates repeated occupations of the site, as in Nahal Issaron, a Neolithic site in the Negev, where archaeologists, by employing traditional typological methods in analysing the typology of the lithic assemblages, could not reach this conclusion (Carmi et al., 1994). The request for many more dates is therefore justified, but not as a wholesale approach (Byrd, 1994, 1998). Unfortunately, dating is not only a matter of quantity. Ten readings that are completely wrong due to introduction of charcoal into older deposits via a rodent hole are not better than one or two that come from a well-preserved campfire. Therefore, in such cases micromorphology may assist archaeologists in evaluating the evident discrepancy between the cultural definition, as based on other dated sites, and the given radiocarbon readings.

In this context, we cannot ignore the difficulties caused by the offhand working habits of most of us, the archaeologists, with the following examples:

1. Sending charcoal samples to be dated without previous knowledge of the kind of trees and/or other perennials that constitute the sample. It is therefore advisable (although not always possible) to submit seeds for dating, or wood charcoal after the initial botanical identification.
2. When the dates are published, the readers often find it difficult to locate the samples within the stratigraphy or the spatial dimensions of the reported site. Recently we provided an example from the Upper Paleolithic layers at Kebara Cave of a more informative way of conveying the results (Bar-Yosef et al., 1996). A drawing and a photograph of the section from which most of the dated samples were derived indicate the location of the dates, followed by a detailed discussion in which particular effort was dedicated to explaining the aberrant readings. This practice should be also adopted by earth scientists working with archaeologists, and not only for ^{14}C dates.
3. A better understanding of site formation processes of both open-air sites and cave sites would increase the ability of the excavators to interpret the available dates and to recognize the "good" and "bad" ones. The presence of older dates in more recent layers is a common phenomenon in mounds as well as in open-air alluvial sites. Younger dates in older layers are commonly noticed but in most cases can be explained by one or more of several mechanisms such as borrowing animals, infiltration through soil cracks, which are common in the clayey soils of the Mediterranean basin charcoal that is an integral component of a later occupation but was removed by wind or water (often downslope) to then accumulate in an area which was abandoned earlier; and finally, a later campfire introduced into an older occupation layer by people who did not leave behind identifiable elements of their material culture.
4. Archaeologists tend to keep the "bad" and the "good" dates in summary lists and conclude (in some cases) that only a certain percentage of the readings are of value (e.g., (Byrd, 1994). There is no doubt that the full lists of dates must be reported. However, when writing overviews, it is sufficient to cite the dates that are considered meaningful and refer the readers to the full lists as published (often in site reports or lists of radiocarbon labs). By repeating inaccurate data, archaeologists confuse the uninitiated and those who are not familiar with the intricacies of the particular sites or will never have the time to check the original reports for the evidence from stratigraphic and material culture.

4.2. The Preradiocarbon Techniques

Dating Tertiary and Quaternary evolutionary events is undoubtedly an important issue. Diachronic morphological changes among hominoid and hominid fossils are often interpreted as the emergence of new species or subspecies. However, the reconstruction of the global story, as expected from the fragmentary nature of the fossil record, is still stained with variable definitions and contradictory interpretations. Therefore, sound dating is of crucial importance, and in recent years, especially for Middle Paleolithic layers and assemblages, and in particular for the human fossils (e.g., Klein, 1999).

The surge of interest in the Middle Paleolithic was propelled by the renewed debate concerning the emergence of modern humans—as proposed by the genetic studies—and the demise of Neanderthals (e.g., Stringer and McKie, 1996; Wolpoff and Caspari, 1996). The dispersal of modern humans from sub-Saharan Africa, once dubbed as the "African Eve hypothesis", and the possibility of these "replacing" the older populations, has caused endless arguments that rely on the dates of the human fossils and their morphological classification. None of these endeavors is an easy task, as described here with examples from the Levant, where both archaic modern humans and the so-called Western Asian Neanderthals were the bearers of Mousterian industries and practised intentional burials (e.g., Arensburg and Belfer-Cohen, 1998; Belfer-Cohen and Hovers, 1992; Mann, 1995; Rak, 1986, 1990, 1998; Stringer, 1994, 1995, 1998; Stringer and McKie 1996; Trinkaus, 1998, 1995; Trinkaus et al., 1998; Vandermeersch, 1981, 1985, 1989, 1992, 1995; Wolpoff, 1996; Wolpoff and Caspari, 1996).

In accordance with my opening remarks, I illustrate this issue by citing a recent example. It concerns the isolated human jaw from layer C in Tabun cave (Israel), which holds a key position. On the basis of its morphology, the jaw was classified as modern human (Quam and Smith, 1998) a view that was shared by the excavator (see Garrod and Bate, 1937). However, a different study has suggested that it belongs to the Neanderthal population (Stefan and Trinkaus, 1998), as was also thought by others in the 1930s. Arensburg and Belfer-Cohen (1998), similar to the conclusion of McCown and Keith (1939), view the suite of Levantine Mousterian fossils as members of a local but morphologically variable population.

Due to this type of controversy and the need to test the genetic hypothesis of modern human dispersal from sub-Saharan Africa, possibly through the Levant, the dating of assemblages, layers, and fossils have become of prime importance. The main dating techniques are thermoluminescence (TL), Electron Spin Resonance (ESR), and related U-series techniques (Wagner, 1998; Rink, Chapter 14, this volume).

Series of dates are now available (for a full list see Bar-Yosef, 1998; Clark et al., 1997; Schwarcz and Rink, 1998; Valladas et al., 1998), but not all the results correlate with each other. Sometimes the results produced by one technique contradict those of another, and the archaeologists do not possess the knowledge to select between them. This is nicely detailed in papers published in recent years (Jelinek, 1992; Clark, 1997) in contrast to the selective view expressed by the author of this chapter (Bar-Yosef, 1998). Others try to demonstrate why certain dates would not fit either the stratigraphic evidence or the overall geological interpretation of the site sequence (e.g., Farrand, 1994). Without making the presumption that I know something the geochronologists don't, it seems that the most consistent results were provided by TL. ESR readings corroborate mainly for the Late Mousterian ("Tabun B-type") and the previous "phase" known also as "Tabun C-type" assemblages (Fig. 17.2). The current resolution correlates the dates among similar assemblages that were deposited in the same stratigraphic

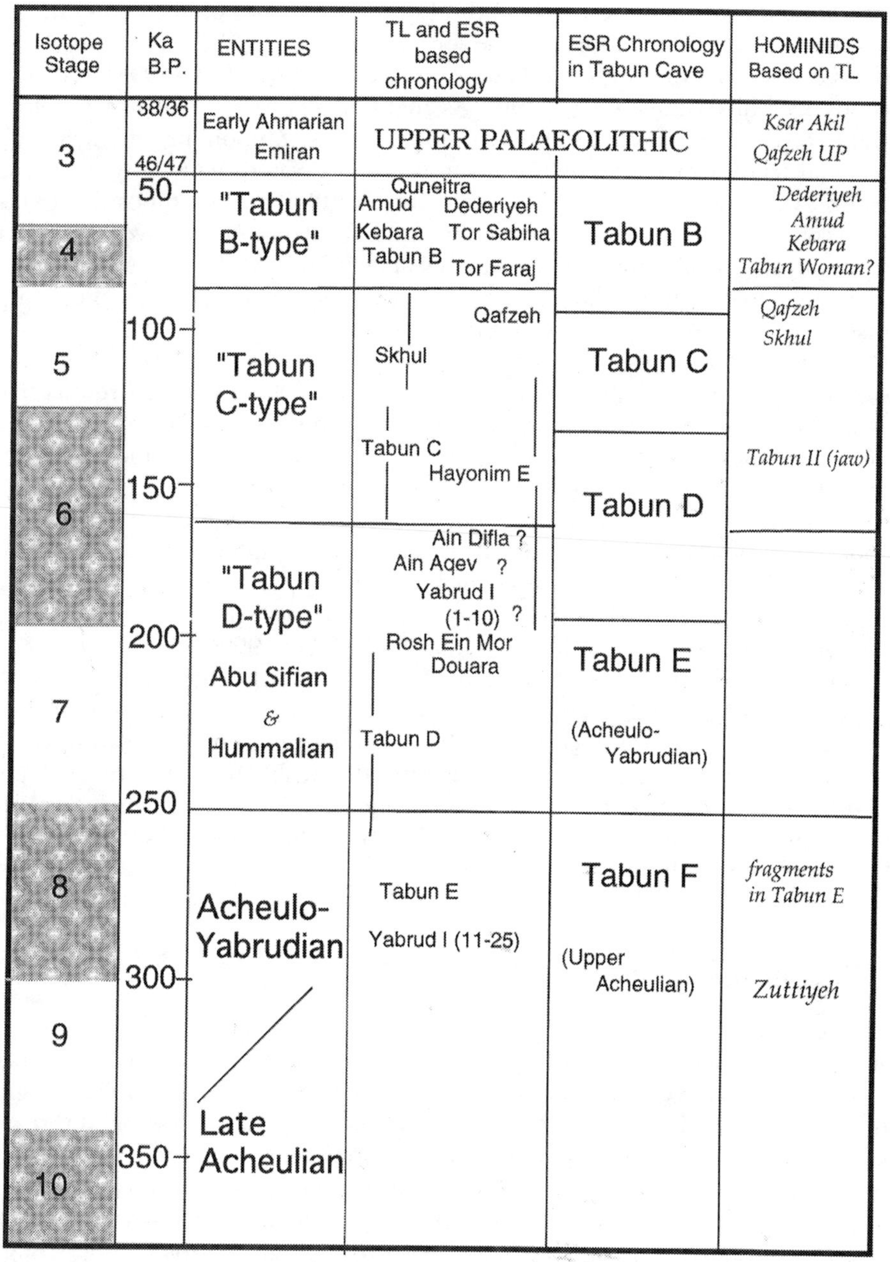

Isotope Stage	Ka B.P.	ENTITIES	TL and ESR based chronology	ESR Chronology in Tabun Cave	HOMINIDS Based on TL
3	38/36 — 46/47 — 50	Early Ahmarian Emiran	UPPER PALAEOLITHIC		Ksar Akil Qafzeh UP
4		"Tabun B-type"	Quneitra, Amud, Dederiyeh, Kebara, Tor Sabiha, Tabun B, Tor Faraj	Tabun B	Dederiyeh Amud Kebara Tabun Woman?
5	100	"Tabun C-type"	Qafzeh, Skhul, Tabun C, Hayonim E	Tabun C	Qafzeh Skhul
6	150			Tabun D	Tabun II (jaw)
7	200 — 250	"Tabun D-type" Abu Sifian & Hummalian	Ain Difla ?, Ain Aqev ?, Yabrud I (1-10) ?, Rosh Ein Mor, Douara, Tabun D	Tabun E (Acheulo-Yabrudian)	
8	300	Acheulo-Yabrudian	Tabun E, Yabrud I (11-25)	Tabun F (Upper Acheulian)	fragments in Tabun E Zuttiyeh
9					
10	350	Late Acheulian			

Figure 17.2. Chronological chart essentially based on TL and ESR dates with prehistoric entities, dated sites and undated (?) sites compared to ESR dates from Tabun Cave. The last column mentions the main hominid remains in relation to the prehistoric entities of the first column. Contemporaneity between the Acheulo-Yabrudian in the central northern Levant and late Acheulian in the southern Levant is based on archaeological considerations and recent TL and ESR dates (Bar-Yosef, 1998; Porat et al., 1999).

order, which in a small area such as the Levant, were supposedly contemporary (Bar-Yosef, 1998). The reasons for the somewhat aberrant ESR dates from Tabun Cave (Grün et al., 1990) are explained by Schwarcz and Rink (1998) and more recently by Millard and Pike (1999). As mentioned previously, the role of site formation processes is often so poorly known that the taphonomy of burnt lithics or animal teeth, especially when the sites were excavated many years ago (as is the case of Garrod's excavations in Tabun and Skhul), is almost impossible to take into account. Even in current excavations such as at Hayonim Cave, the role of burrows in site promotion is not always clear. Undoubtedly, more investigations are needed to clarify site formation processes, especially in those sites where biogenic agencies were of major significance.

A new and experimental technique is the U-series dating by gamma-ray spectroscopy, which is noninvasive (Schwarcz et al., 1998). The first results from this technique propose new dates for Tabun woman (that came from either layer B or C) and the isolated jaw from layer C. The obtained readings indicate an EU age of 19 ± 2 ka for the femur of the woman and an LU age of 33 ± 4 ka. The mandible produced an EU age of 34 ± 5 ka and an LU age of 70 ± 25 ka. It is important to note that none of these readings is acceptable at face value (given that noncalibrated radiocarbon ages for the early Upper Paleolithic are 45/43 ka BP (e.g., Bar-Yosef, et al., 1996) and given that there is still much to study concerning U-series dates of bones (Millard and Pike, 1999).

Surprisingly, the lesson taught by W. Libby in the first attempt at ^{14}C dating—when he dated calendrically known Egyptian wood samples in order to assess the validity of the technique—have been forgotten. The dating of the Tabun woman with a new technique illustrates the problem of why archaeologists are reluctant to give full credence to geochronologists, especially when the latter do not take various data sets into account but trust their machines to be a source of the unbiased truth.

5. Conclusions

Can we learn something from this fragmentary and incomplete discussion of the positive and negative contributions of earth scientists involved in archaeological investigations? My lesson after these 40 years is that in the course of all research we must first phrase the archaeological questions and the various hypothetical resolutions and their implications for paleoanthropological or historical interpretations. We must always follow the same procedure and ask ourselves what we wish to learn in resolving an archaeological–anthropological problem by using a particular scientific technique. Is the proposed technique the only one that can provide the answer? Is the scientific resolution only a "scientific assurance" that the archaeological or anthropological "common sense" is actually right? These questions should be in the mind of all parties. Archaeologists need to know how to explain to a scientist why they want to know a certain thing and the scientists need to explain how they reach their conclusions. Most important is to demonstrate how a certain technique was decided on, how it operates in a given situation, and why therefore we may assume that when an older material sample or landscape is tested, there is some assurance of the obtained results. Science is not magic and scientists, like archaeologists, can be wrong. Only stable, close cooperation will allow both parties to produce durable results. There is no escape from the social aspect of working together and both parties need to know the other's way of thinking. There is no primacy in scientific self-interest in respect to the ongoing discussion that would bring fruitful and solid resolutions to the intriguing problems of the story of human evolution. Joint work and publication would make this story a more interesting one. Doesn't all this sound just like an old teacher?

6. References

Aitken, M. J. 1990, *Science-Based Dating in Archaeology*. Longman, New York.
Aitken, M. J., Stringer, C. B., and Mellars, P. A. (eds). 1993, *The Origin of Modern Humans and the Impact of Chronometric Dating*, Princeton University Press, Princeton, NJ.
Arensburg, B., and Belfer-Cohen, A. 1998, Sapiens and Neandertals: Rethinking the Levantine Middle Paleolithic Hominids. In *Neandertals and Modern Humans in Western Asia*, edited by T. Akazawa, K. Aoki, and O. Bar-Yosef, pp.. 311–322. Plenum Press, New York.
Bar-Yosef, O., 1989, Geochronology of the Levantine Middle Palaeolithic. In *The Human Revolution: Behavioural and Biological Perspectives on the Origins of Modern Humans*, edited by P. Mellars and C. Stringer, pp. 589–610. Edinburgh University Press, Edinburgh.
Bar-Yosef, O. 1993, Site formation processes from a Levantine viewpoint. In *Formation Processes in Archaeological Context*, edited by P. Goldberg, D. T. Nash, and M. D. Petraglia, pp. 11–32. Prehistory Press, Madison, WI.
Bar-Yosef, O. 1998b, The Chronology of the Middle Paleolithic of the Levant. In *Neanderthals and Modern Humans in Western Asia*, edited by T. Akazawa, K. Aoki, and O. Bar-Yosef, pp. 39–56, Plenum Press, New York.
Bar-Yosef, O., 1998a, Jordan Prehistory: A View from the West. In *The Prehistoric Archaeology of Jordan*, edited by D. O. Henry, pp. 162–178. BAR S705, Oxford.
Bar-Yosef, O., Vandermeersch, B., Arensburg,B., Belfer-Cohen, A., Goldberg, P., Laville, H., Meignen, H. Rak, Y., Speth, J. D., Tchernov, E., Tillier, E. M., and Weiner, S., 1992, The excavations in Kebara Cave, Mount Carmel, *Current Anthropology* 33(5):497–550.
Bar-Yosef, O., Arnold, M. Belfer-Cohen, A., Goldberg, P., Housley, R., Laville, H., Meignen, L., Mercier, N., Vogel, J. C., and Vandermeersch, B., 1996, The Dating of the Upper Paleolithic Layers in Kebara Cave, Mount Carmel, *Journal of Archaeological Science* 23:297–306.
Belfer-Cohen, A. and Hovers, E., 1992, In the Eye of the Beholder: Mousterian and Natufian burials in the Levant, *Current Anthropology* 33(4):463–471.
Berglund, B. E. (ed.), 1991, The Cultural Landscape during 6000 Years in Southern Sweden—The Ystad Project, *Ecological Bulletins 41* Munkshaard International, Copenhagen.
Blumenschine, R. J., and Peters, C. R., 1998, Archaeological Predictions for Hominid Land Use in the Paleo-Olduvai Basin, Tanzania, during Lowermost Bed II Times, *Journal of Human Evolution* 34(6):565–607.
Butzer, K. W., 1982, *Archaeology as Human Ecology: Method and Theory for Contextual Approach*, Cambridge University Press, Cambridge.
Butzer, K. W. 1997, Sociopolitical discontinuity in the Near East c. 2200 BCE: Scenarios from Palestine and Egypt. In *Third Millennium BC Climate Change and Old World Collapse*, edited by H. Nüzhet Dalfes, G. Kukla, and H. Weiss, pp. 245–296, Springer-Verlag, Berlin.
Byrd, B. F. 1994, Late Quaternary Hunter–Gatherer Complexes in the Levant between 20,000 and 10,000 BP. In *Late Quaternary Chronology and Paleoclimates of the Eastern Mediterranean*, edited by O. Bar-Yosef and R. Kra, pp. 205–226, Radiocarbon and the Peabody Museum of Archaeology and Ethnology, Harvard University, Tucson, AZ, and Cambridge, MA.

Byrd, B. F., 1998, Spanning the Gap between the Upper Paleolithic and the Natufian: The Early and Middle Epipaleolithic. In *The Prehistoric Archaeology of Jordan*, edited by D. O. Henry, pp. 64–82. Archaeopress BAR S705, Oxford.

Carmi, I. Segal, D., Goring-Morris, A. N., and Gopher, A., 1994, Dating the Prehistoric Site Nahal Issaron in the Southern Negev, Israel, *Radiocarbon* 36(3):391–398.

Clark, G. A., 1997, The Middle-Upper Paleolithic Transition in Europe: An American Perspective, *Norwegian Archaeology Review* 30:25–53.

Clark, G. A., Schuldenrein, J., Donaldson, M. L., Schwarcz, H. P., Rink, W. J., and Fish, S. K., 1997, Chronostratigraphic Contexts of Middle Paleolithic Horizons at the 'Ain Difla Rockshelter (WHS 634), West-Cental Jordan. In *The Prehistory of Jordan, II. Perspectives from 1997*, edited by H-G. Gebel, Z. Kafafi, and G. O. Rollefson, pp. 77–100. Ex Oriente, Berlin.

Courty, M. A., Goldberg, P., and Macphail, R., 1989, *Soils and Micromorphology in Archaeology*, Cambridge University Press, Cambridge, UK.

Farrand, W. R. 1979, Chronology and Paleoenvironment of Levantine Prehistoric Sites as Seen from Sediment Studies, *Journal of Archaeological Science* 6:369–392.

Farrand, W. R., 1994, Confrontation of Geological Stratigraphy and Radiometric Dates from Upper Pleistocene Sites in the Levant. In *Late Quaternary Chronology and Paleoclimates of the Eastern Mediterranean*, edited by O. Bar-Yosef and R. Kra, pp. 21–31. Radiocarbon and the Peabody Museum of Archaeology and Ethnology, Harvard University, Tucson AZ, Cambridge, MA.

Feibel, C. S., 1997, Debating the environmental factors in hominid evolution. *GSA Today* 7(3):1–7.

Garrod, D. A. E., and Bate, D. M., 1937, *The Stone Age of Mount Carmel*. Clarendon Press, Oxford.

Goldberg, P., 1992,Micromorphology, Soils, and Archaeological Sites. In *Soils in Archaeology*, edited by V. Holliday, pp. 145–18. Smithsonian Institution, Washington, DC.

Goldberg, P., and Bar-Yosef, O., 1998, Site Formation Processes in Kebara and Hayonim Caves and Their Significance in Levantine Prehistoric Caves. In *Neanderthals and Modern Humans in Western Asia*, edited by T. Akazawa, K. Aoki, and O. Bar-Yosef. New York, Plenum Press: 107–125.

Goldberg, P., and Laville, H. 1991, Etude Géologique Dépôts de la Grotte de Kebara (Mont Carmel): Campagne 1982–1984. In *Le Sequelette Mousterien de Kebara 2*, edited by O. Bar-Yosef and B. Vandermeersch, CNRS, Paris, pp. 29–42.

Grun, R., Beaumont, P. B., and Stringer, C. B., 1990, ESR Dating Evidence for Early Modern Humans qat Border Cave in South Africa, *Nature* 344:537–539.

Hay, R. L., 1990, *Geology of the Olduvai Gorge: A Study of Sedimentation in a Semiarid Basin*, University of California Press, Berkeley.

Isaac, G. L., 1984, The archaeology of human origins: Studies of the Lower Pleistocene in East Africa 1971–1981. In *Advances in World Archaeology*, edited by F. Wendorf and A. Close, pp. 3:1–87. Academic Press, New York.

Jarman, M. R., Bailey, G. N., and Jarman., H. N. (eds.), 1982, *Early European Agriculture: Its Foundation and Development*, Cambridge University Press, Cambridge.

Jelinek, A. J. 1992, Problems in the Chronology of the Middle Paleolithic and the First Appearance of Early Modern *Homo sapiens* in Southwest Asia. In *The Evolution and Dispersal of Modern Humans in Asia*, edited by T. Akazawa, K. Aoki, and T. Kimura, pp. 253–275. Tokyo University Press, Tokyo.

Klein, R. G., 1999, *The Human Career: Human Biological and Cultural Origins*, University of Chicago Press, Chicago.

Laden, G., 1992, *Ethnoarchaeology and Land Use Ecology of the Efe (Pygmies) of the Ituri Rain Forest, Zaire: A Behavioral Ecological Study of Land Use Paterns and Foraging Behavior*. Harvard University Press, Cambridge.

Larsson, L., Callmer, J., and Stjernquist, B. (eds.), 1992, *The Archaeology of the Cultural Landscape*, Almqvist & Wiksell International, Stockholm.

Laville, H., Rigaud, J.-P., and Sackett, J., 1980, *Rock Shelters of the Perigord*, Academic Press, New York.

Lev, E., and Kislev, M. E., 1993, *The Subsistence and the Diet of the "Neanderthal" Man in Kebara Cave, Mt. Carmel*, edited by A. Perebolotsky. Ramat Ha Nadiv. Project Report, Series Pub., 9, Society for the Protection of Nature, Tel Aviv.

Mann, A., 1995, Moldern Human Origins: Evidence from the Near East, *Paleorient* 21(2):35–46.

Matthews, W., French, C. A. I., Lawrence, T., Cutler, D. F., and Jones, M. K. 1997, Microstratigraphic Traces of Site Formation Processes and Human Activities, *World Arch.* 29(2 — *High Definition Archaeology*):281–308.

McCown, T. D., and Keith, A., 1939, *The Stone Age of Mount Carmel II: The Fossil Human Remains from the Levallois-Mousterian*, Clarendon Press, Oxford.
Meignen, L., 1995, Levallois Lithic Production Systems in the Middle Paleolithic of the Near East. In *The Case of the Unidirectional Method. The Definition and Interpretation of Levallois Technology*, edited by H. Dibble and O. Bar-Yosef, pp. 361–380. Prehistory Press, Madison, WI.
Meignen, L., Bar-Yosef, O., and Goldberg, P., 1989, Les Structures de Combustion moustériennes de la grotte de Kebara (Mont Carmel, Israël). In *Nature et Fonctions des Foyers Prehistoriques*, edited by M. Olive and Y. Taborin. pp. 2: 141-146. APRAIF, Nemours.
Meignen, L., Beyries, S., Speth, J., and Bar-Yosef, O., 1998, Acquisition, Traitement des Matières Animales et Fonction du Site and Paléolithique Moyen dans la Grotte de Kebara (Israël): Approche Interdisciplinarire. In *Economie Préhistorique: Les Comportements de Subsistance au Paléolithique*, edited by J.-P. Brugal, L. Meignen, and M. Patou-Matis, pp. 227–241. Editions APDCA, Sophia Antipolis.
Millard, A. R., and Pike, A. W. G., 1999, Uranium-Series Dating of the Tabun Neanderthal: A Cautionary Note, *Journal of Human Evolution* 36(5):581–586.
Quam, R. M., and Smith, F. H., 1998, A Reassessment of the Tabun C_2 Mandile. In *Neandertals and Modern Humans in Western Asia*, edited by T. Akazawa, K. Aoki, and O. Bar-Yosef, pp. 405–421. Plenum Press, New York.
Rak, Y., 1986, The Neanderthal: A New Look at an Old Face, *Journal of Human Evolution* 15:151–164.
Rak, Y., 1990, On the Differences Between Two Pelvises of Mousterian Context from the Qafzeh and Kebara Caves, Israel, *American Journal of Physical Anthropology* 81:323–332.
Rak, Y., 1998, Does Any Mouterian Cave Present Evidence of Two Hominid Species? In *Neandertals and Modern Humans in Western Asia*, edited by T. Akazawa, K. Aoki, and O. Bar-Yosef, pp. 353–366. Plenum Press, New York.
Roberts, N., Eastwood, W. J., Lamb, H. F., and Tibby, J. C., 1997, The Age and Causes of Mid-Late Holocene Enviromental Change in Southwest Turkey. In *Third Millennium BC Climate Change and Old World Collapse*, edited by H. Nüzhet Dalfes, G. Kukla, and H. Weiss, pp. 409–429. Springer-Verlag, Berlin.
Rosen, A. M., 1986, *Cities of Clay: The Geoarchaeology of Tells*. University of Chicago Press, Chicago.
Sacket, J. R., 1991, Straight Archaeology French Style: The Phylogenetic Paradigm in Historic Perpsective. In *Perspectives on the Past: Theoretical Biases in Mediterranean Hunter–Gatherer Research*, edited by G. A. Clark, pp. 109–139. University of Pennsylvania Press, Philadelphia.
Schiegl, S., Lev-Yadun, S., Bar-Yosef, O., E. Goresy, A., and Weiner, A., 1994, Siliceous Aggregates from Prehistoric Wood Ash: A Major Component of Sediments in Kebara and Hayonim Caves (Israel), *Israel Journal of Earth Sciences* 43:267–278.
Schiegl, S., Goldberg, P., Bar-Yosef, O., and Weiner, 1996, Ash Deposits in Hayonim and Kebara Caves, Israel: Macroscopic, Microscopic and Mineralogical Observations, and their Archaeological Implications, *Journal of Archaeological Science* 23:763–781.
Schiffer, M. B., 1987, *Formation Processes of the Archaeological Record*, University of New Mexico Press, Albuquerque.
Schwarcz, H. P., and Rink, W. J., 1998, Progress in ESR and U-Series Chronology of the Levantine Paleolithic. In *Neandertals and Modern Humans in Western Asia*, edited by T. Akazawa, K. Aoki, and O. Bar-Yosef, pp. 57–67. Plenum Press, New York.
Schwarcz, H. P., Simpson, J. J., and Stringer, C. B., 1998, Neanderthal Skeleton from Tabun: U-Series Data by Gamma-Ray Spectrometry, *Journal of Human Evolution* 35:635–645.
Stefan, V. H., and Trinkaus, E., 1998, Discrete Trait and Morphometric Affinities of the Tabun 2 Mandible (hu970210), *Journal of Human Evolution* 34(5):443–468.
Stern, N., 1993, The Structure of the Lower Pleistocene Archaeological Record, *Current Anthropology* 34(3):201–225.
Stiner, M. C., Munro, N. D., Surrovell, T. A., Tchernov, E., and Bar-Yosef, O., 1999, Paleolithic Population Growth Pulses Evidenced by Small Animal Exploitation, *Science* 283:190–194.
Stringer, C. B., 1994, Out of Africa: A personal history. In *Origins of Anatomically Modern Humans*, edited by M. H. Nitecki and D. V. Nitecki, pp. 149–174. Plenum Press, New York.
Stringer, C. B., 1995, The Evolution and Distribution of Later Pleistocene Human Populations. In *Paleoclimate and Evolution, with Emphasis on Human Origins*, edited by E. Vrba, G. Denton, T. Partridge, and L. Burckle, pp. 524–431. Yale University Press, New Haven.
Stringer, C. B., 1998, Chronological and Biogeographic Perspectives on Later Human Evolution. In

Neanderthals and Modern Humans in Western Asia, edited by T. Akazawa, K. Aoki, and O. Bar-Yosef, pp. 29–37. Plenum Press, New York.

Stringer, C. B., and McKie, R., 1996, *African Exodus: The Origins of Modern Humanity*. Pimlico, London.

Stuiver, M., and Braziunas, T. F., 1993, Modeling Atmospheric ^{14}C Influences and ^{14}C Ages of Marine Samples to 10,000 B.C., *Radiocarbon* 35(1):137–189.

Stuiver, M., Reimer, P. J., Bard, E., Beck, J. W., Burr, G. S., Hughen, K. A., Kromer, B., McCormac, G., van der Plicht, J., and Spurk, M., 1998, INTCAL98 Radiocarbon Age Calibration, 24,000-0 cal BP, *Radiocarbon* 40(3):041–1084.

Taylor, C. B., 1997, On the Isotopic Composition of Dissolved Inorganic Carbon in Rivers and Shallow Groundwater: A Diagrammatic Approach to Process Identification and a More Realistic Model of the Open System, *Radiocarbon* 39(3):251–268.

Trinkaus, E. (ed.), 1989, *The Emergence of Modern Humans: Biocultural Adaptations in the Later Pleistocene. School of American Research Seminar Series*. Cambridge University Press, Cambridge, UK.

Trinkaus, E., 1995, Near Eastern Late Archaic Humans, *Paléorient* 21(2):9–24.

Trinkaus, E., Ruff, C. B., and Churchill, 1998, Upper Limb versus Lower Limb Loading Patterns among Near Eastern Middle Paleolithic Hominids. In *Neandertals and Modern Humans in Western Asia*, edited by T. Akazawa, K. Aoki, and O. Bar-Yosef, pp. 391–404. Plenum Press, New York.

Valladas, H., Mercier, N., Joron, J.-L., and Reyss, J.-L., 1998, GIF Laboratory Dates for Middle Paleolithic Levant. In *Neandertals and Modern Humans in Western Asia*, edited by T. Akazawa, K. Aoki, and O. Bar-Yosef. Plenum Press, New York.

van Andel, T. H., 1998, Middle and Upper Palaeolithic Environments and the Calibration of ^{14}C Dates Beyond 10,000 BP, *Antiquity* 72(275):26–33.

van der Plicht, J., 1999, Radiocarbon Calibration for the Middle/Upper Palaeolithic: A Comment, *Antiquity* 73(279):119–123.

Vandermeersch, B., 1981, *Les Hommes Fossiles de Qafzeh (Israel)*. CNRS, Paris.

Vandermeersch, B., 1985, *The Origin of Neandertals. Ancestors: The Hard Evidence*, pp. 306–309. Alan R. Liss, New York.

Vandermeersch, B., 1989, L'extinction des Neandertaliens. In *L'Homme de Néandertal*, edited by M. Otte, p. 11–21. Etudes et Recherches Archèologiques de l'Université de Liège.

Vandermeersch, B., 1992, The Near Eastern Hominids and the Origins of Modern Humans in Eurasia. In *The Evolution and Dispersal of Modern Humans in Asia*, edited by T. Akazawa, K. Aoki, and T. Kimura, pp. 29–38. Hokusen-Sha, Tokyo.

Vandermeersch, B., 1995, Le Role du Levant dans l'Evolution de l'Humanit, au Pleistocene Supèrieur, *Paléorient* 21(2):25–34.

Vita-Finzi, C., and Higgs, E. S., 1970, Prehistoric Economy in the Mount Carmel Area of Palestine: Site Catchment Analysis, *Proceedings of the Prehistoric Society* 36(1):1–37.

Waelkens, M., Paulissen, E., Vermoere, M., Degryse, P., Celis, D., Schroyen, K., De Cupere, B., Librecht, I., Nackearts, K., Vanhaverbeke, H., Viaene, W., Muchez, P., Ottenburgs, R., Deckers, S., Van Neer, W., Smets, E., Govers, G., Verstraeten, G., Steegen, A., and Cauwenberhs, K., 1999, Man and Environment in the Territory of Sagalassos, a Classical City in SW Turkey, *Quaternary Science Reviews* 18(4–5):697–709.

Wagner, G. A., 1998, *Age Determination of Young Rocks and Artifacts: Physical and Chemical Clocks in Quaternary Geology and Archaeology*. Springer-Verlag, Berlin.

Walter, R. C., Manega, P. C., Hay, R. L., Drake, R. E., and Curtis, G. H., 1991, Laser-Fusion ^{40}Ar/^{39}Ar Dating of Bed I, Olduvai Gorge, Tanzania, *Nature* 354:145–149.

Weiner, S., Goldberg, P., and Bar-Yosef, O., 1993, Bone Preservation in Kebara Cave, Israel Using On-Site Fourier Transform Infrared Spectrometry, *Journal of Archaeological Science* 20:613–627.

Weiner, S., Schiegl,S., Goldberg, P., and Bar-Yosef, O., 1995, Mineral Assemblages in Kebara and Hayonim Caves, Israel: Excavation Strategies, Bone Preservation and Wood Ash Remnants, *Israel Journal of Chemistry* 35:143–154.

Wolpoff, M., 1996, Multiregional Evolution and Modern Human Origins. International Institute for Advanced Studies. The Origins and Past of Modern Humans: Towards Reconciliation, World Scientific, Kyoto.

Wolpoff, M., and Caspari, R., 1996, *Why Aren't Neandertals Modern Humans?* XIII International Congress of Prehistoric and Protohistoric Sciences, Forli, Italy, ABACO Edizioni, Italy.

Index

Ablation, 14
Accretion, discontinuous, 233
Accumulation rate, *see* Sedimentation rate
Acheulean, 483
   Southwest France, 221
Acheuleo-Yabrudian, 483
Acropolis limestone, 450
Activity areas, 56, 210, 338
Aegean, 450
Air photos, 110, 116, 119, 122, 366
   and vegetation patterns, 379
African "Eve" hypothesis, 481
African-Jordan Rift System, 455
"Age of Man," 10
"Age of Humanity," 10
Aggradation, *see also* Alluviation
   aggradational terrace, 118
   climatically driven, 79
   landscape, 132
Agriculture
   canals, 98
   early sites, Peru, 120
   early societies, 234
   effects on soils, 214
   and erosion, 98, 135
   fields, 282
   and landscapes, 116
   maize, 370
   origins of, 475
   prehistoric, 64
   raised fields, 98, 122
   ridged fields, 63, 122
Agricultural settlement systems, 98
Alfisols, 257
   in early-middle Holocene alluvium, 182
   on terraces, 88
Allogenic rivers, 83

Allostratigraphy, Alloformations, *see* Stratigraphy
Alluvial architecture, 57
Alluvial chronologies, 58–62, 66
   and dendrochronology, 71
   precision, 60
Alluvial environments
   backswamp, 191
   discharge, 63, 70
   floodplain, 79
      occupation of, 232
   levee, 83, 191
   point bars, 83
   at Pleistocene and Holocene scales, 79
Alluvial fans, 61, 85, 118–119, 121
   bajadas, 85
   coalesced, 85
   early Holocene Texas, 435
   and formational processes, 187
   middle Holocene, Kansas, 89
   and monumental architecture, 125
   Pleistocene, 123
   and soil formation, 187
   and soils stratigraphy, 91
Alluvial morphogenesis, 86–88
Alluvial stratigraphic sequence, 57–58
   materials used in radiocarbon dating, 59
Alluvial terraces, *see* Terraces
Alluviation, 56, 65–67
   anthropogenic, 65–66
   and base level, 56–57, 62, 64–68
   and climate, 71, 78, 437
   and discharge, 70
   East Africa, 475
   eustacy and 78
   factors of, 78
   Holocene, 62
   and land use, 62

Alluviation (cont.)
  and population variation, 66
  and preservation potential, 444
  and settlement patterns, 64
  and soil formation, 85–88, 139, 175, 435–438
  and tectonics, 62, 67–68
Alluvium, 186, 277
  dating, 393
  faunas in, 263
  floodplain, 85–87, 182
  and intercalated eolian sand, 118
  provenance, 67
  and soils stratigraphy, 91
  and terraces, 79
Altithermal, see Stratigraphy
Amazon Basin, 86
American Bottom culture, 313, 319
Amino acid racemization, 149, 387
Amphibole species, 464
Andes Mountains, 115, 454
Andesite, 455
Antarctica, 15
Anglo-Danish Period, 248
Anomalies, 356, 380
  archaeological testing of, 356, 379
Anhysteretic remenant magnetism, (ARM), 91
Apatite, and fission-track dating, 395
  neoformed, 244
Aplastics, 299
^{40}Ar/^{39}Ar dating, 387, 391–392, 476, 481, 484; see also Geochronology, SCLF
  human bone, 393
  incremental heating, 392
  materials, 387, 393
  total fusion, 392
  sampling requirements, 393
Archaeobotanical objects, 46
Archaeogeophysics, 359–381
Archaeological associations,
  and changing environments, 154
  and changes in site use, 154
  and piercing features, 154
Archaeological excavations,
  geomorphology and, 57
  buried soils and, 183
  strategies and dating potentials, 386
Archaeological features, see Features
Archaeological sites
  Africa
    Thebes (Egypt), 458,
    Colossi of Memnon, 458
    Zimbabwe, 455
  Asia
    Great Wall of China, 155
    Shama Gully, 157
  North America
    3D Ranch, KS, 377–378

Archaeological sites (cont.)
  North America (cont.)
    Alum Creek Site, KS, 178
    Aubrey Clovis Site, TX, 92
    Azatlan WI, 315
    Breed's Hill, MA, 373
    Buckner Creek, KS, 184
    Bunker Hill National Monument, MA, 372
    Cahokia Mounds State Park, IL, 315, 318, 363
    Chaco Canyon NM, 310
    Cherokee Sewer Site, IA, 187
    Chinaberry Site (31HT285), 278, 286
    Delaware Canyon, OK, 96
    Flaming Arrow Site, OK, 370
    Fleur de Lys Quarry (Labrador), 456
    Fort Hood, TX, 60
    Fort Ross, CA, 155
    Fred Edwards WI, 300, 313–315, 320
    Gemma Site, TX, 95
    Hartley Fort IL, 315, 320
    Koster, IL, 97
    Loy Site, TN, 340–347
    Lubbock Lake TX, 92
    MAD Site, IA, 192–193
    Main Site, KY, 191
    Menoken Village, ND, 370
    Mississippi Valley, lower, 90
    Moundville AL, 310
    Pecos Pueblo NM, 319
    Rench IL, 315
    Seal Cove Site, CA, 158–165
    Site 13HA385, IA 177
    Site 38GE261, SC, 282–283
    Sluss Cabin, KS, 372
    Whistling Elk Village, SD, 366–370
  Europe
    Arene Candide (IT), 248
    Barbas (FR), 221
    Bilzingsleben II (Germany), 90
    Boxgrove (GB), 263
    Butser (GB), 247–248
    Carrera Quarry (IT), 457
    Colchester House, London (GB), 258
    Easton Down, GB, 259
    El Castillo (Spain), 410
    Ephesos, 457
    Epirus, Greece, 146
    Folly Lane (GB), 244
    Franchthi Cave (Greece), 450
    Grotte du Lazaret (FR), 223
    Guild Hall (GB), 248
    Haynes Park (GB), 258–260
    Kenchreai, Greece, 150
    Lazaret (FR), 223
    Mons Claudianus (IT), 456
    Naven Fort, (Northern Ireland), 375–377
    Overton Experimental Earthwork (GB), 245

Index

Archaeological sites (*cont.*)
  Europe (*cont.*)
    Port Marianne (FR), 228
    Potterne (GB), 258–259
    Ribemont-sur-Ancre (FR), 230
    Rome (IT), 451
    Rounds (GB), 255, 257
    Stonehenge (GB), 451
    Vaufrey (FR), 223
  Near East
    Hayonim Cave (Israel), 477
    Kebara Cave (Israel), 477–482
    Lagash, (Iraq), 98
    Nahal Issaron (Israel), 480
    Skuhl Cave (Israel), 482
    Tabun Cave (Israel), 411, 481–483
    Tell Arqa (Lebanon), 226
    Tell Beydar, 223
    Tell Brak, 223
    Tell Dja'de, 210, 222
    Tell Leilan, 222–223
    Ubeidiya, 95
    Umm el Tlel (Syria), 225
    Ur, (Iraq), 98
    Vadum Jacob (Israel), 154–155
  South America,
    Huaca de la Luna, Peru, 122
    Huaca del Sol, Peru, 122
    Lomas las Altas, Peru, 119
    Manchan, Peru, 122
Archaeological site evaluation, 186, 195, 210
  area-focused geophysical survey, 356
  auger, 160
  and buried soils, 181,185
  in CRM archaeology, 185, 354
  deep testing, 185
  microartifacts and, 330
  microscreening, 184, 333
  sampling off site, 331
  shovel probes, 94, 160
  site integrity, 187, 210, 215–218
    horizontal and vertical, 185
  siting stratigraphic excavations, 57
  spatial patterning and, 477
  stratigraphic context, 185
  stratigraphy and, 190, 210
  surface collections, 185
  surface soils and, 181
  test pits, 160, 185
  trenches, 94, 160
Archaeological survey (regional)
  air photos, 69, 110–112, 379
  photomaps, photomosaics, 182
    interpretations, 355
  and alluviation, 64
  in arid lands, 57, 94
  backhoe trenches, 94

Archaeological survey (regional) (*cont.*)
  and buried sites, 64, 93
  and cognitive geography, 115
  coring and microscreening, 94, 113, 332–338
  data collection strategies, 115, 195
  in East African rift valleys, 476
  erosion and, 475–476
  geoarchaeological, 116, 184
  geophysical,
    in CRM, 354
    high precision, 357
  geomorphic, 108–109, 181
  Greece, 55
  ground penetrating radar (GPR), 113
  hand auger, 186, 349
    using GIS, 112, 127, 347
    in CRM, 181
  and landscape stability, 175
  locating sites, 175
  magnetic, 357–358
  mechanical coring, 184
  patterned geometries on landscape, 355
  protocol, 126
  and relational data base, 127
  resistivity profiles, 113
  sampling,
    biases, 40
    and discovery methods, 183
  seismic profiling, 345
  shovel testing, 113, 184
  "siteless," 109
  and site discovery methods, 93–94
  site potential, 182
  site prediction models, 93
  soils 175
    buried soils, 344
    soil surveys, 184, 272
  stratigraphy, and 57
  submerged sites, 94, 147
  survey data, use of, 64
    survey area and feature detection, 355
  transects, 110
  trenching, 186, 344
  vibracoring, 345
Archaeological stratigraphy, *see* Stratigraphy
  micromorphology as indispensible companion of, 235
Archaelogy
  behavioral, 40–41, 208
  costs, 353
  ethnoarchaeology, 347
    experimental, 247
  historical, 282, 377
  interdisciplinary, 474
  landscape, 344, 475
  the New Archaeology, 38
  Paleolithic, 410

Archaeology (cont.)
  regional, 330, 473
  research strategies, 211
  stratigraphic, 90, 217
  textbooks, 5
Archaeoseismology, 143–170
  and stratigraphy, 150
Archaeostratigraphy, 45
Archaic sites, 23, 85, 91–92, 97, 155, 183–184, 187, 191, 281, 283, 286, 434, 440
Architecture
  monumental, Peru, 124
Argillic horizon, see Soils
Arroyos, 57
  the arroyo problem, 83
  and paleodischarge, 70–71
Artifacts, see also Chert, Ceramics, Obsidian
  abrasion of, 94
  amber, 450, 463
  artifact stability, 339
  assemblages, 42
    composition, 338
    and site preservation, 215
  bricks, 228, 230
  chemical analysis and sourcing, 464
  copper, 130, 450
  detrital, 121
  fire-cracked rock (FCR), 56, 160
  forgeries, 451
  iron, and induced magentic fields, 357
  lithic, 478
    microflakes, 221
    reduction strategies, 335, 338, 479
  Pleistocene, 123
  as marker fossils, 114
  metal, petrographic description, 451
  procurement, 40; see also Quarries
  reference collections, 247
  size classes, 334
  soapstone, 450
  and stratigraphy, 89, 114
Artifact densities
  and geomorphology, 134
  and plowing, 134
  and surface stability, 95, 132–133
  prediction, 134
Artifact distributions
  across landscape, 132
  of archaeological sites, 40
  on surfaces, 114, 221, 379
  and systemic use, 40
  vertical, 284
Artifact patterns, 46
  behaviors, use and manufacture, 45
Asia Minor, 454
Ash, 411
  calcitic, 262

Ash (cont.)
  fused, 258, 262
Aswan, 460
Athenian acropolis, 449
Atmospheric CO_2, 21
Atomic absorbtion spectometry (AAS), 452
Atomic emission spectrometer, 452
Aurignacian, 221
Autofluorescence, 258
Australia
  early peopling of, 386, 408
Australopithecines, 23
Avifauna, exploitation, 130
Avulsion
  meanderbelt, Trinity River, 86
  rates, 85
  and site preservation, 85

Bajadas, 85
Balkans, 454
Baltic, 450, 463
Bargone (IT), 261
Barrows, 255
Basalt, 455
Base level, see also Alluviation
  and channel entrenchment, 68
Basin of Mexico, 66
Beach ridges,
  Peru, 117
Bedrock lithology and structures, 90
Behavioral laws, 48; see also Archaeology
Behavioral process, 44
Biomantle, 192
Biomass, 423
  and soils, 437–438
Biostratigraphic sequence, 10
Biostratigraphy, see Stratigraphy
Bioturbation, 69, 178–179, 185, 227, 270, 280, 481; see also Faunalturbation, Formational processes, Turbation
  anthropic pedoturbation, 185
  by ants, 185, 274, 281
  and burial depth, 283
  and dating, 387
  by earthworms, 185, 254, 281
    and the worm's fecal pellets, 284
  versus eolian sedimentation, 272, 286
  faunalturbation, 269
  floralturbation, 69
  krotovina, 187–224
  and large artifacts, 160
  and soil formation, 177
  by termites, 185
  by tortoises, 281
  by gophers, 281
Bogs, 56
  ores, 254

Index

Bogs (cont.)
  and soils, 175
Bones, see also Faunal remains
  bone beds, 44
  dating, 393
  taphonomy, 95
Borrow pits, 58
Brazil, 455
Bronze age, 257
  chert use, 459
  mining, 130
  Scandanavia, 460
  subsistence, 130
Buckner Creek Paleosol, 183–184
Burned rock, 56
Buried soils, 69, 121, 175–180, 194, 259, 275, 287, 428, 435–437
  A horizons, 274–275, 287–288, 290, 344
  alluvial, 439
  evidence for human occupation, 183
  keys to identifying buried cultural deposits, 183
  sterile of archaeological materials, 183
  superposed, 188
^{14}C, see Radiocarbon
Cairo, quartzite quarry, 458–459
Calciustolls, 193
Calcrete, see Soil carbonates
CAM plants, 422
Canals, 116
Canton Ware, 313
Carbon analyzer, 275
Carbonate nodules
  kithogenic, pedogenic, detrital, groundwater, 439–430
Carbonates, see Groundwater, Limestone, Soil carbonates
  microfabrics and cave configuration, 231
  sourcing by C, O and Sr isotopes, 453
  trace element analysis, 462
Carrara marble quarries, 457
Cascadia Subduction Zone, CA, 147
  land level change, 147
  tsunami, 147
Casma period, Peru, 120
Cathodoluminescence (CL), 452, 462–464
  data indexed by colors, 464
  sourcing marble, 457
CL ultraviolet (UV) spectra database, 457
"Cathodomicrofacies" database, 457
Caves
  Eurasia, 45
  and ESR dating, 478
  Europe, 223, 231, 476–477
  formation processes, 45, 231
  Greece 450
  Levant, 411, 477–483
  paintings, dating, 408

493

Caves (cont.)
  speleothems, 392
Celtic, 230
Cemeteries,
  Peru, 121
Cenozoic era, 7
Central place foraging, 479
Central Plains, 89, 93, 97
Ceramics
  body versus paste, 304, 321
  Chaco, 310
  classification, 307
  cord-decorated, 313
  detrital, as age constraint, 118, 157
  elemental composition, 322
  exchange, 318, 322; see also Trade
  fission-track dating, 395
  groundmass, 299
  levigation, 304
  matrix, 299
  microattributes, 331
  microsherds, 331–339
  Mimbres, 304
  Mississippi Valley, 308
  Modal analysis, 305
  Neutron activation analysis (NAA), 452
  paste, 304
  paste index, 315
  phasing with pottery analysis, 244
  petrography, 297–323
  point counting, 305–307
    Delesse (area-volume) relation, 306
    Glagolev-Chayes method, 306
    sampling interval, 306
  seriation, 157
  shell temper, 303, 310–311, 316
  Southern Plains, 316
  southwestern, 308
  temper
    grog, 311
    grit, 301–302, 308, 311, 313
    sand, 301–302, 311
    shell, 303, 310–311, 316
  utilitarian, 311
Cesium-vapor gradiometer, 358
Chalcedony,
  dating, 405
  biogenic opal, 406
  sourcing, 459
Chalcolithic,
  Cyprus, 130, 134
Chalk, see Limestone, Carbonates
Chalk downlands, GB, 257
Channels,
  braided, 82
  classification, 82, 84
  chute, 95

Channels (cont.)
  cut-off and site formation, 95
  entrenchment and architecture, 58
    and stratigraphic variation, 58
  meandering, 82
  patterns, 81
  stability, 81
Charcoal,
  and radiocarbon dating, 119
  in soils, 245
Chert,
  burned, 404, 478
  in ceramics, 318
  as chemical siliceous sediment, 459
  dating, 404
  petrography, 459
  sourcing by O isotopes, 453, 461
Chihuahuan Desert NM, 443
Chimu Period (Peru, Late Intermediate Period), 123
China, 400–444
Chronology,
  archaeological, 119
  cultural, 5
  Cyprus, 131
Chronometric scale, 7
Chronostratigraphic scale, 7
Chronostratigraphy, 13; *see also* Stratigraphy
  columns, 79
CL, *see* Cathodoluminescence
Class, ceramic, 307
Clay, 299
  coatings, 245
    impure, 255
  mineralogy and soil age, 87
  and floodplain soil formation, 87
  and trace element composition of carbonates, 462
  smectite, 87
  subsoil, 321
Climate,
  and alluviation, 71
  and archaeological record, 27
  effects on environment, 3
  Eastern Mediterranean, 129
  El Niño, 116
  and fluvial response, 63, 84, 88
  frequency, 24
  global, 234
  high precision proxy records, 72
  and human adaptations, 27
  Israel, 129
  macroclimatic, 79
  and magnetic susceptibility, 475
  Neoglacial, Wyoming, 193
  and ocean circulation, 24
  paleoclimates, 234

Climate (cont.)
  and pollen data, 24, 129
  post-Pleistocene, 79
  and tectonism, 63–65
Climate-response models, 20–22, 81
Climatic forcing, 63
  on fluvial systems, 80
Climatic models,
  air masses, 425
  archaeoclimatic, 22, 25
  computer simulations, 20
  COHMAP, 20–22
  and geologic data, 24
  GCM, 20, 25
Climatic optimum, 130
Clovis culture, 92
Coastal deposits, *see also* Sediments
  dating, 389
Coasts,
  Holocene progradation, 117, 124
  coastal plain SE United States, 270
Colluvium, 69, 81, 85, 111, 377
  episodic colluviation, 231
Colorado Plateau, 88
Compilation of Historical Materials of Chinese Earthquakes, 156
Complex social systems,
  and trade, 126
Context
  archaeological, 39
  cultural, 39–42
  environmental, 135
  systematic, 39, 47
Contextual archaeology, 4
Contextual relationships of grains, 44,
Cooperative Holocene Mapping Project (COHMAP), 20
Copan Lake, OK, 181
Copan Paleosol, 188
Copper,
  mining, 130
  native, 130
  trade networks, 132
Coprolites, mineralized, 244
Cordilleran ice sheet, 24
Cores from glaciers, 15
Cosmogenic isotopes, 149
Cosmogenic nuclide analysis, 114
Coupled ESR/U series dating, 388
Crack propagation, 310
Cretaceous rocks, 111
  White chalk of Western Europe, 459
Critical power, 79
CRM, *see* Cultural resources management
Crusader castles, Israel, 154
Cryoturbation, 187
  and sand wedges, 187

Cryturbation (*cont.*)
  and stone polygons, 187
  and solifluction lobes, 187
Cryptocrystalline quartz, 459
Crystallitic, 262
Cultural activities and deposits, 44, 173
"Cultural determinism," 309
Cultural particles, 329
Cultural patterns, 39
Cultural resources management (CRM), *see also* Archaeological surveys, Archaeological site evaluations
  in fluvial settings, 78
  geophysical surveys, 354
  National Register eligibility, 185
  site mitigation, 190
Cumberland River Valley, KY, 191
Cyprus, 112–115, 126–136

Dating, *see* Geochronology
Daub, 222, 329
Day length, 12
Dead Sea transform fault (Israel), 146
Debris flow, 175
  events, 118
Deep sea cores, 14–15
Deforestation, Cyprus, 126, 132
Delaware River Valley PA, 459
Delesse relation, 307
Dendrochronology, 60, 149; *see also* Geochronology
  dating historic deposits, 70
  Illinois, 69–71
  precision, 60
  Southwestern sites, 60–61
Dendrohydrology, 63, 70
Densities, *see* Artifact densities
Denudation
  chemical and mechanical, 80
Deposition, *see* Sedimentation
  complex modes, 387
  co-seismic deposits, 148
  faulted deposits, 164
Depositional geomorphic features, 86
Depositional history of artifacts, 38, 329
Deposition rates, *see* Sedimentation rates
Deposits, *see also* Sediments
  lithology and sedimentary history, 44
  occupation, 254
  as organizational units, 44
  vertical accretion, 87
Desert pavement, 118
Des Moines River Valley, IA, 181–182
Dabase, 455
Diagenesis,
  ash, 262–263
  and petrography, 323

Diatoms, 244–245, 261
  semiquantitative analysis, 259
Dielectric coefficient, 363
Dielectric permetivity
  relative, 363, 381
Disaggregation, 332
Discharge,
  mean annual, 63
  reconstructed, 70
Discriminant function analysis, 455
Disturbance, *see* Formational processes
Displacement, artifacts, 283
Dolomite, dolostone, 461
Domestication, corn, 419
Dordogne Valley (FR), 221, 223
Dosimeters, 397
Drainages,
  allogenic versus ephemeral, 83
  and allometric change, 88
  Holocene, Peru, 79, 117–119, 121
  hydrogeomorphic evolution of, 79
Drumlins, and site formation, 372–375
Duck River Basin, TN, 90
Dunes, 274; *see also* Eolian
  barchanoid, 275
  Southern High Plains, 85
Dung, 261
  organic herbivore, 244

Early Metal Age, 459
Earthquakes, 156; *see also* Seismicity, Tsunami
  archaeological documentation, 148
  artifact seriation and, 148
  characteristic earthquake concept, 156
  clustered earthquake concept, 157
  dating, 145
  effects, 144
  event chronology, 156
  historical, 148, 157
  Holocene, 165
  Khait, 151
  Lisbon 1755, 143
  New Madrid, 152, 157
  recurrence interval, 145, 156
  San Francisco 1906, 155
    and offset of Archaic Period features, 155
  surface-rupturing, 164
  Yiuchuaun-Pingluo (1739 A.D.), China, 155
Earth sciences and archaeology, 4
Earthworks, 58
East Africa, 94, 475
Easter Island, 454
Eccentricity, 17
Effective miosture,
  and alluviation, 71
Egypt, 450, 479–480
  calendrical chronology, 479

Egyptian Sahara, 94
Elbe-Saale region, Germany, 90
El Kowm Basin (Syria), 225
El Niño
  cultural, technological change, 126
  ecological and environmental stress, 126
  deposit, 118
  loss of arable land, 123
Electrical resistivity, 94–95, 113, 354, 380
  depth of prospection, 359
  and ground moisture, 361
  meter, twin probe array, 361, 381
  sampling intervals, 356
  Wenner configuration, 359, 381
Electromagnetic conductivity (EM), 354, 362–365, 380
  buried metal, 362
  sampling intervals, 356
  sensitivity, 362
  spectrum, 451
  survey results, Whistling Elk Village, 357–358
  compared with resistivity data, 369
Electron microscope, 452
Electron spin resonance (ESR), 149, 396–404, 478–483
  age uncertainty, 388
  calcite speleothems, 404
  coupled ESR/U series, 388, 403
  costs, 402
  dating range, 388, 402
  dosimetry, 397–398
  equivalent dose, 397
  ESR spectra and artifact sourcing, 457–461
  Gamma spectrometer, 398
  in karst-hosted deposits, 399
  isochron method, 402
  materials, 388
  Optical ESR method, 403
  paleodose, 397
  sampling tooth enamel, 400, 402
  shell dating, 403–405
  teeth, kinds dated, 402
Electronic gamma spectrometer, 398
Element map, 262
Elluviation, 179; see also Soils
EM, see Electromagnetic conductivity
Emergent Period (California), 155
Environmental change
  continental scale, 474
  and correlated culture change, 125
  and human prehistory, 3
  late Cenozoic, 26
  rate of change, 475
  and rates of evolution, 24
  and technological readjustments, 27
Environmental radioactivity, 399
Environmental reconstruction, 5, 193, 211, 434

Eocene biosparite, 450; see also Limestone
Eolian, see also Sand
  endemic airborne dust input, 233
Ephemeral streams, 111; see also Arroyo, Korongo, Wadi
Epipaleolithic, 78
Epoxy impregnation, 298
Equilibrium state, 84
Erosion, 56–57, 475; see also sheetwash
  agriculture versus pastoralism, 135
  anthropogenic, 135
  and aggradation, 58, 222
  climates and, 65–66, 69
  cycles, 222
  and deforestation, 135
  dendrogeomorphological criteria for, 70
  and differential preservation, 435
  erosional contacts, 227
  erosional surfaces,
    bedrock, 121
    Pleistocene, 129
  erosion factor (K), 286
  and internal factors, 68
  and mining, 135
  potential, 278
  rates, 98, 135, 435, 494
  by sea cliff retreat, 161
  and temporal-spatial variation in settlement, 64
  and stratigraphic record, 432
  and vegetation, 440
Eskimo, serpentine use, 456
ESR, see Electron spin resonance
ESR spectroscopy, for sourcing and dating cherts, 461
Estuaries, 123
Ethiopia, 391
Ethnoarchaeological research and archaeological practice, 347
Ethnoarchaeological settings, 41
Ethnostratigraphic units, see Stratigraphy
Eustacy
  and alluvial aggradation, 68
  and base level, 64, 68
  and channel entrenchment, 68
  and site formation, 46, 56, 263
  and tectonic activity, 64
  variation, 62
Evolution, cultural, 109, 130
Excavation, see Archaeological excavation
Exchange, see Trade
Excrement, soil animal, 245

Fabrics, depositional, 162
Facies, 207
  analysis, 214
  anthropogenic, 226
  classification, 215
  contacts between, 208

Index 497

Facies (*cont.*)
  horizontal associations, 208
  microstratified, 222
  pedosedimentary, 228
  and sampling, 433
  shell-rich, 164
Fan deltas, 117–118
Fanglomerates, 129, 134
Faults
  cumulative offset, 165
  Dead Sea transform, Israel, 155
  dextral slip, 158
  hanging wall and footwall, 154
  Hierapolis, Turkey, 150, 157
  Honggouzigou, China, 155
  Korinth, Greece, 150
  normal faults, 152
  rates of fault slip, 152
  rotated shells and, 164
  San Andreas, 150
  San Gregorio 158
  Scarps
    collapse, 151
    and occupation sites, 144
  strand, 164
  strike-slip faults, 153–154
    and artifact isoconcentrations, 154–156
    and associated archaeological materials, 154
    and colluviation, 154
    surface fault rupture, 163
    thrust faults, 153
      Weihe Graben, China, 157
Faunal remains
  archaeological study, 20
  microfauna, 478
  of migrating animals, 163
  taphonomy, 477
  and paleoclimatic reconstruction, 441
FAUNMAP Working Group, 24
Features, *see also* Middens, Phosphates
  burials Qafseh (Israel), 479
  cess pits, 258
  dating, 59
  detection by air photos, 366
  earth platform, 229
  hearths, 209, 371, 477; *see also* Ash
    and charcoal, 60
    relict cooking hearths, 162
    and magnetic susceptibility, 357
  floors, 247
    burned, 357
    chipping, 263
    and ground penetrating radar (GPR), 363
    living, 220, 232
    occupation, 254
    stabling versus domestic, 248–254
  fortifications, 355, 366

Features (*cont.*)
  funerary platform, 230
  houses
    burned, 370–371
    and geophysical survey data, 354–355
    mud brick, 210–230
  kilns, 357
  magnetic and dipole fields, 357
  palisades, and GPR survey, 370
  phosphates and 174, 282
  plastered walls, 222–226
  post-holes, and GPR survey, 370
  resistant features, 362
  response to remote sensing technologies, 380
  seismogenic, 150
  shallow burial and visibility of, 370
  stratigraphic distribution of, 245
  visibility with high and low frequency GPR antennae, 364
Feldspars, and dating, 392
  luminescence, 407
Fission-track dating, 394, 454
  age uncertainty, 388
  dating range, 388
  materials, 388
  physical basis, 394
  samples required, 395
  Toba eruption, 395
Floods, 277
  catastrophic, 24
  and El Niño, 116
  frequency and magnitude, 80
Flood drape, 177
Floodplain abandonment, 86
Flint, *see* Chert
Fluvial deposits, *see* Alluvium
Fluvial environments,
  Pleistocene, 79
Fluvial systems,
  human modification of, 78, 97
Fluvial records,
  Pleistocene versus Holocene, 79
Fluxgate gradiometer, 358
Formational processes, 37, 45, 192, 232, 476; *see also* Bioturbation, Tree throws
  actualistic studies, 339
  artifact zones, 271
  ash, 258, 262–263, 477
    diagenesis, 411
  collapse (walls), 222–227
  complicated, 258
  construction, 93
  context
    archaeological, 39
    cultural, 39–42, 208
    early horticulture, 95
    systematic, 39, 47

Formational processes (cont.)
  and ^{14}C dating, 481–482
  cultural sorting (pickup cleaning), 339
  deposit as unit of analysis, 38, 43
  dumping, 227
  and facies analysis, 95
  faunalturbation and floralturbation, 270
  in fluvial environments, 94–95
  granulometry and, 47
  geogenic versus anthrogenic, 477
  history of, 38
  induration, 220
  and landscape development, 173
  natural factors, 208
    versus cultural factors, 94
  natural forcing versus sociocultural dynamics, 233
  overprinting via soil formation, 95
  Paleolithic, 478
  plow zones, 282, 339, 366–368
  post-depositional processes, 478
  postoccupational alluvial histories, 93
  refuse, 40, 208
    de facto, 40
    primary, 40
    secondary, 40
  rodent activity and, 367
  soil-artifact context model, 189
  soil-stratigraphic context, 195
  and stratigraphy, 217
  and surface stability, 95
  three-dimensional reconstruction, 229–232
  tillage, 338–339
  trampling, 221, 226, 231, 246, 255, 338–342
  transformations, 218
    effects on artifact assemblages, 227
Formative Period, 66
Fort Bliss TX, 439
Fort Bragg Military Reservation, 256, 278, 286
Fort Hood, TX, 439
Fort Stewart, GA, 280
Full-glacial-postglacial transition, 12
Fulvic acid, 257

Galisteo, 319
Gamma ray spectroscopy, 484
General circulation models (GCM), 20–22
  and archaeological sites, 25
Geoarchaeological records, 77
  completeness, 80
Geoarchaeology, 4, 43–45, 47–48, 55, 58, 60, 242, 434, 439
  contexts, 174
  in fluvial environments, 77–106
  and formational processes, 38
  in hunter-gatherer sites, 97
  and site excavations, 94

Geoarchaeology (cont.)
  and soils stratigraphy, 91
Geochemical fingerprints, 450
  techniques, 451
Geochronologic scale, 7
Geochronology of deposits, see also ^{40}Ar/^{39}Ar, Electron spin resonance
  age uncertainties, 387
  ante quem age estimate, 150
  archaeological, 229
  in archaeoseismology, 149
    ground cracks, 151
  artifact seriation, 148
  dating range of isotopes, 387
  correlative dating, see Isotopes, Stratigraphy
  fission track, 454
  high precision, 72
  ^{40}K/^{40}Ar, 449
  materials used, 387
  numerical dating, see ^{40}Ar/^{39}Ar, ^{14}C, Dendrochronology, Electron spin resonance, Thermoluminescence dating
  organic carbon ratio (OCR), 188
  and paleoclimates, 389
  and paleoenvironments, 389
  radiation exposure, 386
  radiogenic isotopes, 386
  sideral, 148–149
  soils, 174; see also Soils chronosequences, Soil stratigraphy
  tephrochronology, 149
Geochronology of surfaces, 90
  dendrochronology, 135, 151
  elevation above thalweg, 118
  geomorphic and soils methods, 90
  lichenometry, 135, 151
  morphostratigraphy, 57, 112
  rock varnish, 118
  surface roughness, 118
  soil development, 113, 180–181, 190, 427; see also Soils chronosequences, Soil stratigraphy
  and site formation, 474
  terraces, 90
Geoecology, 4
Geographic information systems (GIS), 337
  and geomorphological maps, 127–128
  and morphostratigraphy, 112
  and remote sensing, 94
Geography, 4
  cognitive, 115
  textbooks, 23
Geological Society of America, 11
Geologic time scale, 7
Geomagnetic polarity time scale
  Bruhnes-Matuyama paleomagnetic reversal, 13
  Gauss chron, 13

Geomagnetic polarity time scale (*cont.*)
  Gauss-Matuyama paleomagnetic reversal, 11
  Matuyama chron, 11–13
  Olduvai subchron, 11
  reversals, 11–13
  time scale, 8
Geomorphic change, 86
Geomorphic position, 149
Geomorphological history, 124
Geomorphology,
  and artifact densities, 134
  geomorphic maps, 109
  geomorphic thresholds, 57
  geomorphic units, 113
  reconnaissance, 116
  and site contexts, 126
  and site distribution, 122
  surveys, 109
Geophysical prospection, *see also*
    Electromagnetic conductivity, Electrical
    resistivity, Ground penetrating radar,
    Magnetic surveys
  active versus passive sensing methods, 355, 379
  anomalies, 356, 380
  archaeological features, 356
  attenuation of signal, 38-
  characterisitcs of principal methods, 360
  testing, 356, 379
Geophysical data processing
  amplitude variance analysis, 378
  analytical surface shading, 376
  clipping, 380
  concatenation, 380
  contrast enhancement, 380
  data logger, 380
  despiking, 380
  edge matching, 380
  high-pass filter, 378
  image processing, 380
  interpolation of time slices, 378
  least squares plane, 377
  regional trend analysis, 378
  residuals analysis, 377
  deserts, problems in, 362
  high precision surveys, 357
  in-field analysis and survey strategies, 365
  magnetic surveys, 357–358
  and microtopography, 379
  software, 365
Germany, 61
Gila River Valley, AZ, 83, 98
GIS, *see* Geographic information systems
Glacial-interglacial cycles, 13, 234
  dating of, 389
  and environments, 443
  Bond cycles, 17
  Dansgaard-Oeschger events, 17

Glacial-interglacial cycles (*cont.*)
  Heinrich events, 17
  periodicity and nature of climatic changes, 234
Glacial-interglacial stages, 9
  Alpine system, 13
  Midwestern US system, 13
Glacial to post-glacial environments, 78
  time-transgressive character, 78
Glacial stratigraphy,
  and intercalated ashes, 13–14
Glaciation
  astronomic theory, 16
  Milankovitch, 18
  orbital forcing, 17
  and oxygen isotope curve, 15
  penultimate, 15
Glaciers, 10, 13–15
Glacio-eustatic rebound, 80
Glaze I Period, 319
Gley, gleyed, *see* Soils
  and floodplain soils, 87, 194
Global cooling, 11
GPR, *see* Ground penetrating radar
Graded stream, 80
Grain size, mean, 305
Granite, 455
  Egypt, 455
  Megalithic sites, 455
  Roman sites, 455
  Zimbabwe, 455
Granodiorite, 446
Granulometry, 277–279
  and correlation of deposits, 91
  definition of lithologic discontinuities, 275
  and depositional events, 47
  disaggregation and dispersal of sample
    using sodium hexametaphosphate, 332
  electronic particle size analyzer, 277
  hydrometer technique, 277, 332
  laser technique, 277
  and microartifacts, 330
  Wentworth scale, 277, 335
  phi scale, 277, 335
  pipette technique, 277, 332
  sedimentological laws and grain distributions, 329
  and soils analysis, 242
Grasses, cool versus warm season, 422
Gravel, matrix supported, 118
Graywacke, 458
Great Britain, 61, 354
Great Plains, 91
Greece, 55, 487
Greenland, 15
Gregorio fault (CA), 145
Grids and geophysical surveys, 356
Grit temper, 301–302, 308, 311, 313, 318

Grog temper, 311
Groundmass, amorphous, 306
Ground penetrating radar (GPR), 94, 113, 363–365, 380
 antennas, 357
 buried feature detection, 363
 compared to resistivity profiling, 366
 depth of prospection versus frequency, 364
 history of use, 354
 and organic content, 363
 parallel transects and three dimensional interpretations, 364, 378
 plow zone effects, 366–367
 real time results, 364
 sandy drumlins, 373
 signal variance and horizontal feature definition, 365
 and stratigraphy, 363
 transducer, 380
 travel times and estimated depths of anomalies, 363–364
  calibrating by feature excavation, 364
Groundwater, 430
 calcretes, 179
 carbonate nodules, 434
 and alluvial pedogenesis, 87
 supersaturated, 430
 uranium solubility, 392
Gulf of Mexico shelf, FL, 94
Günz glacial stage, 13
Gypsum, sourcing by Sr and S isotopes, 453

Hackberry Creek Paleosol, 183–184
Half lives,
 isotopes used in dating, 387, 390
Hamlets, 193
Hand lens, 451
Harris matrix, *see* Stratigraphy
Hazard assessment, 145, 164
Hawaii, 455
Heavy minerals, 278–281
Hematite, and geophysical surveys, 357
Hertz (Hz), 380
Hiatus, depositional, 175
 between El Niño events, 118
High Plains TX, 87, 91, 438
Hillslopes,
 colluvial sedimentation, 271
 components, 272
Historical sites and geophysical surveys, 357, 377
Hohokam culture, 83, 98
Holocene epoch, 4, 7, 78
 abrupt climatic fluctuations, 234
 alluvial change, 98
 climatic stability, 234
 environments, 85, 129, 438
  Middle East, 130

Holocene epoch (*cont.*)
 faulting, 129
 floodplain aggradation, 438
 geomorphic change in, 476
 sea level rise, 68
 soils, 91, 436
 terraces, 135
 transgression, 117, 123
Hominids, 10, 441
 Asia, 391
 Australopithecines, 23
 bipedality, 441
 dating, 391
 Ethiopia, 391
 evolution of, 26, 389, 442, 481
 first, 10
 Kenya, 391
 modern humans, 3
 migration patterns, 389
 sites, 449, 475–476
 taxonomic and adaptive diversity, 26
Hominoids, 481
 Miocene, 441–442
Homo (genus),
 appearance in early Pleistocene, 13
Homo erectus,
 age, 386
 China, 443–444
 Europe, 90, 443
 very late, 400
Homo sapiens, adaptations, 26
Horizontal dipole mode, 380
Human evolution, 4, 26, 379, 399, 406, 476, 481
 and climate, 442
 vegetation changes, 444
Hungary, 451, 453
Hunter-gatherers,
 foraging, 475
 settlement patterns, 440
 settlement systems, 97
Hydrologic gradient, 111
Hydrology; *see* Groundwater
Hydrometer analysis, 332; *see also* Granulometry
Hydroxyapatite, 248

Icebergs, 19
Ice cores, 14–15
Iceland, 454
Ignimbrite, 454
Illinoian glacial stage, 13
Illinois River Valleys, 97, 188
Image processing, 380
Inceptisols, 182
India, 455
Indian cultures, serpentine, 456
Indian Ocean, 443
Induced magnetic fields, 357

Infrared stimulated luminescence (IRSL), 149, 284
Inorganic carbon, 430
Interglacial, 9; *see also* Glacial
  dating of, 389
International Association for the Study of Marble and Other Stones in Antiquity (ASMOSIA), 456
International Geological Congress (IGC), 10
International Congress for Quaternary Research (INQUA), 12
International Union of Geological Sciences, 10
Interstadial, 9; *see also* Glacial
  Late Pleistocene, 15
Iowa, 320
Iron
  depletion zones, 430
  secondary, 245
Iron age, 375
  agriculture, 247
Iron phosphate, 261
IRSL, *see* Infrared stimulated luminescence
Isothermal plateau fission-track (ITPFT) dating, 395; *see also* Fission-track dating
Isotope ratio mass spectrometer (IRMS), 433
Isotopic fingerprints, 450
Isthmus of Panama, 16
Italian granite
  source, 455

Jasper, 459, 461
  source, 455
  XRF analysis, 461
Jordan River Valley, (Israel), 95, 155

Kansan glacial stage, 13
KBS tuff,
  fission-track dating, 395
Kenya, 391, 395, 440–442
  Lake Turkana, 395
  Malawi Rift, 95
  Pleistocene rift valleys, 97
KHz, *see* Hertz
Killpecker Dune Field, WY, 275
Kill sites, 45
Knox Black Chert, 341
Korongo, 476
Krotovina, 187, 274; *see also* Bioturbation

Labrador, serpentine, 456
Lacustrine deposits, 56, 224, 475
  biogenic, 225
  Late Pleistocene, 438
  sequences, 72
Lamps, serpentine, Eskimo, 456
Landforms
  relict fluvial landforms, 345

Landforms (*cont.*)
  tectonic, 144
Landscapes
  alluvial, 97
  analysis with pedology, 175
  aggradational, 132
  changes, 97, 210–211
  cultural, 107–108
  desert, 97
  elements, 115
  evolution of, 81, 108–109, 116, 330
    climatic controls, 123
  and formational processes, 173
  human transformations, 78, 97, 132, 234
  hyperarid, 109
  past climates, 116
  past enviornments and, 62
  reconstructions, 257
  and sea level, 116
  stability,
    and soil development, 175
    and site locations, 121
  tectonics and, 116
Land use,
  and alluvial chronologies, 66
  effects on fluvial systems, 81
  intensification, 132
Landslides, 111
  and seismic activity, 67
Langbein-Schuum sediment yield model, 81
La Plata Valley, CO, 310
Lake Creek Valley, TX, 181
Lake Eire, Ontario, 95
Last interglacial period, 13
Late Cenozoic record, 4
  environmental cooling, 10
  magnetic polarity, 8
  oxygen isotope stratigraphy, 8
  time stratigraphy, 8
Laurentide ice sheet, 24
Lead in sediments from mining, 67
Leucite-Phyric lava, 455
Levant, southern, 97
Levigation, 304
Levsen Punctated, 308
Levsen Stamped, 308
Lichenometry, 135, 149, 151
Limestone, *see also* Carbonates, Dolomite
  petrography, 462
  sourcing, 457
  by NAA and SIRA, 462
Limestone species,
  oolitic, crinoidal, micritic, pisolitic, 462
Limnic deposits, 475
  dating, 389
Linear discriminant analysis, 462
Linn Ware, 308

Liquefaction features, 147
  and archaeological deposits, 152
  and blow sands, 151
  New Madrid Seismic Zone, 147
  sand beds, 157
Lithic, *see* Artifacts
Lithostratigraphy, 5, 13; *see also* Stratigraphy
Litter, plant, 428
Littoral drift deposits, 117, 123
Living floors, 221
Living surfaces, 221; *see also* Occupation surfaces
Llano Estacado, TX, *see also* Southern High Plains
  alluvial record, 65
  soils genesis and chronosequence, 189
Loess, 69
  eastern Europe, 14
  OSL dating, 407
  and stratigraphy, 11, 14
  and TL dating, 407
  upland, 85
  and welded soils on terraces, 88
Loessal plains, 86
Loss on ignition (LOI), 248
Lower Paleolithic, 90, 94, 263, 483
Lower-Middle Pleistocene
  East African rift, 95
Ludowici, GA, 276
Luminescence dating, *see* IRSL, OSL, TL
Lyell, Charles, 10

Macroartifact assemblages, 331
Macrofossils, 242
Macrogeology, 242
Macroremains, 262
Maghaemite, and geophysical surveys, 357
Magnetic dipole, 380
Magnetic field strength,
  and nanotesla, picotesla, 357
  vertical gradient, 358
Magnetic gradiometer, 354, 380
  and data loggers, 354
  sampling intervals, 356
  survey results, 368
Magnetic stratigraphy, 8; *see* Geomagnetic polarity time scale
Magnetic surveys
  depth of penetration, 358
  and diurnal variation, 358
  and earth resistivity, 361
    and large area surveys, 361
  and iron compounds, 357
  and magnetic storms, 358
  probe separation and configuration, 361
  Wenner configuration, 361
Magnetic susceptibility, 248, 381
  and buried occupation surfaces, 286

Magnetic susceptibility (*cont.*)
  and climatic fluctuations, 475
  of granite, 455
  of soils, 242, 261, 357
  sourcing obsidian, 453–454
Magnetite and geophysical surveys, 357
Mammal communities
  at last glacial maximum, 24
  decline and human predation, 130
Manganese nodules, 245
Maps
  anthropogenic features, 354
  archaeological, 160
  architectural, 354
  density contour, 337
  geomorphic, 109, 160
  from stereo-aerial photographs, 128
  Holocene landform sediment assemblages, 182
  in GIS layers, 128
  from geophysical prospection, 356
  morphostratigraphic, 110–112
  paleointensity, 145
  photomosaic/photomap, 182
  planimetric, 116
  topographic, 111
Marble, 417
  Carrara, 457
  CL sourcing of white marble, 457
  CL untraviolet (UV) spectra, 457
  Italian and French, 457
  Lepsius' quarry descriptions, 456
  ^{18}O-^{13}C isotopic plot, 457
  portrait of Livia, 457
  Quarries on Ephesos, 457
  Quarries on Paros, 457
  Roman, 459
  Sourcing by ESR spectra of Mn^{2+}, 457
Marine stratigraphic sections, 10
  boundary stratotype, 11
Mass spectrographic techniques, 453
  paleodiet, 453
  sourcing artifacts, 453
Mediterranean Basin, 455, 477, 481
Medieval, 376
Megafauna,
  Cyprus, 129–130
Megalithic, 455
Megascopic examination of sherds, 308
Melos, 450, 454
Meltwater, and deglaciation, 19
Mesoamerica,
  Basin of Mexico, 66
Mesolithic, 23, 78, 450, 454
Mesopotamia, 450
Metal detector survey, 371
Metallurgy, Cypriot
  resources, 126

Index

Metallurgy, Cypriot (*cont.*)
  and trade, 126
mHz, *see* Hertz
Metaquartzite, 459
Meteoric water
  isotopic composition, 425
Micrite, *see also* Limestone
  included allochems, 461
  as microcrystalline calcite, 461
Microanalysis, 451
  of middens, 328
Microartifacts, 220–221, 225, 327–329
  analysis protocol, 328
  comparative collections, 331
  data representation, 337
  descriptive statistics, 336
  natural and cultural, 329
  point counting, 336
  quantification, 335–327
  recovery procedures, 333
  reference collections, 247
  region-scale research, 330
  size classes, 334
  spatial distributions of, 332
  vertical distributions of, 225
Microdebitage, 345
  attributes, 331
Microfabric, 244, 247
Microfacies, 207
  stratified, 222
Microfossils, 242
Micropans, 255
Microprobe analysis, 244, 256, 262
Microsherds, 331, 335, 339
Microstratigraphy, 95, 215, 229, 243, 259
  three-dimensional, 228–231
  in multidisciplinary studies, 248
  sequences, 226–227
Microstructures and human activity, 284
Microtopography, 231
  surface expressions of buried features, 379
Middens, 45, 189; *see also* Shell middens
  and geophysical surveys, 355, 363
  microanalysis of, 328
Middle Holocene
  alluviation, 440
  buried soils, 437
  Chihuahuan Desert, 443
  dry period, Texas, 438
  East Africa, 441
  environments, 436–437
  geomorphic change during, 193
  soils, 436
  water tables, 439
Middle-Late Pleistocene boundary, 13
Middle Paleolithic
  bone taphonomy and, 477

Middle Paleolithic (*cont.*)
  burials, 479
  and TL dating, 406
  Europe, 406
  France, 223, 231
  Greece, 92
  hearths and, 477
  Levant, 225, 410, 475, 479–482
  and U-series dating, 92
  western European caves, 231
Middle Pleistocene, 402
Middle Ucayali River, 86
Midwestern U.S., 91, 93, 97
Milankovitch forcing, 16–19
Millisiemens/meter, 381
Millstones, 455
Mimbres ceramics, 310
Mindel glacial stage, 13
Mineral inclusions, 305, 322
Mining,
  copper, 130
  landscape impacts, 132
  Medieval-Roman, 132
  and sediment provenance, 67
Miocene,
  East Africa, 441–442
Mississippi River Valley, 68, 97, 188, 308
Mississippian Period, 97, 281
  Dallas Phase TN, 340–343, 347
  ceramics, 313, 318, 320
Missouri River Valley, 366–370, 378
Modern humans, 26, 386, 481
  Africa, 400
  China, 400
  Eve theory and, 481
Mollisols, on terraces, 88
  in early-middle Holocene alluvium, 182
Morphostratigraphic analysis, 136
Morphostratigraphy,
Morphostratigraphy, 90, 108–110, 116, 150; *see also*
  Scarp morphology, Stratigraphy
Mossbauer spectra and sourcing obsidian, 454
Mounds
  burial, 362
  geophysical surveys of, 363
Mount Kreatos, 112
Mousterian, *see* Middle Paleolithic
Mud Creek IA, 69
Multianalytical approaches, 243

Nanosecond (nS), 381
National Register of Historic Places (NRHP), *see*
  Cultural resources management
National Resource Conservation Service (NRCS,
  formerly Soil Conservation Service), 182
Natriargids, 193
Natural levels, 44

Neanderthals, 26, 386, 400, 481–482
Nebrascan glacial stage, 13
Nene River Valley, GB, 257
Neolithic, *see also* Pre-Pottery Neolithic
  Aegean, 451, 457
  Africa, 441
  Britain, 257
  Chassean (Middle Neolithic), France, 228
  Europe, 450–451, 456
  Greece, 450
  Italy, 248
  Negev Desert, 480
  Russia, 459
  Southwest Asia, 78, 463, 475
Netherlands, 61
Neutron activation analysis (NAA), 450, 459
  of ceramics, 305
  intra- and inter-quarry variation, 457
  large databases available, 456
  of Na and Mn in Greek marble, 456
  of steatite, 456
New Mexico, 310
Nice round numbers, 12
Nile Valley, 83
Norman, 260
North American Committee on Stratigraphic Nomenclature (NACSN), 7, 57, 110
North Texas, 89, 95
North Troodos Foothills, Cyprus, 126
Northwest Coast, 45
Novaculite, 459
Numerical data presentation, 243

Obliquity, 17–18
Obsidian
  central Hungary, 451
  color, 454
  composition, 453
  eastern Mediterranean, 450
  fission-track dating, 395
  hydration band measurement, 149, 160
  luster, 454
  sourcing, 160
    by Mossbauer spectra, 454
    by specific gravity, 454
    by Sr isotopes, 453–454
  translucency, 454
  use, 160
  vesicles, 454
Occupation patterns, 39
  duration, 56, 224, 233
  intensity, 56, 193
  phosphorus concentration, 174
  and landscapes, 114
  Middle Paleolithic, 479
  multiple, 95, 120, 163
  periodicity, 56, 283

Occupation patterns (*cont.*)
  phases, 229
  short-term, 187
  transformations, 40
    and deposition, 44
Occupation surfaces, 151, 220–221, 284
  evidenced by artifact orientations and conjoined pieces, 220
  inframillimetric layers, 220
  and subhorizontally layered microflakes, 221
Oceans, thermohaline circulation, 17
Ochre, sourcing in Australia, 461
Ohio River Valley, 344
Ohm's law, 381
Ohm-meter, 381
Oklahoma, 313
Old-fashioned generalists, 235
Oldowan, 449
Olmec, 455
Oolitic microsparite, 450; *see also* Limestone
Opal, 459
Optical emission spectroscopy (OES), 452
Optically stimulated luminescence dating (OSL), 60, 149, 284, 406
  of alluvium, 61
  of eolian sediments, 389
  European sites, 61
  of feldspars, 407
  inadequate bleaching and erroneously old ages, 61
  of loess, 407
  of quartz, 407
  sampling for, 408
  and sedimentary chronologies, 61
  single-grain technique, 286
Organic carbon residence time, 188, 420
Organic matter, detrital, 428, 435
Osaka and Lake Biwa, Japan, 151
OSL, *see* Optically stimulated luminescence
Ostrich egg shell and U-series dating, 392
Ouse Valley, GB, 257
Oxidized root channels, 430
Oxygen isotopes, *see* Isotopes

Paleoanthropology, 477
  East Africa, 475
Paleoclimate reconstructions
  anomalies and alluvial events, 65
  chronologies, 389
  proxy data, 20–21
  and synchronous aggradation or incision, 65
Paleodemography, 64, 71
Paleodischarge, 63
Paleoenvironment reconstruction, 211
  dating, 389
  and lake levels, 129
  and modern analogues, 24

Paleoenvironment reconstruction (*cont.*)
  and paleosol morphology, 441
  and pollen data, 129
  and regional research, 330
  and stable isotopes, 434
Paleogeography, 4
Paleohydrology, 129
Paleoindian, 23, 78, 281, 283, 435, 440
  Clovis, 92
  Southern High Plains, 91
Paleolandscapes, 135
Paleolithic, western Asia, 474
Paleomagnetism, 149
  and dating, *see* Magnetic stratigraphy
Paleoprecipitation, 63
Paleoseismic settings, archaeological sites, 146–147
  and occupation horizons, 151
Paleoseismic trench, 160
Paleoshorelines of ancestral Great Lakes, 189
Paleosols, 188, 207, 257, 344, 428–430
  buried, 431
  cumulic, 431
  in East Africa, 442–444
  isotopic analysis of, 429, 435
  morphology, 441
  and paleoenvironments, 441
Paleosurface topography and GPR reflection data, 362
Palynology, 71, 201, 244, 475; *see also* Pollen
  in archaeology, 20
  of peat bogs, 261
Pan sculpture, 449
Paros marble quarries, 457
Pastoralism and erosion, 135
Parthenon Museum, 450
Particle size analysis, *see* Granulometry
Pawnee River Watershed, KS, 183
Peat bogs, 261
Pedofacies, 207
Pedofeatures, 244–245
Pedimentation, 271
Pedogenesis, *see also* Soils
  biomechanical processes, 185; *see also* Bioturbation
  and artifact burial, 185
  and soil upbuilding, 185
  and clay mineralogy, 87
  horizonation versus homogenization, 224, 233
  and parent materials, 87
  post-depositional, 435
  progressive and regressive, 27
  rate of, 87, 189
  and soil-stratigraphic positioning of archaeological materials, 188
  and texture, 87, 189
  theory of soil genesis, 233

Pedogenesis (*cont.*)
  and translocation, 189
Pedology and landscape/climate reconstructions, 174
Pedostratigraphy, *see* Soils stratigraphy
Pedoturbation
  and artifact burial, 270
  categories, 270
  effects on occupation zones, 187
  equifinality of, 271
  progressive (proanisotropic), 271
  regressive (proisotropic), 271
  versus sedimentation, 272, 280
    and burial processes, 286
  and shrink-swell, 87, 177
Pee Dee Belemnite (PDB) limestone, 433
Penultimate glaciation, 15
Peru, coast, 83, 109
Petrographic microscope, 298, 451
Petrography, 450–451; *see also* Ceramics, Soil micromorphology
  ceramic, 297–323
  chert, 459
  data presentation, 246
  digitized images, 245
  granite, 455
  limestone, 462
  modal analysis, 305
  point counting,
    ceramics, 305, 314
    interval, 306
    soils, 245
    versus area counting, 245
    sampling interval, 306
Petrology, 451
Phenocrysts in obsidian, 453
Phosphate minerals, 174, 244, 254, 260
  nodules, 254
  organic, 260
  "P-ratio," 244
  post-depositional change of ash, 478
  staining of fused ash, 262
Phosphorus
  and buried occupation surfaces, 286
  mapping, 256
  and occupation intensity, 174
  translocation in soil, 189
Photo-stimulated luminescence dating (PSL), 411; *see also* Geochronology, Thermoluminescence dating
Photosynthetic pathways, 63; *see also* CAM, Carbon, Isotopes
Phytoliths, 245, 261, 287
  and archaeological research, 20, 24
  to define buried soils, 276
Piedmont drape, 129
Pipette analysis, 332; *see also* Granulometry

Pixel, 381
Plains Woodland cultures, 96, 183
Plains Village cultures, 96
　initial coalescent variant, 366–370
Plant pseudomorphs, 262
Pleistocene epoch, 4, 7–9
Pleistocene-Holocene boundary, 12
　diachronous, 12
　environments, eastern Mediterranean, 9, 129–130
　transition and archaeological record, 12
Plio-Pleistocene boundary, 11–12, 94
Plowing, see Formational processes
Plummets, Eskimo, serpentine, 456
Poaceae phytoliths, 262
Point counting, see Petrography
Pollen, see also Palynology
　in alluvial sequences, 62
　hydrodynamic sorting of, 62
　Pleistocene-Holocene, 129
　preservation, 24, 62
Population,
　density and climate, 438, 440
　increase, 130
Porcellanite, 459
Porosity analysis, 412
Postglacial environments, 9
Postulate, Local Products Match, 316
Potosi Plain Series, 313–316
Potterne's components and microfabrics, 262
Pottery, see Ceramics
Powell Plain, 313, 316
PPL, plane polarized light (in petrography), 262, 299
Preceramic period, 124
Precession, 17–18
Precipitation, 80
　and alluvial aggradation and incision, 63
　gastropods and micromammals, 194
　mean annual, 63
　monsoonal, 443
Pre-Pottery Neolithic, Syria, 222
Proactinium (Pa), see Uranium series dating
Prospection, see Geophysical prospection
Proton induced gamma ray emission (PIGME), 452
Proton induced X-ray emission (PIXE), 452
Proton precession magnetometer, 354, 358, 376, 381
Proxy records for environmental reconstruction, 61
　and alluviation, 57
　defined, 20–21
　lack of modern analogues, 24
Pueblo II Period, 311
Puebloan cultures, 310–311, 319
Pumice, and dating, 392

Pyramid, Peru, 121
Pyroxenes, 464

Quarries, 450–452, 455–456
　Quartzite, 458–460
　Colossi of Memnon, 458
　ferruginous, 458
　quarry near Cairo, 458
　sourcing by Eu versus Fe plot, 459
Quaternary environmental change and human prehistory, 3, 5
Quaternary geology and geomorphology, 4, 109
Quaternary period, 4, 7
Quaternary Research, basic principles, 4

Radiation exposure techniques, 385, 394
　key aspects of, 388
Radioactive decay, 390
Radiocarbon (^{14}C) dating
　accelerator mass spectrometry (AMS), 391, 433
　bulk sediment samples, 60, 91
　calibration, 27, 387, 480
　charcoal, 163
　contamination, 409
　with allochthonous organic matter, 63
　"hard-water error," 59
　hearths and maximum ages, 59
　humate fractions, 91, 427
　lignite contamination, 157
　materials, 59–60, 160
　mussel shell, 160, 163
　and organic carbon residence time, 188
　organic sediments, 59
　paleosol organic matter, 59
　　mean residence error, 59
　peat bogs, 261
　reservoir ages, 163
　sediments and soils, 63
　seismic events, 157
　and soil humates, 59, 188
　and soil organic matter, 425
　and soil stratigraphy, 91, 431, 436
　and stratigraphy, 58, 427
　tilled soil, 135
Radiogenic dating methods, 149
Ramey Incised ware, 313, 318
Rare earth elements (REE), 456
　chondrite-normalized distribution, and sourcing, 456
Rate of deformation, 149
Rate of deposition, see Sedimentation rate
Raw materials, see Artifacts
Reelfoot Fault, 146
Refractive index and obsidian sourcing, 453–454
Refuse: see Formational processes
Relative dielectric permitivity, 381

Index

Remote sensing, 94
Residues, cooking, 322
Resistivity surveys and stratigraphy, 361
Resistivity tomography, 361–362, 381
Resonance techniques, 452
Revised Universal Soil Loss Equation (RUSLE), 273
Rhyolite, 453
Rio Casma Valley, 116–118
Riss glacial stage, 13
Rock color chart, 461
Rock and mineral weathering, 149
Rockshelters, 146
 formation processes, 45
 cultural and natural transportation agents, 43
 Cyprus, 130
Rock varnish development, 115, 118, 149
Rolling Plains, TX, 88
Roman, 133, 135, 260
 columns, granite source, 455
 granodiorite, 456
 magnetic susceptibility, 455–456
 millstones, 455
 Mons Claudius quarry, 456
 reuse of, 456
Rooting depth and radiocarbon age, 427
Rhyolite, 453
Rift valleys,
 Kenya, 440–442
 Pleistocene, 97
 Tanzania, 442
Sabine River Valley, TX, 345
Salinas, Peru, 117, 123
Sampling
 bulk soil, 332
 in ceramic petrography, 305–306
 demands by archaeochronologists, 386
 design and geomorphology, 136
 for fission-track dating, 395
 for flotation, 333
 and geophysical survey detail, 354
 Kubiena boxes, 243
 of landscape, 114–115
 for micromorphology, 212–213
 off-site control samples, 331
 for OSL dating, 408
 for radiocarbon dating, 92
 random, 136
 for soils analysis, 432
 soil monoliths, 243–244
 for stable isotope analysis, 429
 strategies in geochronology, 385
 stratified, 115, 134, 136
 subsurface, 113
 tooth enamel, 400–402
 vertical distributions, 340
San Andreas fault system, USA, 145–146

San Gregario fault, USA, 146
San Juan Basin, NM, 194
Sand, 299
 Ridges and swales, 363; *see also* Dunes
Sanders Plain ceramics, 317
Sandhills of North Carolina, 273
Sandstone,
 argillaceous, 458
 arkose, 458
 fabric, 458
 feldspathic/volcanolithic, 458
 sourcing, 458
Sand temper, *see* Ceramics
Santa Fe, 319
Sardinia, and granite quarrying, 455
Scanning electron microscopy (SEM), 244, 450–452
 and ceramic analysis, 298
 of soils, 241
Scarp morphology and landform development, 149
SCLF, *see* Single crystal laser fusion
Sea faring trade, 450
Seasonality, and O isotopes, 420
 and stream processes, 80
Sedimentary environments, *see also* Alluvial, Lacustrine, Littoral, Eolian
 and prehistoric site settings, 234
Sediment cores, 129
Sediment-landform assemblages, 182
Sedimentary deposits, *see* Sediments
Sedimentary facies, 207; *see also* Microfacies
 alluvial, 84–85
Sedimentary sequence, 6
Sedimentary structures, 85, 274–275
Sedimentation, 61; *see also* Alluviation
 anthropogenic, 56, 215, 192, 477
 and artifact burial, 113
 biogenic, 479
 colluvial, 272
 as complex response, 68
 and differential preservation, 435
 discontinuous, 227
 equifinality of, 271–272
 hiatus in, 177
 marine, 345
 and paleodischarge, 70
 post-depositional processes, 387
 spatial patterns and occupations, 193
Sedimentation rates, 56–57, 66, 129, 149, 192, 283
 alluvial, 58, 80, 90
 and biomass, 440
 in caves, 231, 478
 and depositional events, 44
 and landscape stability, 438
 and pedogenesis, 87, 207, 231

Sedimentation rates (*cont.*)
  rapid and punctuated, 193
  versus sea level rise, 124
  versus soil formation, 174, 223–224, 432, 438
  in tells, 222–224
Sedimentology, global approach, 214
Sediments, *see also* Alluvium, Sand, Clay
  anthrogenic, 111, 115, 477
  ashes, 477
  coastal, 345; *see also* Littoral
  colluvium, 56, 272
  control samples, 281
  cross-bedded, 275
  eolian, 194, 272, 277–278, 289–290, 476
    dating, 388, 407
    dunes, 274, 278
    dune belts, 278
    dust, 233
    loess, 14
  granulometry and, 277–279
  heavy minerals in, 281
  playas, 194
  porosity, 399
    and moisture content, 412
  provenance studies, 67
  primary features, 176, 274
  sandy, 304
  and soil like properties, 175
  upward-fining sequences, 186
  yield and archaeological preservation, 81
Seismicity, *see also* Earthquakes
  artifacts dating event, 157
  damage, 145
  landslides, 67
  liquefaction sand beds, 157
  paleointensity, 145
  piercing feature displacement, 153, 156
    and dating, 153
    stratigraphy of, 153
  risk, 155
  sand blows, 148
Seismic hazard assessment (SHA), 145, 148
Series, ceramic, 307
Serpentine, 456
Settlement patterns, 136, 195, 479
  Bronze age, 130
  and evolution of physical landscape, 136
  and GIS, 97, 337
  hunter-gatherer versus agricultural, 440
  and isotopic data, 444
  reconstruction of, 108
  and resource distributions, 136
Sheetwash, 186–187; *see also* Erosion
  and microartifacts, 335
Shell, dating, 403
Shell temper, 303, 310–311, 316, 318
Shell middens, 45–46

Shell middens (*cont.*)
  matrix color, 45
  Peruvian coast, 120, 136
Shepard, Anna, 308
Sherds, powdered, 305
Shock, thermal, 310
Shuttle imaging radar (SIR),
  site surveys, 94
Siberia, early occupations, 408
Silex, *see* Chert
Silt, 299
Single crystal laser fusion (SCLF), 392, 395
Site discovery methods, 93
Site distributions and chronology, 120
Site prediction models, 93
Site structure, 330, 379
Slag, 244
Slopes
  catastrophic failure, 66
  components of, 273
Slumping, 192
Soils, *see also* Buried soils, Paleosols, Soil profiles,
    Soils stratigraphy
  alluvial, 86–88, 139, 436, 439
    and facies, 87
  on alluvial fans, 187
  analysis, 181
  anthropogenic, 207
  arable, 247
  archaeological, 242
  and archaeological potential, 182–183
  and archaeological surveys, 174, 181–185; *see also* Archaeological surveys
  argillans, 182, 245; *see also* Cutans
  arid regions, 88, 221
  and biological mixing, 433
  and biomass, 420
  calcic horizons, see Soil horizons 88, 179
  catenas, 114, 128, 259–260
  chemistry, 149, 174
  chronosequences, 88, 114, 188
    dating archaeological materials, 195
    Southern High Plains, 189
    and terrace correlation, 88, 128
  Cienega soils, 87
  clay coats, *see* Argillans
    dark clay coatings, 254–255
  clay mineralogy, 189
  compaction, 222
  compaction, and increased ground resistivity, 373
  concretions, 176
    and stable isotopes, 63
  conservation, 135
  cumulative/cumulic soils, 87, 188, 431; *see also* Overthickened horizons
    and near-surface processes, 187

Index

Soils stratigraphy (*cont.*)
  cumulative/cumulic soils (*cont.*)
    and site formational processes, 187
    role of sedimentary input, 233
  cumulative anthrosol, 260
  cutans, 176, 181; *see also* Argillans
  development,
    indices of, 88
    and sand content, 82
  diatoms, 261
  electromagnetic conductivity, 362
  elluviation, 179
  epipedon, 20
    isotopic variation of, 63
  fabrics, 177, 206
  fine-grained soils, 427
  floodplain soils, 85, 87, 175, 192
    radiocarbon dating of, 91
  formation, *see* Pedogenesis
  groundwater calcretes and, 179, 430; *see also* Groundwater
  heterogeneous mineralogic, 248
  horizonation, 224, 233; *see also* Soil Horizons
  illuviation, 88
  inclusions,
    anthropogenic, 244, 260
  indices of development, 88
  isotopes in, 419, 421
  landforms and, 18
  and loess, 86
  magnetism, 357
  microhorizons, 224; *see also* Soil micromorphology
  morphology, 193, 215
    and climate, 87
  nodules, 245, 424
    lithogenic vs pedogenic, 429
  organic carbon source and carbon isotopes, 63
  organic content and GPR penetration, 363; *see also* Soil organic matter
  overthickened, 87, 178
  pedofacies, 85, 207
  pedosedimentary properties, 227–228
  periglacial, 224
  and phytoliths, 245, 276, 287
  polders, 263
  and pollen preservation, 46, 62, 174
  "polygenetic" soils, 88
  post-Altithermal soils, 193
  post-depositional transformations, 88, 207
  post-glacial, 224
  rates of formation, 87, 189
  regolith weathering and erosion, 125
  Rendzina soils, 255
  Romano-British, 259
  ripened soils, 263
  sapropels, 129

Soils stratigraphy (*cont.*)
  and site evaluations, 174, 185–190; *see also* Archaeological site evaluation
  and site excavations, 190–194; *see also* Archaeological excavations
  shrink-swell, 87, 177, 433
  silt coats, 87
  solum, 274
  soil fauna, 254
  soil organic matter (SOM), 59
  soil sediment, 263
  and stable isotopes, 63, 420–424
Soils stratigraphy
  structure, 176–177, 180
  soil surface dynamics, 233
  surfaces and microconditions during occupations, 220
  soil surveys, 182
  soil taxa, 175, 182, 193
  on terraces, 87, 192
  terra rosa soils, 231
  and texture, 427
  urban soils, 255
  soil velocity analysis, 375
  welded soils, 88, 191
Soil carbonates, 111, 179, 193–194
  allogenic (inherited) carbonate, 88
  and ceramic diagenesis, 323
  and chronosequences, 88
  depth and climate, 430
  depth to leaching, 193
  filaments, 421
  and groundwater, 194, 430
  nodules, 421
    microsampling concentric increments, 430
  rate of development, 88
  root casts, 430
  and stable isotopes, 420–425, 441
    in buried soils, 437
  stages of carbonate morphology, 194
Soil conductivity, 362
Soil Conservation Service (SCS), *see* Natural Resource Conservation Service
Soil horizons
  A, 177, 186, 421
  albic, 179
  argillic, 245
  B, 186, 428, 432, 435
  boundary characteristics, 176
  Bk, 194, 256, 421
  Bt, 180, 271
  truncated Bt, 179–182
  Bw, 182
  Calcic, see Soil carbonates
  C, 179
  E, 179–180
  Eb, 164

Soil horizons (*cont.*)
  incipient, 118
  mixed, 192
  O, 421
  Spodic, 189
  and time, 114
Soil micromorphology, 174, 180, 206, 248, 262, 274
  applications, 209
  archaeological soil micromorphology working group, 246
  data presentation, 244
  distinguishing soils and sediments, 177
  East Africa, 476
  and depositional events, 47
  microstratified pedogenic facies, 221
  and pedogenic carbonates, 430
  pedoturbation and, 47, 284
  plasma, 181
  research strategies, 211
  skeletal grains, 181
  voids, 181
Soil organic matter, 63
  basinwide inputs, 431
  detrital sources, 428
  isotopic fractionation, 423
  conversion by microbial respiration, 423
  pedogenic sources, 428, 431
Soil profiles
  A-Bw, 187
  A-C, 96, 182–192
  Ab-Eb-Bwb, 276
  Ap-A-Bw, 191
  Ap-Bt-B-C, 190
  Ak-Bk, 183
  A/E/Bs, 164
  A-E-Bt-C, 271
  A-E-Bt-E′-B′t, 286
  development of, 149, 174–177
  inverted, 186
  and pedofacies, 85, 207
  rates of development, 189
Soil stratigraphy, 174, 186, 191–192, 430–432
  and geoarchaeology, 91
  radiocarbon dating of, 431, 436
  and paleoseismic deposits, 150
  sections, 431
  and archaeological surveys, 195
Soil Survey of England, 244
South Carolina coastal plain, 280
South Carolina Sand Hills, 283
South Platte River, CO, 94
Southern High Plains, 65, 88–89, 189; *see also* Llano Estacado
Southwestern U.S., 308
Spain, 61, 410
Spatial patterning
  activity areas, 337–338

Spatial patterning (*cont.*)
  flintknapping areas, 341
  public and private areas, 340–341,
  microartifacts versus macroartifacts in, 343
Specialized ceramic production, 312, 317, 319
Specific gravity and sourcing artifacts, 454
Spectrometry
  atomic absorbtion (AAS), 452
  inductively coupled plasma emission (ICP), 452–454
Speleothems, 392
  dating, 403
Sphene, and fission-track dating, 394
Spodic horizon, 189
Spodosols, 257
Spores, fungal, 245
Spring Hollow Incised, 308
Spring Hollow Cordmarked, 308
Spring Hollow Plain, 308
Springs
  and Cienega soils, 87
  paleo-spring deposits, 124
  and prehistoric occupations, 439
Stable isotopes, 421
  activity, 390
  and atmospheric CO_2, 423
  carbon, 421
    and biomass, 440
    C_3-C_4 plants, 42, 63
  depth of rooting and, 426
  $\delta^{13}C$ as paleoclimatic proxy, 424
  and pedogenic carbonates, 63, 423–425
  contamination from lithogenic sources, 433
  daughter, 390
  floodbasin settings, 429
  fractionation, 421–423
  middle-late Holocene isotopic boundary, 440
  mixing lines, 424
  $\delta^{18}O$ and soil organic matter, 425
    and meteoric water composition, 425
  open and closed systems, 390
  ^{18}O, 14
  Oxygen isotope stages, 14–15
    and temperature, 8
  and preservation potentials, 444
  radiogenic, 386, 389
  Rayleigh distillation, 425
  riparian settings, 429
  sampling, 432
  soil organic matter, 63
  sourcing artifacts, 453
  strontium (Sr), 453–454
  sulfur (S), 453
  and past vegetation communities, 423
Stable isotope ratio analysis (SIRA), 63
  and meteoric waters, 425
Stadial, 9; *see also* Glacial

Index

Standard error, 413
Statuary, Greek, 457
Steatite, 456
　NAA for sourcing, 456
　New York to Virginia, 456
　sourcing with transition metals, 456
Stone lines, 162, 271
Stones, allochthonous, 244
Strata
　anthropogenic, 144
　archaeological, 217–219, 232
Stratigraphic relationships and three-dimensional reconstruction, 229–232
Stratigraphic units, 6
Stratigraphy, see also Microstratigraphy, Morphostratigraphy, Soil stratigraphy
　allostratigraphic sequence, 57, 90–91
　allostratigraphy, 57, 90, 109, 111
　archaeological, 149, 207, 215, 232
　　and paleoenvironmental reconstructions, 234
　biostratigraphy, 5,10, 90
　boundaries, 6, 177, 227
　　diachronous, 7
　　middle-late Holocene, 440
　　mythological, 12
　　Plio-Pleistocene, 11, 23
　　Pleistocene-Holocene, 23, 130
　　stratotype, 10–12
　　synchroneity, 6–7, 12
　chronostratigraphy, 5, 113–114
　contacts and geophysical prospecting, 367
　correlations, 91, 149, 229
　decompressing, 479
　discontinuities, 274
　　bounding, 90, 113
　　breaks, 190
　　and soils, 174
　　lithologic, 287
　ethnostratigraphic units, 160
　event stratigraphy, 150; see also Earthquakes
　GPR survey line and stratigraphy, 363
　Harris matrix, 113, 150
　inverted, 192
　isotope stratigraphy, 8, 14–15
　　and deposits, 477
　　and ice cores, 14–15
　　and marine regression, 477
　lithostratigraphy, 5, 57
　loess stratigraphy, 14
　magnetic, 449
　nomenclature, 7
　pedostratigraphy, see Soil stratigraphy
　principles of, 7
　nitty-gritty of, 246
　regional framework, 114
　sclerochronology, 149

Stratigraphy (cont.)
　unconformities, 430
　urban stratigraphy, 260
Streams,
　bedload, 82
　classes, 83
　ingrown meandering, 90
　order, 58
　offset, 165
　properties, 81
　suspended load, 82
Structural logic, 216
Structures, geologic,
　and hydrology, 144, 163
Subneolithic Combed Ware Period,
　chert traded from Russia, 459
Subsistence strategies, 195, 435
　and isotopic data, 444
　prehistoric, 437
Surface soils, 114, 435, 440, 443
　in Holocene alluvium, 182
　organic matter isotopic values, 428
　strongly developed, 192
Surface stability,
　and artifact contexts, 95, 136
　and artifact densities, 132–133
　measures, 133
　and relative age, 136
　and soil formation, 95
　and vegetation, 62
Surfaces, see also Surface soils
　aggradational, 115
　dating, see Dating
　eroded, 115, 135
　occupation, 95
　stability and settlement patterns, 64
　sterile, 108
　from non-occupation versus post-occupation erosion, 136
Survey, see Archaeological survey
Susquehanna River Valley, PA, 95
Switzerland, 61, 458
Sydney-Cyprus-Survey-Project (SCSP), 109

Tabun woman, 482–484
Taphonomy, 482
Tanzania, 441–442
Tectonic regions,
　depressions, 144
　niches for human occupation, 144
Tectonism
　and alluvial response, 67–68, 83
　and climate, 22
　closing of Isthmus of Panama, 16
　and continental drift, 16
　and landscape change, 116
　Neogene, in Cyprus, 128–129

Tectonism (*cont.*)
  and soils, 144
  and strath terraces, 67
  and uplift in Tibetan Plateau, 16
Teeth, dating, 393, 399
  enamel sampling, 400
Temper, *see also* Ceramics, Grit, Grog, Sand
  hematite, 313–314
  media, 329
  non-shell, 311
Tephra hydration, 149
Tephrochronology, 149
Terraces, 57
  agricultural, 126
  aggradational, 118
  artificial, 135
  asynchronous, 67–68
  correlation, 128
  dating, 389
  genesis, 84
  hillslope, 135
  and loess deposition, 88
  morphogenesis, 86
  Pleistocene, 272
  strath, 67, 90, 118
  unpaired, as complex response, 67
Terrain, erosional, 108
Tertiary Period, 10
Tesla
  nanotesla, picotesla, 357, 381
  and background magnetic field strength, 357
Test excavations, 379
Test tiles, 305
Texas, central, 45, 93, 97
Trinity River Valley, TX, 92, 437
Thermal ionization mass spectrometry (TIMS), 392
  and cave speleothems, 392
  and precision of U-series dating, 392
Thermoluminescence (TL) dating, 58–61, 149, 284, 406
  age uncertainty, 388
  Australia, 61
  burned flint, 478
  cave deposits, 479
  dating range, 388
  eolian deposits, 389
  high cost, 285
  materials, 388
  and paleoseismic deposits, 148
  Tabun cave, 483
  zeroing, 389
Thermoremanant magnetism, 381
  and response, 357
Thin sections, 321; *see also* Petrography, Micromorphology
  soil, 241, 245

Thin sections (*cont.*)
  impregnation, 244
TL, *see* Thermoluminescence dating
Tibetan Plateau, 16
Tigris-Euphrates River Valley, 83
Time series photography, 66
Toba, 385
Total field magnetic survey, 381
Trace element analysis, 454, 456, 462
  by discriminant function analysis, 455
Tree throws, 186, 192, 271–274
Trade patterns,
  ceramics and, 310, 319
  copper, 132
  Neolithic, 450
Transgression,
  Holocene, 117
Transition metals,
  sourcing steatite, 456
Tsunami, 147
  radiocarbon dated deposits, 148
Turbation, *see also* Bioturbation, Cryoturbation, Pedogenesis, Pedoturbation, Tree-throws
Turf mound, 244
Turkey,
  granite sources, 455
Twin electrode array, 381
Type, ceramic, 307

$^{234}U+^{230}Th$, *see* Uranium series
$^{234}U+^{230}Pa$, *see* Uranium series
U.S. Geological Survey, 11
United Kingdom; *see* Great Britain
Universal Soil Loss Equation (USLE); *see* RUSLE
Upper Paleolithic
  Aurignacian, 221
  Europe, 78
  Kebara Cave, 481
  Levant, 475
  early, radiocarbon dates, 484
Upper Peninsula, MI, 189
Uranium concentrations, 392
Uranium series (U-series) dating, 149, 391, 480, 482
  Alpha-spectrometric methods, 393
  burial age, 393
  calcite, 392
  closed and open systems, 387, 392
  coupled with ESR, 390
  groundwater carbonates, 92
  human bone, 393
  mass-spectrometric methods, 393
  materials, 387, 393
  open system, 393
  pedogenic carbonates, 92
  sediment porosity, 399, 412
  practical limit, 389

Uranium series dating (*cont.*)
  sediment moisture content, 412
  tooth enamel burial, 390
  and travertines, 90
Uranium-thorium dating (U/Th)
  archaeological applications, 392
  of calcite, 392
  of Fe/Mn oxyhydroxides/oxides, 92
Urban sites
  origins of, 475
  proto-urban sites, 222
UVL (ultra-violet light) in petrography, 262

Vegetation
  and buried features, 379
  cryptogenic, 224
  and erosion, 440
  lomas (fog-drip), 119
  macroplant remains, 258
  paleovegetation, 420
  proxy data, *see* Palynology, Carbon isotopes
  stable isotopes and, 426
Veracruz, Mexico, 98
Vertical and horizontal dipole modes, 381
Vertisols
  on floodplains, 87
Vesiculated pumice, 453
Viewshed, 115
Village layout and geophysical survey technology, 354, 366
  bastioned fortification ditch, 366
Virgin River, UT, 70
Visual percentage charts, 336
Vivianite, 244, 261
  neoformed, 245
Volcanic glass, and dating, 392
  and fission-track dating, 395
Volcanism, *see also* Ash, Basalt, Obsidian, Rhyolite
  and climate, 14, 17
  and radiogenic dating, 391
  shield volcanoes and basalt, 455
  ash, as temper, 301
  and ice cores, 14–15
Vrica section as Pleistocene boundary stratotype, 11

Wadis
  Late Quaternary aggradation along Nile, 84

Wabash Valley seismic zone, 151
Wales, 244
Wares
  ceramic, 307
  culinary, 310
  nonculinary, 310
Waste, ceral processing, 258
Water
  glacial meltwaters, 19
  groundwater, 194, 322, 430
  and stable isotopes, 14, 434
  meteoric, 425
Watershed, and archaeological survey, 115
Water table, 194
  and climate, 439
  and sea level, 46
Weak acid extraction, 305
Weathering, 113, 175
Wenner configuration, *see* Magnetometry
Wetland sites, 242
Wildfires, 222–223
Wisconsin, 308, 313
Wisconsin glacial stage, 13
Witness section, 406
Woodland Period, 85, 97, 189–193, 281, 320
  ceramics, 308
  late, 313, 318
  middens, 192
  middle, 308
  Plains Woodland, 96
Worms, *see* Bioturbation
Würm glacial stage, 13
Wyoming, 193

XPL (cross-polarized light) in petrography, 262, 299
X-radiography, 301
X-ray diffraction (XRD), 248, 451–452
  for amber sourcing, 463
X-ray fluorescence (XRF), 452

Younger Dryas
  and archaeological records, 19
  and paleoenvironments, 19, 27, 130

Zinc in sediments from mining, 67
Zion National Park, UT, 70
Zircon, and fission-track dating, 388, 394–395